37 -

The Analysis of Variance

THE ANALYSIS OF VARIANCE

Fixed, Random and Mixed Models

Hardeo Sahai
Mohammed I. Ageel

Birkhäuser
Boston • Basel • Berlin

Hardeo Sahai
University of Puerto Rico
San Juan, Puerto Rico
USA

Mohammed I. Ageel
King Saud University
Abha Campus, Abha
Saudi Arabia

University of Veracruz
Xalapa, Veracruz
Mexico

Library of Congress Cataloging-in-Publication Data

Sahai, Hardeo.
   The analysis of variance : fixed, random and mixed models /
Hardeo Sahai and Mohammed I. Ageel.
     p. cm.
   Includes bibliographical references.
   ISBN 0-8176-4012-6. — ISBN 3-7643-4012-6
   1. Analysis of variance. I. Ageel, Mohammed I., 1959–   .
II. Title.
QA279.S22   2000
519.5'38—DC21   98-2788
                                                                                                  CIP

AMS Subject Classifications: 62H, 62J

Printed on acid-free paper.
© 2000 Birkhäuser Boston   ***Birkhäuser***

This book contains information obtained from authentic and highly regarded sources. Reprinted material is quoted with permission, and sources are indicated. A wide variety of references are listed. Reasonable efforts have been made to publish reliable data and information, but the author and the publisher cannot assume responsibility for the validity of all materials or for the consequences of their use.

All rights reserved. This work may not be translated or copied in whole or in part without the written permission of the publisher (Birkhäuser Boston, c/o Springer-Verlag New York, Inc., 175 Fifth Avenue, New York, NY 10010, USA), except for brief excerpts in connection with reviews or scholarly analysis. Use in connection with any form of information storage and retrieval, electronic adaptation, computer software, or by similar or dissimilar methodology now known or hereafter developed is forbidden.
The use of general descriptive names, trade names, trademarks, etc., in this publication, even if the former are not especially identified, is not to be taken as a sign that such names, as understood by the Trade Marks and Merchandise Marks Act, may accordingly be used freely by anyone.

ISBN 0-8176-4012-6
ISBN 3-7643-4012-6   SPIN 19901449

Formatted from the authors' Microsoft Word files.
Printed and bound by Sheridan Books, Inc., Ann Arbor, MI.
Printed in the United States of America.

9 8 7 6 5 4 3 2 1

To My Children: Amogh, Mrisa, Pankaj
and
to the Memory of My Father and Mother

H.S.

To My Family: Lyla, Ibrahim, Yagoub, Khalid and their mother
and
to My Father and Mother who contributed to this book by teaching me
that hard work can be satisfying if the task is worthwhile

M.I.A.

"What model are you considering?" "I am not considering one — I am using analysis of variance."

N. R. Draper and H. Smith

The analysis of variance is (not a mathematical theorem but) a simple method of arranging arithmetical facts so as to isolate and display the essential features of a body of data with the utmost simplicity.

Sir Ronald A. Fisher

No aphorism is more frequently repeated in connection with field trials, than that we must ask Nature few questions or, ideally, one question, at a time. The writer is convinced that this view is wholly mistaken. Nature, he suggests, will best respond to a logical and carefully thought out questionnaire; indeed, if we ask her a single question, she will often refuse to answer until some other topic has been discussed.

Sir Ronald A. Fisher

The statistician is no longer an alchemist expected to produce gold from any worthless material offered him. He is more like a chemist capable of assaying exactly how much of value it contains, and capable also of extracting this amount, and no more. In these circumstances, it would be foolish to commend a statistician because his results are precise or to reprove because they are not. If he is competent in his craft, the value of the result follows solely from the value of the material given him. It contains so much information and no more. His job is only to produce what it contains.

Sir Ronald A. Fisher

The new methods occupy an altogether higher plane than that in which ordinary statistics and simple averages move and have their being. Unfortunately, the ideas of which they treat, and still more the many technical phrases employed in them, are as yet unfamiliar. The arithmetic they require is laborious, and the mathematical investigations on which the arithmetic rests are difficult reading even for experts . . . this new departure in science makes its appearance under conditions that are unfavourable to its speedy recognition, and those who labour in it must abide for some time in patience before they can receive sympathy from the outside world.

Sir Francis F. Galton

When statistical data are collected as natural observations, the most sensible assumptions about the relevant statistical model have to be inserted. In controlled experimentation, however, randomness could be introduced deliberately into the design, so that any systematic variability other than [that] due to imposed treatments could be eliminated.

The second principle Fisher introduced naturally went with the first. With statistical analysis geared to the design, all variability not ascribed to the influence of treatments did not have to inflate the random error. With equal numbers of replications for the treatment, each replication could be contained in a distinct block, and only variability among plots in the same block were a source of error — that between blocks could be removed.

Sir Maurice S. Bartlett

Preface

The analysis of variance (ANOVA) models have become one of the most widely used tools of modern statistics for analyzing multifactor data. The ANOVA models provide versatile statistical tools for studying the relationship between a dependent variable and one or more independent variables. The ANOVA models are employed to determine whether different variables interact and which factors or factor combinations are most important. They are appealing because they provide a conceptually simple technique for investigating statistical relationships among different independent variables known as factors.

Currently there are several texts and monographs available on the subject. However, some of them such as those of Scheffé (1959) and Fisher and McDonald (1978), are written for mathematically advanced readers, requiring a good background in calculus, matrix algebra, and statistical theory; whereas others such as Guenther (1964), Huitson (1971), and Dunn and Clark (1987), although they assume only a background in elementary algebra and statistics, treat the subject somewhat scantily and provide only a superficial discussion of the random and mixed effects analysis of variance.

This book has been designed to bridge this gap. It provides a thorough and elementary discussion of the commonly employed analysis of variance models with emphasis on intelligent applications of the methods. We have tried to present a logical development of the subject covering both the assumptions made and methodological and computational details of the techniques involved. Most of the important results on estimation and hypothesis testing related to analysis of variance models have been included. An attempt has been made to present as many necessary concepts, principles, and techniques as possible without resorting to the use of advanced mathematics and statistical theory. In addition, the book contains complete citations of most of the important and related works, including an up-to-date and comprehensive bibliography of the field. A notable feature of the presentation is that the fixed, random, and mixed effect analysis of variance models are treated in tandem.

No attempt has been made to present the theoretical derivations of the results and techniques employed in the text. In such cases the reader is referred to the appropriate sources where such discussions can be found. It is hoped that the inquisitive reader will go through these sources to get a thorough grounding in the theory involved. However, whenever considered appropriate, some elementary derivations involving results on expectations of mean squares have been included.

The computational formulae needed to perform analysis of variance calculation are presented in full detail in order to facilitate the requisite computations using a handheld scientific calculator. Many modern electronic calculators are sufficiently powerful to handle complex arithmetic and algebraic computations and can be readily employed for this task. We are of the opinion that researchers and scientists should clearly understand a procedure before using it and manual computations provide a better understanding of the working of a procedure as well as any limitations of the experimental data. In addition, a separate chapter has been included to describe the use of some well-known statistical packages to perform a computer-assisted analysis of variance and specific commands relevant to individual analysis of variance models are included. Each chapter contains a number of worked examples where both manual and computer-assisted analysis of variance computations are illustrated in complete detail.

The only prerequisite for the understanding of the material is a preparation in a precalculus introductory course in statistical inference with special emphasis on the principles of estimation and hypothesis testing. Although some of the statistical concepts and principles used in the text (e.g., maximum likelihood and minimum variance unbiased estimation, and other related concepts and principles) may not be familiar to students who have not taken any intermediate and advanced level courses in statistical inference, these concepts have been included for the sake of completeness and to enhance the reference value of the text. However, the use of such results is generally incidental, without any mathematical and technical formality, and can be skipped without any loss in continuity. In addition, most of the results of this nature are usually kept out of the main body of the text and are generally indicated under a *remark* or following a *footnote*. Moreover, many important results in probability and statistics useful for understanding the analysis of variance models have been included in the appendices.

The book can be employed as a textbook in an introductory analysis of variance course for students whose specialization is not statistics, such as those in biological, social, engineering, and management sciences but nevertheless use analysis of variance quite extensively in their work. It can also be used as a supplement for a theoretical course in analysis of variance to balance and complement the theoretical aspects of the subject with practical applications. The book contains ample discussion and references to many theoretical results and will be immensely useful to students with advanced training in statistics. The investigators concerned with the analysis of variance techniques of data analysis will also find it a useful source of reference for many important results and applications.

Inasmuch as the principles and procedures of the analysis of variance are fairly general and common to all academic disciplines, the book can be employed as a text in all curricula. Although the examples and exercises are drawn primarily from behavioral, biological, engineering, and management sciences, the illustrations and interpretations are relevant to all disciplines. This underscores the interdisciplinary nature of research problems from all substantive fields and the absolute generality and discipline-free nature of statistical theory and methodology. Examples and exercises, in addition to assessing basic definitional and computational skills, are designed to illustrate basic conceptual understanding including applications and interpretations.

The textbook contains an abundance of *footnotes* and *remarks*. They are intended for statistically sophisticated readers who wish to pursue the subject matter in greater depth and it is not necessary that the student beginning an analysis of variance course read them. They often expand and elaborate on a particular theme, point the way to generalization and to other techniques, and make historical comments and remarks. In addition, they contain literature citations for further exploration of the topic and refer to finer points of theory and methods. We are confident that this approach will be pedagogically appealing and useful to readers with a higher degree of scholarly interest.

Finally, there is something to be said concerning the use of real-life data. We certainly believe that real-life examples are important as motivational devices for students to convince them that the techniques are indeed used in substantive fields of research; and, whenever possible, we have tried to make use of data from actual experiments and studies reported in books and papers by other authors. However, real-life data that are easy to describe, without requiring too much time and space, and are helpful in illustrating a particular technique are not always easy to find. Thus, we have included many examples and exercises that are realistically constructed using hypothetical data in order to fit the illustration of a particular statistical model under consideration. We believe that in many instances the use of such "artificial" data is just as instructive and motivating as the "real" data. The reader interested in working through some more examples and exercises involving real-data sets is referred to the books by Andrews and Herzberg (1985) and Hand et al. (1993).

Acknowledgments

No book is ever written in isolation and we would like to express our appreciation and gratitude to all the individuals and organizations who directly or indirectly have contributed to this book.

The first author wants to thank Professor Richard L. Anderson for introducing him to analysis of variance models and encouraging him to do further work in this field. He would also like to acknowledge two sabbatical leaves (1978–1979 and 1993–1994) granted by the Administrative Board of the University of Puerto Rico which provided the time to write this book.

Over a period of ten years (1972–1982), the first author taught a course in analysis of variance at the University of Puerto Rico. It was taken primarily by students in substantive fields such as biology, business, psychology, and engineering, among others. An initial set of lectures was developed during this period and although this book drastically expands and updates those notes, it never would have been possible to embark upon a project of this magnitude without that groundwork.

Parts of the manuscript were used as lecture notes during a short course on analysis of variance offered in the autumn of 1993 at the Division of Biostatistics of the University of Granada (Spain) Medical and Dental Faculties and we greatly benefited from the comments and criticisms received. We would particularly like to thank Professor Antonio Martin Andrés for an invitation to offer the course.

Several portions of the manuscript were written and revised during the first author's successive visits (Spring 1994, Summer 1995, Winter 1996, Autumn 1997, Summer 1998) to the University of Veracruz (Mexico) and he is especially grateful to Dr. Mario Miguel Ojeda, and other University authorities for organizing and hosting the visits and providing a stimulating environment for research and study.

A preliminary draft of the manuscript was used as a principal reference source in a course on analysis of variance offered at the Department of Mathematics and Statistics of the National University of Colombia in Santafé de Bogota during the summer of 1994 and we received much useful feedback. We would like to thank Dr. Jorge Martinez for organizing the visit and an invitation to offer the course.

Several parts of the manuscript were used as primary reference materials for a short course on analysis of variance offered at the National University of Trujillo (Peru) and we greatly benefited from suggestions that improved the text. We are especially indebted to Professor Segundo M. Chuquilin Terán and other University authorities for an invitation to offer the course.

Selected portions of the manuscript were also used during a short course on analysis of variance offered at the Fifth Meeting of the International Biometric Society Network Meeting for Central America, Mexico, Colombia, and Venezuela, held in Xalapa, Mexico in August, 1997, and we received many useful ideas from the participants that included students and professors from substantive fields of research and a diverse group of professionals from industry and government.

The authors would like to thank all the individuals who have participated in our analysis of variance courses, both on and off campus and in a variety of forums and settings, and have worked through "preprinted" versions of the text for their kind comments and sufferance. Undoubtedly, all their suggestions and remarks have had a positive effect on the book. We are also greatly indebted to authors of textbooks and journal articles which we have generously consulted and who are cited throughout the book.

We have received several excellent comments and suggestions on various parts of the book from many of our friends and colleagues. We would especially like to mention our appreciation to Dr. Shahariar Huda of the King Saud University (Saudi Arabia), Dr. Mario M. Ojeda of the University of Veracruz (Mexico), and Dr. Satish C. Misra of the Food and Drug Administration (USA) who have commented upon various parts of the manuscript.

Of course, we did not address all the comments and suggestions received and any remaining errors are the authors' sole responsibility.

We would also like to record a note of appreciation to Dr. Russel D. Wolfinger of the SAS Institute, Inc. (USA) and to Dr. David Nichols of SPSS, Inc. (USA) for reviewing contents of the manuscript on performing analysis of variance using statistical packages and pointing out some omissions and inaccuracies.

Dr. Raul Micchiavelli of the University of Puerto Rico, Mr. Guadalupe Hernandez Lira of the University of Veracruz (Mexico), and Mr. Victor Alvarez of the University of Guatemala in Guatemala City assisted us in running worked examples using statistical packages and their helpful support is greatly appreciated.

Dr. Jayanta Banarjee, Professor of Mechanical Engineering at the University of Puerto Rico, deserves special recognition for kindly helping us to construct a number of realistic exercises using hypothetical data.

The first author wishes to extend a warm appreciation to members and staff of the Puerto Rico Center for Addiction Research, especially Dr. Rafaela R. Robles, Dr. Hector M. Colón, Ms. Carmen A. Marrero, M. P. H., Mr. Tomás L. Matos, and Mr. Juan C. Reyes, M. P. H. who as an innovative research group, for well over a decade provided an intellectually stimulating environment and a lively research forum to discuss and debate the role of analysis of variance models in social and behavioral research.

Our grateful and special thanks go to our publisher, especially Mr. Wayne Yuhasz, Executive Director of Computational Sciences and Engineering, for

Acknowledgments

his encouragement and support of the project. Equally, we would like to record our thanks to the editorial and production staff at Birkhäuser, especially Mr. Michael Koy, Production Editor, for all their help and cooperation in bringing the project to its fruition.

The authors and Birkhäuser would like to thank many authors, publishers, and other organizations for their kind permission to use the data and to reprint whole or parts of statistical tables and charts from their previously published copyrighted materials, and the acknowledgments are made in the book where they appear.

Finally, we must make a special acknowledgment of gratitude to our families, who were patient during the many hours of daily work devoted to the book, in what seemed like an endless process of revisions for finalizing the manuscript, and we are greatly indebted for their continued help and support.

The authors welcome any suggestions and criticisms of the book in regard to omissions, inaccuracies, corrections, additions, or ways of presentation that would be rectified in any further revision of this work.

Contents

Preface ix
Acknowledgments xiii
List of Tables xxix
List of Figures xxxiii

1. **Introduction** 1
   1.0 Preview 1
   1.1 Historical Developments 3
   1.2 Analysis of Variance Models 4
   1.3 Concept of Fixed and Random Effects 6
   1.4 Finite and Infinite Populations 7
   1.5 General and Generalized Linear Models 8
   1.6 Scope of the Book 8

2. **One-Way Classification** 11
   2.0 Preview 11
   2.1 Mathematical Model 11
   2.2 Assumptions of the Model 11
   2.3 Partition of the Total Sum of Squares 14
   2.4 The Concept of Degrees of Freedom 15
   2.5 Mean Squares and Their Expectations 17
   2.6 Sampling Distribution of Mean Squares 20
   2.7 Test of Hypothesis: The Analysis of Variance F Test 22
       Model I (Fixed Effects) 22
       Model II (Random Effects) 24
   2.8 Analysis of Variance Table 26
   2.9 Point Estimation: Estimation of Treatment Effects and
       Variance Components 26
   2.10 Confidence Intervals for Variance Components 31
   2.11 Computational Formulae and Procedure 35
   2.12 Analysis of Variance for Unequal Number of Observations . . . 36
   2.13 Worked Examples for Model I 39
   2.14 Worked Examples for Model II 43
   2.15 Use of Statistical Computing Packages 52
   2.16 Worked Examples Using Statistical Packages 52
   2.17 Power of the Analysis of Variance F Test 52
        Model I (Fixed Effects) 57
        Model II (Random Effects) 60

2.18 Power and Determination of Sample Size 61
     Sample Size Determination Using Smallest
     Detectable Difference 63
2.19 Inference About the Difference Between Treatment Means:
     Multiple Comparisons 64
     Linear Combination of Means, Contrast and
     Orthogonal Contrasts 65
     Test of Hypothesis Involving a Contrast 67
     The Use of Multiple Comparisons 70
         Tukey's method 70
         Scheffé's method 73
         Interpretation of Tukey's and Scheffé's methods 76
         Comparison of Tukey's and Scheffé's methods 77
     Other Multiple Comparison Methods 77
         Least significant difference test 77
         Bonferroni's test 78
         Dunn-Šidák's test 79
         Newman-Keuls's test 80
         Duncan's multiple range test 81
         Dunnett's test 81
     Multiple Comparisons for Unequal Sample Sizes
     and Variances 82
         Unequal sample sizes 82
         Unequal population variances 84
2.20 Effects of Departures from Assumptions Underlying the
     Analysis of Variance Model 84
     Departures from Normality 85
     Departures from Equal Variances 86
     Departures from Independence of Error Terms 88
2.21 Tests for Departures from Assumptions of the Model 89
     Tests for Normality 89
         Chi-square goodness-of-fit test 89
         Test for skewness 90
         Test for kurtosis 91
     Other Tests for Normality 93
         Shapiro-Wilk's W test 93
         Shapiro-Francia's test 94
         D'Agostino's D test 96
     Tests for Homoscedasticity 97
         Bartlett's test 98
         Hartley's test 104
         Cochran's test 105
         Comments on Bartlett's, Hartley's and Cochran's tests 106
         Other tests of homoscedasticity 107

Contents   xix

    2.22 Corrections for Departures from Assumptions of the Model . . 108
            Transformations to Correct Lack of Normality 109
                   Logarithmic transformation 109
                   Square-root transformation 109
                   Arcsine transformation 110
            Transformations to Correct Lack of Homoscedasticity 110
                   Logarithmic transformation 111
                   Square-root transformation 111
                   Reciprocal transformation 111
                   Arcsine transformation 112
                   Square transformation 112
                   Power transformation 112
   Exercises 113

3. **Two-Way Crossed Classification Without Interaction** 125
   3.0 Preview 125
   3.1 Mathematical Model 126
   3.2 Assumptions of the Model 127
   3.3 Partition of the Total Sum of Squares 128
   3.4 Mean Squares and Their Expectations 130
         Model I (Fixed Effects) 131
         Model II (Random Effects) 132
         Model III (Mixed Effects) 133
   3.5 Sampling Distribution of Mean Squares 135
         Model I (Fixed Effects) 135
         Model II (Random Effects) 136
         Model III (Mixed Effects) 136
   3.6 Tests of Hypotheses: The Analysis of Variance F Tests 137
         Model I (Fixed Effects) 137
         Model II (Random Effects) 139
         Model III (Mixed Effects) 139
   3.7 Point Estimation 139
         Model I (Fixed Effects) 139
         Model II (Random Effects) 141
         Model III (Mixed Effects) 141
   3.8 Interval Estimation 142
         Model I (Fixed Effects) 142
         Model II (Random Effects) 143
         Model III (Mixed Effects) 143
   3.9 Computational Formulae and Procedure 144
   3.10 Missing Observations 145
   3.11 Power of the Analysis of Variance F Tests 148
   3.12 Multiple Comparison Methods 149
   3.13 Worked Example for Model I 151

3.14 Worked Example for Model II 154
3.15 Worked Example for Model III 157
3.16 Worked Example for Missing Value Analysis 161
3.17 Use of Statistical Computing Packages 164
3.18 Worked Examples Using Statistical Packages 164
3.19 Effects of Violations of Assumptions of the Model 168
Exercises 168

4. **Two-Way Crossed Classification With Interaction** 177
   4.0 Preview 177
   4.1 Mathematical Model 177
   4.2 Assumptions of the Model 180
   4.3 Partition of the Total Sum of Squares 182
   4.4 Mean Squares and Their Expectations 184
       Model I (Fixed Effects) 186
       Model II (Random Effects) 188
       Model III (Mixed Effects) 190
   4.5 Sampling Distribution of Mean Squares 193
       Model I (Fixed Effects) 193
       Model II (Random Effects) 196
       Model III (Mixed Effects) 196
   4.6 Tests of Hypotheses: The Analysis of Variance F Tests 197
       Model I (Fixed Effects) 198
           Test for AB interactions 198
           Test for factor B effects 198
           Test for factor A effects 199
       Model II (Random Effects) 200
           Test for AB interactions 200
           Test for factor B effects 201
           Test for factor A effects 201
       Model III (Mixed Effects) 202
           Test for AB interactions 202
           Test for factor B effects 202
           Test for factor A effects 202
       Summary of Models and Tests 203
   4.7 Point Estimation 203
       Model I (Fixed Effects) 203
       Model II (Random Effects) 205
       Model III (Mixed Effects) 206
   4.8 Interval Estimation 207
       Model I (Fixed Effects) 207
       Model II (Random Effects) 208
       Model III (Mixed Effects) 209
   4.9 Computational Formulae and Procedure 210

Contents   xxi

    4.10 Analysis of Variance with Unequal Sample Sizes Per Cell . . 212
           Fixed Effects Analysis 215
                Proportional frequencies 215
                General case of unequal frequencies 217
           Random Effects Analysis 223
                Proportional frequencies 223
                General case of unequal frequencies 223
           Mixed Effects Analysis 226
    4.11 Power of the Analysis of Variance F Tests 227
           Model I (Fixed Effects)....................... 227
                Test for AB interactions 227
                Test for factor B effects 227
                Test for factor A effects 227
           Model II (Random Effects)..................... 228
                Test for AB interactions 228
                Test for factor B effects 228
                Test for factor A effects 228
           Model III (Mixed Effects) 229
                Test for AB interactions 229
                Test for factor B effects 229
                Test for factor A effects 229
    4.12 Multiple Comparison Methods 230
    4.13 Worked Example for Model I 233
    4.14 Worked Example for Model I: Unequal Sample
           Sizes Per Cell 237
    4.15 Worked Example for Model II 244
    4.16 Worked Example for Model III 247
    4.17 Use of Statistical Computing Packages............. 252
    4.18 Worked Examples Using Statistical Packages 253
    4.19 The Meaning and Interpretation of Interaction 253
    4.20 Interaction With One Observation Per Cell 259
    4.21 Alternate Mixed Models 264
    4.22 Effects of Violations of Assumptions of the Model 268
           Model I (Fixed Effects)....................... 269
           Model II (Random Effects)..................... 269
           Model III (Mixed Effects) 270
    Exercises................................... 270

5. **Three-Way and Higher-Order Crossed Classifications** 281
    5.0 Preview 281
    5.1 Mathematical Model 281
    5.2 Assumptions of the Model 284
    5.3 Partition of the Total Sum of Squares............... 285
    5.4 Mean Squares and Their Expectations 286

5.5 Tests of Hypotheses: The Analysis of Variance F Tests 286
       Model I (Fixed Effects) 288
       Model II (Random Effects) 288
       Model III (Mixed Effects) 291
5.6 Point and Interval Estimation 292
5.7 Computational Formulae and Procedure 297
5.8 Power of the Analysis of Variance F Tests 298
5.9 Multiple Comparison Methods 299
5.10 Three-Way Classification with One Observation Per Cell ... 301
5.11 Four-Way Crossed Classification 302
5.12 Higher-Order Crossed Classifications 307
5.13 Unequal Sample Sizes in Three- and Higher-Order
     Classifications 311
5.14 Worked Example for Model I 314
5.15 Worked Example for Model II 322
5.16 Worked Example for Model III 328
5.17 Use of Statistical Computing Packages 333
5.18 Worked Examples Using Statistical Packages 334
Exercises 338

6. Two-Way Nested (Hierarchical) Classification 347
6.0 Preview 347
6.1 Mathematical Model 349
6.2 Assumptions of the Model 350
6.3 Analysis of Variance 350
6.4 Tests of Hypotheses: The Analysis of Variance F Tests 351
6.5 Point Estimation 354
       Model I (Fixed Effects) 354
       Model II (Random Effects) 355
       Model III (Mixed Effects) 356
6.6 Interval Estimation 356
       Model I (Fixed Effects) 356
       Model II (Random Effects) 358
       Model III (Mixed Effects) 359
6.7 Computational Formulae and Procedure 359
6.8 Power of the Analysis of Variance F Tests 360
6.9 Multiple Comparison Methods 360
6.10 Unequal Numbers in the Subclasses 362
       Tests of Hypotheses 363
       Point and Interval Estimation 365
6.11 Worked Example for Model I 368
6.12 Worked Example for Model II 371
6.13 Worked Example for Model II: Unequal Numbers
     in the Subclasses 374
6.14 Worked Example for Model III 378

Contents   xxiii

    6.15 Use of Statistical Computing Packages 381
    6.16 Worked Examples Using Statistical Packages 381
    Exercises 386

7. **Three-Way and Higher-Order Nested Classifications** 395
    7.0 Preview 395
    7.1 Mathematical Model 395
    7.2 Analysis of Variance 396
    7.3 Tests of Hypotheses and Estimation 399
    7.4 Unequal Numbers in the Subclasses 400
    7.5 Four-Way Nested Classification 403
    7.6 General q-Way Nested Classification 406
    7.7 Worked Example for Model II 407
    7.8 Worked Example for Model II: Unequal Numbers
         in the Subclasses 411
    7.9 Worked Example for Model III 417
    7.10 Use of Statistical Computing Packages 421
    7.11 Worked Examples Using Statistical Packages 421
    Exercises 425

8. **Partially Nested Classifications** 431
    8.0 Preview 431
    8.1 Mathematical Model 431
    8.2 Analysis of Variance 433
    8.3 Computational Formulae and Procedure 437
    8.4 A Four-Factor Partially Nested Classification 438
    8.5 Worked Example for Model II 439
    8.6 Worked Example for Model III 444
    8.7 Use of Statistical Computing Packages 448
    8.8 Worked Example Using Statistical Packages 451
    Exercises 451

9. **Finite Population and Other Models** 461
    9.0 Preview 461
    9.1 One-Way Finite Population Model 461
    9.2 Two-Way Crossed Finite Population Model 462
         Tests of Hypotheses 465
         F Tests 466
         Point Estimation 467
         Interval Estimation 468
    9.3 Three-Way Crossed Finite Population Model 470
    9.4 Four-Way Crossed Finite Population Model 472
    9.5 Nested Finite Population Models 474
    9.6 Unbalanced Finite Population Models 475
    9.7 Worked Example for a Finite Population Model 475

	9.8 Other Models 481
	9.9 Use of Statistical Computing Packages 481
	Exercises 481
10.	**Some Simple Experimental Designs** **483**
	10.0 Preview 483
	10.1 Principles of Experimental Design 483
	Replication 483
	Randomization 483
	Control 484
	10.2 Completely Randomized Design 485
	Model and Analysis 485
	Worked Example 487
	10.3 Randomized Block Design 488
	Model and Analysis 490
	Both blocks and treatments fixed 490
	Both blocks and treatments random 492
	Blocks random and treatments fixed 492
	Blocks fixed and treatments random 493
	Missing Observations 493
	Relative Efficiency of the Design 493
	Replications 494
	Worked Example 494
	10.4 Latin Square Design 495
	Model and Analysis 498
	Point and Interval Estimation 500
	Power of the F Test 501
	Multiple Comparisons 501
	Computational Formulae 502
	Missing Observations 502
	Tests for Interaction 503
	Relative Efficiency of the Design 503
	Replications 504
	Worked Example 505
	10.5 Graeco-Latin Square Design 507
	Model and Analysis 507
	Worked Example 510
	10.6 Split-Plot Design 512
	Model and Analysis 513
	Worked Example 516
	10.7 Other Designs 516
	Incomplete Block Designs 516
	Lattice Designs 519
	Youden Squares 520
	Cross-Over Designs 520

Contents   xxv

        Repeated Measures Designs 521
        Hyper-Graeco-Latin and Hyper Squares 522
        Magic and Super Magic Latin Squares 522
        Split-Split-Plot Design 523
        2^p Design and Fractional Replications 523
   10.8 Use of Statistical Computing Packages 524
   Exercises 525

11. Analysis of Variance Using Statistical Computing Packages . . 543
   11.0 Preview 543
   11.1 Analysis of Variance Using SAS 543
   11.2 Analysis of Variance Using SPSS 550
   11.3 Analysis of Variance Using BMDP 558
   11.4 Use of Statistical Packages for Computing Power 560
   11.5 Use of Statistical Packages for Multiple
        Comparison Procedures 561
   11.6 Use of Statistical Packages for Tests of Homoscedasticity . . 565
   11.7 Use of Statistical Packages for Tests of Normality 567

Appendices
   A   Student's t Distribution 569
   B   Chi-Square Distribution 570
   C   Sampling Distribution of $(n-1)S^2/\sigma^2$ 571
   D F Distribution 572
   E   Noncentral Chi-Square Distribution 573
   F   Noncentral and Doubly Noncentral t Distributions 574
   G   Noncentral F Distribution 575
   H   Doubly Noncentral F Distribution 576
   I   Studentized Range Distribution 576
   J   Studentized Maximum Modulus Distribution 577
   K   Satterthwaite Procedure and Its Application to
        Analysis of Variance 578
   L   Components of Variance 580
   M   Intraclass Correlation 580
   N   Analysis of Covariance 581
   O   Equivalence of the Anova F and Two-Sample t Tests 582
   P   Equivalence of the Anova F and Paired t Tests 584
   Q   Expected Value and Variance 586
   R   Covariance and Correlation 587
   S   Rules for Determining the Analysis of Variance Model 588
   T   Rules for Calculating Sums of Squares
        and Degrees of Freedom 590
   U   Rules for Finding Expected Mean Squares 591
   V   Samples and Sampling Distribution 595
   W   Methods of Statistical Inference 596

X	Some Selected Latin Squares	601
Y	Some Selected Graeco-Latin Squares	602
Z	PROC MIXED Outputs for Some Selected Worked Examples	603

Statistical Tables and Charts 605

Tables

I	Cumulative Standard Normal Distribution	605
II	Percentage Points of the Standard Normal Distribution	607
III	Critical Values of the Student's t Distribution	608
IV	Critical Values of the Chi-Square Distribution	610
V	Critical Values of the F Distribution	612
VI	Power of the Student's t Test	618
VII	Power of the Analysis of Variance F Test	621
VIII	Power Values and Optimum Number of Levels for Total Number of Observations in the One-Way Random Effects Analysis of Variance F Test	630
IX	Minimum Sample Size Per Treatment Group Needed for a Given Value of p, α, $1 - \beta$ and Effect Size (C) in Sigma Units	634
X	Critical Values of the Studentized Range Distribution	636
XI	Critical Values of the Dunnett's Test	639
XII	Critical Values of the Duncan's Multiple Range Test	642
XIII	Critical Values of the Bonferroni t Statistic and Dunn's Multiple Comparison Test	645
XIV	Critical Values of the Dunn-Šidák's Multiple Comparison Test	648
XV	Critical Values of the Studentized Maximum Modulus Distribution	652
XVI	Critical Values of the Studentized Augmented Range Distribution	654
XVII(a)	Critical Values of the Distribution of $\hat{\gamma}_1$ for Testing Skewness	655
XVII(b)	Critical Values of the Distribution of $\hat{\gamma}_2$ for Testing Kurtosis	655
XVIII	Coefficients of Order Statistics for the Shapiro-Wilk's W Test for Normality	656
XIX	Critical Values of the Shapiro-Wilk's W Test for Normality	657
XX	Critical Values of the D'Agostino's D Test for Normality	658
XXI	Critical Values of the Bartlett's Test for Homogeneity of Variances	659

XXII	Critical Values of the Hartley's Maximum F Ratio Test for Homogeneity of Variances 663
XXIII	Critical Values of the Cochran's C Test for Homogeneity of Variances 664
XXIV	Random Numbers 665

Charts

I	Power Functions of the Two-Sided Student's t Test 669
II	Power Functions of the Analysis of Variance F Tests (Fixed Effects Model): Pearson-Hartley Charts 672
III	Operating Characteristic Curves for the Analysis of Variance F Tests (Random Effects Model) 681
IV	Curves of Constant Power for Determination of Sample Size in a One-Way Analysis of Variance (Fixed Effects Model): Feldt-Mahmoud Charts 686

References **689**
Author Index **717**
Subject Index **725**

List of Tables

Table 2.1	Analysis of Variance for Model (2.1.1)	26
Table 2.2	Analysis of Variance for Model (2.1.1) with Unequal Sample Sizes	37
Table 2.3	Data on Yields of Four Different Varieties of Wheat (in bushels per acre)	39
Table 2.4	Analysis of Variance for the Yields Data of Table 2.3	41
Table 2.5	Data on Blood Analysis of Animals Injected with Five Drugs	41
Table 2.6	Analysis of Variance for the Blood Analysis Data of Table 2.5	43
Table 2.7	Interview Ratings by Five Staff Members	43
Table 2.8	Analysis of Variance for the Interview Ratings Data of Table 2.7	45
Table 2.9	Data on Yields of Six Varieties of Corn (in bushels per acre)	48
Table 2.10	Analysis of Variance for the Yields Data of Table 2.9	50
Table 2.11	Pairwise Differences of Sample Means $\bar{y}_{i.} - \bar{y}_{i'}$	72
Table 2.12	Calculations for Shapiro-Francia's Test	95
Table 2.13	Calculations for Bartlett's Test	101
Table 2.14	Data on Log-bids of Five Texas Offshore Oil and Gas Leases	103
Table 3.1	Data for a Two-Way Experimental Layout	126
Table 3.2	Analysis of Variance for Model (3.1.1)	134
Table 3.3	Loss in Weights Due to Wear Testing of Four Materials (in mg)	151
Table 3.4	Analysis of Variance for the Weight Loss Data of Table 3.3	153
Table 3.5	Number of Minutes Observed in Grazing	154
Table 3.6	Analysis of Variance for the Grazing Data of Table 3.5	156
Table 3.7	Breaking Strength of the Plastics (in lbs.)	158
Table 3.8	Analysis of Variance for the Breaking Strength Data of Table 3.7	160
Table 3.9	Analysis of Variance for the Weight Loss Data of Table 3.3 with one Missing Value	163
Table 4.1	Data for a Two-Way Crossed Classification with n Replications per Cell	178
Table 4.2	Analysis of Variance for Model (4.1.1)	194
Table 4.3	Test Statistics for Models I, II, and III	203

Table 4.4	Data for a Two-Way Crossed Classification with n_{ij} Replications per Cell	214
Table 4.5	Analysis of Variance for the Unbalanced Fixed Effects Model in (4.1.1) with Proportional Frequencies	216
Table 4.6	Unweighted Means Analysis for the Unbalanced Fixed Effects Model in (4.1.1) with Disproportional Frequencies	218
Table 4.7	Weighted Means Analysis for the Unbalanced Fixed Effects Model in (4.1.1) with Disproportional Frequencies	221
Table 4.8	Analysis of Variance for Model (4.10.11)	226
Table 4.9	Running Time (in Seconds) to Complete a 1.5 Mile Course	234
Table 4.10	Analysis of Variance for the Running Time Data of Table 4.9	236
Table 4.11	Weight Gains (in grams) of Rats under Different Diets (Data Made Unbalanced by Deleting Observations)	238
Table 4.12	Analysis of Variance Using Unweighted and Weighted Squares-of-Means Analysis of the Unbalanced Data on Weight Gains (in grams) of Table 4.11	243
Table 4.13	Screen Lengths (in inches) from a Quality Control Experiment	245
Table 4.14	Analysis of Variance for the Screen Lengths Data of Table 4.13	246
Table 4.15	Yield Loads for Cement Specimens	248
Table 4.16	Analysis of Variance for the Yield Loads Data of Table 4.15	250
Table 4.17	Analysis of Variance for Model (4.20.1)	261
Table 4.18	Analysis of Variance for an Alternate Mixed Model	265
Table 5.1	Data for a Three-Way Crossed Classification with n Replications per Cell	282
Table 5.2	Analysis of Variance for Model (5.1.1)	287
Table 5.3	Tests of Hypotheses for Model (5.1.1) under Model I	289
Table 5.4	Tests of Hypotheses for Model (5.1.1) under Model II	292
Table 5.5	Tests of Hypotheses for Model (5.1.1) under Model III (A Fixed, B and C Random)	293
Table 5.6	Tests of Hypotheses for Model (5.1.1) under Model III (A Random, B and C Fixed)	294
Table 5.7	Estimates of Parameters and Their Variances under Model I	295
Table 5.8	Analysis of Variance for Model (5.10.1)	303
Table 5.9	Analysis of Variance for the Unbalanced Fixed Effects Model in (5.1.1) with Proportional Frequencies	314
Table 5.10	Data on Average Resistivities (in m-ohms/cm^3) for Electrolytic Chromium Plate Example	316
Table 5.11	Cell Totals $y_{ijk.}$	317
Table 5.12	Sums over Levels of Degrasing $y_{ij..}$	317

List of Tables

Table 5.13	Sums over Levels of Time $y_{i.k.}$	317
Table 5.14	Sums over Levels of Temperature $y_{.jk.}$	318
Table 5.15	Analysis of Variance for the Resistivity Data of Table 5.10	320
Table 5.16	Data on the Melting Points of a Homogeneous Sample of Hydroquinone	323
Table 5.17	Sums over Analysts $y_{ij.}$	323
Table 5.18	Sums over Weeks $y_{i.k}$	324
Table 5.19	Sums over Thermometer $y_{.jk}$	324
Table 5.20	Analysis of Variance for the Melting Point Data of Table 5.16	326
Table 5.21	Data on Diamond Pyramid Hardness Number of Dental Fillings Made from Two Alloys of Gold	328
Table 5.22	Sums over Dentists $y_{ij.}$	329
Table 5.23	Sums over Methods $y_{i.k}$	330
Table 5.24	Sums over Alloys $y_{.jk}$	330
Table 5.25	Analysis of Variance for the Gold Hardness Data of Table 5.21	332
Table 6.1	Data for a Two-Way Nested Classification	349
Table 6.2	Analysis of Variance for Model (6.1.1)	352
Table 6.3	Point Estimates and Their Variances under Model III	357
Table 6.4	Analysis of Variance for Model (6.1.1) Involving Unequal Numbers in the Subclasses	364
Table 6.5	Weight Gains of Chickens Placed on Four Feeding Treatments	368
Table 6.6	Analysis of Variance for the Weight Gains Data of Table 6.5	370
Table 6.7	Moisture Content from Two Analyses on Two Samples of 15 Batches of Pigment Paste	372
Table 6.8	Analysis of Variance for the Moisture Content of Pigment Paste Data of Table 6.7	373
Table 6.9	Data on Weight Gains from a Breeding Experiment	375
Table 6.10	Analysis of Variance for the Weight Gains Data of Table 6.9	377
Table 6.11	Average Daily Weight Gains of Two Pigs of Each Litter	379
Table 6.12	Analysis of Variance for the Weight Gains Data of Table 6.11	380
Table 7.1	Analysis of Variance for Model (7.1.1)	398
Table 7.2	Analysis of Variance for Model (7.1.1) Involving Unequal Numbers in the Subclasses	402
Table 7.3	Analysis of Variance for Model (7.5.1)	405
Table 7.4	Percentage of Ingredient of a Batch of Material	408
Table 7.5	Analysis of Variance for the Material Homogeneity Data of Table 7.4	410
Table 7.6	Number of Diatoms per Square Centimeter Colonizing Glass Slides	412
Table 7.7	Analysis of Variance for the Diatom Data of Table 7.6	415

Table 7.8	Glycogen Content of Rat Livers in Arbitrary Units	418
Table 7.9	Analysis of Variance for the Glycogen Data of Table 7.8	420
Table 8.1	Analysis of Variance for Model (8.1.1)	436
Table 8.2	Analysis of Variance for Model (8.4.1)	440
Table 8.3	Analysis of Variance for the Calcium Content of Turnip Leaves Data	441
Table 8.4	Measured Strengths of Tire Cords from Two Plants Using Different Production Processes	444
Table 8.5	Calculation of Cell and Marginal Totals	446
Table 8.6	Analysis of Variance for the Tire Cords Strength Data of Table 8.4	449
Table 9.1	Analysis of Variance for Model (9.1.1)	462
Table 9.2	Analysis of Variance for Model (9.2.1)	464
Table 9.3	Expected Mean Squares for Model (9.3.1)	472
Table 9.4	Analysis of Variance for Model (9.5.1)	475
Table 9.5	Production Output from an Industrial Experiment	476
Table 9.6	Analysis of Variance for the Production Output Data of Table 9.5	477
Table 10.1	Analysis of Variance for the Completely Randomized Design with Equal Sample Sizes	486
Table 10.2	Analysis of Variance for the Completely Randomized Design with Unequal Sample Sizes	487
Table 10.3	Data on Weights of Mangold Roots in a Uniformity Trial	488
Table 10.4	Analysis of Variance for the Data on Weights of Mangold Roots	488
Table 10.5	Analysis of Variance for the Randomized Block Design	491
Table 10.6	Data on Yields of Wheat Straw from a Randomized Block Experiment	494
Table 10.7	Analysis of Variance for the Data on Yields of Wheat Straw	495
Table 10.8	Analysis of Variance for the Latin Square Design	499
Table 10.9	Data on Responses of Monkeys to Different Stimulus Conditions	505
Table 10.10	Analysis of Variance for the Data on Responses of Monkeys to Different Stimulus Conditions	505
Table 10.11	Analysis of Variance for the Graeco-Latin Square Design	509
Table 10.12	Data on Photographic Density for Different Brands of Flash Bulbs	510
Table 10.13	Analysis of Variance for the Data on Photographic Density for Different Brands of Flash Bulbs	510
Table 10.14	Analysis of Variance for the Split-Plot Design	515
Table 10.15	Data on Yields of Two Varieties of Soybean	517
Table 10.16	Analysis of Variance for the Data on Yields of Two Varieties of Soybean	517

List of Figures

Figure 2.1	Schematic Representation of Model I and Model II	13
Figure 2.2	Program Instructions and Output for the One-Way Fixed Effects Analysis of Variance: Data on Yields of Four Different Varieties of Wheat (Table 2.3)	53
Figure 2.3	Program Instructions and Output for the One-Way Fixed Effects Analysis of Variance with Unequal Numbers of Observations: Data on Blood Analysis of Animals Injected with Five Drugs (Table 2.5)	54
Figure 2.4	Program Instructions and Output for the One-Way Random Effects Analysis of Variance: Data on Interview Ratings by Five Staff Members (Table 2.7)	55
Figure 2.5	Program Instructions and Output for the One-Way Random Effects Analysis of Variance with Unequal Numbers of Observations: Data on Yields of Six Varieties of Corn (Table 2.9)	56
Figure 2.6	Curves Exhibiting Positive and Negative Skewness and Symmetrical Distribution	91
Figure 2.7	Curves Exhibiting Positive and Negative Kurtosis and the Normal Distribution	92
Figure 3.1	Program Instructions and Output for the Two-Way Fixed Effects Analysis of Variance with One Observation Per Cell: Data on Loss in Weights Due to Wear Testing of Four Materials (Table 3.3)	165
Figure 3.2	Program Instructions and Output for the Two-Way Random Effects Analysis of Variance with One Observation per Cell: Data on Number of Minutes Observed in Grazing (Table 3.5)	166
Figure 3.3	Program Instructions and Output for the Two-Way Mixed Effects Analysis of Variance with One Observation per Cell: Data on Breaking Strength of the Plastics (Table 3.7)	167
Figure 4.1	Program Instructions and Output for the Two-Way Fixed Effects Analysis of Variance with Two Observations per Cell: Data on Running Time (in seconds) to Complete a 1.5 Mile Course (Table 4.9)	254
Figure 4.2	Program Instructions and Output for the Two-Way Fixed Effects Analysis of Variance with Unequal Numbers of	

	Observations per Cell: Data on Weight Gains of Rats under Different Diets (Table 4.11)	255
Figure 4.3	Program Instructions and Output for the Two-Way Random Effects Analysis of Variance with Two Observations per Cell: Data on Screen Lengths from a Quality Control Experiment (Table 4.13)	256
Figure 4.4	Program Instructions and Output for the Two-Way Mixed Effects Analysis of Variance with Three Observations per Cell: Data on Yield Loads for Cement Specimens (Table 4.15)	258
Figure 4.5	Patterns of Observed Cell Means and Existence or Nonexistence of Interaction Effects	260
Figure 5.1	Program Instructions and Output for the Three-Way Fixed Effects Analysis of Variance: Data on Average Resistivities (in m-ohms/cm^3) for Electrolytic Chromium Plate Example (Table 5.10)	334
Figure 5.2	Program Instructions and Output for the Three-Way Random Effects Analysis of Variance: Data on the Melting Points of a Homogeneous Sample of Hydroquinone (Table 5.16)	335
Figure 5.3	Program Instructions and Output for the Three-Way Mixed Effects Analysis of Variance: Data on Diamond Pyramid Hardness Number of Dental Fillings Made from Two Alloys of Gold (Table 5.21)	337
Figure 6.1	A Layout for the Two-Way Nested Design Where Barrels Are Nested within Locations	348
Figure 6.2	A Layout for the Two-Way Nested Design Where Spindles Are Nested within Machines	348
Figure 6.3	Program Instructions and Output for the Two-Way Fixed Effects Nested Analysis of Variance: Weight Gains Data for Example of Section 6.11 (Table 6.5)	382
Figure 6.4	Program Instructions and Output for the Two-Way Random Effects Nested Analysis of Variance: Moisture Content of Pigment Paste Data for Example of Section 6.12 (Table 6.7)	383
Figure 6.5	Program Instructions and Output for the Two-Way Random Effects Nested Analysis of Variance with Unequal Numbers in the Subclasses: Breeding Data for Example of Section 6.13 (Table 6.9)	384
Figure 6.6	Program Instructions and Output for the Two-Way Mixed Effects Nested Analysis of Variance: Average Daily Weight Gains Data for Example of Section 6.14 (Table 6.11)	385
Figure 7.1	A Layout for the Three-Way Nested Design Where Barrels Are Nested within Vats and Samples Are Nested within Barrels	396

List of Figures

Figure 7.2	Program Instructions and Output for the Three-Way Random Effects Nested Analysis of Variance: Material Homogeneity Data for Example of Section 7.7 (Table 7.4)	421
Figure 7.3	Program Instructions and Output for the Three-Way Random Effects Nested Analysis of Variance with Unequal Numbers in the Subclasses: Diatom Data for Example of Section 7.8 (Table 7.6)	423
Figure 7.4	Program Instructions and Output for the Three-Way Mixed Effects Nested Analysis of Variance: Glycogen Data for Example of Section 7.9 (Table 7.8)	424
Figure 8.1	A Layout for the Partially Nested Design Where Days Are Crossed with Methods and Operators Are Nested within Methods	432
Figure 8.2	Program Instructions and Outputs for the Partially Nested Mixed Effects Analysis of Variance: Measured Strengths of Tire Cords from Two Plants Using Different Production Processes (Table 8.4)	450
Figure 10.1	Program Instructions and Output for the Completely Randomized Design: Data on Weights of Mangold Roots in a Uniformity Trial (Table 10.3)	489
Figure 10.2	A Layout of a Randomized Block Design	490
Figure 10.3	Program Instructions and Output for the Randomized Block Design: Data on Yields of Wheat Straw from a Randomized Block Design (Table 10.6)	496
Figure 10.4	Some Selected Latin Squares	497
Figure 10.5	Program Instructions and Output for the Latin Square Design: Data on Responses of Monkeys to Different Stimulus Conditions (Table 10.9)	506
Figure 10.6	Some Selected Graeco-Latin Squares	507
Figure 10.7	Program Instructions and Output for the Graeco-Latin Square Design: Data on Photographic Density for Different Brands of Flash Bulbs (Table 10.12)	511
Figure 10.8	A Layout of a Split-Plot Design	512
Figure 10.9	Program Instructions and Output for the Split-Plot Design: Data on Yields of Two Varieties of Soybean (Table 10.15)	518

1 Introduction

1.0 PREVIEW

The variation among physical observations is a common characteristic of all scientific measurements. This property of observations, that is, their failure to reproduce themselves exactly, arises from the necessity of taking the observations under different conditions. Thus, in a given experiment, readings may have to be taken by different persons at different periods of time or under different operating or experimental conditions. For example, there may be a large number of external conditions over which the experimenter has no control. Many of these uncontrolled external conditions may not affect the results of the experiment to any significant degree. However, some of them may change the outcome of the experiment appreciably. Such external conditions are commonly known as the factors.

The analysis of variance methodology is concerned with the investigation of the factors likely to contribute significant effects, by suitable choice of experiments. It is a technique by which variations associated with different factors or defined sources may be isolated and estimated. The procedure involves the division of total observed variation in the data into individual components attributable to various factors and those due to random or chance fluctuation, and performing tests of significance to determine which factors influence the experiment. The methodology was originally developed by Sir Ronald A. Fisher (1918, 1925, 1935) who gave it the name of "analysis of variance." The analysis of variance is the most widely used tool of modern (post-1950) statistics by research workers in the substantive fields of biology, psychology, sociology, education, agriculture, engineering, and so forth. The demand for knowledge and development of this topic has largely come from the aforementioned substantive fields. The development of analysis of variance methodology has in turn affected and influenced the types of experimental research being carried out in many fields. For example, in quantitative genetics which relies extensively on separating the total variation into environmental and genetic components, many of the concepts are directly linked to the principles and procedures of the analysis of variance. Nowadays, the analysis of variance models are widely used to analyze the effects of the independent variables under study on the dependent variable or response measure of interest.

The general synthesis of the analysis of variance procedure can be summarized as follows. Given a collection of n observations y_i's, we define the aggregate variation, called the total sum of squares, by

$$\sum_{i=1}^{n}(y_i - \bar{y})^2,$$

where

$$\bar{y} = \frac{\sum_{i=1}^{n} y_i}{n}.$$

Then the technique consists of partitioning the total sum of squares into component variations due to different factors also called sums of squares. For example, suppose there are Q such factors. Then the total sum of squares (SS_T) is partitioned as

$$SS_T = SS_A + SS_B + \cdots + SS_Q,$$

where $SS_A, SS_B, \ldots,$ and SS_Q represent the sums of squares associated with factors $A, B, \ldots,$ and Q, respectively and which account in some sense for the variation that can be attributed to these factors or sources of variation. The experiment is so designed that each sum of squares reflects the effect of just one factor or that of the random error attributed to chance. Furthermore, all such sums of squares are made comparable by dividing each by an appropriate number known as the degrees of freedom.[1] The quantities obtained by dividing each sum of squares by the corresponding degrees of freedom are called mean squares. The mean squares and other relevant quantities provide the basis for making statistical inference either in terms of a test of hypothesis or point and interval estimation. For example, the random distribution of the appropriate ratio of a pair of mean squares often permits certain tests to be made about the effects of the factors involved. In this way, we can decide whether the apparent effect of any one factor is readily explainable purely by chance.

The use of the term "analysis of variance" to describe the statistical methodology dealing with the means of a variable across groups of observations is not wholly satisfactory. The term seems to have its origin in the work of Fisher (1918) when he partitioned the total variance of a human attribute into components attributed to heredity, environment, and other factors; this led to an equation

$$\sigma^2 = \sigma_1^2 + \sigma_2^2 + \cdots + \sigma_k^2,$$

where σ^2 is the total variance and σ_i^2's are variances associated with different

---

[1] The degrees of freedom designate the number of statistically independent quantities (response variables) that comprise a sum of square (see Section 2.1 for further details).

factors. In such a case, the term "analysis of variance" is entirely appropriate. However, as Kempthorne (1976) states, this is rather a limited view of the entire statistical methodology falling under the nomen of "analysis of variance."[2]

1.1 HISTORICAL DEVELOPMENTS

Credit for much of the early developments in the field of the analysis of variance goes to Sir Ronald Aylmer Fisher[3] (1890–1962) who originally developed it in the 1920's.[4] He was the pioneer and innovator of the uses and applications of statistical methods in experimental design. Shortly after the end of World War I, Fisher resigned a public school teaching job and accepted a position at the Statistical Laboratory of the Rothamsted Agricultural Experimental Station in Harpenden, England. The center was heavily engaged in agricultural research and for many years Fisher was in charge of statistical data analysis there. He developed and employed the analysis of variance as the principal method of data analysis in experimental design. Frank Yates was Fisher's coworker at Rothamsted, and both of them collaborated on many important research projects. Yates also was primarily responsible for making many early contributions to the literature on analysis of variance and experimental design. Moreover, both Fisher and Yates made numerous indirect contributions through the staff of the Rothmsted Experimental Station.

On the retirement of Karl Pearson in 1933, Fisher was appointed Galton Professor at the University of London. Later, he moved to the faculty of Cambridge University to accept the Arthur Balfour Chair created by an endowment from the great founder of the science of eugenics. He also traveled abroad widely and held visiting professorships at several universities throughout the world. In 1936, Fisher visited the United States and received an honorary degree from the Harvard University on the occasion of its Tercentenary Celebrations. Early in 1937, he accepted the Honorary Fellowship of the Indian Statistical Institute and an invitation to preside over the first session of the Indian Statistical Congress held in January, 1938. In 1935, Fisher published a book, *The Design of Experiments*, in which logical and theoretical principles of experimental design were developed and expounded with a great variety of illustrative examples. The book has gone through numerous editions and has become a classic of statistical literature.

Fisher developed and used analysis of variance mainly during the 1920s and 1930s (see Fisher (1918, 1925, 1935)) for the study of data from agricultural experiments. From the beginning, Fisher employed both fixed effects and random

---

[2] Kendall et al. (1983, p. 2) point out that it would be more appropriate to call it "analysis of sum of squares," "but history and brevity are against this logical usage."

[3] The brief biographical sketch of Fisher given here has been drawn, plagiaristically in some ways, from Mahalanobis (1962).

[4] There had been some early work on analysis of variance carried out by W. H. R. A. Lexis and T. N. Thiele in the late nineteenth century (Kendal et al. (1983, p. 2)).

effects models (the latter, at least, when he treated intraclass correlations), and alluded to what has later been designated a "mixed model" in his book *The Design of Experiments*. The useful "analysis of variance table," including sum of squares, degrees of freedom, and mean squares for various sources of variation, was first published in a paper by Fisher and Mackenzie (1923). Tippett (1931) appears to have been the first to include a column of expected mean squares for variance components and to estimate these components.

Eisenhart (1947a) seems to have originated the terms "Model I" and "Model II" for fixed effects and random effects models, and he also mentions the possibility of a mixed model. Mixed models have been designated as Model III (see, e.g., Ostle and Malone (1988)). Some authors use Model III to denote random effects models in which random effects are drawn from a finite population of possible values (see, e.g., Dunn and Clark (1974, Chapter 9)). We have referred to these models as finite population models (see Chapter 9). Prior to Eisenhart (1947a), more formal treatments of the models had appeared in the papers by Daniels (1939) and Crump (1946) regarding random effects models, and by Jackson (1939) regarding the mixed model. However, a more complete treatment of the mixed model did not appear until Scheffé (1956a). The development of various models was very intensive in the 1940s and 1950s. These early developments in this field can be found in expository articles by Crump (1951) and Plackett (1960).

Most of the early applications of analysis of variance were in the field of agricultural sciences. Fisher's application of statistical theory in agricultural experiments brought many new developments and advances in the field. In fact, much of modern statistics originated to meet the research needs of agriculture experimental stations, and it is to this legacy that much of the terminology in the field is derived from agricultural experimentations. However, many of the experimental design terms such as "treatment," "plot," and "block" used earlier in an agricultural context, have lost their original meaning and are nowadays used in all areas of research.

Today, the methods of experimental design and analysis of variance are commonly used in nearly all fields of study and scientific investigation. Some of the disciplines where statistical design and analysis of experiments are routinely used include agriculture, biology, medicine, health, physical sciences, engineering, education, and social and behavioral sciences.

1.2 ANALYSIS OF VARIANCE MODELS

It is assumed that an analysis of variance model for the observations can be approximated by linear combinations (functions) of certain unobservable quantities known as "effects" corresponding to each factor of the experiment. The effects may be of two kinds: systematic or random. If the effect is systematic, it is called a fixed effect or Model I effect; otherwise it is called a random effect or Model II effect. The equation expressing the observations as a linear combination of the effects is known as a linear model.

Introduction

In every linear model there is at least one set of random effects, equal in number to the number of observations, a different one of which appears in every observation. This is called the residual effect or the error term. Furthermore, there is usually one fixed effect that appears in every model equation and is known as the general constant and is the mean, in some sense, of all the observations. Thus, a general linear model is represented as[5]

$$y_{ijk...q} = \mu + \alpha_i + \beta_j + \gamma_k + \cdots + e_{ijk...q}, \quad (1.2.1)$$

where $y_{ijk...q}$ is the observed score, $-\infty < \mu < \infty$ is the overall mean common to all the observations, $\alpha_i, \beta_j, \gamma_k, \ldots$ are unobservable effects attributable to different factors or sources of variation, and $e_{ijk...q}$ is an unobservable random error associated with the observation $y_{ijk...q}$ and is assumed to be independently distributed with mean zero and variance σ_e^2. Model (1.2.1) is called a fixed effects model, or Model I if the random effects in the model equation are only the error terms. Thus, under a fixed effects model, the quantities $\alpha_i, \beta_j, \gamma_k, \ldots$ are assumed to be constants. The objective in a fixed effects model is to make inferences about the unknown parameters $\mu, \alpha_i, \beta_j, \gamma_k, \ldots$, and σ_e^2. It is called a random effects model or a variance components model or Model II if all the effects in the model equation except the additive constant are random effects. Thus, under a random effects model, the quantities $\alpha_i, \beta_j, \gamma_k, \ldots$ are assumed to be random variables with means of zero and variances $\sigma_\alpha^2, \sigma_\beta^2, \sigma_\gamma^2, \ldots$, respectively. The objective in a random effect model is to make inferences about the variances $\sigma_\alpha^2, \sigma_\beta^2, \sigma_\gamma^2, \ldots, \sigma_e^2$ and/or certain functions of them. A case falling under none of these categories is called a mixed model or Model III. Thus, in a mixed model, some of the effects in the model equation are fixed and some are random. Mixed models contain a mixture of fixed and random effects and therefore represent a blend of the fixed and the random effects models. Mixed models include, as special cases, the fixed effects model in which all effects are assumed fixed except the error terms, and the random effects model in which all effects are assumed random except the general constant. The objective in a mixed model is to make inferences about the fixed effect parameters and the variances of the random effects. There are widespread applications of such models in a variety of substantive fields including genetics, animal husbandry, social sciences, and engineering.

Throughout this volume we consider analysis of variance based on Models I, II, and III. These are the most widely applicable models although other models have also been proposed (Tukey (1949a)).

---

[5] In a linear model it is customary to use a lower or an upper case Roman letter to represent a random effect and a Greek letter to designate a fixed effect. However, the practice is not universal, and some authors do just the opposite; that is, they use Greek letters to represent random effects and Roman letters to designate fixed effects (see, e.g., Kempthorne and Folks (1971, pp. 456–470)). In order to keep our notations simple and uniform, we use Greek letters to represent both fixed and random effects except the error or residual term which is denoted by the lower case Roman letter (e).

1.3 CONCEPT OF FIXED AND RANDOM EFFECTS

Whether an effect should be considered fixed or random will depend on the way the experimental treatments (levels of a factor) are selected and the kind of inferences one wishes to make from the analysis. In the fixed effects case, the experimenter is working with a systematically chosen set of treatments or levels of factors that are of particular intrinsic interest, and the inferences are to be made only about differences among the treatments actually being studied and about no other treatments that might have been included.

Thus, in the case of fixed effects, in advance of the actual experiment, the experimenter decides that he wants to see if differences in effects exist among some predetermined set of treatments or treatment combinations. His interest lies in these treatments or treatment combinations and no others. That is, treatments are chosen a priori by the researcher because they are of special interest to him. Each treatment of practicable interest to the experimenter has been included in the experiment, and the set of treatments or treatment combinations included in the experiment cover the entire set of treatments about which the experimenter wants to make inferences. The effect of any treatment is "fixed" in the sense that it must appear in any new trial of the experiment on other subjects or experimental units. Thus, model terms that represent treatments, blocks, and interactions are parameters.

An example of fixed effects may be an experiment in which the effects of different drugs on groups of animals are examined. Here, the treatments (drugs) are fixed and determined by the experimenter, and the interest being in the results of the treatments and the differences between them. This is also the case when we test the effects of different levels (doses) of a given factor such as a chemical or the amount of light to which a plant has been exposed. Another example of fixed effects would be a study of body weights for several age groups of animals. The treatments (levels) would be age groups that are fixed. Other sets of factors or variables that are usually considered fixed are types of disease, treatment therapy, gender, marital and socioeconomic status, and so forth.

In the random effects case, the experimenter is working with a randomly selected subset of treatments from a much larger population of all possible treatments about which the experimenter wants to make inferences. The subset of treatments included in the experiment do not exhaust the set of all possible treatments of interest. Thus, the treatment levels at which the experiment is conducted are not of interest in themselves, but rather they represent some of the many treatments on which the experiment could have been performed. Here, the effect of the treatment is not regarded as fixed, since any particular treatment itself need not be included each time the experiment is carried out. In such a case, in any repetition of the experiment a new sample of treatments is to be included. The experimenter may not actually plan to repeat the experiment, but conceptually each repetition involves a fresh sample of treatments. The interest of the researcher in such situations lies in determining whether different treatment levels yield different responses.

For an example of random effects, suppose that in a psychological experiment the personality of the experimenter herself may have an effect on the results. There may be a large group of people, each presumably having a distinct personality, who might possibly serve as experimenters. Trying out each such person in the experiment would not be practically feasible. So, instead, one draws a random sample from some chosen population of potential experimenters. In this case, each experimenter constitutes an experimental treatment given to one group of subjects assigned to her at random. Since the experimental treatments employed are themselves a random sample and inferences are to be made about experimenter effects extending to the population of potential experimenters, the effects are regarded as random. Other examples of factors or variables that are usually considered random are animals, days, subjects, plots, and so forth.

It is sometimes difficult to decide whether a given factor is fixed or random. The main distinction depends on whether the levels of the factor can be considered a random sample of a large collection of such levels of interest, or are fixed treatments whose differences the investigator wishes to investigate. Moreover, in many experimental studies involving the levels of a random factor, the researcher does not actually select a random sample from a large population of all the levels of interest. For example, animals, subjects, days, and the like, being studied are the ones that happen to be conveniently available. The usual assumption in such a situation is that the natural process leading to the availability of such cases is a random one involving no systematic bias and the cases being studied are sufficiently representative of their class type. However, if there are reasons to doubt the representativeness of the sample being studied, the estimation process will be biased, raising serious questions about the validity of the results.

1.4 FINITE AND INFINITE POPULATIONS

As mentioned earlier, in the case of random effects, treatments included in the experiment are usually assumed to be a random sample from a population of all possible treatments of interest. Such populations are generally considered to be of infinite size. However, in the definition of random effects it is not necessary to require that the population be of infinite size; it also could be finite. For example, a set of similar machines may be used for certain operations but measurements are obtained from only a limited number of them. In the case of a finite population, however, the population may be very large so that for all practical purposes the population may be considered to be infinite.

Thus, while considering random effects, one must distinguish between two cases: when the population is finite and when it is infinite — either because it is really so or because the finite population is sufficiently large so that for all practical purposes, it can be considered as infinite. In this volume we are primarily concerned with random effects arising out of infinite populations. These are probably the most common situations in many real-life applications.

The results dealing with the so-called finite population theory are treated only briefly. In the case of the so-called finite population theory, the analysis of variance is performed in the standard way but the results on expected mean squares are different. Knowing the expected mean squares, one can then decide which of the ratios should be used to test for the statistical significance of the factors or sources of variation of interest. The details on analyses and results covering finite population situations can be found in the works of Bennett and Franklin (1954, p. 404), McHugh and Mielke (1968), Gaylor and Hartwell (1969), Searle and Fawcett (1970), and Sahai (1974a).

1.5 GENERAL AND GENERALIZED LINEAR MODELS

In this volume and in many other books dealing with the analysis of variance and regression models, much of the theory and methodology being described makes the fundamental assumption of the normality of the error terms. The analysis of variance models along with the linear regression models are commonly known as the linear (statistical) models. The linear (statistical) models considered in this book are just one particular formulation of a model called the general linear model which presents a unified treatment of regression and analysis of variance models. A comprehensive treatment of statistical analysis based on general linear models can be found in Graybill (1961, 1976), Searle (1971b, 1987), Hocking (1985, 1996), Littell et al. (1991), Wang and Chow (1994), Rao and Toutenburg (1995), Christensen (1996), and Neter et al. (1990, 1996). Furthermore, in recent years, considerable effort has been devoted to the development of a wider class of linear models encompassing other probability distributions for their error structure. For example, in many medical and epidemiological works, a common type of outcome measure is a binary response and the mean response entails the binomial parameter. Similarly, in many health studies, the response variable is the observed number of cases from a certain rare disease, and, thus, the error structure has a Poisson distribution. Furthermore, in many environmental studies, the response variable often follows a gamma or inverse Gaussian distribution. The linear (statistical) models that allow for theory and methodology to be applicable to a much more general class of linear models, of which the normal theory is a special case, are known as generalized linear models. A considerable body of literature has been developed for these models and the interested reader is referred to the books by McCullagh and Nelder (1983, 1989), Dobson (1990), and Hinkley et al. (1991).

1.6 SCOPE OF THE BOOK

In this volume, we consider univariate analysis of variance models, that is, models with a single response variate. However, many applications of analysis of variance models involve simultaneous measurements on several correlated response variates and for statistical analysis of multivariate response data one uses multivariate analysis of variance, which generalizes the analysis of variance procedure for univariate normal populations to multivariate normal populations. The analysis of variance models with multivariate response are discussed in the works of Krishnaiah (1980), Anderson (1984), Morrison (1990), and Lindman (1992), among others. Similarly, statistical methods presented in this book are based on normal theory assumption for inference problems. The nonparametric analysis of variance procedures based on ranks, which do not require normal theory assumption, are not considered. There is a significant body of literature on nonparametric analysis of variance and the interested reader is referred to Conover (1971), Lehmann (1975), Daniel (1990), Sprent (1997), and Hollander and Wolfe (1998), among others. Finally, the main focus of this book is on classical analysis of variance procedures, where parameters are assumed to be unknown constants, and the accepted statistical techniques are based on the frequentist theory of hypothesis testing developed by Neyman and Pearson. The impact of the frequentist approach is reflected in the established use of type I and type II errors, and p-values and confidence intervals for statistical analysis. In contrast, there is a growing body of literature on the use of Bayesian theory of statistical inference in linear modes. In the Bayesian approach, all parameters are regarded as "random" in the sense that all uncertainty about them should be expressed in terms of a probability distribution. The basic paradigm of Bayesian statistics involves a choice of a joint prior distribution of all parameters of interest that could be based on objective evidence or subjective judgment or a combination of both. Evidence from experimental data is summarized by a likelihood function, and the joint prior distribution multiplied by the likelihood function is the (unnormalized) joint posterior density. The final (normalized) joint posterior distribution and its marginals form the basis of all Bayesian inference (see, e.g., Lee (1997)). The Bayesian method frequently provides more information and makes inference more readily attainable than the traditional frequentist approach. A drawback of the Bayesian approach is that it tends to be computationally intensive and, even for simple unbalanced mixed models, the computation of the joint posterior density of the parameters and its marginals involves high-dimensional integrals. Fortunately, as a result of recent advances in computing hardware and numerical algorithms, Bayesian methods have gained wide applicability, and the scope and complexity of Bayesian applications have greatly increased. Readers interested in the Bayesian approach to analysis of variance problems are referred to Box and Tiao (1973), Broemeling (1985), Schervish (1992), and Searle et al. (1992), among others.

2 One-Way Classification

2.0 PREVIEW

In this chapter we consider the analysis of variance associated with experiments having only one factor or experimental variable. Such an experimental layout is commonly known as one-way classification in which sample observations are classified (grouped) by only a single criterion. It provides the simplest data structure containing one or more observations at every level of a single factor. One-way classification is a very useful model in statistics. Many complex experimental situations can often be considered as one-way classification. In the succeeding chapters, we discuss situations involving two or more experimental variables.

2.1 MATHEMATICAL MODEL

Consider an experiment having a treatment groups or a different levels of a single factor. Suppose n observations have been made at each level giving a total of $N = an$ observations. Let y_{ij} be the observed score corresponding to the j-th observation at the i-th level or treatment group. The analysis of variance model for this experiment is given as

$$y_{ij} = \mu + \alpha_i + e_{ij} \quad (i = 1, \ldots, a; j = 1, \ldots, n), \qquad (2.1.1)$$

where $-\infty < \mu < \infty$ is the general or overall mean (true grand mean) common to all the observations, α_i is the effect due to the i-th level of the factor and e_{ij} is the random error associated with the j-th observation, at the i-th level or treatment group.

Model (2.1.1) states that the score for observation j at level i is based on the sum of three components: the true grand mean μ of all the treatment populations, the effect α_i associated with the particular treatment i, and a third part e_{ij} which is strictly peculiar to the j-th observation made under the i-th level or treatment group. The term e_{ij} takes into account all those factors that have not been included in model (2.1.1).

2.2 ASSUMPTIONS OF THE MODEL

Before one can use model (2.1.1) to make inferences about the existence of effects, certain assumptions must be made:

(i) The errors e_{ij}'s are assumed to be randomly distributed with mean zero and common variance σ_e^2.
(ii) The errors associated with any pair of observations are assumed to be uncorrelated; that is,

$$E(e_{ij}\ e_{i'j'}) = 0 \quad \begin{cases} i \neq i', j \neq j' \\ i = i', j \neq j' \end{cases} \quad (2.2.1)$$

(iii) Under Model I, the effects α_i's are assumed to be fixed constants subject to the constraint that $\sum_{i=1}^{a} \alpha_i = 0$. This implies that the observations y_{ij}'s are distributed with mean $\mu + \alpha_i$ and common variance σ_e^2.
(iv) Under Model II, the effects α_i's are also assumed to be randomly distributed with mean zero and variance σ_α^2. Furthermore, α_i's are uncorrelated with each other and each of the α_i's and e_{ij}'s are also uncorrelated; that is,

$$E(\alpha_i\ \alpha_{i'}) = 0, \quad i \neq i' \quad (2.2.2)$$

and

$$E(\alpha_i e_{ij}) = 0, \quad \text{for all } i\text{'s and } j\text{'s.}$$

Then from model (2.1.1), we have $\sigma_y^2 = \sigma_\alpha^2 + \sigma_e^2$ and so σ_α^2 and σ_e^2 are components of σ_y^2, the variance of an observation. Hence, σ_α^2 and σ_e^2 are called "components of variance" (see Appendix L). This implies that the observations y_{ij}'s are distributed with mean μ and common variance $\sigma_\alpha^2 + \sigma_e^2$.

Remarks: (i) Under Model I, the assumption that $\sum_{i=1}^{a} \alpha_i = 0$ can be made without any loss of generality since if $\sum_{i=1}^{a} \alpha_i$ were equal to some nonzero constant c, μ could be replaced by $\mu + c$ and $\sum_{i=1}^{a} \alpha_i$ would then be equal to zero. Similarly, e_{ij}'s could be assumed to have zero mean. Under Model II, as in the case of the fixed effects model, the assumption of a zero mean can also be made without any loss of generality. In this case, however, we do not assume that $\sum_{i=1}^{a} \alpha_i = 0$ since the α_i's are randomly selected from a population theoretically of infinite size, and their sum need not be zero. Instead we have $E(\alpha_i^2) = \sigma_\alpha^2$ and $E(\alpha_i \alpha_{i'}) = 0$ for $i \neq i'$.

(ii) Figure 2.1 is a schematic representation of the levels of a factor with a levels, under the assumptions of the fixed and random effects one-way analysis of variance model. Under the fixed effects (Model I) case, $\mu_1, \mu_2, \ldots, \mu_a$ are the means of preselected subpopulations fixed by the design. Under the random effects (Model II) case, $\mu_1, \mu_2, \ldots, \mu_a$ are the means of a randomly selected subpopulations from a population with mean μ and variance σ_α^2.

One-Way Classification

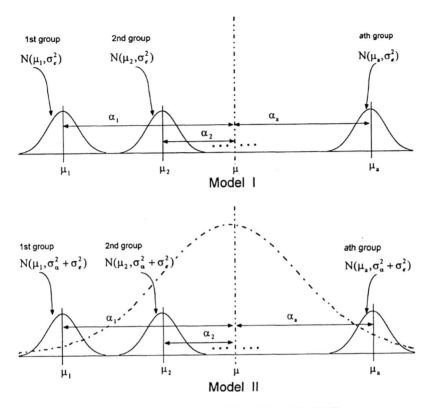

FIGURE 2.1 Schematic Representation of Model I and Model II.

(iii) One-way classification can be regarded as a one-way nested classification where a factor corresponding to "replications" or "samples" is nested within the levels of the treatment factor. Such a layout is often termed a two-stage nested design since it involves random sampling performed in two stages. One-way random effects models frequently arise in experiments involving a two-stage nested design. Some examples are as follows:

(a) A sample a of batches of toothpaste is selected at random from a production process comprising a large number of batches and a chemical analysis to determine its composition is performed on n samples taken from each batch.

(b) A sample a of tobacco leaves is selected from a shipment batch and a chemical analysis is performed on n samples taken from each leaf to determine the nicotine and tar content.

(c) A sample a of blocks is selected from a city having a large number of blocks and an interview is conducted on a sample of n individuals from each block to determine their voting preferences.

(d) A sample a of bales is selected from a shipment of bales and an analysis to determine its content purity is performed on n cores taken from each bale.

2.3 PARTITION OF THE TOTAL SUM OF SQUARES

Using the notation that a bar on the top and a dot in place of a suffix means that the particular suffix has been averaged out over the appropriate observations, we write the following identity

$$y_{ij} - \bar{y}_{..} = (\bar{y}_{i.} - \bar{y}_{..}) + (y_{ij} - \bar{y}_{i.}), \qquad (2.3.1)$$

where

$$\bar{y}_{i.} = \sum_{j=1}^{n} y_{ij}/n = y_{i.}/n \qquad (2.3.2)$$

and

$$\bar{y}_{..} = \sum_{i=1}^{a}\sum_{j=1}^{n} y_{ij}/an = y_{..}/an. \qquad (2.3.3)$$

Since it is an algebraic identity, equation (2.3.1) must hold for any value of y_{ij}. Squaring both sides of (2.3.1) and summing over i and j, we have

$$\sum_{i=1}^{a}\sum_{j=1}^{n}(y_{ij}-\bar{y}_{..})^2 = \sum_{i=1}^{a}\sum_{j=1}^{n}(\bar{y}_{i.}-\bar{y}_{..})^2 + \sum_{i=1}^{a}\sum_{j=1}^{n}(y_{ij}-\bar{y}_{i.})^2$$

$$+ 2\sum_{i=1}^{a}\sum_{j=1}^{n}(\bar{y}_{i.}-\bar{y}_{..})(y_{ij}-\bar{y}_{i.}). \qquad (2.3.4)$$

Now, note that the cross-product term vanishes; that is,

$$\sum_{i=1}^{a}\sum_{j=1}^{n}(\bar{y}_{i.}-\bar{y}_{..})(y_{ij}-\bar{y}_{i.}) = \sum_{i=1}^{a}(\bar{y}_{i.}-\bar{y}_{..})\sum_{j=1}^{n}(y_{ij}-\bar{y}_{i.}) = 0 \qquad (2.3.5)$$

and

$$\sum_{i=1}^{a}\sum_{j=1}^{n}(\bar{y}_{i.}-\bar{y}_{..})^2 = n\sum_{i=1}^{n}(\bar{y}_{i.}-\bar{y}_{..})^2. \qquad (2.3.6)$$

Then, equation (2.3.4) simplifies to the following identity[1]

$$\sum_{i=1}^{a}\sum_{j=1}^{n}(y_{ij}-\bar{y}_{..})^2 = n\sum_{i=1}^{a}(\bar{y}_{i.}-\bar{y}_{..})^2 + \sum_{i=1}^{a}\sum_{j=1}^{n}(y_{ij}-\bar{y}_{i.})^2. \qquad (2.3.7)$$

---

[1] For a geometric interpretation of the algebraic partition of the sum of squares given by identity (2.3.7), see Kendall et al. (1983, p.11).

One-Way Classification

Equation (2.3.7) states that the sum of the squared deviations of individual observations from the overall mean is equal to the sum of squared deviations of the group means from the overall or grand mean plus the sum of squared deviations of the observations from the group means. The term on the left of (2.3.7) is called the total sum of squares and will be abbreviated as SS_T. The first term to the right of (2.3.7) is called the sum of squares "between groups" or "due to treatments" and is abbreviated as SS_B; and the second term is called the sum of squares "within groups" and is abbreviated as SS_W. The second term represents the variation of the individual observations about their own sample means and it is sometimes called the error or residual sum of squares.

Remark: The meaning of the partition of the total sum of squares into between and within group sums of squares can be explained as follows. Individual observations in any sample will differ from each other or exhibit variability. These observed differences among individual observations can be ascribed to specific sources. First, some pairs of observations are in different treatment groups and their differences are due either to the different treatments, or to chance variation, or to both. The sum of squares between groups reflects the contribution of different treatments to intergroup differences as well as chance. On the other hand, observations in the same treatment group can differ only because of chance variation, since each observation within the group received exactly the same treatment. The sum of squares within groups reflects these intragroup differences due only to chance variation. Thus, in any sample, two kinds of variability — the sum of squares between groups, reflecting the variability due to treatments and chance, and the sum of squares within groups reflecting chance variation alone — can be isolated.

2.4 THE CONCEPT OF DEGREES OF FREEDOM

Before we proceed further, it is time to examine the notion of the degrees of freedom. Recall the basic definition of the sample variance S^2 of a random sample $y_1, y_2, \ldots, y_n$ as

$$S^2 = \frac{\sum_{i=1}^{n}(y_i - \bar{y})^2}{n-1} = \frac{\sum_{i=1}^{n} d_i^2}{n-1}, \qquad (2.4.1)$$

where

$$\bar{y} = \sum_{i=1}^{n} y_i/n \qquad (2.4.2)$$

and

$$d_i = y_i - \bar{y}. \qquad (2.4.3)$$

From (2.4.1) we note that the quantity S^2 is based upon a sum of squared deviations from the sample mean. However, we know that the sum of the deviations about the mean, defined by (2.4.3), must be zero; that is,

$$\sum_{i=1}^{n} d_i = 0. \tag{2.4.4}$$

The property (2.4.4) has a very important consequence. For example, suppose that in a random sample of $n = 4$, one is to choose all the deviations from the mean. For the first deviation, one can guess any number, say, $d_1 = 6$. Similarly, quite arbitrarily, one can assign two more deviations, say, $d_2 = -9$ and $d_3 = -7$. However, when one comes to choose the value of the fourth deviation, one is no longer free to take any number one pleases. The value of d_4 must be given by

$$d_4 = 0 - (d_1 + d_2 + d_3)$$
$$= 0 - (6 - 9 - 7)$$
$$= 10.$$

In short, given the values of $n - 1$ deviations from the mean, which could be any arbitrarily assigned numbers, the value of the last deviation is completely determined. Thus, we say that there are $n - 1$ degrees of freedom for a sample variance, reflecting the fact that only $n - 1$ deviations are "free" to be any number. Given the value of these "free" numbers, the last value is automatically determined.[2]

To obtain the degrees of freedom associated with different sums of squares, we note that since the total sum of squares (SS_T) is based on an deviations $y_{ij} - \bar{y}_{..}$'s of the an observations with one constraint on the deviations, namely,

$$\sum_{i=1}^{a} \sum_{j=1}^{n} (y_{ij} - \bar{y}_{..}) = 0, \tag{2.4.5}$$

it has $an - 1$ degrees of freedom. Similarly, since the between group sum of squares (SS_B) is computed from the deviations of the a independent class means $\bar{y}_{i.}$'s from the overall mean $\bar{y}_{..}$, but with one constraint, namely,

$$\sum_{i=1}^{a} (\bar{y}_{i.} - \bar{y}_{..}) = 0, \tag{2.4.6}$$

hence it has $a - 1$ degrees of freedom.

Finally, since the within sum of squares (SS_W) involves the deviations of the an observations from the a sample means, it has $an - a = a(n - 1)$ degrees

---

[2] For a geometric interpretation of degrees of freedom, see Walker (1940).

One-Way Classification

of freedom. Alternatively, the degrees of freedom corresponding to the within groups can be argued as follows. Consider the component of the within sum of squares corresponding to the i-th factor level, namely,

$$\sum_{j=1}^{n}(y_{ij} - \bar{y}_{i.})^2. \tag{2.4.7}$$

The expression (2.4.7) is the equivalent of a total sum of squares considering only the i-th factor level. Hence, there are $n - 1$ degrees of freedom associated with this sum of squares. Since SS_W is a sum of squares comprising component sums of squares, the i-th component being given by (2.4.7), the degrees of freedom associated with SS_W is the sum of the a component degrees of freedom, namely,

$$\sum_{i=1}^{a}(n-1) = a(n-1). \tag{2.4.8}$$

2.5 MEAN SQUARES AND THEIR EXPECTATIONS

The next question is how to use the partition of the total sum of squares and the corresponding degrees of freedom in making inferences about the existence of treatment effects. In the analysis of variance, ordinarily, it is convenient to deal with the quantities known as mean squares instead of sums of squares. The mean squares are obtained by dividing each sum of squares by the corresponding degrees of freedom. We denote the two mean squares, namely, between and within by MS_B and MS_W, respectively.

Next, we examine the expectations of within and between mean squares. From model equation (2.1.1), we obtain

$$\bar{y}_{i.} = \sum_{j=1}^{n}(\mu + \alpha_i + e_{ij})/n = \mu + \alpha_i + \bar{e}_{i.} \tag{2.5.1}$$

and

$$\bar{y}_{..} = \sum_{i=1}^{a}(\mu + \alpha_i + \bar{e}_{i.})/a = \mu + \bar{\alpha}_{.} + \bar{e}_{..}. \tag{2.5.2}$$

Substituting the values of y_{ij}, $\bar{y}_{i.}$, and $\bar{y}_{..}$ from (2.1.1), (2.5.1), and (2.5.2), respectively, into the expressions for SS_W and SS_B defined in (2.3.7), we find

$$SS_W = \sum_{i=1}^{a}\sum_{j=1}^{n}(e_{ij} - \bar{e}_{i.})^2 \tag{2.5.3}$$

and

$$SS_B = n \sum_{i=1}^{a} (\alpha_i - \bar{\alpha}_. + \bar{e}_{i.} - \bar{e}_{..})^2. \qquad (2.5.4)$$

Now, because the e_{ij}'s are uncorrelated and identically distributed with mean zero and variance σ_e^2, it follows, using the formulae of the variance of the sampling distribution of e_{ij}'s, that

$$E(e_{ij}^2) = \sigma_e^2, \qquad (2.5.5)$$

$$E(\bar{e}_{i.}^2) = \sigma_e^2/n, \qquad (2.5.6)$$

and

$$E(\bar{e}_{..}^2) = \sigma_e^2/an. \qquad (2.5.7)$$

It is then a matter of straightforward simplification to derive the expectations of mean squares. First, taking the expectation of (2.5.3), we get

$$E(SS_W) = \sum_{i=1}^{a} E\left[\sum_{j=1}^{n} e_{ij}^2 - 2\bar{e}_{i.} \sum_{j=1}^{n} e_{ij} + n\bar{e}_{i.}^2\right]$$

$$= \sum_{i=1}^{a} \left[\sum_{j=1}^{n} E(e_{ij}^2) - nE(\bar{e}_{i.}^2)\right]. \qquad (2.5.8)$$

On substituting the values of $E(e_{ij}^2)$ and $E(\bar{e}_{i.}^2)$ from (2.5.5) and (2.5.6), respectively, into (2.5.8), we find

$$E(SS_W) = \sum_{i=1}^{a} \left(n\sigma_e^2 - n\frac{\sigma_e^2}{n}\right)$$

$$= a(n-1)\sigma_e^2 \qquad (2.5.9)$$

The expectation of MS_W is, therefore, given by

$$E(MS_W) = E\left(\frac{SS_W}{a(n-1)}\right) = \sigma_e^2. \qquad (2.5.10)$$

Note that result (2.5.10) is true under both Models I and II. To derive the expectation of MS_B, we first consider the case of Model I and then that of Model II. First, if we restrict ourselves to Model I, then the α_i's are fixed quantities depending on the particular levels (treatments) selected in the experiment with the restriction that $\bar{\alpha}_. = 0$. Then, on taking the expectation of (2.5.4), we

One-Way Classification

get

$$E(SS_B) = n\left[\sum_{i=1}^{a} \alpha_i^2 + E\sum_{i=1}^{a}(\bar{e}_{i.} - \bar{e}_{..})^2\right], \quad (2.5.11)$$

by virtue of the fact that the α_i's are constants and the expectation of the cross-product term vanishes.

Now, using results (2.5.6) and (2.5.7), we find that

$$E\sum_{i=1}^{a}(\bar{e}_{i.} - \bar{e}_{..})^2 = \sum_{i=1}^{a} E(\bar{e}_{i.}^2) - aE(\bar{e}_{..}^2)$$

$$= a\frac{\sigma_e^2}{n} - a\frac{\sigma_e^2}{an}$$

$$= (a-1)\frac{\sigma_e^2}{n}. \quad (2.5.12)$$

Furthermore, substituting (2.5.12) into (2.5.11), we obtain

$$E(SS_B) = n\sum_{i=1}^{a} \alpha_i^2 + (a-1)\sigma_e^2. \quad (2.5.13)$$

Finally, the expectation of MS_B is given by

$$E(MS_B) = \frac{n}{a-1}\sum_{i=1}^{a}\alpha_i^2 + \sigma_e^2. \quad (2.5.14)$$

Next, consider the case of Model II when the α_i's are also randomly distributed with mean zero and variance σ_α^2, and the α_i's are uncorrelated with each other and with the e_{ij}'s. It then follows, using the formula for the variance of the sampling distribution of the mean of the α_i's, that

$$E(\alpha_i^2) = \sigma_\alpha^2 \quad (2.5.15)$$

and

$$E(\bar{\alpha}_.)^2 = \sigma_\alpha^2/a. \quad (2.5.16)$$

Now, on taking the expectation of (2.5.4), we get

$$E(SS_B) = n\left[E\sum_{i=1}^{a}(\alpha_i - \bar{\alpha}_.)^2 + E\sum_{i=1}^{a}(\bar{e}_{i.} - \bar{e}_{..})^2\right]. \quad (2.5.17)$$

Using results (2.5.15) and (2.5.16), we find that

$$E \sum_{i=1}^{a}(\alpha_i - \bar{\alpha}.)^2 = \sum_{i=1}^{a} E(\alpha_i^2) - a E(\bar{\alpha}.)^2$$

$$= a\sigma_\alpha^2 - a\frac{\sigma_\alpha^2}{a}$$

$$= (a-1)\sigma_\alpha^2. \tag{2.5.18}$$

On substituting (2.5.12) and (2.5.18) into (2.5.17), we get

$$E(\text{SS}_B) = n\left[(a-1)\sigma_\alpha^2 + (a-1)\frac{\sigma_e^2}{n}\right]$$

$$= n(a-1)\sigma_\alpha^2 + (a-1)\sigma_e^2. \tag{2.5.19}$$

Finally, the expectation of MS_B is given by

$$E(\text{MS}_B) = n\sigma_\alpha^2 + \sigma_e^2. \tag{2.5.20}$$

It is important to recognize that in the derivation of the results on expected sums of squares and mean squares, we have not made any distribution assumptions for e_{ij}'s under Model I and for e_{ij}'s as well as α_i's under Model II. This fact has important implications while procuring unbiased estimates for the parameters under both Models I and II.

2.6 SAMPLING DISTRIBUTION OF MEAN SQUARES

The between and within sums of squares or mean squares are functions of the sample observations, and thus must have a sampling distribution. However, to derive the form of their distributions, we require the assumption of normality for the random components of model (2.1.1). Thus, we assume that under Model I, the e_{ij}'s are completely independent and are normally distributed with mean zero and variance σ_e^2. Furthermore, under Model II, the α_i's are also completely independent of each other and of the e_{ij}'s, and are normally distributed with mean zero and variance σ_α^2.

Now, first of all, we note that for a normal parent population with variance σ^2

$$\frac{\hat{\sigma}^2}{\sigma^2} \sim \frac{\chi^2[\nu]}{\nu}, \tag{2.6.1}$$

where $\hat{\sigma}^2$ is an unbiased estimator of σ^2 based on ν degrees of freedom (see Appendix C). Furthermore, under both Models I and II, we have from (2.5.10) that

$$E(\text{MS}_W) = \sigma_e^2. \tag{2.6.2}$$

One-Way Classification

From (2.6.1) and (2.6.2), it follows that

$$\frac{\text{MS}_W}{\sigma_e^2} \sim \frac{\chi^2[a(n-1)]}{a(n-1)}; \tag{2.6.3}$$

that is, the ratio of MS_W to σ_e^2 is a $\chi^2[a(n-1)]$ variable divided by $a(n-1)$.

To derive the sampling distribution of MS_B, however, we have to distinguish between the cases of Models I and II. Under Model I, we have from (2.5.14) that

$$E(\text{MS}_B) = \frac{n}{a-1} \sum_{i=1}^{a} \alpha_i^2 + \sigma_e^2. \tag{2.6.4}$$

Therefore, MS_B is an unbiased estimator of σ_e^2 only when $\alpha_i = 0, i = 1, 2, \ldots, a$; that is, the effects of all the levels are the same. Hence, from (2.6.1) and (2.6.4), it follows that when $\alpha_i = 0$ $(i = 1, 2 \ldots, a)$ we have

$$\frac{\text{MS}_B}{\sigma_e^2} \sim \frac{\chi^2[a-1]}{a-1}; \tag{2.6.5}$$

that is, the ratio of MS_B to σ_e^2 is a $\chi^2[a-1]$ variable divided by $a-1$. It is important to note at this point that when the effects of all the levels are not the same,

$$\frac{\text{MS}_B}{\sigma_e^2} \sim \frac{\chi^{2'}[a-1, \lambda]}{a-1}, \tag{2.6.6}$$

where $\chi^{2'}[a-1, \lambda]$ represents a noncentral $\chi^2[a-1]$ variable with the noncentrality parameter λ given by

$$\lambda = \frac{n}{2\sigma_e^2} \sum_{i=1}^{a} \alpha_i^2. \tag{2.6.7}$$

The proof of this result is beyond the scope of this volume and is not presented here. However, the interested reader is referred to Appendix E for a definition of the noncentral chi-square variable.

Under Model II, we have from (2.5.20) that

$$E(\text{MS}_B) = \sigma_e^2 + n\sigma_\alpha^2. \tag{2.6.8}$$

It can further be shown that

$$\frac{\text{MS}_B}{\sigma_e^2 + n\sigma_\alpha^2} \sim \frac{\chi^2[a-1]}{a-1}; \tag{2.6.9}$$

that is, the ratio of MS_B to $\sigma_e^2 + n\sigma_\alpha^2$ is a $\chi^2[a-1]$ variable divided by $a-1$. The result (2.6.9) is true irrespective of whether $\sigma_\alpha^2 = 0$ or not. A proof of this result can be found in Graybill (1961, pp. 344–345; 1976, pp. 609–610) and Kempthorne and Folks (1971, pp. 467–470). Furthermore, it can be proven that under both Models I and II, the two statistics MS_B and MS_W are statistically independent. In other words, the value of MS_W, whether particularly large or small, gives us no information about whether the value of MS_B is particularly large or small. The mathematical proof that MS_B and MS_W are independent is rather involved and is not presented here (see, e.g., Graybill (1961, pp. 345–346; 1976, pp. 609–610); Scheffé (1959, Chapter 2)). Intuitively, one can argue as follows. Since SS_B is based solely on the group mean values, it has nothing to do with the individual variation within any group. Similarly, SS_W is based solely on the individual variation within groups (i.e., measured from their respective group means) and is, therefore, not affected whatever the group means happen to be.

2.7 TEST OF HYPOTHESIS: THE ANALYSIS OF VARIANCE F TEST

In this section, we present the usual hypothesis about the treatment effects and the appropriate F test for fixed and random effects models.

MODEL I (FIXED EFFECTS)

Under Model I, the usual null hypothesis of interest is that all the treatments (factor levels) have the same effect; that is,

$$H_0 : \alpha_1 = \alpha_2 = \cdots = \alpha_a = 0. \quad (2.7.1)$$

The alternative is

$$H_1 : \text{not all } \alpha_i\text{'s are zero.}$$

In order to develop a test statistic for the hypothesis (2.7.1), we note from (2.5.10) and (2.5.14) that when H_0 is true

$$E(MS_W) = \sigma_e^2 \quad (2.7.2)$$

and

$$E(MS_B) = \sigma_e^2; \quad (2.7.3)$$

that is, both the mean squares MS_W and MS_B are unbiased estimates of the same quantity σ_e^2. On the other hand, when (2.7.1) is false,

$$E(MS_B) > E(MS_W). \quad (2.7.4)$$

One-Way Classification

Furthermore, since under H_0 both of these mean squares divided by σ_e^2 are independently distributed as chi-square variables divided by their respective degrees of freedom, it follows that their ratio is distributed as Snedecor's F. Thus, the ratio

$$F = \frac{\text{MS}_B/\sigma_e^2}{\text{MS}_W/\sigma_e^2} = \frac{\text{MS}_B}{\text{MS}_W} \qquad (2.7.5)$$

is distributed as an F variable with $a-1$ and $a(n-1)$ degrees of freedom (see Appendix D). Hence, the null hypothesis (2.7.1) can be tested by computing the ratio (2.7.5) and comparing it directly with the one-tailed values in the tables of the F distribution with $a-1$ and $a(n-1)$ degrees of freedom.[3] An α-level is chosen in advance and if the calculated value is greater than $100(1-\alpha)$th percentage point of the F distribution with $a-1$ and $a(n-1)$ degrees of freedom, we may conclude that the hypothesis is false at the α-level of significance.

The formal hypothesis testing procedure involving fixed α as described is very useful and has led to many important developments in statistical theory. However, an alternative way to test the hypothesis (2.7.1) is in terms of the p-value. The p-value for a sample outcome is the probability of obtaining a result equal to or more extreme than the observed one. In this case, $p = P[F[a-1, a(n-1)] > F_0]$, where $F[a-1, a(n-1)]$ has an F distribution with $a-1$ and $a(n-1)$ degrees of freedom and F_0 is the observed value of the statistic (2.7.5). Larger p-values support H_0 and smaller p-values support H_1. A fixed α-level test can be carried out by comparing the p-value with the specified α-level. If the p-value is greater than the specified α, H_0 is concluded, otherwise not.[4] For a further discussion of p-value, see Gibbons and Pratt (1975) and Pratt and Gibbons (1981, pp. 23–32).

It should be observed that the F statistic defined in (2.7.5) always provides a one-tailed test of H_0 in terms of the sampling distribution of F. This is the case since under H_1

$$E(\text{MS}_B) > E(\text{MS}_W),$$

and thus the F statistic must show a value greater than one. The value of an F statistic less than one can signify nothing except the sampling error,[5] or perhaps nonrandomness of the sample observations, or violation of the assumptions.

---

[3] The F test considered here is designed for alternatives $\alpha_i \neq \alpha_{i'}$ for some pair (i, i') in all possible directions. For a discussion of monotone alternatives $\alpha_1 \leq \alpha_2 \leq \cdots \leq \alpha_a$, see Miller (1986, Section 3.1.3) and references cited therein.

[4] Hodges and Lehmann (1970, p. 317) suggest that one can consider the p-value as a "measure of the degree of surprise" which the experimental data should cause in the belief-of the null hypothesis. Miller (1986, p. 2) has termed the p-value a "measure of the credibility" of the null hypothesis. The smaller the value, the less likely one feels about the veracity of the null hypothesis.

[5] When the null hypothesis is true, one can expect a value of the F ratio less than one at least 50 percent of the time.

For example, if the measurements are not made in a random order, then any uncontrolled factor may have some varying effect on the sequence of experiments. This could cause an increase in within-group variance but may leave the between-group variance unaffected. Thus, if the value of the statistic (2.7.5) is obtained as significantly less than one, then it is possible that an important uncontrolled factor has not been randomized during the course of the experiment and much of the usefulness of the experimental results has been invalidated.

Remarks: (i) It should be noted that

$$E(F) \neq \frac{E(MS_B)}{E(MS_W)},$$

since, in general, the expected value of a ratio of two random variables is not equal to the ratio of the expected values of the random variables, even though the latter may be equal. Actually, it can be shown that when the null hypothesis is true

$$E(F) = \frac{a(n-1)}{a(n-1)-2}.$$

Thus, under H_0,

$$\frac{E(MS_B)}{E(MS_W)} = 1, \quad \text{but } E(F) > 1.$$

(ii) The analysis of variance model (2.1.1) may also be looked upon as a way to explain the variation in the dependent variable. The ratio of the between-group sum of squares to the total sum of squares gives the proportion of the total sum of squares accounted for by the linear model being posited and provides a measure of how well the model fits the data. A very low value indicates that the model fails to explain a lot of variation in the dependent variable and one may want to look for additional factors that may help to account for a higher proportion of the variation in the dependent varible.

MODEL II (RANDOM EFFECTS)

Under Model II if all the factor levels have the same effect in the population of random effects α_i's, then $\sigma_\alpha^2 = 0$. Hence, the Model II analogue of the null hypothesis (2.7.1) is

$$H_0 : \sigma_\alpha^2 = 0, \tag{2.7.6}$$

against the alternative

$$H_1 : \sigma_\alpha^2 > 0. \tag{2.7.7}$$

Again, from (2.5.10) and (2.5.20), it follows that when H_0 is true,

$$E(MS_W) = \sigma_e^2 \tag{2.7.8}$$

One-Way Classification

and

$$E(\text{MS}_B) = \sigma_e^2; \qquad (2.7.9)$$

that is, both the mean squares MS_B and MS_W are unbiased for the error variance σ_e^2. Furthermore, as stated in the preceding section, $\text{MS}_B/(\sigma_e^2 + n\sigma_\alpha^2)$ is distributed as a chi-square variable divided by $a - 1$ and MS_B is statistically independent of MS_W. Thus, the ratio

$$\frac{\text{MS}_B/(\sigma_e^2 + n\sigma_\alpha^2)}{\text{MS}_W/\sigma_e^2} = \left(1 + n\frac{\sigma_\alpha^2}{\sigma_e^2}\right)^{-1} \frac{\text{MS}_B}{\text{MS}_W} \qquad (2.7.10)$$

is distributed as an F variable with $a - 1$ and $a(n - 1)$ degrees of freedom and, when H_0 is true, the statistic MS_B/MS_W has an F distribution. Hence, the same test statistic, as used in the case of Model I, can be employed to test the hypothesis (2.7.6).

Remarks: (i) Sometimes, it is quite likely that the hypothesis (2.7.6) versus (2.7.7) may not be a realistic choice. For example, it may be thought that some differences between the factor levels (groups) are almost certain to exist, and then it makes little sense to test the hypothesis of no difference. However, the researcher may want to see whether $\sigma_\alpha^2 \leq \sigma_e^2/2$; that is, if the variability between groups is half or less than the variability within groups. In other words, a researcher may want to test a hypothesis of the type:

$$H_0': \sigma_\alpha^2/\sigma_e^2 \leq \rho_o \quad \text{vs.} \quad H_1': \sigma_\alpha^2/\sigma_e^2 > \rho_o, \qquad (2.7.11)$$

where ρ_o is a specific value of $\sigma_\alpha^2/\sigma_e^2$. In this case, the statistic (2.7.10) with $\sigma_\alpha^2/\sigma_e^2 = \rho_o$ provides the proper F test with large values of the statistic providing a ground for rejection. Thus, the test consists of rejecting H_0' if $F_{obs} > (1 + n\rho_o) F[a - 1, a(n - 1); 1 - \alpha]$.[6]

(ii) One is occasionally interested in testing the hypothesis that the overall mean (μ) is equal to some given constant μ_o. The test can be performed by considering the quantity $\text{MS}_o = an(\bar{y}_{..} - \mu_o)^2$, which has one degree of freedom. Furthermore, it can be shown that

$$E(\text{MS}_o) = \begin{cases} \sigma_e^2 + an(\mu - \mu_o)^2, & \text{for Model I} \\ \sigma_e^2 + an(\mu - \mu_o)^2 + n\sigma_\alpha^2, & \text{for Model II.} \end{cases}$$

Thus, under Model I, the hypothesis can be based on the ratio MS_o/MS_W which has an F distribution with 1 and $a(n - 1)$ degrees of freedom. While, under Model II, the hypothesis can be based on the ratio MS_o/MS_B which has an F distribution with 1 and $a - 1$ degrees of freedom. It should be noticed that in this example Models I and II have led to different significance tests.

---

[6] Spjotvoll (1967) has studied the structure of optimum tests of the hypothesis of the type (2.7.11).

TABLE 2.1
Analysis of Variance for Model (2.1.1)

Source of Variation	Degrees of Freedom	Sum of Squares	Mean Square	Expected Mean Square Model I	Expected Mean Square Model II	F Value
Between	$a-1$	SS_B	MS_B	$\sigma_e^2 + \dfrac{n\sum_{i=1}^{a}\alpha_i^2}{a-1}$	$\sigma_e^2 + n\sigma_\alpha^2$	MS_B/MS_W
Within	$a(n-1)$	SS_W	MS_W	σ_e^2	σ_e^2	
Total	$an-1$	SS_T				

2.8 ANALYSIS OF VARIANCE TABLE

The results on the partition of the total sum of squares, degrees of freedom, expected mean squares, and the analysis of variance F test are usually summarized in the form of a table commonly referred to as the analysis of variance table. The table shows in a certain order the sums of squares and other related quantities used in the computation of the F test. Such a table greatly simplifies the arithmetic and algebraic details of the analysis, which tend to become rather complicated in more complex designs.

In an analysis of variance table, the first column, designated as the *source of variation*, represents the partitioning of the total response variation into the various components included in the linear model. The second column, designated as *degrees of freedom*, partitions the sample size into various components that relate the amount of information corresponding to each source of variation of the model. The third column, designated as *sum of squares*, contains the sums of squares associated with various components or sources of variation of the model. The fourth column, designated as *mean square*, lists the respective sums of squares divided by the corresponding degrees of freedom. The fifth column, designated as *expected mean square*, contains expected mean squares that represent expected values of the mean squares derived under the assumption of an analysis of variance model. The sixth column, designated as *F value*, contains the values of the F ratios which are generally formed by taking ratios of two mean squares. In most of the worked examples presented in this volume, we have generally added a seventh column, designated as *p-value*, which contains probabilities of obtaining a result equal to or more extreme than the observed F ratios if the null hypothesis were true.

Table 2.1 shows the general form of the analysis of variance table for the one-way classification model (2.1.1).

2.9 POINT ESTIMATION: ESTIMATION OF TREATMENT EFFECTS AND VARIANCE COMPONENTS

It should be recognized that performance of the F test and construction of the analysis of variance table by no means complete all the inferences the investigator may want to draw. The experimenter's main objective is not always

One-Way Classification

to test the equality of all the treatment means. In many instances, he may want to estimate various parameters or functions of parameters. Under fixed effects, the parameters of the model (2.1.1) are $\mu, \alpha_1, \ldots, \alpha_a$ and σ_e^2 which can be estimated from the sample data. The random effects version of the model (2.1.1) involves the parameters μ, σ_e^2, and σ_α^2 which also can be estimated.

As estimators of the parameters, we consider the best linear and best quadratic unbiased estimators. Under Model I, we have

$$E(y_{ij}) = \mu + \alpha_i. \tag{2.9.1}$$

Then it is straightforward to see that

$$E(\bar{y}_{i.}) = \frac{(\mu + \alpha_i) + \cdots + (\mu + \alpha_i)}{n}$$

$$= \mu + \alpha_i \tag{2.9.2}$$

and

$$E(\bar{y}_{..}) = \frac{(\mu + \alpha_1) + \cdots + (\mu + \alpha_1) + \cdots + (\mu + \alpha_a) + \cdots + (\mu + \alpha_a)}{an}$$

$$= \mu, \tag{2.9.3}$$

since generally the α_i's are chosen such that $\sum_{i=1}^{a} \alpha_i = 0$. From (2.9.2) and (2.9.3), the unbiased estimates of μ and α_i are

$$\hat{\mu} = \bar{y}_{..} \tag{2.9.4}$$

and

$$\hat{\alpha}_i = \bar{y}_{i.} - \bar{y}_{..}. \tag{2.9.5}$$

It can be shown that (2.9.4) and (2.9.5) are the so-called best linear unbiased estimates (BLUE) for μ and α_i, respectively. With the additional assumption of normality they are the best unbiased estimates. Furthermore, from (2.5.10), it follows that MS_W is an unbiased estimate for σ_e^2. In addition, it can be shown that MS_W is the best quadratic unbiased estimate of σ_e^2 and if the e_{ij}'s are normally distributed, MS_W is the best unbiased estimate (see Graybill (1954)).

Under Model II, instead of estimating the effects directly by taking differences of the treatment means from the grand mean as in the case of Model I, the problem of major interest is the estimation of the components of variance σ_α^2 and σ_e^2. One set of estimators of σ_α^2 and σ_e^2 are immediately obtained using the standard method of moment estimation based on the expected mean squares appearing in the analysis of variance table. Thus, from Table 2.1, we obtain that

$$E(\mathrm{MS}_W) = \sigma_e^2 \qquad (2.9.6)$$

and

$$E(\mathrm{MS}_B) = \sigma_e^2 + n\sigma_\alpha^2. \qquad (2.9.7)$$

Hence, it follows that

$$\hat{\sigma}_e^2 = \mathrm{MS}_W \qquad (2.9.8)$$

and

$$\hat{\sigma}_\alpha^2 = (\mathrm{MS}_B - \mathrm{MS}_W)/n \qquad (2.9.9)$$

are unbiased estimators of σ_e^2 and σ_α^2, respectively.

Remarks: (i) It can be shown that the estimators (2.9.8) and (2.9.9) are also the maximum likelihood estimators[7] (corrected for bias) of the corresponding parameters (see, e.g., Graybill (1961, pp. 338–344)). Furthermore, it can be proven that in the class of all quadratic unbiased estimators (quadratic functions of the observations), the estimators (2.9.8) and (2.9.9) have minimum variance (see Graybill and Hultquist (1961); Graybill (1976, pp. 614–615)).

(ii) The aforesaid property of the 'minimum variance quadratic unbiased estimation' of the above estimators of variance components holds irrespective of the form of the distribution of the random effects α_i's and e_{ij}'s. Moreover, with the additional assumption of normality, the estimators (2.9.8) and (2.9.9) can be shown to have minimum variance within the class of all unbiased estimators (see Graybill (1954); Graybill and Wortham (1956)). Thus, the estimators (2.9.8) and (2.9.9) have certain optimal properties.[8] Nevertheless, the estimate of σ_α^2 can be negative.

(iii) It is clearly embarrassing to estimate a variance component as a negative number, which, by definition, is a nonnegative quantity. Several courses of action, however, are available. One procedure is to accept the negative estimate as an indication that the true population value of the variance component is close to zero, assuming of course that the sampling variability produced a negative estimate. This seems to have some intuitive appeal but the replacement of a negative estimate by zero affects some statistical properties such as unbiasedness of the estimates. Alternatively, one may reestimate the variance components using methods of estimation that always produce nonnegative estimates. Still, another alternative is to interpret the negative estimate as an indication that the assumed model is wrong. A full discussion of these results is beyond the scope of

---

[7] For the nonnegative maximum likelihood estimators, see Sahai and Thompson (1973).
[8] There are biased estimators that have more desirable properties in terms of the mean squared error criterion.

One-Way Classification

this volume. The interested reader is referred to the survey papers by Harville (1969, 1977), Searle (1971a), and Khuri and Sahai (1985) and books by Searle (1971b) and Searle et al. (1992) which contain ample discussions of this and other related topics.

Knowing the estimates of σ_e^2 and σ_α^2, the estimate of the total variance σ_y^2 is obtained as

$$\hat{\sigma}_y^2 = \hat{\sigma}_e^2 + \hat{\sigma}_\alpha^2. \tag{2.9.10}$$

The fact that the total variance consists of σ_e^2 and σ_α^2 permits one to make a somewhat more informative use of the estimates of σ_e^2 and σ_α^2. We can take the ratio of the estimated σ_α^2 to the estimated total variance ($\hat{\sigma}_y^2$) to find the estimated proportion of variance accounted for by the factor levels. It is highly informative to estimate variance components individually and to employ them in evaluating proportions of variance accounted for by different factors. Such proportions give one of the best ways to decide if a factor is a predictably important one. For example, it is entirely possible for a given factor to give statistically significant results in a study, even though only a very small percentage of variance is attributable to that factor. This is most likely to happen, of course, if the sample n is very large. On the other hand, when there is significant evidence for effects of a factor and the factor also accounts for a relatively large percentage of variance, then this information may be an important instrument in interpreting the experimental results or in deciding how the experimental findings might be applied. Thus, in Model II experiments, when the levels of a factor are sampled, it is a good practice to estimate the components of variance, and to judge the significance of the factor on the basis of the explained variation in addition to the results of the F test.

In many areas of research, particularly in genetics and industrial work, interest may center on the estimation of the variance ratio[9] $\theta = \sigma_\alpha^2/\sigma_e^2$, or the intraclass correlation defined by $\rho = \sigma_\alpha^2/(\sigma_e^2 + \sigma_\alpha^2)$ (see Appendix M for a definition of the intraclass correlation). It can be shown that the uniformly minimum variance unbiased (UMVU) estimator of θ is given by (see, e.g., Graybill (1961, pp. 378–379); Winer et al. (1991, p. 97)).

$$\begin{aligned}\hat{\theta} &= \frac{1}{n}\left[\frac{\mathrm{MS}_B}{\mathrm{MS}_W} \cdot \frac{a(n-1)-2}{a(n-1)} - 1\right]\\ &= \frac{[a(n-1)-2]\mathrm{MS}_B - a(n-1)\mathrm{MS}_W}{an(n-1)\mathrm{MS}_W}\\ &= \frac{\mathrm{MS}_B - m\,\mathrm{MS}_W}{m\,n\,\mathrm{MS}_W},\end{aligned} \tag{2.9.11}$$

---

[9] The ratio $\theta = \sigma_\alpha^2/\sigma_e^2$ measures the size of the population variability relative to the error variability present in the data.

where

$$m = \frac{a(n-1)}{a(n-1)-2}.$$

By way of contrast

$$\frac{\hat{\sigma}_\alpha^2}{\hat{\sigma}_e^2} = \frac{MS_B - MS_W}{n MS_W}. \tag{2.9.12}$$

To get an idea of the order of the magnitude of the bias in the estimator (2.9.12), consider the following data (Winer et al. 1991, pp. 97–98):

$$a = 4, \quad n = 10,$$
$$MS_B = 250, \quad MS_W = 50.$$

For this example,

$$\hat{\sigma}_e^2 = 50 \quad \text{and} \quad \hat{\sigma}_\alpha^2 = \frac{1}{10}(250 - 50) = 20.$$

Hence,

$$\frac{\hat{\sigma}_\alpha^2}{\hat{\sigma}_e^2} = \frac{20}{50} = 0.40.$$

Whereas, from (2.9.11), we have

$$\hat{\theta} = \frac{250 - \frac{36}{34}(50)}{\left(\frac{360}{34}\right)(50)} = 0.372.$$

Thus, the estimator (2.9.12) is slightly positively biased in relation to the estimator (2.9.11).

The UMVU estimator of the intraclass correlation cannot be expressed in closed form (Olkin and Pratt 1958). A computer program to calculate the UMVU estimator is given by Donoghue and Collins (1990). A biased estimator, however, can be obtained by substituting the estimates of the individual components in the formula for the intraclass correlation. Thus,

$$\hat{\rho} = \frac{\hat{\sigma}_\alpha^2}{\hat{\sigma}_e^2 + \hat{\sigma}_\alpha^2} = \frac{MS_B - MS_W}{MS_B + (n-1)MS_W}. \tag{2.9.13}$$

One-Way Classification

The estimator (2.9.13), however, can produce a negative estimate. Alternatively, an estimate of the intraclass correlation in terms of the unbiased estimate of θ is

$$\hat{\rho}' = \frac{\hat{\theta}}{1+\hat{\theta}}. \qquad (2.9.14)$$

For the numerical example in the preceding paragraph, the estimator (2.9.13) gives

$$\hat{\rho} = \frac{20}{50+20} = 0.286,$$

whereas the alternative estimator (2.9.14) in terms of the estimator of θ yields

$$\hat{\rho}' = \frac{0.372}{1+0.372} = 0.271.$$

Hence, the latter estimator ($\hat{\rho}'$) slightly underestimates ρ in relation to the estimate provided by $\hat{\rho}$. For a review of other estimators of ρ, see Donner (1986).

2.10 CONFIDENCE INTERVALS FOR VARIANCE COMPONENTS

An examination of the mean square and the expected mean square columns of the analysis of variance Table 2.1 suggests which quantities can be readily estimated by a confidence interval. First, note that each of the entries of the mean square column can be transformed into a chi-square variable by multiplying it by the corresponding degrees of freedom and then dividing by the corresponding expected mean square value. This chi-square variable can then be used to obtain a confidence interval for the quantity appearing in the expected mean squares column. Thus, a $100(1-\alpha)$ percent confidence interval for σ_e^2 can be obtained by noting that

$$\frac{a(n-1)\mathrm{MS}_W}{\sigma_e^2} \sim \chi^2[a(n-1)]. \qquad (2.10.1)$$

The desired confidence interval is then given by

$$P\left[\frac{a(n-1)\mathrm{MS}_W}{\chi^2[a(n-1),1-\alpha/2]} < \sigma_e^2 < \frac{a(n-1)\mathrm{MS}_W}{\chi^2[a(n-1),\alpha/2]}\right] = 1-\alpha, \qquad (2.10.2)$$

where $\chi^2[a(n-1), 1-\alpha/2]$ and $\chi^2[a(n-1), \alpha/2]$ denote the $100(1-\alpha/2)$th and $100(\alpha/2)$th percentage points of the chi-square distribution with $a(n-1)$ degrees of freedom.[10] Similarly, a $100(1-\alpha)$ percent confidence interval for $\sigma_e^2 + n\sigma_\alpha^2$ can be obtained by noting that

$$\frac{(a-1)\text{MS}_B}{\sigma_e^2 + n\sigma_\alpha^2} \sim \chi^2[a-1]. \tag{2.10.3}$$

The desired confidence interval is then given by

$$P\left[\frac{(a-1)\text{MS}_B}{\chi^2[a-1, 1-\alpha/2]} < \sigma_e^2 + n\sigma_\alpha^2 < \frac{(a-1)\text{MS}_B}{\chi^2[a-1, \alpha/2]}\right] = 1-\alpha, \tag{2.10.4}$$

where again $\chi^2[a-1, 1-\alpha/2]$ and $\chi^2[a-1, \alpha/2]$ represent the $100(1-\alpha/2)$th and $100(\alpha/2)$th percentage points of the chi-square distribution with $(a-1)$ degrees of freedom.

Unfortunately, the expressions in the expected mean square column are the only quantities for which one can obtain exact confidence intervals by the procedure just described. In particular, an exact confidence interval for σ_α^2 does not exist. Various approximate confidence intervals have been proposed in the literature. For a detailed discussion of these procedures and their relative merits, the reader is referred to Boardman (1974) and Burdick and Graybill (1988, 1992, pp. 60–63). Here, we briefly describe some procedures that have been recommended for the problem. A conservative $100(1-2\alpha)$ percent confidence interval based on the distributions of MS_B and MS_B/MS_W is obtained as (see, e.g., Williams (1962); Graybill (1976, pp. 618–620)):

$$P\left[\frac{(a-1)\text{MS}_B}{n\chi^2[a-1, 1-\alpha/2]}\left(1 - \frac{F[a-1, a(n-1); 1-\alpha/2]}{F^*}\right)\right.$$
$$\left. < \sigma_\alpha^2 < \frac{(a-1)\text{MS}_B}{n\chi^2[a-1, \alpha/2]}\left(1 - \frac{F[a-1, a(n-1); \alpha/2]}{F^*}\right)\right] \geq 1-2\alpha, \tag{2.10.5}$$

where $F^* = \text{MS}_B/\text{MS}_W$. The empirical evidence seems to indicate that the probability $1-2\alpha$ in (2.10.5) can be replaced by $1-\alpha$. Similarly, two approximate $100(1-\alpha)$ percent confidence intervals based on the distribution of the ratio of mean squares are obtained as (see, e.g., Bulmer (1957); Scheffé (1959,

---

[10] A slightly shorter confidence interval could be obtained by considering unequal probabilities in each tail. Tate and Klett (1959) and Murdock and Williford (1977) provide tables of chi-square values that provide shortest two-sided intervals.

One-Way Classification

p. 235); Searle (1971b, p. 414)).[11]:

$$P\left[\frac{\text{MS}_W}{nF[a-1,\infty;1-\alpha/2]}\left(\frac{\text{MS}_B}{\text{MS}_W}-F[a-1,a(n-1);1-\alpha/2]\right)\right.$$
$$\left.<\sigma_\alpha^2<\frac{\text{MS}_W}{nF[a-1,\infty;\alpha/2]}\left(\frac{\text{MS}_B}{\text{MS}_W}-F[a-1,a(n-1);\alpha/2]\right)\right]\doteq 1-\alpha.$$
(2.10.6)

and

$$P\left[\frac{\text{MS}_W(F^*-F_2)(F^*+F_2-F_2')}{nF^*F_2'}\right.$$
$$\left.<\sigma_\alpha^2<\frac{\text{MS}_W(F^*-F_1)(F^*+F_1-F_1')}{nF^*F_1'}\right]\doteq 1-\alpha. \quad (2.10.7)$$

where $F^* = \text{MS}_B/\text{MS}_W$, $F_1 = F[a-1, a(n-1); \alpha/2]$, $F_2 = F[a-1, a(n-1); 1-\alpha/2]$, $F_1' = F[a-1, \infty; \alpha/2]$, and $F_2' = F[a-1, \infty, 1-\alpha/2]$.

Finally, an approximate procedure that seems to provide a shorter interval and has better coverage property is given by (Ting et al. (1990); Burdick and Graybill (1992, pp. 60–61)):

$$P\left\{\frac{1}{n}(\text{MS}_B - \text{MS}_W - \sqrt{V_L}) \leq \sigma_\alpha^2 \leq \frac{1}{n}(\text{MS}_B - \text{MS}_W + \sqrt{V_U})\right\} \doteq 1-\alpha,$$
(2.10.8)

where

$$V_L = G_1^2\text{MS}_B^2 + H_2^2\text{MS}_W^2 + G_{12}\text{MS}_B\text{MS}_W,$$
$$V_U = H_1^2\text{MS}_B^2 + G_2^2\text{MS}_W^2 + H_{12}\text{MS}_B\text{MS}_W,$$

with

$$G_1 = 1 - F^{-1}[a-1, \infty; 1-\alpha/2], \quad G_2 = 1 - F^{-1}[a(n-1), \infty; 1-\alpha/2],$$
$$H_1 = F^{-1}[a-1, \infty; \alpha/2] - 1, \quad H_2 = F^{-1}[a(n-1), \infty; \alpha/2] - 1,$$
$$G_{12} = \frac{(F[a-1, a(n-1); 1-\alpha/2] - 1)^2 - G_1^2 F^2[a-1, a(n-1); 1-\alpha/2] - H_2^2}{F[a-1, a(n-1); 1-\alpha/2]},$$

---

[11] A confidence interval for σ_α^2 that is robust to departures from normality can be obtained using the jackknife technique (Arvesen and Schmitz (1970); Miller (1986, Section 3.6.3)).

and

$$H_{12} = \frac{(1 - F[a-1, a(n-1); \alpha/2])^2 - H_1^2 F^2[a-1, a(n-1); \alpha/2] - G_2^2}{F[a-1, a(n-1); \alpha/2]}.$$

Although we have only approximate procedures for confidence limits for σ_α^2, we are able to obtain exact confidence limits for the ratio $\sigma_\alpha^2/\sigma_e^2$, the intraclass correlation $\sigma_\alpha^2/(\sigma_e^2 + \sigma_\alpha^2)$, and $\sigma_e^2/(\sigma_e^2 + \sigma_\alpha^2)$. Using the distribution result in (2.7.10); that is,

$$\left(1 + n\frac{\sigma_\alpha^2}{\sigma_e^2}\right)^{-1} \frac{\text{MS}_B}{\text{MS}_W} \sim F[a-1, a(n-1)], \qquad (2.10.9)$$

it can be shown that the probability is $1 - \alpha$ that

$$\frac{1}{n}\left\{\frac{\text{MS}_B}{\text{MS}_W} \cdot \frac{1}{F[a-1, a(n-1); 1-\alpha/2]} - 1\right\}$$
$$< \frac{\sigma_\alpha^2}{\sigma_e^2} < \frac{1}{n}\left\{\frac{\text{MS}_B}{\text{MS}_W} \cdot \frac{1}{F[a-1, a(n-1); \alpha/2]} - 1\right\}. \qquad (2.10.10)$$

Furthermore, on rearranging the inequalities in (2.10.10), it readily follows that with probability $1 - \alpha$, the following relations hold:

$$\frac{F^* - F[a-1, a(n-1); 1-\alpha/2]}{F^* + (n-1)F[a-1, a(n-1); 1-\alpha/2]}$$
$$< \frac{\sigma_\alpha^2}{\sigma_e^2 + \sigma_\alpha^2} < \frac{F^* - F[a-1, a(n-1); \alpha/2]}{F^* + (n-1)F[a-1, a(n-1); \alpha/2]} \qquad (2.10.11)$$

and

$$\frac{nF[a-1, a(n-1); \alpha/2]}{F^* + (n-1)F[a-1, a(n-1); \alpha/2]}$$
$$< \frac{\sigma_e^2}{\sigma_e^2 + \sigma_\alpha^2} < \frac{nF[a-1, a(n-1); 1-\alpha/2]}{F^* + (n-1)F[a-1, a(n-1); 1-\alpha/2]}, \qquad (2.10.12)$$

where $F^* = \text{MS}_B/\text{MS}_W$. The inequalities (2.10.11) and (2.10.12) provide exact confidence limits for the intraclass correlation (ρ) and $1 - \rho$, respectively. Since $\rho \geq 0$, negative limits are defined to be zero.

Remark: Singhal (1987) and Groggel et al. (1988) provide methods for determining approximate confidence intervals for ρ under the assumption of nonnormality.

One-Way Classification

2.11 COMPUTATIONAL FORMULAE AND PROCEDURE

For performing analysis of variance, packaged computer programs are widely available for handling calculations that would have been highly tedious or simply not feasible in the precomputer age. It is assumed that computer software is used in the handling of analysis of variance computations for all but the simplest data sets. For hand calculations, however, the definitional formulae for SS_T, SS_B, and SS_W given in Section 2.3 are usually not very convenient. In the following we give useful computational formulae, which are algebraically identical to the definitional formulae. Thus,

$$\text{Total Sum of Squares } (SS_T) = \sum_{i=1}^{a} \sum_{j=1}^{n} (y_{ij} - \bar{y}_{..})^2$$

$$= \sum_{i=1}^{a} \sum_{j=1}^{n} y_{ij}^2 - \frac{y_{..}^2}{an}, \quad (2.11.1)$$

$$\text{Between Group Sum of Squares } (SS_B) = n \sum_{i=1}^{a} (\bar{y}_{i.} - \bar{y}_{..})^2$$

$$= \frac{1}{n} \sum_{i=1}^{a} y_{i.}^2 - \frac{y_{..}^2}{an}, \quad (2.11.2)$$

and

$$\text{Within Group Sum of Squares } (SS_W) = \sum_{i=1}^{a} \sum_{j=1}^{n} (y_{ij} - \bar{y}_{i.})^2$$

$$= \sum_{i=1}^{a} \sum_{j=1}^{n} y_{ij}^2 - \frac{1}{n} \sum_{i=1}^{a} y_{i.}^2. \quad (2.11.3)$$

It should be noted that the within-group sum of squares (SS_W) can also be computed by making use of the identity (2.3.7), giving

$$SS_W = SS_T - SS_B. \quad (2.11.4)$$

The relation (2.11.4) can further be used to check the validity of the earlier computations.

Now, the steps in the analysis of variance computations can be summarized as follows:

(i) Sum the observations for each level to form $y_{i.}$ for all i, and then obtain the grand total $y_{..}$.

(ii) Form the sum of the squares of the individual observations to yield $\sum_{i=1}^{a} \sum_{j=1}^{n} y_{ij}^2$.

(iii) Form the sum of squares of the totals for each level and divide it by n to yield $\sum_{i=1}^{a} y_{i.}^2/n$.
(iv) Square the grand total and divide it by an to yield a $y_{..}^2/an$ term, which is known as the correction factor.
(v) The three sums of squares can now be obtained by using the computational formulae (2.11.1) through (2.11.3).

Remark: The computational formulae given in this section are very convenient to use. But a word of caution must be included for individuals who use a computer or an electronic calculator with an eight digit capacity. If we sum the squares of numbers containing three or more digits, we can exceed their capacity easily and thereby get erroneous results. In that case it would be better to use their definitional formulae to calculate the appropriate sums of squares.

2.12 ANALYSIS OF VARIANCE FOR UNEQUAL NUMBERS OF OBSERVATIONS

Equal numbers of observations for each treatment or at each factor level are desirable because of the simplicity of organizing the experiment and subsequent data analysis. Furthermore, for a given sample size, the analysis of variance procedure is most powerful; that is, it provides the smallest value of the probability for committing a type II error, when the number of observations for each level of a factor is the same. In addition, it has been found that the F test is relatively insensitive to the violation of the assumption of the homogeneity of variances when the samples are of equal size. However, due to a variety of reasons, it may happen that it is impossible to collect an equal number of observations at each level of the factor. Part of the data may have been lost, or certain treatment or factor levels, which are important for some other reasons, may have been emphasized by taking more observations at these levels. Thus, if more data are available at some levels than at others, we must take them all into the analysis.

The analysis of variance for the one-way classification model with unequal numbers of observations is essentially the same as for the balanced case. One needs to make only minor changes in the formulae to account for unequal sample sizes. To avoid unnecessary repetition, we simply indicate the necessary changes in notation and present a summary table of the analysis of variance. Thus, suppose that the factor has a different levels and that n_i ($i = 1, 2, \ldots, a$) observations have been made at each level, giving a total of $N = \sum_{i=1}^{a} n_i$ observations in all. The analysis of variance models and their assumptions remain the same except that under Model I in place of $\sum_{i=1}^{a} \alpha_i = 0$, we now need the restriction that $\sum_{i=1}^{a} n_i \alpha_i = 0$. Also, remember that the class mean $\bar{y}_{i.}$ is now based on n_i observations. With this basic change, an identical analysis can be carried out as in the balanced case without any conceptual difficulty. The details of the analysis are summarized in the form of an analysis of variance table as given in Table 2.2. The derivation of expected between mean squares under Models I and II is somewhat involved and can be found in Graybill (1961, pp.

TABLE 2.2
Analysis of Variance for Model (2.1.1) with Unequal Sample Sizes

Source of Variation	Degrees of Freedom	Sum of[1] Squares	Mean Square	Expected Mean Square[2] Model I	Expected Mean Square[2] Model II	F Value
Between	$a - 1$	SS_B	MS_B	$\sigma_e^2 + \dfrac{\sum_{i=1}^{a} n_i \alpha_i^2}{a - 1}$	$\sigma_e^2 + n_o \sigma_\alpha^2$	MS_B/MS_W
Within	$N - a$	SS_W	MS_W	σ_e^2	σ_e^2	
Total	$N - 1$	SS_T				

[1] The sums of squares in this case are defined as follows:

$$SS_B = \sum_{i=1}^{a} n_i (\bar{y}_{i.} - \bar{y}_{..})^2, \quad SS_W = \sum_{i=1}^{a}\sum_{j=1}^{n_i} (y_{ij} - \bar{y}_{i.})^2, \quad \text{and} \quad SS_T = \sum_{i=1}^{a}\sum_{j=1}^{n_i} (y_{ij} - \bar{y}_{..})^2,$$

with

$$\bar{y}_{i.} = \frac{\sum_{j=1}^{n_i} y_{ij}}{n_i} \quad \text{and} \quad \bar{y}_{..} = \frac{\sum_{i=1}^{a}\sum_{j=1}^{n_i} y_{ij}}{N} = \frac{\sum_{i=1}^{a} n_i \bar{y}_{i.}}{N}.$$

[2] $n_o = (N^2 - \sum_{i=1}^{a} n_i^2)/N(a-1)$. If the number of observations is the same for each group, that is, $n_1 = n_2 = \cdots = n_a = n$, then it follows that $n_o = (a^2 n^2 - an^2)/an(a-1) = n$. Thus, the results of the analysis of variance for the unbalanced design reduce to that for the balanced case.

351–354; 1976, pp. 517–518). The analysis of variance F test and estimation of variance components can be accomplished as before. For example, unbiased estimators of σ_e^2 and σ_α^2 are given by

$$\hat{\sigma}_e^2 = MS_W \tag{2.12.1}$$

and

$$\hat{\sigma}_\alpha^2 = (MS_B - MS_W)/n_o, \tag{2.12.2}$$

where n_o is defined following Table 2.2. The estimator (2.12.1) is still the minimum variance unbiased for σ_e^2. However, the estimator (2.12.2) is not the best estimator of σ_α^2 especially when the n_i's differ greatly among them.[12]

It should be remarked that in the case of the unbalanced design, the between-mean square (MS_B) does not have a chi-distribution when $\sigma_\alpha^2 > 0$, but instead a weighted combination of chi-square distributions. Under the null hypothesis

---

[12] For some further discussions on this point, see Robertson (1962) and Kendall et al. (1983, Section 36.26). For some alternative estimators of variance components for an unbalanced design see Searle (1971b, Chapter 10).

$H_0 : \sigma_\alpha^2 = 0$, the statistic MS_B/MS_W has an F distribution with $a-1$ and $N-a$ degrees of freedom, and can be used to test the corresponding null hypothesis; but its distribution under the alternative ($\sigma_\alpha^2 > 0$) is much more complicated than the corresponding balanced case. For some additional results on this topic see Singh (1987) and Donner and Koval (1989). Similarly, the normal theory confidence interval for σ_e^2 can be obtained in the usual way but the determination of confidence intervals for σ_α^2, $\sigma_\alpha^2/\sigma_e^2$, and $\sigma_\alpha^2/(\sigma_e^2 + \sigma_\alpha^2)$ is much more complicated. The interested reader is referred to the book by Burdick and Graybill (1992, pp. 68–77) for a concise discussion of methods of constructing confidence intervals in the case of unbalanced design.[13] In passing, we may note that an exact $100(1-\alpha)$ percent confidence interval for σ_e^2 and approximate $100(1-\alpha)$ percent confidence intervals for $\sigma_\alpha^2, \sigma_\alpha^2/\sigma_e^2$, and the intraclass correlation $\rho = \sigma_\alpha^2/(\sigma_e^2 + \sigma_\alpha^2)$, based on the distribution of the mean squares, are given by

$$P\left[\frac{(N-a)\text{MS}_W}{\chi^2[N-a, 1-\alpha/2]} < \sigma_e^2 < \frac{(N-a)\text{MS}_W}{\chi^2[N-a, \alpha/2]}\right] = 1-\alpha, \quad (2.12.3)$$

$$P\left[\frac{L\text{MS}_B^*}{(1+n_o^*L)\,F[a-1,\infty;1-\alpha/2]} < \sigma_\alpha^2 < \frac{U\text{MS}_B^*}{(1+n_o^*U)\,F[a-1,\infty;\alpha/2]}\right]$$
$$\doteq 1-\alpha, \quad (2.12.4)$$

$$P\left[\frac{\text{MS}_B^*}{n_o^*\text{MS}_W\,F[a-1,N-a;1-\alpha/2]} - \frac{1}{n_{\min}}\right.$$
$$\left. < \frac{\sigma_\alpha^2}{\sigma_e^2} < \frac{\text{MS}_B^*}{n_o^*\text{MS}_W\,F[a-1,N-a;\alpha/2]} - \frac{1}{n_{\max}}\right] \geq 1-\alpha, \quad (2.12.5)$$

and

$$P\left[\frac{F^*/F[a-1,N-a;1-\alpha/2]-1}{F^*/F[a-1,N-a;1-\alpha/2]+(n_o-1)}\right.$$
$$\left. \leq \frac{\sigma_\alpha^2}{\sigma_e^2+\sigma_\alpha^2} \leq \frac{F^*/F[a-1,N-a;\alpha/2]-1}{F^*/F[a-1,N-a;\alpha/2]+(n_o-1)}\right] \doteq 1-\alpha, \quad (2.12.6)$$

where

$$n_o = \frac{N^2 - \sum_{i=1}^{a} n_i^2}{N(a-1)}, \quad n_o^* = \frac{a}{\sum_{i=1}^{a} 1/n_i},$$

---

[13] Methods for constructing confidence intervals for $\sigma_\alpha^2, \sigma_\alpha^2/\sigma_e^2, \sigma_e^2 + \sigma_\alpha^2$, and $\sigma_\alpha^2/(\sigma_e^2 + \sigma_\alpha^2)$ have been discussed by Thomas and Hultquist (1978), Burdick and Graybill (1984), Burdick and Eickman (1986), Burdick et al. (1986b), and Donner and Wells (1986).

One-Way Classification

TABLE 2.3
Data on Yields of Four Different Varieties of Wheat (in bushels per acre)

	Varieties		
I	II	III	IV
96	93	60	76
37	81	54	89
58	79	78	88
69	101	56	84
73	96	61	75
81	102	69	68

$$F^* = \frac{MS_B}{MS_W}, \quad MS_B^* = n_o^* \sum_{i=1}^{a} \frac{(\bar{y}_{i.}^* - \bar{y}_{..}^*)^2}{a - 1},$$

$$L = \frac{MS_B^*}{n_o^* MS_W F[a - 1, N - a; 1 - \alpha/2]} - \frac{1}{n_{\min}},$$

$$U = \frac{MS_B^*}{n_o^* MS_W F[a - 1, N - a; \alpha/2]} - \frac{1}{n_{\max}},$$

$$n_{\min} = \min(n_1, n_2, \ldots, n_a), \quad n_{\max} = \max(n_1, n_2, \ldots, n_a),$$

with

$$\bar{y}_{i.}^* = \sum_{j=1}^{n_i} y_{ij}/n_i \quad \text{and} \quad \bar{y}_{..}^* = \sum_{i=1}^{a} \bar{y}_{i.}^*/a.$$

2.13 WORKED EXAMPLES FOR MODEL I

Suppose a crop scientist wishes to test the effect of four varieties of wheat on the resultant yield. She designs an experiment with 24 plots of the same size and shape and sows each variety at random in 6 of the 24 plots. The yields from these 24 plots provide the data for a one-way classification with equal sample sizes and are presented in Table 2.3.

The data from the experiment just described must be analyzed under Model I since the four varieties are specially selected by the experimenter to be of particular interest to her. Hence, the factor under investigation (varieties of wheat) will have a fixed effect. In this example, $a = 4$, $n = 6$, and the resultant calculations for the sums of squares, using the computational formulae given in Section 2.11, are summarized in the following.

The marginal totals corresponding to the different varieties are

$$y_{1.} = 414, \quad y_{2.} = 552, \quad y_{3.} = 378, \quad y_{4.} = 480;$$

and the grand total is

$$y_{..} = 1{,}824.$$

The other quantities needed in the calculations of the sums of squares are:

$$\frac{y_{..}^2}{an} = \frac{(1{,}824)^2}{24} = 138{,}624,$$

$$\frac{1}{n}\sum_{i=1}^{a} y_{i.}^2 = \frac{1}{6}[(414)^2 + (552)^2 + (378)^2 + (480)^2] = 141{,}564,$$

and

$$\sum_{i=1}^{a}\sum_{j=1}^{n} y_{ij}^2 = (96)^2 + (37)^2 + \cdots + (75)^2 + (68)^2 = 144{,}836.$$

The resultant sums of squares are, therefore, obtained as follows.

$$SS_T = \sum_{i=1}^{a}\sum_{j=1}^{n} y_{ij}^2 - \frac{y_{..}^2}{an} = 144{,}836 - 138{,}624$$
$$= 6{,}212,$$

$$SS_B = \frac{1}{n}\sum_{i=1}^{a} y_{i.}^2 - \frac{y_{..}^2}{an} = 141{,}564 - 138{,}624$$
$$= 2{,}940,$$

and

$$SS_W = \sum_{i=1}^{a}\sum_{j=1}^{n} y_{ij}^2 - \frac{1}{n}\sum_{i=1}^{a} y_{i.}^2 = 144{,}836 - 141{,}564$$
$$= 3{,}272.$$

Finally, the results of the analysis of variance calculations are summarized in Table 2.4. If we choose the significance level of $\alpha = 0.05$, we find from Appendix Table V that the 95th percentage point of the F distribution with 3 and 20 degrees of freedom is 3.10. Since the value of the F statistic from Table 2.4 is 5.99, which is greater than 3.10 ($p = 0.004$), we may conclude that the effects of the four varieties of wheat are significantly different. Stated another way, the response variability attributable to the means of varieties is significantly greater

TABLE 2.4
Analysis of Variance for the Yields Data of Table 2.3

Source of Variation	Degrees of Freedom	Sum of Squares	Mean Square	Expected Mean Square	F Value	p-Value
Between Varieties	3	2,940	980.000	$\sigma_e^2 + \frac{6}{4-1}\sum_{i=1}^{4}\alpha_i^2$	5.99	0.004
Within Varieties	20	3,272	163.600	σ_e^2		
Total	23	6,212				

TABLE 2.5
Data on Blood Analysis of Animals Injected With Five Drugs

		Drugs		
A	B	C	D	E
19.0	7.0	4.0	6.0	6.0
11.0	1.0	7.0	6.0	4.0
15.0	4.0	7.0	6.0	2.0
	4.0		10.0	

than the variability due to uncontrolled experimental error. The conclusion of the F test from Table 2.4 may not have surprised the agricultural investigator. In the first place, she conducted the study because she expected the four varieties of wheat to have different effects on yield and was interested in finding which varieties lead to higher yield. We discuss this problem, namely, how to study the nature of the factor level effects when differences exist, in Section 2.19.

For another example, involving unequal sample sizes, suppose a pharmaceutical research company conducts an experiment to compare efficacy of five drugs. There are 20 animals available for the trial and each drug is injected into 4 randomly selected animals. Three animals die during the course of the experiment. The blood samples from the remaining animals are taken and analyzed. The data on blood pH reading from each blood analysis in certain standardized units are presented in Table 2.5.

The data of Table 2.5 should be analyzed again under Model I since the five drugs are specially chosen by the company to be of particular interest. Hence, the factor under investigation (drugs) will have a systematic effect. In this example, $a = 5$, $n_1 = 3$, $n_2 = 4$, $n_3 = 3$, $n_4 = 4$, and $n_5 = 3$; and the resultant calculations for the sums of squares are summarized in the following.

The marginal totals corresponding to the five different drugs are

$$y_{1.} = 45.0, \quad y_{2.} = 16.0, \quad y_{3.} = 18.0, \quad y_{4.} = 28.0, \quad y_{5.} = 12.0;$$

and the grand total is

$$y_{..} = 119.0.$$

The other quantities needed in the calculations of the sums of squares are obtained as

$$\frac{y_{..}^2}{N} = \frac{(119)^2}{17} = 833.000,$$

$$\sum_{i=1}^{a} \frac{y_{i.}^2}{n_i} = \frac{(45)^2}{3} + \frac{(16)^2}{4} + \frac{(18)^2}{3} + \frac{(28)^2}{4} + \frac{(12)^2}{3} = 1{,}091.000,$$

and

$$\sum_{i=1}^{a} \sum_{j=1}^{n_i} y_{ij}^2 = (19)^2 + (11)^2 + \cdots + (2)^2 = 1{,}167.000.$$

The resulting sums of squares are, therefore, given as

$$SS_T = \sum_{i=1}^{a} \sum_{j=1}^{n_i} y_{ij}^2 - \frac{y_{..}^2}{N} = 1{,}167.000 - 833.000$$

$$= 334.000,$$

$$SS_B = \sum_{i=1}^{a} \frac{y_{i.}^2}{n_i} - \frac{y_{..}^2}{N} = 1{,}091.000 - 833.000$$

$$= 258.000,$$

and

$$SS_W = \sum_{i=1}^{a} \sum_{j=1}^{n_i} y_{ij}^2 - \sum_{i=1}^{a} \frac{y_{i.}^2}{n_i} = 1{,}167.000 - 1{,}091.000$$

$$= 76.000.$$

Finally, the results of the analysis of variance calculations are summarized in Table 2.6. If we choose the level of significance $\alpha = 0.05$, we find from Appendix Table V that the 95th percentage point of the F distribution with 4 and 12 degrees of freedom is 3.26 ($p < 0.001$). Since the value of the F statistic from Table 2.6 is 10.18, which is greater than 3.26, we may conclude that the

TABLE 2.6
Analysis of Variance for the Blood Analysis Data of Table 2.5

Source of Variation	Degrees of Freedom	Sum of Squares	Mean Square	Expected Mean Square	F Value	p-Value
Between Drugs	4	258.000	64.500	$\sigma_e^2 + \dfrac{1}{5-1}\sum_{i=1}^{5} n_i \alpha_i^2$	10.18	<0.001
Within Drugs	12	76.000	6.333	σ_e^2		
Total	16	334.000				

TABLE 2.7
Interview Ratings by Five Staff Members

	Staff Members			
I	II	III	IV	V
86	67	57	83	85
75	86	74	80	84
94	90	71	96	92
86	76	55	99	91

different drugs do not lead to the same mean response; that is, there is a relation between the injected drug and the pH reading from the blood analysis.

This conclusion may not have surprised the drug company. In the first place, it conducted the study because it was suspected that five drugs would have different reactions and the company was interested in finding out the nature of these differences. In Section 2.19, we discuss the second stage of the analysis, namely, how to study the nature of the factor level effects when differences exist.

2.14 WORKED EXAMPLES FOR MODEL II

Suppose a college admissions office wishes to study the results of the interview ratings of prospective students by its staff members. Five staff members are selected at random and four prospective students are assigned randomly to each. The results provide the data for a one-way classification with equal sample sizes and are given in Table 2.7.

The data of Table 2.7 should be analyzed under Model II since the five staff members are randomly selected from the list of college staff and the results of the analysis are to be valid for the entire pool of college staff. Hence, the factor under study should be regarded as having random effects. Here, $a = 5$

and $n = 4$ and the resultant calculations for the sums of squares using the computational formulae are given in the following.

The marginal totals for ratings corresponding to the five staff members are

$$y_{1.} = 341, \quad y_{2.} = 319, \quad y_{3.} = 257, \quad y_{4.} = 35, \quad y_{5.} = 352;$$

and the grand total is

$$y_{..} = 1{,}627.$$

The other quantities needed in the calculations of the sums of squares are obtained as

$$\frac{y_{..}^2}{an} = \frac{(1{,}627)^2}{20} = 132{,}356.450,$$

$$\frac{1}{n}\sum_{i=1}^{a} y_{i.}^2 = \frac{1}{4}[(341)^2 + (319)^2 + (257)^2 + (358)^2 + (352)^2] = 134{,}039.750,$$

and

$$\sum_{i=1}^{a}\sum_{j=1}^{n} y_{ij}^2 = (86)^2 + (75)^2 + \cdots + (92)^2 + (91)^2 = 135{,}137.$$

The resultant sums of squares are, therefore, given as

$$SS_T = \sum_{i=1}^{a}\sum_{j=1}^{n} y_{ij}^2 - \frac{y_{..}^2}{an} = 135{,}137 - 132{,}356.450$$

$$= 2{,}780.550,$$

$$SS_B = \frac{1}{n}\sum_{i=1}^{a} y_{i.}^2 - \frac{y_{..}^2}{an} = 134{,}039.750 - 132{,}356.450$$

$$= 1{,}683.300,$$

and

$$SS_W = \sum_{i=1}^{a}\sum_{j=1}^{n} y_{ij}^2 - \frac{1}{n}\sum_{i=1}^{a} y_{i.}^2 = 135{,}137 - 134{,}039.750$$

$$= 1097.250.$$

Finally, the results of the analysis of variance calculations are summarized in Table 2.8. Here, the null hypothesis states that the variability in interview rating among staff members is due entirely to the natural variability among students, whereas the alternative hypothesis states that there is an additional

One-Way Classification

TABLE 2.8
Analysis of Variance for the Interview Ratings Data of Table 2.7

Source of Variation	Degrees of Freedom	Sum of Squares	Mean Square	Expected Mean Square	F Value	p-Value
Between Staff	4	1,683.300	420.825	$\sigma_e^2 + 4\sigma_\alpha^2$	5.753	0.005
Within Staff	15	1,097.250	73.150	σ_e^2		
Total	19	2,780.550				

variability among staff members due primarily to differences in staff members' rating schemes. If we choose the level of significance $\alpha = 0.05$, we find from Appendix Table V that the 95th percentage point of the F distribution with 4 and 15 degrees of freedom is 3.06. Since the computed F value of 5.753 from Table 2.8 is greater than 3.06 ($p = 0.005$), we may conclude that $\sigma_\alpha^2 > 0$, or that the mean ratings of the staff members differ significantly.

Furthermore, if the experimenter is interested in estimating the magnitudes of the components of variance σ_e^2 and σ_α^2, we may obtain their unbiased estimates using the formulae (2.9.8) and (2.9.9). Hence, from (2.9.8) and (2.9.9), we find that

$$\hat{\sigma}_e^2 = 73.150$$

and

$$\hat{\sigma}_\alpha^2 = \frac{420.825 - 73.150}{4} = 86.919.$$

The estimate of the total variance σ_y^2 is then given by

$$\hat{\sigma}_y^2 = \hat{\sigma}_e^2 + \hat{\sigma}_\alpha^2 = 73.150 + 86.919$$
$$= 160.069,$$

and the estimated proportion of the total variance accounted for by the staff members is

$$\frac{\hat{\sigma}_\alpha^2}{\hat{\sigma}_y^2} = \frac{86.919}{160.069} = 0.543.$$

Thus, we observe that about 54 percent of the variance among interview ratings seems to be due to differences among staff members. This would be a most important finding in such an experiment, as it would suggest that repetitions of this experiment involving different staff members would not be comparable.

A change in experimental procedure or some better control over staff ratings would clearly be advisable.

To obtain a 95 percent confidence interval for σ_e^2, we have

$$MS_W = 73.150, \quad \chi^2[15, 0.025] = 6.262, \quad \text{and} \quad \chi^2[15, 0.975] = 27.488.$$

Substituting these values in (2.10.2), the desired 95 percent confidence interval for σ_e^2 is given by

$$P\left[\frac{15 \times 73.150}{27.488} < \sigma_e^2 < \frac{15 \times 73.150}{6.262}\right] = 0.95$$

or

$$P[39.917 < \sigma_e^2 < 175.224] = 0.95.$$

Similarly, to obtain a 95 percent conservative confidence interval for σ_α^2 from (2.10.5), we have

$$\chi^2[4, 0.025] = 0.484, \quad \chi^2[4, 0.975] = 11.143,$$
$$F[4, 15; 0.025] = 0.116, \quad \text{and} \quad F[4, 15; 0.975] = 3.804.$$

Substituting appropriate values in (2.10.5), the desired 95 percent confidence interval for σ_α^2 is given by

$$P\left[\frac{4 \times 420.825}{4 \times 11.143}\left(1 - \frac{3.804}{5.753}\right) < \sigma_\alpha^2 < \frac{4 \times 420.825}{4 \times 0.484}\left(1 - \frac{0.116}{5.753}\right)\right] \geq 0.95$$

or

$$P[12.794 < \sigma_\alpha^2 < 851.942] \geq 0.95.$$

Furthermore, to obtain a 95 percent approximate confidence interval for σ_α^2 from (2.10.6) and (2.10.7), we have

$$F[4, \infty; 0.025] = 0.121, \quad \text{and} \quad F[4, \infty; 0.975] = 2.790.$$

Substituting appropriate values in (2.10.6), the desired 95 percent confidence interval for σ_α^2 is given by

$$P\left[\frac{73.150}{4 \times 2.790}(5.753 - 3.804) < \sigma_\alpha^2 < \frac{73.150}{4 \times 0.121}(5.753 - 0.116)\right] \doteq 0.95$$

or

$$P[12.775 < \sigma_\alpha^2 < 851.956] \doteq 0.95.$$

One-Way Classification

Similarly, substituting the appropriate values in (2.10.7), the desired 95 percent confidence interval for σ_α^2 is given by

$$P\left\{\frac{73.150(5.753 - 3.804)(5.753 + 3.804 - 2.790)}{4 \times 5.753 \times 2.790} < \sigma_\alpha^2 < \frac{73.150(5.753 - 0.116)(5.753 + 0.116 - 0.121)}{4 \times 5.753 \times 0.121}\right\} \doteq 0.95$$

or

$$P\{15.027 < \sigma_\alpha^2 < 851.215\} \doteq 0.95.$$

Likewise, to obtain a 95 percent confidence interval for σ_α^2 from (2.10.8), we compute the following quantities:

$$G_1 = 0.6416, \quad G_2 = 0.4536, \quad H_1 = 7.2645, \quad H_2 = 1.3981,$$
$$G_{12} = -0.1289, \quad H_{12} = -1.1587,$$
$$V_L = 79{,}392.0996 \quad \text{and} \quad V_U = 9{,}311{,}190.0700.$$

Substituting appropriate values in (2.10.8), we obtain

$$P\{16.477 < \sigma_\alpha^2 < 849.775\} \doteq 0.95.$$

Note that this interval results in a slightly shorter interval than the intervals for σ_α^2 reported previously.

Finally, substituting the appropriate values in (2.10.10) through (2.10.12), the desired 95 percent confidence intervals for $\sigma_\alpha^2/\sigma_e^2$, $\sigma_\alpha^2/(\sigma_e^2 + \sigma_\alpha^2)$, and $\sigma_e^2/(\sigma_e^2 + \sigma_\alpha^2)$ are given by

$$P\left[\frac{1}{4}\left(\frac{5.753}{3.804} - 1\right) < \frac{\sigma_\alpha^2}{\sigma_e^2} < \frac{1}{4}\left(\frac{5.753}{0.116} - 1\right)\right] \doteq 0.95$$

or

$$P\left[0.128 < \frac{\sigma_\alpha^2}{\sigma_e^2} < 12.149\right] \doteq 0.95,$$

$$P\left[\frac{5.753 - 3.804}{5.753 + (4-1)3.804} < \frac{\sigma_\alpha^2}{\sigma_e^2 + \sigma_\alpha^2} < \frac{5.753 - 0.116}{5.753 + (4-1)0.116}\right] \doteq 0.95$$

or

$$P\left[0.114 < \frac{\sigma_\alpha^2}{\sigma_e^2 + \sigma_\alpha^2} < 0.924\right] \doteq 0.95,$$

TABLE 2.9
Data on Yields of Six Varieties of Corn (in bushels per acre)

		Varieties			
Four Country	Silver King	Iodent	Lancaster	Osterland	Clark
7.3	7.7	6.9	9.6	4.8	4.3
4.5	5.4	6.8	7.8	9.2	8.4
7.4	5.2	7.6	9.6	8.5	6.6
7.4	4.0	8.1	7.7	8.8	4.9
5.0		9.4	8.2	7.9	5.8
5.9		12.0	7.3	5.9	7.6
6.4		15.9	11.3	9.2	3.7
6.3		7.4	9.5		
5.0		9.0	8.8		
6.1		5.2	8.4		
7.9		9.2	6.8		
5.7		8.6			

Source: Snedecor (1934). Used with permission.

and

$$P\left[\frac{4 \times 0.116}{5.753 + (4-1)0.116} < \frac{\sigma_e^2}{\sigma_e^2 + \sigma_\alpha^2} < \frac{4 \times 3.804}{5.753 + (4-1)3.804}\right] = 0.95$$

or

$$P\left[0.076 < \frac{\sigma_e^2}{\sigma_e^2 + \sigma_\alpha^2} < 0.887\right] = 0.95.$$

For another example involving unequal sample sizes, consider the data from an experiment reported by Snedecor (1934) who compared the yields of a number of varieties of corn, each variety being represented by several inbred lines. The data on yields (in bushels per acre) for six varieties of their inbred lines are given in Table 2.9.

The data in Table 2.9 should again be analyzed under Model II since each variety of corn is being represented by several inbred lines and the results of the analysis are to be applicable for all the varieties. Here, $a = 6$, $n_1 = 12$, $n_2 = 4$, $n_3 = 12$, $n_4 = 11$, $n_5 = 7$, $n_6 = 7$, $N = \sum_{i=1}^{a} n_i = 53$, $n_0 = (N^2 - \sum_{i=1}^{a} n_i^2)/N(a-1) = 8.626$; and the resultant calculations for the sums of squares are given in the following.

The marginal totals for yields corresponding to six varieties of corn are

$y_{1.} = 74.9$, $y_{2.} = 22.3$, $y_{3.} = 106.1$, $y_{4.} = 95.0$, $y_{5.} = 54.3$, $y_{6.} = 41.3$;

One-Way Classification

and the grand total is

$$y_{..} = 393.9.$$

The other quantities needed in the calculations of the sums of squares are obtained as

$$\frac{y_{..}^2}{N} = \frac{(393.9)^2}{53} = 2{,}927.495,$$

$$\sum_{i=1}^{a} \frac{y_{i.}^2}{n_i} = \frac{(74.9)^2}{12} + \frac{(22.3)^2}{4} + \frac{(106.1)^2}{12} + \frac{(95.0)^2}{11} + \frac{(54.3)^2}{7} + \frac{(41.3)^2}{7}$$

$$= 3{,}015.262,$$

and

$$\sum_{i=1}^{a} \sum_{j=1}^{n_i} y_{ij}^2 = (7.3)^2 + (4.5)^2 + \cdots + (3.7)^2 = 3{,}174.010.$$

The resulting sums of squares are, therefore, given as

$$SS_T = \sum_{i=1}^{a} \sum_{j=1}^{n_i} y_{ij}^2 - \frac{y_{..}^2}{N} = 3{,}174.010 - 2{,}927.495$$

$$= 246.515,$$

$$SS_B = \sum_{i=1}^{a} \frac{y_{i.}^2}{n_i} - \frac{y_{..}^2}{N} = 3{,}015.262 - 2{,}927.495$$

$$= 87.767,$$

and

$$SS_W = \sum_{i=1}^{a} \sum_{j=1}^{n_i} y_{ij}^2 - \sum_{i=1}^{a} \frac{y_{i.}^2}{n_i} = 3{,}174.01 - 3{,}015.262$$

$$= 158.748.$$

Finally, the results of the analysis of variance calculations are summarized in Table 2.10. Here, the null hypothesis states that the variation in yields among varieties of corn is due entirely to the natural variability among replicates, whereas the alternative hypothesis states that there is an additional variability among varieties of corn. If we choose the level of significance $\alpha = 0.05$, we find from Appendix Table V that the 95th percentage point of the F distribution with 5 and 47 degrees of freedom is 2.41. Since the computed F value of 5.196

TABLE 2.10
Analysis of Variance for the Yields Data of Table 2.9

Source of Variation	Degrees of Freedom	Sum of Squares	Mean Square	Expected Mean Square	F Value	p-Value
Between Varieties	5	87.767	17.553	$\sigma_e^2 + 8.626\sigma_\alpha^2$	5.196	<0.001
Within Varieties	47	158.748	3.378	σ_e^2		
Total	52	246.515				

from Table 2.10 is greater than 2.41 ($p < 0.001$), we may conclude that $\sigma_\alpha^2 > 0$, or that the mean yields of the varieties differ significantly.

Furthermore, if the experimenter is interested in estimating the magnitudes of the components of variance σ_e^2 and σ_α^2, we may obtain their unbiased estimates using the formulae (2.12.1) and (2.12.2). Hence, from (2.12.1) and (2.12.2), we find that

$$\hat{\sigma}_e^2 = 3.378$$

and

$$\hat{\sigma}_\alpha^2 = \frac{17.553 - 3.378}{8.626} = 1.643.$$

The estimate of the total variance σ_y^2 is then given by

$$\hat{\sigma}_y^2 = \hat{\sigma}_e^2 + \hat{\sigma}_\alpha^2 = 3.378 + 1.643$$
$$= 5.021,$$

and the estimated proportion of the total variance accounted for by the varieties is

$$\frac{\hat{\sigma}_\alpha^2}{\hat{\sigma}_y^2} = \frac{1.643}{5.021} = 0.327.$$

Thus, we observe that about 33 percent of the variance among yields seems to be due to differences among varieties. The remaining 67 percent of variance can be attributed to within-variety variation.

To obtain an exact 95 percent confidence interval for σ_e^2, we have

$$MS_W = 3.378, \quad \chi^2[47, 0.025] = 29.956, \quad \text{and} \quad \chi^2[47, 0.975] = 67.821.$$

Substituting these values in (2.12.3), the desired 95 percent confidence interval

One-Way Classification

for σ_e^2 is given by

$$P\left[\frac{47 \times 3.378}{67.821} < \sigma_e^2 < \frac{47 \times 3.378}{29.956}\right] \doteq 0.95$$

or

$$P[2.341 < \sigma_e^2 < 5.300] = 0.95.$$

Now, to obtain an approximate confidence interval for σ_α^2, we have

$$F[5, 47; 0.975] = 2.851, \quad F[5, 47; 0.025] = 0.163,$$
$$F[5, \infty; 0.975] = 2.570, \quad F[5, \infty; 0.025] = 0.166,$$
$$n_0 = 8.626, \quad n_0^* = 7.563,$$
$$F^* = 5.196, \quad MS_B^* = 15.591,$$
$$L = -0.036, \quad \text{and} \quad U = 3.661.$$

Substituting these values in (2.12.4), the desired 95 percent confidence interval for σ_α^2 is given by

$$P[-0.030 < \sigma_\alpha^2 < 11.985] \doteq 0.95.$$

Furthermore, to obtain an approximate confidence interval for $\sigma_\alpha^2/\sigma_e^2$, we substitute the required quantities in (2.12.5), which yields the desired interval as

$$P\left[-0.036 < \frac{\sigma_\alpha^2}{\sigma_e^2} < 3.661\right] \doteq 0.95.$$

It is to be understood that the negative limits are defined to be zero. It is, however, informative to leave them with negative signs.

Similarly, to obtain an approximate 95 percent confidence interval for the intraclass correlation, we have

$$F^* = 5.195, \quad F[5, 47; 0.025] = 0.163, \quad \text{and} \quad F[5, 47; 0.975] = 2.851.$$

Substituting these values in (2.12.6), the desired 95 percent confidence interval for the intraclass correlation is given by

$$P\left[\frac{(5.196/2.851) - 1}{(5.196/2.851) + (8.626 - 1)} < \frac{\sigma_\alpha^2}{\sigma_e^2 + \sigma_\alpha^2} < \frac{(5.196/0.163) - 1}{(5.196/0.163) + (8.626 - 1)}\right]$$
$$\doteq 0.95$$

or

$$P\left[0.087 < \frac{\sigma_\alpha^2}{\sigma_e^2 + \sigma_\alpha^2} < 0.782\right] \doteq 0.95.$$

2.15 USE OF STATISTICAL COMPUTING PACKAGES

One-way analysis of variance can be performed by a number of statistical packages using either a mainframe or a microcomputer. SAS, SPSS, and BMDP each contains various procedures to perform one-way analysis of variance. However, PROC ANOVA of SAS, ONEWAY of SPSS, and BMDP 7D are more suited for simple one-way designs including both balanced and unbalanced data sets. For the random effects model involving variance component estimation, one may prefer to use SAS GLM, SPSS GLM, and BMDP 8V or 3V. The output from these procedures provides an analysis of variance table, treatment means, and their standard errors. BMDP 7D has an extra feature of printing comparative histograms and all descriptive measures of location and variability for each group and for combined data. This provides an effective visual aid in making comparisons between different group means. To obtain similar descriptive measures using SAS, one can use the MEANS statement in GLM or UNIVARIATE procedure for each group and for combined data. For an introduction to SAS, SPSS, and BMDP procedures for performing analysis of variance, see Chapter 11.

2.16 WORKED EXAMPLES USING STATISTICAL PACKAGES

In this section, we illustrate the applications of statistical packages to perform one-way analysis of variance for the data sets employed in examples presented in Sections 2.13 and 2.14. Figures 2.2, 2.3, 2.4, and 2.5 illustrate the program instructions and the output results for analyzing data in Tables 2.3, 2.5, 2.7, and 2.9 using SAS ANOVA/GLM, SPSS ONEWAY/GLM, and BMDP 7D/8V/3V procedures. The typical output provides the data format listed at the top followed by the number of observations for each factor level, estimates of the factor level means, and the entries of the analysis of variance table. Note that in each case the results are the same as those obtained using manual computations in Sections 2.13 and 2.14.

2.17 POWER OF THE ANALYSIS OF VARIANCE F TEST

The power of the F test of the analysis of variance is important in evaluating the sensitivity of the test and also in determining sample size needed to attain a given value of the power. We recall that the power of a test refers to the probability that the decision procedure will reject the null hypothesis when in

One-Way Classification

```
DATA WHEATYLD;                      The SAS System
INPUT VARIETY YIELD;           Analysis of Variance Procedure
DATALINES;                     Dependent Variable: YIELD
1 96
1 37 .                                  Sum of      Mean
. .                            Source    DF  Squares   Square   F Value  Pr > F
4 68                           Model      3  2940.0000 980.0000  5.99   0.0044
;                              Error     20  3272.0000 163.6000
PROC ANOVA;
CLASSES VARIETY;               Corrected 23  6212.0000
MODEL YIELD = VARIETY;         Total
RUN;                           R-Square    C.V.      Root MSE    YIELD Mean
CLASS   LEVELS   VALUES        0.473278  16.82977    12.791       76.000
VARIETY    4      1 2 3 4
NUMBER OF OBS. IN DATA         Source  DF  Anova SS  Mean Square F Value Pr > F
SET=24                         VARIETY  3  2940.0000  980.0000    5.99   0.0044
```

(i) SAS application: SAS ANOVA instructions and output for the one-way fixed effects analysis of variance.

```
DATA LIST                           Test of Homogeneity of Variances
/VARIETY 1
 YIELD 3-5.                             Levene     df1   df2    Sig.
BEGIN DATA.                             Statistic
1 96                            YIELD    1.402      3    20    .271
1 37
1 58                                              ANOVA
1 69
. .                                     Sum of   df    Mean       F     Sig.
4 68 .                                  Squares        Square
END DATA.
ONEWAY YIELD BY  | YIELD Between Groups 2940.000  3   980.000    5.990  .004
VARIETY (1,4)    |       Within Groups  3272.000 20   163.600
/STATISTICS=ALL. |       Total          6212.000 23
```

(ii) SPSS application: SPSS ONEWAY instructions and output for the one-way fixed effects analysis of variance.

```
/INPUT     FILE='C:\SAHAI    BMDP7D - ONE- AND TWO-WAY ANALYSIS OF VARIANCE WITH
           \TEXTO\EJE1.TXT'.         DATA SCREENING Release: 7.0 (BMDP/DYNAMIC)
           FORMAT=FREE.       -----------------------------------------------
           VARIABLES=2.      | ANALYSIS OF VARIANCE TABLE FOR MEANS      TAIL |
/VARIABLE  NAMES=VART,YIELD. |SOURCE    SUM OF    DF   MEAN    F VALUE  PROB.|
/GROUP     CODES(VART)=1,2,  |-------- -SQUARE------    SQUARE---- ------ -------|
                    3,4.     |VARIETY   2940.0000  3  980.0000   5.99   0.0044|
           NAMES(VART)=I,II, |ERROR     3272.0000 20  163.6000               |
                    III,IV.  |-----------------------  ----------------------|
/HISTOGRAM GROUPING=VART.               EQUALITY OF MEANS TESTS;
           VARIABLE=YIELD.         VARIANCES ARE NOT ASSUMED TO BE EQUAL
/END                         | WELCH            3, 11       8.94    0.0028|
1 96                         | BROWN-FORSYTHE   3, 11       5.99    0.0113|
1 37                         |------------------------------------------|
. .                          |LEVENE'S TEST FOR VARIANCES 3, 20   1.40   0.2714|
4 68
```

(iii) BMDP application: BMDP 7D instructions and output for the one-way fixed effects analysis of variance.

FIGURE 2.2 Program Instructions and Output for the One-Way Fixed Effects Analysis of Variance: Data on Yields of Four Different Varieties of Wheat (Table 2.3).

```
DATA BLOODANA;                          The SAS System
INPUT DRUG BLOODPH;              Analysis of Variance Procedure
DATALINES;
1 19                         Dependent Variable: BLOODPH
1 11                                      Sum of      Mean
. .                          Source    DF  Squares    Square   F Value  Pr > F
5 2                          Model      4 258.00000  64.50000   10.18   0.0008
;                            Error     12  76.00000   6.33333
PROC ANOVA;
CLASSES DRUG;                Corrected 16 334.00000
MODEL BLOODPH=DRUG;          Total
RUN;                                   R-Square    C.V.    Root MSE   BLOODPH Mean
CLASS    LEVELS   VALUES               0.772455  35.95159   2.5166      7.0000
DRUG       5      1 2 3 4 5
NUMBER OF OBS. IN DATA       Source    DF  Anova SS  Mean Square F Value Pr > F
SET=17                       DRUG       4 258.00000   64.50000   10.18   0.0008
```

(i) SAS application: SAS ANOVA instructions and output for the one-way fixed effects analysis of variance with unequal numbers of observations.

```
DATA LIST                          Test of Homogeneity of Variances
 /DRUG 1
  BLOOPH 3-4.                           Levene      df1     df2     Sig.
BEGIN DATA.                             Statistic
1 19                         BLOODPH     .448        4       12     .772
1 11
1 15                                             ANOVA
2 7
. .                                      Sum of    df    Mean         F     Sig.
5 2                                      Squares        Square
END DATA.
ONEWAY BLOODPH BY  |BLOODPH  Between Groups  258.000  4   64.500    10.184  .001
DRUG (1,5)                   Within Groups    76.000 12    6.333
/STATISTICS=ALL.             Total           334.000 16
```

(ii) SPSS application: SPSS ONEWAY instructions and output for the one-way fixed effects analysis of variance with unequal numbers of observations.

```
/INPUT    FILE ='C:\SAHAI     BMDP7D - ONE- AND TWO-WAY ANALYSIS OF VARIANCE WITH
          \TEXTO\EJE2.TXT'.            DATA SCREENING Release: 7.0 (BMDP/DYNAMIC)
          FORMAT = FREE.     -----------------------------------------------------
          VARIABLES=2.       | ANALYSIS OF VARIANCE TABLE FOR MEANS         TAIL |
/VARIABLE NAMES=DRUG,BLOOPH. |SOURCE    SUM OF    DF    MEAN      F VALUE  PROB. |
/GROUP    CODES(DRUG)=1,2,3, |-------   SQUARES-------  SQUARE- - ----------------|
                  4,5.       |DRUG      258.0000   4    64.5000    10.18  0.0008 |
          NAMES(DRUG)=A,B,C, |ERROR      76.0000  12     6.3333                  |
                  D,E.       |----------------------------------------------------|
/HISTOGRAM GROUPING=DRUG.              QUALITY OF MEANS TESTS;
          VARIABLE=BLOOPH.       VARIANCES ARE NOT ASSUMED TO BE EQUAL
/END                         |WELCH             4, 6               4.03   0.0635 |
1 19                         |BROWN-FORSYTHE    4, 7               9.70   0.0055 |
1 11                         |----------------------------------------------------|
. .                          |LEVENE'S TEST FOR VARIANCES 4, 12    0.45   0.7721 |
5 2                          -----------------------------------------------------
```

(iii) BMDP application: BMDP 7D instructions and output for the one-way fixed effects analysis of variance with unequal numbers of observations.

FIGURE 2.3 Program Instructions and Output for the One-Way Fixed Effects Analysis of Variance with Unequal Numbers of Observations: Data on Blood Analysis of Animals Injected with Five Drugs (Table 2.5).

One-Way Classification

```
DATA INTERRAT;
INPUT STAFF RESP;
DATALINES;
1 86
1 75
1 94
1 86
2 67
 . .
5 91
;
PROC GLM;
CLASSES STAFF;
MODEL RESP=STAFF;
RANDOM STAFF;
RUN;
CLASS    LEVELS    VALUES
STAFF      5       1 2 3 4 5
NUMBER OF OBS. IN DATA SET=20
```

```
                The SAS System
           General Linear Models Procedure
Dependent Variable: RESP
                  Sum of        Mean
Source      DF    Squares       Square    F Value   Pr > F
Model        4    1683.3000    420.8250    5.75     0.0052
Error       15    1097.2500     73.1500
Corrected   19    2780.5500
Total

         R-Square    C.V.       Root MSE    RESP Mean
         0.605384   10.51356     8.5528      81.350

Source   DF   Type I SS   Mean Square   F Value   Pr > F
STAFF     4   1683.3000    420.8250      5.75     0.0052
Source   DF  Type III SS  Mean Square   F Value   Pr > F
STAFF     4   1683.3000    420.8250      5.75     0.0052

Source   Type III Expected Mean Square
STAFF    Var(Error)+4 Var(STAFF)
```

(i) SAS application: SAS GLM instructions and output for the one-way random effects analysis of variance.

```
DATA LIST
/STAFF 1
 RESP 3-4.
BEGIN DATA.
1 86
1 75
1 94
1 86
1 67
1 86
 . .
5 91 .
END DATA.
GLM RESP BY
STAFF
/DESIGN STAFF
/RANDOM STAFF.
```

```
Tests of Between-Subjects Effects     Dependent Variable: RESP

Source               Type III SS    df    Mean Square     F       Sig.
STAFF   Hypothesis   1683.300        4      420.825     5.753    .005
        Error        1097.250       15       73.150(a)
a  MS(Error)

                    Expected Mean Squares(a,b)
                       Variance Component
Source       Var(STAFF)                Var(ERROR)
STAFF         4.000                      1.000
ERROR          .000                      1.000
a  For each source, the expected mean square equals the sum of the
   coefficients in the cells times the variance components, plus
   aquadratic term involving effects in the Quadratic Term cell.
b  Expected Mean Squares are based on the Type III Sums of Squares.
```

(ii) SPSS application: SPSS GLM instructions and output for the one-way random effects analysis of variance.

```
/INPUT    FILE='C:\SAHAI
          \TEXTO\EJE3.TXT'.
          FORMAT=FREE.
          VARIABLES=4.
/VARIABLE NAMES=R1,R2,R3,R4.
/DESIGN   NAMES=STAFF, RESP.
          LEVELS=5, 4.
          RANDOM=STAFF,
                 RESP.
          MODEL='S, R(S)'.
/END
86   75    94    86
 .    .     .     .
85   84    92    91
ANALYSIS OF VARIANCE DESIGN
INDEX              STAFF   RESP
NUMBER OF LEVELS     5       4
POPULATION SIZE    INF      INF
MODEL    'S, R(S)'
```

```
BMDP8V - GENERAL MIXED MODEL ANALYSIS OF VARIANCE -
         EQUAL CELL SIZES Release: 7.0 (BMDP/DYNAMIC)

ANALYSIS OF VARIANCE FOR DEPENDENT VARIABLE  1

SOURCE   ERROR   SUM OF         D.F.    MEAN         F        PROB.
         TERM    SQUARES                SQUARE
1 MEAN   STAFF   132356.450      1    132356.5    314.52    0.0001
2 STAFF  R(S)      1683.300      4       420.8      5.75    0.0052
3 R(S)                1097.250  15        73.2

SOURCE          EXPECTED MEAN               ESTIMATES OF
                  SQUARE                VARIANCE COMPONENTS
1 MEAN        20(1)+4(2)+(3)                6596.78125
2 STAFF         4(2) + (3)                    86.91875
3 R(S)             (3)                        73.15000
```

(iii) BMDP application: BMDP 8V instructions and output for the one-way random effects analysis of variance.

FIGURE 2.4 Program Instructions and Output for the One-Way Random Effects Analysis of Variance: Data on Interview Ratings by Five Staff Members (Table 2.7).

56   The Analysis of Variance

```
DATA CORNVARI;                              The SAS System
INPUT VARIETY YIELD;              General Linear Models Procedure
DATALINES;                    Dependent Variable: YIELD
1 7.3                                         Sum of        Mean
1 4.5                         Source    DF    Squares       Square     F Value  Pr > F
1 7.4                         Model      5    87.767041    17.553408    5.20    0.0007
...                           Error     47   158.748431     3.377626
6 3.7                         Corrected 52   246.515472
;                             Total
PROC GLM;                     R-Square         C.V.      Root MSE  YIELD Mean
CLASSES VARIETY;              0.356031       24.72838     1.8378      7.4321
MODEL YIELD = VARIETY;
RANDOM VARIETY;               Source    DF    Type I SS   Mean Square  F Value  Pr > F
RUN;                          VARIETY    5    87.767041   17.553408     5.20    0.0007
CLASS   LEVELS   VALUES       Source    DF    Type III SS Mean Square  F Value  Pr > F
VARIETY    6    1 2 3 4 5     VARIETY    5    87.767041   17.553408     5.20    0.0007
6
NUMBER OF OBS. IN DATA        Source          Type III Expected Mean Square
SET=53                        VARIETY         Var(Error) + 8.6264 Var(VARIETY)
```

(i) SAS application: SAS GLM instructions and output for the one-way random effects analysis of variance with unequal numbers of observations.

```
DATA LIST            Tests of Between-Subjects Effects    Dependent Variable: YIELD
/VARIETY 1
  YIELD 3-6(1)       Source              Type III SS  df  Mean Square    F      Sig.
BEGIN DATA.          VARIETY Hypothesis    87.767      5   17.553      5.197   .001
1 7.3                        Error        158.748     47    3.378(a)
1 4.5                a MS(Error)
1 7.4
1 7.4                                    Expected Mean Squares(a,b)
1 5.0                                         Variance Component
...                  Source       Var(VARIETY)              Var(ERROR)
6 3.7                VARIETY         8.626                    1.000
END DATA.            ERROR            .000                    1.000
GLM YIELD BY         a For each source, the expected mean square equals the sum of the
VARIETY              coefficients in the cells times the variance components, plus a
/DESIGN VARIETY      quadratic term involving effects in the Quadratic Term cell.
/RANDOM VARIETY.     b Expected Mean Squares are based on the Type III Sums of Squares.
```

(ii) SPSS application: SPSS GLM instructions and output for the one-way random effects analysis of variance with unequal numbers of observations.

```
/INPUT     FILE='C:\SAHAI     BMDP3V - GENERAL MIXED MODEL ANALYSIS OF VARIANCE
           \TEXTO\EJE4.TXT'.           Release: 7.0       (BMDP/DYNAMIC)
           FORMAT=FREE.                DEPENDENT VARIABLE YIELD
           VARIABLES=2.
/VARIABLE  NAMES=VAR,YIELD.   PARAMETER  ESTIMATE  STANDARD    EST/     TWO-TAIL PROB.
/GROUP     CODES(VAR)=1,2,3,                       ERROR      ST.DEV.   (ASYM. THEORY)
                  4,5,6.     ERR.VAR.    3.377     0.696
           NAMES(VAR)=F,S,I,  CONSTANT    7.231     0.584     12.376      0.000
                  L,O,C.      RAND( 1)    1.619     1.294
/DESIGN    DEPENDENT=YIELD.
           RANDOM=VAR.        TESTS OF FIXED EFFECTS BASED ON ASYMPTOTIC VARIANCE
           METHOD=REML.       -COVARIANCE MATRIX
/END
1 7.3                        SOURCE    F-STATISTIC   DEGREES OF     PROBABILITY
...                                                   FREEDOM
6 3.7                        CONSTANT    153.18        1    52        0.00000
```

(iii) BMDP application: BMDP 3V instructions and output for the one-way random effects analysis of variance with unequal numbers of observations.

FIGURE 2.5 Program Instructions and Output for the One-Way Random Effects Analysis of Variance with Unequal Numbers of Observations: Data on Yields of Six Varieties of Corn (Table 2.9).

One-Way Classification

fact the null hypothesis is false. We now illustrate the calculation of the power of the F test for Models I and II separately.

MODEL I (FIXED EFFECTS)

It follows from (2.6.3) and (2.6.6) that the power of the F test for the hypothesis $H_0: \alpha_i = 0$ ($i = 1, \ldots, a$) is given by

$$1 - \beta = P\{F'[a - 1, a(n - 1); \lambda] > F[a - 1, a(n - 1); 1 - \alpha]\}, \quad (2.17.1)$$

where $F[a - 1, a(n - 1); 1 - \alpha]$ is the $100(1 - \alpha)$-th percentage point of the F distribution with $a - 1$ and $a(n - 1)$ degrees of freedom and $F'[a - 1, a(n - 1); \lambda]$ is a statistic having a noncentral F distribution (see Appendix G) with $a - 1$ and $a(n - 1)$ degrees of freedom and the noncentrality parameter λ given by

$$\lambda = \frac{n}{2\sigma_e^2} \sum_{i=1}^{a} \alpha_i^2. \quad (2.17.2)$$

When there are unequal numbers of observations (n_i) at different factor levels, the noncentrality parameter takes the form

$$\lambda = \frac{1}{2\sigma_e^2} \sum_{i=1}^{a} n_i \alpha_i^2. \quad (2.17.3)$$

Remark: To evaluate the probability expression given by (2.17.1), one needs to employ noncentral F tables which involve evaluation of the integrals for the noncentral F distributions. Such tables of power or the probability expression (2.17.1) have been calculated by Tang (1938) and by Tiku (1967, 1972). To tabulate the power for each α, it would require a triple-entry table consisting of v_1, v_2, and λ. Tang's tables give the probability of type II error, β, corresponding to the degrees of freedom $v_1 = 1$ (1) 8, $v_2 = 2, 4, 6, 7$ (1) 30, 60, ∞; normalized noncentrality parameter $\phi = \{2\lambda/(v_1 + 1)\}^{1/2} = 1(0.5) \ 3(1)8$; and level of significance $\alpha = 0.05$ and 0.01. The computations of these probabilities are based upon the incomplete noncentral beta distribution. These tables are reproduced in Graybill (1961, pp. 444–459). Tiku's tables give the power corresponding to $v_1 = 1$ (1) 10, 12; $v_2 = 2$ (2) 30, 40, 60, 120, ∞; $\phi = 0.5, 1.0 \ (0.2) \ 2.2 \ (0.4) \ (3.0)$; and $\alpha = 0.005, 0.01, 0.025, 0.05$, and 0.10 and are also reproduced in Graybill (1976, pp. 672–686). An abriged version of these tables, $\alpha = 0.01, 0.05$, and 0.10, is also reprinted in this volume as Appendix Table VII. Among other tables, Lehmer (1944) calculated ϕ as a function of $(\alpha, 1 - \beta, v_1, v_2)$ for $\alpha = 0.01, 0.05$; $1 - \beta = 0.7, 0.8$. More extensive tables, which also include Lehmer tables, were published by the National Bureau of Standards, Washington, DC in 1960, for $\alpha = 0.01, 0.02, 0.05, 0.10$; $1 - \beta = 0.10, 0.50, 0.90, 0.95, 0.99$; except $(\alpha, \beta) = (0.10, 0.10), (0.20, 0.10)$; $v_1 = 1$ (1) (10), 12, 15, 20, 24, 30, 40, 60, 120, ∞; and $v_2 = 2$ (2) 12, 20, 24, 30, 40, 60, ∞.

In addition to the tables previously described, the computation of the power of the F test is facilitated by the availability of power charts, prepared by Pearson and Hartley (1951) and Fox (1956), which make calculation of the probabilities (2.17.1) quite simple. Pearson-Hartley charts, reproduced in Appendix Chart II, are used as follows:

(i) The charts are given for $v_1 = 1, 2, \ldots, 8$, the value of which is shown in the upper left-hand corner of each chart.
(ii) Two levels of significance, namely, $\alpha = 0.05$ and $\alpha = 0.01$, are given in the charts.
(iii) There are two X-scales (abscissas) corresponding to the two significance levels used. The left set of curves for each chart corresponds to $\alpha = 0.05$ and the right set refers to $\alpha = 0.01$.
(iv) Separate curves are provided for different values of v_2. For each chart, the values of v_2 are given at the top of the chart. Because only selected values of v_2 are provided in the charts, an interpolation is generally required for intermediate values of v_2. There are eight curves corresponding to $v_2 = 6, 7, 8, 9, 10, 12, 15, 20, 30, 60, \infty$.
(v) The X-scale (abscissa) represents ϕ, the normalized noncentrality parameter, and the Y-scale (ordinate) represents the desired power $(1 - \beta)$ as defined in (2.17.1).

Remarks: (i) For the calculation of the power $(1 - \beta)$, we need to know the value of ϕ beforehand (whose exact value is unknown). To estimate the value of ϕ, one requires the knowledge of α_i's and the error variance σ_e^2. An estimate of σ_e^2 can be obtained based on previous experimentation or a pilot study. If several estimates are available, the largest one should be taken. The larger the value of σ_e^2, the larger will be the value of n required to achieve a given level of power.

(ii) A special condensation of the Pearson and Hartley charts was published by Duncan (1957), who plotted on a single set of axes the values of ϕ corresponding to $1 - \beta = 0.5$ and 0.90 for various values of v_1 and v_2. Separate charts were presented for $\alpha = 0.05$ and 0.01. Having v_1 and v_2 on the same chart facilitates the computations involving both of these degrees of freedom.

(iii) The Fox charts, constructed using the Tang and Lehmer tables, are useful in determining the design parameters (combination of v_1 and v_2 values) in order to obtain a desired power against a specified alternative. The charts consist of the following:

(a) In a (v_1, v_2)-plane and for fixed values of α and β, the charts give the contours of ϕ. These are curves on which ϕ has certain constant values.
(b) Because of the choice of a reciprocal scale for v_1 and v_2, these curves appear to be nearly straight lines.
(c) The curves are arranged in eight separate charts for $\alpha = 0.01$ and 0.05, and $1 - \beta = 0.5, 0.7, 0.8,$ and 0.9.
(d) There are two monograms included among the Fox charts. They are designed to facilitate the interpolation to values of $1 - \beta$ different from $0.5, 0.7, 0.8,$ and 0.9.

The Fox charts, reprinted in Scheffé (1959, pp. 446–455), are used as follows. For a given pair of values (β, ϕ), the point corresponding to this pair is located in each of the two grids and the straight line drawn through these two points is then the (approximate) contour of ϕ.

One-Way Classification

(iv) The Pearson and Hartley charts and the Fox charts serve a somewhat complementary role: the former being designed for $v_1 = 1, 2, \ldots, 8$; and the latter for $v_1 = 3, 4, 5, \ldots, \infty$.

Example 1. Consider the pharmaceutical research example described in Section 2.13 and suppose that the company wishes to know the power of the F test of the experiment when there are substantial differences between mean blood pH readings for different drugs. More specifically, suppose one wishes to consider the case when $\alpha_1 = 8$, $\alpha_2 = -3$, $\alpha_3 = -1$, $\alpha_4 = 0$, and $\alpha_5 = -3$. From (2.17.3), the value of the noncentrality parameter λ is

$$\lambda = \frac{1}{2\sigma_e^2}[\{3(8)^2 + 4(-3)^2 + 3(-1)^2 + 4(0)^2 + 3(-3)^2\}]$$

$$= \frac{1}{2\sigma_e^2}(258.0), \qquad (2.17.4)$$

where an estimate of σ_e^2 is obtained from $MS_W = 6.333$. Substituting the value of σ_e^2 in (2.17.4), we get

$$\lambda = \frac{1}{2(6.333)}(258.0) = 20.37$$

and the normalized noncentrality parameter is

$$\phi = \sqrt{\frac{2(20.37)}{5}} = 2.85.$$

Furthermore, for this example, we have $v_1 = 4$ and $v_2 = 12$. Hence, for $\alpha = 0.01$, we find from the Pearson-Hartley charts given in Appendix Chart II that the power is approximately 0.94. The use of Tiku's tables given in Appendix Table VII with appropriate interpolation gives essentially the same result. Thus, there are about 94 chances in 100 that the F test will detect the differences in the mean blood pH readings of the five drugs given the specified differences in their magnitude.

In addition to tables and charts described above one can use a normal approximation to the distribution of the square root of the noncentral F variate to evaluate the power expression (2.17.1). For example, it can be shown that (see, e.g., Johnson et al. (1995, pp. 491–492)).

$$Z = \frac{\sqrt{v_1 v_2^{-1}(2v_2 - 1) F[v_1, v_2; \lambda]} - \sqrt{2(v_1 + 2\lambda) - (v_1 + 2\lambda)^{-1}(v_1 + 4\lambda)}}{\sqrt{v_1 v_2^{-1} F[v_1, v_2; \lambda] + (v_1 + 2\lambda)^{-1}(v_1 + 4\lambda)}}$$

$$(2.17.5)$$

is approximately normally distributed with mean zero and standard deviation one.[14] Applying the normal approximation (2.17.5) to equation (2.17.1), it follows that (see, e.g., Fleiss (1986, pp. 372–373); Day and Graham (1991))

$$1 - \beta \simeq P(Z \leq z_{1-\beta}),$$

where

$$z_{1-\beta} = \frac{\sqrt{v_2 [2(v_1 + 2\lambda)^2 - (v_1 + 4\lambda)]} - \sqrt{v_1(v_1 + 2\lambda)(2v_2 - 1) F[v_1, v_2; 1 - \alpha]}}{\sqrt{v_1 (v_1 + 2\lambda) F[v_1, v_2; 1 - \alpha] + v_2(v_1 + 4\lambda)}}.$$

Example 2. For Example 1 considered previously, $v_1 = 4$, $v_2 = 12$, $\lambda = 20.37$, $\alpha = 0.01$, $F[4, 12; 0.99] = 5.41$, and $z_{1-\beta}$ is determined by

$$z_{1-\beta} = \frac{\sqrt{12[2(4 + 2 \times 20.37)^2 - (4 + 4 \times 20.37)]} - \sqrt{4(4 + 2 \times 20.37)(2 \times 12 - 1) \times 5.41}}{\sqrt{4(4 + 2 \times 20.37) \times 5.41 + 12(4 + 4 \times 20.37)}}$$

$$= 1.51.$$

Now, from Appendix Table I, the power is given by

$$1 - \beta \simeq P(Z \leq 1.51) = 0.93,$$

which gives nearly the same value as that obtained earlier using Pearson Hartley charts and Tiku's tables.

MODEL II (RANDOM EFFECTS)

Under Model II, it follows from (2.7.10) that the power of the F test for the null hypothesis $H_0 : \sigma_\alpha^2 = 0$ is given by

$$= P \left\{ F[a - 1, a(n - 1)] > \frac{F[a - 1, a(n - 1); 1 - \alpha]}{1 + n\sigma_\alpha^2/\sigma_e^2} \right\}. \quad (2.17.6)$$

Thus, in the case of Model II, the power of the F test depends only upon the (central) F distribution and is, therefore, more readily calculable than the power under Model I, which involves the noncentral F distribution. Furthermore, it is readily seen that the power of the F test for the more general hypothesis

---

[14] More accurate approximations of the noncentral F variate to evaluate the power expression (2.17.1) can be based on the central F distribution. For a discussion of these and some other approximations to the noncentral F distribution, see Johnson et al. (1995, pp. 491–495).

(2.7.11) will be given by

$$P\left\{F[a-1, a[n-1]] > \frac{1+n\rho_o}{1+n\sigma_\alpha^2/\sigma_e^2} F[a-1, a(n-1); 1-\alpha]\right\}, \quad (2.17.7)$$

which again involves only the central F distribution.[15]

To simplify the computation of probabilities (2.17.6) and (2.17.7), curves giving 1−power have been drawn (see, e.g., Bowker and Lieberman (1972, pp. 309–313)). These are similar to the Pearson-Hartley charts for the fixed effects case described earlier. These curves are reproduced in Appendix Chart III.

Example 3. Suppose that $a = 4, n = 6, \sigma_\alpha^2/\sigma_e^2 = 2.5$, and α is taken to be 0.05. From Appendix Table V, we then find that $F[3, 20; 0.95] = 3.10$, and using (2.17.6) the power of the test is given by

$$= P\{F[3, 20] > 3.10/(1 + 6 \times 2.5)\}$$

$$= P\{F[3, 20] > 0.19\}$$

$$= 0.90.$$

2.18 POWER AND DETERMINATION OF SAMPLE SIZE

We have seen in the preceding section that under Model I, once $\sum_{i=1}^{a} \alpha_i^2$ and σ_e^2 have been specified, the power of the F test becomes merely a function of n. By making n suitably large, the power of the F test can be made accordingly large for any nonzero specified value of $\sum_{i=1}^{a} \alpha_i^2/\sigma_e^2$. That is, for a fixed value of $\sum_{i=1}^{a} \alpha_i^2/\sigma_e^2$, as n increases, ϕ also increases. The larger ϕ, for a fixed level of significance α, the larger is the power. Therefore, the sample size n can be determined so as to make the power of the F test sufficiently large, say, 0.80 or 0.90 with respect to the specified values of $\sum_{i=1}^{a} \alpha_i^2$ and σ_e^2. Equivalently, the sample size can be determined so as to make the experiment sensitive enough to detect differences in the parameters α_i's that are considered large enough to be of practical importance. As mentioned earlier, an estimate of σ_e^2 can be obtained either from a pilot survey or from previous experimentation. However, the problem of specifying $\sum_{i=1}^{a} \alpha_i^2$ is rather a difficult one. Generally, $\sum_{i=1}^{a} \alpha_i^2$ should be taken as the smallest value that is considered to be of practical importance. An expert judgment supported by a reasonable rationale is generally required.

---

[15] For discussions of the power function under nonnormality, see Tan and Wong (1980) and Singhal and Sahai (1994). For discussion of a general approach for making power calculations in the most frequently encountered statistical applications, including a fixed effect analysis of variance model, without reference to any tables and charts, see Wheeler (1974).

Example 1. Consider an example with $a = 3$ and the overall F test to be made at the level of significance $\alpha = 0.05$. From tables of the noncentral F distribution, the power of the overall test may be determined for various combinations of ϕ and n. Suppose the minimum value of α_i thought to be of practical importance is 3.0. If σ_e^2 is estimated to be 25, then

$$\phi = \sqrt{\frac{n}{a\sigma_e^2}\sum_{i=1}^{a}\alpha_i^2} = \sqrt{\frac{n(3^2 + 3^2 + 3^2)}{3 \times 25}} = 0.6\sqrt{n}.$$

Hence, for this special case, using Appendix Table VII, the values of ϕ associated with various values of n and the corresponding power are determined as follows:

n:	3	5	7	9	11	21
ϕ:	1.0	1.3	1.6	1.8	2.0	2.7
$1-\beta$:	0.21	0.31	0.62	0.75	0.85	0.99

If the power of 0.80 is considered to be appropriate, then letting $n = 10$ will provide a test having approximately this power. If the power of 0.90 is desired, then n should be between 11 and 21. It can be seen that when $n = 13$, the power is approximately 0.90.

To facilitate the computation of the sample size in the absence of noncentral F distribution tables, special tables and charts have been prepared by Feldt and Mahmoud (1958a, b), Kastenbaum et al. (1970a), Bowman and Kastenbaum (1975), Cohen (1988, Chapter 8), and Day and Graham (1991). Here, we briefly describe the use of Feldt-Mahmoud charts. The charts, reproduced in Appendix Chart IV, give the values of n (Y-scale) as a function of $\phi' = \frac{1}{\sigma_e}\sqrt{\sum_{i=1}^{a}\alpha_i^2/a}$ (X-scale) for specified values of the number of factors (a), the level of significance (α), and the power $(1 - \beta)$. The only difference between ϕ and ϕ' is that ϕ' does not involve n, because we wish to determine it. Thus, the relation of ϕ' to ϕ is simply

$$\phi' = \phi/\sqrt{n}.$$

Two levels of significance are used in the charts, namely, $\alpha = 0.05$ and $\alpha = 0.01$. The charts are given for $r = a = 2, 3, 4, 5$ and the values of $1 - \beta$ employed in the charts are: 0.5, 0.7, 0.8, 0.9, and 0.95. There are two X-scales depending on which level of significance is employed. Furthermore, the left set of curves on each chart refers to $\alpha = 0.05$ and the right set to $\alpha = 0.01$. There are separate curves for different values of $P = 1 - \beta$ and the curves are indexed according to the value of P at the top of the chart. Since only the selected values of P are used in the chart, one needs to interpolate for intermediate values of $1 - \beta$. The sample size n may be read from the ordinate of the curve.

One-Way Classification

Example 2. Suppose an engineer wishes to determine whether four leading brand names of light bulbs have the same mean life. The overall F test is to be made at $\alpha = 0.05$ with a power of $1 - \beta = 0.9$ and the value of ϕ' is found to be 0.8. To determine the required sample size, we refer to the chart for $r = 4$, locate the curve for $P = 0.9$ in the left set of curves corresponding to $\alpha = 0.05$, and read the value of the ordinate at $\phi' = 0.8$ on the X-scale. We find the value of n to be approximately equal to 7. Hence, 7 bulbs of each brand should be tested to meet the given specifications.

SAMPLE SIZE DETERMINATION USING SMALLEST DETECTABLE DIFFERENCE

The problem of sample size determination considered previously makes use of the specified values of $\sum_{i=1}^{a} \alpha_i^2$ and σ_e^2. However, as indicated there, the task of specifying $\sum_{i=1}^{a} \alpha_i^2$ is a difficult one. Moreover, for fixed effects, $\sum_{i=1}^{a} \alpha_i^2$ is difficult to interpret as a meaningful measure. Instead, one can define the sensitivity in terms of the magnitude of the difference between any pair of the α_i's, say, Δ, which is meaningful to detect with a reasonably high probability. Suppose that for at least one pair of treatments $|\alpha_i - \alpha_j| \geq \Delta$ ($i \neq j$). Now, the minimum value of λ ($\lambda_{\min}$) in (2.17.2), subject to the condition that at least two of the α_i's differ by Δ or more, occurs when the α_i's are such that $|\alpha_j - \alpha_k| \geq \Delta$ and $\alpha_i = 0$ for all $i \neq j, i \neq k$. That is, when only two of the α_i's differ by Δ and the remaining $a - 2$ α_i's are zero.[16] If the specified power of the test, $1 - \beta$, is determined for $\lambda_{\min}$, then since power increases with λ, the power will be at least as large for all sets of α_i's satisfying the previous condition.

Now, from equation (2.17.2), it follows that

$$\lambda_{\min} = \frac{n \times 2 \times \dfrac{\Delta^2}{4}}{2\sigma_e^2} = \frac{n\Delta^2}{4\sigma_e^2}.$$

Again, for a given value of Δ, knowledge of σ_e is necessary to determine the required value of the sample size n. Since σ_e is often not known precisely, the sensitivity parameter Δ is expressed as Δ/σ_e rather than Δ itself. On using equation (2.17.1) for the power of the test with $\lambda = n\Delta^2/4\sigma_e^2$, the smallest value of n can be determined such that $1 - \beta \geq 1 - \beta_0$, where β is the actual value of the type II error and β_0 is the specified one. This implies that the actual power is at least as large as the specified value. Note that when there are only

---

[16] Let the difference between the largest treatment effect, $\alpha_{\max}$, and the smallest treatment effect, $\alpha_{\min}$, be denoted by $\Delta_{\max} = \alpha_{\max} - \alpha_{\min}$. For any set of α_i ($i = 1, 2, \ldots, a$) satisfying this condition, the smallest value of λ given by (2.17.2) is obtained when the remaining $a - 2$ treatment effects $\alpha_i = (\alpha_{\max} + \alpha_{\min})/2$. Since α_i's satisfy the constraint that $\sum_{i=1}^{a} \alpha_i = 0$, this implies that $\alpha_{\max} = \Delta/2$, $\alpha_{\min} = -\Delta/2$, and $\alpha_i = 0$, otherwise.

two treatments, the problem is equivalent to that of the two-sample, two-sided Student's t test. Furthermore, it has been shown by Nelson (1983) that the sample sizes obtained for a fixed effects analysis are generally comparable to those obtained using the analysis of means method.

Remark: Tables of sample sizes using this approach were prepared by Bratcher et al. (1970) for $\alpha = 0.5, 0.4, 0.3, 0.25, 0.2, 0.1, 0.05, 0.01$; $1 - \beta = 0.7, 0.8, 0.9, 0.95$; $\Delta/\sigma_e = 1.0\ (0.25)\ 2, 2.5, 3$; and $a = 2\ (1)\ 11, 13, 16, 21, 25$, and 31. Nelson (1985) extended these tables for some additional values; that is, $\alpha = 0.1, 0.05, 0.01$; $1-\beta = 0.5, 0.8, 0.9, 0.95$; $\Delta/\sigma_e = 0.4\ (0.1)\ 1\ (0.2)\ 2\ (0.5)\ 3.0$; and $a = 2\ (1)\ 9$. Some of these tables are reprinted in Appendix Table IX. A similar but more comprehensive set of tables has been developed by Bowman (1972) and Bowman and Kastenbaum (1975).

> **Example 3.** Consider a one-way layout involving three treatments ($a = 3$), a significance level of $\alpha = 0.05$, and a type II error of $\beta = 0.2$. For an effect size of $\Delta/\sigma_e = 1.5$, Appendix Table IX shows that a sample size of 10 for each treatment will be required.

2.19 INFERENCE ABOUT THE DIFFERENCE BETWEEN TREATMENT MEANS: MULTIPLE COMPARISONS

In an analysis of variance problem involving the comparison of a group of treatment means, simply stating that the group means are significantly different may not be sufficient. In addition, the investigator probably also wants to know which particular means differ significantly from others, or if there is some relation among them. For example, in many controlled experiments, the investigator plans the experiment in order to estimate and test hypotheses regarding a limited number of specific quantities. The analysis of variance F test does not directly provide answers to these questions. New test procedures, known as multiple comparisons, have been developed to answer questions such as these. A full discussion of these procedures is beyond the scope of this volume. We present only a brief introduction of some of these procedures.

Remark: For detailed discussions of the topic, the interested reader is referred to the survey papers by Kurtz et al. (1965), O'Neil and Wetherill (1971), Chew (1976a, b, c), Miller (1977, 1985), Krishnaiah (1979), Stoline (1981), and Tukey (1991), including books by Miller (1966, 1981), Rosenthal and Rosnow (1985), Hochberg and Tamhane (1987), Toothaker (1991), and Hsu (1996). Several standard textbooks on statistics also contain references to many of these procedures. The survey paper by O'Neill and Wetherill (1971) also provides a selected and classified bibliography of the subject.

We begin with a definition of a linear combination of means, a contrast and orthogonal contrasts.

Linear Combination of Means, Contrast and Orthogonal Contrasts

Any expression of the form

$$L = \ell_1\mu_1 + \ell_2\mu_2 + \cdots + \ell_a\mu_a, \tag{2.19.1}$$

where ℓ_i's are arbitrary constants is called a linear combination of means. If one adds the constraint that $\sum_{i=1}^{a} \ell_i = 0$, then the linear combination is called a contrast of a means μ_i's, $i = 1, 2, \ldots, a$. The expressions $\mu_1 - \mu_3$ and $\mu_1 - 2\mu_2 + \mu_3$ are examples of contrasts. Two contrasts L_1 and L_2 defined by

$$L_1 = \ell_1\mu_1 + \ell_2\mu_2 + \cdots + \ell_a\mu_a$$

and

$$L_2 = \ell'_1\mu_1 + \ell'_2\mu_2 + \cdots + \ell'_a\mu_a$$

are said to be orthogonal if

$$\ell_1\ell'_1 + \ell_2\ell'_2 + \cdots + \ell_a\ell'_a = 0.$$

The expressions $\mu_1 - \mu_2$ and $\mu_1 + \mu_2 - \mu_3 - \mu_4$ are examples of orthogonal contrasts.

Given a linear combination or a contrast of factor level means defined in (2.19.1), we can estimate it unbiasedly by

$$\hat{L} = \ell_1\bar{y}_1. + \ell_2\bar{y}_2. + \cdots + \ell_a\bar{y}_a. \tag{2.19.2}$$

The sum of squares associated with $\hat{L}$ is defined by

$$\text{SS}_{\hat{L}} = \frac{\left(\sum_{i=1}^{a} \ell_i \bar{y}_i.\right)^2}{\sum_{i=1}^{a} (\ell_i^2/n_i)}. \tag{2.19.3}$$

If $n_i = n$, $i = 1, 2, \ldots, a$, then (2.19.3) reduces to

$$\text{SS}_{\hat{L}} = \frac{n\left(\sum_{i=1}^{a} \ell_i \bar{y}_i.\right)^2}{\sum_{i=1}^{a} \ell_i^2}. \tag{2.19.4}$$

It is easy to show that the expressions (2.19.3) and (2.19.4) are general formulae for any sum of squares distributed as a chi-square with one degree of freedom.

If we assume that each of the sample means is based on the same number of observations, then it can be shown that $a - 1$ orthogonal contrasts can be formed using a sample means. These $a - 1$ contrasts form a set of mutually orthogonal contrasts. In addition, it can be shown that the sums of squares of the $a - 1$ orthogonal contrasts will add up to the between-group sum of squares. In other words, the between-group sum of squares can be partitioned into $a - 1$ sums of squares each having one degree of freedom corresponding to the $a - 1$ orthogonal contrasts. Furthermore, mutual orthogonality is desirable because it leads to independence of the $a - 1$ sums of squares associated with the orthogonal contrasts.

Remark: When the levels of a factor are quantitative rather than qualitative in nature, the investigator often wants to know whether the means follow a systematic pattern or trend; say, linear, quadratic, cubic, and the like. In such a case, the particular contrasts of interest involve those measuring linear, quadratic, cubic, or other higher-order trends in a series of means. As usual, with a means, we have $a - 1$ orthogonal contrasts, each having one degree of freedom. Thus, for three means, we have only two orthogonal contrasts: the linear and quadratic. With four means, there are three orthogonal contrasts: the linear, quadratic, and cubic, and so on. Many books on statistical design provide coefficients for linear, quadratic, cubic, and other contrasts (see, e.g., Fisher and Yates, (1963)). The most complete set of coefficients is given by Anderson and Houseman (1942). This type of analysis can be used to measure the trend of factor-level means associated with equally spaced values of a quantitative variable. It can also be used to assess the trend components of means obtained at fixed intervals of times; for example, in ongoing surveys of population characteristics and other time series data.

Example 1. For the data on the yields of four different varieties of wheat given in Table 2.3, consider the following set of three mutually orthogonal contrasts:

$$L_1 = \mu_1 - \mu_2,$$
$$L_2 = \mu_3 - \mu_4,$$

and

$$L_3 = \mu_1 + \mu_2 - \mu_3 - \mu_4.$$

The unbiased estimates of the preceding contrasts are obtained as

$$\hat{L}_1 = \bar{y}_{1.} - \bar{y}_{2.} = 69 - 92 = -23,$$
$$\hat{L}_2 = \bar{y}_{3.} - \bar{y}_{4.} = 63 - 80 = -17,$$

One-Way Classification

Example 1 (*continued*)

and

$$\hat{L}_3 = \bar{y}_{1.} + \bar{y}_{2.} - \bar{y}_{3.} - \bar{y}_{4.} = 69 + 92 - 63 - 80 = 18.$$

The corresponding sums of squares are calculated as

$$SS_{\hat{L}_1} = \frac{6(-23)^2}{(1+1)} = 1{,}587,$$

$$SS_{\hat{L}_2} = \frac{6(-17)^2}{(1+1)} = 867,$$

and

$$SS_{\hat{L}_3} = \frac{6(18)^2}{(1+1+1+1)} = 486.$$

Now, it is readily verified that

$$SS_B = SS_{\hat{L}_1} + SS_{\hat{L}_2} + SS_{\hat{L}_3};$$

that is, $2940 = 1587 + 867 + 486$.

TEST OF HYPOTHESIS INVOLVING A CONTRAST

Since the $\bar{y}_{i.}$'s are independent, the variance of (2.19.2) is given by

$$\text{Var}(\hat{L}) = \sum_{i=1}^{a} \ell_i^2 \, \text{Var}(\bar{y}_{i.})$$

$$= \sigma_e^2 \sum_{i=1}^{a} \ell_i^2 / n_i.$$

An unbiased estimate of this variance is

$$\widehat{\text{Var}}(\hat{L}) = MS_W \sum_{i=1}^{a} \ell_i^2 / n_i. \qquad (2.19.5)$$

Inasmuch as $\hat{L}$ is a linear combination of independent normal random variables, it is also normally distributed. It then follows that the statistic

$$(\hat{L} - L)/\sqrt{\widehat{\text{Var}}(\hat{L})} \qquad (2.19.6)$$

has a Student's t distribution with $N - a$ degrees of freedom. Therefore, a suitable test statistic for testing the null hypothesis

$$H_0 : L = L_0 \qquad (2.19.7)$$

versus

$$H_1 : L \neq L_0$$

is

$$t[N - a] = \frac{\hat{L} - L_0}{\sqrt{\text{MS}_W \sum_{i=1}^{a} \ell_i^2 / n_i}} \qquad (2.19.8)$$

or, equivalently,

$$F[1, N - a] = \frac{(\hat{L} - L_0)^2}{\text{MS}_W \sum_{i=1}^{a} \ell_i^2 / n_i}. \qquad (2.19.9)$$

A two-sided critical region is used with the t test given by (2.19.8) whereas the critical region for the F test given by (2.19.9) is determined by the right tail.

Finally, a $100(1 - \alpha)$ percent confidence interval for L is given by

$$P[\hat{L} - \psi < L < \hat{L} + \psi] = 1 - \alpha, \qquad (2.19.10)$$

where

$$\psi = t[N - a, 1 - \alpha/2] \sqrt{\text{MS}_W \sum_{i=1}^{a} \ell_i^2 / n_i} \qquad (2.19.11)$$

and $t[N - a, 1 - \alpha/2]$ denotes the $100(1 - \alpha/2)$th percentage point of the t distribution with $N - a$ degrees of freedom.

One-Way Classification

Example 2. For the data on blood analysis of animals injected with five drugs given in Table 2.5, consider the contrast L defined by

$$L = \mu_1 + 2\mu_2 - \mu_3 - \mu_4 - \mu_5.$$

An unbiased estimate of this contrast is

$$\hat{L} = \bar{y}_{1.} + 2\bar{y}_{2.} - \bar{y}_{3.} - \bar{y}_{4.} - \bar{y}_{5.} = 15 + 2(4) - 6 - 7 - 4 = 6$$

A test for the hypothesis

$$H_0: L = 0$$

versus

$$H_1: L \neq 0$$

can be carried out by calculating the value of statistic (2.19.8) with $L_0 = 0$. In this example, $a = 5, n_1 = 3, n_2 = 4, n_3 = 3, n_4 = 4, n_5 = 3$, and $MS_W = 6.333$. Then, on substituting in (2.19.8), we obtain

$$t[6] = \frac{6}{\sqrt{\{6.333 \left(\frac{1}{3} + \frac{4}{4} + \frac{1}{3} + \frac{1}{4} + \frac{1}{3}\right)\}}} = \frac{6}{\sqrt{14.249}} = 1.59.$$

From Appendix Table III, we find that $t[12, 0.975] = 2.179$ so that H_0 is sustained at $\alpha = 0.05$ level of significance.

Finally, on substituting the appropriate quantities in (2.19.10), a 95 percent confidence interval for L is given by

$$P\left[6 - 2.179\sqrt{14.249} < L < 6 + 2.179\sqrt{14.249}\right] = 0.95,$$

or

$$P[-2.23 < L < 14.23] = 0.95.$$

Since the interval includes the value zero, we conclude that it is not significantly different from zero. Thus, the results of the confidence interval are in agreement with that of the t test given above.

THE USE OF MULTIPLE COMPARISONS

After rejecting the null hypothesis (2.7.1), one might be tempted to make a comparison between each pair of factor level means, that is, 1 versus 2, ..., 1 versus a, 2 versus 3, ..., $a-1$ versus a by using the test procedures (2.19.8) or (2.19.9). But how many such comparisons need to be made? For a factor levels there are $\binom{a}{2} = \frac{1}{2}a(a-1)$ pairs to be compared, although there are only $a-1$ degrees of freedom for factor levels. Clearly, not all such comparisons are independent. Thus, it is not proper to use tests on more than $a-1$ such comparisons. Further suppose that an experimenter wishes to compare a factor levels using c independent (orthogonal) contrasts. If each one of the comparisons is tested with the same significance level, say, α, and if we assume that MS_W has an infinite number of degrees of freedom (so the tests are independent), then when all the null hypotheses involving c comparisons are true, the probability of falsely rejecting at least one of them is equal to $1-(1-\alpha)^c$. For $\alpha = 0.05$ and $c = 5$, this probability is $1-(0.95)^5 = 0.2262$, and for $c = 10$, the probability increases to $1-(0.95)^{10} = 0.4013$. Thus, if we do not reject the null hypothesis with the initial F test, and if we then perform tests based on contrasts, we increase the overall probability of committing a type I error.[17] Moreover, it is difficult to obtain an expression equivalent to $1-(1-\alpha)^c$ for comparisons made with nonindependent (nonorthogonal) contrasts. The difficulties just described in connection with the test procedures (2.19.8) or (2.19.9); or the confidence interval (2.19.10), are resolved by using a multiple comparison method. Several multiple comparison procedures are available in the literature. In the following, we consider two such widely employed procedures known as Tukey's method and Scheffé's method, respectively.

Tukey's method

According to this method, if L is any contrast estimated by $\hat{L}$, then a $100(1-\alpha)$ percent simultaneous confidence interval for L is given by

$$\hat{L} - T\sqrt{n^{-1}MS_W}\left(\frac{1}{2}\sum_{i=1}^{a}|\ell_i|\right) < L < \hat{L} + T\sqrt{n^{-1}MS_W}\left(\frac{1}{2}\sum_{i=1}^{a}|\ell_i|\right),$$

(2.19.12)

where $T = q[a, a(n-1); 1-\alpha]$ is the $100(1-\alpha)$th percentage point of the Studentized range distribution with parameters a and $a(n-1)$. (For a definition of the Studentized range distribution, see Appendix I.) Some selected

---

[17] Since we are considering individual comparisons as well as sets of such comparisons, there are two different types of error rates or significance levels at issue. When considering individual comparisons, the significance level is referred to as comparisonwise error rate. When considering an entire set of comparisons, the significance level associated with all the comparisons in the set is called the experimentwise error rate.

One-Way Classification

percentage points of the Studentized range distribution are given in Appendix Table X. Tukey (1953) has shown that for a given value of α, the intervals given by (2.19.12) hold simultaneously for every possible contrast that may be constructed (see also Scheffé (1959, p.74)). If the interval contains the value zero, the contrast is said to be not significantly different from zero; whereas if the interval does not contain the value zero, the contrast is said to be significantly different from zero. Thus, to test the null hypothesis (2.19.7), we note whether

$$|\hat{L}| \Big/ \left\{ \sqrt{n^{-1} MS_W} \left(\frac{1}{2} \sum_{i=1}^{a} |\ell_i| \right) \right\} > q[a, a(n-1); 1-\alpha]. \qquad (2.19.13)$$

Tukey's method was originally designed for contrasts comparing two means; that is, $L = \mu_1 - \mu_2$, and so on. It is seldom used in practice except for this special contrast. For this situation,

$$\frac{1}{2} \sum_{i=1}^{a} |\ell_i| = \frac{1}{2}(|1| + |-1|) = 1,$$

and the intervals given by (2.19.12) reduce to

$$\bar{y}_{i.} - \bar{y}_{i'.} - T\sqrt{n^{-1} MS_W} < \mu_i - \mu'_i < \bar{y}_{i.} - \bar{y}_{i'.} + T\sqrt{n^{-1} MS_W}. \qquad (2.19.14)$$

Thus, according to the Tukey's method, the probability is $1-\alpha$ that the intervals (2.19.14) contain all $a(a-1)/2$ pairwise differences of the type $\mu_i - \mu'_i$, $i \neq i'$.

Example 3. To illustrate this procedure, consider the data on the yield of four different varieties of wheat given in Table 2.3. Here, $a = 4$, $n = 6$, $N - a = 20$, and $MS_W = 163.600$. If we let $\alpha = 0.05$, then from the values of the Studentized range distribution given in Appendix Table X, $q[4, 20; 0.95] = 3.58$. Now, the pairwise differences of sample means will be compared to

$$q[4, 20; 0.95]\sqrt{n^{-1} MS_W} = 3.58\sqrt{163.600/6} = 18.69.$$

Note that there are $4(3)/2 = 6$ pairs of differences to be compared. The four sample means are

$$\bar{y}_{1.} = 69, \quad \bar{y}_{2.} = 92, \quad \bar{y}_{3.} = 63, \quad \text{and} \quad \bar{y}_{4.} = 80;$$

and the 6 pairs of differences can be arranged systematically as in Table 2.11.

Example 3 *(continued)*

TABLE 2.11
Pairwise Differences of Sample Means
$\bar{y}_{i\cdot} - \bar{y}_{i'\cdot}$.

	$\bar{y}_{i\cdot}$	$\bar{y}_{i\cdot} - \bar{y}_{3\cdot}$	$\bar{y}_{i\cdot} - \bar{y}_{1\cdot}$	$\bar{y}_{i\cdot} - \bar{y}_{4\cdot}$
$\bar{y}_{2\cdot}$	92	29	23	12
$\bar{y}_{4\cdot}$	80	17	11	
$\bar{y}_{1\cdot}$	69	6		
$\bar{y}_{3\cdot}$	63			

The differences above the dotted line exceed 18.69. The conclusions are that the variety II is significantly better than I and III. There is no significant difference between the varieties II and IV. The probability that we have made one or more incorrect statements is 0.05.

Example 4. To illustrate Tukey's method for a more complex contrast, say, $L = \mu_1 + \mu_4 - \mu_2 - \mu_3$, we use the same data as in Example 3. Here,

$$\frac{1}{2}\sum_{i=1}^{a}|\ell_i| = \frac{1}{2}(1+1+1+1) = 2, \quad T = q[4, 20; 0.95] = 3.58,$$

$$T\sqrt{n^{-1}\,\text{MS}_W}\left(\frac{1}{2}\sum_{i=1}^{a}|\ell_i|\right) = 3.58\sqrt{(163.600/6)}(2) = 37.39,$$

and

$$\hat{L} = 69 + 80 - 92 - 63 = -6.$$

Now, substituting the appropriate quantities in (2.19.12), we get a 95 percent simultaneous confidence interval for L as

$$-6 - 37.39 < L < -6 + 37.39$$

or

$$-43.39 < L < 31.39.$$

Since the interval includes the value zero, we conclude that L is not significantly different from zero.

Scheffé's method

According to this method if L is any contrast estimated by $\hat{L}$, then a $100(1-\alpha)$ percent simultaneous confidence interval for L is given by

$$\hat{L} - S\sqrt{(a-1)\text{MS}_W \sum_{i=1}^{a} \ell_i^2/n_i} < L < \hat{L} + S\sqrt{(a-1)\text{MS}_W \sum_{i=1}^{a} \ell_i^2/n_i},$$
(2.19.15)

where $S^2 = F[a-1, N-a; 1-\alpha]$ is the $100(1-\alpha)$th percentage point of the F distribution with $a-1$ and $N-a$ degrees of freedom. For a given value of α, Scheffé (1953) has shown that the intervals given by (2.19.15) hold simultaneously for every possible contrast that can be constructed (see also Scheffé (1959, p. 69)).[18] Again, as in the Tukey's method, to test the null hypothesis (2.19.7), we note whether[19]

$$|\hat{L}| \bigg/ \left\{\sqrt{(a-1)\text{MS}_W \sum_{i=1}^{a} \ell_i^2/n_i}\right\} > \{F[a-1, N-a; 1-\alpha]\}^{1/2}.$$
(2.19.16)

Example 5. To illustrate this procedure, we again use the data of Table 2.3. As before, $a = 4, n = 6, N - a = 20$, and $\text{MS}_W = 163.600$. If we again let $\alpha = 0.05$, then from Appendix Table V, we find that $S^2 = F[3, 20; 0.95] = 3.10$. Furthermore, for the contrasts consisting of the differences of two means, we have

$$\sum_{i=1}^{a} \ell_i^2/n_i = \frac{1}{6} + \frac{1}{6} = \frac{1}{3}.$$

Now, the differences between the sample means will be compared to

$$\sqrt{\left\{(4-1)(163.600)\left(\frac{1}{3}\right)(3.10)\right\}} = 22.52,$$

instead of 18.69 as in Tukey's method. Hence, in Table 2.11, again the sample differences $\bar{y}_{2.} - \bar{y}_{3.}$ and $\bar{y}_{2.} - \bar{y}_{1.}$ are significant by the Scheffé's

---

[18] For a simple proof of the result (2.19.15) using elementary calculus, see Klotz (1969).
[19] For an extension of Scheffé's method that tests all possible sets of comparisons encompassing all pairs of means, all possible contrasts between groups of means, and all possible partitions of the means, see Gabriel (1964).

Example 5 (*continued*)

method. It should, however, be observed that the critical value for mean differences for Scheffé's method is larger than for Tukey's method. In general, for simple contrasts of this type, Tukey's method gives shorter intervals and consequently finds more differences significant.

Example 6. To illustrate the computation of a more complex contrast, again, as in Tukey's method, consider $L = \mu_1 + \mu_4 - \mu_2 - \mu_3$. Then

$$\hat{L} = 69 + 80 - 92 - 63 = -6,$$

$$\sum_{i=1}^{a} \ell_i^2/n_i = (1 + 1 + 1 + 1)/6 = \frac{2}{3},$$

and

$$S\left\{\sqrt{(a-1)\mathrm{MS}_W \sum_{i=1}^{a} \ell_i^2/n_i}\right\} = \sqrt{(3.10)}\sqrt{\left\{(4-1)(163.600)\left(\frac{2}{3}\right)\right\}}$$

$$= 31.85.$$

On substituting the appropriate quantities in (2.19.15), we get a 95 percent simultaneous confidence interval for L as

$$-6 - 31.85 < L < -6 + 31.85$$

or

$$-37.85 < L < 25.85.$$

So in this case the interval given by Scheffé's method is shorter than that provided by Tukey's method. In general, for more complex contrasts, Scheffé's method gives shorter intervals than Tukey's method.

Example 7. To illustrate further, suppose that the experimenter had decided before conducting the experiment that the only contrast of interest is $L = \mu_1 + \mu_4 - \mu_2 - \mu_3$. Then, we can get an interval for L using (2.19.10), which is based upon the usual t distribution. Now, from

One-Way Classification

> **Example 7** (*continued*)
>
> Appendix Table III, $t[20, 0.975] = 2.086$ and, on substituting the appropriate quantities in (2.19.10), the 95 percent confidence interval for L is obtained as
>
> $$-6 - 2.086\sqrt{(163.600)\left(\frac{2}{3}\right)} < L < -6 + 2.086\sqrt{(163.600)\left(\frac{2}{3}\right)}$$
>
> or
>
> $$-27.78 < L < 15.68.$$
>
> Thus, the interval constructed from (2.19.10) is much shorter than the one given by either Tukey's or Scheffé's method.

Naturally, we would expect to get a shorter interval from a procedure designed to capture one prechosen contrast than from procedures that try to catch all possible contrasts. However, it is quite unlikely that the experimenter would specify a contrast in advance. Usually, we first test the hypothesis of equal factor level means. If this is rejected, we attempt to discover the contrasts that are significantly different from zero. As has been discussed earlier in this section, it is usually difficult to calculate the significance level associated with the several intervals of the type (2.19.10). Thus, we would use the intervals (2.19.12) and (2.19.15), given by Tukey's method and Scheffé's method, respectively, which may be computed for contrasts that are selected after the experiment is conducted and for which we can make an exact probability statement.

> **Example 8.** Finally, we illustrate Scheffé's method for the case of model (2.1.1) with unequal sample sizes by using the data on blood analysis of animals injected with five drugs given in Table 2.5. Let the contrast of interest be defined by
>
> $$L = \mu_1 + 2\mu_2 - \mu_3 - \mu_4 - \mu_5$$
>
> which is estimated by
>
> $$\hat{L} = 15 + 2(4) - 6 - 7 - 4 = 6.$$
>
> Furthermore, in this case, we have $a = 5$, $n_1 = 3$, $n_2 = 4$, $n_3 = 3$, $n_4 = 4$, $n_5 = 3$, and $MS_W = 6.333$. For $\alpha = 0.05$, we find from Appendix Table V

Example 8 *(continued)*

that $S^2 = F[4, 12; 0.95] = 3.26$. Now, for the given contrast, we get

$$\sum_{i=1}^{a} \ell_i^2/n_i = \frac{1}{3} + \frac{4}{4} + \frac{1}{3} + \frac{1}{4} + \frac{1}{3} = \frac{9}{4},$$

and

$$S\sqrt{(a-1)\,\text{MS}_W \sum_{i=1}^{a} \ell_i^2/n_i} = \sqrt{(3.26)}\sqrt{\left\{(5-1)(6.333)\left(\frac{9}{4}\right)\right\}} = 13.63.$$

On substituting the appropriate quantities in (2.19.15), the 95 percent simultaneous confidence interval for L is obtained as

$$6.00 - 13.63 < L < 6.00 + 13.63$$

or

$$-7.63 < L < 19.63.$$

Note from Example 2 that this interval is larger than the interval based on the usual t distribution. Again, the interval includes zero, and hence the contrast is not significantly different from zero.

Interpretation of Tukey's and Scheffé's methods

Suppose one were to calculate the confidence intervals for all conceivable contrasts by using (2.19.12) or (2.19.15). Then, according to Tukey's and Scheffé's methods, the entire set of confidence intervals would be correct in $100(1 - \alpha)$ percent of repetitions of the experiment. Note that this is a somewhat different interpretation than one from the ordinary confidence interval. When we make a 95 percent confidence interval for, say, a single parameter θ, we are correct in saying that if we took all possible random samples of size n and calculated the interval for each, the interval would cover the true value of θ in 95 percent of the cases. For the multiple comparisons, however, we are referring to all possible comparisons that might be made on a given set of data, and the probability statement is about the event of all such intervals computed from a set of data covering the corresponding true values. It is, therefore, important that the initial F test be significant giving us prior reason to believe that reliable departures from the hypothesis exist. These differences are to be found among the possible comparisons.

One-Way Classification

Remark: Researchers who have used Tukey's and Scheffé's methods have occasionally been somewhat surprised to find that a significance of the overall analysis of variance F test has not led to at least one significant contrast. We expect that if the overall test is significant at the α-level, then at least the maximum possible contrast will also be significant at the α-level. Unfortunately, the maximum possible contrast may have been of little interest, and, therefore, may not have been computed. There is no guarantee that the obvious contrasts (i.e., the differences within pairs of means) or the contrasts most interesting to the experimenter will be significant when the overall F test is significant.

Comparison of Tukey's and Scheffé's methods

In the following, we provide relative merits and drawbacks of Tukey's and Scheffé's methods of multiple comparisons.

1. The Tukey's method can be used only with equal sample sizes for all factor levels, but the Scheffé's method is applicable whether the sample sizes are equal or not.
2. Although the Tukey's method is applicable for any general contrast, the procedure is most powerful when comparing simple pairwise differences and not when making more complex comparisons.
3. If only pairwise comparisons are of interest, and all factor levels have equal sample sizes, Tukey's method gives shorter confidence intervals and thus is more powerful.
4. In the case of comparisons involving general contrasts, Scheffé's method tends to give narrower confidence limits, and thus provides a more powerful significance test.
5. The Scheffé's method has the property that if the F test produces significant results, then the corresponding Scheffé's multiple comparison will detect at least one statistically significant contrast from all possible contrasts. Thus, we are able to draw more conclusions than merely that all factor level means are different.
6. The Scheffé's method requires the use of the tables of the F distribution which are more readily available than the tables of the Studentized range distribution used by the Tukey's method.
7. The Scheffé's method is less sensitive to violations of normality and homogeneity of variance assumptions than is the Tukey's method.

OTHER MULTIPLE COMPARISON METHODS

In addition to Tukey's and Scheffé's methods described previously, there are a number of other multiple comparison procedures that are used widely in many substantive fields of research. In the following, we briefly outline some other common procedures for making a post hoc comparison.

Least significant difference test

The test also commonly referred as the protected least significant difference (LSD) was originally proposed by Fisher (1935). The test is carried out in steps as follows:

1. First, an overall F test at a given level of significance (α) is carried out to determine whether there are significant differences among the treatment groups.
2. Only if the F test in step 1 is significant, pairwise comparisons among treatments are performed using a t test at level α.

Assuming a two-sided alternative, the pair of means μ_i and μ'_i would be declared significant if

$$|\bar{y}_{i.} - \bar{y}_{i'.}| > t[N - a, 1 - \alpha/2]\sqrt{\mathrm{MS}_W\left(\frac{1}{n_i} + \frac{1}{n_{i'}}\right)}. \qquad (2.19.17)$$

The quantity to the right of the inequality in (2.19.17) is called the least significant difference. If the design is balanced, that is, $n_1 = n_2 = \cdots = n_a = n$, then it reduces to $t[a(n - 1), 1 - \alpha/2]\sqrt{2\mathrm{MS}_W/n}$. To use the LSD procedure, one need simply compare the observed differences between each pair of sample means to the corresponding least significant difference. If the sample difference exceeds this quantity, one may conclude that the pair of means are significantly different.

Remark: The use of the preliminary F test in the LSD procedure helps to protect the overall error rate under the null hypothesis of no differences in the set of treatment effects. However, even if a single comparison differs significantly from zero, the LSD approach does not protect against finding chance differences among other comparisons.

Bonferroni's test

The test is based on the principle that if there are k null hypotheses to be tested, then a desired overall error rate of at most α can be achieved by testing each null hypothesis at level α/k. Equivalently, if there are k confidence intervals each constructed at confidence level $100(1 - \alpha/k)$ percent, then they all hold simultaneously with confidence level of at least $100(1 - \alpha)$ percent. To see how this works, suppose that each null hypothesis is tested at level α^* and let E_i denote the event that the i-th null hypothesis is rejected. Then, the overall probability of a type I error (α) is given by

$$\begin{aligned} \alpha &= P\{E_1 U E_2 U \cdots U E_k\} \\ &\leq P(E_1) + P(E_2) + \cdots + P(E_k) \\ &= k\alpha^*. \end{aligned}$$

Thus, if each one of the k null hypotheses is rejected at level α/k, the overall error rate is at most α. The procedure is known as the Bonferroni's method, since it is based on the Bonferroni or Boole inequality. Sometimes it is also referred to as Dunn's multiple comparison procedure following Dunn (1961) who examined the properties of the procedure in detail and prepared tables to facilitate its use. The method is fairly simple and versatile and gives reasonably good results if k is not very large. The procedure tends to be somewhat conservative, that is, true confidence levels tend to be greater than $1 - \alpha$. The method should be used when there are only few comparisons to be made and none of the other procedures are appropriate.[20] Special percentage points of the t distribution for very small values of α are usually required to determine Bonferroni intervals. Specially designed tables for this purpose are given in Dunn (1961), Pearson and Hartley (1970, Table 9), Bailey (1977), Miller (1981, p. 238), and Kafadar and Tukey (1988). Moses (1978) provides charts for finding upper percentage points for α in the range of 0.01 to 0.00001. Koehler (1983) gives a fairly simple and accurate approximation for the extreme percentiles of the t distribution. Many statistical packages and other computer programs have standard routines for calculating percentage points. Some selected percentage points of the t distribution to determine Bonferroni intervals are given in Appendix Table XIII. For some further discussions and details about the Bonferroni statistic, see Dunn (1959, 1961) and Dunn and Massey (1965).

Remark: Holm (1979) introduced a modified Bonferroni test that consists of a class of sequentially rejective Bonferroni (SRB) procedures which results in greater power than the Bonferroni's test. Under Holm's SRB criterion, if any hypothesis is rejected at the level $\alpha^* = \alpha/k$, then the denominator of α^* for the next test is $k - 1$, and the criterion continues to be modified in a stepwise manner, with the denominator of α^* decreased by 1 each time a hypothesis is rejected, so that tests can be conducted at successively higher significance levels. The experimentwise error rate of the SRB procedures is $\leq \alpha$ as is that of the standard Bonferroni procedure. Shaffer (1986) introduced a refinement of Holm's SRB test known as the modified sequentially rejective Bonferroni (MSRB) test which is at least as powerful as the Holm's test while maintaining an experimentwise error rate $\leq \alpha$.

Dunn-Šidák's test

According to Bonferroni's or Dunn's test, given k comparisons or contrasts each to be tested at the level α^*, the overall error rate (α) cannot exceed $k\alpha^*$. For small values of α, it provides an excellent approximation to the upper bound. However, an even better approximation to the upper bound can be obtained by

---

[20] Fleiss (1986, pp. 106–107) made the following recommendations regarding the use of the Bonferroni method. It should be preferred to Scheffé if the number of comparisons is less than a^2; it should be preferred to Tukey if fewer than all $a(a - 1)/2$ comparisons are needed or if a relatively small number of other comparisons are to be made; and it should be preferred over the Dunnett if the comparisons of interest are other than or in addition to those between each of several treatments and a control.

a multiplicative inequality given by Šidák (1967). It can be shown that[21]

$$\alpha \le 1 - (1 - \alpha^*)^k \le k\alpha^*.$$

Thus, instead of testing each contrast at the α/k level of significance, as in Bonferroni's or Dunn's method, each contrast can be tested at the $1-(1-\alpha)^{1/k}$ level of significance[22]. In general, the use of the Dunn-Šidák's method requires a slightly smaller critical value and hence leads to a more powerful test and a narrower confidence interval than the Bonferroni's method. For example, let $\alpha = 0.05$ and suppose there are five contrasts to be tested. Now, $\alpha/k = 0.05/5 = 0.01$ and $1 - (1 - \alpha)^{1/k} = 1 - (1 - 0.05)^{1/5} = 0.0102$. Thus, the difference between the two significance levels is negligible. Games (1977) developed tables of critical values of the t statistic for use with the Dunn-Šidák's method. The values are reprinted in Appendix Table XIV. Dunn (1961) and Games (1977) made comparisons of Bonferroni and Dunn-Šidák procedures with Tukey and Scheffé procedures and found that when there are many means in an experiment and the number of comparisons of interest is relatively small compared to the number of means in the experiment, the Bonferroni and Dunn-Šidák procedures yield shorter confidence intervals than either the Tukey or Scheffé procedure.

Newman-Keuls's test

The test also known as the Student-Newman-Keuls's test was first proposed by Newman (1939) and subsequently popularized by Keuls (1952). The procedure follows a predetermined criterion for grouping means into subsets and adjusts the overall error rate α according to the number of means to be tested. It uses the Studentized critical range values determined by

$$W_{(a)} = q[a, a(n-1); 1-\alpha]\sqrt{\frac{MS_W}{n}},$$

where $q[a, a(n-1); 1-\alpha]$ is the $100(1-\alpha)$th percentage point of the Studentized range distribution with parameters a and $a(n-1)$. The procedure consists of arranging the a means $\bar{y}_{1.}, \bar{y}_{2.}, \ldots, \bar{y}_{a.}$ in ascending order as $\bar{y}_{(1)} < \bar{y}_{(2)} < \cdots < \bar{y}_{(a)}$. It then divides the group means into mutually exclusive subsets so that means within a subset are not significantly different and the means from distinct subsets are different. For this purpose, the quantity $\bar{y}_{(a)} - \bar{y}_{(1)}$, called the range of the set of means, is calculated. Now the observed range $\bar{y}_{(a)} - \bar{y}_{(1)}$ is compared with $W_{(a)}$. If it is less than $W_{(a)}$, the procedure stops and we conclude that the a means are not significantly different. If $\bar{y}_{(a)} - \bar{y}_{(1)}$ is greater than $W_{(a)}$, we

---

[21] This result was earlier proved by Dunn (1958) for certain special cases.
[22] Assuming k independent significance tests, each using a significance level of α^*, the probability of not committing type I error in any of the k tests is $(1-\alpha^*)^k$. The overall or experimentwise error rate (i.e., the probability of committing at least one type I error rate) is then given by $\alpha = 1 - (1 - \alpha^*)^k$. The solution for α^* yields $\alpha^* = 1 - (1 - \alpha)^{1/k}$.

divide $\bar{y}_{(1)}, \bar{y}_{(2)}, \ldots, \bar{y}_{(a)}$ into two groups, one containing $\bar{y}_{(a)}, \bar{y}_{(a-1)}, \ldots, \bar{y}_{(2)}$, and the other containing $\bar{y}_{(a-1)}, \bar{y}_{(a-2)}, \ldots, \bar{y}_{(1)}$. Next, each of the ranges in two subgroups, viz., $\bar{y}_{(a)} - \bar{y}_{(2)}$ and $\bar{y}_{(a-1)} - \bar{y}_{(1)}$, is compared with $W_{(a-1)}$ determined by

$$W_{(a-1)} = q[a - 1, a(n - 1); 1 - \alpha]\sqrt{\frac{MS_W}{n}}.$$

If either range does not exceed $W_{(a-1)}$, then the means in each of the two groups are not significantly different and the procedure stops. If either or both ranges exceed $W_{(a-1)}$, then the $a - 1$ means in the corresponding group(s) are further divided into two groups of $a - 2$ means each and the ranges for these groups are compared with $W_{(a-2)}$. The procedure is continued until a group of i means is found whose range does not exceed $W_{(i)}$ or all the means have been compared.

Duncan's Multiple Range test

This test developed by Duncan (1952, 1955) also ranks the group means by magnitude and then obtains subsets of means that are not significantly different. The method adjusts its overall error rate for each comparison rather than a prechosen level based on the total number of group means to be determined. The procedure is carried out exactly the same way as the Newman-Keuls's test except that the observed ranges are now compared with Duncan's critical range values determined by

$$D_{(a)} = R[a, a(n - 1); 1 - \alpha]\sqrt{\frac{MS_W}{n}},$$

where $R[a, a(n - 1); 1 - \alpha]$ denotes the $100(1 - \alpha)$th percentage point based on Duncan's multiple range distribution with parameters a and $a(n - 1)$. Some selected percentage points of Duncan's multiple range distribution are given in Appendix Table XII. The procedure has been found to be somewhat less conservative than Newman-Keuls's test.

Dunnett's test

The test developed by Dunnett (1955) is especially designed for experiments that include a control group and the researcher wishes to compare all the remaining group means with the control. The procedure is a simple modification of the usual t test with the change that the differences between means involving the control group $|\bar{y}_{i.} - \bar{y}_c|, i = 1, 2, \ldots, a-1$ are compared with the critical ranges determined by

$$D[a - 1, a(n - 1); 1 - \alpha]\sqrt{MS_W\left(\frac{1}{n_i} + \frac{1}{n_c}\right)},$$

where $\bar{y}_c$ is the mean of the control group and $D[a-1, a(n-1); 1-\alpha]$ denotes the $100(1-\alpha)$th percentage point of the Dunnett distribution with parameters $a-1$ and $a(n-1)$. Some selected percentage points of the Dunnett distribution are given in Appendix Table XI. Since the test does not make any comparison among the noncontrol groups, it is generally more powerful than other procedures for comparing a control group with other groups. It is important to point out that the Dunnett procedure should be used when the only comparisons of interests are the individual treatments against the control.[23]

Remark: The SAS PROBMC function computes probabilities or quantiles from the one-sided or two-sided distribution of the Dunnett's statistic with finite and infinite degrees of freedom for the variance estimate. For futher information and numerical examples of PROBMC function, see SAS Institute (1997, Chapter 28).

MULTIPLE COMPARISONS FOR UNEQUAL SAMPLE SIZES AND VARIANCES

Most of the multiple comparison procedures presented so far are appropriate for designs involving equal sample sizes and assuming equal population variances. We now summarize some of the procedures designed to be used with unequal sample sizes and variances.

Unequal sample sizes

For pairwise comparisons, Tukey (1953) and Kramer (1956, 1957) proposed a modification of Tukey's method where a harmonic mean of n_i and $n_{i'}$ is inserted for n in equation (2.19.14). The resulting intervals known as Tukey-Kramer intervals are given by

$$\bar{y}_{i.} - \bar{y}_{i'.} - q[a, N-a; 1-\alpha]\sqrt{\frac{1}{2}\left(\frac{1}{n_i} + \frac{1}{n_{i'}}\right)\text{MS}_W} < \mu_i - \mu_{i'}$$

$$< \bar{y}_{i.} - \bar{y}_{i'.} + q[a, N-a; 1-\alpha]\sqrt{\frac{1}{2}\left(\frac{1}{n_i} + \frac{1}{n_{i'}}\right)\text{MS}_W}. \quad (2.19.18)$$

Remark: Winer (1962; 1971, p. 216) and Miller (1966, p. 43; 1981, p. 43) proposed the idea for general type contrasts where the harmonic mean of unequal n_i's ($i = 1, 2, \ldots, a$) is substituted for n in equation (2.19.12). Simulation studies by Dunnett (1980a) showed that for pairwise comparisons Tukey-Kramer intervals (2.19.18) provide approximate probability coverage. Later, Hayter (1984) proved analytically that the probability coverage is always conservative ($\geq 1-\alpha$). However, Tukey-Kramer-Miller-Winer procedures are not robust to unequal variances (see, e.g., Howell and Games (1973); Keselman et al. (1975); Keselman and Rogan (1978)).

---

[23] If the researcher is interested in comparing the combination of groups to the control group, Scheffé's test or a generalization of Dunnett's test proposed by Shaffer (1977) may be used.

One-Way Classification

For pairwise comparisons involving k simultaneous intervals, the Bonferroni intervals are given by

$$\bar{y}_{i.} - \bar{y}_{i'.} - t[N - a, 1 - \alpha/2k]\sqrt{\text{MS}_W \left(\frac{1}{n_i} + \frac{1}{n_{i'}}\right)} < \mu_i - \mu_{i'}$$

$$< \bar{y}_{i.} - \bar{y}_{i'.} + t[N - a, 1 - \alpha/2k]\sqrt{\text{MS}_W \left(\frac{1}{n_i} + \frac{1}{n_{i'}}\right)}.$$

In the one-way classification involving all pairwise comparisons k is $a(a-1)/2$. However, occasionally k would be less if some mean comparisons *a priori* are not of interest.

For pairwise comparisons, Hochberg (1974) proposed a modification of Tukey-Kramer intervals given by

$$\bar{y}_{i.} - \bar{y}_{i'.} - m[a(a-1)/2, N - a; 1 - \alpha]\sqrt{\text{MS}_W \left(\frac{1}{n_i} + \frac{1}{n_{i'}}\right)} < \mu_i - \mu_{i'}$$

$$< \bar{y}_{i.} - \bar{y}_{i'.} + m[a(a-1)/2, N - a; 1 - \alpha]\sqrt{\text{MS}_W \left(\frac{1}{n_i} + \frac{1}{n_{i'}}\right)},$$

where $m[p, v, 1 - \alpha]$ is the $100(1 - \alpha)$th percentage point of the Studentized maximum modulus distribution for $p = a(a - 1)/2$ pairwise means with $v = N - a$ degrees of freedom for the error. (For a definition of the Studentized maximum modulus distribution, see Appendix J.) Some selected percentage points of the Studentized maximum modulus distribution are given in Appendix Table XV. The studies by Tamhane (1979) and Dunnett (1980a) have shown that the procedure is not robust to unequal variances involving unequal sample sizes.

Similarly, Spjotvoll and Stoline (1973) proposed the intervals

$$\bar{y}_{i.} - \bar{y}_{i'.} - q'[a, N - a; 1 - \alpha]\sqrt{\text{MS}_W}\, \max\left\{\frac{1}{\sqrt{n_i}}, \frac{1}{\sqrt{n_{i'}}}\right\}$$

$$< \mu_i - \mu_{i'} < \bar{y}_{i.} - \bar{y}_{i'.} + q'[a, N - a; 1 - \alpha]\sqrt{\text{MS}_W}\, \max\left\{\frac{1}{\sqrt{n_i}}, \frac{1}{\sqrt{n_{i'}}}\right\},$$

where $q'[p, v; 1-\alpha]$ is the $100(1-\alpha)$th percentage point of the Studentized augmented range distribution for p means with v degrees of freedom for the error. Tables of $q'[p, v; 1 - \alpha]$ have been prepared by Stoline (1978). Some selected percentage points of the Studentized augmented range distribution are given in Appendix Table XVI. These intervals, however, are conservative in the sense that the true coverage probability is greater than or equal to $1 - \alpha$.

Remark: Ury (1976) studied some of the foregoing procedures including Scheffé and Dunn-Šidák intervals and found that the choice of the "best interval" depends upon the particular combination of sample sizes, significance level, number of groups, and the error degrees of freedom. Stoline (1981) made a detailed comparison of all the foregoing procedures and recommended the general use of the Tukey-Kramer intervals.

Unequal population variances

For pairwise comparisons, Games and Howell (1976) proposed a procedure where MS_W in equation (2.19.18) is replaced by $\sqrt{S_i^2/n_i + S_{i'}^2/n_{i'}}$ for the difference involving the (i, i')-pair of means. The error degrees of freedom $N - a$ in $q[a, N - a; 1 - \alpha]$ is further replaced by

$$v_{i,i'} = \frac{\left(S_i^2/n_i + S_{i'}^2/n_{i'}\right)^2}{\left[\left(S_i^2/n_i\right)^2/(n_i - 1)\right] + \left[\left(S_{i'}^2/n_{i'}\right)^2/(n_{i'} - 1)\right]}.$$

Extensive Monte Carlo studies by Tamhane (1979) and Dunnett (1980b) have shown that the procedure can give nonconservative α values, as high as 0.84, with unequal variances. For a modification of (2.19.18) based on the Studentized augmented range distribution, see Hochberg (1976).

Dunnett (1980b) suggested a modification of the foregoing procedure in which the critical value $q[a, v_{i,i'}; 1 - \alpha]$ is replaced by

$$\frac{q[a, n_i - 1; 1 - \alpha]\left(S_i^2/n_i\right) + q[a, n_{i'} - 1; 1 - \alpha]\left(S_{i'}^2/n_{i'}\right)}{\left(S_i^2/n_i\right) + \left(S_{i'}^2/n_{i'}\right)}$$

It should be noted that the critical value given above corresponds to Cochran's (1964) approximate solution to the Behrens-Fisher problem and thus the procedure is expected to be conservative (Dunnett (1980b)).

Dunnett (1980b) proposed another modification that utilizes the same statistic as in Games and Howell but the critical value $q[a, v_{i,i'}; 1 - \alpha]$ is replaced by the critical value $m[a(a - 1)/2, v_{i,i'}; 1 - \alpha]$ of the Studentized maximum modulus distribution. Dunnett (1980b) found that the procedure is also conservative with unequal variances.

Remark: Weerahandi (1995) proposed a modification of the Scheffé's procedure of multiple comparison given by (2.19.16) to the case of unequal variances. The procedure is, however, too complicated and mathematically intractable for the practitioner to use in routine work.

2.20 EFFECTS OF DEPARTURES FROM ASSUMPTIONS UNDERLYING THE ANALYSIS OF VARIANCE MODEL

In making inferences from the analysis of variance model (2.1.1), we have made the following assumptions:

(i) e_{ij}'s are normally distributed;

(ii) e_{ij}'s have same variance σ_e^2; and
(iii) e_{ij}'s are independently distributed.

It stands to reason that in any real-life applications none of the preceding assumptions can be expected to be completely satisfied. One rarely draws independent random samples from populations that are exactly normally distributed with precisely equal variances. The question naturally arises: What are the effects of any departure from the assumptions of the model on the inferences made? For a thorough discussion of the topic, the reader is referred to Scheffé (1959, pp. 331–369), Miller (1986, Chapter 3), and Snedecor and Cochran (1989, Chapter 15). Here, we briefly summarize some of the main findings.

DEPARTURES FROM NORMALITY

For Model I, many investigations have been made to study the effect of nonnormality on both the level of significance and the power of the F test employed in the analysis of variance. Both analytic results (see, e.g., Scheffé (1959, pp. 345–351)) and empirical studies by Pearson (1931), Geary (1947), Gayen (1950), Box and Anderson (1955), Boneau (1960, 1962), Srivastava (1959), Bradley (1964), Tiku (1964, 1971), and Donaldson (1968) attest to the fact that the failure to satisfy this assumption has little effect on the F test. Thus, if departure from normality is not too extreme, the lack of normality does not present any serious problem, since the means will follow the normal distribution more closely than the variates themselves. Both the level of significance and the power of the F test are only slightly affected by any departure from normality. However, extreme nonnormality may result in a biased test. In this connection, it is important to mention that any departure from the kurtosis of the normal distribution (either more or less peaked) is much more serious than the skewness of the distribution in terms of the effects on inferences. Also, platykurtic (flat) and leptokurtic (peaked) distributions have little effect on the significance level but can have a marked effect on power, particularly when the sample sizes are small. Furthermore, only highly skewed distributions would have any marked effect either on the level of significance or the power of the F test.

The point estimates of the factor level means and their contrasts are unbiased irrespective of whether populations are normal or not. Hence, the F test is generally robust against any departures from normality (in skewness and/or kurtosis) if sample sizes are large or even if moderately large. For instance, the nominal level of significance might be 0.05 whereas the actual level for a nonnormal population might vary from .044 to .052 depending on the sample size and the magnitude of the kurtosis (Box and Anderson (1955)). Generally, the actual level of significance in the presence of positive kurtosis (platykurtic) is slightly higher than the specified one and the real power of the test for positive kurtosis is slightly higher than the normal one. If the underlying population has negative kurtosis (leptokurtic), the actual power of the test will be slightly lower than the normal one (Glass et al. (1972)). Single interval estimates of the factor level means and contrasts and some of the multiple comparison methods are also not much affected by the lack of normality provided the sample sizes are not too small. The robustness of multiple comparison tests in general has not been as thoroughly studied. Among the few studies in this area is that of Brown (1974). A number of studies, however, have investigated the robustness of several multiple comparison procedures, including Tukey and Scheffé, for exponential and chi-square distributions and found little effect on both significance level and power (see, e.g., Petrinovich and Hardyck (1969); Keselman and Rogan (1978)). Dunnett (1982) reported that Tukey is conservative both with respect to significance level and power for long-tailed distributions and to outliers. Similarly, Ringland (1983) found that the Scheffé was conservative for distributions with influence to outliers.

For Model II, the lack of normality has more serious implications than Model I. The estimates of the variance components are still unbiased, but their variances depend on the kurtosis of the distribution and the actual confidence coefficients for interval estimates of $\sigma_e^2, \sigma_\alpha^2, \sigma_\alpha^2/\sigma_e^2$ may be substantially different from the specified one (Singhal and Sahai (1992)). Furthermore, when testing the null hypothesis that the variance of a random effect is some specified value different from zero, the test is not robust to the assumption of normality. For some illustrations and numerical results, the reader is referred to Arvesen and Schmitz (1970) and Arvesen and Layard (1975). However, if one is concerned only with a test of the hypothesis $\sigma_\alpha^2 = 0$, then slight departures from normality have only minor consequences for the conclusions reached when the sample size is reasonably large (see, e.g., Tan and Wong (1980); Singhal and Singh (1984); Singhal et al. (1988)).

DEPARTURES FROM EQUAL VARIANCES

Both the analytical derivations by Box (1954a) and the empirical studies cited earlier indicate that if the variances are unequal, the F test for the equality of means under Model I is only slightly affected with respect to moderate violations of this assumption provided the sample sizes do not differ greatly and the parent populations are approximately normally distributed[24] (Glass et al. (1972)). Generally, unequal error variances increase the actual level of significance slightly higher than the specified level and result in a slight elevation of the power function to a degree related to the magnitude of differences among

---

[24] When the variances are unequal, an approximate test similar to the approximate t test when two group variances are unequal may be used (Welch (1956)). For a description of the test and some illustrative examples, see Zar (1996, pp. 189–190). The method has been shown to perform rather well when population variances are unequal (Kohr and Games (1974); Levy (1978a); Dijkstra and Werter (1981)). For some other approaches to analysis of variance involving heterogeneous variances, see James (1951), Brown and Forsythe (1974a,b), Bishop and Dudewicz (1978), Clinch and Keselman (1982), Krutchkoff (1988), Wilcox (1988, 1993), and Alexander and Govern (1994). For a survey and comparisons of traditional ANOVA alternatives with other alternative procedures, see Coombs et al. (1996).

variances (Box (1954a)). If larger variances are associated with larger sample sizes, the level of significance will be slightly less than the nominal value, and if they are associated with smaller sample sizes, it will be slightly greater than the nominal value (Horsnell (1953); Kohr and Games (1974)). Similarly, if the sample sizes do not differ greatly, the Scheffé's method for multiple comparison is only slightly affected due to any lack of homogeneity of error variances. Thus, the F test and related procedures are fairly robust against any violation from equal error variances provided the sample sizes are nearly equal. Comparisons of factor level means based on a single contrast, however, are significantly affected by unequal variances even when samples sizes are equal.

On the other hand, when different numbers of cases appear in various samples, even relatively small departures from the assumption of homogeneous variances can have very serious consequences for the validity of the final inference (see, e.g., Scheffé (1959, p. 351); Welch (1956); James (1951); Box (1954a); Brown and Forsythe (1974a); Bishop and Dudewicz (1978); Tan and Tabatabai (1986)). According to Box (1954a), for samples of unequal sizes, even a small violation of this assumption can have a marked effect on the level of significance. The actual significance level will exceed the nominal level when smaller samples are drawn from more heterogenous populations and will be less than the nomimal value when the smaller samples are drawn from more homogeneous populations. Furthermore, Rogan and Keselman (1977) found that the actual significance level may be appreciably larger when the variances are quite heterogenous. Moreover, the effect of unequal variances is not appreciably reduced simply by increasing the samples sizes, as long as the ratios of the sample sizes remain unchanged.

Krutchkoff (1988) made an extensive simulation study in order to determine the size and power of several analysis of variance procedures, including the F test, Kruskal-Wallis test and a new procedure called the K test. It was found that both the F test and the Kruskal-Wallis test are highly sensitive whereas the K test is relatively insensitive to the heterogeneity of variances. The Kruskal-Wallis test, however, is not as sensitive to the unequal error variances as the F test; and was found to be more robust to nonnormality (when the error variances are equal) than either the F test or the K test. A more recent study by Lix et al. (1996) seems to indicate that violations of the variance homogeneity assumptions can have serious consequences for control of the type I error rate regardless of whether group sizes are equal or unequal, but particularly in the latter case. The study also found that all the parametric alternatives of the analysis of variance test had superior performance when the variance homogeneity assumption was violated. Furthermore, when the group sizes were equal, the effect of nonnormality on the type I error rate of the F test was no different when variances were equal than when they were unequal. The error rates remained close to the nominal level regardless of the degree of nonnormality when variances were equal and were always inflated across the nonnormal distributions when variances were unequal. The pattern was also evident when group sizes were unequal. Thus, whenever possible, the experimenter should try to achieve a nearly equal number of cases in each factor level unless the assumption of equal population variances can reasonably be assured in the experimental context. It should be observed that the use of equal sample sizes for all factor levels not only tends to minimize the effects of unequal variances using the F test, but also simplifies the computational procedure.

For Model II, the effect on the robustness of the F test is the same as for the fixed effects model. For balanced designs the effects are minimal but can have serious effects for unbalanced designs. However, the lack of homoscedasticity or unequal error variances can have serious effects on inferences about the estimation of variance components even when all factor levels contain equal sample sizes.

Departures from Independence of Error Terms

Lack of independence can result from biased measurements or possibly from a poor allocation of treatments to experimental units. Departure from independence could also arise in an experiment in which experimental units or plots are laid out in a field so that adjacent plots give similar yields. Lack of independence can likewise result from correlation in time rather than in space. Thus, the most frequent violation of independence assumption occurs when the observations are recorded over some time-space coordinate in which adjacent observations tend to be correlated. Nonindependence of the error terms can have important effects on inferences for both Models I and II. If this assumption is not met, both the level of significance and the power of the F test may be strongly affected and very serious errors in inferences can be made (Scheffé (1959, p. 945)). The direction of the effect depends on the nature of the dependence of the error terms. In most cases encountered in practice, the dependence tends to make the value of the ratio too large and consequently the significance level will be smaller than it should be (although the opposite can also be the case). Thus, positive correlations among the error variances within a factor level may cause too many significant results based on the F test and the effect on the t test may be even greater.

Since the violation of this assumption is often difficult to remedy, every possible effort should be made to obtain independent random samples. The use of ramdon sampling or randomization in various stages of the study can be a most important protection against independence of error terms. In general, great care should be taken to see that the data are based on independent observations, both between and within groups; that is, each observation is in no way related to any of the other observations. Although dependency among the error terms creates a special problem in any analysis of variance, it is not required that the observations themselves be completely independent for Model II to apply. However, just as Model I is not robust to the assumption of independence, Model II is also not robust to this assumption. Violation of this assumption generally results in declaring too many significant results in the F test. The effects on various point and interval estimates of σ_α^2, however, are unknown.

2.21 TESTS FOR DEPARTURES FROM ASSUMPTIONS OF THE MODEL

As we have seen in the preceding section, the analysis of variance procedure is robust and can tolerate certain departures from the specified assumptions. It is, nevertheless, recommended that whenever a departure is suspected it should be investigated. In this section, we briefly discuss the tests for normality and homoscedasticity.

Remark: Before carrying out the formal statistical procedures for testing normality and homogeneity described here, it may be fairly informative and useful to explore the data graphically. For example, one can use box-plots for different groups and/or within group histograms to see if the distribution of values in each group is symmetric and free of any gross outliers and other anomalies in the data; and if the spread of the data across groups is fairly constant. Ideally, if the analysis of variance assumptions are satisfied, box-plots should be symmetric and the spreads across the groups should be nearly the same. However, in many practical problems involving small sample sizes, the skewness and homogeneity may be difficult to evaluate in this way. Box-plots are discussed in most introductory statistics textbooks, or one may refer to a book on exploratory data analysis (see, e.g., Tukey (1977); Chambers et al. (1983); Hoaglin et al. (1983); Cleveland (1985)). Box-plots and related graphical techniques are also available in most statistical packages currently being used for data analysis. For a discussion of analysis of variance from the viewpoint of exploratory data analysis, see Hoaglin et al. (1991).

TESTS FOR NORMALITY

A relatively simple technique to determine the appropriateness of the assumption of normality is to graph the data points on a normal probability paper. If a straight line can be drawn through the plotted points, the assumption of normality is considered to be reasonable.

We now consider some formal tests for normality. They are the chi-square goodness-of-fit test, and the tests for skewness and kurtosis which are often used as supplements to the chi-square test.

Chi-square goodness-of-fit test

In this test the data are grouped into classes to form a frequency distribution and the sample mean and standard deviation are calculated. From these quantities a normal distribution is fitted and expected frequencies in each class are obtained. Let o_i and e_i represent the observed and expected frequencies for the i-th class.

Then the test criterion is based on the quantity

$$X^2 = \sum_i (o_i - e_i)^2 / e_i, \qquad (2.21.1)$$

where the summation is taken over all the classes. If the data actually come from a normal distribution, then the quantity (2.21.1) follows approximately a chi-square distribution with $k - 3$ degrees of freedom, where k is the number of classes used in the calculation of X^2. If the data come from some other distribution, the observed o_i will tend to agree poorly with the values of e_i that are expected on the assumption of normality and the calculated value of X^2 will become large. Consequently, large values of X^2 lead to the rejection of the hypothesis of normality. Thus, if the calculated value of the statistic X^2 exceeds $\chi^2[k - 3, 1 - \alpha]$, the $100(1 - \alpha)$th percentage point of the chi-square distribution with $k - 3$ degrees of freedom, we reject the null hypothesis that the sample is selected from a normal population.

For the validity of the chi-square test, it is required that the expected frequencies e_i should not be too small. Small expected values are likely to occur only in the extreme classes. A working rule is that two extreme expectations may be each as low as 1, provided that most of the other expected values exceed 5. If the expected values are lower than 1, classes are combined to give an expectation of at least 1. For a more detailed discussion of these questions refer to Cochran (1954), Larntz (1978), and Koehler and Larnzt (1980).

Test for skewness

One indication of nonnormality occurs when the relative frequency histogram for the sample data is highly skewed to either the left or right. A measure of amount of skewness is given by μ_3, the third moment about the population mean, which is the average value of $(x - \mu)^3$ taken over the population. The skewness is positive or negative according to the sign of μ_3. If low values are clustered close to the mean μ but high values extend far above the mean, μ_3 will be positive since the large positive contributions of $(x - \mu)^3$ when x exceeds μ will predominate over the smaller negative contributions of $(x - \mu)^3$ obtained when x is less than μ. Similarly, μ_3 will be negative when the lower tail is the extended one. The meaning of positive and negative values of μ_3 is illustrated in Figure 2.6

The actual measure of skewness is given by the coefficient of skewness defined as

$$\gamma_1 = \frac{\mu_3}{\mu_2^{3/2}} = \frac{\mu_3}{\sigma^3}. \qquad (2.21.2)$$

The quantity (2.21.2) is independent of the measurement scale and can be

One-Way Classification

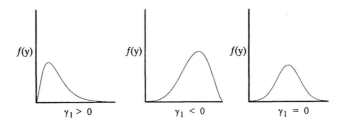

FIGURE 2.6 Curves Exhibiting Positive and Negative Skewness and Symmetrical Distribution.

estimated from the sample data by

$$\hat{\gamma}_1 = \frac{m_3}{m_3^{3/2}}, \qquad (2.21.3)$$

where

$$m_3 = \sum_{i=1}^{n}(y_i - \bar{y})^3/n \quad \text{and} \quad m_2 = \sum_{i=1}^{n}(y_i - \bar{y})^2/n.$$

A test of the null hypothesis that the sample data are selected from a normal population can be based on the statistic

$$Z = \frac{\hat{\gamma}_1}{\sqrt{6/n}}, \qquad (2.21.4)$$

where Z is a standard normal variate. The assumption that Z has approximately a standard normal distribution is accurate enough for this test if n exceeds 150. For sample sizes between 25 and 200 the one-tailed 5 percent and 10 percent significance values of $\hat{\gamma}_1$ have been determined from a more accurate approximation and appear in Pearson and Hartley (1970). Some of these values are reprinted in Appendix Table XVII(a).

Test for kurtosis
A second kind of departure from normality can be detected by examining the kurtosis of the distribution. The kurtosis of a distribution is measured by the quantity

$$\gamma_2 = \frac{\mu_4}{\mu_2^2} = \frac{\mu_4}{\sigma^4}, \qquad (2.21.5)$$

where γ_2 is called the coefficient of kurtosis. Unlike the coefficient of skewness (γ_1), γ_2 measures the heaviness of the tail of a distribution. For the normal population $\mu_4 = 3\mu_2^2$ so that $\gamma_2 = 3$. The lighter-tailed distributions will have a large pile-up near μ and so $\gamma_2 > 3$. The heavier-tailed distributions, such as a t distribution, will have less pile-up about μ and so $\gamma_2 < 3$. This is illustrated in Figure 2.7.

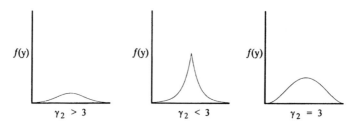

FIGURE 2.7 Curves Exhibiting Positive and Negative Kurtosis and the Normal Distribution.

The quantity (2.21.5) can be estimated by

$$\hat{\gamma}_2 = \frac{m_4}{m_2^2}, \tag{2.21.6}$$

where

$$m_4 = \sum_{i=1}^{n}(y_i - \bar{y})^4/n \quad \text{and} \quad m_2 = \sum_{i=1}^{n}(y_i - \bar{y})^2/n.$$

For large sample sizes ($n \geq 1{,}000$) a test of the null hypothesis that $\gamma_2 = 3$ can be based on the statistic

$$Z = \frac{\hat{\gamma}_2 - 3}{\sqrt{24/n}}, \tag{2.21.7}$$

where again Z has approximately a standard normal distribution. Unfortunately, one seldom encounters a sample size with 1,000 or more observations and the test statistic (2.21.7) has very little practical utility. For smaller sample sizes, however, upper and lower percentage points of the distribution of $\hat{\gamma}_2$ have been tabulated and can be used to establish the veracity of the null hypothesis. Tables of critical values are given in Pearson and Hartley (1970). Some of these values are reprinted in Appendix Table XVII(b).

Geary (1935, 1936) developed an alternative test criterion for kurtosis based on the statistic

$$G = \text{Mean deviation/Standard deviation}$$

$$= \frac{\sum_{i=1}^{n}|y_i - \bar{y}|/n}{\sqrt{m_2}}. \tag{2.21.8}$$

One-Way Classification

The significance values of the statistic (2.21.8) have been tabulated for sample sizes down to $n = 11$. If y is a normal deviate, the value of G when determined for the whole population is 0.7979. Positive kurtosis yields higher values and negative kurtosis lower values of G. When applied to the same data, the statistics $\hat{\gamma}_2$ and G usually agree well in their conclusions. The advantages of G are that tables are available for smaller sample sizes and that G is relatively easier to compute.

OTHER TESTS FOR NORMALITY

The foregoing procedures are some of the classical tests of normality. Over the years a large number of other techniques have been developed for testing for departures from normality. In the following we describe some powerful omnibus tests proposed for the problem. For further information on tests of normality, see Royston (1983, 1991, 1993a,b,c).

Shapiro-Wilk's W test

Shapiro and Wilk (1965) proposed a relatively powerful procedure based on the statistic

$$W = \frac{\left(\sum_{i=1}^{n} a_i y_{(i)}\right)^2}{\sum_{i=1}^{n}(y_i - \bar{y})^2}, \qquad (2.21.9)$$

where $y_{(1)} \leq y_{(2)} \leq \cdots \leq y_{(n)}$ represent the order statistics and the coefficients a_i's are the optimal weights for the weighted least squares estimator of the standard deviation for a normal population. Inasmuch as $a_{n-i+1} = -a_i$, the expression $\sum_{i=1}^{n} a_i y_{(i)}$ can be written as $\sum_{i=1}^{k} a_{n-i+1}(y_{(n-i+1)} - y_{(i)})$ where $k = n/2$, if n is even, or $(n-1)/2$, if n is odd. For n odd, the middle observation is used in the calculation of $\sum_{i=1}^{n}(y_i - \bar{y})^2$, but is not used in the calculation of $\sum_{i=1}^{k} a_{n-i+1}(y_{(n-i+1)} - y_{(i)})$. Thus, for n odd, $a_{(n+1)/2} = a_{k+1}$ appears as zero in Appendix Table XVIII. Also note that the W test is two-sided because the test statistic (2.21.9) is in a quadratic form. The hypothesis of normality is rejected at the α significance level if W is less than the $(1-\alpha)$th quantile of the null distribution of W. The coefficients a_i, for $2 \leq n \leq 50$, were given by the authors and some selected values are given in Appendix Table XVIII. A short table of critical values of the statistic (2.21.9) originally given by Shapiro and Wilk (1965) is also reprinted in Appendix Table XIX. Royston (1982a, b) has provided an approximation to the null distribution of W and a FORTRAN algorithm for $n \leq 2000$. The Shapiro and Wilk W test is one of the most powerful omnibus tests for testing normality. Extensive empirical Monte Carlo simulation studies by Shapiro et al. (1968) and Pearson et al. (1977) have shown that W is more powerful against a wide range of alternative distributions. The test is found to be good against short or very long-tailed alternatives even for a sample as small as 10.

Example 1. To illustrate the procedure, consider a sample of 10 observations given by 2.4, 2.7, 2.6, 3.4, 3.2, 3.5, 3.2, 3.4, 3.6, and 3.5. The ordered statistics are determined as

$$2.4, 2.6, 2.7, 3.2, 3.2, 3.4, 3.4, 3.5, 3.5, 3.6;$$

Since $n = 10$, we have $k = 5$. Using the 5 coefficients a_1, a_2, a_3, a_4, a_5 from Appendix Table XVIII, we obtain

$$\sum_{i=1}^{k} a_{n-i+1}(y_{(n-i+1)} - y_{(i)})$$

$$= 0.5739(3.6 - 2.4) + 0.3291(3.5 - 2.6)$$
$$+ 0.2141(3.5 - 2.7) + 0.1224(3.4 - 3.2) + 0.0399(3.4 - 3.2)$$
$$= 1.18861.$$

Furthermore, for the given set of observations, $\sum_{i=1}^{n}(y_i - \bar{y})^2 = 1.645$. Hence, $W = (1.18861)^2/1.645 = 0.859$. From Appendix Table XIX, the 5 percent critical value of the W statistic is $W(10, 0.05) = 0.842$. Since $W > W(10, 0.05)$, we cannot reject the hypothesis of normality and conclude that it is reasonable to assume that the data are normally distributed.

Shapiro-Francia's test

Shapiro and Francia (1972) proposed a modification of the W statistic defined by

$$W' = \frac{\left(\sum_{i=1}^{n} b_i y_{(i)}\right)^2}{\sum_{i=1}^{n}(y_i - \bar{y})^2}, \qquad (2.21.10)$$

where the coefficients b_i are determined by

$$b_i = \frac{m_i}{\sqrt{\sum_{i=1}^{n} m_i^2}},$$

with m_i representing the expected values of the order statistics from a unit normal distribution. Inasmuch as $b_{n-i+1} = -b_i$, the expression $\sum_{i=1}^{n} b_i y_{(i)}$ can be written as $\sum_{i=1}^{k} b_{n-i+1}(y_{(n-i+1)} - y_{(i)})$ where $k = n/2$, if n is even, or $(n-1)/2$, if n is odd. For n odd, the middle observation is used in the calculation of

One-Way Classification

$\sum_{i=1}^{n}(y_i - \bar{y})^2$, but is not used in the calculation of $\sum_{i=1}^{k} b_{n-i+1}(y_{(n-i+1)} - y_{(i)})$. Again, note that W' test is two-sided because the test statistic (2.21.10) is in a quadratic form. The hypothesis of normality is rejected at the α significance level if W' is less than $(1-\alpha)$th quantile of the null distribution of W'. Extensive tables of m_i are given in Harter (1961, 1969b). A small table of critical values of the statistic (2.21.10) is given by Shapiro and Francia (1972).

Example 2. To illustrate the procedure, we use the data on birthweights of twelve piglets in a particular litter from an experiment reported by Royston et al. (1982). The data have also been referred to and analyzed by Royston (1993c). It is widely believed that piglets provide a good model for the human neonate, especially in studies involving turnover of glucose. The order statistics of birthweight data and the corresponding expected values of the order statistics from a unit normal distribution are given in Table 2.12.

TABLE 2.12
Calculations for Shapiro Francia's Test

i	1	2	3	4	5	6	7	8	9	10	11	12
$y_{(i)}$	605	858	862	992	1006	1018	1020	1079	1088	1110	1120	1166
m_i	−1.6292	−1.1157	−0.7929	−0.5368	−0.3122	−0.1025	0.1025	0.3122	0.5368	0.7929	1.1157	1.6292

For $n = 12, k = 6$, the coefficients $b_i, i = 1, 2, \ldots, 6$, are determined as

$$b_1 = \frac{0.1025}{\sqrt{9.84778}} = 0.0327, \quad b_2 = \frac{0.3122}{\sqrt{9.84778}} = 0.0995,$$

$$b_3 = \frac{0.5368}{\sqrt{9.84778}} = 0.1711, \quad b_4 = \frac{0.7929}{\sqrt{9.84778}} = 0.2527,$$

$$b_5 = \frac{1.1157}{\sqrt{9.84778}} = 0.3555, \quad \text{and} \quad b_6 = \frac{1.6296}{\sqrt{9.84778}} = 0.5192.$$

Now,

$$\sum_{i=1}^{k} b_{n-i+1}(y_{(n-i+1)} - y_{(i)})$$
$$= 0.5192(1166 - 605) + 0.3555(1120 - 858)$$
$$+ 0.2527(1110 - 862) + 0.1711(1088 - 992)$$
$$+ 0.0995(1079 - 1006) + 0.0327(1020 - 1018)$$
$$= 470.8363;$$

Example 2 *(continued)*

and

$$\sum_{i=1}^{n}(y_i - \bar{y})^2 = 263{,}616.6667.$$

Hence, the Shapiro-Francia Statistic W' is given by

$$W' = (470.8363)^2/263{,}616.6667 = 0.841.$$

Since the critical values of the W' statistic are not readily available, we employ a normal approximation due to Royston (1993c). It can be shown that the statistic $\log_e(1 - W')$ is approximately normally distributed with mean $\hat{\mu} = -1.2725 + 1.052(v - u)$ and standard deviation $\hat{\sigma} = 1.0308 - 0.26758(v + 2/u)$ where $u = \log_e(n)$ and $v = \log_e(u)$. The values of the normal deviate $Z' = \{\log_e(1 - W') - \hat{\mu}\}/\hat{\sigma}$ are referred to the upper-tail critical values of the standard normal distribution. Values of $Z' > 1.645$ indicate departures from normality at the 5 percent significance level. For the birthweight data considered previously,

$$\hat{\mu} = -2.929, \quad \hat{\sigma} = 0.572,$$

and

$$Z' = \{\log_e(1 - 0.841) - (-2.929)\}/(0.572) = 1.91.$$

Since $Z' > 1.645$, the hypothesis of normality of the birthweight data is rejected ($p = 0.028$).

D'Agostino's D test
D'Agostino (1971) proposed a test statistic

$$D = \frac{\sum_{i=1}^{n}\left\{i - \frac{1}{2}(n+1)\right\}y_{(i)}}{n\sqrt{n\sum_{i=1}^{n}(y_i - \bar{y})^2}}, \qquad (2.21.11)$$

which is also a modification of the W statistic where the coefficients a_i are replaced by $W_i = i - \frac{1}{2}(n+1)$ and, thus, no tables of coefficients are needed.

One-Way Classification

Note that in contrast to W and W' tests, the D test is two-sided since the statistic (2.21.11) is in a linear form. The hypothesis of normality is rejected at the α significance level if D is less than the $\alpha/2$th quantile or greater than $(1 - \alpha/2)$th quantile of the null distribution of D. The test, originally proposed for moderate sample sizes, is also an omnibus test and can detect deviations from normality both for skewness or kurtosis. Tables of percentage points of the approximate standardized D distribution of the statistic (2.21.11) were given by D'Agostino (1972). The test is computationally much simpler than the Shapiro-Wilk's test. The studies by Theune (1973) have shown that the Shapiro-Wilk's test is preferable over D'Agostino's test for sample sizes up to 50 for lognormal, chi-square, uniform, and U-shaped alternatives.

Example 3. To illustrate the procedure, we again consider the data of Example 1. For the given data set,

$$\sum_{i=1}^{n}\left\{i - \frac{1}{2}(n+1)\right\}y_{(i)} = \sum_{i=1}^{n} iy_{(i)} - \frac{1}{2}(n+1)\sum_{i=1}^{n} y_{(i)}$$

$$= 1(2.4) + 2(2.6) + \cdots + 10(3.6)$$

$$- \frac{1}{2}(10+1)(2.4 + 2.6 + \cdots + 3.6)$$

$$= 184.2 - (5.5)(31.5)$$

$$= 10.95$$

and, as before, $\sum_{i=1}^{n}(y_i - \bar{y})^2 = 1.645$. Hence, $D = 10.95/\{10\sqrt{10(1.645)}\} = 0.26998$. From Appendix Table XX, the 5 percent critical value of the D statistic is $D(10, 0.05) = (0.2513, 0.2849)$. Since the calculated value of D lies in this interval, the hypothesis of normality is not rejected and we may conclude that it is reasonable to assume that the data are normally distributed.

For discussions of tests especially designed for detecting outliers, see Hawkins (1980), Beckman and Cook (1983), and Barnett and Lewis (1994). Robust estimation procedures have also been employed in detecting extreme observations. The procedures give less weight to data values that are extreme in comparison to the rest of the data. Robust estimation techniques have been reviewed by Huber (1981) and Hampel et al. (1986).

TESTS FOR HOMOSCEDASTICITY

If there are just two populations (i.e., $a = 2$), the equality of two population variances can be tested by using the usual F test from the fact that the statistic

$$F = \frac{S_1^2/\sigma_1^2}{S_2^2/\sigma_2^2}$$

has the Snedecor's F distribution with $n_1 - 1$ and $n_2 - 1$ degrees of freedom. Here, σ_1^2 and σ_2^2 are population variances and S_1^2 and S_2^2 are the corresponding sample estimators based on independent samples of sizes n_1 and n_2, respectively. However, with $a > 2$, rather than making all pairwise F tests, we want a single test that can be used to verify the assumption of equality of population variances. There are several tests available for this purpose. The three most commonly used tests are Bartlett's, Hartley's, and Cochran's tests.[25] The Bartlett's test compares the weighted arithmetic and geometric means of the sample variances. The Hartley's test compares the ratio of the largest to the smallest variance. The Cochran's test compares the largest sample variance to the average of all the sample variances. We now describe these procedures and illustrate their applications with examples. They, however, have lower power than is desired for most applications and are adversely affected by nonnormality.

In the following, we are concerned with testing the hypothesis:

$$H_0: \sigma_1^2 = \sigma_2^2 = \cdots = \sigma_a^2$$

versus (2.21.12)

$$H_1: \sigma_i^2 \ne \sigma_j^2 \quad \text{for at least one } (i, j) \text{ pair.}$$

Bartlett's test

The basic idea under the Bartlett's (1937a, b) test is as follows. Given the observations y_{ij}'s from model (2.1.1), let

$$T_A = \frac{1}{\sum_{i=1}^{a}(n_i - 1)} \sum_{i=1}^{a}(n_i - 1)S_i^2$$

and

$$T_G = \left\{ \prod_{i=1}^{a} (S_i^2)^{n_i - 1} \right\}^{1/\sum_{i=1}^{a}(n_i - 1)},$$

---

[25] For a discussion of an exact test based on the generalized likelihood ratio principle, which is asymptotically equivalent to the Bartlett's test, see Weerahandi (1995). For a discussion of a general class of tests for homogeneity of variances and their properties, see Cohen and Strawderman (1971).

where

$$S_i^2 = \frac{\sum_{i=1}^{n_i}(y_{ij} - \bar{y}_{i.})^2}{n_i - 1}.$$

It should be noted that T_A and T_G are weighted arithmetic and geometric averages of the S_i^2's which are the usual sample variances of the observations at different factor levels. It is well known that

$$T_G \leq T_A$$

and the two averages are equal if all S_i^2's are equal. Thus, the greater the variation among the S_i^2's the farther apart the two averages will be. Hence, if the ratio

$$R = T_A/T_G$$

is close to 1, we have the evidence that the population variances are equal. If R is large, it would indicate that the population variances are unequal. The same conclusion would follow if we use $\log_e(R) = \log_e(T_A) - \log_e(T_G)$ instead of R. Thus, Bartlett's test is based on the statistic R or $\log_e(R)$; rejecting the null hypothesis if the statistic is significantly greater than unity.[26]

Inasmuch as the sampling distribution of R or $\log_e(R)$ is not readily available, Bartlett considered two approximations of R. First, for large sample sizes, a function of $\log_e(R)$ has approximately a chi-square distribution with $a - 1$ degrees of freedom under the hypothesis that the population variances are equal. More specifically, if each $n_i > 5$, the statistic

$$B = \frac{K}{1 + L}, \qquad (2.21.13)$$

where

$$K = \sum_{i=1}^{a}(n_i - 1)\log_e(T_A) - \sum_{i=1}^{a}(n_i - 1)\log_e(S_i^2) \qquad (2.21.14)$$

and

$$L = \frac{1}{3(a-1)}\left[\sum_{i=1}^{a}\frac{1}{n_i - 1} - \frac{1}{\sum_{i=1}^{a}(n_i - 1)}\right] \qquad (2.21.15)$$

---

[26] Equivalently, Bartlett's test can be based on the statistic $R^{-1} = T_G/T_A$, rejecting the null hypothesis if the statistic is significantly smaller than unity.

has approximately a chi-square distribution with $a-1$ degrees of freedom (see also Nagasenkar (1984)). Thus, if the calculated value of the statistic B exceeds $\chi^2[a-1, 1-\alpha]$, the $100(1-\alpha)$th percentage point of the chi-square distribution with $a-1$ degrees of freedom, we reject the null hypothesis that the population variances are equal. The accuracy of this approximation has been considered by Bishop and Nair (1939), Hartley (1940), and Barnett (1962).

The chi-square approximation to the distribution of the Bartlett's test statistic (2.21.13) is not appropriate when any of the n_i's are less than five. An approximation which is more accurate when some of the n_i's are small is based on the F distribution. The approximation consists of considering the statistic

$$B' = \frac{v_2 K}{v_1(M-K)}, \qquad (2.21.16)$$

where

$$v_1 = a - 1, \qquad (2.21.17)$$
$$v_2 = (a+1)/L^2, \qquad (2.21.18)$$

and

$$M = v_2/\{1 - L + 2/v_2\}, \qquad (2.21.19)$$

which has a sampling distribution approximated by an F distribution with v_1 and v_2 degrees of freedom. The values of v_2 will usually not be an integer and it may be necessary to interpolate in the F table. Good accuracy can be achieved by the method of two-way harmonic interpolation (see, e.g., Laubscher (1965)) based on the reciprocals of the degrees of freedom. Usually, however, the observed value of B' will differ significantly from the tabulated value and in that case an interpolation may not be required. When $k=2$, and for equal sample sizes, Bartlett's test reduces to the two-sided variance ratio F test. When two sample sizes are unequal, the two methods, however, may give different results (Maurais and Quimet (1986)).

Remark: Exact critical values obtained from the null distributions of Bartlett's statistic for the case involving equal sample sizes have been given by Harsaae (1969), Glaser (1976), and Dyer and Keating (1980). For very small values of n_i's tables are given in Hartley (1940) and Pearson and Hartley (1970). For equal sample sizes, some selected percentage points of the distribution are given in Appendix Table XXI. Algebraic expressions for determining exact critical values for Bartlett's test for unequal sample sizes have been derived by Chao and Glaser (1978) and Dyer and Keating (1980).

One-Way Classification

Example 4. To illustrate the procedure, consider the data given in Table 2.5. The calculations needed for Bartlett's test are summarized in Table 2.13. On substituting the appropriate quantities into (2.21.14) and (2.21.15), and then into (2.21.13), we obtain

$$K = 2.1005,$$
$$L = 0.1736,$$

and

$$B = 2.1005/(1 + 0.1736) = 1.7898.$$

Since from Appendix Table IV, $\chi^2[4, 0.95] = 9.49$ with a p-value of 0.774, we do not reject the null hypothesis that the five variances are all equal.

TABLE 2.13
Calculations for Bartlett's Test

Treatment	$n_i - 1$	S_i^2	$\log_e S_i^2$	$(n_i - 1)S_i^2$	$(n_i - 1)\log_e S_i^2$
1	2	16.0000	2.7726	32.0000	5.5452
2	3	6.0000	1.7918	18.0000	5.3754
3	2	3.0000	1.0986	6.0000	2.1972
4	3	4.0000	1.3863	12.0000	4.1589
5	2	4.0000	1.3863	8.0000	2.7726

To use the F approximation given by (2.21.16), on substituting the appropriate quantities into (2.21.17), (2.21.18), and (2.21.19), and then into (2.21.16), we have

$$v_1 = 4$$
$$v_2 = 6/(0.1736)^2 = 199.1,$$

$$M = 199.1/(1 - 0.1736 + 2/199.1) = 238.0311,$$

and

$$B' = \frac{(199.1)(2.1005)}{4(238.0311 - 2.1005)} = 0.44.$$

Example 4 *(continued)*

Furthermore, for $\alpha = 0.05$, we obtain from Appendix Table V that $F[4, 120; 0.95] = 2.45$ and $F[4, \infty; 0.95] = 2.37$. Using the harmonic interpolation, based on the reciprocals of the degrees of freedom, we have

$$F[4, 199.1; 0.95] = 2.45 + \frac{\frac{1}{199.1} - \frac{1}{120}}{\frac{1}{\infty} - \frac{1}{120}}(2.37 - 2.45) = 2.41.$$

Since $B' = 0.44 < 2.41$ with a p-value of $P\{F[4, 199.1] > 0.44\} = 0.779$, we may conclude that the five variances are all equal. Thus, the conclusion from the Bartlett's test using the F approximation is the same as using the chi-square approximation.

Example 5. In this example, we simultaneously illustrate the Shapiro-Wilk's W test for normality followed by the Bartlett's test for homogeneity of variances. We further describe the use of exact critical values for the Bartlett's test statistic given in Appendix Table XXI. Dyer and Keating (1980) reported and analyzed data on the sealed bids on each of five Texas offshore oil and gas leases selected from 110 leases issued on May 21, 1968. Using the probability plots it was shown that the bids on each of the five leases are lognormally distributed. The logarithmic scores of the sealed bids on each of five leases are given in Table 2.14.

Proceeding as in Example 1, the Shapiro-Wilk's W statistic for each of the five groups of leases is determined as:

$$W_1(8) = \{0.6052(16.269 - 13.521) + \cdots$$
$$+ 0.0561(15.035 - 14.847)\}^2 / 7(0.842) = 0.982,$$
$$W_2(10) = \{0.5739(16.292 - 12.597) + \cdots$$
$$+ 0.0399(14.430 - 14.307)\}^2 / 9(1.282) = 0.970,$$
$$W_3(5) = \{0.6646(13.980 - 11.629) + \cdots$$
$$+ 0.2413(13.273 - 12.134)\}^2 / 4(0.859) = 0.982,$$
$$W_4(12) = \{0.5475(17.589 - 13.003) + \cdots$$
$$+ 0.0922(15.539 - 15.370)\}^2 / 11(1.883) = 0.960,$$

One-Way Classification

Example 5 *(continued)*

TABLE 2.14
Data on Log-bids of Five Texas Offshore Oil and Gas Leases

		Lease No.		
I	II	III	IV	V
$16.269	$16.292	$13.980	$17.859	$17.188
15.733	15.223	13.273	16.557	16.712
15.256	15.100	12.616	16.264	16.259
15.035	14.995	12.134	15.957	16.128
14.847	14.430	11.629	15.910	15.463
14.223	14.307		15.539	15.100
13.987	13.520		15.370	14.565
13.521	13.463		14.847	14.519
	13.129		14.785	13.521
	12.597		13.521	13.014
			13.503	13.003
			13.003	12.622
				12.530

$n_1 = 8$, $S_1^2 = 0.842$, $n_2 = 10$, $S_2^2 = 1.282$,
$n_3 = 5$, $S_3^2 = 0.859$, $n_4 = 12$, $S_4^2 = 1.883$, $n_5 = 13$, $S_5^2 = 2.635$.

Source: Dyer and Keating (1980). Used with permission.

and

$$W_5(13) = \{0.5359(17.188 - 12.530) + \cdots \\ + 0.0539(15.100 - 14.519)\}^2 / 12(2.635) = 0.928.$$

From Appendix Table XIX, the 5 percent critical values for the W statistic in each of the five groups are:

$$W_1(8, 0.05) = 0.818, \quad W_2(10, 0.05) = 0.842,$$
$$W_3(5, 0.05) = 0.762, \quad W_4(12, 0.05) = 0.859,$$

and

$$W_5(13, 0.05) = 0.866.$$

Since in each group, $W_i(n_i) > W_i(n_i, 0.05)$, the lognormality of the bids data is not rejected at the 5 percent significance level. In fact, it can be verified that the hypothesis of lognormality is sustained at a significance level of 0.5 or lower.

Example 5 (*continued*)

We now test for homogeneity of variances using Bartlett's test. For the log-bids data, the weighted arithmetic and geometric means of S_i^2, T_A and T_G, are determined as

$$T_A = \frac{\sum_{i=1}^{5}(n_i - 1)S_i^2}{\sum_{i=1}^{5}(n_i - 1)} = \frac{(7 \times 0.842) + \cdots + (12 \times 2.635)}{7 + \cdots + 12} = 1.7023$$

and

$$T_G = \prod_{i=1}^{5}(S_i^2)^{(n_i-1)/\sum_{i=1}^{5}(n_i-1)} = (0.842)^{\left(\frac{7}{7+\cdots+12}\right)} \cdots (2.635)^{\left(\frac{12}{7+\cdots+12}\right)}$$
$$= 1.5560.$$

Hence, the Bartlett's test statistic B is given by

$$B = T_G/T_A = 1.5560/1.7023 = 0.9141.$$

From Appendix Table XXI, the 5 percent critical value is approximately determined as

$$B(8, 10, 5, 12, 13; 0.05)$$
$$\cong \left(\frac{8}{48}\right)(0.7512) + \left(\frac{10}{48}\right)(0.8025) + \left(\frac{5}{48}\right)(0.5952)$$
$$+ \left(\frac{12}{48}\right)(0.8364) + \left(\frac{13}{48}\right)(0.8493)$$
$$= 0.7935.$$

Since $B > 0.7935$, the hypothesis of homogeneity is not rejected at the 5 percent significance level. As a matter of fact, it can be verified that the approximate 25 percent critical value is 0.8757 and the hypothesis of homogeneity is sustained at a significance level of 0.25 or lower.

Hartley's test

Hartley (1950) developed a test for the hypothesis (2.21.12) when the sample sizes are all equal; that is, $n_i = n$, $i = 1, 2, \ldots, a$. The test represents a natural extension to the F test for the case with $a = 2$. If the S_i^2's denote the sample

One-Way Classification

variances, then the test statistic is defined by

$$H = \frac{\max(S_i^2)}{\min(S_i^2)}, \qquad (2.21.20)$$

where $\max(S_i^2)$ and $\min(S_i^2)$ denote the largest and smallest sample variances, respectively. Naturally, when the population variances are all equal, the value of H would be expected near 1 and greater the variation between S_i^2's, the larger the value of H. The decision rule consists of rejecting the null hypothesis (2.21.12) if the calculated value of H exceeds $H[a, v; 1-\alpha]$, the $100(1-\alpha)$th percentage point of the distribution of H.

Remark: The distribution of the statistic (2.21.20) depends on a and n and initial tables for 1 and 5 percent critical values were originally given by Hartley (1950). Later, David (1952) gave tables for $\alpha = 0.05, 0.01, a = 2(1)12$, and $v = n - 1 = 2(1)10, 12, 15, 20, 30, 60, \infty$. These tables are also given in Owen (1962) and Pearson and Hartley (1970). Some selected percentage points of the H distribution are given in Appendix Table XXII.

Example 6. To illustrate the procedure, we consider the data given in Table 2.3. The sample variances are as follows:

$$S_1^2 = 406.8, \quad S_2^2 = 97.6, \quad S_3^2 = 80.8, \quad \text{and} \quad S_4^2 = 69.2$$

Now, we have

$$\max(S_i^2) = 406.8, \quad \min(S_i^2) = 69.2,$$

and the statistic (2.21.20) is given by

$$H = \frac{406.8}{69.2} = 5.88.$$

From Appendix Table XXII, we have $H[4, 5; 0.95] = 13.70$ and so we do not reject the null hypothesis that the four variances are all equal.

Cochran's test

Cochran (1941) developed a test for homoscedasticity especially designed for the case when one variance is very much larger than the others and the sample sizes are all equal. The test statistic is given by

$$C = \frac{\max(S_i^2)}{\sum_{i=1}^{a} S_i^2}. \qquad (2.21.21)$$

The decision rule consists of rejecting the null hypothesis (2.21.12) if the calculated value of C exceeds $C[a, v; 1 - \alpha]$, the $100(1 - \alpha)$th percentage point of the distribution of C.

Remark: The distribution of the statistic (2.21.21) depends on a and n, and initial tables for the upper 5 percentage points for $a = 3(1)10$ and $v = n-1 = 1(1)6(2)10$ were given by Cochran (1941). Later, Eisenhart and Solomon (1947) gave tables for $\alpha = 0.05, 0.01$, $a = 2(1)$ 12, 15, 20, 24, 30, 40, 60, 120, ∞, and $v = 1(1)$ 10, 16, 36, 144, ∞. These tables are also given in Pearson and Hartley (1973). A more comprehensive tabulation of the statistic C appears in a publication by Japanese Standards Association (1972). The latter publication presents the percentage points of C for $a = 2(1)20$; $n = 2(1)31, 41, 61, 121, \infty$; and $\alpha = 0.05, 0.01$. Some selected percentage points for the distribution of C are reprinted in Appendix Table XXIII.

> **Example 7.** To illustrate the procedure, we again consider the data of Table 2.3 as in the case of the Hartley's test. The sample variances lead to the value of the test statistic (2.21.21) given by
>
> $$C = \frac{406.8}{406.8 + 97.6 + 80.8 + 69.2} = 0.6216.$$
>
> From Appendix Table XXIII, we have $C[4, 5; 0.95] = 0.5895$ and therefore we reject the null hypothesis (2.21.12) that the variances are all equal. Note that for the same data Cochran's test leads to the rejection of the null hypothesis whereas Hartley's test fails to reach the critical value.

Comments on Bartlett's, Hartley's and Cochran's tests

(i) In most practical situations, the Hartley's and Cochran's tests will lead to similar conclusions. Since Cochran's test utilizes more information in the sample data, it is generally more sensitive than Hartley's test. When the normality assumption can be relied upon, Bartlett's test is more powerful than other tests (Gartside (1972)).

(ii) Both Hartley's and Cochran's tests require that all sample sizes be equal. If the sample sizes are unequal, but do not differ greatly, they may still be used as approximate tests. In this case, the value of n would be the average sample size for the determination of the percentage points of the test statistics. Some statisticians recommend the use of the largest n for this purpose. The procedure will result in the probability of type I error being slightly larger than the prescribed value.

(iii) All the test procedures are sensitive to departures from normality (Box (1953); Box and Anderson (1955)). That is, if the populations from which samples are taken are not normally distributed, the actual level of significance may differ greatly from the specified one. They all tend to mask existing differences in variances if the kurtosis is smaller than zero, or to exhibit nonexistent differences if the kurtosis is greater than zero. Thus, the values of the test statistics may lead to an erroneous rejection of the null hypothesis. Therefore, it is not advisable to test for homoscedasticity unless there is sufficient evidence to assume that the distributions are at least approximately normal. It is recommended that any homoscedasticity test be used only when preceded by a preliminary test which does not reject normality. However, if the test is performed as a check before using an analysis of variance procedure, then the rejection of the null hypothesis indicates that at least one of the two underlying assumptions is violated. For example, it has been found that Bartlett's test is a good one for testing departures from normality.

(iv) As noted in the preceding section, the analysis of variance F test is not much affected by the unequal variances as long as the differences in the variances are not too large and the sample sizes are nearly equal. Hence, a fairly low level of α may be justified in conducting the test for the equality of variances when the sample sizes are nearly equal. This would be appropriate in determining the aptness of the analysis of variance model (2.1.1) since only large differences between variances need to be detected.

Other tests of homoscedasticity

The preceding tests of homoscedasticity are traditional tests based on normal theory for testing the null hypothesis of equal variances. However, they all are very sensitive to the assumption of normality and give too many significant results for data coming from a long-tailed distribution. In recent years, a number of tests have appeared in the literature that are less sensitive to normality in the data and are found to have a good power for a variety of population distributions. Levene (1960) proposed a test that considers the scores $z_{ij} = (y_{ij} - \bar{y}_{i.})^2$ as identically distributed normal variates and applies the usual F test on these scores. A significant difference between means of the transformed scores is considered as evidence of significant differences in variances of the groups. Levene (1960) also proposed using F tests based on the scores $z_{ij} = |y_{ij} - \bar{y}_{i.}|$, $z_{ij} = \log_e |y_{ij} - \bar{y}_{i.}|$, and $z_{ij} = |y_{ij} - \bar{y}_{i.}|^{1/2}$.

Following Levene (1960), a number of other robust procedures have been proposed that are essentially based on techniques of applying analysis of variance to transformed scores. For example, Brown and Forsythe (1974c) proposed using the transformed scores based on the absolute deviations from the median. In order to increase power when sample sizes are odd, Ramsey and Brailsford (1989) suggested that the median be replaced by the pseudo median equal to the midpoint of the scores just above and below the median. A somewhat different approach known as the jackknife was proposed by Miller (1968) where the original scores in each group were replaced by the contribution of that observation to the group variance.[27] O'Brien (1979, 1981) proposed a procedure that is a blend of Levene's squared deviation scores and the jackknife. It performs analysis of variance using

$$\frac{(n_i - 1.5)n_i(y_{ij} - \bar{y}_{i.})^2 - 0.5 s_i^2(n_i - 1)}{(n_i - 1)(n_i - 2)},$$

where $\bar{y}_{i.}$ and s_i^2 represent mean and variance, respectively, for the i-th factor level.

In recent years, there have been a number of studies investigating the robustness of these procedures and they point toward the robustness of Brown-Forsythe and O'Brien procedures. More recently, Algina et al. (1995) have proposed a procedure, called *maximum test for scale*, in which the test statistic is the more extreme of the Brown-Forsythe and O'Brien test statistics. Some limited simulation work for the two-sample case indicates better properties for type I and type II error rates than either the Brown-Forsythe or O'Brien procedure. For further discussions and details, the reader is referred to Games et al. (1972), Hall (1972), Layard (1973), Levy (1978b), Keselman et al. (1979), Conover et al. (1981), Olejnik and Algina (1987), Micceri (1989), Ramsey (1994), and Algina et al. (1995).

2.22 CORRECTIONS FOR DEPARTURES FROM ASSUMPTIONS OF THE MODEL

If the data set in a given problem violates the assumptions of the analysis of variance model (2.1.1), a choice of possible corrective measures is available. One is to modify the model. However, this approach has the disadvantage that more often than not the modified model involves fairly complex analysis. Another approach may be to consider using some nonparametric tests which do not make the normal theory assumption for inference problems. A third approach to be discussed in this section is to use transformations on the data. Sometimes it is possible to make an algebraic transformation of the data to make them appear more nearly normally distributed, or to make the variances of the error terms constant. Conclusions derived from the statistical analyses performed on the transformed data are also applicable to the original data. In this section, we briefly discuss some commonly used transformations to correct for the lack of

---

[27] The jackknife procedure computes sample variances within each group by deleting one observation at a time. Thus, in the i-th group, n_i variances are computed as follows:

$$s_{i(\ell)}^2 = \frac{1}{n_i - 1} \sum_{\substack{j=1 \\ j \neq \ell}}^{n_i} (y_{ij} - \bar{y}_{i(\ell)})^2 \quad \text{where} \quad \bar{y}_{i(\ell)} = \frac{1}{n_i - 1} \sum_{\substack{j=1 \\ j \neq \ell}}^{n_i} y_{ij}.$$

The analysis of variance is performed on the transformed scores $z_{i\ell} = n_i \log_e(s_i^2) - (n_i - 1) \log_e(s_{i(\ell)}^2)$ ($i = 1, 2, \ldots, a; \ell = 1, 2, \ldots, n_i$) and the test statistic is the usual F statistic with $a - 1$ and $N - a$ degrees of freedom.

One-Way Classification

normality and homoscedasticity. Tukey (1955) discussed the use of transformations such that effects in the transformed scale are additive. Although individual transformations that will correct for lack of normality, homoscedasticity, and nonadditivity may be different, Box and Cox (1964, 1982) found that often a single transformation will simultaneously rectify all the problems.

Remark: For further discussions of transformations, the reader may refer to Bartlett (1936, 1947), Cochran (1940), Curtiss (1943), Bartlett and Kendal (1946), Eisenhart (1947b), Freeman and Tukey (1950), Tukey (1957), Draper and Hunter (1969), Cox (1977), Draper and Smith (1981, pp. 220–221), Efron (1982), Berry (1987), and Hoaglin (1988). Natrella (1963, Chapter 20) provides a detailed and thorough discussion of the use of transformations. An extremely thorough and detailed monograph on transformation methodology has been prepared by Thöni (1967). An excellent and thorough introduction and a bibliography of the topic can be found in a review paper by Hoyle (1973). For a more recent bibliography of articles on transformations, see Draper and Smith (1981, pp. 683–684).

TRANSFORMATIONS TO CORRECT LACK OF NORMALITY

Here, we discuss some transformations to correct for the departures from normality.

Logarithmic transformation
Suppose the data are distributed according to the relationship

$$y_{ij} = \beta(\alpha_i + e_{ij}), \tag{2.22.1}$$

where the e_{ij}'s are normally and independently distributed, each with mean zero and variance σ_e^2. Then, on making a logarithmic transformation of (2.22.1), we get

$$\log_e(y_{ij}) = \log_e(\beta) + \alpha_i + e_{ij},$$

which can be rewritten as

$$y'_{ij} = \mu + \alpha_i + e_{ij}. \tag{2.22.2}$$

From (2.22.2), we notice that although the y_{ij}'s are not normally distributed, the transformed variables y'_{ij}'s are. This may be the case when the distribution of y_{ij}'s is skewed.

Square-root transformation
Suppose the sample observations are given by the relationship

$$y_{ij} = (\mu + \alpha_i + e_{ij})^2, \tag{2.22.3}$$

where, as in (2.22.1), the e_{ij}'s are normally and independently distributed with mean zero and variance σ_e^2. Then, on making the square-root transformation, we get

$$y'_{ij} = \sqrt{y_{ij}} = \mu + \alpha_i + e_{ij}. \qquad (2.22.4)$$

From (2.22.4), we notice that although the y_{ij}'s are not normally distributed, the transformed variables y'_{ij}'s are. This may be the case when the y_{ij}'s are nonnegative real numbers and their distribution is skewed to the right.

Arcsine transformation

Suppose the sample scores y_{ij}'s are binomial proportions with mean μ based on samples of size n. Then, the transformed scores[28]

$$y'_{ij} = 2 \arcsin \sqrt{y_{ij}} \qquad (2.22.5)$$

are approximately normally distributed with approximate mean $\mu' = 2 \arcsin \sqrt{\mu}$ and variance $1/n$. The transformation (2.22.5) does not perform as well at the extreme ends of the possible values (near 0 and n). Anscombe (1948) and Freeman and Tukey (1950) proposed some improved arcsine transformations given by

$$y'_{ij} = \arcsin \sqrt{(ny_{ij} + 3/8)/(n + 3/4)}$$

and

$$y'_{ij} = \frac{1}{2} \left[\arcsin \sqrt{\frac{ny_{ij}}{n+1}} + \arcsin \sqrt{\frac{ny_{ij} + 1}{n+1}} \right].$$

TRANSFORMATIONS TO CORRECT LACK OF HOMOSCEDASTICITY

There are several types of data in which the variances of the error terms are not constant. If there is evidence of some systematic relationship between treatment mean and variance, homogeneity of the error variance may be achieved through an appropriate transformation of the data. Bartlett (1936) has given a formula for deriving such transformations provided the relationship between μ_i and σ_e^2 is known. In many cases where the nature of the relationship is not clear, the experimenter can, through trial and error, find a transformation that will stabilize the variance. We now consider some commonly employed transformations to stabilize the variance.

---

[28] If the data come from a population having the so-called negative binomial distribution, then the use of inverse hyperbolic sines may be more appropriate (Beall (1942); Bartlett (1947); Anscombe (1948)).

Logarithmic transformation

This transformation is applicable when $\sigma_i^2 \propto \mu_i^2$ or $\sigma_i \propto \mu_i$, that is, when the factor level standard deviation is proportional to the corresponding mean. This type of situation arises when the distribution of scores is markedly skewed. The transformation is also applicable when the scores are standard deviations. In this case $s_i/\bar{y}_{i.}$ tends to be constant and so a logarithmic transformation, that is,

$$y'_{ij} = \log_e(y_{ij}), \qquad (2.22.6)$$

would stabilize the variance. If some of the measurements are small (particularly zero), the recommended transformation is (Bartlett 1947)

$$y'_{ij} = \log_e(y_{ij} + 1). \qquad (2.22.7)$$

Square-root transformation

This transformation is applicable when $\sigma_i^2 \propto \mu_i$, that is, when the means and variances are proportional for each factor level. This type of situation is often found when the observed variable y_{ij} is a count, such as the number of auto accidents in a given year. In this case, the sample statistic $s_i^2/\bar{y}_{i.}$ tends to be constant and so a square-root transformation such as

$$y'_{ij} = \sqrt{y_{ij}} \qquad (2.22.8)$$

would stabilize the variance. If some of the observations y_{ij}'s are very small (particularly zero), homogeneity of variance is more likely to be achieved by the transformation[29] (Bartlett (1936))

$$y'_{ij} = \sqrt{y_{ij} + 0.5}. \qquad (2.22.9)$$

The square-root transformation is usually applied to all data assumed to follow a Poisson distribution. For a discussion of the use of square-root transformation to perform analysis of variance for Poisson data, see Budescu and Applebaum (1981).

Reciprocal transformation

This type of transformation is applicable when $\sigma_i \propto \mu_i^2$, that is, when the factor level standard deviation is proportional to the square of the corresponding mean. In this case $s_i/\bar{y}_{i.}^2$ tends to be constant and an appropriate transformation to

---

[29] The transformation $y'_{ij} = \sqrt{y_{ij} + 3/8}$ has an even better variance stabilizing property than equation (2.22.9) (Anscombe (1948); Kihlberg et al., (1972)). Freeman and Tukey (1950) showed that the transformation $y'_{ij} = \sqrt{y_{ij}} + \sqrt{y_{ij} + 1}$ will yield similar results as (2.22.9) but is preferable for $y_{ij} \leq 2$.

stabilize the variance is the reciprocal transformation;[30] that is,

$$y'_{ij} = 1/y_{ij}. \tag{2.22.10}$$

The transformation (2.22.10) is generally used when $y'_{ij} = y_{ij}^{-1}$ has a definite physical meaning and where the possibility of the random variable being less than or equal to zero is negligible. For example, data on the failures of a machine may be collected as either "intervals between failures," or "the number of failures per unit time." Similarly, the reciprocal of the survival data is related to the death rate and the reciprocal of the waiting time unit when some phenomenon occurs is related to the speed with which the phenomenon occurs.

Arcsine transformation

This transformation is applicable when $\sigma_i^2 \propto \mu_i(1 - \mu_i)$, that is, when scores are proportions. For example, the factor levels may be different treatment procedures, the unit of observation is a clinical center, and the observed variable y_{ij} is the proportion of patients in the i-th treatment group for the j-th clinical center who benefited by the treatment. In this case an appropriate transformation to stabilize the variance is the arcsine transformation; that is,

$$y'_{ij} = \arcsin \sqrt{y_{ij}}. \tag{2.22.11}$$

The transformed score using (2.22.11) is the angle whose sine is equal to the square root of the original score. Tables to facilitate this transformation have been prepared (see, e.g., Fisher and Yates (1963); Owen (1962)).

Square transformation

If the standard deviation decreases as the corresponding factor level mean increases, then the transformation

$$y'_{ij} = y_{ij}^2 \tag{2.22.12}$$

would stablize the variance. The transformation (2.22.12) is generally useful when the distribution is skewed to the left.

Power transformation

When there does not exist a theoretical basis to select a transformation, or the transformations described fail to achieve normality or homoscedasticity, a class of transformations proposed by Box and Cox (1964) can be used to achieve the desired objective. The general form of the transformation is given by

$$f(y_{ij}) = \begin{cases} y_{ij}^{\lambda}, & \lambda \neq 0 \\ \log_e(y_{ij}) & \text{for } \lambda = 0, \end{cases} \tag{2.22.13}$$

---

[30] If y_{ij} represents counts, then $y'_{ij} = 1/(y_{ij} + 1)$ may be used to avoid a possibility of division by zero.

One-Way Classification

where λ is a parameter to be determined from the data. The analyst tries different values of λ in (2.22.13) until the transformed scores conform to the assumption in question. It should be noted that the transformation (2.22.13) includes the following simple transformations as special cases.

$$\lambda = -1, \quad f(y_{ij}) = \frac{1}{y_{ij}}$$

$$\lambda = -0.5, \quad f(y_{ij}) = \frac{1}{\sqrt{y_{ij}}}$$

$$\lambda = 0, \quad f(y_{ij}) = \log_e(y_{ij}) \quad \text{(by definition)}$$

$$\lambda = 0.5, \quad f(y_{ij}) = \sqrt{y_{ij}}$$

$$\lambda = 2, \quad f(y_{ij}) = y_{ij}^2.$$

These are some of the more commonly used transformations. Still other transformations can be found to be applicable for various other relationships between the means and variances. Furthermore, the transformations to stabilize the variance also often make the population distribution nearly normal. However, the use of such transformations may often result in different group means. It is possible that the means of the original scores are equal but the means of the transformed scores are not, and vice versa. Moreover, the means of transformed scores are often changed in ways that are not intuitively meaningful or are difficult to interpret.

EXERCISES

1. In an effort to increase the service life of a handbrake, an automobile manufacturing company has developed three new designs. To assess their performance against a standard design of a handbrake, 12 automobiles of a certain make were randomly chosen and assigned to four different groups with 3 cars in each group. The handbrakes of four different designs were then randomly assigned to each group with each of the 3 cars in every group using a handbrake of one of the four designs. The relevant data on service life, measured in months, for each handbrake are given as follows.

Standard Design	21.2	13.4	17.0
New Design-I	21.4	12.0	13.0
New Design-II	5.2	9.1	4.2
New Design-III	8.7	35.8	39.0

   (a) Describe the model and the assumptions for the experiment.
   (b) Analyze the data and report the analysis of variance table.
   (c) Perform an appropriate F test to determine if there are significant differences in the average service life in the four groups of handbrakes. Use $\alpha = 0.05$.

(d) Carry out the test for homoscedasticity at $\alpha = 0.01$ by employing
   (i) Bartlett's test,
   (ii) Hartley's test,
   (iii) Cochran's test.
(e) Would you consider using appropriate contrasts? If so, perform the following, and interpret your results:
   (i) Orthogonal contrasts,
   (ii) Tukey's procedure,
   (iii) Scheffé's procedure.
(f) If it is found that the measures of service life for handbrakes have a distribution skewed to the right, what transformation would be appropriate to correct it? Make the required transformation on the data and repeat the analyses carried out in parts (b), (c), and (d).
(g) Why were all the automobiles included in the experiment of a certain preselected model? Is it possible to generalize the results of this study to automobiles of any other model?

2. A study was carried out to determine if different types of savings institutions attract similar amounts of savings after adjusting for factors such as advertising, years in operation, and size of the neighborhoods of the branches, and so on. A research analyst randomly selected 5 out of a large number of savings institutions included in the study and from each of these 5 institutions, 5 branches were selected at random. The total savings, in millions of dollars, in the 25 branches included in the study are given as follows.

Types of Savings Institutions				
A	B	C	D	E
37.2	33.4	37.5	31.0	30.9
38.4	37.7	36.6	33.4	37.0
36.0	38.8	35.8	36.7	36.2
31.3	32.8	37.0	39.0	38.1
32.4	33.7	35.6	37.1	36.8

(a) Describe the mathematical model and the assumptions involved. Would you use Model I or Model II? Explain.
(b) Analyze the data and report the analysis of variance table.
(c) Determine if there is significant evidence to conclude that the average accumulated savings are not the same among the different types of savings institutions under study. Use $\alpha = 0.01$.
(d) Would you consider using contrasts? Explain.

3. Calculations of the sums of squares for a one-way analysis of variance

from certain experimental data yielded the following results:

$$SS_B = 570.23,$$
$$SS_W = -19.72,$$

and

$$SS_T = 550.51.$$

What can you say about the correctness of the results? Explain.

4. A consumer organization carried out a study to determine whether the price being offered for a used car differed with the personality of the owner of the car. Four individuals were selected for the study, and each, pretending to be the owner, was sent to 5 different dealers. From each of the 20 dealers selected in the study the price quotes were obtained on a five-year old medium price car. The amounts offered, in hundreds of dollars, by each of the 20 dealers in the study are given as follows.

\multicolumn{4}{c}{Owners}			
A	B	C	D
40	40	34	35
38	43	37	40
40	41	38	37
41	42	40	36
37	43	35	34

(a) Describe the mathematical model and the assumptions involved. Would you consider using a Model I or Model II? Explain.
(b) Analyze the data and report the analysis of variance table.
(c) Perform an appropriate F test to determine whether the price quotes differ according to the personality traits of the owner. Use $\alpha = 0.05$.
(d) What are the variance components associated with the assumptions of the Model II?
(e) Obtain appropriate point and interval estimates of the variance components identified in part (d).

5. Out of three different textbooks published by three leading publishers, a statistics professor is trying to choose one for adoption for his basic statistics class. He designed an experiment with 30 students of his class, whom he randomly assigned into three different groups, placing 10 in each group. The three textbooks, from John Wiley, Prentice-Hall, and Wadsworth, were then randomly assigned to each group. After the end of the course, all the students who completed the course took the same examination. The scores of the examination are given in the following.

Textbooks

John Wiley	Prentice-Hall	Wadsworth
80	55	65
80	62	55
81	80	62
71	70	67
81	70	58
75	66	72
82	77	70
78	75	60
86		52
84		

(a) Describe the mathematical model and the assumptions involved.
(b) Analyze the data and report the analysis of variance table.
(c) Perform an appropriate F test for the hypothesis that the average scores using three different textbooks are the same. Use $\alpha = 0.05$.
(d) Carry out the test for homoscedasticity at $\alpha = 0.01$ by employing
  (i) Bartlett's test,
  (ii) Hartley's test,
  (iii) Cochran's test.

6. An automobile company wants to know the length of time during which the premiums are given by the different agents employed by the company. A study was conducted in which four agents were chosen at random and the number of transactions completed by each agent in a given week were recorded. The delay, in days, for completing the transaction was noted for each sample case and the relevant data are given as follows.

Agents

I	II	III	IV
9	19	27	22
7	17	30	28
9	22	32	23
13	23	26	19
10	28	29	20
19	21	30	19
17	21	27	24
12	27	28	26
6	25	33	27
11	16	20	
10		35	
13		28	

(a) Describe the mathematical model and the assumptions involved. Would you consider using Model I or Model II? Explain.
(b) Analyze the data and report the analysis of variance table.
(c) Perform an appropriate F test to determine if the mean delay time varies from agent to agent. Use $\alpha = 0.01$.
(d) What are the variance components associated with the assumptions of the Model II?
(e) Obtain appropriate point and interval estimates of the variance components identified in part (d).

7. Consider an experiment designed to investigate differences in blood counts in three groups of monkeys randomly administered to two drugs and a control. The data on blood counts are given as follows.

Type of Drug		
A	B	Control
11.8	14.8	9.4
10.9	11.7	10.5
9.7	14.2	9.2
11.4	11.2	10.2
	12.6	11.8
		10.3

(a) Describe the mathematical model and the assumptions for the experiment.
(b) Analyze the data and report the analysis of variance table.
(c) Perform an appropriate F test to determine if the average blood count varies for three types of drugs. Use $\alpha = 0.05$.
(d) Carry out the test for homoscedasticity at $\alpha = 0.05$ by employing
  (i) Bartlett's test,
  (ii) Hartley's test,
  (iii) Cochran's test.
(e) Determine 95% one-sided and two-sided confidence intervals for the single contrast comparing control with the mean of the other two drugs using the Dunnett's statistic and interpret your results.

8. A zoologist studying the structural traits of certain species of mammals classifies them into three groups: small, medium, or large according to the size of the vertebrate. He selects three random samples of size 8 from each group and then records the length of each in the sample. The relevant data on length measurements in certain standard units are given as follows.

Mammal Groups		
Small	Medium	Large
8.1	11.4	8.6
8.8	11.2	7.1
10.5	10.6	7.4
7.2	7.7	9.0
9.6	9.5	8.6
9.8	8.1	9.1
10.1	9.5	10.3
7.2	12.1	9.5

(a) Describe the mathematical model and the assumptions involved.

(b) Analyze the data and report the appropriate analysis of variance table.

(c) Perform an appropriate F test for the hypothesis that the mean length of each group is the same. Use $\alpha = 0.01$.

(d) Carry out the test for homoscedasticity at $\alpha = 0.01$ by employing

  (i) Bartlett's test,

  (ii) Hartley's test,

  (iii) Cochran's test.

(e) Would you consider using contrasts? If so, perform the following and interpret your results

  (i) Orthogonal contrasts,

  (ii) Tukey's procedure,

  (iii) Scheffé's procedure.

9. Consider the null hypothesis $H_0 : \alpha_1 = \alpha_2 = \alpha_3 = \alpha_4 = 0$ versus the alternative H_A : not all α_i's are zero.

  (a) Determine three orthogonal contrasts.

  (b) Are the three orthogonal contrasts given in part (a) unique; that is, can you construct two or more separate sets of three orthogonal contrasts?

  (c) Can you construct four orthogonal contrasts?

10. A manufacturing company employs a large number of presses that are used to produce certain automobile parts. A study was conducted to assess the performance of the presses. A sample of four presses was selected at random from the entire plant and then 10 parts were taken at random from the production line of each press. The measures on the length of the 40 parts were determined and the calculations on the sums of squares yielded the following results:

$$SS_B = 0.0264 \text{ and } SS_A = 0.0380$$

(a) Describe the mathematical model and the assumptions involved.

One-Way Classification 119

   (b) Prepare the pertinent analysis of variance table.
   (c) Perform an appropriate F test to determine if the presses are similar in their average performance. Use $\alpha = 0.05$.
   (d) Obtain appropriate point and interval estimates of the variance components associated with the model assumed in part (a).
   (e) Test the hypothesis that the between and within component ratio is equal to or less than $1/2$. Use $\alpha = 0.05$.
   (f) Find an interval estimate for the between and within component ratio and the intraclass correlation using the confidence coefficient of 0.95.
11. A study was performed to determine the effect of different varieties of fertilizers upon potato yields. Three fertilizers, designated by N, P, and K, were used. Each fertilizer was randomly assigned to 10 plots and the yields were determined for each of the 30 plots. The yield totals corresponding to the three fertilizer groups are

$$Y_N = 50, \quad Y_p = 70, \quad Y_k = 100;$$

and the total sum of squares is calculated to be 580.
   (a) Describe the mathematical model and the assumptions involved.
   (b) Prepare the pertinent analysis of variance table.
   (c) Perform an appropriate F test for the hypothesis that the average yield for each fertilizer group is the same. Use $\alpha = 0.01$.
   (d) Consider the following hypothesis of interest: the average for N equals the average of the other two fertilizers. Carry out the corresponding test of the hypothesis and state your conclusions. Use $\alpha = 0.01$.
12. In a study involving health and nutrition survey, 15 families each spending comparable amount in their grocery bills were administered a survey questionnaire regarding their dietary habits. The families were classified according to whether they lived in a rural, urban or suburban district and the data on average daily protein consumption are given as follows.

District		
Urban	Suburban	Rural
371	365	491
334	352	421
358	362	441
300	321	461
343	342	
302		

   (a) Describe the mathematical model and the assumptions involved.
   (b) Analyze the data and report the analysis of variance table.

(c) Perform an appropriate F test to determine if the average daily protein consumptions are equal for the three districts. Use $\alpha = 0.05$.

(d) Carry out the test for homoscedasticity at $\alpha = 0.05$ by employing

    (i) Bartlett's test,

    (ii) Hartley's test,

    (iii) Cochran's test.

(e) Would you consider using contrasts? If so, perform the following and interpret your results:

    (i) Orthogonal contrasts,

    (ii) Tukey's procedure,

    (iii) Scheffé's procedure.

13. A study was conducted to study the relationship between intelligence and ability to concentrate. Thirty students were randomly selected from a large psychology class and were administered tests of intelligence and concentration ability. The students were classified into five groups according to their concentration ability and the data on IQ score are given as follows.

Concentration Ability				
I	II	III	IV	V
121	115	130	74	96
129	132	118	105	94
140	114	132	106	88
	118	106	104	
	123	116	97	
	111	99	103	
		121	108	
		113	96	
		111		
		121		

(a) Describe the mathematical model and the assumptions involved.

(b) Analyze the data and report the analysis of variance table.

(c) Perform an appropriate F test to determine if the average IQ scores are equal for the five groups. Use $\alpha = 0.05$.

(d) Carry out the test for homoscedasticity at $\alpha = 0.05$ by employing

    (i) Bartlett's test,

    (ii) Hartley's test,

    (iii) Cochran's test.

(e) Would you consider using contrasts? If so, perform the following and interpret your results:

    (i) Orthogonal contrasts,

(ii) Tukey's procedure,

(iii) Scheffé's procedure.

14. Hendy and Charles (1970) reported data on the silver content (% Ag) of a number of Byzantine coins discovered in Cyprus. There were nine coins from the first coinage of the reign of King Manuel I, Commenus (1143–1180); seven of the coins came from the second coinage minted several years later and four from the third coinage (still later); another seven were from a fourth coinage. The question of interest is whether there were significant differences in the silver content of coins minted early and late in King Manuel's reign. The data are given as follow.

First Coinage	Second Coinage	Third Coinage	Fourth Coinage
5.9	6.9	4.9	5.3
6.8	9.0	5.5	5.6
6.4	6.6	4.6	5.5
7.0	8.1	4.5	5.1
6.6	9.3		6.2
7.7	9.2		5.8
7.2	8.6		5.8
6.9			
6.2			

Source: Hendy and Charles (1970). Used with permission.

(a) Describe the mathematical model and the assumptions involved.
(b) Analyze the data and report the analysis of variance table.
(c) Perform an appropriate F test to determine if the average silver contents are equal for the four coinage. Use $\alpha = 0.05$.
(d) Carry out the test for homoscedasticity at $\alpha = 0.05$ by employing

(i) Bartlett's test,

(ii) Hartley's test,

(iii) Cochran's test.

(e) Would you consider using contrasts? If so, perform the following and interpret your results:

(i) Orthogonal contrasts,

(ii) Tukey's procedure,

(iii) Scheffé's procedure.

15. Anionwu et al. (1981) reported data on steady-state haemoglobin levels for patients with different types of sickle cell disease. The question of interest is whether the steady-state haemoglobin levels differ significantly between patients with different types. The date are given as follows.

Type of Sickle Cell Disease		
HB SS	HB S/-thalassaemia	HB SC
7.2	8.1	10.7
7.7	9.2	11.3
8.0	10.0	11.5
8.1	10.4	11.6
8.3	10.6	11.7
8.4	10.9	11.8
8.4	11.1	12.0
8.5	11.9	12.1
8.6	12.0	12.3
8.7	12.1	12.6
9.1		12.6
9.1		13.3
9.1		13.3
9.8		13.8
10.1		13.9
10.3		

Source: Anionwu et al. (1981). Used with permission.

(a) Describe the mathematical model and the assumptions involved.
(b) Analyze the data and report the analysis of variance table.
(c) Perform an appropriate F test to determine if the average levels of haemoglobin are equal for the three types of sickle cell disease. Use $\alpha = 0.05$.
(d) Carry out the test for homoscedasticity at $\alpha = 0.05$ by employing
   (i) Bartlett's test,
   (ii) Hartley's test,
   (iii) Cochran's test.
(e) Would you consider using contrasts? If so, perform the following and interpret your results:
   (i) Orthogonal contrasts,
   (ii) Tukey's procedure,
   (iii) Scheffé's procedure.

16. Sokal and Rohlf (1994, p. 237) reported data on the number of eggs laid per female per day for the first 14 days of life (per diem fecundity) for 25 females of each of three genetic lines of the fruitfly *Drosophila melanogaster*. The genetic lines to be labelled RS and SS were selectively bred for resistance and the susceptibility to DDT, respectively, and the line NS is a nonselected control strain. The purpose of the study was to investigate whether the two selected lines (RS and SS) differ in fecundity from the nonselected line; and whether the RS line

One-Way Classification

differs in fecundity from the SS line. The data are given as follows.

Genetic Lines		
Resistant (RS)	Susceptible (SS)	Nonselected (NS)
12.8	38.4	35.4
21.6	32.9	27.4
14.8	48.5	19.3
23.1	20.9	41.8
34.6	11.6	20.3
19.7	22.3	37.6
22.6	30.2	36.9
29.6	33.4	37.3
16.4	26.7	28.2
20.3	39.0	23.4
29.3	12.8	33.7
14.9	14.6	29.2
27.3	12.2	41.7
22.4	23.1	22.6
27.5	29.4	40.4
20.3	16.0	34.4
38.7	20.1	30.4
26.4	23.3	14.9
23.7	22.9	51.8
26.1	22.5	33.8
29.5	15.1	37.9
38.6	31.0	29.5
44.4	16.9	42.4
23.2	16.1	36.6
23.6	10.8	47.4

Source: Sokal and Rohlf (1994, p. 237). Used with permission.

(a) Describe the mathematical model and the assumptions involved.
(b) Analyze the data and report the analysis of variance table.
(c) Perform an appropriate F test to determine if the average fecundity levels are equal for the three genetic lines. Use $\alpha = 0.05$.
(d) Carry out the test for homoscedasticity at $\alpha = 0.05$ by employing
   (i) Bartlett's test,
   (ii) Hartley's test,
   (iii) Cochran's test.
(e) Would you consider using contrasts? If so, test the following contrasts and interpret your results: (i) RS + SS − 2NS (ii) RS − SS.

3 Two-Way Crossed Classification Without Interaction

3.0 PREVIEW

The major advantage of the one-way classification (one-factor design) discussed in the preceding chapter is its simplicity, which extends to the experimental layout, the model and assumptions underlying the analysis of variance, and the computations involved in the analysis. The major disadvantage of such a design is its relative inefficiency. The error variance will usually be large compared to that resulting from other designs. This is in part offset by the fact that no other design yields as many degrees of freedom for the error variance as does this design.

In many investigations, however, it is desirable to measure response at combinations of levels of two or more factors considered simultaneously. For example, we might desire to investigate blood pressure for different gender and ethnic groups, or to investigate weight loss comparing four diets among urban, suburban, and rural subjects according to their gender, or to investigate miles per gallon among five makes of automobiles for both city and country driving. In investigations involving many factors, the effect of each factor on the response variable may be analyzed using one-way classification. Such an analysis, however, will not be economical or efficient with respect to time, effort, and money. Moreover, such a procedure would give no information about the possible interactions that may exist among different factors.

The theory of analysis of variance permits the investigation of several factors or independent variables within the same experiment. Such a procedure is efficient, time saving, and, equally important, it permits the investigation of the joint effects of several factors or interactions between them. This and the next chapter deal with the statistical model and analysis of variance involving two factors such that every level of one factor included in the experiment occurs with every level of the second factor and vice versa. Such a layout is termed two-way crossed classification. Two-way crossed layout allows a researcher to examine fully the main effects of both factors and their interactions. The term crossed classification or classification comes from the fact that in many fields of investigation, the measurements or observations can be classified in the form of

TABLE 3.1
Data for a Two-Way Experimental Layout

			Factor B			
A \ B	B_1	B_2	...	B_j	...	B_b
A_1	y_{11}	y_{12}	...	y_{1j}	...	y_{1b}
A_2	y_{21}	y_{22}	...	y_{2j}	...	y_{2b}
.	.	.	.	.	.	.
.	.	.	.	.	.	.
A_i	y_{i1}	y_{i2}	...	y_{ij}	...	y_{ib}
.	.	.	.	.	.	.
.	.	.	.	.	.	.
A_a	y_{a1}	y_{a2}	...	y_{aj}	...	y_{ab}

(Factor A labels the rows.)

a two-way table where the rows of the table correspond to the levels of a factor and the columns to the levels of another factor. In a crossed classification it is customary to refer to the combinations of levels of the factors as cells rather than treatments.

3.1 MATHEMATICAL MODEL

Two factors are said to be crossed if the data contain observations at each combination of a level of one factor with a level of the other factor. Consider two factors A and B having a and b levels, respectively, and let there be exactly one observation in each of the $a \times b$ cells of the two-way layout. Let y_{ij} be the observed score corresponding to the i-th level of factor A and the j-th level of factor B. The data involving the total of $N = a \times b$ scores y_{ij}'s can then be schematically represented as in Table 3.1.

The analysis of variance model for this type of experimental layout is given as

$$y_{ij} = \mu + \alpha_i + \beta_j + e_{ij} \quad (i = 1, 2, \ldots, a; \ j = 1, \ldots, b), \quad (3.1.1)$$

where $-\infty < \mu < \infty$ is the overall mean, α_i is the effect due to the i-th level of the factor A, β_j is the effect due to the j-th level of the factor B, and e_{ij} is the error term that takes into account the random variation within a particular cell. The model (3.1.1) states that the observed score y_{ij} corresponding to the (i, j)-th cell consists of the sum of the components: (i) the grand mean μ, (ii) the effect α_i associated with the i-th level of factor A, (iii) the effect β_j associated with the j-th level of factor B, and (iv) an error term e_{ij} which is strictly peculiar to the (i, j)-th cell.

Two-Way Crossed Classification Without Interaction

3.2 ASSUMPTIONS OF THE MODEL

Similar to the one-way classification model (2.1.1), the following assumptions are made in order to make inferences about the existence of effects in model (3.1.1).

(i) The errors e_{ij}'s are assumed to be randomly distributed with mean zero and common variance σ_e^2.
(ii) The errors associated with any pair of observations are assumed to be uncorrelated; that is,

$$E(e_{ij}e_{i'j'}) = \begin{cases} 0, & i \neq i', j \neq j'; \\ 0, & i \neq i', j = j'; \\ 0, & i = i', j \neq j'; \\ \sigma_e^2, & i = i', j = j'. \end{cases} \quad (3.2.1)$$

(iii) Under Model I, the effects α_i's and β_j's are assumed to be fixed constants subject to the constraints

$$\sum_{i=1}^{a} \alpha_i = \sum_{j=1}^{b} \beta_j = 0.$$

This implies that the observations y_{ij}'s are distributed with mean $\mu + \alpha_i + \beta_j$ and common variance σ_e^2.

(iv) Under Model II, the effects α_i's and β_j's are assumed to be randomly distributed with zero means and variances σ_α^2 and σ_β^2, respectively. Furthermore, α_i's, β_j's, and e_{ij}'s are mutually and completely uncorrelated; that is, in addition to (3.2.1), the following relations hold

$$\begin{aligned} E(\alpha_i \alpha_{i'}) &= 0, \quad i \neq i'; \\ E(\beta_j \beta_{j'}) &= 0, \quad j \neq j'; \\ E(\alpha_i \beta_j) &= 0, \quad \text{all } (i, j)\text{'s}; \\ E(\alpha_i e_{ij}) &= 0, \quad \text{all } (i, j)\text{'s}; \end{aligned} \quad (3.2.2)$$

and

$$E(\beta_j e_{ij}) = 0, \quad \text{all } (i, j)\text{'s}.$$

Then, from the model equation (3.1.1), we have $\sigma_y^2 = \sigma_\alpha^2 + \sigma_\beta^2 + \sigma_e^2$; and thus σ_α^2, σ_β^2, and σ_e^2 are components of σ_y^2, the variance of an observation. This implies that the observations y_{ij}'s are distributed with mean μ and common variance $\sigma_\alpha^2 + \sigma_\beta^2 + \sigma_e^2$.

(v) Under Model III, the effects α_i's are assumed to be fixed subject to the constraint $\sum_{i=1}^{a} \alpha_i = 0$ and the effects β_j's are assumed to be randomly distributed with mean zero and variance σ_β^2. Furthermore, as before, the β_j's are uncorrelated with each other and each of the β_j's and e_{ij}'s are also uncorrelated; that is,

$$E(\beta_j \beta'_j) = 0, \quad j \neq j';$$
and (3.2.3)
$$E(\beta_j e_{ij}) = 0, \quad \text{all } (i, j)\text{'s}.$$

In this case, $\sigma_y^2 = \sigma_\beta^2 + \sigma_e^2$ and so σ_β^2 and σ_e^2 are components of σ_y^2; and the y_{ij}'s are distributed with mean $\mu + \alpha_i$ and common variance $\sigma_\beta^2 + \sigma_e^2$.

Remark: Under the assumptions of the random effects model in (3.1.1), the observations within the same level of the factor A have the correlation given by $\rho_\alpha = \sigma_\alpha^2/(\sigma_e^2 + \sigma_\beta^2 + \sigma_\alpha^2)$. Similarly, the observations within the same level of factor B have the correlation given by $\rho_\beta = \sigma_\beta^2/(\sigma_e^2 + \sigma_\beta^2 + \sigma_\alpha^2)$. Under the assumptions of the mixed model in (3.1.1), the observations within the same level of the random factor B have the correlation given by $\rho_\beta = \sigma_\beta^2/(\sigma_e^2 + \sigma_\beta^2)$. These correlations are referred to as the intraclass correlations.

3.3 PARTITION OF THE TOTAL SUM OF SQUARES

The total variation or total sum of squares with respect to model (3.1.1) is $\sum_{i=1}^{a} \sum_{j=1}^{b} (y_{ij} - \bar{y}_{..})^2$, which can be partitioned as follows:

$$\sum_{i=1}^{a} \sum_{j=1}^{b} (y_{ij} - \bar{y}_{..})^2 = \sum_{i=1}^{a} \sum_{j=1}^{b} [(\bar{y}_{i.} - \bar{y}_{..}) + (\bar{y}_{.j} - \bar{y}_{..})$$
$$+ (y_{ij} - \bar{y}_{i.} - \bar{y}_{.j} + \bar{y}_{..})]^2$$
$$= \sum_{i=1}^{a} \sum_{j=1}^{b} (\bar{y}_{i.} - \bar{y}_{..})^2 + \sum_{i=1}^{a} \sum_{j=1}^{b} (\bar{y}_{.j} - \bar{y}_{..})^2$$
$$+ \sum_{i=1}^{a} \sum_{j=1}^{b} (y_{ij} - \bar{y}_{i.} - \bar{y}_{.j} + \bar{y}_{..})^2$$
$$= b \sum_{i=1}^{a} (\bar{y}_{i.} - \bar{y}_{..})^2 + a \sum_{j=1}^{b} (\bar{y}_{.j} - \bar{y}_{..})^2$$
$$+ \sum_{i=1}^{a} \sum_{j=1}^{b} (y_{ij} - \bar{y}_{i.} - \bar{y}_{.j} + \bar{y}_{..})^2, \quad (3.3.1)$$

where

$$\bar{y}_{i.} = \frac{\sum_{j=1}^{b} y_{ij}}{b}, \quad \bar{y}_{.j} = \frac{\sum_{i=1}^{a} y_{ij}}{a},$$

Two-Way Crossed Classification Without Interaction

and

$$\bar{y}_{..} = \frac{\sum_{i=1}^{a} \bar{y}_{i.}}{a} = \frac{\sum_{j=1}^{b} \bar{y}_{.j}}{b} = \frac{\sum_{i=1}^{a} \sum_{j=1}^{b} y_{ij}}{ab}.$$

The identity (3.3.1) is valid since all the cross-product terms are equal to zero.

The first two sums of squares to the right of (3.3.1) measure the variation due to the α_i's and β_j's, respectively, and the last one corresponds to the error e_{ij}'s. We use the notation SS_A, SS_B, and SS_E to denote the sums of squares due to the α_i's, β_j's and e_{ij}'s, respectively. The corresponding mean squares, obtained by dividing SS_A, SS_B, and SS_E by $(a-1)$, $(b-1)$, and $(a-1)(b-1)$, respectively, are denoted by MS_A, MS_B, and MS_E, respectively. Here, $(a-1)$, $(b-1)$, and $(a-1)(b-1)$ are obtained by partitioning the total degrees of freedom $ab - 1$ into three components: due to the α_i's, β_j's, and e_{ij}'s.

Remark: The three cross-product terms arising in equation (3.3.1) are:

$$2 \sum_{i=1}^{a} \sum_{j=1}^{b} (\bar{y}_{i.} - \bar{y}_{..})(\bar{y}_{.j} - \bar{y}_{..}) = 2 \sum_{i=1}^{a} (\bar{y}_{i.} - \bar{y}_{..}) \sum_{j=1}^{b} (\bar{y}_{.j} - \bar{y}_{..})$$
$$= 2a(\bar{y}_{..} - \bar{y}_{..})b(\bar{y}_{..} - \bar{y}_{..})$$
$$= 0;$$

$$2 \sum_{i=1}^{a} \sum_{j=1}^{b} (\bar{y}_{i.} - \bar{y}_{..})(y_{ij} - \bar{y}_{i.} - \bar{y}_{.j} + \bar{y}_{..})$$
$$= 2 \sum_{i=1}^{a} (\bar{y}_{i.} - \bar{y}_{..}) \sum_{j=1}^{b} (y_{ij} - \bar{y}_{i.} - \bar{y}_{.j} + \bar{y}_{..})$$
$$= 2 \sum_{i=1}^{a} (\bar{y}_{i.} - \bar{y}_{..}) b(\bar{y}_{i.} - \bar{y}_{i..} - \bar{y}_{..} + \bar{y}_{..})$$
$$= 0;$$

and

$$2 \sum_{i=1}^{a} \sum_{j=1}^{b} (\bar{y}_{.j} - \bar{y}_{..})(y_{ij} - \bar{y}_{i.} - \bar{y}_{.j} + \bar{y}_{..})$$
$$= 2 \sum_{j=1}^{b} (\bar{y}_{.j} - \bar{y}_{..}) \sum_{i=1}^{a} (y_{ij} - \bar{y}_{i.} - \bar{y}_{.j} + \bar{y}_{..})$$
$$= 2 \sum_{j=1}^{b} (\bar{y}_{.j} - \bar{y}_{..}) a(\bar{y}_{.j} - \bar{y}_{..} - \bar{y}_{.j} + \bar{y}_{..})$$
$$= 0.$$

Thus, all cross-product terms are equal to zero.

3.4 MEAN SQUARES AND THEIR EXPECTATIONS

Next, we examine the expectations of the mean squares. On taking successive averages of the model equation (3.1.1), we obtain

$$\bar{y}_{i.} = \mu + \alpha_i + \bar{\beta}_. + \bar{e}_{i.}, \tag{3.4.1}$$

$$\bar{y}_{.j} = \mu + \bar{\alpha}_. + \beta_j + \bar{e}_{.j}, \tag{3.4.2}$$

and

$$\bar{y}_{..} = \mu + \bar{\alpha}_. + \bar{\beta}_. + \bar{e}_{..}. \tag{3.4.3}$$

Substituting the values of y_{ij}, $\bar{y}_{i.}$, $\bar{y}_{.j}$, and $\bar{y}_{..}$, from (3.1.1), (3.4.1), (3.4.2), and (3.4.3), respectively, into the expressions for SS_A, SS_B, and SS_E defined in (3.3.1), we find that

$$SS_E = \sum_{i=1}^{a} \sum_{j=1}^{b} (e_{ij} - \bar{e}_{i.} - \bar{e}_{.j} + \bar{e}_{..})^2, \tag{3.4.4}$$

$$SS_B = a \sum_{j=1}^{b} (\beta_j - \bar{\beta}_. + \bar{e}_{.j} - \bar{e}_{..})^2, \tag{3.4.5}$$

and

$$SS_A = b \sum_{i=1}^{a} (\alpha_i - \bar{\alpha}_. + \bar{e}_{i.} - \bar{e}_{..})^2. \tag{3.4.6}$$

Now, because the e_{ij}'s are uncorrelated and identically distributed with mean zero and variance σ_e^2, it follows that

$$E(e_{ij}^2) = \sigma_e^2, \tag{3.4.7}$$

$$E(\bar{e}_{i.}^2) = \sigma_e^2/b, \tag{3.4.8}$$

$$E(\bar{e}_{.j}^2) = \sigma_e^2/a, \tag{3.4.9}$$

and

$$E(\bar{e}_{..}^2) = \sigma_e^2/ab. \tag{3.4.10}$$

It is then a matter of straightforward computations to derive the expectations of

Two-Way Crossed Classification Without Interaction

mean squares. First, taking the expectation of (3.4.4.), we obtain

$$E(SS_E) = \sum_{i=1}^{a} \sum_{j=1}^{b} E(e_{ij} - \bar{e}_{i.} - \bar{e}_{.j} + \bar{e}_{..})^2$$

$$= \sum_{i=1}^{a} \sum_{j=1}^{b} [E(e_{ij}^2) + E(\bar{e}_{i.}^2) + E(\bar{e}_{.j}^2) + E(\bar{e}_{..}^2)$$
$$- 2E(e_{ij}\bar{e}_{i.}) - 2E(e_{ij}\bar{e}_{.j}) + 2E(e_{ij}\bar{e}_{..})$$
$$+ 2E(\bar{e}_{i.}\bar{e}_{.j}) - 2E(\bar{e}_{i.}\bar{e}_{..}) - 2E(\bar{e}_{.j}\bar{e}_{..})]$$

$$= \sum_{i=1}^{a} \sum_{j=1}^{b} \left[\sigma_e^2 + \frac{1}{b}\sigma_e^2 + \frac{1}{a}\sigma_e^2 + \frac{1}{ab}\sigma_e^2 - \frac{2}{b}\sigma_e^2 - \frac{2}{a}\sigma_e^2 \right.$$
$$\left. + \frac{2}{ab}\sigma_e^2 + \frac{2}{ab}\sigma_e^2 - \frac{2}{ab}\sigma_e^2 - \frac{2}{ab}\sigma_e^2 \right]$$

$$= ab\left[\sigma_e^2 - \frac{1}{b}\sigma_e^2 - \frac{1}{a}\sigma_e^2 + \frac{1}{ab}\sigma_e^2 \right]$$

$$= \sigma_e^2(ab)\left[1 - \frac{1}{b} - \frac{1}{a} + \frac{1}{ab} \right]$$

$$= (ab - a - b + 1)\sigma_e^2$$

$$= (a-1)(b-1)\sigma_e^2. \qquad (3.4.11)$$

The expectation of MS_E is, therefore, given by

$$E(MS_E) = E\left[\frac{SS_E}{(a-1)(b-1)} \right] = \sigma_e^2. \qquad (3.4.12)$$

Note that the result (3.4.12) is true under the assumptions of fixed and random as well as mixed effects models.

Now, to derive the expectations of MS_B and MS_A, we consider the cases of Models I, II, and III separately.

MODEL I (FIXED EFFECTS)

Under Model I, the α_i's and β_j's are fixed quantities depending on the particular levels included in the experiment with the restriction that $\bar{\alpha}_. = \bar{\beta}_. = 0$. First, on taking the expectation of (3.4.5), we obtain

$$E(SS_B) = a\left[\sum_{j=1}^{b} \beta_j^2 + E\sum_{j=1}^{b} (\bar{e}_{.j} - \bar{e}_{..})^2 \right], \qquad (3.4.13)$$

by virtue of the fact that the β_j's are constant and the expectation of the cross-product term is zero. Now, using the results (3.4.9) and (3.4.10), we find that

$$E \sum_{j=1}^{b} (\bar{e}_{.j} - \bar{e}_{..})^2 = \sum_{j=1}^{b} E(\bar{e}_{.j}^2) - bE(\bar{e}_{..}^2)$$

$$= b\frac{1}{a}\sigma_e^2 - b\frac{1}{ab}\sigma_e^2$$

$$= \frac{(b-1)}{a}\sigma_e^2. \qquad (3.4.14)$$

Furthermore, on substituting (3.4.14) into (3.4.13), we obtain

$$E(SS_B) = a \sum_{j=1}^{b} \beta_j^2 + (b-1)\sigma_e^2. \qquad (3.4.15)$$

Therefore, the expectation of MS_B is given by

$$E(MS_B) = E\left(\frac{SS_B}{b-1}\right) = \frac{a}{b-1} \sum_{j=1}^{b} \beta_j^2 + \sigma_e^2. \qquad (3.4.16)$$

Similarly, from symmetry, it follows that the expectation of MS_A is given by

$$E(MS_A) = \frac{b}{a-1} \sum_{i=1}^{a} \alpha_i^2 + \sigma_e^2. \qquad (3.4.17)$$

MODEL II (RANDOM EFFECTS)

Under Model II, the α_i's and β_j's are also randomly distributed with mean zero and variances σ_α^2 and σ_β^2, respectively. It then follows, using the formulae for the variances of the sampling distribution of the means of the α_i's and β_j's, that

$$E(\alpha_i^2) = \sigma_\alpha^2, \qquad (3.4.18)$$

$$E(\bar{\alpha}^2) = \sigma_\alpha^2/a, \qquad (3.4.19)$$

$$E(\beta_j^2) = \sigma_\beta^2, \qquad (3.4.20)$$

and

$$E(\bar{\beta}^2) = \sigma_\beta^2/b. \qquad (3.4.21)$$

Now, on taking the expectation of (3.4.5), we get

$$E(SS_B) = a\left[E \sum_{j=1}^{b} (\beta_j - \bar{\beta}_.)^2 + E \sum_{j=1}^{b} (\bar{e}_{.j} - \bar{e}_{..})^2 \right], \qquad (3.4.22)$$

since the expectation of the cross-product term is zero. Furthermore, using the results (3.4.20) and (3.4.21), we find that

$$E \sum_{j=1}^{b} (\beta_j - \bar{\beta}_.)^2 = \sum_{j=1}^{b} E(\beta_j^2) - bE(\bar{\beta}_.^2)$$

$$= b\sigma_\beta^2 - b\frac{1}{b}\sigma_\beta^2$$

$$= (b-1)\sigma_\beta^2. \tag{3.4.23}$$

On substituting (3.4.14) and (3.4.23) into (3.4.22), we obtain

$$E(SS_B) = a\left[(b-1)\sigma_\beta^2 + (b-1)\frac{1}{a}\sigma_e^2\right]$$

$$= a(b-1)\sigma_\beta^2 + (b-1)\sigma_e^2. \tag{3.4.24}$$

Therefore, the expectation of MS_B is given by

$$E(MS_B) = E\left(\frac{SS_B}{b-1}\right) = a\sigma_\beta^2 + \sigma_e^2. \tag{3.4.25}$$

Similarly, from symmetry, it follows that the expectation of MS_A is given by

$$E(MS_A) = E\left(\frac{SS_A}{a-1}\right) = b\sigma_\alpha^2 + \sigma_e^2. \tag{3.4.26}$$

MODEL III (MIXED EFFECTS)

Under Model III, the α_i's are fixed quantities and the β_j's are randomly distributed with mean zero and variance σ_β^2. Furthermore, the β_j's are uncorrelated with each other and each of the β_j's and e_{ij}'s are also uncorrelated. Therefore, using the results (3.4.17) and (3.4.25), it follows that the expectations of MS_B and MS_A are given by

$$E(MS_B) = a\sigma_\beta^2 + \sigma_e^2$$

and

$$E(MS_A) = \frac{b}{a-1}\sum_{i=1}^{a}\alpha_i^2 + \sigma_e^2.$$

The foregoing results of Sections 3.3 and 3.4 can now be summarized in a tabular form as the analysis of variance table shown in Table 3.2.

TABLE 3.2
Analysis of Variance for Model (3.1.1)

Source of Variation	Degrees of Freedom	Sum of Squares	Mean Square	Expected Mean Square		
				Model I	Model II	Model III
Due to A	$a - 1$	SS_A	MS_A	$\sigma_e^2 + \dfrac{b}{a-1}\sum_{i=1}^{a}\alpha_i^2$	$\sigma_e^2 + b\sigma_\alpha^2$	$\sigma_e^2 + \dfrac{b}{a-1}\sum_{i=1}^{a}\alpha_i^2$
Due to B	$b - 1$	SS_B	MS_B	$\sigma_e^2 + \dfrac{a}{b-1}\sum_{j=1}^{b}\beta_j^2$	$\sigma_e^2 + a\sigma_\beta^2$	$\sigma_e^2 + a\sigma_\beta^2$
Error	$(a-1)(b-1)$	SS_E	MS_E	σ_e^2	σ_e^2	σ_e^2
Total	$ab - 1$	SS_T				

3.5 SAMPLING DISTRIBUTION OF MEAN SQUARES

It is important to recognize that in the derivations of the results on expected mean squares given in the preceding section, we have not made any distribution assumptions for e_{ij}'s under Model I; for α_i's, β_j's, and e_{ij}'s under Model II; and for β_j's and e_{ij}'s under Model III. However, to derive the form of their sampling distributions, we require the assumption of normality for the random components of model (3.1.1). Thus, under Model I, we assume that the e_{ij}'s are independent and normal random variables with mean zero and variance σ_e^2. Under Model II, all α_i's, β_j's, and e_{ij}'s are mutually and completely independent normal random variables with mean zero and variances $\sigma_\alpha^2, \sigma_\beta^2$, and σ_e^2, respectively. Finally, under Model III, the α_i's are constants subject to the restriction that $\sum_{i=1}^{a} \alpha_i = 0$, and the β_j's and e_{ij}'s are mutually and completely independent normal random variables with mean zero and variances σ_β^2 and σ_e^2, respectively.

In the following we give the results on sampling distributions of mean squares for fixed, random, and mixed effects models. The derivation of these results is beyond the scope of this volume and can be found in Scheffé (1959), Graybill (1961), and Searle (1971b).

MODEL I (FIXED EFFECTS)

Under the distribution assumptions of Model I, it can be shown that:

(a) The quantities MS_E, MS_B, and MS_A are statistically independent.
(b) The following results are true:

(i)
$$\frac{\text{MS}_E}{\sigma_e^2} \sim \frac{\chi^2[(a-1)(b-1)]}{(a-1)(b-1)}, \qquad (3.5.1)$$

(ii)
$$\frac{\text{MS}_B}{\sigma_e^2} \sim \frac{\chi^{2\prime}[b-1, \lambda_B]}{b-1}, \qquad (3.5.2)$$

and

(iii)
$$\frac{\text{MS}_A}{\sigma_e^2} \sim \frac{\chi^{2\prime}[a-1, \lambda_A]}{a-1}, \qquad (3.5.3)$$

where, as usual, $\chi^2[.]$ denotes a central and $\chi^{2\prime}[.,.]$ denotes a noncentral chi-square variable with respective degrees of freedom, and the noncentrality parameters λ_B and λ_A are defined by

$$\lambda_B = \frac{a}{2\sigma_e^2} \sum_{j=1}^{b} \beta_j^2$$

and

$$\lambda_A = \frac{b}{2\sigma_e^2} \sum_{i=1}^{a} \alpha_i^2.$$

It, therefore, follows from (3.5.2) and (3.5.3) that
(ii)' If $\beta_j = 0$, for all j, then

$$\frac{MS_B}{\sigma_e^2} \sim \frac{\chi^2[b-1]}{b-1}. \tag{3.5.4}$$

(iii)' If $\alpha_i = 0$, for all i, then

$$\frac{MS_A}{\sigma_e^2} \sim \frac{\chi^2[a-1]}{a-1}. \tag{3.5.5}$$

MODEL II (RANDOM EFFECTS)

Under the distribution assumptions of Model II, it can be shown that:

(a) The quantities MS_E, MS_B, and MS_A are statistically independent.
(b) The following results are true:
(i)

$$\frac{MS_E}{\sigma_e^2} \sim \frac{\chi^2[(a-1)(b-1)]}{(a-1)(b-1)}, \tag{3.5.6}$$

(ii)

$$\frac{MS_B}{\sigma_e^2 + a\sigma_\beta^2} \sim \frac{\chi^2[b-1]}{b-1}, \tag{3.5.7}$$

and
(iii)

$$\frac{MS_A}{\sigma_e^2 + b\sigma_\alpha^2} \sim \frac{\chi^2[a-1]}{a-1}. \tag{3.5.8}$$

That is, the ratio of MS_E to σ_e^2 is a $\chi^2[(a-1)(b-1)]$ variable divided by $(a-1)(b-1)$, the ratio of MS_B to $\sigma_e^2 + a\sigma_\beta^2$ is a $\chi^2[b-1]$ variable divided by $b-1$, and the ratio of MS_A to $\sigma_e^2 + b\sigma_\alpha^2$ is a $\chi^2[a-1]$ variable divided by $a-1$.

MODEL III (MIXED EFFECTS)

Under the distribution assumptions of Model III, it can be shown that:

(a) The quantities MS_E, MS_B, and MS_A are statistically independent.

(b) The following results are true:
   (i)
$$\frac{\mathrm{MS}_E}{\sigma_e^2} \sim \frac{\chi^2[(a-1)(b-1)]}{(a-1)(b-1)}, \qquad (3.5.9)$$

   (ii)
$$\frac{\mathrm{MS}_B}{\sigma_e^2 + a\sigma_\beta^2} \sim \frac{\chi^2[b-1]}{(b-1)}, \qquad (3.5.10)$$

and
   (iii)
$$\frac{\mathrm{MS}_A}{\sigma_e^2} \sim \frac{\chi^{2\prime}[a-1, \lambda_A]}{a-1}, \qquad (3.5.11)$$

where
$$\lambda_A = \frac{b}{2\sigma_e^2} \sum_{i=1}^{a} \alpha_i^2.$$

It follows from (3.5.11) that if $\alpha_i = 0$, for all i, then
   (iii)$'$
$$\frac{\mathrm{MS}_A}{\sigma_e^2} \sim \frac{\chi^2[a-1]}{a-1}. \qquad (3.5.12)$$

3.6 TESTS OF HYPOTHESES: THE ANALYSIS OF VARIANCE F TESTS

In this section, we present the usual hypotheses about the effects of A and B factors and the appropriate F tests for fixed, random, and mixed effects models.

MODEL I (FIXED EFFECTS)

Under Model I, the usual hypotheses of interests are:
$$H_0^B : \text{all } \beta_j\text{'s} = 0$$
versus
$$H_1^B : \text{not all } \beta_j\text{'s are zero}; \qquad (3.6.1)$$

and
$$H_0^A : \text{all } \alpha_i\text{'s} = 0$$
versus
$$H_1^A : \text{not all } \alpha_i\text{'s are zero}. \qquad (3.6.2)$$

In order to develop test procedures for the hypotheses (3.6.1) and (3.6.2), we note from (3.4.12), (3.4.16), and (3.4.17) that when the null hypotheses H_0^B and H_0^A are true, we have

$$E(\text{MS}_E) = \sigma_e^2,$$
$$E(\text{MS}_B) = \sigma_e^2,$$

and

$$E(\text{MS}_A) = \sigma_e^2,$$

that is, $\text{MS}_E, \text{MS}_B,$ and MS_A are unbiased estimates of the same quantity σ_e^2. It then follows, from (3.5.1), (3.5.4), and (3.5.5), that

$$F_B = \frac{\text{MS}_B/\sigma_e^2}{\text{MS}_E/\sigma_e^2} = \frac{\text{MS}_B}{\text{MS}_E} \sim F\,[b-1,(a-1)(b-1)] \qquad (3.6.3)$$

and

$$F_A = \frac{\text{MS}_A/\sigma_e^2}{\text{MS}_E/\sigma_e^2} = \frac{\text{MS}_A}{\text{MS}_E} \sim F\,[a-1,(a-1)(b-1)]. \qquad (3.6.4)$$

Therefore, the statistics (3.6.3) and (3.6.4) provide suitable test procedures for testing hypotheses (3.6.1) and (3.6.2), respectively. Thus, H_0^B is rejected at the α-level of significance if

$$F_B > F\,[b-1,(a-1)(b-1);1-\alpha]. \qquad (3.6.5)$$

Similarly, H_0^A is rejected at the α-level of significance if

$$F_A > F\,[a-1,(a-1)(b-1);1-\alpha]. \qquad (3.6.6)$$

It should be noted, however, that when the null hypotheses H_0^B and H_0^A are not true, it follows from (3.5.1) through (3.5.3) that

$$\frac{\text{MS}_B}{\text{MS}_E} \sim F'[b-1,(a-1)(b-1);\lambda_B] \qquad (3.6.7)$$

and

$$\frac{\text{MS}_A}{\text{MS}_E} \sim F'[a-1,(a-1)(b-1);\lambda_A], \qquad (3.6.8)$$

where $F'[.,.;.]$ denotes a statistic having a noncentral F distribution with respective degrees of freedom and the noncentrality parameters λ_B and λ_A defined by

$$\lambda_B = \frac{a}{2\sigma_e^2} \sum_{j=1}^{b} \beta_j^2$$

and

$$\lambda_A = \frac{b}{2\sigma_e^2} \sum_{i=1}^{a} \alpha_i^2.$$

The results (3.6.7) and (3.6.8) are employed in evaluating the power of these F tests in Section 3.11.

MODEL II (RANDOM EFFECTS)

Under Model II, testing significance of the effects of a factor is equivalent to testing the hypothesis that the corresponding variance component is zero. Thus, the usual analogues of hypotheses (3.6.1) and (3.6.2) are:

$$H_0^B : \sigma_\beta^2 = 0 \quad \text{versus} \quad H_1^B : \sigma_\beta^2 > 0 \qquad (3.6.9)$$

and

$$H_0^A : \sigma_\alpha^2 = 0 \quad \text{versus} \quad H_1^A : \sigma_\alpha^2 > 0. \qquad (3.6.10)$$

It can be readily seen that the statistics (3.6.3) and (3.6.4), obtained in the case of Model I, also provide suitable test procedures for the hypotheses (3.6.9) and (3.6.10), respectively. However, under the alternative hypotheses, the aforesaid statistics have a (central) F distribution rather than a noncentral F as in the case of Model I, a fact which greatly simplifies the computation of power under Model II.

MODEL III (MIXED EFFECTS)

Under Model III, the hypotheses of interest are: $\sigma_\beta^2 = 0$ and α_i's $= 0$. Again, it can be seen that the test statistics (3.6.3) and (3.6.4) developed earlier are also applicable for these hypotheses.

3.7 POINT ESTIMATION

In this section, we present results on point estimation for parameters of interest under fixed, random, and mixed effects models.

MODEL I (FIXED EFFECTS)

In the case of the fixed effects model, the least squares estimators[1] of the parameters μ, α_i's, and β_j's are obtained by minimizing the residual sum of squares

$$Q = \sum_{i=1}^{a} \sum_{j=1}^{b} (y_{ij} - \mu - \alpha_i - \beta_j)^2, \qquad (3.7.1)$$

---

[1] The least squares estimators in this case are the same as those obtained by the maximum likelihood method under the assumption of normality.

with respect to μ, α_i's, and β_j's; and subject to the restrictions:

$$\sum_{i=1}^{a} \alpha_i = \sum_{j=1}^{b} \beta_j = 0. \tag{3.7.2}$$

The resultant estimators are obtained to be:

$$\hat{\mu} = \bar{y}_{..}, \tag{3.7.3}$$
$$\hat{\alpha}_i = \bar{y}_{i.} - \bar{y}_{..}, \quad i = 1, 2, \ldots, a, \tag{3.7.4}$$

and

$$\hat{\beta}_j = \bar{y}_{.j} - \bar{y}_{..}, \quad j = 1, 2, \ldots, b. \tag{3.7.5}$$

These are the so-called best linear unbiased estimators (BLUE). The variances of the estimators (3.7.3) through (3.7.5) are:

$$\text{Var}(\hat{\mu}) = \sigma_e^2/ab, \tag{3.7.6}$$
$$\text{Var}(\hat{\alpha}_i) = (a-1)\sigma_e^2/ab, \tag{3.7.7}$$

and

$$\text{Var}(\hat{\beta}_j) = (b-1)\sigma_e^2/ab. \tag{3.7.8}$$

The other parameters of interest are: $\mu + \alpha_i$ (mean levels of the factor A), $\mu + \beta_j$ (mean levels of the factor B), pairwise differences $\alpha_i - \alpha_{i'}$ and $\beta_j - \beta_{j'}$, and the contrasts of the type $\sum_{i=1}^{a} \ell_i \alpha_i$ ($\sum_{i=1}^{a} \ell_i = 0$) and $\sum_{j=1}^{b} \ell'_j \beta_j$ ($\sum_{j=1}^{b} \ell'_j = 0$). Their respective estimates together with the variances are given by

$$\widehat{\mu + \alpha_i} = \bar{y}_{i.}, \qquad \text{Var}(\widehat{\mu + \alpha_i}) = \sigma_e^2/b; \tag{3.7.9}$$
$$\widehat{\mu + \beta_j} = \bar{y}_{.j}, \qquad \text{Var}(\widehat{\mu + \beta_j}) = \sigma_e^2/a; \tag{3.7.10}$$
$$\widehat{\alpha_i - \alpha_{i'}} = \bar{y}_{i.} - \bar{y}_{i'.}, \qquad \text{Var}(\widehat{\alpha_i - \alpha_{i'}}) = 2\sigma_e^2/b; \tag{3.7.11}$$
$$\widehat{\beta_j - \beta_{j'}} = \bar{y}_{.j} - \bar{y}_{.j'}, \qquad \text{Var}(\widehat{\beta_j - \beta_{j'}}) = 2\sigma_e^2/a; \tag{3.7.12}$$
$$\sum_{i=1}^{a} \widehat{\ell_i \alpha_i} = \sum_{i=1}^{a} \ell_i \bar{y}_{i.}, \quad \text{Var}\left(\sum_{i=1}^{a} \widehat{\ell_i \alpha_i}\right) = \sum_{i=1}^{a} \ell_i^2 \sigma_e^2/b; \tag{3.7.13}$$

and

$$\sum_{j=1}^{b} \widehat{\ell'_j \beta_j} = \sum_{j=1}^{b} \ell'_j \bar{y}_{.j}, \quad \text{Var}\left(\sum_{j=1}^{b} \widehat{\ell'_j \beta_j}\right) = \sum_{j=1}^{b} \ell'^2_j \sigma_e^2/a. \tag{3.7.14}$$

Two-Way Crossed Classification Without Interaction

The variance σ_e^2 is, of course, estimated unbiasedly by

$$\hat{\sigma}_e^2 = \text{MS}_E. \tag{3.7.15}$$

MODEL II (RANDOM EFFECTS)

In the case of the random effects model, the variance components may be estimated by the analysis of variance method, that is, by equating the observed mean squares in the lines of the analysis of variance table to their respective expected values and solving the equations for the variance components. The resulting estimators are:

$$\hat{\sigma}_e^2 = \text{MS}_E, \tag{3.7.16}$$

$$\hat{\sigma}_\beta^2 = \frac{1}{a}(\text{MS}_B - \text{MS}_E), \tag{3.7.17}$$

and

$$\hat{\sigma}_\alpha^2 = \frac{1}{b}(\text{MS}_A - \text{MS}_E). \tag{3.7.18}$$

It can be shown that these estimators are also the maximum likelihood estimators (corrected for bias) of the corresponding parameters. The parameter μ is, of course, estimated by

$$\hat{\mu} = \bar{y}_{..}, \tag{3.7.19}$$

as in the case of the fixed effects model. The remarks concerning the negative estimates and the optimal properties of the analysis of variance estimators made in Section 2.9 also apply here.[2]

MODEL III (MIXED EFFECTS)

In the case of a mixed effects model, with A fixed and B random, the usual parameters of interest are μ, α_i's, σ_β^2, and σ_e^2. The corresponding estimators are:

$$\hat{\mu} = \bar{y}_{..}, \tag{3.7.20}$$

$$\hat{\alpha}_i = \bar{y}_{i.} - \bar{y}_{..}, \quad i = 1, 2, \ldots, a, \tag{3.7.21}$$

$$\hat{\sigma}_\beta^2 = \frac{1}{a}(\text{MS}_B - \text{MS}_E), \tag{3.7.22}$$

and

$$\hat{\sigma}_e^2 = \text{MS}_E. \tag{3.7.23}$$

---

[2] For a discussion of the nonnegative maximum likelihood as well as other nonnegative estimation procedures and their properties, the reader is referred to Sahai (1974b) and Sahai and Khurshid (1992).

3.8 INTERVAL ESTIMATION

In this section, we present results on confidence intervals for parameters of interest under fixed, random, and mixed effects models.

MODEL I (FIXED EFFECTS)

Using the distribution theory of the error mean square; that is,

$$\frac{\mathrm{MS}_E}{\sigma_e^2} \sim \frac{\chi^2[(a-1)(b-1)]}{(a-1)(b-1)}, \qquad (3.8.1)$$

a $100(1-\alpha)$ percent confidence interval for σ_e^2 is obtained as

$$\frac{(a-1)(b-1)\mathrm{MS}_E}{\chi^2[(a-1)(b-1), 1-\alpha/2]} < \sigma_e^2 < \frac{(a-1)(b-1)\mathrm{MS}_E}{\chi^2[(a-1)(b-1), \alpha/2]}. \qquad (3.8.2)$$

Also, it is possible to construct confidence intervals using the t distribution for a pairwise difference $\alpha_i - \alpha_{i'}$ or the contrast $\sum_{i=1}^{a} \ell_i \alpha_i$, where $\sum_{i=1}^{a} \ell_i = 0$. To obtain the confidence limits for $\alpha_i - \alpha_{i'}$, we note from (3.7.11) that

$$\widehat{\alpha_i - \alpha_{i'}} = \bar{y}_{i.} - \bar{y}_{i'.}$$

with

$$\mathrm{Var}(\bar{y}_{i.} - \bar{y}_{i'.}) = 2\sigma_e^2/b.$$

The confidence limits on $\alpha_i - \alpha_i'$ can, therefore, be derived from the relation

$$\frac{(\bar{y}_{i.} - \bar{y}_{i'.}) - (\alpha_i - \alpha_{i'})}{\sqrt{2\mathrm{MS}_E/b}} \sim t[(a-1)(b-1)].$$

Thus, the corresponding $100(1-\alpha)$ percent confidence limits for $\alpha_i - \alpha_i'$ are given by

$$(\bar{y}_{i.} - \bar{y}_{i'.}) \pm t[(a-1)(b-1), 1-\alpha/2]\sqrt{2\mathrm{MS}_E/b}.$$

Similarly, confidence limits for the contrast $\sum_{i=1}^{a} \ell_i \alpha_i$ can be obtained from the relation

$$\frac{\sum_{i=1}^{a} \ell_i \bar{y}_{i.} - \sum_{i=1}^{a} \ell_i \alpha_i}{\sqrt{\mathrm{MS}_E \sum_{i=1}^{a} \ell_i^2/b}} \sim t[(a-1)(b-1)],$$

which yields

$$P\left\{\sum_{i=1}^{a} \ell_i \bar{y}_{i.} - M < \sum_{i=1}^{a} \ell_i \alpha_i < \sum_{i=1}^{a} \ell_i \bar{y}_{i.} + M\right\} = 1 - \alpha,$$

where

$$M = t\left[(a-1)(b-1), 1 - \alpha/2\right] \sqrt{MS_E \sum_{i=1}^{a} \ell_i^2 / b}.$$

Similar results hold for any pairwise difference $\beta_j - \beta_{j'}$, or the contrast $\sum_{j=1}^{b} \ell'_j \beta_j (\sum_{j=1}^{b} \ell'_j = 0)$. However, for the reasons given in Section 2.19, the multiple comparison procedures discussed in Section 3.12 should be preferred.

MODEL II (RANDOM EFFECTS)

Exact confidence intervals for σ_e^2, $\sigma_e^2 + a\sigma_\beta^2$, $\sigma_e^2 + b\sigma_\alpha^2$, variance ratios $\sigma_\beta^2/\sigma_e^2$ and $\sigma_\alpha^2/\sigma_e^2$, and proportions of variances $\sigma_e^2/(\sigma_e^2+\sigma_\beta^2)$, $\sigma_e^2/(\sigma_e^2+\sigma_\alpha^2)$, $\sigma_\beta^2/(\sigma_e^2+\sigma_\beta^2)$, and $\sigma_\alpha^2/(\sigma_e^2+\sigma_\alpha^2)$ can be obtained by using the results on the sampling distribution for mean squares. In particular, the probability is $1 - \alpha$ that the interval

$$\left[\frac{1}{a}\left(\frac{MS_B}{MS_E} \cdot \frac{1}{F[b-1,(a-1)(b-1); 1-\alpha/2]} - 1\right),\right.$$
$$\left.\frac{1}{a}\left(\frac{MS_B}{MS_E} \cdot \frac{1}{F[b-1,(a-1)(b-1); \alpha/2]} - 1\right)\right]$$

captures $\sigma_\beta^2/\sigma_e^2$. However, as before, exact confidence intervals for σ_α^2 and σ_β^2 do not exist. The approximate procedures available in the case of one-way classification are also applicable here.[3]

MODEL III (MIXED EFFECTS)

The objective in this case is to set confidence limits on the variance components σ_e^2, σ_β^2, and on fixed effects α_i's. The limits on σ_e^2 are the same as given in (3.8.2). Again, an exact interval for σ_β^2 is not available, but one can set exact limits on $\sigma_e^2 + a\sigma_\beta^2$ and $\sigma_\beta^2/\sigma_e^2$. Similarly, one can obtain confidence intervals for a pairwise difference $\alpha_i - \alpha_{i'}$ or the contrast $\sum_{i=1}^{a} \ell_i \alpha_i (\sum_{i=1}^{a} \ell_i = 0)$. Approximate

---

[3] For some results on approximate confidence intervals for the variance components σ_β^2 and σ_α^2, and the total variance $\sigma_e^2 + \sigma_\beta^2 + \sigma_\alpha^2$, including a numerical example, see Burdick and Graybill (1992, pp. 126–128). The problem of setting confidence intervals on the proportions of variability $\sigma_e^2/(\sigma_e^2+\sigma_\beta^2+\sigma_\alpha^2)$, $\sigma_\beta^2/(\sigma_e^2+\sigma_\beta^2+\sigma_\alpha^2)$, and $\sigma_\alpha^2/(\sigma_e^2+\sigma_\beta^2+\sigma_\alpha^2)$ has been considered by Arteaga et al. (1982). For a concise summary of the results, including a numerical example, see Burdick and Graybill (1992, pp. 129–132).

confidence intervals for σ_β^2 and a single factor A level mean $\mu + \alpha_i$ can be determined using the Satterthwaite procedure (see Appendix K). For some additional results including numerical examples, see Burdick and Graybill (1992, pp. 154–156).

3.9 COMPUTATIONAL FORMULAE AND PROCEDURE

As in the case of one-way classification, the sums of squares SS_T, SS_A, SS_B, and SS_E can be expressed in more computationally suitable forms. These are:

$$SS_T = \sum_{i=1}^{a} \sum_{j=1}^{b} y_{ij}^2 - \frac{y_{..}^2}{ab}, \quad (3.9.1)$$

$$SS_A = \sum_{i=1}^{a} \frac{y_{i.}^2}{b} - \frac{y_{..}^2}{ab}, \quad (3.9.2)$$

$$SS_B = \sum_{j=1}^{b} \frac{y_{.j}^2}{a} - \frac{y_{..}^2}{ab}, \quad (3.9.3)$$

and

$$SS_E = \sum_{i=1}^{a} \sum_{j=1}^{b} y_{ij}^2 - \sum_{i=1}^{a} \frac{y_{i.}^2}{b} - \sum_{j=1}^{b} \frac{y_{.j}^2}{a} + \frac{y_{..}^2}{ab}, \quad (3.9.4)$$

where

$$y_{i.} = \sum_{j=1}^{b} y_{ij}, \quad y_{.j} = \sum_{i=1}^{a} y_{ij},$$

and

$$y_{..} = \sum_{i=1}^{a} y_{i.} = \sum_{j=1}^{b} y_{.j} = \sum_{i=1}^{a} \sum_{j=1}^{b} y_{ij}.$$

The error sum of squares SS_E is usually calculated by subtracting $SS_A + SS_B$ from SS_T; that is,

$$SS_E = SS_T - SS_A - SS_B. \quad (3.9.5)$$

The computational procedure for the sums of squares can be performed in the following sequence of steps:

(i) Sum the observations for each row to form the row totals:

$$y_{1.}, y_{2.}, \ldots, y_{a.}.$$

(ii) Sum the observations for each column to form the column totals:

$$y_{.1}, y_{.2}, \ldots, y_{.b}.$$

(iii) Sum all the observations to obtain the overall or grand total:

$$y_{..} = \sum_{i=1}^{a} y_{i.} = \sum_{j=1}^{b} y_{.j}.$$

(iv) Form the sum of squares of the individual observations to yield:

$$\sum_{i=1}^{a} \sum_{j=1}^{b} y_{ij}^2 = y_{11}^2 + y_{12}^2 + \cdots + y_{ab}^2.$$

(v) Form the sum of squares of the totals for each row and divide it by b to yield:

$$\sum_{i=1}^{a} y_{i.}^2 / b.$$

(vi) Form the sum of squares of the totals for each column and divide it by a to yield:

$$\sum_{j=1}^{b} y_{.j}^2 / a.$$

(vii) Square the grand total and divide it by ab to obtain the correction factor:

$$\frac{y_{..}^2}{ab}.$$

Now, the required sums of squares, SS_T, SS_A, SS_B, and SS_E are obtained by using the computational formulae (3.9.1), (3.9.2), (3.9.3), and (3.9.4) or (3.9.5), respectively.

It is expected that most investigators would use computers in the handling of analysis of variance calculations. Otherwise, an electronic calculator is highly recommended, especially for a large data set. Such calculators have the additional advantage that the totals and sums of squares can be determined at the same time providing a check on the previous calculation.

3.10 MISSING OBSERVATIONS

In the analysis of variance discussed in this chapter, it is assumed that there is exactly one observation in each cell of the two-way layout as shown in Table 3.1. However, in the process of conducting an experiment, some of the observations may be lost. For example, the experimenter may fail to record an observation, animals or plants may die during the course of the experiment, or a subject may withdraw before the completion of the experiment. In such cases an approximate analysis discussed here may be used. The method consists of inserting estimates of the missing values and then carrying out the usual analysis as if no observations were missing. The estimates are obtained so as to minimize the residual or error sum of squares. However, care should be exercised in not including these estimates when computing the relevant degrees of freedom. Thus, for every missing value being estimated, the degrees of freedom for the residual mean square are reduced by one.

Suppose the observation corresponding to the i-th row and the j-th column is missing and let it be denoted by y_{ij}. Then all the sums of squares are computed in the usual way except, of course, that they all involve y_{ij}. It is then an elementary calculus problem to show that the value of y_{ij} which minimizes the error sum of squares is given by

$$\hat{y}_{ij} = \frac{by'_{.j} + ay'_{i.} - y'_{..}}{(a-1)(b-1)}, \qquad (3.10.1)$$

where $y'_{i.}$ denotes the total of $b-1$ observations in the i-th row, $y'_{.j}$ denotes the total of $a-1$ observations in the j-th column, and $y'_{..}$ denotes the sum of all $ab-1$ observations. The mathematical derivation of the formula (3.10.1) may be found in an intermediate or advanced level text (see, e.g., Peng (1967, pp. 109–110); Montgomery (1991, pp. 148–151); Hinkelmann and Kempthorne (1994, pp. 266–267)). If $\hat{y}_{ij}$ obtained from (3.10.1) is substituted for the missing value, then SS_A, SS_B, and SS_E can be computed in the usual way.

The formula (3.10.1) was first discussed by Allen and Wishart (1930). The F test will be slightly biased and the reader is referred to the paper by Yates (1933) for a discussion on this point. Also, it can be shown that when there is a missing observation in the i-th row,

$$\text{Var}(\bar{y}_{i.}) = \left[\frac{1}{b} + \frac{a}{b(a-1)(b-1)}\right]\sigma_e^2 \qquad (3.10.2)$$

and

$$\text{Var}(\bar{y}_{i.} - \bar{y}_{i'.}) = \left[\frac{2}{b} + \frac{a}{b(a-1)(b-1)}\right]\sigma_e^2. \qquad (3.10.3)$$

The expression (3.10.2) can also be written as

$$\text{Var}(\bar{y}_{i.}) = \left[1 + \frac{1}{b(a-1)}\right]\frac{\sigma_e^2}{b-1}. \qquad (3.10.4)$$

Note that the expression (3.10.4) is slightly greater than $\sigma_e^2/(b-1)$ and the variance of a row mean with no missing value is σ_e^2/b. Furthermore, the error mean square is estimated correctly but the mean squares for factors A and B are somewhat inflated. To correct for this bias the quantity

$$\frac{[y'_{.j} - (a-1)\hat{y}_{ij}]^2}{a(a-1)^2} \tag{3.10.5}$$

is subtracted from the factor A mean square. Similarly, to test for the factor B effects, the quantity

$$\frac{[y'_{i.} - (b-1)\hat{y}_{ij}]^2}{b(b-1)^2} \tag{3.10.6}$$

is subtracted from the factor B mean square.

If there are two missing values, one can either repeat the foregoing procedure with two simultaneous equations obtained by minimizing the error sum of squares with respect to two missing values, or one can obtain an iterative procedure of guessing one value, and fitting the other by formula (3.10.1), then going back and fitting the first value, and so on. When there are several missing values, one first guesses values for all units except the first one. Formula (3.10.1) is then used to find an initial estimate of the first missing value. With this initial estimate for the first one and the values guessed for the others, the formula (3.10.1) is again used to obtain an estimate of the second value. The process is continued in this manner to obtain estimates for the remaining values. After completing the first cycle of the initial estimates, a second set of estimated values is found and the entire process is repeated several times until the estimated values are not different from those obtained in the previous cycle. The details may be found in Tocher (1952), Bennett and Franklin (1954, p. 382), Cochran and Cox (1957, pp. 110–112), and Steel and Torrie (1980, pp. 211–213). Healy and Westmacott (1956) gave a more general iterative method using a program that analyzes complete data rather rapidly, and Rubin (1972) presented a noniterative method. For m missing values, a computer program is usually required that will invert an $m \times m$ matrix. The general problem of missing data can usually be dealt with much more efficiently using an algorithm developed by Dempster et al. (1977). For further discussion of this topic the reader is referred to Anderson (1946), Dodge (1985), and Snedecor and Cochran (1989, pp. 273–278).[4] For a discussion of correction for bias in mean squares for factors A and B when two or more observations in a row or column are missing, see Glen and Kramer (1958).

It should be remarked that the use of estimates for missing values does not in any way recover the information that is lost through the missing data. It is merely a computational procedure to enable the experimenter to make an approximate

---

[4] Hoyle (1971) gives an introduction to spoilt data (missing, extra, and mixed up observations) with an extensive bibliography. Afifi and Elashoff (1966, 1967) in a two-part article have considered the problem of missing data in multivariate statistics.

analysis. It is important that the investigator examine carefully the nature of the missing values. If the reasons for missing values can be attributed to chance, the treatment comparisons based on the remaining values will be unbiased, and the methods described previously may be generally applied. It is worthwhile to remember that the analysis of variance does not take lightly to missing data and utmost caution should be exercised to ensure that no observation is lost. The situation is best summed up by Cochran and Cox (1957, p. 82) when they state: "... the only complete solution to the 'missing data' problem is not to have them...."

3.11 POWER OF THE ANALYSIS OF VARIANCE F TESTS

The discussion on the power of the analysis of variance F test for the one-way classification given in Section 2.17 also applies here. Thus, under Model I, it follows from (3.6.5) and (3.6.7) that the power of the F test for the hypothesis on β_j's is given by

$$\begin{aligned}\text{Power} &= P\left\{\frac{\text{MS}_B}{\text{MS}_A} > F[b-1,(a-1)(b-1);1-\alpha] \mid \beta_j > 0 \right.\\ &\qquad \left. \text{for at least one } j \right\} \\ &= P\{F'[b-1,(a-1)(b-1);\phi_B] \\ &\qquad > F[b-1,(a-1)(b-1);1-\alpha]\},\end{aligned} \qquad (3.11.1)$$

where

$$\phi_B = \frac{1}{\sigma_e}\sqrt{\frac{a}{b}\sum_{j=1}^{b}\beta_j^2}.$$

Similarly, for the hypothesis on α_i's, we obtain

$$\text{Power} = P\{F'[a-1,(a-1)(b-1);\phi_A] \\ > F[(a-1),(a-1)(b-1);1-\alpha]\}, \qquad (3.11.2)$$

where

$$\phi_A = \frac{1}{\sigma_e}\sqrt{\frac{b}{a}\sum_{i=1}^{a}\alpha_i^2}.$$

The expressions (3.11.1) and (3.11.2) can be evaluated by using noncentral F tables or Pearson-Hartley charts as described in Section 2.17.

Under Model II, the power of the test for the hypothesis

$$H_0 : \sigma_\beta^2 = 0 \quad \text{versus} \quad H_1 : \sigma_\beta^2 > 0$$

is given by

$$P\left\{ \frac{\mathrm{MS}_B}{\mathrm{MS}_E} > F[b-1, (a-1)(b-1); 1-\alpha] \mid \sigma_\beta^2 > 0 \right\}$$

$$= P\left\{ F[b-1, (a-1)(b-1)] \right.$$

$$\left. > \left(1 + a\frac{\sigma_\beta^2}{\sigma_e^2}\right)^{-1} F[b-1, (a-1)(b-1); 1-\alpha] \right\}. \quad (3.11.3)$$

Likewise, the power of the test for the hypothesis

$$H_0 : \sigma_\alpha^2 = 0 \quad \text{versus} \quad H_1 : \sigma_\alpha^2 > 0$$

is given by

$$P\left\{ F[a-1, (a-1)(b-1)] \right.$$

$$\left. > \left(1 + b\frac{\sigma_\alpha^2}{\sigma_e^2}\right)^{-1} F[a-1, (a-1)(b-1); 1-\alpha] \right\}. \quad (3.11.4)$$

Powers of the tests corresponding to the more general hypotheses of the type $\sigma_\beta^2 / \sigma_e^2 \le \rho$ can also be obtained similarly.

Under Model III, the power of the test for β_j effects involves the central F distribution and for α_i effects involves the noncentral F distribution. The power results for σ_β^2 are then the same as given in (3.11.3) and for α_i's the results are given by (3.11.2).

3.12 MULTIPLE COMPARISON METHODS

The results on multiple comparisons discussed in Section 2.19 are also applicable here with a few minor modifications. The procedures can be utilized for the fixed as well as the mixed effects models. For the fixed effects case, as we have seen in Section 3.8, the contrasts of interest may involve α_i's or β_j's and will be of the form

$$L = \ell_1 \alpha_1 + \ell_2 \alpha_2 + \cdots + \ell_a \alpha_a$$

or

$$L' = \ell'_1\beta_1 + \ell'_2\beta_2 + \cdots + \ell'_b\beta_b,$$

where

$$\sum_{i=1}^{a} \ell_i = \sum_{j=1}^{b} \ell'_j = 0.$$

The estimates of L and L' are given by

$$\hat{L} = \ell_1 \bar{y}_{1.} + \ell_2 \bar{y}_{2.} + \cdots + \ell_a \bar{y}_{a.}$$

and

$$\hat{L}' = \ell'_1 \bar{y}_{.1} + \ell'_2 \bar{y}_{.2} + \cdots + \ell'_b \bar{y}_{.b}.$$

The results on Tukey's and Scheffé's methods are the same as given in Section 2.19 except that now MS_E replaces MS_W and $(a-1)(b-1)$ replaces $a(n-1)$ in degrees of freedom entries. Furthermore, MS_B will be replaced by MS_A or MS_B depending upon whether the inferences are sought on α_i's or β_j's. Thus, for example, if Tukey's method is used, L is significant at the α-level if

$$\hat{L} \Big/ \left\{ \sqrt{b^{-1}MS_E}\left(\frac{1}{2}\sum_{i=1}^{a}|\ell_i|\right)\right\} > q\,[a,(a-1)(b-1);1-\alpha].$$

Similarly, if Scheffé's method is used, L is significant at the α-level if

$$\hat{L} \Big/ \left\{ (a-1)MS_E \sum_{i=1}^{a} \ell_i^2/b \right\}^{1/2}$$
$$> \{F\,[a-1,(a-1)(b-1);1-\alpha]\}^{1/2}.$$

Likewise, the significance of the contrast L' can be tested. The other multiple comparison procedures can also be similarly modified.

For a single pairwise comparison, one can use $\sqrt{2}t\,[(a-1)(b-1),\,1-\alpha/2]$ instead of $T = q[a,(a-1)(b-1);1-\alpha]$. For a limited number of pairwise comparisons, the Bonferroni method can be employed by using $\sqrt{2}t\,[(a-1)(b-1),\,1-\alpha/2k]$ instead of T.

Under Model III, the contrasts of interest involve only the α_i's and the results are identical to those given previously.[5]

---

[5] For a general discussion of multiple comparison methods in a two-way layout, see Hirotsu (1973).

TABLE 3.3
Loss in Weights Due to Wear Testing of Four Materials (in mg)

Material	Position		
	1	2	3
1	241	270	274
2	195	241	218
3	235	273	230
4	234	236	227

3.13 WORKED EXAMPLE FOR MODEL I

Consider the following example described in Davies (1954). An experiment was carried out for wear testing of four materials. A test piece of each material was extracted from each of the three positions of a testing machine. The reduction in weight due to wear was determined on each piece of material in milligrams and the data are given in Table 3.3.

It is desired to test whether there are significant differences due to different materials and machine positions. Clearly, the data of Table 3.3 should be analyzed under Model I, since the four materials and the three positions of testing machines are especially chosen by the experimenter to be of particular interest to her and thus will both have systematic effects. The mathematical model would be

$$y_{ij} = \mu + \alpha_i + \beta_j + e_{ij} \quad (i = 1, 2, 3, 4; \ j = 1, 2, 3),$$

where μ is the grand mean, α_i is the effect of the i-th material, β_j is the effect of the j-th position, and the e_{ij}'s are random errors, with $\sum_{i=1}^{4} \alpha_i = 0$, $\sum_{j=1}^{3} \beta_j = 0$, and $e_{ij} \sim N(0, \sigma_e^2)$. It is further assumed that no interaction between the material and position is likely to exist.

To perform the analysis of variance computations, we first obtain the row and column totals as

$$y_{1.} = 785, \quad y_{2.} = 654, \quad y_{3.} = 738, \quad y_{4.} = 697;$$
$$y_{.1} = 905, \quad y_{.2} = 1{,}020, \quad y_{.3} = 949;$$

and the grand total is

$$y_{..} = 2{,}874.$$

The other quantities required in the calculations of the sums of squares are:

$$\frac{y_{..}^2}{ab} = \frac{(2874)^2}{4 \times 3} = 688,323,$$

$$\frac{1}{b}\sum_{i=1}^{a} y_{i.}^2 = \frac{1}{3}[(785)^2 + (654)^2 + (738)^2 + (697)^2] = 691,464.667,$$

$$\frac{1}{a}\sum_{j=1}^{b} y_{.j}^2 = \frac{1}{4}[(905)^2 + (1,020)^2 + (949)^2] = 690,006.500,$$

and

$$\sum_{i=1}^{a}\sum_{j=1}^{b} y_{ij}^2 = (241)^2 + (270)^2 + \cdots + (227)^2 = 694,322.$$

The resultant sums of squares are, therefore, given by

$$SS_T = \sum_{i=1}^{a}\sum_{j=1}^{b} y_{ij}^2 - \frac{y_{..}^2}{ab} = 694,322 - 688,323 = 5,999.000,$$

$$SS_A = \frac{1}{b}\sum_{i=1}^{a} y_{i.}^2 - \frac{y_{..}^2}{ab} = 691,464.667 - 688,323 = 3,141.667,$$

$$SS_B = \frac{1}{a}\sum_{j=1}^{b} y_{.j}^2 - \frac{y_{..}^2}{ab} = 690,006.500 - 688,323 = 1,683.500,$$

and

$$SS_E = SS_T - SS_A - SS_B = 5,999.000 - 3,141.667 - 1,683.500$$
$$= 1,173.833.$$

Finally, the results of the analysis of variance calculations are summarized in Table 3.4.

If we choose the level of significance $\alpha = 0.05$, we find from Appendix Table V that $F[3, 6; 0.95] = 4.76$ and $F[2, 6; 0.95] = 5.14$. Comparing these values with the computed F values given in Table 3.4, we do not reject the hypothesis of no "position" effects ($p = 0.069$), but reject the hypothesis of no "material" effects ($p = 0.039$). That is, we may conclude that there is probably a significant difference due to the materials but not due to positions.

To determine which materials differ, we use Tukey's and Scheffé's procedures for pairwise comparisons. For the Tukey's procedure, we find from Appendix Table X that

$$q[a, (a-1)(b-1); 1-\alpha] = q[4, 6; 0.95] = 4.90.$$

TABLE 3.4
Analysis of Variance for the Weight Loss Data of Table 3.3

Source of Variation	Degrees of Freedom	Sum of Squares	Mean Square	Expected Mean Square	F Value	p-Value
Material	3	3,141.667	1,047.222	$\sigma_e^2 + \dfrac{3}{4-1}\sum_{i=1}^{4}\alpha_i^2$	5.35	0.039
Position	2	1,683.500	841.750	$\sigma_e^2 + \dfrac{4}{3-1}\sum_{j=1}^{3}\beta_j^2$	4.30	0.069
Error	6	1,173.833	195.639	σ_e^2		
Total	11	5,999.000				

Now, the pairwise differences of sample means for materials would be compared to

$$q[a, (a-1)(b-1); 1-\alpha]\sqrt{\frac{MS_E}{b}} = 4.90\sqrt{\frac{195.639}{3}} = 39.57.$$

The four sample means for the materials are

$$\bar{y}_{1.} = 261.67, \quad \bar{y}_{2.} = 218.00, \quad \bar{y}_{3.} = 246.00, \quad \bar{y}_{4.} = 232.33,$$

and there are six pairs of differences to be compared. Furthermore,

$$|\bar{y}_{1.} - \bar{y}_{2.}| = 43.67 > 39.57, \quad |\bar{y}_{1.} - \bar{y}_{3.}| = 15.67 < 39.57,$$
$$|\bar{y}_{1.} - \bar{y}_{4.}| = 29.34 < 39.57, \quad |\bar{y}_{2.} - \bar{y}_{3.}| = 28.00 < 39.57,$$
$$|\bar{y}_{2.} - \bar{y}_{4.}| = 14.33 < 39.57, \quad |\bar{y}_{3.} - \bar{y}_{4.}| = 13.67 < 39.57.$$

Hence, we may conclude that materials one and two are probably significantly different but not the others.

For the Scheffé's procedure, we find from Appendix Table V that

$$S^2 = F[a-1, (a-1)(b-1); 1-\alpha] = F[3, 6; 0.95] = 4.76.$$

Furthermore, for the contrasts consisting of the differences between two means, we have

$$\sum_{i=1}^{a} \frac{\ell_i^2}{b} = \frac{1}{3} + \frac{1}{3} = \frac{2}{3}.$$

TABLE 3.5
Number of Minutes Observed in Grazing

	Animal									
Observer	1	2	3	4	5	6	7	8	9	10
1	34	76	75	31	61	82	82	67	72	38
2	33	76	72	29	60	82	84	67	72	36
3	35	78	76	30	65	86	88	66	76	37
4	34	77	71	29	60	78	83	67	72	37
5	33	77	70	27	59	81	82	67	70	33

Source: John (1971, p. 68). Used with permission.

So that the differences among the sample means will now be compared to

$$\sqrt{S^2(a-1) \text{MS}_E \sum_{i=1}^{a} \ell_i^2 / b} = \sqrt{4.76\,(3)\,(195.639)\left(\frac{2}{3}\right)} = 43.16.$$

Here, again, we may conclude that materials one and two are probably significantly different but not the others. However, evidence from the Scheffé's method is not as strong as that from the Tukey's method. The significance of any other contrasts of interest can also similarly be evaluated.

3.14 WORKED EXAMPLE FOR MODEL II

The following example is based on a study on the grazing habits of Zebu cattle in Entebbe, Uganda. A group of 10 cattle was observed and recorded every minute they were grazing. A group of five observers was chosen and the same group was used during the entire experiment. They followed a group of cattle in the same paddock for 88 minutes during one afternoon. The data given in Table 3.5 are taken from John (1971, p. 68) and represent the number of minutes in which observer i ($i = 1, \ldots, 5$) reported animal j ($j = 1, \ldots, 10$) grazing.

We now proceed to analyze the data of Table 3.5 under Model II since the group of observers and animals in the study can be regarded as random samples from the respective populations of observers and cattle and the results of the analysis are to be valid for the entire populations. Thus, the factors of observers and animals will both have random effects. The mathematical model would be

$$y_{ij} = \mu + \alpha_i + \beta_j + e_{ij} \quad (i = 1, \ldots, 5; \; j = 1, \ldots, 10),$$

where μ is the general mean, α_i is the effect of the i-th observer, β_j is the effect

Two-Way Crossed Classification Without Interaction

of the j-th cattle, and the e_{ij}'s are random errors with

$$\alpha_i \sim N(0, \sigma_\alpha^2), \quad \beta_j \sim N(0, \sigma_\beta^2), \quad \text{and} \quad e_{ij} \sim N(0, \sigma_e^2).$$

In addition, the α_i's, β_j's, and e_{ij}'s are mutually and completely independent.

To perform the analysis of variance computations, we first obtain the row and column totals as

$$y_{1.} = 618, \quad y_{2.} = 611, \quad y_{3.} = 637, \quad y_{4.} = 608, \quad y_{5.} = 599;$$
$$y_{.1} = 169, \quad y_{.2} = 384, \quad y_{.3} = 364, \quad y_{.4} = 146, \quad y_{.5} = 305,$$
$$y_{.6} = 409, \quad y_{.7} = 419, \quad y_{.8} = 334, \quad y_{.9} = 362, \quad y_{.10} = 181;$$

and the grand total is

$$y_{..} = 3{,}073.$$

The other quantities needed in the calculations of the sums of squares are:

$$\frac{y_{..}^2}{ab} = \frac{(3{,}073)^2}{50} = 188{,}866.580,$$

$$\frac{1}{b}\sum_{i=1}^{a} y_{i.}^2 = \frac{1}{10}[(618)^2 + (611)^2 + \cdots + (599)^2] = 188{,}947.900,$$

$$\frac{1}{a}\sum_{j=1}^{b} y_{.j}^2 = \frac{1}{5}[(169)^2 + (384)^2 + \cdots + (181)^2] = 208{,}211.400,$$

and

$$\sum_{i=1}^{a}\sum_{j=1}^{b} y_{ij}^2 = (34)^2 + (76)^2 + \cdots + (33)^2 = 208{,}367.$$

The resultant sums of squares are, therefore, given by

$$\text{SS}_T = \sum_{i=1}^{a}\sum_{j=1}^{b} y_{ij}^2 - \frac{y_{..}^2}{ab} = 208{,}367 - 188{,}866.580 = 19{,}500.420,$$

$$\text{SS}_A = \frac{1}{b}\sum_{i=1}^{a} y_{i.}^2 - \frac{y_{..}^2}{ab} = 188{,}947.900 - 188{,}866.580 = 81.320,$$

$$\text{SS}_B = \frac{1}{a}\sum_{j=1}^{b} y_{.j}^2 - \frac{y_{..}^2}{ab} = 208{,}211.400 - 188{,}866.580 = 19{,}344.820,$$

TABLE 3.6
Analysis of Variance for the Grazing Data of Table 3.5

Source of Variation	Degrees of Freedom	Sum of Squares	Mean Square	Expected Mean Square	F Value	p-Value
Observer	4	81.320	20.330	$\sigma_e^2 + 10\sigma_\alpha^2$	9.85	<0.001
Animal	9	19,344.820	2,149.424	$\sigma_e^2 + 5\sigma_\beta^2$	1,041.89	<0.001
Error	36	74.280	2.063	σ_e^2		
Total	49	19,500.420				

and

$$SS_E = SS_T - SS_A - SS_B = 19{,}500.420 - 81.320 - 19{,}344.820 = 74.280.$$

Finally, the results of the analysis of variance calculations are summarized in Table 3.6.

If we choose the level of significance $\alpha = 0.05$, we find from Appendix Table V that $F[4, 36; 0.95] = 2.63$ and $F[9, 36; 0.95] = 2.15$. Comparing these values with the computed F values given in Table 3.6, we may conclude that $\sigma_\alpha^2 > 0$ and $\sigma_\beta^2 > 0$, and there are strong significant differences among observers ($p < 0.001$) as well as among animals ($p < 0.001$).

Furthermore, to evaluate the relative magnitude of the variance components σ_e^2, σ_β^2, and σ_α^2, we may obtain their unbiased estimates using the formulae (3.7.16), (3.7.17), and (3.7.18), respectively. Hence, we find that

$$\hat{\sigma}_e^2 = 2.063,$$

$$\hat{\sigma}_\beta^2 = \frac{1}{5}(2149.424 - 2.063) = 429.472,$$

$$\hat{\sigma}_\alpha^2 = \frac{1}{10}(20.330 - 2.063) = 1.827,$$

and the best estimate of the total variance σ_y^2 is given by

$$\hat{\sigma}_y^2 = \hat{\sigma}_e^2 + \hat{\sigma}_\beta^2 + \hat{\sigma}_\alpha^2$$
$$= 2.063 + 429.472 + 1.827,$$
$$= 433.362.$$

Now, the estimated proportions of the relative contribution of the variance components to the total variance are:

$$\frac{\hat{\sigma}_e^2}{\hat{\sigma}_y^2} = \frac{2.063}{433.362} = 0.005,$$

$$\frac{\hat{\sigma}_\beta^2}{\hat{\sigma}_y^2} = \frac{429.472}{433.362} = 0.991,$$

and

$$\frac{\hat{\sigma}_\alpha^2}{\hat{\sigma}_y^2} = \frac{1.827}{433.362} = 0.004.$$

Thus, we note that about 99 percent of the variation in the observations is attributable to animals. This would probably be the most important finding in the experiment suggesting that the cattle vary vastly in their habits of grazing.

To obtain a 95 percent confidence interval for σ_e^2, we have

$$MS_E = 2.063, \quad \chi^2[36, 0.025] = 21.34, \quad \text{and} \quad \chi^2[36, 0.975] = 54.44.$$

Substituting these values in (3.8.2), the desired 95 percent confidence interval for σ_e^2 is given by

$$P\left[\frac{36 \times 2.063}{54.44} < \sigma_e^2 < \frac{36 \times 2.063}{21.34}\right] = 0.95$$

or

$$P\left[1.364 < \sigma_e^2 < 3.480\right] = 0.95.$$

3.15 WORKED EXAMPLE FOR MODEL III

In a plastics manufacturing factory, it is discovered that there is considerable variation in the breaking strength of the plastics produced by three different machines. The raw material is considered to be uniform and hence can be discarded as a possible source of variability. An experiment was performed to determine the effects of the machine and the operator on the breaking strength. Four operators were randomly selected and each assigned to a machine. The data are given in Table 3.7.

It is desired to test whether there are significant differences among machines and operators. Since four operators were selected at random from a large pool of operators, who in turn were assigned to three specific machines, the experiment fits the assumptions of the mixed effects model. The mathematical model would

**TABLE 3.7
Breaking Strength of the
Plastics (in lbs.)**

	Operator			
Machine	1	2	3	4
1	106	110	106	104
2	107	111	108	110
3	109	113	112	111

be

$$y_{ij} = \mu + \alpha_i + \beta_j + e_{ij} \quad (i = 1,\ldots,3;\ j = 1,\ldots,4),$$

where μ is the general mean, α_i is the effect of the i-th machine, β_j is the effect of the j-th operator, and e_{ij}'s are random errors with

$$\sum_{i=1}^{3} \alpha_i = 0, \quad \beta_j \sim N(0, \sigma_\beta^2), \quad \text{and} \quad e_{ij} \sim N(0, \sigma_e^2).$$

It is further assumed that no interaction between the machine and the operator is likely to exist and the β_j's and the e_{ij}'s are mutually and completely independent.

For the validity of the preceding assumptions, it is, of course, necessary that the experimenter must take appropriate measures to ensure that operators are randomly selected from the large pool of operators available. Moreover, systematic errors due to other factors should be avoided including possible sources of variation in the working conditions and in measuring the breaking strength. Random assignment of raw materials is also important.

To perform the analysis of variance computations, we first obtain the row and column totals as

$$y_{1.} = 426, \quad y_{2.} = 436, \quad y_{3.} = 445;$$
$$y_{.1} = 322, \quad y_{.2} = 334, \quad y_{.3} = 326, \quad y_{.4} = 325;$$

and the grand total is

$$y_{..} = 1{,}307.$$

Two-Way Crossed Classification Without Interaction

The other quantities needed in the calculations of the sums of squares are:

$$\frac{y_{..}^2}{ab} = \frac{(1,307)^2}{12} = 142,354.083,$$

$$\frac{1}{b}\sum_{i=1}^{a} y_{i.}^2 = \frac{1}{4}[(426)^2 + (436)^2 + (445)^2] = 142,399.250,$$

$$\frac{1}{a}\sum_{j=1}^{b} y_{.j}^2 = \frac{1}{3}[(322)^2 + (334)^2 + (326)^2 + (325)^2] = 142,380.333,$$

and

$$\sum_{i=1}^{a}\sum_{j=1}^{b} y_{ij}^2 = (106)^2 + (110)^2 + \cdots + (111)^2 = 142,437.$$

The resultant sums of squares are, therefore, given by

$$SS_T = \sum_{i=1}^{a}\sum_{j=1}^{b} y_{ij}^2 - \frac{y_{..}^2}{ab} = 142,437 - 142,354.083 = 82.917,$$

$$SS_A = \frac{1}{b}\sum_{i=1}^{a} y_{i.}^2 - \frac{y_{..}^2}{ab} = 142,399.250 - 142,354.083 = 45.167,$$

$$SS_B = \frac{1}{a}\sum_{j=1}^{b} y_{.j}^2 - \frac{y_{..}^2}{ab} = 142,380.333 - 142,354.083 = 26.250,$$

and

$$SS_E = SS_T - SS_A - SS_B = 82.917 - 45.167 - 26.250 = 11.500.$$

Finally, the results of the analysis of variance calculations are summarized in Table 3.8.

If we choose the level of significance $\alpha = 0.05$, we find from Appendix Table V that $F[2, 6; 0.95] = 5.14$. Since the calculated F value of 11.78 exceeds 5.14 ($p = 0.008$), we may conclude that the machines differ significantly. Similarly, we find that $F[3, 6; 0.95] = 4.76$, which is greater than the calculated value of 4.57 ($p = 0.054$) and so we do not reject the hypothesis of no significant operator effects.

To determine which machines differ, we use Tukey's and Scheffé's procedures for paired comparisons. For the Tukey's procedure, we find from Appendix Table X that

$$q[a, (a-1)(b-1); 1-\alpha] = q[3, 6; 0.95] = 4.34.$$

TABLE 3.8
Analysis of Variance for the Breaking Strength Data of Table 3.7

Source of Variation	Degrees of Freedom	Sum of Squares	Mean Square	Expected Mean Square	F Value	p-Value
Machine	2	45.167	22.583	$\sigma_e^2 + \frac{4}{3-1}\sum_{i=1}^{3}\alpha_i^2$	11.78	0.008
Operator	3	26.250	8.750	$\sigma_e^2 + 3\sigma_\beta^2$	4.57	0.054
Error	6	11.500	1.917	σ_e^2		
Total	11	82.917				

Now, the pairwise differences of sample means for machines are compared to

$$q[a, (a-1)(b-1); 1-\alpha]\sqrt{\frac{MS_E}{b}} = 4.34\sqrt{\frac{1.917}{4}} = 3.01.$$

The three sample means for machines are

$$\bar{y}_{1.} = 106.50, \quad \bar{y}_{2.} = 109.00, \quad \text{and} \quad \bar{y}_{3.} = 111.25,$$

and there are three pairs of differences to be compared. Furthermore,

$$|\bar{y}_{1.} - \bar{y}_{2.}| = 2.50 < 3.01,$$
$$|\bar{y}_{1.} - \bar{y}_{3.}| = 4.75 > 3.01,$$

and

$$|\bar{y}_{2.} - \bar{y}_{3.}| = 2.25 < 3.01.$$

Hence, we may conclude that machines one and three are probably significantly different but not the others.

For the Scheffé's procedure, we find from Appendix Table V that

$$S^2 = F[a-1, (a-1)(b-1); 1-\alpha] = F[2, 6; 0.95] = 5.14.$$

Furthermore, for the contrasts consisting of the differences between the two means, we have

$$\sum_{i=1}^{a} \frac{\ell_i^2}{b} = \frac{1}{4} + \frac{1}{4} = \frac{1}{2}.$$

Two-Way Crossed Classification Without Interaction

So that the differences among the sample means are now compared to

$$\sqrt{S^2(a-1)\mathrm{MS}_E \sum_{i=1}^{a} \ell_i^2 / b} = \sqrt{5.14(2)(1.917)\left(\frac{1}{2}\right)} = 3.14.$$

Here, again we conclude that machines one and three are probably significantly different but not the others.

Furthermore, if the experimenter is interested in estimating the magnitudes of the variance components σ_e^2 and σ_β^2, these may be estimated unbiasedly by using the formulae (3.7.22) and (3.7.23) as

$$\hat{\sigma}_e^2 = 1.917$$

and

$$\hat{\sigma}_\beta^2 = \frac{1}{3}(8.750 - 1.917) = 2.278.$$

Finally, suppose we want to calculate the power of the test when the effect of one machine is higher than the other two by 3 lbs. Thus, we have,

$$\alpha_3 = 3 + \alpha_2 = 3 + \alpha_1,$$

which gives $\alpha_1 = \alpha_2 = -1$, and $\alpha_3 = 2$.

So that

$$\phi = \frac{1}{\sigma_e}\sqrt{\frac{4}{3}\sum_{i=1}^{3}\alpha_i^2} = \sqrt{\frac{4\{(-1)^2 + (-1)^2 + (2)^2\}}{3 \times 1.917}} = 2.04,$$

where an estimate of σ_e^2 is obtained from $\mathrm{MS}_E = 1.917$. Now, using the Pearson-Hartley chart, given in Appendix Chart II, with $v_1 = 2$, $v_2 = 6$, $\alpha = 0.05$, and $\phi = 2.04$, the power is found to be about 0.66.

3.16 WORKED EXAMPLE FOR MISSING VALUE ANALYSIS

Consider the loss in weight due to wear testing data given in Table 3.3 and suppose that the observation corresponding to Material 1 and Position 1 is missing. We then have:

$$a = 4, \quad b = 3, \quad y'_{..} = 2{,}633, \quad y'_{1.} = 544, \quad \text{and} \quad y'_{.1} = 664.$$

Therefore, on substituting in the formula (3.10.1), we obtain

$$\hat{y}_{11} = \frac{3(664) + 4(544) - 2{,}633}{(4-1)(3-1)} = 255.8.$$

This value is then entered in Table 3.3 in place of 241 and all sums of squares are computed as usual. The relevant computations are given in the following.

The row and column totals, after substituting for the missing value, are:

$$y_{1.} = 799.8, \quad y_{2.} = 654, \quad y_{3.} = 738, \quad y_{4.} = 697.$$
$$y_{.1} = 919.8, \quad y_{.2} = 1{,}020, \quad y_{.3} = 949;$$

and the grand total is

$$y_{..} = 2{,}888.8.$$

The other quantities needed in the calculations of the sums of squares are:

$$\frac{y_{..}^2}{ab} = \frac{(2{,}888.8)^2}{4 \times 3} = 695{,}430.453,$$

$$\frac{1}{b}\sum_{i=1}^{a} y_{i.}^2 = \frac{1}{3}[(799.8)^2 + (654)^2 + (738)^2 + (697)^2] = 699{,}283.013,$$

$$\frac{1}{a}\sum_{j=1}^{b} y_{.j}^2 = \frac{1}{4}[(919.8)^2 + (1{,}020)^2 + (949)^2] = 696{,}758.260,$$

and

$$\sum_{i=1}^{a}\sum_{j=1}^{b} y_{ij}^2 = (255.8)^2 + (270)^2 + \cdots + (227)^2 = 701{,}674.640.$$

The resultant sums of squares are, therefore, given by

$$SS_T = \sum_{i=1}^{a}\sum_{j=1}^{b} y_{ij}^2 - \frac{y_{..}^2}{ab} = 701{,}674.640 - 695{,}430.453 = 6{,}244.187,$$

$$SS_A = \frac{1}{b}\sum_{i=1}^{a} y_{i.}^2 - \frac{y_{..}^2}{ab} = 699{,}283.013 - 695{,}430.453 = 3{,}852.560,$$

$$SS_B = \frac{1}{a}\sum_{j=1}^{b} y_{.j}^2 - \frac{y_{..}^2}{ab} = 696{,}758.260 - 695{,}430.453 = 1{,}327.807,$$

TABLE 3.9
Analysis of Variance for the Weight Loss Data of Table 3.3 with One Missing Value

Source of Variation	Degrees of Freedom	Sum of Squares	Mean Square (Biased)	Mean Square (Unbiased)	F Value	p-Value
Material	3	3,852.560	1,284.187	987.199	4.64	0.066
Position	2	1,327.807	663.904	576.424	2.71	0.159
Error	6-1	1,063.820	212.764	212.764		
Total	11-1	6,244.187				

and

$$SS_E = SS_T - SS_A - SS_B = 6,244.187 - 3,852.560 - 1,327.807$$
$$= 1,063.820.$$

Finally, the results of the analysis of variance calculations are summarized in Table 3.9. Notice that the degrees of freedom in the total and error sums of square are reduced by one.

To correct for the bias, from equation (3.10.5), the quantity to be subtracted from the material mean square is

$$\frac{[664 - (4-1)(255.8)]^2}{4(4-1)^2} = 296.988.$$

This gives $1,284.187 - 296.988 = 987.199$ for the correct mean square. Similarly, from equation (3.10.6), the quantity to be subtracted from the position mean square is

$$\frac{[544 - (3-1)(255.8)]^2}{3(3-1)^2} = 87.480.$$

This gives $663.904 - 87.480 = 576.424$ for the correct mean square. The corrected mean squares, the variance ratios, and the associated p-values are also shown in Table 3.9. Note that the conclusions drawn in the Worked Example in Section 3.13 regarding differences in material and position effects are slightly affected.

From equation (3.10.2), the estimated standard error of the sample mean of the material (with the missing value) is

$$\widehat{SE}(\bar{y}_{1.}) = \sqrt{\left[\frac{1}{3} + \frac{4}{3(4-1)(3-1)}\right](212.764)} = 10.872.$$

Similarly, from equation (3.10.3), the estimated standard error of the difference between the means for materials 1 and 2, is

$$\widehat{SE}(\bar{y}_{1.} - \bar{y}_{2.}) = \sqrt{\left[\frac{2}{3} + \frac{4}{3(4-1)(3-1)}\right](212.764)} = 13.752.$$

The same standard error applies for the comparison of the mean of material 1 with means of materials 3 and 4. In contrast, the estimated error for the comparison of means of a pair materials with no missing values (i.e., 2 vs. 3, 2 vs. 4, and 3 vs. 4), is given by $\sqrt{2(212.764)/3} = 11.910$.

3.17 USE OF STATISTICAL COMPUTING PACKAGES

Two-way fixed effects analysis of variance with one observation per cell and no missing values can be performed using the SAS ANOVA procedure. The missing observations could be estimated and then the ANOVA procedure could be employed to perform the modified analysis as described in Section 3.10. For random and mixed model analysis of variance, the F tests remain unchanged, and no special analysis is required. The moment estimates of variance components can easily be computed from the entries of the analysis of variance table. For estimating variance components, using other methods, PROC MIXED or VARCOMP can be used. The details of SAS commands for executing these procedures are given in Section 11.1.

Among SPSS procedures, either ANOVA, MANOVA, or GLM could be used, although ANOVA would be simpler. The analysis of data with missing values could be handled as indicated previoulsy. As before, for random and mixed effects analysis, no special tests are required. Further, SPSS Release 7.5 now includes a VARCOMP procedure which provides for three methods for the estimation of variance components. For instructions regarding SPSS commands, see Section 11.2.

In using the BMDP package, two programs suited for this model are 7D and 2V if the analysis involves only fixed effects in the model. For the analysis involving random and mixed effects models 3V and 8V are recommended.

3.18 WORKED EXAMPLES USING STATISTICAL PACKAGES

In this section, we illustrate the applications of statistical packages to perform two-way analysis of variance with one observation per cell for the data sets employed in examples presented in Sections 3.13 through 3.15. Figures 3.1, 3.2, and 3.3 illustrate the program instructions and the output results for analyzing data in Tables 3.3, 3.5, and 3.7 using SAS ANOVA/GLM, SPSS ANOVA/GLM and BMDP 2V/8V procedures. The typical output provides the data format listed at the top, cell means, and the entries of the analysis of variance table. Note that in each case the results are the same as those provided using manual computations in Sections 3.13 through 3.15.

Two-Way Crossed Classification Without Interaction 165

```
DATA WEARTEST;
INPUT MATERIAL POSITION
WEIGHT;
DATALINES;
1 1 241
. . .
4 3 227
;
PROC ANOVA;
CLASSES MATERIAL POSITION;
MODEL WEIGHT=MATERIAL
POSITION;
RUN;
CLASS    LEVELS   VALUES
MATERIAL    4     1 2 3 4
POSITION    3     1 2 3
NUMBER OF OBS. IN DATA
SET=12
```

```
                    The SAS System
              Analysis of Variance Procedure

Dependent Variable: WEIGHT
              Sum of           Mean
Source   DF   Squares         Square     F Value   Pr > F
Model    5    4825.1667       965.0333   4.93      0.0388
Error    6    1173.8333       195.6389
Corrected 11  5999.0000
Total

R-Square         C.V.        Root MSE    WEIGHT Mean
0.804328         5.840124    13.987      239.50

Source     DF   Anova SS    Mean Square   F Value   Pr > F
MATERIAL   3    3141.6667   1047.2222     5.35      0.0393
POSITION   2    1683.5000   841.7500      4.30      0.0693
```

(i) SAS application: SAS ANOVA instructions and output for the two-way fixed effects analysis of variance with one observation per cell.

```
DATA LIST
/MATERIAL 1
 POSITION 3
 WEIGHT 5-7.
BEGIN DATA.
1 1 241
1 2 270
. . .
4 3 227
END DATA.
ANOVA WEIGHT BY
MATERIAL(1,4)
POSITION(1,3)
/MAXORDER=NONE
/STATISTICS=ALL.
```

```
                         ANOVA(a,b)
                        Unique Method
                  Sum of     df   Mean         F        Sig.
                  Squares         Square
WEIGHT  Main  (Combined) 4825.167  5   965.033    4.933   .039
        Effects
              MATERIAL   3141.667  3   1047.222   5.353   .039
              POSITION   1683.500  2   841.750    4.303   .069
        Model            4825.167  5   965.033    4.933   .039
        Residual         1173.833  6   195.639
        Total            5999.000 11   545.364
a WEIGHT by MATERIAL, POSITION  b All effects entered simultaneously
```

(ii) SPSS application: SPSS ANOVA instructions and output for the two-way fixed effects analysis of variance with one observation per cell.

```
/INPUT    FILE ='C:\SAHAI
          \TEXTO\EJE5.TXT'.
          FORMAT=FREE.
          VARIABLES=3.
/VARIABLE NAMES=MAT,POS,
          WEIGHT.
/GROUP    VARIABLE=MAT,POS.
          CODES(MAT)=1,2,3,4.
          NAMES(MAT)=M1,M2,
          M3,M4.
          CODES(POS)=1,2,3.
          NAMES(POS)=P1,P2,P3.
/DESIGN   DEPENDENT=WEIGHT.
/END.
1 1 241
. . .
4 3 227
```

```
BMDP2V - ANALYSIS OF VARIANCE AND COVARIANCE WITH
         REPEATED MEASURES Release: 7.0 (BMDP/DYNAMIC)

ANALYSIS OF VARIANCE FOR THE 1-ST DEPENDENT VARIABLE
THE TRIALS ARE REPRESENTED BY THE VARIABLES:WEIGHT

THE HIGHEST ORDER INTERACTION IN EACH TABLE HAS BEEN
REMOVED FROM THE MODEL SINCE THERE IS ONE SUBJECT PER
CELL

SOURCE     SUM OF        D.F.   MEAN           F        TAIL
           SQUARES              SQUARE                  PROB.

MEAN       688323.00000  1      688323.00000   3518.33  0.0000
MATERIAL   3141.66667    3      1047.22222     5.35     0.0393
POSITION   1683.50000    2      841.75000      4.30     0.0693
ERROR      1173.83333    6      195.63889
```

(iii) BMDP application: BMDP 2V instructions and output for the two-way fixed effects analysis of variance with one observation per cell.

FIGURE 3.1 Program Instructions and Output for the Two-Way Fixed Effects Analysis of Variance with One Observation per Cell: Data on Loss in Weights Due to Wear Testing of Four Materials (Table 3.3)

```
DATA ANMLGRAZ;                      The SAS System
INPUT OBSERVER ANIMAL      General Linear Models Procedure
GRAZING;                   Dependent Variable: GRAZING
DATALINES;                              Sum of        Mean
1 1 34                     Source    DF  Squares     Square    F Value   Pr > F
. . .                      Model     13  19426.140   1494.318  724.23    0.0001
5 10 33                    Error     36  74.280      2.063
;                          Corrected 49  19500.420
PROC GLM;                  Total
CLASSES OBSERVER ANIMAL;
MODEL GRAZING=OBSERVER                R-Square    C.V.     Root MSE   GRAZING Mean
ANIMAL;                               0.996191   2.337180   1.4364      61.460
RANDOM OBSERVER ANIMAL;    Source    DF  Type I SS   Mean Square  F Value   Pr > F
RUN;                       OBSERVER  4   81.320      20.330       9.85     0.0001
CLASS   LEVELS   VALUES    ANIMAL    9   19344.820   2149.424     1041.72  0.0001
OBSERVER  5    1 2 3 4 5   Source    DF  Type III SS Mean Square  F Value   Pr > F
ANIMAL    10   1 2 3 4 5   OBSERVER  4   81.320      20.330       9.85     0.0001
               6 7 8 9 10  ANIMAL    9   19344.820   2149.424     1041.72  0.0001
NUMBER OF OBS. IN DATA     Source    Type III Expected Mean Square
SET=50                     OBSERVER  Var(Error) + 10 Var(OBSERVER)
                           ANIMAL    Var(Error) + 5 Var(ANIMAL)
```

(i) SAS Application: SAS GLM instructions and output for the two-way random effects analysis of variance with one observation per cell.

```
DATA LIST           Tests of Between-Subjects Effects   Dependent Variable: GRAZING
/OBSERVER 1
 ANIMAL 3-4         Source                  Type III SS  df   Mean Square     F       Sig.
 GRAZING 6-7.       OBSERVER Hypothesis     81.320       4    20.330          9.853   .000
BEGIN DATA.                  Error          74.280       36   2.063(a)
1 1  34             ANIMAL   Hypothesis     19344.820    9    2149.424        1041.724 .000
1 2  76                      Error          74.280       36   2.063(a)
1 3  75             a  MS(Error)
. . .
5 10 33                                        Expected Mean Squares(a,b)
END DATA.                                          Variance Component
GLM GRAZING BY      Source         Var(OBSERVER)    Var(ANIMAL)        Var(Error)
 OBSERVER ANIMAL    OBSERVER       10.000           .000               1.000
 /DESIGN            ANIMAL         .000             5.000              1.000
 OBSERVER           Error          .000             .000               1.000
 ANIMAL             a For each source, the expected mean square equals the sum of the
 /RANDOM            coefficients in the cells times the variance components, plus a
 OBSERVER           quadratic term involving effects in the Quadratic Term cell.
 ANIMAL.            b Expected Mean Squares are based on the Type III Sums of Squares.
```

(ii) SPSS Application: SPSS GLM instructions and output for the two-way random effects analysis of variance with one observation per cell.

```
/INPUT  * FILE='C:\SAHAI    BMDP8V - GENERAL MIXED MODEL ANALYSIS OF VARIANCE
          \TEXTO\EJE6.TXT'.        - EQUAL CELL SIZES  Release: 7.0 (BMDP/DYNAMIC)
        FORMAT=FREE.
        VARIABLES=10.       ANALYSIS OF VARIANCE FOR DEPENDENT VARIABLE  1
/VARIABLE NAMES=A1,...,A10.
/DESIGN NAMES=OBSR,ANIM.    SOURCE    ERROR  SUM OF     D.F.  MEAN        F      TAIL
        LEVELS=5, 10.                 TERM   SQUARES          SQUARE             PROB.
        RANDOM=OBSR,ANIM.   1 MEAN           188866.5800  1   188866.580
        MODEL='O,A'.        2 OBSERVER OA    81.3200      4   20.330      9.85   0.0000
/END                        3 ANIMAL   OA    19344.8200   9   2149.424    1041.72 0.0000
34 76 75 31 61 82 82 67 72  4 OA             74.2800      36  2.063
38
. . .                       SOURCE         EXPECTED MEAN              ESTIMATES OF
33 77 70 27 59 81 82 67 70                    SQUARE               VARIANCE COMPONENTS
33                          1 MEAN       50(1)+10(2)+5(3)+(4)         3733.97778
ANALYSIS OF VARIANCE DESIGN 2 OBSERVER   10(2) + (4)                  1.82667
INDEX           OBSR  ANIM  3 ANIMAL     5(3) + (4)                   429.47222
NUMBER OF LEVELS  5    10   4 OA         (4)                          2.06333
POPULATION SIZE  INF   INF
MODEL   O, A
```

(iii) BMDP Application: BMDP 8V instructions and output for the two-way random effects analysis of variance with one observation per cell.

FIGURE 3.2 Program Instructions and Output for the Two-Way Random Effects Analysis of Variance with One Observation per Cell: Data on Number of Minutes Observed in Grazing (Table 3.5).

```
DATA BREAKSTR;                     The SAS System
INPUT MACHINE OPERATOR        General Linear Models Procedure
    BREAKING;                 Dependent Variable: BREAKING
DATALINES;                                  Sum of        Mean
1 1 106                       Source    DF  Squares      Square    F Value   Pr > F
1 2 110                       Model      5  71.4167     14.2833      7.45    0.0149
. . .                         Error      6  11.5000      1.9167
3 4 111                       Corrected 11  82.9167
;                             Total
PROC GLM;                     R-Square      C.V.    Root MSE  BREAKING Mean
CLASSES MACHINE OPERATOR;     0.861307    1.271098   1.3844       108.92
MODEL BREAKING=MACHINE        Source    DF  Type I SS  Mean Square  F Value  Pr > F
    OPERATOR;                 MACHINE    2  45.1667     22.5833      11.78   0.0084
RANDOM OPERATOR;              OPERATOR   3  26.2500      8.7500       4.57   0.0543
RUN;                          Source    DF Type III SS  Mean Square  F Value  Pr > F
CLASS    LEVELS   VALUES      MACHINE    2  45.1667     22.5833      11.78   0.0084
MACHINE    3       1 2 3      OPERATOR   3  26.2500      8.7500       4.57   0.0543
OPERATOR   4       1 2 3 4    Source        Type III Expected Mean Square
NUMBER OF OBS. IN DATA        MACHINE       Var(Error) + Q(MACHINE)
    SET=12                    OPERATOR      Var(Error) + 3 Var(OPERATOR)
```

(i) SAS application: SAS GLM instructions and output for the two-way mixed effects analysis of variance with one observation per cell.

```
DATA LIST              Tests of Between-Subjects Effects  Dependent Variable: BREAKING
/MACHINE 1
  OPERETOR 3           Source                  Type III SS   df   Mean Square    F      Sig.
  BREAKING 5-7.        MACHINE  Hypothesis       45.167       2     22.583    11.783    .008
BEGIN DATA.                     Error            11.500       6      1.917(a)
1 1 106                OPERATOR Hypothesis       26.250       3      8.750     4.565    .054
1 2 110                         Error            11.500       6      1.917(a)
1 3 106                a  MS(ERROR)
. . .
3 4 111                                      Expected Mean Squares (a, b)
END DATA.                                          Variance Component
GLM BREAKING BY        Source           Var(OPERETOR)   Var(Error)   Quadratic Term
  MACHINE              MACHINE              .000           1.000        Machine
  OPERATOR             OPERATOR            3.000           1.000
/DESIGN                Error                .000           1.000
  MACHINE              a For each source, the expected mean square equals the sum of the
  OPERETOR             coefficients in the cells times the variance components, plus a
/RANDOM                quadratic term involving effects in the Quadratic Term cell.
  OPERETOR             b Expected Mean Squares are based on the Type III Sums of Squares.
```

(ii) SPSS application: SPSS GLM instructions and output for the two-way mixed effects analysis of variance with one observation per cell.

```
/INPUT      FILE='C:\SAHAI       BMDP8V - GENERAL MIXED MODEL ANALYSIS OF VARIANCE
            \TEXTO\EJE7.TXT'.             - EQUAL CELL SIZES  Release: 7.0 (BMDP/DYNAMIC)
            FORMAT=FREE.
            VARIABLES=4.         ANALYSIS OF VARIANCE FOR DEPENDENT VARIABLE  1
/VARIABLE   NAMES=O1,...,O4.
/DESIGN     NAMES=M,O.           SOURCE      ERROR     SUM OF    D.F.    MEAN        F      PROB.
            LEVELS=3, 4.                     TERM      SQUARES           SQUARE
            RANDOM=O.            1 MEAN                142354.0833  1  142354.083
            MODEL='M, O'.        2 MACHINE    MO         45.1667    2      22.583   11.78  0.0084
/END                             3 OPERATOR   MO         26.2500    3       8.750    4.57  0.0543
106 110 106 104                  4 MO                    11.5000    6       1.917
107 111 108 110
109 113 112 111                  SOURCE      EXPECTED MEAN          ESTIMATES OF
ANALYSIS OF VARIANCE DESIGN                  SQUARE                 VARIANCE COMPONENTS
 INDEX           M   O           1 MEAN      12(1)+4(2)+3(3)+(4)    11860.38889
 NUMBER OF LEVELS 3  4            2 MACHINE  4(2)+(4)                   5.16667
 POPULATION SIZE INF INF          3 OPERATOR 3(3)+(4)                   2.27778
 MODEL     M, O                   4 MO       (4)                        1.91667
```

(iii) BMDP application: BMDP 8V instructions and output for the two-way mixed effects analysis of variance with one observation per cell.

FIGURE 3.3 Program Instructions and Output for the Two-Way Mixed Effects Analysis of Variance with One Observation per Cell: Data on Breaking Strength of the Plastics (Table 3.7)

3.19 EFFECTS OF VIOLATIONS OF ASSUMPTIONS OF THE MODEL

For the two-way crossed model (3.1.1), with a single observation and no interaction, Welch (1937) and Pitman (1938) compared the moments of the beta statistic, related to the usual F statistic, both under the assumption of normality and under permutation theory. There was close agreement between the two results which lends support to the robust nature of the F test as in the case of one-way classification. For a further discussion of this point, the reader is referred to Hinkelman and Kempthorne (1994, Chapter 8).

The problem of the effect of unequal error variances on the inferences of the model (3.1.1) was studied by Box (1954b). The results showed that for minor departures from homoscedasticity, the effects are not large. If the variances differ row-wise but are constant over columns, then the actual α-level for the test of row effects is slightly greater than the nominal value. For the test of the column effects, the reverse is true.

Box (1954b) also studied the effect of a first-order serial correlation between rows within columns. It was found that treatment (row) comparisons are not appreciably affected by serial correlation between treatment measurements within a block (column). However, serial correlations among the measurements on each treatment can seriously affect the validity of treatment comparisons.

For a discussion of effects of violation of assumptions under Models II and III, see Section 4.19.

EXERCISES

1. The drained weights of frozen oranges were measured for various compositions and concentrations of a drink. The original weights in each case were the same. Any observed differences in drained weights can thus be attributed to differences in concentration or composition of the drink. The relevant data on weights (oz.) are as follows.

	Composition			
Concentration (%)	C_1	C_2	C_3	C_4
20	21.52	21.32	22.19	22.19
30	22.32	21.52	22.32	23.15
40	22.56	23.12	22.42	21.52
50	23.31	22.15	21.32	22.16

   (a) Describe the mathematical model and the assumptions for the experiment.
   (b) Analyze the data and report the analysis of variance table.
   (c) Does the concentration of the drink have a significant effect on the drained weight? Use $\alpha = 0.05$.
   (d) Does the composition of the drink have a significant effect on the drained weight? Use $\alpha = 0.05$.

(e) If there are significant differences in drained weights due to drink concentration, use a suitable multiple comparison method to determine which concentrations differ. Use $\alpha = 0.01$.
(f) Same as part (e) but for drink composition.

2. Three levels of fertilizer in combination with two levels of irrigation were employed in a field experiment. The six treatment combinations were randomly assigned to plots. The relevant data on yields are as follows.

Level of Irrigation	Level of Fertilizer		
	High	Medium	Low
Yes	380	340	305
No	330	360	340

(a) Describe the appropriate mathematical model and the assumptions for the experiment.
(b) Analyze the data and report the analysis of variance table.
(c) Are there significant differences among the levels of the fertilizer? Use $\alpha = 0.05$.
(d) Obtain point and interval estimates of the overall difference in yield due to irrigation.
(e) Obtain point and interval estimates of the mean difference in yield between high and low fertilizer levels.

3. An experiment was conducted to assess the effect of four brands of cutting fluids on the abrasive wear of four types of cutting tools. The measure of wear was reported in terms of the logarithm of loss of tool flank weight (in grams times 100) in a 1-hour test run. The relevant results are summarized as follows.

Cutting Tool	Cutting Fluid Brand			
	CF1	CF2	CF3	CF4
CT1	1.171	1.057	1.061	1.011
CT2	0.705	0.612	0.631	0.598
CT3	0.538	0.418	0.457	0.412
CT4	0.414	1.308	1.371	1.251

(a) Describe the mathematical model and the assumptions for the experiment.
(b) Analyze the data and report the analysis of variance table.
(c) Do the fluid brands have a significant effect on the measure of abrasive wear? Use $\alpha = 0.05$.
(d) Does the type of cutting tool have a significant effect on the measure of abrasive wear? Use $\alpha = 0.05$.
(e) If there are significant differences in the measure of abrasive wear due to the fluid brand, use a suitable multiple comparison method to determine which fluid brands differ. Use $\alpha = 0.01$.

(f) Same as part (e) but for the type of cutting tool.
4. During the manufacturing process of a certain component, its breaking strength was measured for three operating temperatures and for each temperature there was influence of seven furnace pressures. The relevant data in certain standard units are given as follows.

	Pressure						
Temperature	P1	P2	P3	P4	P5	P6	P7
T1	0.803	0.836	1.303	1.276	1.161	1.054	1.307
T2	0.705	0.630	1.005	1.062	0.616	0.803	0.618
T3	1.321	0.815	0.771	1.110	0.710	1.022	0.717

(a) Describe the mathematical model and the assumptions for the experiment.
(b) Analyze the data and report the analysis of variance table.
(c) Do the operating temperatures have a significant effect on the breaking strength? Use $\alpha = 0.05$.
(d) Do the furnace pressures have a significant effect on the breaking strength? Use $\alpha = 0.05$.
(e) If there are significant differences in breaking strength due to temperatures, use a suitable multiple comparison method to determine which temperatures differ. Use $\alpha = 0.01$.
(f) Same as part (e) but for the furnace pressure.

5. A university computing department manages four resource centers on the campus. Each center houses one timesharing terminal and two types of personal computers. During a given week, the numbers of hours a certain type of computing machine was being used were recorded and the relevant data are as follows.

	Resource Center			
Equipment	1	2	4	4
Time-Sharing Terminal	70	70	50	60
Apple Computer	40	40	20	40
IBM Computer	30	30	10	30

(a) Describe the mathematical model you will employ to analyze the effects of resource center location and the type of computing machine.
(b) Analyze the data and report the analysis of variance table.
(c) Perform appropriate F tests to determine whether the two factors have main effects. Use $\alpha = 0.05$.

6. A factorial experiment was performed to study the effect of the level of pressure and the level of temperature on the compressive strength of thermoplastics. The relevant data in certain standard units are given

as follows.

Pressure (lb/in.^{2})	Temperature (°F)		
	250	260	270
120	8.00	10.57	8.30
130	8.01	9.40	8.86
140	7.72	10.30	8.32
150	8.14	9.73	8.01

(a) Describe the mathematical model and the assumptions for the experiment considering both factors to be fixed.
(b) Analyze the data and report the analysis of variance table.
(c) Does the level of temperature have a significant effect on the compressive strength? Use $\alpha = 0.05$.
(d) Does the level of pressure have a significant effect on the compressive strength? Use $\alpha = 0.05$.
(e) If there are significant differences in the compressive strength due to the temperature, use a suitable multiple comparison method to determine which temperatures differ. Use $\alpha = 0.01$.
(f) Same as part (e) but for the pressure.
(g) Suppose the temperature and pressure levels are selected at random. Assuming that there is no interaction between pressure and temperature, state the analysis of variance model and report appropriate conclusions. How do your conclusions differ from those obtained in parts (c) and (d).

7. A study was conducted to determine the effect of the size of a group on the results of a brainstorming session. Three different types of company executives were used, one for each group size. Each group was assigned a problem and was given an hour to generate ideas. The variable of interest was the number of new ideas proposed. The relevant data are given as follows.

Type of Group	Size of Group			
	2	3	4	5
Sales executives	22	30	38	34
Advertising executives	19	25	32	31
Marketing executives	16	20	26	28

(a) Describe the mathematical model you will employ to analyze the effects of group size and the type of group on the number of new ideas being proposed.
(b) Perform appropriate F tests to determine whether the factors have any main effects. Use $\alpha = 0.05$.
(c) If there are significant differences between the types of groups, use a suitable multiple comparison method to determine which group types differ. Use $\alpha = 0.01$.

8. Consider a two-factor experiment designed to investigate the breaking strength of a bond of pieces of material. There are 5 ingots of a composition material that are used with one of the three metals as the bonding agent. The data on amount of pressure required to break a bond from an ingot that uses one of the metals as the bonding agent are given as follows.

	Type of Metal		
Ingot	Copper	Iron	Nickel
1	83.3	83.0	78.1
2	77.5	79.9	78.6
3	85.6	93.8	87.1
4	78.4	89.2	83.8
5	84.3	85.3	84.2

(a) Describe the mathematical model and the assumptions for the experiment. It is assumed that metals are fixed and ingots are random.
(b) Analyze the data and report the analysis of variance table.
(c) Do the metals have a significant effect on the breaking strength? Use $\alpha = 0.05$.
(d) Do the ingots have a significant effect on the breaking strength? Use $\alpha = 0.05$.
(e) If there are significant differences in the breaking strength due to metals, use a suitable multiple comparison method to determine which metals differ. Use $\alpha = 0.01$.
(f) Obtain point and interval estimates of the variance components associated with the assumptions of the model given in part (a).

9. Mosteller and Tukey (1977, p. 503) reported data from a study where six experimenters measured the specific heat of water at various temperatures. The interest lies in investigating the reliability of the measurements and in determining an accurate estimate of the specific heat. The data are given as follows.

	Temperature (°C)					
Investigator	5	10	15	20	25	30
Liidin	1.0027	1.0010	1.0000	0.9994	0.9993	0.9996
Dieterici	1.0050	1.0021	1.0000	0.9987	0.9983	0.9984
Bonsfreld	1.0039	1.0016	1.0000	0.9991	0.9989	0.9990
Ronland	1.0054	1.0019	1.0000	0.9979	0.9972	0.9969
Bartollis	1.0041	1.0017	1.0000	0.9994	1.0000	1.0016
Janke	1.0040	1.0016	1.0000	0.9991	0.9987	0.9988

Source: Mosteller and Tukey (1977, p. 503). Used with permission.

(a) Describe the mathematical model and the assumptions for the experiment. Would you use Model I, Model II, or Model III? Explain.
(b) Analyze the data and report the analysis of variance table.
(c) Does the level of the temperature have a significant effect on the measurement of specific heat. Use $\alpha = 0.05$.
(d) Does the investigator have a significant effect on the measurement of specific heat. Use $\alpha = 0.05$.
(e) If there are significant differences in specific heats due to temperature, use a suitable multiple comparison method to determine which levels of the temperature differ. Use $\alpha = 0.01$.
(f) Same as in part (e) but for the investigator.
(g) If you assumed the investigator to be a random factor, determine the point and interval estimates of the variance components of the model.

10. Weekes (1983, Table 1.1) reported data of Michelson and Morley on the speed of light. The data came from 5 experiments, each consisting of 20 consecutive runs. The results are given below where reported measurement is the speed of light in suitable units.

Run	Experiment				
	1	2	3	4	5
1	850	960	880	890	890
2	740	940	880	810	840
3	900	960	880	810	780
4	1070	940	860	820	810
5	930	880	720	800	760
6	850	800	720	770	810
7	950	850	620	760	790
8	980	880	860	740	810
9	980	900	970	750	820
10	880	840	950	760	850
11	1000	830	880	910	870
12	980	790	910	920	870
13	930	810	850	890	810
14	650	880	870	860	740
15	760	880	840	880	810
16	810	830	840	720	940
17	1000	800	850	840	950
18	1000	790	840	850	800
19	960	760	840	850	810
20	960	800	840	780	870

Source: Weekes (1983, Table 1.1). Used with permission.

(a) Describe the mathematical model and the assumptions for the experiment. Would you use Model I, Model II, or Model III? Explain.
(b) Analyze the data and report the analysis of variance table.
(c) Does the experiment have a significant effect on the measurement of the speed of light? Use $\alpha = 0.05$.
(d) Does the run have a significant effect on the measurement of the speed of light? Use $\alpha = 0.05$.
(e) Assuming that the experiment and run effects are random, estimate the variance components of the model and determine their relative importance.

11. Berry (1987) reported data from an experiment designed to investigate the performance of different types of electrodes. Five different types of electrodes were applied to the arms of 16 subjects and the readings for resistance were taken. The data are given below where the measures of resistance are given in the original units of kilohms.

	Electrode Type				
Subject	1	2	3	4	5
1	500	400	98	200	250
2	600	600	600	75	310
3	250	370	220	250	220
4	72	140	240	33	54
5	135	300	450	430	70
6	27	84	135	190	180
7	100	50	82	73	78
8	105	180	32	58	32
9	90	180	220	34	64
10	200	290	320	280	135
11	15	45	75	88	80
12	160	200	300	300	220
13	250	400	50	50	92
14	170	310	230	20	150
15	66	1000	1050	280	220
16	107	48	26	45	51

Source: Berry (1987). Used with permission.

(a) Describe the mathematical model and the assumptions for the experiment. Would you use Model I, Model II, or Model III? Explain.
(b) Analyze the data and report the analysis of variance table.
(c) Does the electrode type have a significant effect on the measures of resistance? Use $\alpha = 0.05$.

(d) Does the subject have a significant effect on the measures of resistance? Use $\alpha = 0.05$.
(e) If there are significant differences in resistance due to electrode type and it is considered to be a fixed effect, use a suitable multiple comparison method to determine which electrode types differ. Use $\alpha = 0.01$.
(f) If you assumed any of the effects to be random, determine the point and interval estimates of the variance components of the model.
(g) It is found that the measures of resistance have a skewed distribution to the right. Make a logarithmic transformation on the data and repeat the analyses carried out in parts (b) through (f).

4 Two-Way Crossed Classification with Interaction

4.0 PREVIEW

Suppose that we relax the requirement of model (3.1.1) that there be exactly one observation in each of the $a \times b$ cells of the two-way layout. The model remains the same except that we could now use y_{ijk} to designate the k-th observation at the i-th level of A and the j-th level of B, that is, in the (i,j)-th cell. We now suppose that there are n ($n \geq 1$) observations in each cell. With $n = 1$, the model (3.1.1) will be a special case of the model being considered here. With an arbitrary integer value of n, the analysis of variance will be a simple extension of that described in Chapter 3. However, an important and somewhat restrictive implication of the simple additive model discussed in Chapter III is that the value of the difference between the mean responses at two levels of A is the same at each level of B. However, in many cases, this simple additive model may not be appropriate. The failure of the differences between the mean responses at the different levels of A to remain constant over the different levels of B is attributed to interaction between the two factors. Having more than one observation per cell allows a researcher to investigate the main effects of both factors and their interaction. In this chapter, we study the model involving two factors with interaction terms.

4.1 MATHEMATICAL MODEL

Consider two factors A and B having a and b levels, respectively, and let there be n observations at each combination of levels of A and B (i.e., n observations in each of the $a \times b$ cells). Let y_{ijk} be the k-th observation at the i-th level of A and the j-th level of B. The data involving a total of $N = abn$ scores y_{ijk}'s can then be schematically represented as in Table 4.1.

The notations employed here are a straightforward extension of the notations for the preceding chapter. A dot in the subscript indicates aggregation or total over the variable represented by the index, and a dot and a bar represent the corresponding mean. Thus, the sum of the observations corresponding to the i-th

TABLE 4.1
Data for a Two-Way Crossed Classification with n Replications per Cell

				Factor B			
B_1	B_2	B_3	$\cdots$	B_j	$\cdots$	B_b

Factor A							
A_1	$y_{111}, y_{112}, \ldots, y_{11n}$	$y_{121}, y_{122}, \ldots, y_{12n}$	$y_{131}, y_{132}, \ldots, y_{13n}$	$\cdots$	$y_{1j1}, y_{1j2}, \ldots, y_{1jn}$	$\cdots$	$y_{1b1}, y_{1b2}, \ldots, y_{1bn}$
A_2	$y_{211}, y_{212}, \ldots, y_{21n}$	$y_{221}, y_{222}, \ldots, y_{22n}$	$y_{231}, y_{232}, \ldots, y_{23n}$	$\cdots$	$y_{2j1}, y_{2j2}, \ldots, y_{2jn}$	$\cdots$	$y_{2b1}, y_{2b2}, \ldots, y_{2bn}$
A_3	$y_{311}, y_{312}, \ldots, y_{31n}$	$y_{321}, y_{322}, \ldots, y_{32n}$	$y_{331}, y_{332}, \ldots, y_{33n}$	$\cdots$	$y_{3j1}, y_{3j2}, \ldots, y_{3jn}$	$\cdots$	$y_{3b1}, y_{3b2}, \ldots, y_{3bn}$
$\vdots$							
A_i	$y_{i11}, y_{i12}, \ldots, y_{i1n}$	$y_{i21}, y_{i22}, \ldots, y_{i2n}$	$y_{i31}, y_{i32}, \ldots, y_{i3n}$	$\cdots$	$y_{ij1}, y_{ij2}, \ldots, y_{ijn}$	$\cdots$	$y_{ib1}, y_{ib2}, \ldots, y_{ibn}$
$\vdots$							
A_a	$y_{a11}, y_{a12}, \ldots, y_{a1n}$	$y_{a21}, y_{a22}, \ldots, y_{a2n}$	$y_{a31}, y_{a32}, \ldots, y_{a3n}$	$\cdots$	$y_{aj1}, y_{aj2}, \ldots, y_{ajn}$	$\cdots$	$y_{ab1}, y_{ab2}, \ldots, y_{abn}$

Two-Way Crossed Classification with Interaction

level of factor A and the j-th level of factor B is

$$y_{ij.} = \sum_{k=1}^{n} y_{ijk}$$

The corresponding mean is

$$\bar{y}_{ij.} = \frac{y_{ij.}}{n} = \frac{\sum_{k=1}^{n} y_{ijk}}{n}.$$

The total of all the observations for the i-th level of factor A is

$$y_{i..} = \sum_{j=1}^{b} \sum_{k=1}^{n} y_{ijk},$$

and the corresponding mean is

$$\bar{y}_{i..} = \frac{y_{i..}}{bn} = \frac{\sum_{j=1}^{b} \sum_{k=1}^{n} y_{ijk}}{bn}.$$

Similarly, for the j-th level of factor B, the sum of all the observations and the corresponding mean are denoted by

$$y_{.j.} = \sum_{i=1}^{a} \sum_{k=1}^{n} y_{ijk},$$

and

$$\bar{y}_{.j.} = \frac{y_{.j.}}{an} = \frac{\sum_{i=1}^{a} \sum_{k=1}^{n} y_{ijk}}{an}.$$

Finally, the sum of all the observations, the grand total, is

$$y_{...} = \sum_{i=1}^{a} \sum_{j=1}^{b} \sum_{k=1}^{n} y_{ijk},$$

and the grand or overall mean is

$$\bar{y}_{...} = \frac{y_{...}}{abn} = \frac{\sum_{i=1}^{a} \sum_{j=1}^{b} \sum_{k=1}^{n} y_{ijk}}{abn}.$$

The analysis of variance model for this type of experimental layout is given as

$$y_{ijk} = \mu + \alpha_i + \beta_j + (\alpha\beta)_{ij} + e_{ijk} \quad \begin{cases} i = 1, 2, \ldots, a \\ j = 1, 2, \ldots, b \\ k = 1, 2, \ldots, n, \end{cases} \quad (4.1.1)$$

where $-\infty < \mu < \infty$ is the overall mean, α_i is the effect due to the i-th level of factor A, β_j is the effect due to the j-th level of factor B, $(\alpha\beta)_{ij}$ is the interaction effect representing the departure of the mean of the observations in the (i, j)-th cell, denoted by μ_{ij}, from the sum of the first three terms of (4.1.1), and e_{ijk} is the customary error term accounting for the random variation from cell to cell. The terms α_i and β_j are known as main effects. They are average effects corresponding to each level of factor A and each level of factor B. The term $(\alpha\beta)_{ij}$ is called an interaction effect. If the levels of factor A and factor B behave in a strictly additive manner, that is, if a level of factor A contributes a certain amount to the average yield, irrespective of the level of factor B, we say that the $(\alpha\beta)_{ij}$'s are all zero. On the other hand, if a level of factor A, say, 3, increases yield more with, say, level 1 than with level 2 of factor B, we say that $(\alpha\beta)_{31}$ is positive and $(\alpha\beta)_{32}$ is negative. The model (4.1.1) states that an observation y_{ijk} consists of these components:

  (i) the overall mean μ,
  (ii) the main effect α_i for factor A at the i-th level,
  (iii) the main effect β_j for factor B at the j-th level,
  (iv) the interaction effect $(\alpha\beta)_{ij}$ when factor A is at the i-th level and factor B is at the j-th level, and
  (v) an error or residual term e_{ijk} which is the deviation of a particular observation from the cell mean μ_{ij}.

Remark: Two-way crossed classifications are widely used in many areas of scientific research and applications. Some examples are as follows:

  (a) A city has a sources of a pollutant, n samples are taken from each source, and are sent to b laboratories for analysis of its chemical composition.
  (b) An organism has a species, n females are taken from each species, and each one is used in b experiments in order to measure variation in progeny.
  (c) An industrial production involves a machines, b workers, and n samples of the product are taken from each machine × worker combination.

4.2 ASSUMPTIONS OF THE MODEL

As in the case of model (3.1.1), the assumptions of the model (4.1.1) can be summarized as follows:

  (i) The errors e_{ijk}'s are randomly distributed with mean zero and common variance σ_e^2.

(ii) The errors associated with any pair of observations are uncorrelated.
(iii) Under Model I, the α_i's, β_j's, and $(\alpha\beta)_{ij}$'s are assumed to be fixed constants subject to the constraints

$$\sum_{i=1}^{a} \alpha_i = \sum_{j=1}^{b} \beta_j = \sum_{i=1}^{a} (\alpha\beta)_{ij} = \sum_{j=1}^{b} (\alpha\beta)_{ij} = 0.$$

(iv) Under Model II, the α_i's, β_j's, and $(\alpha\beta)_{ij}$'s are assumed to be randomly distributed with zero means and variances σ_α^2, σ_β^2, and $\sigma_{\alpha\beta}^2$, respectively. Furthermore, the α_i's, β_j's, and $(\alpha\beta)_{ij}$'s are mutually and completely uncorrelated.[1] In this case, σ_α^2, σ_β^2, $\sigma_{\alpha\beta}^2$, and σ_e^2 are the variance components of the model (4.1.1) and the inferences are sought about them.

(v) Although the distribution properties for Models I and II were fairly straightforward to enumerate, this is not the case for Model III. Generally, if any element in an interaction term is considered random, it may be appropriate to assume that the interaction term has a Model II effect. However, in that case, we must assume the corresponding distribution properties of the interaction terms. The proper error term used to test certain hypotheses and to construct certain confidence intervals will depend on the distribution properties being assumed.

There are several types of distribution properties that have been proposed as realistic for various experimental situations and their full discussion is beyond the scope of this volume. The interested reader is referred to the works of Wilk (1955), Wilk and Kemthorne (1955, 1956), Scheffé (1956a,b), and Harville (1978). The articles by Hocking (1973) and Sahai (1988) provide an excellent summary of various mixed models. We assume the following distribution properties and an analysis of variance for this situation is presented in succeeding sections. If an experimenter desires distribution properties other than those given here, tests and confidence intervals must be modified accordingly. One such alternate mixed model is discussed briefly in Section 4.19.

We assume that α_i's are constants subject to the constraint $\sum_{i=1}^{a} \alpha_i = 0$; β_j's are uncorrelated random variables with mean zero and variance σ_β^2; $(\alpha\beta)_{ij}$'s are random variables with mean zero and variance $[(a-1)/a]\sigma_{\alpha\beta}^2$ and subject to the constraints $\sum_{i=1}^{a} (\alpha\beta)_{ij} = 0$ for all j.[2] This introduces dependence between

---
[1] Some statisticians have expressed concern over the fact that the interaction terms $(\alpha\beta)_{ij}$'s are assumed to be uncorrelated to those of α_i's and β_j's. However, the assumption is consistent with the results from the finite population models that define the interaction to be a function of the main effects (see, e.g., Cornfield and Tukey (1956); Scheffé (1959, Section 7.4)).
[2] Since the α_i's are assumed to be fixed subject to the constraint that $\sum_{i=1}^{a} \alpha_i = 0$, it is felt that it is reasonable to assume that the summation of the interaction terms, over all the levels of factor A within a given level of factor B, should be equal to zero.

certain interaction terms at different levels of the fixed factor. In fact, one can show that (Graybill (1961, pp. 396–397))[3]

$$E[(\alpha\beta)_{ij}(\alpha\beta)_{i'j'}] = \begin{cases} -\dfrac{1}{a}\sigma^2_{\alpha\beta}, & i \neq i'; \\ 0, & i = i', j \neq j'. \end{cases} \quad (4.2.1)$$

Note that in this model the variance of $(\alpha\beta)_{ij}$ is defined as $[(a-1)/a]\sigma^2_{\alpha\beta}$ instead of $\sigma^2_{\alpha\beta}$ in order to simplify the expressions for expected mean squares. Furthermore, the β_j's, $(\alpha\beta)_{ij}$'s, and e_{ijk}'s are uncorrelated with each other, and e_{ijk}'s have mean zero and variance σ^2_e.

The objective in this model is to test the hypotheses: $\sigma^2_\beta = 0, \sigma^2_{\alpha\beta} = 0, \alpha_i$'s $= 0$; and to find point and interval estimates for the variance components $\sigma^2_e, \sigma^2_{\alpha\beta}, \sigma^2_\beta$; fixed effects α_i's and their contrasts.

Remarks: (i) Under Model I, the assumptions that $\sum_{i=1}^{a}(\alpha\beta)_{ij} = 0 = \sum_{j=1}^{b}(\alpha\beta)_{ij}$ can be made without any loss of generality because, by definition, $(\alpha\beta)_{ij}$ is a differential effect and if the sum, say, over i, were equal to some non-zero constant c, we could replace β_j with $\beta_j - c/a$ and the resultant $(\alpha\beta)_{ij}$ will then sum to zero. Similarly, the assumptions that $\sum_{i=1}^{a}\alpha_i = 0 = \sum_{j=1}^{b}\beta_j$ can be made without any loss of generality (see Remark at the end of Section 2.2).

(ii) As indicated in the Remark of Section 3.2, under the assumptions of the random effects model in (4.1.1), there are three intraclass correlations defined as follows: The correlation between the observations within the same level of factor A; within the same level of factor B; and between the observations within the same cell. Denoting these correlations by ρ_α, ρ_β, and $\rho_{\alpha\beta}$, we have $\rho_\alpha = \sigma^2_\alpha/(\sigma^2_e + \sigma^2_{\alpha\beta} + \sigma^2_\beta + \sigma^2_\alpha)$, $\rho_\beta = \sigma^2_\beta/(\sigma^2_e + \sigma^2_{\alpha\beta} + \sigma^2_\beta + \sigma^2_\alpha)$, and $\rho_{\alpha\beta} = \sigma^2_{\alpha\beta}/(\sigma^2_e + \sigma^2_{\alpha\beta} + \sigma^2_\beta + \sigma^2_\alpha)$. Under the assumptions of the mixed model in (4.1.1), the correlation between the observations within the same cell is given by $\rho_{\alpha\beta} = \sigma^2_{\alpha\beta}/(\sigma^2_e + \sigma^2_{\alpha\beta} + \sigma^2_\beta)$.

4.3 PARTITION OF THE TOTAL SUM OF SQUARES

To partition the total sum of squares, we start with the identity

$$y_{ijk} - \bar{y}_{...} = (\bar{y}_{i..} - \bar{y}_{...}) + (\bar{y}_{.j.} - \bar{y}_{...}) + (\bar{y}_{ij.} - \bar{y}_{i..} - \bar{y}_{.j.} + \bar{y}_{...}) + (y_{ijk} - \bar{y}_{ij.}),$$

and square and sum over i, j, and k to yield:

$$\sum_{i=1}^{a}\sum_{j=1}^{b}\sum_{k=1}^{n}(y_{ijk} - \bar{y}_{...})^2$$

---

[3] This implies that the interaction terms are correlated with a covariance equal to $-(1/a)\sigma^2_{\alpha\beta}$ and that the covariance decreases as the number of levels of the factor A increases.

Two-Way Crossed Classification with Interaction

$$= \sum_{i=1}^{a}\sum_{j=1}^{b}\sum_{k=1}^{n}[(\bar{y}_{i..} - \bar{y}_{...}) + (\bar{y}_{.j.} - \bar{y}_{...}) \quad (4.3.1)$$
$$+ (\bar{y}_{ij.} - \bar{y}_{i..} - \bar{y}_{.j.} + \bar{y}_{...}) + (y_{ijk} - \bar{y}_{ij.})]^2$$
$$= \sum_{i=1}^{a}\sum_{j=1}^{b}\sum_{k=1}^{n}(\bar{y}_{i..} - \bar{y}_{...})^2 + \sum_{i=1}^{a}\sum_{j=1}^{b}\sum_{k=1}^{n}(\bar{y}_{.j.} - \bar{y}_{...})^2$$
$$+ \sum_{i=1}^{a}\sum_{j=1}^{b}\sum_{k=1}^{n}(\bar{y}_{ij.} - \bar{y}_{i..} - \bar{y}_{.j.} + \bar{y}_{...})^2 + \sum_{i=1}^{a}\sum_{j=1}^{b}\sum_{k=1}^{n}(y_{ijk} - \bar{y}_{ij.})^2$$
$$= bn\sum_{i=1}^{a}(\bar{y}_{i..} - \bar{y}_{...})^2 + an\sum_{j=1}^{b}(\bar{y}_{.j.} - \bar{y}_{...})^2$$
$$+ n\sum_{i=1}^{a}\sum_{j=1}^{b}(\bar{y}_{ij.} - \bar{y}_{i..} - \bar{y}_{.j.} + \bar{y}_{...})^2 + \sum_{i=1}^{a}\sum_{j=1}^{b}\sum_{k=1}^{n}(y_{ijk} - \bar{y}_{ij.})^2,$$
$$(4.3.2)$$

where

$$\bar{y}_{ij.} = \frac{\sum_{k=1}^{n} y_{ijk}}{n}, \quad \bar{y}_{i..} = \frac{\sum_{j=1}^{b}\sum_{k=1}^{n} y_{ijk}}{bn},$$

$$\bar{y}_{.j.} = \frac{\sum_{i=1}^{a}\sum_{k=1}^{n} y_{ijk}}{an}, \quad \text{and} \quad \bar{y}_{...} = \frac{\sum_{i=1}^{a}\sum_{j=1}^{b}\sum_{k=1}^{n} y_{ijk}}{abn}.$$

The identity (4.3.1) is valid since all cross-product terms are equal to zero.

The terms to the right of (4.3.1) are denoted by SS_A, SS_B, SS_{AB}, and SS_E in respective order and measure the variation due to the α_i's, β_j's, $(\alpha\beta)_{ij}$'s, and e_{ijk}'s, respectively. The identity states that we have partitioned the total variation SS_T into the following components:

(i) SS_A, called the A-factor sum of squares, representing the variation in the y_{jjk} due to the A-factor effects;
(ii) SS_B, called the B-factor sum of squares, representing variation in the y_{ijk} due to the B-factor effects;
(iii) SS_{AB}, called the interaction sum of squares, representing variation in the y_{jjk} due to the interaction effects; and
(iv) SS_E, called the error sum of squares, representing variation in the y_{jjk} after removing A-factor effects, B-factor effects, and interaction effects.

Remark: In a two-way crossed classification, one can compute ab separate variances corresponding to each cell as $\sum_{k=1}^{n}(y_{ijk} - \bar{y}_{ij.})^2$, which can then be tested for homogeneity of variances (see Section 2.22).

4.4 MEAN SQUARES AND THEIR EXPECTATIONS

Mean squares are obtained in the usual way by dividing the sums of squares by their corresponding degrees of freedom. For SS_A and SS_B terms, we have the condition that

$$\sum_{i=1}^{a}(\bar{y}_{i..} - \bar{y}_{...}) = \sum_{j=1}^{b}(\bar{y}_{.j.} - \bar{y}_{...}) = 0 \tag{4.4.1}$$

and so the degrees of freedom are $a - 1$ and $b - 1$, respectively. For the SS_{AB} term, note that the random quantities θ_{ij}'s defined by

$$\theta_{ij} = \bar{y}_{ij.} - \bar{y}_{i..} - \bar{y}_{.j.} + \bar{y}_{...}$$

are subject to the conditions:

$$\sum_{i=1}^{a} \theta_{ij} = 0, \quad \text{for each } j \ (b \text{ relations}), \tag{4.4.2}$$

$$\sum_{j=1}^{b} \theta_{ij} = 0, \quad \text{for each } i \ (a \text{ relations}). \tag{4.4.3}$$

However, in effect, there are only $a + b - 1$ independent restrictions on the θ_{ij}'s. This is so since b restrictions (4.4.2), when summed, determine the relation

$$\sum_{j=1}^{b}\left(\sum_{i=1}^{a} \theta_{ij}\right) = 0.$$

Similarly, restrictions (4.4.3), when summed, must also be zero; that is,

$$\sum_{i=1}^{a}\left(\sum_{j=1}^{b} \theta_{ij}\right) = 0.$$

Thus, only $a - 1$ of the a relations (4.4.3) will be independent and the total number of independent restrictions on θ_{ij}'s is $a + b - 1$. Since the number of θ_{ij}'s is ab, the number of degrees of freedom associated with the SS_{AB} term is

$$ab - (a + b - 1) = (a - 1)(b - 1).$$

By subtraction, the number of degrees of freedom associated with the error term SS_E is

$$abn - (a - 1) - (b - 1) - (a - 1)(b - 1) = ab(n - 1).$$

Two-Way Crossed Classification with Interaction

The corresponding mean squares MS_A, MS_B, MS_{AB}, and MS_E are, therefore, defined by

$$MS_A = \frac{SS_A}{a-1}, \quad MS_B = \frac{SS_B}{b-1},$$

$$MS_{AB} = \frac{SS_{AB}}{(a-1)(b-1)}, \quad \text{and} \quad MS_E = \frac{SS_E}{ab(n-1)}.$$

Next, we examine the expectations of mean squares, which can be readily derived from the assumptions of the model (4.1.1) and the usual laws of expectations. On taking successive averages of the model equation (4.1.1), we obtain

$$\bar{y}_{ij.} = \mu + \alpha_i + \beta_j + (\alpha\beta)_{ij} + \bar{e}_{ij.}, \tag{4.4.4}$$

$$\bar{y}_{i..} = \mu + \alpha_i + \bar{\beta}_. + \overline{(\alpha\beta)}_{i.} + \bar{e}_{i..}, \tag{4.4.5}$$

$$\bar{y}_{.j.} = \mu + \bar{\alpha}_. + \beta_j + \overline{(\alpha\beta)}_{.j} + \bar{e}_{.j.}, \tag{4.4.6}$$

and

$$\bar{y}_{...} = \mu + \bar{\alpha}_. + \bar{\beta}_. + \overline{(\alpha\beta)}_{..} + \bar{e}_{...}, \tag{4.4.7}$$

where, as usual, the bars indicate means over the subscripts shown by dots. Substituting the values of y_{ijk}, $\bar{y}_{ij.}$, $\bar{y}_{i..}$, $\bar{y}_{.j.}$, and $\bar{y}_{...}$ given by (4.1.1), (4.4.4), (4.4.5), (4.4.6), and (4.4.7), respectively, into the expressions for the SS_A, SS_B, SS_{AB}, and SS_E terms defined in (4.3.1), we obtain the following expressions:

$$SS_E = \sum_{i=1}^{a}\sum_{j=1}^{b}\sum_{k=1}^{n}(e_{ijk} - \bar{e}_{ij.})^2, \tag{4.4.8}$$

$$SS_{AB} = n\sum_{i=1}^{a}\sum_{j=1}^{b}[(\alpha\beta)_{ij} - \overline{(\alpha\beta)}_{i.} - \overline{(\alpha\beta)}_{.j} + \overline{(\alpha\beta)}_{..}$$
$$+ \bar{e}_{ij.} - \bar{e}_{i..} - \bar{e}_{.j.} + \bar{e}_{...}]^2, \tag{4.4.9}$$

$$SS_B = an\sum_{j=1}^{b}[\beta_j - \bar{\beta}_. + \overline{(\alpha\beta)}_{.j} - \overline{(\alpha\beta)}_{..} + \bar{e}_{.j.} - \bar{e}_{...}]^2, \tag{4.4.10}$$

and

$$SS_A = bn\sum_{i=1}^{a}[\alpha_i - \bar{\alpha}_. + \overline{(\alpha\beta)}_{i.} - \overline{(\alpha\beta)}_{..} + \bar{e}_{i..} - \bar{e}_{...}]^2. \tag{4.4.11}$$

Now, because the e_{ijk}'s are uncorrelated and identically distributed with mean zero and variance σ_e^2, it follows that

$$E(e_{ijk}^2) = \sigma_e^2, \tag{4.4.12}$$

$$E(\bar{e}_{ij.}^2) = \sigma_e^2/n, \tag{4.4.13}$$

$$E(\bar{e}_{i..}^2) = \sigma_e^2/bn, \tag{4.4.14}$$

$$E(\bar{e}_{.j.}^2) = \sigma_e^2/an, \tag{4.4.15}$$

and

$$E(\bar{e}_{...}^2) = \sigma_e^2/abn. \tag{4.4.16}$$

It is then a matter of straightforward computation to derive the expectations of mean squares. First, taking the expectation of (4.4.8), we obtain

$$E(\text{SS}_E) = \sum_{i=1}^{a} \sum_{j=1}^{b} \sum_{k=1}^{n} E(e_{ijk} - \bar{e}_{ij.})^2$$

$$= \sum_{i=1}^{a} \sum_{j=1}^{b} \sum_{k=1}^{n} \left[E(e_{ijk}^2) - 2E(e_{ijk}\bar{e}_{ij.}) + E(\bar{e}_{ij.}^2) \right]$$

$$= abn\left[\sigma_e^2 - 2\frac{\sigma_e^2}{n} + \frac{\sigma_e^2}{n} \right]$$

$$= ab(n-1)\sigma_e^2. \tag{4.4.17}$$

The expectation of MS_E is, therefore, given by

$$E(\text{MS}_E) = E\left(\frac{\text{SS}_E}{ab(n-1)} \right) = \sigma_e^2. \tag{4.4.18}$$

Note that the result (4.4.18) is true under the assumptions of the fixed and random, as well as mixed effects models.

Now, to derive the expectations of MS_{AB}, MS_B, and MS_A, we consider the cases of Models I, II, and III separately.

MODEL I (FIXED EFFECTS)

Under Model I, the α_i's, β_j's, and $(\alpha\beta)_{ij}$'s are fixed quantities with the restrictions that $\bar{\alpha}_. = \bar{\beta}_. = \overline{(\alpha\beta)}_{i.} = \overline{(\alpha\beta)}_{.j} = \overline{(\alpha\beta)}_{..} = 0$. Therefore, the expressions

Two-Way Crossed Classification with Interaction

(4.4.9) through (4.4.11) reduce to

$$SS_{AB} = n \sum_{i=1}^{a} \sum_{j=1}^{b} [(\alpha\beta)_{ij} + \bar{e}_{ij.} - \bar{e}_{i..} - \bar{e}_{.j.} + \bar{e}_{...}]^2, \quad (4.4.19)$$

$$SS_B = an \sum_{j=1}^{b} [\beta_j + \bar{e}_{.j.} - \bar{e}_{...}]^2, \quad (4.4.20)$$

and

$$SS_A = bn \sum_{i=1}^{a} [\alpha_i + \bar{e}_{i..} - \bar{e}_{...}]^2. \quad (4.4.21)$$

On taking the expectation of (4.4.19), we obtain

$$E(SS_{AB}) = n \sum_{i=1}^{a} \sum_{j=1}^{b} E[(\alpha\beta)_{ij} + \bar{e}_{ij.} - \bar{e}_{i..} - \bar{e}_{.j.} + \bar{e}_{...}]^2$$

$$= n \sum_{i=1}^{a} \sum_{j=1}^{b} (\alpha\beta)_{ij}^2 + n \sum_{i=1}^{a} \sum_{j=1}^{b} E(\bar{e}_{ij.} - \bar{e}_{i..} - \bar{e}_{.j.} + \bar{e}_{...})^2, \quad (4.4.22)$$

since the $(\alpha\beta)_{ij}$'s are constants and the expectation of the cross-product term is zero. By proceeding as in the derivation of the result (3.4.11), it can be readily shown that

$$\sum_{i=1}^{a} \sum_{j=1}^{b} E(\bar{e}_{ij.} - \bar{e}_{i..} - \bar{e}_{.j.} + \bar{e}_{...})^2 = (a-1)(b-1)\frac{\sigma_e^2}{n}. \quad (4.4.23)$$

Then, on substituting (4.4.23) into (4.4.22), we obtain

$$E(SS_{AB}) = n \sum_{i=1}^{a} \sum_{j=1}^{b} (\alpha\beta)_{ij}^2 + (a-1)(b-1)\sigma_e^2. \quad (4.4.24)$$

Therefore, the expectation of MS_{AB} is given by

$$E(MS_{AB}) = E\left(\frac{SS_{AB}}{(a-1)(b-1)}\right) = \sigma_e^2 + \frac{n \sum_{i=1}^{a} \sum_{j=1}^{b} (\alpha\beta)_{ij}^2}{(a-1)(b-1)}. \quad (4.4.25)$$

To find the expectation of MS_B, we note from (4.4.20) that

$$E(SS_B) = an\left[\sum_{j=1}^{b} \beta_j^2 + E\sum_{j=1}^{b}(\bar{e}_{.j.} - \bar{e}_{...})^2\right], \quad (4.4.26)$$

since β_j's are constants and the expectation of the cross-product term is zero. Now, using the results (4.4.15) and (4.4.16), we find that

$$E\sum_{j=1}^{b}(\bar{e}_{.j.} - \bar{e}_{...})^2 = \sum_{j=1}^{b} E(\bar{e}_{.j.}^2) - bE(\bar{e}_{...}^2)$$

$$= b\frac{\sigma_e^2}{an} - b\frac{\sigma_e^2}{abn}$$

$$= \frac{b-1}{an}\sigma_e^2. \quad (4.4.27)$$

Then, on substituting (4.4.27) into (4.4.26), we obtain

$$E(SS_B) = an\sum_{j=1}^{b}\beta_j^2 + (b-1)\sigma_e^2. \quad (4.4.28)$$

Therefore, the expectation of MS_B is given by

$$E(MS_B) = E\left(\frac{SS_B}{b-1}\right) = \sigma_e^2 + \frac{an}{b-1}\sum_{j=1}^{b}\beta_j^2. \quad (4.4.29)$$

Finally, from symmetry, it follows that the expectation of MS_A is given by

$$E(MS_A) = \sigma_e^2 + \frac{bn}{a-1}\sum_{i=1}^{a}\alpha_i^2. \quad (4.4.30)$$

MODEL II (RANDOM EFFECTS)

Under Model II, the α_i's, β_j's, and $(\alpha\beta)_{ij}$'s are mutually and completely uncorrelated random variables with mean zero and variances σ_α^2, σ_β^2, and $\sigma_{\alpha\beta}^2$ respectively. It then follows, using the formulae for the variances of the sampling distribution of the means of α_i's, β_j's, and $(\alpha\beta)_{ij}$'s, that

$$E(\alpha_i^2) = \sigma_\alpha^2, \quad (4.4.31)$$

$$E(\bar{\alpha}_.^2) = \sigma_\alpha^2/a, \quad (4.4.32)$$

$$E(\beta_j^2) = \sigma_\beta^2, \quad (4.4.33)$$

Two-Way Crossed Classification with Interaction

$$E(\bar{\beta}^2) = \sigma_\beta^2/b, \qquad (4.4.34)$$

$$E[(\alpha\beta)_{ij}^2] = \sigma_{\alpha\beta}^2, \qquad (4.4.35)$$

$$E[\overline{(\alpha\beta)}_{i.}^2] = \sigma_{\alpha\beta}^2/b, \qquad (4.4.36)$$

$$E[\overline{(\alpha\beta)}_{.j}^2] = \sigma_{\alpha\beta}^2/a, \qquad (4.4.37)$$

and

$$E[\overline{(\alpha\beta)}_{..}^2] = \sigma_{\alpha\beta}^2/ab. \qquad (4.4.38)$$

First, taking the expectation of (4.4.9), we obtain

$$E(\text{SS}_{AB}) = n \sum_{i=1}^{a} \sum_{j=1}^{b} E[(\alpha\beta)_{ij} - \overline{(\alpha\beta)}_{i.} - \overline{(\alpha\beta)}_{.j} + \overline{(\alpha\beta)}_{..}]^2$$

$$+ n \sum_{i=1}^{a} \sum_{j=1}^{b} E[\bar{e}_{ij.} - \bar{e}_{i..} - \bar{e}_{.j.} + \bar{e}_{...}]^2, \qquad (4.4.39)$$

since the expectation of the cross-product term is zero. By proceeding as in the derivation of the result (3.4.11), it can be shown that

$$\sum_{i=1}^{a} \sum_{j=1}^{b} E[(\alpha\beta)_{ij} - \overline{(\alpha\beta)}_{i.} - \overline{(\alpha\beta)}_{.j} + \overline{(\alpha\beta)}_{..}] = (a-1)(b-1)\sigma_{\alpha\beta}^2 \qquad (4.4.40)$$

and

$$\sum_{i=1}^{a} \sum_{j=1}^{b} E[\bar{e}_{ij.} - \bar{e}_{i..} - \bar{e}_{.j.} + \bar{e}_{...}]^2 = (a-1)(b-1)\frac{\sigma_e^2}{n}. \qquad (4.4.41)$$

On substituting (4.4.40) and (4.4.41) into (4.4.39), we obtain

$$E(\text{SS}_{AB}) = (a-1)(b-1)[\sigma_e^2 + n\sigma_{\alpha\beta}^2]. \qquad (4.4.42)$$

Therefore, the expectation of MS_{AB} is given by

$$E(\text{MS}_{AB}) = E\left(\frac{\text{SS}_{AB}}{(a-1)(b-1)}\right) = \sigma_e^2 + n\sigma_{\alpha\beta}^2. \qquad (4.4.43)$$

Next, taking the expectation of (4.4.10), we obtain

$$E(SS_B) = an\left[E\sum_{j=1}^{b}(\bar{\beta}_j - \bar{\beta}_.)^2 + E\sum_{j=1}^{b}(\overline{(\alpha\beta)}_{.j} - \overline{(\alpha\beta)}_{..})^2 \right.$$
$$\left. + E\sum_{j=1}^{b}(\bar{e}_{.j.} - \bar{e}_{...})^2\right], \qquad (4.4.44)$$

since again the expectations of the cross-product terms are zero. Using the results (4.4.33) and (4.4.34), it follows that

$$E\sum_{j=1}^{b}(\bar{\beta}_j - \bar{\beta}_.)^2 = (b-1)\sigma_\beta^2. \qquad (4.4.45)$$

Similarly, using (4.4.37) and (4.4.38), we have

$$E\sum_{j=1}^{b}[\overline{(\alpha\beta)}_{.j} - \overline{(\alpha\beta)}_{..}]^2 = (b-1)\frac{\sigma_{\alpha\beta}^2}{a}; \qquad (4.4.46)$$

and, finally, using (4.4.15) and (4.4.16), we have

$$E\sum_{j=1}^{b}(\bar{e}_{.j.} - \bar{e}_{...})^2 = (b-1)\frac{\sigma_e^2}{an}. \qquad (4.4.47)$$

Substituting (4.4.45), (4.4.46), and (4.4.47) into (4.4.44), we obtain

$$E(SS_B) = (b-1)[\sigma_e^2 + n\sigma_{\alpha\beta}^2 + an\sigma_\beta^2]. \qquad (4.4.48)$$

The expectation of MS_B is, therefore, given by

$$E(MS_B) = E\left(\frac{SS_B}{b-1}\right) = \sigma_e^2 + n\sigma_{\alpha\beta}^2 + an\sigma_\beta^2. \qquad (4.4.49)$$

Finally, from symmetry, it follows that

$$E(MS_A) = \sigma_e^2 + n\sigma_{\alpha\beta}^2 + bn\sigma_\alpha^2. \qquad (4.4.50)$$

MODEL III (MIXED EFFECTS)

Under Model III, the α_i's are constants with the restriction that $\sum_{i=1}^{a}\alpha_i = 0$; β_j's are uncorrelated random variables with mean zero and variance σ_β^2, and $(\alpha\beta)_{ij}$'s are random variables with mean zero and variance-covariance structure given by (4.2.1), and subject to the constraints $\sum_{i=1}^{a}(\alpha\beta)_{ij} = 0$ for all j. Furthermore, the β_j's, $(\alpha\beta)_{ij}$'s, and e_{ijk}'s are uncorrelated with each other. It then follows,

Two-Way Crossed Classification with Interaction

using the formulae for the variances of the sampling distributions of the means of the β_j's and $(\alpha\beta)_{ij}$'s that

$$E(\beta_j^2) = \sigma_\beta^2, \qquad (4.4.51)$$

$$E(\bar{\beta}_.^2) = \sigma_\beta^2/b, \qquad (4.4.52)$$

$$E[(\alpha\beta)_{ij}^2] = \frac{a-1}{a}\sigma_{\alpha\beta}^2, \qquad (4.4.53)$$

and

$$E[\overline{(\alpha\beta)}_{i.}^2] = \frac{a-1}{ab}\sigma_{\alpha\beta}^2. \qquad (4.4.54)$$

Now, under the restrictions that $\bar{\alpha}_. = \overline{(\alpha\beta)}_{.j} = \overline{(\alpha\beta)}_{..} = 0$, the expressions (4.4.9) through (4.4.11) reduce to

$$SS_{AB} = n\sum_{i=1}^{a}\sum_{j=1}^{b}[(\alpha\beta)_{ij} - \overline{(\alpha\beta)}_{i.} + \bar{e}_{ij.} - \bar{e}_{i..} - \bar{e}_{.j.} + \bar{e}_{...}]^2, \qquad (4.4.55)$$

$$SS_B = an\sum_{j=1}^{b}[\beta_j - \bar{\beta}_. + \bar{e}_{.j.} - \bar{e}_{...}]^2, \qquad (4.4.56)$$

and

$$SS_A = bn\sum_{i=1}^{a}[\alpha_i + \overline{(\alpha\beta)}_{i.} + \bar{e}_{i..} - \bar{e}_{...}]^2. \qquad (4.4.57)$$

First, taking the expectation of (4.4.55), we obtain

$$E(SS_{AB}) = n\sum_{i=1}^{a}\sum_{j=1}^{b}E[(\alpha\beta)_{ij} - \overline{(\alpha\beta)}_{i.}]^2 + n\sum_{i=1}^{a}\sum_{j=1}^{b}E[\bar{e}_{ij.} - \bar{e}_{i..} - \bar{e}_{.j.} + \bar{e}_{...}]^2, \qquad (4.4.58)$$

since the expectation of the cross-product term is zero. Using the results (4.4.53) and (4.4.54), we obtain

$$\sum_{i=1}^{a}\sum_{j=1}^{b}E[(\alpha\beta)_{ij} - \overline{(\alpha\beta)}_{i.}]^2 = \sum_{i=1}^{a}\left[\sum_{j=1}^{b}E(\alpha\beta)_{ij}^2 - bE\overline{(\alpha\beta)}_{i.}^2\right]$$

$$= \sum_{i=1}^{a}\left[b\frac{a-1}{a}\sigma_{\alpha\beta}^2 - b\frac{a-1}{ab}\sigma_{\alpha\beta}^2\right]$$

$$= (a-1)(b-1)\sigma_{\alpha\beta}^2. \qquad (4.4.59)$$

Furthermore, as shown in (3.4.11), we have

$$\sum_{i=1}^{a}\sum_{j=1}^{b} E(\bar{e}_{ij.} - \bar{e}_{i..} - \bar{e}_{.j.} + \bar{e}_{...})^2 = (a-1)(b-1)\frac{\sigma_e^2}{n}. \qquad (4.4.60)$$

Substituting (4.4.59) and (4.4.60) into (4.4.58), we obtain

$$E(SS_{AB}) = (a-1)(b-1)\left[\sigma_e^2 + n\sigma_{\alpha\beta}^2\right]. \qquad (4.4.61)$$

Therefore, the expectation of MS_{AB} is given by

$$E(MS_{AB}) = E\left(\frac{SS_{AB}}{(a-1)(b-1)}\right) = \sigma_e^2 + n\sigma_{\alpha\beta}^2. \qquad (4.4.62)$$

Next, taking the expectation of (4.4.56), we have

$$E(SS_B) = an\left[E\sum_{j=1}^{b}(\bar{\beta}_j - \bar{\beta}_.)^2 + E\sum_{j=1}^{b}(\bar{e}_{.j.} - \bar{e}_{...})^2\right]. \qquad (4.4.63)$$

As in (4.4.45) and (4.4.47), it is easy to verify that

$$E\sum_{j=1}^{b}(\bar{\beta}_j - \bar{\beta}_.)^2 = (b-1)\sigma_\beta^2 \qquad (4.4.64)$$

and

$$E\sum_{j=1}^{b}(\bar{e}_{.j.} - \bar{e}_{...})^2 = (b-1)\frac{\sigma_e^2}{an}. \qquad (4.4.65)$$

Substituting (4.4.64) and (4.4.65) into (4.4.63), we obtain

$$E(SS_B) = (b-1)\left[\sigma_e^2 + an\sigma_\beta^2\right]. \qquad (4.4.66)$$

So that the expectation of MS_B is given by

$$E(MS_B) = E\left(\frac{SS_B}{b-1}\right) = \sigma_e^2 + an\sigma_\beta^2. \qquad (4.4.67)$$

Finally, taking the expectation of (4.4.57), we obtain

$$E(SS_A) = bn\left[\sum_{i=1}^{a}\alpha_i^2 + E\sum_{i=1}^{a}\overline{(\alpha\beta)}_{i.}^2 + E\sum_{i=1}^{a}(\bar{e}_{i..} - \bar{e}_{...})^2\right], \qquad (4.4.68)$$

Two-Way Crossed Classification with Interaction

since the α_i's are constants and the expectations of the cross-product terms are zero. From (4.4.14), (4.4.16), and (4.4.54), it readily follows that

$$E \sum_{i=1}^{a} \overline{(\alpha\beta)}_{i.}^2 = \frac{a-1}{b} \sigma_{\alpha\beta}^2 \tag{4.4.69}$$

and

$$E \sum_{i=1}^{a} (\bar{e}_{i..} - \bar{e}_{...})^2 = (a-1)\frac{\sigma_e^2}{bn}. \tag{4.4.70}$$

Substituting (4.4.69) and (4.4.70) into (4.4.68), we obtain

$$E(SS_A) = (a-1)\sigma_e^2 + (a-1)n\sigma_{\alpha\beta}^2 + bn\sum_{i=1}^{a}\alpha_i^2.$$

Hence, the expectation of MS_A is given by

$$E(MS_A) = E\left(\frac{SS_A}{a-1}\right) = \sigma_e^2 + n\sigma_{\alpha\beta}^2 + \frac{bn}{a-1}\sum_{i=1}^{a}\alpha_i^2.$$

The foregoing results of Sections 4.3 and 4.4 can now be summarized in a tabular form as the analysis of variance table shown in Table 4.2.

4.5 SAMPLING DISTRIBUTION OF MEAN SQUARES

In this section, we give the distribution results on mean squares for the fixed, random, and mixed effects models. The derivation of these results is beyond the scope of this volume and can be found in Scheffé (1959, pp. 109–112), Graybill (1961, pp. 397–402; 1976, pp. 630–632), and Searle et al. (1992, pp. 131–132). Note that although in the derivation of the expected mean squares we have not made any distribution assumption about the form of the random components of the model (4.1.1), we do require the assumption of normality to derive their sampling distributions.

MODEL I (FIXED EFFECTS)

Under the distribution assumptions of Model I, it can be shown that:

(a) The quantities MS_E, MS_{AB}, MS_B, and MS_A are statistically independent.
(b) The following results are true:
   (i)

$$\frac{MS_E}{\sigma_e^2} \sim \frac{\chi^2[ab(n-1)]}{ab(n-1)}, \tag{4.5.1}$$

TABLE 4.2
Analysis of Variance for Model (4.1.1)

Source of Variation	Degrees of Freedom	Sum of Squares	Mean Square	Expected Mean Square		
				Model I	Model II	Model III
Due to A	$a-1$	SS_A	MS_A	$\sigma_e^2 + \dfrac{bn}{a-1}\sum_{i=1}^{a}\alpha_i^2$	$\sigma_e^2 + n\sigma_{\alpha\beta}^2 + bn\sigma_\alpha^2$	$\sigma_e^2 + n\sigma_{\alpha\beta}^2 + \dfrac{bn}{a-1}\sum_{i=1}^{a}\alpha_i^2$
Due to B	$b-1$	SS_B	MS_B	$\sigma_e^2 + \dfrac{an}{b-1}\sum_{j=1}^{b}\beta_j^2$	$\sigma_e^2 + n\sigma_{\alpha\beta}^2 + an\sigma_\beta^2$	$\sigma_e^2 + an\sigma_\beta^2$
Interaction $A \times B$	$(a-1)(b-1)$	SS_{AB}	MS_{AB}	$\sigma_e^2 + \dfrac{n}{(a-1)(b-1)}\sum_{i=1}^{a}\sum_{j=1}^{b}(\alpha\beta)_{ij}^2$	$\sigma_e^2 + n\sigma_{\alpha\beta}^2$	$\sigma_e^2 + n\sigma_{\alpha\beta}^2$
Error	$ab(n-1)$	SS_E	MS_E	σ_e^2	σ_e^2	σ_e^2
Total	$abn-1$	SS_T				

Two-Way Crossed Classification with Interaction

(ii)

$$\frac{\text{MS}_{AB}}{\sigma_e^2} \sim \frac{\chi^{2'}[(a-1)(b-1), \lambda_{AB}]}{(a-1)(b-1)}, \qquad (4.5.2)$$

(iii)

$$\frac{\text{MS}_B}{\sigma_e^2} \sim \frac{\chi^{2'}[b-1, \lambda_B]}{b-1}, \qquad (4.5.3)$$

and

(iv)

$$\frac{\text{MS}_A}{\sigma_e^2} \sim \frac{\chi^{2'}[a-1, \lambda_A]}{a-1}, \qquad (4.5.4)$$

where, as usual, $\chi^2[\cdot]$ denotes a central and $\chi^{2'}[\cdot,\cdot]$ denotes a noncentral chi-square variable with respective degrees of freedom and the noncentrality parameters λ_{AB}, λ_B, and λ_A defined by

$$\lambda_{AB} = \frac{n}{2\sigma_e^2} \sum_{i=1}^{a} \sum_{j=1}^{b} (\alpha\beta)_{ij}^2,$$

$$\lambda_B = \frac{an}{2\sigma_e^2} \sum_{j=1}^{b} \beta_j^2,$$

and

$$\lambda_A = \frac{bn}{2\sigma_e^2} \sum_{i=1}^{a} \alpha_i^2.$$

It, therefore, follows from (4.5.2) through (4.5.4) that

(ii)' If $(\alpha\beta)_{ij} = 0$, for all i and j, then

$$\frac{\text{MS}_{AB}}{\sigma_e^2} \sim \frac{\chi^2[(a-1)(b-1)]}{(a-1)(b-1)}; \qquad (4.5.5)$$

(iii)' If $\beta_j = 0$, for all j, then

$$\frac{\text{MS}_B}{\sigma_e^2} \sim \frac{\chi^2[b-1]}{b-1}; \qquad (4.5.6)$$

(iv)′ If $\alpha_i = 0$, for all i, then

$$\frac{\text{MS}_A}{\sigma_e^2} \sim \frac{\chi^2[a-1]}{a-1}. \tag{4.5.7}$$

MODEL II (RANDOM EFFECTS)

Under the distribution assumptions of Model II, it can be shown that:

(a) The quantities MS_E, MS_{AB}, MS_B, and MS_A are statistically independent.
(b) The following results are true:

(i)
$$\frac{\text{MS}_E}{\sigma_e^2} \sim \frac{\chi^2[ab(n-1)]}{ab(n-1)}, \tag{4.5.8}$$

(ii)
$$\frac{\text{MS}_{AB}}{\sigma_e^2 + n\sigma_{\alpha\beta}^2} \sim \frac{\chi^2[(a-1)(b-1)]}{(a-1)(b-1)}, \tag{4.5.9}$$

(iii)
$$\frac{\text{MS}_B}{\sigma_e^2 + n\sigma_{\alpha\beta}^2 + an\sigma_\beta^2} \sim \frac{\chi^2[b-1]}{b-1}, \tag{4.5.10}$$

and

(iv)
$$\frac{\text{MS}_A}{\sigma_e^2 + n\sigma_{\alpha\beta}^2 + bn\sigma_\alpha^2} \sim \frac{\chi^2[a-1]}{a-1}. \tag{4.5.11}$$

That is, the ratio of MS_E to σ_e^2 is a $\chi^2[ab(n-1)]$ variable divided by $ab(n-1)$; the ratio of MS_{AB} to $\sigma_e^2 + n\sigma_{\alpha\beta}^2$ is a $\chi^2[(a-1)(b-1)]$ variable divided by $(a-1)(b-1)$; the ratio of MS_B to $\sigma_e^2 + n\sigma_{\alpha\beta}^2 + an\sigma_\beta^2$ is a $\chi^2[b-1]$ variable divided by $b-1$; and the ratio of MS_A to $\sigma_e^2 + n\sigma_{\alpha\beta}^2 + bn\sigma_\alpha^2$ is a $\chi^2[a-1]$ variable divided by $a-1$.

MODEL III (MIXED EFFECTS)

Under the distribution assumptions of Model III, it can be shown that:

(a) The quantities MS_E, MS_{AB}, MS_B, and MS_A are statistically independent.
(b) The following results are true:

(i)
$$\frac{\text{MS}_E}{\sigma_e^2} \sim \frac{\chi^2[ab(n-1)]}{ab(n-1)}, \qquad (4.5.12)$$

(ii)
$$\frac{\text{MS}_{AB}}{\sigma_e^2 + n\sigma_{\alpha\beta}^2} \sim \frac{\chi^2[(a-1)(b-1)]}{(a-1)(b-1)}, \qquad (4.5.13)$$

(iii)
$$\frac{\text{MS}_B}{\sigma_e^2 + an\sigma_\beta^2} \sim \frac{\chi^2[b-1]}{b-1}, \qquad (4.5.14)$$

and

(iv)
$$\frac{\text{MS}_A}{\sigma_e^2 + n\sigma_{\alpha\beta}^2} \sim \frac{\chi^{2'}[a-1, \lambda_A]}{a-1}, \qquad (4.5.15)$$

where
$$\lambda_A = \frac{bn}{2(\sigma_e^2 + n\sigma_{\alpha\beta}^2)} \sum_{i=1}^a \alpha_i^2.$$

It then follows from (4.5.15) that if $\alpha_i = 0$ for all i, then

(iv)'
$$\frac{\text{MS}_A}{\sigma_e^2 + n\sigma_{\alpha\beta}^2} \sim \frac{\chi^2[a-1]}{a-1}. \qquad (4.5.16)$$

4.6 TESTS OF HYPOTHESES: THE ANALYSIS OF VARIANCE F TESTS

In this section, we present the usual hypotheses of interest and appropriate F tests for fixed, random, and mixed effects models. As usual, the test statistic is constructed by comparing two mean squares that have the same expectation under H_0 and the numerator mean square has a larger expectation than the denominator mean square under H_1.

MODEL I (FIXED EFFECTS)

The usual tests of hypotheses of interest are about AB interactions, factor B effects, and factor A effects.

Test for AB Interactions

Ordinarily the two-way classification study begins with a test to determine whether the two factors interact. The hypothesis is

$$H_0^{AB} : \text{all } (\alpha\beta)_{ij}\text{'s} = 0$$
$$\text{versus} \qquad (4.6.1)$$
$$H_1^{AB} : \text{not all } (\alpha\beta)_{ij}\text{'s are zero.}$$

In order to develop a test procedure for the hypothesis (4.6.1), we note from (4.4.18) and (4.4.25) that under H_0^{AB},

$$E(\text{MS}_E) = \sigma_e^2,$$
$$E(\text{MS}_{AB}) = \sigma_e^2;$$

and under H_1^{AB},

$$E(\text{MS}_{AB}) > E(\text{MS}_E).$$

Furthermore, it follows from (4.5.1) and (4.5.5) that under H_0^{AB},

$$F_{AB} = \frac{\text{MS}_{AB}/\sigma_e^2}{\text{MS}_E/\sigma_e^2} = \frac{\text{MS}_{AB}}{\text{MS}_E} \sim F[(a-1)(b-1), ab(n-1)]. \qquad (4.6.2)$$

Thus, the statistic (4.6.2) provides a suitable test procedure for (4.6.1); H_0^{AB} being rejected if

$$F_{AB} > F[(a-1)(b-1), ab(n-1); 1-\alpha].$$

Test for Factor B Effects
The hypothesis is

$$H_0^B : \text{all } \beta_j = 0$$
$$\text{versus} \qquad (4.6.3)$$
$$H_1^B : \text{not all } \beta_j\text{'s are zero.}$$

Two-Way Crossed Classification with Interaction

In order to develop a test procedure for the hypothesis (4.6.3), we note from (4.4.18) and (4.4.29) that under H_0^B,

$$E(\text{MS}_E) = \sigma_e^2,$$
$$E(\text{MS}_B) = \sigma_e^2;$$

and under H_1^B,

$$E(\text{MS}_B) > E(\text{MS}_E).$$

Furthermore, it follows from (4.5.1) and (4.5.6) that under H_0^B,

$$F_B = \frac{\text{MS}_B/\sigma_e^2}{\text{MS}_E/\sigma_e^2} = \frac{\text{MS}_B}{\text{MS}_E} \sim F[b-1, ab(n-1)]. \tag{4.6.4}$$

Thus, the statistic (4.6.4) provides a suitable test procedure for (4.6.3); H_0^B being rejected if

$$F_B > F[b-1, ab(n-1); 1-\alpha].$$

Test for Factor A Effects

The hypothesis is

$$H_0^A : \text{all } \alpha_i = 0$$
$$\text{versus} \tag{4.6.5}$$
$$H_1^A : \text{not all } \alpha_i\text{'s are zero}.$$

Proceeding as in the test for factor B effects, it readily follows that the statistic

$$F_A = \frac{\text{MS}_A/\sigma_e^2}{\text{MS}_E/\sigma_e^2} = \frac{\text{MS}_A}{\text{MS}_E} \sim F[a-1, ab(n-1)]$$

provides a suitable test procedure for (4.6.5); H_0^A being rejected if

$$F_A > F[a-1, ab(n-1); 1-\alpha].$$

Remarks: (i) If a nonsignificant value of F_{AB} occurs, some authors suggest that the MS_{AB} and MS_E terms of the analysis of variance Table 4.2 be pooled to obtain a better estimate of the error term, namely,

$$\frac{(a-1)(b-1)\text{MS}_{AB} + ab(n-1)\text{MS}_E}{(a-1)(b-1) + ab(n-1)} = \frac{\text{SS}_{AB} + \text{SS}_E}{abn - a - b + 1}.$$

The reason put forth is that when no interactions exist, $E(\text{MS}_{AB}) = \sigma_e^2$ gives the same expectation as for MS_E; so that the new estimator of σ_e^2 would have a large number of degrees of freedom associated with it. However, this practice is not always recommended since a nonsignificant F value does not mean that the hypothesis is true. In other words, there is always the possibility of some interaction being present and not showing up in the F test. Hence, the best estimate of σ_e^2 is always to be taken as MS_E, unless the experimenter has additional information confirming the nonexistence of interaction terms. Moreover, the pooling procedure affects both the level of significance and the power of the tests for factor A and factor B effects, in ways that are not yet fully explored. It is generally recommended that the pooling should not be undertaken unless:

(a) the degrees of freedom associated with MS_E are too small; and
(b) the calculated value of the test statistic $\text{MS}_{AB}/\text{MS}_E$ falls well below the critical value. Some authors recommend that MS_{AB} should be nearly equal to MS_E.

Part (a) of this rule is intended to limit pooling to cases where the gains may indeed be important, and part (b) is meant to ensure that in fact there are no interactions. For some general rules of thumb for deciding when to pool see Paull (1950), Bozivich et al. (1956), Srivastava and Bozivich (1962), and Mead et al. (1975).

(ii) It may be of interest to derive the significance level associated with the experiment as a whole. Let α_1, α_2, and α_3 be the significance levels of the F ratios F_A, F_B, and F_{AB}, respectively, and let α be the overall significance comprising all three tests. Then it can be shown that (Kimball (1951)) $\alpha < 1 - (1 - \alpha_1)(1 - \alpha_2)(1 - \alpha_3)$. For example, if $\alpha_1 = \alpha_2 = \alpha_3 = 0.05$, then $\alpha < 1 - (1 - 0.05)^3 = 0.143$. Similarly, if $\alpha_1 = \alpha_2 = \alpha_3 = 0.01$, then $\alpha < 0.030$.

MODEL II (RANDOM EFFECTS)

In Model II, as in Model I, the usual hypotheses of interest are about AB interaction, factor B effects, and factor A effects.

Test for AB Interactions

The presence for interaction terms is tested by the hypothesis

$$H_0^{AB} : \sigma_{\alpha\beta}^2 = 0$$
$$\text{versus} \qquad (4.6.6)$$
$$H_1^{AB} : \sigma_{\alpha\beta}^2 > 0.$$

To obtain an appropriate test procedure for the hypothesis (4.6.6), we note from (4.4.18) and (4.4.43) that under H_0^{AB},

$$E(\text{MS}_E) = \sigma_e^2,$$
$$E(\text{MS}_{AB}) = \sigma_e^2;$$

and under H_1^{AB},

$$E(\text{MS}_{AB}) > E(\text{MS}_E).$$

Two-Way Crossed Classification with Interaction

Furthermore, it follows from (4.5.8) and (4.5.9) that under H_0^{AB}

$$F_{AB} = \frac{\text{MS}_{AB}/\sigma_e^2}{\text{MS}_E/\sigma_e^2} = \frac{\text{MS}_{AB}}{\text{MS}_E} \sim F[(a-1)(b-1), ab(n-1)]. \quad (4.6.7)$$

Thus, the statistic (4.6.7) provides a suitable test procedure for (4.6.6); H_0^{AB} being rejected if

$$F_{AB} > F[(a-1)(b-1), ab(n-1); 1-\alpha].$$

Test for Factor B Effects
The presence for factor B effects is tested by the hypothesis

$$H_0^B : \sigma_\beta^2 = 0$$
$$\text{versus} \quad (4.6.8)$$
$$H_1^B : \sigma_\beta^2 > 0.$$

Again, we note from (4.4.43) and (4.4.49) that under H_0^B,

$$E(\text{MS}_{AB}) = \sigma_e^2 + n\sigma_{\alpha\beta}^2,$$
$$E(\text{MS}_B) = \sigma_e^2 + n\sigma_{\alpha\beta}^2;$$

and under H_1^B,

$$E(\text{MS}_B) > E(\text{MS}_{AB}).$$

Furthermore, it follows from (4.5.9) and (4.5.10) that under H_0^B,

$$F_B = \frac{\text{MS}_B/(\sigma_e^2 + n\sigma_{\alpha\beta}^2)}{\text{MS}_{AB}/(\sigma_e^2 + n\sigma_{\alpha\beta}^2)} = \frac{\text{MS}_B}{\text{MS}_{AB}} \sim F[b-1, (a-1)(b-1)]. \quad (4.6.9)$$

Thus, the statistic (4.6.9) provides a suitable test procedure for (4.6.8); H_0^B being rejected if

$$F_B > F[b-1, (a-1)(b-1); 1-\alpha].$$

Test for Factor A Effects
The presence for factor A effects is tested by the hypothesis

$$H_0^A : \sigma_\alpha^2 = 0$$
$$\text{versus} \quad (4.6.10)$$
$$H_1^A : \sigma_\alpha^2 > 0.$$

Proceeding as in the test for factor B effects, it readily follows that the statistic

$$F_A = \frac{\text{MS}_A/(\sigma_e^2 + n\sigma_{\alpha\beta}^2)}{\text{MS}_{AB}/(\sigma_e^2 + n\sigma_{\alpha\beta}^2)} = \frac{\text{MS}_A}{\text{MS}_{AB}} \sim F[a-1, (a-1)(b-1)] \quad (4.6.11)$$

provides a suitable test procedure for (4.6.10); H_0^A being rejected if

$$F_A > F[a-1, (a-1)(b-1); 1-\alpha].$$

Remarks: (i) The more general hypotheses of interest may be

$$\frac{\sigma_{\alpha\beta}^2}{\sigma_e^2} \leq \rho_1, \quad \frac{\sigma_\beta^2}{\sigma_e^2 + n\sigma_{\alpha\beta}^2} \leq \rho_2, \quad \frac{\sigma_\alpha^2}{\sigma_e^2 + n\sigma_{\alpha\beta}^2} \leq \rho_3$$

which are tested in the obvious way.

(ii) One of the most important differences between the tests of hypotheses in Models I and II being that when a factor has a random effect, the main effect is tested by using the interaction mean square in the denominator, whereas if it has a fixed effect then one must divide it by the error mean square.

MODEL III (MIXED EFFECTS)

Under Model III, the hypotheses of interest are: $\sigma_{\alpha\beta}^2 = 0$, $\sigma_\beta^2 = 0$, and α_i's $= 0$. The appropriate tests are obtained in the same way as in the case of Models I and II.

Test for AB Interactions

The hypothesis $H_0^{AB} : \sigma_{\alpha\beta}^2 = 0$ versus $H_1^{AB} : \sigma_{\alpha\beta}^2 > 0$ may be tested by the ratio $\text{MS}_{AB}/\text{MS}_E$, which under H_0^{AB} has an F distribution with $(a-1)(b-1)$ and $ab(n-1)$ degrees of freedom. Similarly, the hypothesis $\sigma_{\alpha\beta}^2/\sigma_e^2 \leq \rho_1$ can be tested in the obvious way by using the statistic $(1 + \sigma_{\alpha\beta}^2/\sigma_e^2)^{-1}(\text{MS}_{AB}/\text{MS}_E)$.

Test for Factor B Effects

The hypothesis $H_0^B : \sigma_\beta^2 = 0$ versus $H_1^B : \sigma_\beta^2 > 0$ may be tested by the ratio MS_B/MS_E, which under H_0^B has an F distribution with $b-1$ and $ab(n-1)$ degrees of freedom. Similarly, the hypothesis $\sigma_\beta^2/\sigma_e^2 \leq \rho_2$ may be tested in the obvious way by the test statistic $(1 + an\sigma_\beta^2/\sigma_e^2)^{-1}(\text{MS}_B/\text{MS}_E)$.

Test for Factor A Effects

The hypothesis H_0^A : all $\alpha_i = 0$ versus H_1^A : not all $\alpha_i = 0$ may be tested by the ratio $\text{MS}_A/\text{MS}_{AB}$, which under H_0^A has an F distribution with $a-1$ and $(a-1)(b-1)$ degrees of freedom.

Remarks: (i) The tests for the main effects under Model III work conversely to the tests under Models I and II. Here, the test statistic for factor B with random effects is obtained by dividing MS_B by MS_E, whereas if this factor had occurred under Model II, then the statistic would be obtained by dividing MS_B by MS_{AB}. Similarly, the factor A with fixed effects is tested by dividing MS_A by MS_{AB}, whereas if the factor had occurred under Model I, then the test would have been made by dividing MS_A by MS_E. These results on tests of hypotheses were first developed by Johnson (1948).

(ii) If interaction terms are nonsignificant, one may want to test the hypothesis H_o^A: all α_i's $= 0$ by the statistic MS_A/MS_E which may provide more degrees of freedom for the denominator. The possibility of pooling MS_{AB} and MS_E may also be considered if degrees of freedom are few. The earlier comments on pooling also apply here.

SUMMARY OF MODELS AND TESTS

The appropriate test statistics for Models I, II, and III developed in this section are summarized in Table 4.3.

TABLE 4.3
Test Statistics for Models I, II, and III

Hypothesized Effect	Model I (A and B Fixed)	Model II (A and B Random)	Model III (A Fixed, B Random)
Factor A	MS_A/MS_E	MS_A/MS_{AB}	MS_A/MS_{AB}
Factor B	MS_B/MS_E	MS_B/MS_{AB}	MS_B/MS_E
Interaction AB	MS_{AB}/MS_E	MS_{AB}/MS_E	MS_{AB}/MS_E

4.7 POINT ESTIMATION

In this section, we present results on point estimation for parameters of interest under fixed, random, and mixed effects models.

MODEL I (FIXED EFFECTS)

The least squares estimators[4] of the parameters μ, α_i's, β_j's, and $(\alpha\beta)_{ij}$'s for the model (4.1.1) are obtained by minimizing

$$Q = \sum_{i=1}^{a}\sum_{j=1}^{b}\sum_{k=1}^{n}[y_{ijk} - \mu - \alpha_i - \beta_j - (\alpha\beta)_{ij}]^2, \qquad (4.7.1)$$

---

[4] The least squares estimators in this case are the same as those obtained by the maximum likelihood method under the assumption of normality.

with respect to μ, α_i, β_j, and $(\alpha\beta)_{ij}$; and subject to the restrictions:

$$\sum_{i=1}^{a} \alpha_i = \sum_{j=1}^{b} \beta_j = \sum_{i=1}^{a} (\alpha\beta)_{ij} = \sum_{j=1}^{b} (\alpha\beta)_{ij} = 0. \tag{4.7.2}$$

When one performs this minimization, it can be shown by the method of elementary calculus that the following least squares estimators are obtained:

$$\hat{\mu} = \bar{y}_{...}, \tag{4.7.3}$$

$$\hat{\alpha}_i = \bar{y}_{i..} - \bar{y}_{...}, \quad i = 1, 2, \ldots, a, \tag{4.7.4}$$

$$\hat{\beta}_j = \bar{y}_{.j.} - \bar{y}_{...}, \quad j = 1, 2, \ldots, b, \tag{4.7.5}$$

and

$$\widehat{(\alpha\beta)}_{ij} = \bar{y}_{ij.} - \bar{y}_{i..} - \bar{y}_{.j.} + \bar{y}_{...} \quad (i = 1, 2, \ldots, a; j = 1, 2, \ldots, b). \tag{4.7.6}$$

These are the so-called best linear unbiased estimators (BLUE). The variances of estimators (4.7.3) through (4.7.6) are:

$$\text{Var}(\hat{\mu}) = \sigma_e^2/abn, \tag{4.7.7}$$

$$\text{Var}(\hat{\alpha}_i) = (a-1)\sigma_e^2/abn, \tag{4.7.8}$$

$$\text{Var}(\hat{\beta}_j) = (b-1)\sigma_e^2/abn, \tag{4.7.9}$$

and

$$\text{Var}(\widehat{(\alpha\beta)}_{ij}) = (a-1)(b-1)\sigma_e^2/abn. \tag{4.7.10}$$

The other parameters of interest may include $\mu + \alpha_i$ (mean levels of factor A), $\mu + \beta_j$ (mean levels of factor B), $\mu + \alpha_i + \beta_j + (\alpha\beta)_{ij}$ (the cell means), pairwise differences $\alpha_i - \alpha_{i'}$, $\beta_j - \beta_{j'}$, and the contrasts

$$\sum_{i=1}^{a} \ell_i \alpha_i \left(\sum_{i=1}^{a} \ell_i = 0\right), \quad \sum_{j=1}^{b} \ell'_j \beta_j \left(\sum_{j=1}^{b} \ell'_j = 0\right), \quad \text{and}$$

$$\sum_{i=1}^{a} \sum_{j=1}^{b} \ell_{ij} (\alpha\beta)_{ij} \left(\sum_{i=1}^{a} \sum_{j=1}^{b} \ell_{ij} = 0\right).$$

Their respective estimates together with variances are given by

$$\widehat{\mu + \alpha_i} = \bar{y}_{i..}, \quad \text{Var}\left(\widehat{\mu + \alpha_i}\right) = \sigma_e^2/bn; \tag{4.7.11}$$

$$\widehat{\mu + \beta_i} = \bar{y}_{.j.}, \quad \text{Var}\left(\widehat{\mu + \beta_i}\right) = \sigma_e^2/an; \tag{4.7.12}$$

$$\mu + \alpha_i + \widehat{\beta_j + (\alpha\beta)_{ij}} = \bar{y}_{ij\cdot},$$
$$\text{Var}\left(\mu + \alpha_i + \widehat{\beta_j + (\alpha\beta)_{ij}}\right) = \sigma_e^2/n; \quad (4.7.13)$$

$$\widehat{\alpha_i - \alpha_{i'}} = \bar{y}_{i\cdot\cdot} - \bar{y}_{i'\cdot\cdot}, \quad \text{Var}\left(\widehat{\alpha_i - \alpha_{i'}}\right) = 2\sigma_e^2/bn; \quad (4.7.14)$$

$$\widehat{\beta_j - \beta_{j'}} = \bar{y}_{\cdot j\cdot} - \bar{y}_{\cdot j'\cdot}, \quad \text{Var}\left(\widehat{\beta_j - \beta_{j'}}\right) = 2\sigma_e^2/an; \quad (4.7.15)$$

$$\sum_{i=1}^{a} \widehat{\ell_i \alpha_i} = \sum_{i=1}^{a} \ell_i \bar{y}_{i\cdot\cdot}, \quad \text{Var}\left(\sum_{i=1}^{a} \widehat{\ell_i \alpha_i}\right) = \left(\sum_{i=1}^{a} \ell_i^2\right) \sigma_e^2/bn; \quad (4.7.16)$$

$$\sum_{j=1}^{b} \widehat{\ell'_j \beta_j} = \sum_{j=1}^{b} \ell'_j \bar{y}_{\cdot j\cdot}, \quad \text{Var}\left(\sum_{j=1}^{b} \widehat{\ell'_j \beta_j}\right) = \left(\sum_{j=1}^{b} \ell_j'^2\right) \sigma_e^2/an; \quad (4.7.17)$$

and

$$\sum_{i=1}^{a}\sum_{j=1}^{b} \widehat{\ell_{ij}(\alpha\beta)_{ij}} = \sum_{i=1}^{a}\sum_{j=1}^{b} \ell_{ij}(\bar{y}_{ij\cdot} - \bar{y}_{i\cdot\cdot} - \bar{y}_{\cdot j\cdot} + \bar{y}_{\cdot\cdot\cdot}); \quad (4.7.18)$$

$$\text{Var}\left(\sum_{i=1}^{a}\sum_{j=1}^{b} \widehat{\ell_{ij}(\alpha\beta)_{ij}}\right) = \left(\sum_{i=1}^{a}\sum_{j=1}^{b} \ell_{ij}^2 - \frac{1}{b}\sum_{i=1}^{a} \ell_{i\cdot}^2 - \frac{1}{a}\sum_{j=1}^{b} \ell_{\cdot j}^2\right) \sigma_e^2/n \quad (4.7.19)$$

The best quadratic unbiased estimator of σ_e^2 is, of course, provided by the error mean square.

MODEL II (RANDOM EFFECTS)

In the case of the random effects model, for random factors (that have significant effects), one would often like to estimate the magnitude of the variance components. As before, unbiased point estimators can be readily obtained by using linear combinations of the expected mean squares in the analysis of variance Table 4.2. For instance, σ_α^2 can be estimated by noting that

$$E(\text{MS}_A) - E(\text{MS}_{AB}) = bn\sigma_\alpha^2.$$

Hence, an unbiased estimator of σ_α^2 is given by

$$\hat{\sigma}_\alpha^2 = \frac{\text{MS}_A - \text{MS}_{AB}}{bn}. \quad (4.7.20)$$

Similarly,

$$\hat{\sigma}_\beta^2 = \frac{\text{MS}_B - \text{MS}_{AB}}{an}, \quad (4.7.21)$$

$$\hat{\sigma}_{\alpha\beta}^2 = \frac{\text{MS}_{AB} - \text{MS}_E}{n}, \quad (4.7.22)$$

and

$$\hat{\sigma}_e^2 = MS_E. \tag{4.7.23}$$

As usual, the parameter μ is, of course, estimated by $\hat{\mu} = \bar{y}_{...}$. The estimators (4.7.20) through (4.7.23) are the so-called minimum variance quadratic unbiased estimators or the minimum variance unbiased estimators under the assumption of normality. They may, however, produce negative estimates.[5] If one can assume the lack of interaction terms, it is possible to use MS_E in place of MS_{AB} in the estimators (4.7.20) and (4.7.21). Alternatively, one can pool MS_{AB} and MS_E and use this pooled estimator for MS_{AB} in the expressions for σ_α^2 and σ_β^2.

MODEL III (MIXED EFFECTS)

In the case of a mixed effects model with A fixed and B random, the parameters to be estimated are μ, α_i's, $\sigma_\beta^2, \sigma_{\alpha\beta}^2$, and σ_e^2. From the expected mean squares column under Model III of Table 4.2, the $\sigma_\beta^2, \sigma_{\alpha\beta}^2$, and σ_e^2 are estimated unbiasedly by

$$\hat{\sigma}_\beta^2 = \frac{MS_B - MS_E}{an}, \tag{4.7.24}$$

$$\hat{\sigma}_{\alpha\beta}^2 = \frac{MS_{AB} - MS_E}{n}, \tag{4.7.25}$$

and

$$\hat{\sigma}_e^2 = MS_E. \tag{4.7.26}$$

Thus, althuogh $\sigma_{\alpha\beta}^2$ and σ_e^2 have the same estimates as in the case of the random effects model, the estimate of σ_β^2 is different. This general approach of estimating variance components can be used in any mixed model. After deleting the mean squares containing fixed factors, the remaining set of equations can be solved for variance components.[6]

The fixed effects μ, $\mu + \alpha_i$, and α_i's are, of course, estimated by

$$\hat{\mu} = \bar{y}_{...},$$
$$\widehat{\mu + \alpha_i} = \bar{y}_{i..}, \quad i = 1, 2, \ldots, a,$$
$$\hat{\alpha}_i = \bar{y}_{i..} - \bar{y}_{...}, \quad i = 1, 2, \ldots, a,$$

---

[5] For a discussion of the nonnegative maximum likelihood estimation, see Herbach (1959) and Miller (1977).
[6] For a discussion of the maximum likelihood estimation in the mixed model, see Szatrowski and Miller (1980).

which are the same estimates as in the case of a fixed effects model. Furthermore, comparisons involving pairwise contrasts can be estimated by

$$\widehat{\alpha_i - \alpha_{i'}} = \bar{y}_{i..} - \bar{y}_{i'..}.$$

To evaluate the variances of the means and contrasts, we note that

$$\text{Var}(\bar{y}_{...}) = \frac{\sigma_e^2 + n\frac{(a-1)}{a}\sigma_{\alpha\beta}^2 + an\sigma_\beta^2}{abn},$$

$$\text{Var}(\bar{y}_{i..}) = \frac{\sigma_e^2 + n\frac{(a-1)}{a}\sigma_{\alpha\beta}^2 + n\sigma_\beta^2}{bn}, \quad (4.7.27)$$

$$\text{Var}(\bar{y}_{ij.}) = \frac{\sigma_e^2 + n\frac{(a-1)}{a}\sigma_{\alpha\beta}^2 + n\sigma_\beta^2}{n},$$

$$\text{Cov}(\bar{y}_{i..}, \bar{y}_{i'..}) = -\frac{\sigma_{\alpha\beta}^2}{ab} \quad (4.7.28)$$

and

$$\text{Var}(\bar{y}_{i..} - \bar{y}_{i'..}) = \frac{2(\sigma_e^2 + n\sigma_{\alpha\beta}^2)}{bn}. \quad (4.7.29)$$

4.8 INTERVAL ESTIMATION

In this section, we present results on confidence intervals for parameters of interest under fixed, random, and mixed effects models.

MODEL I (FIXED EFFECTS)

An exact confidence interval for σ_e^2 can be based on the chi-square distribution of $ab(n-1)\text{MS}_E/\sigma_e^2$. Thus, a $1 - \alpha$ level confidence interval for σ_e^2 is

$$\frac{ab(n-1)\text{MS}_E}{\chi^2[ab(n-1), 1-\alpha/2]} < \sigma_e^2 < \frac{ab(n-1)\text{MS}_E}{\chi^2[ab(n-1), \alpha/2]}. \quad (4.8.1)$$

Furthermore, it is possible to obtain confidence intervals based on the t distribution for a particular α_i or a particular difference $\alpha_i - \alpha_{i'}$. For example,

$$(\bar{y}_{i..} - \bar{y}_{...}) - t[ab(n-1), 1-\alpha/2]\sqrt{(a-1)\text{MS}_E/abn}$$
$$< \alpha_i < (\bar{y}_{i..} - \bar{y}_{...}) + t[ab(n-1), 1-\alpha/2]\sqrt{(a-1)\text{MS}_E/abn}, \quad (4.8.2)$$

defines a $1 - \alpha$ level confidence interval for α_i.

Similarly, to obtain confidence limits for $\alpha_i - \alpha_{i'}$, we note from (4.7.14) that

$$E(\bar{y}_{i..} - \bar{y}_{i'..}) = \alpha_i - \alpha_{i'}$$

and

$$\text{Var}(\bar{y}_{i..} - \bar{y}_{i'..}) = 2\sigma_e^2/bn.$$

The confidence limits can, therefore, be derived from the relation:

$$\frac{(\bar{y}_{i..} - \bar{y}_{i'..}) - (\alpha_i - \alpha_{i'})}{\sqrt{2\text{MS}_E/bn}} \sim t[ab(n-1)]. \qquad (4.8.3)$$

Similar results on confidence intervals for the β_j's, $(\alpha\beta)_{ij}$'s, and any pairwise differences on them, using the t distribution, can also be obtained. However, multiple comparison methods, discussed in Section 4.12, are usually preferable.

MODEL II (RANDOM EFFECTS)

An exact confidence interval for the error variance component σ_e^2 is obtained as in (4.8.1). However, as indicated in Section 2.10 for the one-way random model, exact confidence intervals for the variance components $\sigma_{\alpha\beta}^2, \sigma_\beta^2$, and σ_α^2 do not exist. One can, nevertheless, obtain exact confidence intervals for $\sigma_e^2 + n\sigma_{\alpha\beta}^2$, $\sigma_e^2 + n\sigma_{\alpha\beta}^2 + bn\sigma_\alpha^2$, $\sigma_e^2 + n\sigma_{\alpha\beta}^2 + an\sigma_\beta^2$; and ratios of particular combinations of variance components, for example, $\sigma_{\alpha\beta}^2/\sigma_e^2$, $\sigma_\beta^2/(\sigma_e^2 + n\sigma_{\alpha\beta}^2)$, $\sigma_\alpha^2/(\sigma_e^2 + n\sigma_{\alpha\beta}^2)$, and $(\sigma_e^2 + n\sigma_{\alpha\beta}^2)/(\sigma_e^2 + n\sigma_{\alpha\beta}^2 + bn\sigma_\alpha^2)$, by taking the appropriate ratios of mean squares as discussed in Section 2.10. For a discussion of approximate confidence intervals for the variance components $\sigma_{\alpha\beta}^2, \sigma_\beta^2, \sigma_\alpha^2$; for the ratios of variance components $\sigma_\beta^2/\sigma_e^2, \sigma_\alpha^2/\sigma_e^2, \sigma_\alpha^2/\sigma_\beta^2$; and the proportions of variability $\sigma_e^2/(\sigma_e^2 + \sigma_{\alpha\beta}^2 + \sigma_\beta^2 + \sigma_\alpha^2)$, $\sigma_{\alpha\beta}^2/(\sigma_e^2 + \sigma_{\alpha\beta}^2 + \sigma_\beta^2 + \sigma_\alpha^2)$, $\sigma_\beta^2/(\sigma_e^2 + \sigma_{\alpha\beta}^2 + \sigma_\beta^2 + \sigma_\alpha^2)$, and $\sigma_\alpha^2/(\sigma_e^2 + \sigma_{\alpha\beta}^2 + \sigma_\beta^2 + \sigma_\alpha^2)$, including numerical examples, see Burdick and Graybill (1992, pp. 121–124).

To obtain confidence limits for μ, we note from (4.4.7) that

$$E(\bar{y}_{...}) = \mu \qquad (4.8.4)$$

and

$$\text{Var}(\bar{y}_{...}) = \frac{\sigma_e^2 + n\sigma_{\alpha\beta}^2 + an\sigma_\beta^2 + bn\sigma_\alpha^2}{abn}. \qquad (4.8.5)$$

In order to get a mean square with expected value equal to the numerator in

(4.8.5), we use the linear combination $MS_A + MS_B - MS_{AB}$ since it has the expected value given by

$$E(MS_A + MS_B - MS_{AB}) = \sigma_e^2 + n\sigma_{\alpha\beta}^2 + an\sigma_\beta^2 + bn\sigma_\alpha^2. \quad (4.8.6)$$

It can now be shown using the Satterthwaite procedure (see Appendix K) that

$$\frac{\bar{y}_{...} - \mu}{\sqrt{(MS_A + MS_B - MS_{AB})/abn}} \sim \text{approx. } t[\nu], \quad (4.8.7)$$

where

$$\nu = \frac{[MS_A + MS_B - MS_{AB}]^2}{\frac{(MS_A)^2}{a-1} + \frac{(MS_B)^2}{b-1} + \frac{(MS_{AB})^2}{(a-1)(b-1)}}. \quad (4.8.8)$$

The confidence limits for μ can now be determined in the usual way, but these will be imprecise because of the approximation involved in (4.8.7).

MODEL III (MIXED EFFECTS)

An exact confidence interval for σ_e^2 is constructed as in (4.8.1). However, as in Model II, exact confidence intervals for σ_β^2 and $\sigma_{\alpha\beta}^2$ do not exist. One can, nevertheless, obtain exact intervals for $\sigma_{\alpha\beta}^2/\sigma_e^2$ and $\sigma_\beta^2/\sigma_e^2$ by basing the procedure on the statistics MS_{AB}/MS_E and MS_B/MS_E, respectively. Approximate confidence intervals for σ_β^2 and $\sigma_{\alpha\beta}^2$ can be constructed by the method of Satterthwaite and other related procedures (see, e.g., Burdick and Graybill, 1992, p. 153)). Also, as in the case of Model I, it is possible to obtain confidence intervals for μ, α_i, $\mu + \alpha_i$, $\alpha_i - \alpha_{i'}$, or the contrast $\sum_{i=1}^a \ell_i \alpha_i$ $(\sum_{i=1}^a \ell_i = 0)$. For example, an exact confidence interval for $\sum_{i=1}^a \ell_i \alpha_i$, with coefficient $1 - \alpha$, is given by

$$\sum_{i=1}^a \ell_i \bar{y}_{i..} - t[(a-1)(b-1), 1-\alpha/2]\sqrt{MS_{AB} \sum_{i=1}^a \ell_i^2/bn} < \sum_{i=1}^a \ell_i \alpha_i$$

$$< \sum_{i=1}^a \ell_i \bar{y}_{i..} + t[(a-1)(b-1), 1-\alpha/2]\sqrt{MS_{AB} \sum_{i=1}^a \ell_i^2/bn}. \quad (4.8.9)$$

Thus, when dealing with a mixed model, the appropriate mean square to be used in the estimated variance formula is no longer MS_E. A simple rule to determine the appropriate mean square is: use the mean square employed in the denominator of the test statistic for testing the presence of the fixed factor under consideration. For instance, with the mixed model (4.1.1) where A is fixed and B random, MS_{AB} is the appropriate mean square (see Table 4.3). The degrees of freedom in constructing the confidence interval are those associated with the mean square utilized for estimating the variance of the contrast. However, it is not always possible to obtain an appropriate mean square for the desired variance estimate. For example, in order to estimate $\mu + \alpha_i$, we notice from (4.7.27) and Table 4.2 that there is no appropriate mean square to estimate the desired variance. An unbiased estimate of $\text{Var}(\bar{y}_{i..})$, however, can be obtained by using an appropriate linear combination of the mean squares. An approximate $1 - \alpha$ level confidence interval for $\mu + \alpha_i$ can be constructed using

$$\bar{y}_{i..} \pm t[\nu, 1 - \alpha/2] \sqrt{\frac{\hat{\sigma}_e^2 + \frac{n(a-1)}{a}\hat{\sigma}_{\alpha\beta}^2 + n\hat{\sigma}_\beta^2}{bn}},$$

where the degrees of freedom ν will be estimated using Satterthwaite procedure similar to equation (4.8.8).

4.9 COMPUTATIONAL FORMULAE AND PROCEDURE

As in Section 3.9, we use the following computational formulae, which are identical to the definitional formulae given in (4.3.1):

$$SS_T = \sum_{i=1}^{a}\sum_{j=1}^{b}\sum_{k=1}^{n} y_{ijk}^2 - \frac{y_{...}^2}{abn},$$

$$SS_A = \frac{1}{bn}\sum_{i=1}^{a} y_{i..}^2 - \frac{y_{...}^2}{abn},$$

$$SS_B = \frac{1}{an}\sum_{j=1}^{b} y_{.j.}^2 - \frac{y_{...}^2}{abn},$$

$$SS_{AB} = \frac{1}{n}\sum_{i=1}^{a}\sum_{j=1}^{b} y_{ij.}^2 - \frac{1}{bn}\sum_{i=1}^{a} y_{i..}^2 - \frac{1}{an}\sum_{j=1}^{b} y_{.j.}^2 + \frac{y_{...}^2}{abn},$$

and

$$SS_E = \sum_{i=1}^{a}\sum_{j=1}^{b}\sum_{k=1}^{n} y_{ijk}^2 - \frac{1}{n}\sum_{i=1}^{a}\sum_{j=1}^{b} y_{ij.}^2,$$

Two-Way Crossed Classification with Interaction

where as before a dot in the subscript indicates the total over the variable represented by the index. Ordinarily, the interaction sum of squares is obtained by the relation

$$SS_{AB} = SS_T - SS_A - SS_B - SS_E$$

or

$$SS_{AB} = SS_{TC} - SS_A - SS_B,$$

where

$$SS_{TC} = \frac{1}{n} \sum_{i=1}^{a} \sum_{j=1}^{b} y_{ij.}^2 - \frac{y_{...}^2}{abn}.$$

The computational procedure for the sums of squares can thus be performed in a systematic manner by the following sequence of steps:

(i) Compute the cell totals: $y_{11.}, y_{12.}, \ldots, y_{ab.}$
(ii) Compute the row totals: $y_{1..}, y_{2..}, \ldots, y_{a..}$
(iii) Compute the column totals: $y_{.1.}, y_{.2.}, \ldots, y_{.b.}$
(iv) Compute the overall or grand total:

$$y_{...} = \sum_{i=1}^{a} y_{i..} = \sum_{j=1}^{b} y_{.j.} = \sum_{i=1}^{a} \sum_{j=1}^{b} y_{ij.}.$$

(v) Compute the raw sum of squares:

$$\sum_{i=1}^{a} \sum_{j=1}^{b} \sum_{k=1}^{n} y_{ijk}^2 = y_{111}^2 + y_{112}^2 + \cdots + y_{abn}^2.$$

(vi) Compute the correction factor:

$$\frac{y_{...}^2}{abn}.$$

(vii) Compute

$$\frac{1}{bn} \sum_{i=1}^{a} y_{i..}^2.$$

(viii) Compute
$$\frac{1}{an}\sum_{j=1}^{a} y_{.j.}^2.$$

(ix) Compute
$$\frac{1}{n}\sum_{i=1}^{a}\sum_{j=1}^{b} y_{ij.}^2.$$

(x) Compute
$$SS_T = \sum_{i=1}^{a}\sum_{j=1}^{b}\sum_{k=1}^{n} y_{ijk}^2 - \frac{y_{...}^2}{abn}.$$

(xi) Compute
$$SS_{TC} = \frac{1}{n}\sum_{i=1}^{a}\sum_{j=1}^{b} y_{ij.}^2 - \frac{y_{...}^2}{abn}.$$

(xii) Compute
$$SS_A = \frac{1}{bn}\sum_{i=1}^{a} y_{i..}^2 - \frac{y_{...}^2}{abn}.$$

(xiii) Compute
$$SS_B = \frac{1}{an}\sum_{j=1}^{b} y_{.j.}^2 - \frac{y_{...}^2}{abn}.$$

(xiv) Compute $SS_{AB} = SS_T - SS_A - SS_B - SS_E = SS_{TC} - SS_A - SS_B$.
(xv) Compute $SS_E = SS_T - SS_{TC}$.

4.10 ANALYSIS OF VARIANCE WITH UNEQUAL SAMPLE SIZES PER CELL

In the two-way classification model discussed in this chapter, it has been assumed that there are the same number of replications of the experiment in each cell of the two-way Table 4.1. If this is not true, it is not possible to partition the overall sum of squares into independent components due to main effects and interaction terms. Care should always be taken to ensure that the number of observations in each cell is constant; but, even with the utmost care, it may happen for a variety of reasons, such as loss of subjects, incomplete records, and the like, that an experiment terminates with unequal sample sizes per cell. Moreover, unequal sample sizes are fairly common with many survey-type data. For example, an agricultural analyst may wish to study the effect of temperature and precipitation on production of certain agricultural crops from data for certain counties in the country. In this type of uncontrolled study, it may easily happen that the number of counties in the various temperature-precipitation categories are not equal.

The model in this case remains the same, except that the sample size corresponding to the i-th level of factor A and the j-th level of factor B is now denoted by n_{ij}. The data layout for a general two-way crossed classification with unequal subclass numbers is displayed in Table 4.4. Now, the total number of observations for the i-th level of factor A is

$$n_{i.} = \sum_{j=1}^{b} n_{ij},$$

for the j-th level of factor B is

$$n_{.j} = \sum_{i=1}^{b} n_{ij},$$

and the total number of observations is

$$N = \sum_{i=1}^{a} n_{i.} = \sum_{j=1}^{b} n_{.j} = \sum_{i=1}^{a}\sum_{j=1}^{b} n_{ij}.$$

When only a few values are missing, one could replace a missing value by the mean for that cell. The standard analysis of variance can then be performed except that for each missing value being estimated, the error degrees of freedom are reduced by one. If there is a wide disparity between the numbers of observations in different cells, one can no longer use the standard analysis of variance described earlier in this chapter and must resort to some other procedures. When the sample sizes for each cell are unequal, the two-way analysis of variance for factor effects becomes complex. The component sums of squares in the analysis of variance are no longer orthogonal; that is, they do not sum to the total sum of squares. The least squares method for obtaining the best estimates of the parameters is rather complicated in the fixed effects model and the best analysis has not been and probably will not be found for the random effects models. In the following, we consider some common methods of analysis of variance for unequal sample size data. For a concise and readable account of the nonorthogonal two-way analysis of variance, see Herr and Gaebelin (1978).

TABLE 4.4
Data for a Two-Way Crossed Classification with n_{ij} Replications per Cell

	Factor B					
	B_1	B_2	B_3	$\ldots$	B_j	$\ldots$	B_b
A_1	$y_{111}, y_{112}, \ldots, y_{11n_{11}}$	$y_{121}, y_{122}, \ldots, y_{12n_{12}}$	$y_{131}, y_{132}, \ldots, y_{13n_{13}}$	$\ldots$	$y_{1j1}, y_{1j2}, \ldots, y_{1jn_{1j}}$	$\ldots$	$y_{1b1}, y_{1b2}, \ldots, y_{1bn_{1b}}$
A_2	$y_{211}, y_{212}, \ldots, y_{21n_{21}}$	$y_{221}, y_{222}, \ldots, y_{22n_{22}}$	$y_{231}, y_{232}, \ldots, y_{23n_{23}}$	$\ldots$	$y_{2j1}, y_{2j2}, \ldots, y_{2jn_{2j}}$	$\ldots$	$y_{2b1}, y_{2b2}, \ldots, y_{2bn_{2b}}$
A_3	$y_{311}, y_{312}, \ldots, y_{31n_{31}}$	$y_{321}, y_{322}, \ldots, y_{32n_{32}}$	$y_{331}, y_{332}, \ldots, y_{33n_{33}}$	$\ldots$	$y_{3j1}, y_{3j2}, \ldots, y_{3jn_{3j}}$	$\ldots$	$y_{3b1}, y_{3b2}, \ldots, y_{3bn_{3b}}$
Factor A $\vdots$							
A_i	$y_{i11}, y_{i12}, \ldots, y_{i1n_{i1}}$	$y_{i21}, y_{i22}, \ldots, y_{i2n_{i2}}$	$y_{i31}, y_{i32}, \ldots, y_{i3n_{i3}}$	$\ldots$	$y_{ij1}, y_{ij2}, \ldots, y_{ijn_{ij}}$	$\ldots$	$y_{ib1}, y_{ib2}, \ldots, y_{ibn_{ib}}$
$\vdots$							
A_a	$y_{a11}, y_{a12}, \ldots, y_{a1n_{a1}}$	$y_{a21}, y_{a22}, \ldots, y_{a2n_{a2}}$	$y_{a31}, y_{a32}, \ldots, y_{a3n_{a3}}$	$\ldots$	$y_{aj1}, y_{aj2}, \ldots, y_{ajn_{aj}}$	$\ldots$	$y_{ab1}, y_{ab2}, \ldots, y_{abn_{ab}}$

Two-Way Crossed Classification with Interaction

Further details on nonorthogonal two-way analysis of variance models can be found in a special issue of *Communications in Statistics: Part A, Theory and Methods* (Vol. 9, No. 2, 1980).

FIXED EFFECTS ANALYSIS

We discuss two methods for fixed effects analysis; one for the case of proportional frequencies and the other for the general case of unequal frequencies.

Proportional Frequencies

Sometimes, the unequal sample sizes follow a proportional pattern;[7] that is,

$$n_{ij} = \frac{n_{i.} n_{.j}}{N}. \quad (4.10.1)$$

The relation (4.10.1) implies that the sample sizes in any of the rows or columns are proportional. This is called the case of proportional frequencies.

When the frequencies are proportional, the analysis of variance discussed earlier in this chapter can be employed with suitable modifications. For example, the definitional and computational formulae for the sums of squares are:

$$SS_T = \sum_{i=1}^{a} \sum_{j=1}^{b} \sum_{k=1}^{n_{ij}} (y_{ijk} - \bar{y}_{...})^2 = \sum_{i=1}^{a} \sum_{j=1}^{b} \sum_{k=1}^{n_{ij}} y_{ijk}^2 - \frac{y_{...}^2}{N},$$

$$SS_A = \sum_{i=1}^{a} n_{i.} (\bar{y}_{i..} - \bar{y}_{...})^2 = \sum_{i=1}^{a} \frac{y_{i..}^2}{n_{i.}} - \frac{y_{...}^2}{N},$$

$$SS_B = \sum_{j=1}^{b} n_{.j} (\bar{y}_{.j.} - \bar{y}_{...})^2 = \sum_{j=1}^{b} \frac{y_{.j.}^2}{n_{.j}} - \frac{y_{...}^2}{N},$$

$$SS_{AB} = \sum_{i=1}^{a} \sum_{j=1}^{b} n_{ij} (\bar{y}_{ij.} - \bar{y}_{i..} - \bar{y}_{.j.} + \bar{y}_{...})^2 \quad (4.10.2)$$

$$= \sum_{i=1}^{a} \sum_{j=1}^{b} \frac{y_{ij.}^2}{n_{ij}} - \sum_{i=1}^{a} \frac{y_{i..}^2}{n_{i.}} - \sum_{j=1}^{b} \frac{y_{.j.}^2}{n_{.j}} + \frac{y_{...}^2}{N},$$

and

$$SS_E = \sum_{i=1}^{a} \sum_{j=1}^{b} \sum_{k=1}^{n_{ij}} (y_{ijk} - \bar{y}_{ij.})^2,$$

$$= \sum_{i=1}^{a} \sum_{j=1}^{b} \sum_{k=1}^{n_{ij}} y_{ijk}^2 - \sum_{i=1}^{a} \sum_{j=1}^{b} \frac{y_{ij.}^2}{n_{ij}},$$

---

[7] It is not necessary to check that the number of replicates n_{ij} in each of the ab cells follows the relation (4.10.1). Only one need check one cell in each of $a - 1$ levels of factor A and one in each of $b - 1$ levels of factor B (Huck and Layne (1974)).

TABLE 4.5
Analysis of Variance for the Unbalanced Fixed Effects Model in (4.1.1) with Proportional Frequencies

Source of Variation	Degrees of Freedom	Sum of Squares	Mean Square	Expected Mean Square
Due to A	$a-1$	SS_A	MS_A	$\sigma_e^2 + \dfrac{1}{a-1}\sum_{i=1}^{a} n_{i.}\alpha_i^2$
Due to B	$b-1$	SS_B	MS_B	$\sigma_e^2 + \dfrac{1}{b-1}\sum_{j=1}^{b} n_{.j}\beta_j^2$
Interaction $A \times B$	$(a-1)(b-1)$	SS_{AB}	MS_{AB}	$\sigma_e^2 + \dfrac{1}{(a-1)(b-1)}\sum_{i=1}^{a}\sum_{j=1}^{b} n_{ij}(\alpha\beta)_{ij}^2$
Error	$N-ab$	SS_E	MS_E	σ_e^2
Total	$N-1$	SS_T		

where

$$y_{ij.} = \sum_{k=1}^{n_{ij}} y_{ijk}, \quad \bar{y}_{ij.} = y_{ij.}/n_{ij},$$

$$y_{i..} = \sum_{j=1}^{b} y_{ij.}, \quad \bar{y}_{i..} = y_{i..}/n_{i.},$$

$$y_{.j.} = \sum_{i=1}^{a} y_{ij.}, \quad \bar{y}_{.j.} = y_{.j.}/n_{.j},$$

and

$$y_{...} = \sum_{i=1}^{a}\sum_{j=1}^{b}\sum_{k=1}^{n_{ij}} y_{ijk}, \quad \bar{y}_{...} = y_{...}/N.$$

The analysis of variance including the expected mean squares is given in Table 4.5. Tests of hypotheses for main effects and interaction can be carried out as before for the case of an equal number of observations per cell. For example, for testing the interaction effects, the statistic is MS_{AB}/MS_E. Under the null hypothesis $H_0^{AB} : (\alpha\beta)_{ij} = 0$ for all i and j, this ratio has an F distribution with $(a-1)(b-1)$ and $N-ab$ degrees of freedom; and the null hypothesis is rejected for large values of this ratio. If there are no significant interaction effects, the main effect due to factor A is tested by the statistic MS_A/MS_E which, under the null hypothesis $H_0^A : \alpha_i = 0$ for all i, has an F distribution with $a-1$ and $N-ab$ degrees of freedom. Similarly, the main effect due to factor B is tested by the statistic MS_B/MS_E which, under the null hypothesis

$H_0^B : \beta_j = 0$ for all j, has an F distribution with $b - 1$ and $N - ab$ degrees of freedom.

General Case of Unequal Frequencies

If the sample sizes n_{ij}'s do not vary considerably, say, by not more than the ratio of 2 to 1, with most n_{ij} being nearly equal and no n_{ij} equal to zero, an approximate analysis of variance suggested by Yates (1934), called the method of unweighted means, may be used. This approximate method is also used in cases where the n_{ij}'s do differ considerably but the researcher desires to obtain a quick initial approximation to a more exact analysis. Note that since $\text{Var}(\bar{y}_{ij.}) = \sigma_e^2/n_{ij}$, the variances are unequal when n_{ij} is not constant.

The procedure is rather a simple one where an analysis of variance is performed using the $\bar{y}_{ij.}$'s as if there were only one observation for each (i, j)-th cell. The sums of squares for the main effects, interaction, and error are calculated in the usual way. Thus, defining $x_{ij} = \bar{y}_{ij.}$, the expressions for the sums of squares are:

$$SS_{Au} = b \sum_{i=1}^{a} (\bar{x}_{i.} - \bar{x}_{..})^2,$$

$$SS_{Bu} = a \sum_{j=1}^{b} (\bar{x}_{.j} - \bar{x}_{..})^2,$$

$$SS_{ABu} = \sum_{i=1}^{a} \sum_{j=1}^{b} (x_{ij} - \bar{x}_{i.} - \bar{x}_{.j} + \bar{x}_{..})^2, \qquad (4.10.3)$$

and

$$SS_E = \sum_{i=1}^{a} \sum_{j=1}^{b} \sum_{k=1}^{n_{ij}} (y_{ijk} - \bar{y}_{ij.})^2,$$

where

$$\bar{x}_{i.} = \sum_{j=1}^{b} x_{ij}/b, \quad \bar{x}_{.j} = \sum_{i=1}^{a} x_{ij}/a,$$

and

$$\bar{x}_{..} = \sum_{i=1}^{a} \sum_{j=1}^{b} x_{ij}/ab.$$

The analysis of variance is shown in Table 4.6 and the approximate F tests are performed based on the usual ratios of mean squares. It should, however,

TABLE 4.6
Unweighted Means Analysis for the Unbalanced Fixed Effects Model in (4.1.1) with Disproportional Frequencies

Source of Variation	Degrees of Freedom	Sum of Squares	Mean Square
Due to A	$a-1$	SS_{Au}	MS_{Au}
Due to B	$b-1$	SS_{Bu}	MS_{Bu}
Interaction $A \times B$	$(a-1)(b-1)$	SS_{ABu}	MS_{ABu}
Error	$N-ab$	SS_E	MS_E
Total	$N-1$	SS_T	

be noted that the sums of squares SS_{Au}, SS_{Bu}, and SS_{ABu} are computed on a "mean" basis, whereas the SS_E is computed on an "individual" basis. Thus, $MS_E = SS_E/(N - ab)$ is not the correct term with which to test for the main effects and interaction mean squares. It must be modified and expressed on a "mean" basis to be comparable to the mean squares for main effects and interaction.

The expected values of the mean squares are obtained as follows (see, e.g., Searle (1971b, pp. 365–366)):

$$E(MS_{Au}) = \frac{b}{a-1} \sum_{i=1}^{a} [\alpha_i + \overline{(\alpha\beta)}_{i.} - \bar{\alpha}_{.} - \overline{(\alpha\beta)}_{..}]^2 + n_h^{-1} \sigma_e^2,$$

$$E(MS_{Bu}) = \frac{a}{b-1} \sum_{j=1}^{b} [\beta_j + \overline{(\alpha\beta)}_{.j} - \bar{\beta}_{.} - \overline{(\alpha\beta)}_{..}]^2 + n_h^{-1} \sigma_e^2, \quad (4.10.4)$$

$$E(MS_{ABu}) = \frac{1}{(a-1)(b-1)} \sum_{i=1}^{a} \sum_{j=1}^{b} [(\alpha\beta)_{ij} - \overline{(\alpha\beta)}_{i.} - \overline{(\alpha\beta)}_{.j} + \overline{(\alpha\beta)}_{..}]^2$$
$$+ n_h^{-1} \sigma_e^2,$$

and

$$E(MS_E) = \sigma_e^2,$$

where

$$n_h = \left(\frac{1}{ab} \sum_{i=1}^{a} \sum_{j=1}^{b} n_{ij}^{-1} \right)^{-1}. \quad (4.10.5)$$

Note that n_h represents the harmonic mean of all $a \times b$ n_{ij}'s. The following

features of the preceing analysis are worth noting:
   (i) The means of the x_{ij}'s are calculated in the usual manner; that is, $\bar{x}_{i.} = \sum_{j=1}^{b} x_{ij}/b$, and so on.
   (ii) The error sum of squares SS_E is calculated exactly as in the case of proportional frequencies.
   (iii) The sums of squares do not add up to the total sum of squares. The first three sums of squares (i. e., SS_{Au}, SS_{Bu}, and SS_{ABu}) add up to $\sum_{i=1}^{a} \sum_{j=1}^{b} (x_{ij} - \bar{x}_{..})^2$, but all four do not add up to the total sum of squares.
   (iv) The sums of squares SS_{Au}, SS_{Bu}, and SS_{ABu} do not have chi-square type distributions as in the case of model (4.1.1), nor they are independent.
   (v) The sum of squares SS_E is independent of SS_{Au}, SS_{Bu}, and SS_{ABu}, and SS_E/σ_e^2 has an exact chi-square distribution. All other sums of squares (divided by σ_e^2) have only approximate noncentral (or central under H_0) chi-square distributions.

Since the mean squares in Table 4.6 do not have exact chi-square distributions, their ratios do not provide exact F statistics for testing hypotheses of interest. However, Gosslee and Lucas (1965) indicated that they provide reasonably adequate F statistics using modified degrees of freedom for the numerator mean squares. For example, the modified numerator degrees of freedom for MS_{Au}/MS_E is

$$v'_a = \frac{(a-1)^2 \left(\sum_{i=1}^{a} h_i\right)^2}{\left(\sum_{i=1}^{a} h_i\right)^2 + a(a-2) \sum_{i=1}^{a} h_i^2}, \qquad (4.10.6)$$

where

$$h_i = \left(\frac{1}{b} \sum_{j=1}^{b} n_{ij}^{-1}\right)^{-1}.$$

Similarly, Rankin (1974) has shown that the approximate F tests give satisfactory results provided the ratios of sample sizes do not exceed 3. He also investigated the problem of modifying the numerator degrees of freedom to adjust for irregularities in sample sizes. Note that although the amended degrees of freedom (4.10.6) modify MS_{Au}/MS_E to be an approximate F statistic, we observe from (4.10.4) that the hypothesis it tests is the equality of $\alpha_i + \overline{(\alpha\beta)}_{i.}$ for all i. Furthermore, since the observation $x_{ij} = \bar{y}_{ij.}$ has variance σ_e^2/n_{ij}, the average variance of all the "observations" is

$$\frac{1}{ab}\sum_{i=1}^{a}\sum_{j=1}^{b}(\sigma_e^2/n_{ij}) = \frac{\sigma_e^2}{ab}\sum_{i=1}^{a}\sum_{j=1}^{b}n_{ij}^{-1} = \frac{\sigma_e^2}{n_h}$$

and the estimated average variance of the "observations" is

$$\frac{MS_E}{ab}\sum_{i=1}^{a}\sum_{j=1}^{b}n_{ij}^{-1} = \frac{MS_E}{n_h},$$

where n_h is the harmonic mean of the n_{ij}'s defined in (4.10.5).

An alternative to the unweighted-means analysis is the weighted-squares-of-means analysis also proposed by Yates (1934). In this technique, the interaction and error sums of squares are defined as earlier, but SS_{Au} and SS_{Bu} sums of squares are weighted in inverse proportion to their variances according to the number of observations in the cell. Thus, letting $x_{ij} = \bar{y}_{ij.}$, the corresponding weighted sums of squares are:

$$SS_{Aw} = \sum_{i=1}^{a} w_{Ai}(\bar{x}_{i.} - \bar{x}_A)^2$$

and

$$SS_{Bw} = \sum_{j=1}^{b} w_{Bj}(\bar{x}_{.j} - \bar{x}_B)^2,$$

where

$$w_{Ai} = \frac{b^2}{\sum_{j=1}^{b}\frac{1}{n_{ij}}}, \quad \bar{x}_A = \frac{\sum_{i=1}^{a} w_{Ai}\bar{x}_{i.}}{\sum_{i=1}^{a} w_{Ai}},$$

$$w_{Bj} = \frac{a^2}{\sum_{i=1}^{a}\frac{1}{n_{ij}}}, \quad \bar{x}_B = \frac{\sum_{j=1}^{b} w_{Bj}\bar{x}_{.j}}{\sum_{j=1}^{b} w_{Bj}},$$

$$x_{ij} = \bar{y}_{ij.} = \frac{1}{n_{ij}}\sum_{k=1}^{n_{ij}} y_{ijk},$$

TABLE 4.7
Weighted Means Analysis for the Unbalanced Fixed Effects Model in (4.1.1) with Disproportional Frequencies

Source of Variation	Degrees of Freedom	Sum of Squares	Mean Square
Due to A	$a-1$	SS_{Aw}	MS_{Aw}
Due to B	$b-1$	SS_{Bw}	MS_{Bw}
Interaction $A \times B$	$(a-1)(b-1)$	SS_{ABw}	MS_{ABw}
Error	$N-ab$	SS_E	MS_E
Total	$N-1$	SS_T	

and $\bar{x}_{i.}, \bar{x}_{.j}$, and $\bar{x}_{..}$ are defined as earlier in the case of unweighted-means analysis. The interaction and error sums of squares are of course given as in unweighted-means analysis; that is,

$$SS_{ABw} = \sum_{i=1}^{a}\sum_{j=1}^{b}(x_{ij} - \bar{x}_{i.} - \bar{x}_{.j} + \bar{x}_{..})^2$$

and

$$SS_E = \sum_{i=1}^{a}\sum_{j=1}^{b}\sum_{k=1}^{n_{ij}}(y_{ijk} - \bar{y}_{ij.})^2.$$

The complete analysis of variance is shown in Table 4.7 where expected values of the mean squares are obtained as follows (see, e.g., Searle (1971b, pp. 369–371)):

$$E(MS_{Aw}) = \frac{1}{a-1}\sum_{i=1}^{a} w_{Ai}\left[\alpha_i + \overline{(\alpha\beta)}_{i.} - \frac{\sum_{i=1}^{a} w_{Ai}(\alpha_i + \overline{(\alpha\beta)}_{i.})}{\sum_{i=1}^{a} w_{Ai}}\right]^2 + \sigma_e^2,$$

$$E(MS_{Bw}) = \frac{1}{b-1}\sum_{j=1}^{b} w_{Bj}\left[\beta_j + \overline{(\alpha\beta)}_{.j} - \frac{\sum_{j=1}^{b} w_{Bj}(\beta_j + \overline{(\alpha\beta)}_{.j})}{\sum_{j=1}^{b} w_{Bj}}\right]^2 + \sigma_e^2,$$

$$E(\mathrm{MS}_{ABw}) = \frac{1}{(a-1)(b-1)} \sum_{i=1}^{a} \sum_{j=1}^{b} [(\alpha\beta)_{ij} - \overline{(\alpha\beta)}_{i.} - \overline{(\alpha\beta)}_{.j} + \overline{(\alpha\beta)}_{..}]^2$$
$$+ n_h^{-1} \sigma_e^2,$$

and

$$E(\mathrm{MS}_E) = \sigma_e^2.$$

It can be shown that the variance ratios $\mathrm{MS}_{Aw}/\mathrm{MS}_E$, $\mathrm{MS}_{Bw}/\mathrm{MS}_E$, and $\mathrm{MS}_{ABw}/n_h^{-1}\mathrm{MS}_E$ provide exact tests of overall hypotheses concerning α_i's, β_j's, and $(\alpha\beta)_{ij}$'s. When the data are balanced, the null hypotheses being tested are:

$$H_0^A : \text{all } \alpha_i = 0, \quad H_0^B : \text{all } \beta_j = 0, \quad H_0^{AB} : \text{all } (\alpha\beta)_{ij} = 0. \quad (4.10.7)$$

However, when data are unbalanced, the corresponding null hypotheses being tested are:

$$H_0^A : \text{all } \alpha_i + \overline{(\alpha\beta)}_{i.} \text{ are equal}, \quad H_0^B : \text{all } \beta_j + \overline{(\alpha\beta)}_{.j} \text{ are equal},$$
$$H_0^{AB} : \text{all } (\alpha\beta)_{ij} - \overline{(\alpha\beta)}_{i.} - \overline{(\alpha\beta)}_{.j} + \overline{(\alpha\beta)}_{..} \text{ are equal}. \quad (4.10.8)$$

Thus, only with the restrictions

$$\overline{(\alpha\beta)}_{i.} = 0, \quad i = 1, 2, \ldots, a; \quad \overline{(\alpha\beta)}_{.j} = 0, \quad j = 1, 2, \ldots, b$$

the null hypotheses (4.10.7) and (4.10.8) are equivalent.

Federer and Zelen (1966) present another approximate analysis that is more exact, but somewhat more complicated than the unweighted and weighted analyses discussed here. Still, another approximate method called the method of expected subclass numbers can be found in Bancroft (1968, pp. 37–41). In situations when the approximate methods are not applicable, for example, badly balanced designs (with 10 or more observations in some cells and only a few in others) and designs with empty cells, a method based on multiple regression analysis may be used. The method consists of considering the analysis of variance model as a regression model, fitting the model for the data, obtaining sums of squares for main effects and interactions as the regression sums of squares, and using the general inferential techniques for the regression model. There are various methods for carrying out this analysis and different methods may lead to different results. For example, one has to determine whether one will test the SS_A adjusted for $\{(\alpha\beta)_{ij}\}$ and $\{\beta_j\}$ or only for $\{\beta_j\}$ if $\{(\alpha\beta)_{ij}$'s$\}$ are not significant, or test the unadjusted SS_A against SS_E. Furthermore, the sums of squares are no longer orthogonal and the sequence in which hypotheses involving fixed effects $\{\alpha_i\}$, $\{\beta_j\}$, $\{(\alpha\beta)_{ij}\}$ are tested may lead to different results. In addition, a unique partition of sums of squares does not exist and the hypotheses being tested do not always correspond to the case of balanced design. For additional details regarding this approach see Draper and Smith (1981) and Searle (1971b, 1987).

RANDOM EFFECTS ANALYSIS

The problems of testing hypotheses and estimation of variance components encountered in unbalanced designs of random effects models having two or more factors are much more complicated than the corresponding balanced case. We again consider two cases for the random effects analysis, one for the case of proportional frequencies and the other for the general case of unequal frequencies.

Proportional Frequencies

For the case of proportional frequencies, the expected values of mean squares can be obtained by the Wilk and Kempthorne (1955) formula. For example, letting $n_{ij} = (n_{i.}n_{.j})/N$, we obtain (see, e.g., Snedecor and Cochran (1967, pp. 478–483)):

$$E(\text{MS}_A) = \sigma_e^2 + \frac{N}{a-1}\left(1 - \sum_{i=1}^{a}\frac{n_{i.}^2}{N^2}\right)\left(\sum_{j=1}^{b}\frac{n_{.j}^2}{N^2}\sigma_{\alpha\beta}^2 + \sigma_\alpha^2\right),$$

$$E(\text{MS}_B) = \sigma_e^2 + \frac{N}{b-1}\left(1 - \sum_{j=1}^{b}\frac{n_{.j}^2}{N^2}\right)\left(\sum_{i=1}^{a}\frac{n_{i.}^2}{N^2}\sigma_{\alpha\beta}^2 + \sigma_\beta^2\right),$$

$$E(\text{MS}_{AB}) = \sigma_e^2 + \frac{N}{(a-1)(b-1)}\left\{\sum_{i=1}^{a}\frac{n_{i.}}{N}\left(1 - \frac{n_{i.}}{N}\right)\right\}$$
$$\times \left\{\sum_{j=1}^{b}\frac{n_{.j}}{N}\left(1 - \frac{n_{.j}}{N}\right)\right\}\sigma_{\alpha\beta}^2,$$

and

$$E(\text{MS}_E) = \sigma_e^2.$$

Approximate tests of hypotheses and variance components estimates can be constructed as earlier. For detailed discussion and numerical examples, see Bancroft (1968, Section 1.6).

General Case of Unequal Frequencies

For the general case of unequal frequencies, Hirotsu (1968) proposed approximate F tests for testing the hypotheses:

$$\begin{aligned}H_0^A : \sigma_\alpha^2 = 0 \quad &\text{versus} \quad H_1^A : \sigma_\alpha^2 > 0,\\ H_0^B : \sigma_\beta^2 = 0 \quad &\text{versus} \quad H_1^B : \sigma_\beta^2 > 0,\end{aligned} \quad (4.10.9)$$

and

$$H_0^{AB}: \sigma_{\alpha\beta}^2 = 0 \quad \text{versus} \quad H_1^{AB}: \sigma_{\alpha\beta}^2 > 0,$$

by using the test statistics analogous to those in the balanced case where now the mean squares are those obtained in the unweighted-means analysis discussed earlier in this section. Thus, the proposed test statistics are:

$$\begin{aligned} \text{MS}_{Au}/\text{MS}_{ABu}, & \quad \text{for } H_0^A; \\ \text{MS}_{Bu}/\text{MS}_{ABu}, & \quad \text{for } H_0^B; \end{aligned} \qquad (4.10.10)$$

and

$$\text{MS}_{ABu}/n_h^{-1}\,\text{MS}_E, \quad \text{for } H_0^{AB};$$

where

$$n_h = \left(\frac{1}{ab}\sum_{i=1}^{a}\sum_{j=1}^{b} n_{ij}^{-1}\right)^{-1}.$$

The test statistics (4.10.10) are to be compared with the $100(1-\alpha)$th percentage points of the F distribution with the degrees of freedom $[(a-1), (a-1)(b-1)]$, $[(b-1), (a-1)(b-1)]$, and $[(a-1)(b-1), N-ab]$, respectively.

Remark: Hirotsu (1968) gave the expressions for the power functions of the tests (4.10.10) with numerical examples, which, however, tend to be very complex. Spjotvoll (1968) and Thomsen (1975) proposed exact tests for main effects variance components under the assumption that the interaction variance component is zero. Khuri and Littel (1987) developed exact tests of variance components that do not require the assumption of nonexistence of interaction variance component. Hussein and Milliken (1978a) considered tests for main effects variance components in a heteroscedastic situation assuming that the interaction variance component is zero. Similarly, Tan et al. (1988) reported tests for main effects as well as interaction variance components involving a heteroscedastic model.

For the estimation of variance components, three methods of estimation were initially proposed and studied in some detail by Henderson (1953). The methods were reexamined and represented in elegant matrix notations by Searle (1968). Since then a variety of new procedures have been developed and the theory has been extended in a number of different directions. Rao (1971, 1972) introduced the concept of minimum norm quadratic unbiased estimation (MINQUE). Similarly, LaMotte (1973) considered minimum variance quadratic unbiased estimators (MIVQUE) and Pukelsheim (1981) investigated the existence of nonnegative unbiased estimators. For detailed discussions of

Two-Way Crossed Classification with Interaction

these and other developments in the field the reader is referred to Searle et al. (1992) and Rao (1997).

To illustrate the nature of the problem of estimating variance components for the case of unbalanced cell frequencies, consider an experiment for which the following two-way additive model is appropriate:

$$y_{ijk} = \mu + \alpha_i + \beta_j + e_{ijk} \quad \begin{cases} i = 1, 2, \ldots, a \\ j = 1, 2, \ldots, b \\ k = 0, 1, \ldots, n_{ij}, \end{cases} \quad (4.10.11)$$

where the α_i's, β_j's, and e_{ijk}'s are uncorrelated random variables with mean zero and variances σ_α^2, σ_β^2, and σ_e^2, respectively. Let the total sum of squares be partitioned as follows:

$$\sum_{i=1}^{a}\sum_{j=1}^{b}\sum_{k=1}^{n_{ij}}(y_{ijk} - \bar{y}_{\ldots})^2 = \left[\sum_{i=1}^{a}\frac{y_{i..}^2}{n_{i.}} - \frac{y_{\ldots}^2}{N}\right] + \left[\sum_{j=1}^{b}\frac{y_{.j.}^2}{n_{.j}} - \frac{y_{\ldots}^2}{N}\right] + SS_E$$

or

$$SS_T = SS_A + SS_B + SS_E,$$

where

$$SS_E = SS_T - SS_A - SS_B.$$

Note that it is possible that SS_E may be negative. The derivation of expected values of the mean squares is complicated and the results may be shown to be those given in Table 4.8 (see, e.g., Graybill (1961, pp. 360–362)), where the coefficients of variance components are determined as follows:

$$c_{aa} = \frac{1}{a-1}\left(N - \sum_{i=1}^{a}\frac{n_{i.}^2}{N}\right), \quad c_{ab} = \frac{1}{a-1}\left(\sum_{i=1}^{a}\sum_{j=1}^{b}\frac{n_{ij}^2}{n_{i.}} - \sum_{j=1}^{b}\frac{n_{.j}^2}{N}\right),$$

$$c_{bb} = \frac{1}{b-1}\left(N - \sum_{j=1}^{b}\frac{n_{.j}^2}{N}\right), \quad c_{ba} = \frac{1}{b-1}\left(\sum_{i=1}^{a}\sum_{j=1}^{b}\frac{n_{ij}^2}{n_{.j}} - \sum_{i=1}^{a}\frac{n_{i.}^2}{N}\right),$$

$$c_{eb} = \frac{1}{N-a-b+1}\left(\sum_{j=1}^{b}\frac{n_{.j}^2}{N} - \sum_{i=1}^{a}\sum_{j=1}^{b}\frac{n_{ij}^2}{n_{i.}}\right),$$

and

$$c_{ea} = \frac{1}{N-a-b+1}\left(\sum_{i=1}^{a}\frac{n_{i.}^2}{N} - \sum_{i=1}^{a}\sum_{j=1}^{b}\frac{n_{ij}^2}{n_{.j}}\right).$$

TABLE 4.8
Analysis of Variance for Model (4.10.11)

Source of Variation	Degrees of Freedom	Sum of Squares	Mean Square	Expected Mean Square
Due to A	$a-1$	SS_A	MS_A	$\sigma_e^2 + c_{ab}\sigma_\beta^2 + c_{aa}\sigma_\alpha^2$
Due to B	$b-1$	SS_B	MS_B	$\sigma_e^2 + c_{bb}\sigma_\beta^2 + c_{ba}\sigma_\alpha^2$
Error	$N-a-b+1$	SS_E	MS_E	$\sigma_e^2 + c_{eb}\sigma_\beta^2 + c_{ea}\sigma_\alpha^2$

If one desires to estimate variance components by the analysis of variance method, that is, by equating mean squares to their corresponding expected values, one obtains the following system of equations:

$$\begin{aligned} MS_A &= \hat{\sigma}_e^2 + c_{ab}\hat{\sigma}_\beta^2 + c_{aa}\hat{\sigma}_\alpha^2, \\ MS_B &= \hat{\sigma}_e^2 + c_{bb}\hat{\sigma}_\beta^2 + c_{ba}\hat{\sigma}_\alpha^2, \\ MS_E &= \hat{\sigma}_e^2 + c_{eb}\hat{\sigma}_\beta^2 + c_{ea}\hat{\sigma}_\alpha^2. \end{aligned} \quad (4.10.12)$$

where parameters have been replaced by their respective estimators. The resultant solution of the system of equations (4.10.12) provides a set of estimators of the variance components. The evaluation of explicit expressions for the estimators is somewhat involved and the results can be found in Searle (1971b, p. 487) and Searle et al. (1992, p. 439). The estimators obtained will be unbiased and consistent, but other optimum properties are still being explored. Although sampling variances can be obtained, other distribution properties cannot, since even under the normality assumptions, distribution of the estimators is unknown.

The only functions of variance components for which exact intervals can be obtained are σ_e^2 and $\sigma_{\alpha\beta}^2/\sigma_e^2$. For a discussion of the problem of setting confidence intervals for the individual variance components, certain ratios of variance components and proportions of variability, including numerical examples, see Burdick and Graybill (1992, pp. 136–145).

MIXED EFFECTS ANALYSIS

Most of the inferential difficulties that are encountered occur in the mixed effects model. The treatment of the unbalanced mixed model is beyond the scope of this volume. The interested reader is referred to Searle (1971b, pp. 429–431; 1987, Chapter 13; 1988), Stroup (1989), McLean et al. (1991), Hocking (1993), and Khuri et al. (1998). Smith (1951) discusses the tests of hypotheses for the mixed model with proportional frequencies. For a discussion of exact tests for the random and fixed effects in an unbalanced two-way crossed classification model, see Gallo and Khuri (1990). Burdick and Graybill (1992, p. 172) give a

4.11 POWER OF THE ANALYSIS OF VARIANCE F TESTS

The power of the analysis of variance F tests for AB interactions, factor B effects, and factor A effects can be evaluated in a manner similar to the case of one-way classification. The results on power calculations are briefly summarized in the following.

MODEL I (FIXED EFFECTS)

The parameter ϕ and the appropriate degrees of freedom for each of the tests are as follows.

Test for AB Interactions

$$\text{Power} = P\{F'[\nu_1, \nu_2; \phi] > F[\nu_1, \nu_2; 1 - \alpha]\},$$

where

$$\nu_1 = (a-1)(b-1), \quad \nu_2 = ab(n-1),$$

and

$$\phi = \frac{1}{\sigma_e} \sqrt{\frac{n \sum_{i=1}^{a} \sum_{j=1}^{b} (\alpha\beta)_{ij}^2}{(a-1)(b-1)+1}}.$$

Test for Factor B Effects

$$\text{Power} = P\{F'[\nu_1, \nu_2; \phi] > F[\nu_1, \nu_2; 1 - \alpha]\},$$

where

$$\nu_1 = b-1, \quad \nu_2 = ab(n-1),$$

and

$$\phi = \frac{1}{\sigma_e} \sqrt{\frac{an}{b} \sum_{j=1}^{b} \beta_j^2}.$$

Test for Factor A Effects

$$\text{Power} = P\{F'[\nu_1, \nu_2; \phi] > F[\nu_1, \nu_2; 1 - \alpha]\},$$

where

$$\nu_1 = a-1, \quad \nu_2 = ab(n-1),$$

and

$$\phi = \frac{1}{\sigma_e}\sqrt{\frac{bn}{a}\sum_{i=1}^{a}\alpha_i^2}.$$

Remark: Kastenbaum et al. (1970b) gave tables showing how large n must be ($1 \le n \le 5$) for $a = 2(1)6$ and $b = 2(1)5$ in testing for factor A effects with $\alpha = 0.05, 0.01$, and $1 - \beta = 0.7, 0.8, 0.9, 0.95, 0.99, 0.995$, when max $|\alpha_i - \alpha_{i'}|/\sigma_e$ is given. More extensive tables are given by Bowman (1972) and Bowman and Kastenbaum (1975).

MODEL II (RANDOM EFFECTS)

The power calculations involve only the central F distribution. The results for each of the tests are as follows.

Test for AB Interactions

$$\text{Power} = P\{F[v_1, v_2] > \lambda^{-2} F[v_1, v_2; 1 - \alpha]\},$$

where

$$v_1 = (a-1)(b-1), \quad v_2 = ab(n-1),$$

and

$$\lambda = \sqrt{1 + \frac{n\sigma_{\alpha\beta}^2}{\sigma_e^2}}.$$

Test for Factor B Effects

$$\text{Power} = P\{F[v_1, v_2] > \lambda^{-2} F[v_1, v_2; 1 - \alpha]\},$$

where

$$v_1 = b - 1, \quad v_2 = (a-1)(b-1),$$

and

$$\lambda = \sqrt{1 + \frac{an\sigma_\beta^2}{\sigma_e^2 + n\sigma_{\alpha\beta}^2}}.$$

Test for Factor A Effects

$$\text{Power} = P\{F[v_1, v_2] > \lambda^{-2} F[v_1, v_2; 1 - \alpha]\},$$

where

$$v_1 = a - 1, \quad v_2 = (a-1)(b-1),$$

Two-Way Crossed Classification with Interaction

and

$$\lambda = \sqrt{1 + \frac{bn\sigma_\alpha^2}{\sigma_e^2 + n\sigma_{\alpha\beta}^2}}.$$

The power of the tests of the hypotheses of the type $\sigma_{\alpha\beta}^2/\sigma_e^2 \leq \rho_1$, $\sigma_\beta^2/(\sigma_e^2 + n\sigma_{\alpha\beta}^2) \leq \rho_2$, or $\sigma_\alpha^2/(\sigma_e^2 + n\sigma_{\alpha\beta}^2) \leq \rho_3$ can similarly be expressed in terms of the central F distribution.

MODEL III (MIXED EFFECTS)

The power of the tests for AB interactions and B effects involves central F distributions and for A effects involves the noncentral F distribution. The results are as follows.

Test for AB Interactions

$$\text{Power} = P\{F[\nu_1, \nu_2] > \lambda^{-2} F[\nu_1, \nu_2; 1 - \alpha]\},$$

where

$$\nu_1 = (a-1)(b-1), \quad \nu_2 = ab(n-1),$$

and

$$\lambda = \sqrt{1 + \frac{n\sigma_{\alpha\beta}^2}{\sigma_e^2}}.$$

Test for Factor B Effects

$$\text{Power} = P\{F[\nu_1, \nu_2] > \lambda^{-2} F[\nu_1, \nu_2; 1 - \alpha]\},$$

where

$$\nu_1 = b-1, \quad \nu_2 = ab(n-1),$$

and

$$\lambda = \sqrt{1 + \frac{an\sigma_\beta^2}{\sigma_e^2}}.$$

Test for Factor A Effects

$$\text{Power} = P\{F'[\nu_1, \nu_2; \phi] > F[\nu_1, \nu_2; 1 - \alpha]\},$$

where

$$\nu_1 = a-1, \quad \nu_2 = (a-1)(b-1),$$

and

$$\phi = \sqrt{\frac{bn \sum_{i=1}^{a} \alpha_i^2}{a(\sigma_e^2 + n\sigma_{\alpha\beta}^2)}}.$$

The power of the tests of the hypotheses of the type $\sigma_{\alpha\beta}^2/\sigma_e^2 \leq \rho_1$ or $\sigma_\beta^2/\sigma_e^2 \leq \rho_2$ can similarly be expressed in terms of the central F distribution.

4.12 MULTIPLE COMPARISON METHODS

Usually, more than one comparison is of interest and the multiple comparison procedures discussed in Section 2.19 can be employed with only minor modifications. The procedures can be utilized for the fixed as well as the mixed effects models. Most comparisons concern the control of the error rate α for each separate family of F tests, that is, the two main effects and the cell means. One may want to control α for the entire experiment comprising all three families of tests but it is rarely of interest.

For example, under Model I, if H_0^{AB} is rejected, we would be interested in comparing the cell means $\mu_{ij} = \mu + \alpha_i + \beta_j + (\alpha\beta)_{ij}$. Then, the Tukey's or Scheffé's method may be used to investigate the contrasts of the type

$$L = \mu_{ij} - \mu_{i'j'},$$

among all cell means, where L is estimated by

$$\hat{L} = \bar{y}_{ij.} - \bar{y}_{i'j'.}.$$

Now, the procedure is equivalent to the one-way classification model with the total number of treatments here being equal to $r = ab$ and the degrees of freedom for MS_E equal to $ab(n-1)$.

Thus, suppose that $\bar{y}_{ij.}$ is larger than $\bar{y}_{i'j'.}$. Then using the Tukey's procedure $\hat{L}$ is significantly different from zero with confidence coefficient $1 - \alpha$ if

$$\frac{\hat{L}}{\sqrt{MS_E/n}} > q[ab, ab(n-1); 1-\alpha].$$

If the Scheffé's method is applied to these comparisons, then $\hat{L}$ is significantly different from zero with confidence coefficient $1 - \alpha$ if

$$\frac{\hat{L}}{\sqrt{(ab-1)MS_E(2/n)}} > \{F[ab-1, ab(n-1); 1-\alpha]\}^{1/2}.$$

If H_0^{AB} is not rejected, we usually would proceed to test H_0^A and H_0^B. If H_0^A or H_0^B is rejected, the Tukey's or Scheffé's method may be used to study contrasts among the α_i's or β_j's of the form

$$L = \sum_{i=1}^{a} \ell_i \alpha_i \quad \left(\sum_{i=1}^{a} \ell_i = 0 \right)$$

or

$$L' = \sum_{j=1}^{b} \ell'_j \beta_j \quad \left(\sum_{j=1}^{b} \ell'_j = 0 \right),$$

where L and L' are estimated unbiasedly by

$$\hat{L} = \sum_{i=1}^{a} \ell_i \bar{y}_{i..},$$

and

$$\hat{L}' = \sum_{j=1}^{b} \ell'_j \bar{y}_{.j.}.$$

Using the Tukey's method, $\hat{L}$ is significant at the α-level if

$$\frac{\hat{L}}{\sqrt{(bn)^{-1} \text{MS}_E \left(\frac{1}{2} \sum_{i=1}^{a} |\ell_i| \right)}} > q[a, ab(n-1); 1-\alpha].$$

Similarly, if the Scheffé's method is used, $\hat{L}$ is significant at the α-level if

$$\frac{\hat{L}}{\sqrt{(a-1)(bn)^{-1} \text{MS}_E \left(\sum_{i=1}^{a} \ell_i^2 \right)}} > \{F[a-1, ab(n-1); 1-\alpha]\}^{1/2}.$$

Likewise, the significance of the contrast L' can be tested.

If one wishes to construct intervals for L, then using the Tukey's method a $100(1-\alpha)$ percent simultaneous confidence interval for L is given by

$$\hat{L} - T \sqrt{(bn)^{-1} \text{MS}_E \left(\frac{1}{2} \sum_{i=1}^{a} |\ell_i| \right)} < L < \hat{L} + T \sqrt{(bn)^{-1} \text{MS}_E \left(\frac{1}{2} \sum_{i=1}^{a} |\ell_i| \right)},$$

(4.12.1)

where

$$T = q[a, ab(n-1); 1-\alpha].$$

In particular, for a pairwise contrast, the Tukey's interval is

$$\bar{y}_{i..} - \bar{y}_{i'..} - T\sqrt{(bn)^{-1} \text{MS}_E} < \alpha_i - \alpha_{i'} < \bar{y}_{i..} - \bar{y}_{i'..} + T\sqrt{(bn)^{-1} \text{MS}_E}.$$

Using the Scheffé's method, the interval will be

$$\hat{L} - S\sqrt{(a-1)(bn)^{-1} \text{MS}_E \left(\sum_{i=1}^{a} \ell_i^2\right)}$$
$$< L < \hat{L} + S\sqrt{(a-1)(bn)^{-1} \text{MS}_E \left(\sum_{i=1}^{a} \ell_i^2\right)}, \quad (4.12.2)$$

where

$$S^2 = F[a-1, ab(n-1); 1-\alpha].$$

The Bonferroni-type confidence interval based on the t distribution is obtained as

$$\hat{L} - t[ab(n-1), 1-\alpha/2m]\sqrt{(bn)^{-1} \text{MS}_E \sum_{i=1}^{a} \ell_i^2}$$
$$< L < \hat{L} + t[ab(n-1), 1-\alpha/2m]\sqrt{(bn)^{-1} \text{MS}_E \sum_{i=1}^{a} \ell_i^2},$$

where m is the number of intervals made, with an overall level of at least $1-\alpha$. Similar confidence intervals can be given for L'.

When a design is slightly unbalanced and one uses the unweighted-means analysis, then the foregoing Tukey's procedure can be used by replacing n by n_h given by (4.10.5). The coverage probability should be approximately $1-\alpha$; however, as the design becomes more imbalanced, the coverage probability deteriorates.

Under Model III, the contrasts of interest involve only α_i's and if H_0^A is rejected, the Tukey's or Scheffé's method can be employed to investigate contrasts of the type $\sum_{i=1}^{a} \ell_i \alpha_i$. For example, suppose we wish to obtain all pairwise comparisons between α_i's by means of Tukey's method. Then

$$L = \alpha_i - \alpha_{i'},$$
$$\hat{L} = \bar{y}_{i..} - \bar{y}_{i'..},$$

and
$$\widehat{\text{Var}}(\hat{L}) = 2\text{MS}_{AB}/bn.$$

The value of T in this case will be
$$T = q[a, (a-1)(b-1); 1-\alpha],$$

leading to the interval
$$\bar{y}_{i..} - \bar{y}_{i'..} - T\sqrt{(bn)^{-1}\text{MS}_{AB}} < \alpha_i - \alpha_{i'} < \bar{y}_{i..} - \bar{y}_{i'..} + T\sqrt{(bn)^{-1}\text{MS}_{AB}}.$$
(4.12.3)

For the Scheffé's interval, we will have
$$\hat{L} - S\sqrt{(a-1)(bn)^{-1}\text{MS}_{AB}\left(\sum_{i=1}^{a}\ell_i^2\right)} < L < \hat{L} + S\sqrt{(a-1)(bn)^{-1}\text{MS}_{AB}\left(\sum_{i=1}^{a}\ell_i^2\right)},$$
(4.12.4)

where
$$S^2 = F[a-1, (a-1)(b-1); 1-\alpha].$$

For just a single confidence interval, one can use $\sqrt{2}t[(a-1)(b-1); 1-\alpha/2]$ in place of T. For a limited number of comparisons k, the Bonferroni intervals are obtained by using $\sqrt{2}t[(a-1)(b-1); 1-\alpha/2k]$ instead of T. For $k < a(a-1)/2$, the Bonferroni intervals are usually shorter than the Tukey intervals.

4.13 WORKED EXAMPLE FOR MODEL I

Steel and Torrie (1980, pp. 217–218) reported data (courtesy of A.C. Linnerud, North Carolina State University) on times (in seconds) to complete a 1.5-mile course. All the runners were men classified in three age groups and in three fitness categories. The data form a two-way classification and are shown in Table 4.9. The example just described can be regarded as a two-way fixed effects model since three age groups and 3 fitness categories are specially chosen by the researcher to be of particular interest and thus both factors will have systematic effects. Since there are two observations for each combination level, this will enable the experimenter to evaluate for the presence of interaction effects. If there were just one observation for each combination level, either lack of interaction would have to be assumed or its presence would be confounded with the error term. It could not be estimated separately.

TABLE 4.9
Running Time (in seconds) to Complete a 1.5 Mile Course

Age Group	Fitness Category		
	Low	Medium	High
40	669	602	527
	671	603	547
50	775	684	571
	821	687	573
60	1,009	824	688
	1,060	828	713

Source: Steel and Torrie (1980, p. 218). Used with permission.

The mathematical model for this experiment would be

$$y_{ijk} = \mu + \alpha_i + \beta_j + (\alpha\beta)_{ij} + e_{ijk} \quad \begin{cases} i = 1, 2, 3 \\ j = 1, 2, 3 \\ k = 1, 2, \end{cases}$$

where μ is the general mean, α_i is the effect of the i-th age group ($\sum_{i=1}^{3} \alpha_i = 0$), β_j is the effect of the j-th fitness category ($\sum_{j=1}^{3} \beta_j = 0$), $(\alpha\beta)_{ij}$ is the fixed effect interaction of the i-th age group with the j-th fitness category ($\sum_{i=1}^{3} (\alpha\beta)_{ij} = 0 = \sum_{j=1}^{3} (\alpha\beta)_{ij}$), and e_{ijk}'s are experimental errors assumed to be independently and normally distributed each with mean zero and variance σ_e^2.

The following computations will lead to the analysis of variance table.

(i) The cell totals:

$$y_{11.} = 1,340, \quad y_{12.} = 1,205, \quad y_{13.} = 1,074;$$
$$y_{21.} = 1,596, \quad y_{22.} = 1,371, \quad y_{23.} = 1,144;$$
$$y_{31.} = 2,069, \quad y_{32.} = 1,652, \quad y_{33.} = 1,401.$$

(ii) The row (age) totals:

$$y_{1..} = 3,619, \quad y_{2..} = 4,111, \quad y_{3..} = 5,122.$$

(iii) The column (fitness) totals:

$$y_{.1.} = 5,005, \quad y_{.2.} = 4,228, \quad y_{.3.} = 3,619.$$

(iv) The grand total:

$$y_{...} = 3,619 + 4,111 + 5,122 = 12,852.$$

(v)
$$\sum_{i=1}^{3}\sum_{j=1}^{3}\sum_{k=1}^{2} y_{ijk}^2 = (669)^2 + (671)^2 + \cdots + (713)^2 = 9{,}557{,}568.$$

(vi)
$$\frac{y_{...}^2}{3 \times 3 \times 2} = \frac{(12{,}852)^2}{18} = 9{,}176{,}328.$$

(vii)
$$\frac{1}{3 \times 2}\sum_{i=1}^{3} y_{i..}^2 = \frac{(3{,}619)^2 + (4{,}111)^2 + (5{,}122)^2}{6} = 9{,}372{,}061.$$

(viii)
$$\frac{1}{3 \times 2}\sum_{j=1}^{3} y_{.j.}^2 = \frac{(5{,}005)^2 + (4{,}228)^2 + (3{,}619)^2}{6} = 9{,}337{,}195.$$

(ix)
$$\frac{1}{2}\sum_{i=1}^{3}\sum_{j=1}^{3} y_{ij.}^2 = \frac{(1{,}340)^2 + (1{,}205)^2 + \cdots + (1{,}401)^2}{2} = 9{,}554{,}680.$$

(x) $SS_T = 9{,}557{,}568 - 9{,}176{,}328 = 381{,}240.$
(xi) $SS_{TC} = 9{,}554{,}680 - 9{,}176{,}328 = 378{,}352.$
(xii) $SS_A = 9{,}372{,}061 - 9{,}176{,}328 = 195{,}733.$
(xiii) $SS_B = 9{,}337{,}195 - 9{,}176{,}328 = 160{,}867.$
(xiv) $SS_{AB} = 378{,}352 - 195{,}733 - 160{,}867 = 21{,}752.$
(xv) $SS_E = 381{,}240 - 378{,}352 = 2{,}888.$

These results, along with the remaining analysis of variance computations, are summarized in Table 4.10. If we choose the level of significance $\alpha = 0.05$, we find from Appendix Table V that

$$F[2, 9; 0.95] = 4.26,$$

and

$$F[4, 9; 0.95] = 3.63.$$

Comparing these values with the computed F values given in Table 4.10, we may reach the following conclusions:

(a) Reject the hypothesis of no interaction effects and conclude that there is strong evidence of interaction between the different age groups and the different fitness categories ($p < 0.001$).

TABLE 4.10
Analysis of Variance for the Running Time Data of Table 4.9

Source of Variation	Degrees of Freedom	Sum of Squares	Mean Square	Expected Mean Square	F Value	p-Value
Age group	2	195,733	97,866.500	$\sigma_e^2 + \dfrac{3 \times 2}{3-1}\sum_{i=1}^{3}\alpha_i^2$	304.99	<0.001
Fitness category	2	160,867	80,433.500	$\sigma_e^2 + \dfrac{3 \times 2}{3-1}\sum_{j=1}^{3}\beta_j^2$	250.66	<0.001
Interaction	4	21,752	5,438.000	$\sigma_e^2 + \dfrac{2}{(3-1)(3-1)} \times \sum_{i=1}^{3}\sum_{j=1}^{3}(\alpha\beta)_{ij}^2$	16.95	<0.001
Error	9	2,888	320.889	σ_e^2		
Total	17	381,240	22,425.882			

(b) Reject the hypothesis of no age effects and conclude that different age groups result in different mean running time to complete the race ($p < 0.001$).

(c) Reject the hypothesis of no fitness effects and conclude that the mean running times to complete the race are not the same for the three categories ($p < 0.001$).

It should be noted that the presence of interaction between age group and fitness category seems to be more than just a chance occurrence. The presence of interactions makes the interpretation of the main effects more difficult. Although F tests still remain valid, the hypotheses about the main effects cannot be interpreted only in terms of the α_i's and the β_j's. Nevertheless, assuming that the interaction effects are unimportant, we attempt to illustrate the use of orthogonal contrasts to partition the sums of squares for age group and fitness category and perform tests on a contrast.

If the hypothesis of no interaction were true, we could make general comparisons regarding the fitness rather than separate comparisons for each age group. Similarly, we might make general comparisons among the age groups rather than separate comparisons for each fitness. For example, we could compare fitness category low versus high and also low and high versus medium. The contrasts for making these comparisons would be

$$L_1 = \beta_1 - \beta_3$$

and

$$L_2 = \beta_1 + \beta_3 - 2\beta_2,$$

respectively. The single degree of freedom sums of squares associated with L_1 and L_2 are obtained as follows

$$SS_{L_1} = \frac{(5{,}005 - 3{,}619)^2}{6[(1)^2 + (-1)^2]} = 160{,}083$$

and

$$SS_{L_2} = \frac{[5{,}005 + 3{,}619 - 2(4{,}228)]^2}{6[(1)^2 + (1)^2 + (-2)^2]} = 784.$$

Notice that $SS_{L_1} + SS_{L_2} = SS_B$ since L_1 and L_2 are two independent orthogonal contrasts which partition the sum of squares for the fitness into two single degree of freedom sums of squares. The computed F values corresponding to L_1 and L_2 are, respectively,

$$F_1 = \frac{160{,}083}{320.889} = 498.87$$

and

$$F_2 = \frac{784}{320.889} = 2.44.$$

Comparing these values to the critical value $F[1, 9; 0.95] = 5.12$, we find that F_1 is highly significant ($p < 0.001$) but F_2 falls well below the significance level ($p = 0.15$). Thus, the results of the F tests indicate that the hypothesis $H_0: \beta_1 - \beta_3 = 0$ is rejected whereas the hypothesis $H_0: \beta_1 + \beta_3 - 2\beta_2$ is sustained.

Similarly, we could compare age groups 1 and 3 and also age groups 1 and 3 versus 2. The resulting F ratios, each with 1 and 9 degrees of freedom are both highly significant ($p < 0.001$). The results indicate that the low age group requires the least running time whereas the upper age group requires the most running time. However, as indicated previously, the researcher should be cautious in making any general conclusions because of the strong interaction effects between the factors. The running time within each age group varies greatly according to the fitness category. Although, in each age group, the running time decreases dramatically as we move from the low to the high fitness category, the decrease is much greater for the upper age group than for the low and the middle age group.

4.14 WORKED EXAMPLE FOR MODEL I: UNEQUAL SAMPLE SIZES PER CELL

The following example is based on an unbalanced design described in Blackwell et al. (1991, pp. 286–287). The original data came from a balanced factorial

TABLE 4.11
Weight Gains (in grams) of Rats under Different Diets (Data Made Unbalanced by Deleting Observations)

Source of Protein	Quantity of Protein	
	High	Low
Beef	81, 100, 102, 104, 107, 111, 117, 118	51, 64, 72, 76, 78, 86, 95
Pork	79, 91, 94, 96, 98, 102, 102, 108	49, 70, 73, 81, 82, 82, 86, 97, 106
Cereal	56, 74, 77, 82, 86, 88, 92, 95, 98, 111	58, 67, 74, 74, 80, 89, 95, 97, 98, 107

Source: Blackwell et al. (1991, p. 287). Used with permission.

experiment reported by Snedecor and Cochran (1989, p. 304) to test the effectiveness of two factors, source of protein: three levels — beef, pork and cereal, and quantity of protein: two levels — high and low, forming six potential protein feeding treatments. Ten male rats were randomly assigned to each treatment and gains in weight were recorded. The experimental data were made unbalanced by deleting eight observations and the remaining observations are given in Table 4.11.

In the following, we illustrate both unweighted- and weighted-squares of means analysis. In this type of analysis, two models are used. The model for the mean of each subclass is

$$x_{ij} = \bar{y}_{ij.} = \mu + \alpha_i + \beta_j + (\alpha\beta)_{ij} + \bar{e}_{ij.} \quad \begin{cases} i = 1, 2, 3 \\ j = 1, 2, \end{cases}$$

where μ is the general mean, α_i is the effect of the i-th source of protein ($\sum_{i=1}^{3} \alpha_i = 0$), β_j is the effect of the j-th quantity of protein ($\sum_{j=1}^{2} \beta_j = 0$), $(\alpha\beta)_{ij}$ is the fixed effect interaction of the i-th protein source with the j-th protein quantity ($\sum_{i=1}^{3} (\alpha\beta)_{ij} = 0 = \sum_{j=1}^{2} (\alpha\beta)_{ij}$), and $\bar{e}_{ij.} = \sum_{k=1}^{n_{ij}} e_{ijk}/n_{ijk}$ is the experimental error associated with the (i, j)-th cell which is assumed to be independently and normally distributed with mean zero and variance σ^2. The preceding equation represents the model being assumed when the sums of squares for factors A and B and the interaction AB are computed. For computing the error sum of squares, the model being used is

$$y_{ijk} = \mu + \alpha_i + \beta_j + (\alpha\beta)_{ij} + e_{ijk} \quad \begin{cases} i = 1, 2, 3 \\ j = 1, 2 \\ k = 1, 2, \ldots, n_{ij}, \end{cases}$$

where, again, $\sum_{i=1}^{3} \alpha_i = \sum_{j=1}^{2} \beta_j = \sum_{i=1}^{3} (\alpha\beta)_{ij} = \sum_{j=1}^{2} (\alpha\beta)_{ij} = 0$ and e_{ijk} is the experimental error assumed to be independently and normally distributed with mean zero and variance σ_e^2. The expectation of error mean square is σ_e^2/n_h,

Two-Way Crossed Classification with Interaction

which has to be divided by $1/n_h$ so that it yields an unbiased estimate of σ_e^2.

The computations proceed as follows:

(i) The cell counts (n_{ij}):

$$n_{11} = 8, \quad n_{12} = 7,$$
$$n_{21} = 8, \quad n_{22} = 9,$$
$$n_{31} = 10, \quad n_{32} = 10.$$

(ii) The cell means (x_{ij}):

$$x_{11} = 105.0000, \quad x_{12} = 74.5714,$$
$$x_{21} = 96.2500, \quad x_{22} = 80.6667,$$
$$x_{31} = 85.9000, \quad x_{32} = 83.9000.$$

(iii) The row (protein source) means:

$$\bar{x}_{1.} = 89.7857, \quad \bar{x}_{2.} = 88.4584, \quad \bar{x}_{3.} = 84.9000.$$

(iv) The column (protein quantity) means:

$$\bar{x}_{.1} = 95.7167, \quad \bar{x}_{.2} = 79.7127.$$

(v) The grand mean:

$$\bar{x}_{..} = 87.7147.$$

(vi)

$$\sum_{i=1}^{3}\sum_{j=1}^{2} x_{ij}^2 = (105.0000)^2 + (74.5714)^2 + \cdots$$
$$+ (83.9000)^2 = 46{,}775.0927.$$

(vii)

$$3 \times 2\bar{x}_{..}^2 = 6(87.7147)^2 = 46{,}163.2116.$$

(viii)

$$2\sum_{i=1}^{3} \bar{x}_{i.}^2 = 2[(89.7857)^2 + (88.4584)^2 + (84.9000)^2] = 46{,}188.7409.$$

(ix)

$$3\sum_{j=1}^{2}\bar{x}_{\cdot j}^{2} = 3[(95.7167)^{2} + (79.7127)^{2}] = 46{,}547.4036.$$

Now, the unweighted sums of squares for factor A (protein source) and factor B (protein quantity) are:

$$SS_{Au} = 2\sum_{i=1}^{3}(\bar{x}_{i\cdot} - \bar{x}_{\cdot\cdot})^{2} = 2\sum_{i=1}^{3}\bar{x}_{i\cdot}^{2} - 3\times 2\bar{x}_{\cdot\cdot}^{2}$$
$$= 46{,}188.7409 - 46{,}163.2116 = 25.5293$$

and

$$SS_{Bu} = 3\sum_{j=1}^{2}(\bar{x}_{\cdot j} - \bar{x}_{\cdot\cdot})^{2} = 3\sum_{j=1}^{2}\bar{x}_{\cdot j}^{2} - 3\times 2\bar{x}_{\cdot\cdot}^{2}$$
$$= 46{,}547.4036 - 46{,}163.2116 = 384.1920.$$

To calculate the corresponding weighted sums of squares, we have

$$w_{A1} = \frac{b^{2}}{\dfrac{1}{n_{11}} + \dfrac{1}{n_{12}}} = \frac{(2)^{2}}{\dfrac{1}{8} + \dfrac{1}{7}} = 14.9333,$$

$$w_{A2} = \frac{b^{2}}{\dfrac{1}{n_{21}} + \dfrac{1}{n_{22}}} = \frac{(2)^{2}}{\dfrac{1}{8} + \dfrac{1}{9}} = 16.9412,$$

$$w_{A3} = \frac{b^{2}}{\dfrac{1}{n_{31}} + \dfrac{1}{n_{32}}} = \frac{(2)^{2}}{\dfrac{1}{10} + \dfrac{1}{10}} = 20.0000,$$

$$w_{B1} = \frac{a^{2}}{\dfrac{1}{n_{11}} + \dfrac{1}{n_{21}} + \dfrac{1}{n_{31}}} = \frac{(3)^{2}}{\dfrac{1}{8} + \dfrac{1}{8} + \dfrac{1}{10}} = 25.7143,$$

$$w_{B2} = \frac{a^{2}}{\dfrac{1}{n_{12}} + \dfrac{1}{n_{22}} + \dfrac{1}{n_{32}}} = \frac{(3)^{2}}{\dfrac{1}{7} + \dfrac{1}{9} + \dfrac{1}{10}} = 25.4260,$$

Two-Way Crossed Classification with Interaction

$$\bar{x}_A = \frac{\sum_{i=1}^{3} w_{Ai}\bar{x}_{i.}}{\sum_{i=1}^{3} w_{Ai}}$$

$$= \frac{(14.9333)(89.7857) + (16.9412)(88.4584) + (20.0000)(84.90000)}{14.9333 + 16.9412 + 20.0000}$$

$$= 87.4686,$$

and

$$\bar{x}_B = \frac{\sum_{j=1}^{3} w_{Bj}\bar{x}_{.j}}{\sum_{j=1}^{3} w_{Bj}} = \frac{(25.7143)(95.7167) + (25.4260)(79.7127)}{25.7143 + 25.4260}$$

$$= 87.7598.$$

Therefore, the corresponding weighted sums of squares are:

$$SS_{Aw} = \sum_{i=1}^{3} w_{Ai}(\bar{x}_{i.} - \bar{x}_A)^2$$

$$= 14.9333(89.7857 - 87.4686)^2 + 16.9412(88.4584 - 87.4686)^2$$
$$+ 20.0000(84.9000 - 87.4686)^2$$

$$= 228.7277$$

and

$$SS_{Bw} = \sum_{j=1}^{2} w_{Bj}(\bar{x}_{.j} - \bar{x}_B)^2$$

$$= 25.7143(95.7167 - 87.7598)^2 + 25.4260(79.7127 - 87.7598)^2$$

$$= 3274.5118.$$

The interaction and error sum of squares for both unweighted and weighted analyses are:

$$SS_{AB} = \sum_{i=1}^{3}\sum_{j=1}^{2}(x_{ij} - \bar{x}_{i.} - \bar{x}_{.j} + \bar{x}_{..})^2$$

$$= \sum_{i=1}^{3}\sum_{j=1}^{2} x_{ij}^2 - 2\sum_{i=1}^{3}\bar{x}_{i.}^2 - 3\sum_{j=1}^{2}\bar{x}_{.j}^2 + 3 \times 2\bar{x}_{..}^2$$

$$= 46{,}775.0927 - 46{,}188.7409 - 46{,}547.4036 + 46{,}163.2116$$
$$= 202.1598$$

and

$$SS_E = \sum_{i=1}^{3}\sum_{j=1}^{2}\sum_{k=1}^{n_{ij}}(y_{ijk} - \bar{y}_{ij.})^2$$
$$= \sum_{i=1}^{3}\sum_{j=1}^{2}\sum_{k=1}^{n_{ij}} y_{ijk}^2 - \sum_{i=1}^{3}\sum_{j=1}^{2} n_{ij}\bar{y}_{ij.}^2$$
$$= 413{,}088 - 403{,}983.0043$$
$$= 9{,}104.9957.$$

Finally, the total sum of squares is obtained as

$$SS_T = \sum_{i=1}^{3}\sum_{j=1}^{2}\sum_{k=1}^{n_{ij}}(y_{ijk} - \bar{y}_{...})^2$$
$$= \sum_{i=1}^{3}\sum_{j=1}^{2}\sum_{k=1}^{n_{ij}} y_{ijk}^2 - \frac{y_{...}^2}{N}$$
$$= 413{,}088 - (4{,}556)^2/52$$
$$= 13{,}912.3077.$$

These results along with the remaining analysis of variance computations are summarized in Table 4.12. Note that for both weighted and unweighted analyses, the sums of squares do not add up to the total sum of squares. If we choose the level of significance $\alpha = 0.05$, we find from Appendix Table V that

$$F[2, 46, 0.95] = 3.20$$

and

$$F[1, 46, 0.95] = 4.05.$$

Comparing these values with the computed F values given in Table 4.12 for both unweighted and weighted analyses, we may reach the following conclusions

(a) Reject the hypothesis of no interaction effects and conclude that there is some evidence of interaction between protein source and protein quantity.
(b) Do not reject the hypothesis of no protein source effects and conclude that there is a lack of evidence that mean weights (population marginal means) for three sources of protein do not differ significantly.

TABLE 4.12
Analysis of Variance Using Unweighted and Weighted Squares of Means Analysis of the Unbalanced Data on Weight Gains (in grams) of Table 4.11

Source of Variation	Degrees of Freedom	Sum of Squares		Mean Square		F Value		p-Value	
		Unweighted	Weighted	Unweighted	Weighted	Unweighted	Weighted	Unweighted	Weighted
Protein source	2*	25.5293	228.7277	12.7646	114.3639	0.55**	0.58	0.581	0.564
Protein quantity	1*	384.1920	3,274.5118	384.1920	3,274.5118	16.54**	16.54	<0.001	<0.001
Interaction	2	202.1598	202.1598	101.0799	101.0799	4.35**	4.35**	0.019	0.019
Error	46	9,104.9957	9,104.9957	197.9347	197.9347				
Total	51	13,912.3077	13,912.3077						

* The amended degrees of freedom using formula (4.10.6) for the unweighted analysis are nearly the same.

** In the computation of F values, the error mean square is modified by dividing it by the harmonic mean of n_{ij}'s given by $6/(1/8 + 1/7 + 1/8 + 1/9 + 1/10 + 1/10) = 8.5231$.

(c) Reject the hypothesis of no protein quantity effects and conclude that there is strong evidence that mean weights for two quantities (marginal means for high and low levels) of protein differ significantly.

It should be noted that the presence of interaction between protein source and quantity seems to be more than just a chance occurrence. The presence of interactions makes the statements about main effects somewhat difficult to interpret. Although the F tests are still valid, the hypotheses about the main effects cannot be interpreted only in terms of the α_i's and the β_j's.

4.15 WORKED EXAMPLE FOR MODEL II

Burdick and Graybill (1992, pp. 11–12) described a quality control experiment designed to study the sources of variability in the length of window screens. It is desired to determine the contribution of the variability in the final product that is due to operators, machines, and the operator × machine interaction. Three operators and four machines are randomly selected from the available operators and machines in the company and each operator makes two screens on each of the selected machines. The data collected in the experiment are given in Table 4.13. This is an example of a two-way crossed classification with replication. Here, our two factors are operators and machines and the experimental units that provide the replication are the machine-operator duos. Inasmuch as both factors are randomly selected from a rather large population, the data should be analyzed using Model II. Furthermore, since there are two observations for each combination of operator and machine, this will enable the experimenter to test for the presence of any interaction.

The mathematical model for this experiment would be

$$y_{ijk} = \mu + \alpha_i + \beta_j + (\alpha\beta)_{ij} + e_{ijk} \quad \begin{cases} i = 1, 2, 3 \\ j = 1, 2, 3, 4 \\ k = 1, 2, \end{cases}$$

where μ is the general mean, α_i is the effect of the i-th operator, β_j is the effect of the j-th machine, $(\alpha\beta)_{ij}$ is the interaction of the i-th operator with the j-th machine, and e_{jjk}'s are experimental errors. It is further assumed that $\alpha_i \sim N(0, \sigma_\alpha^2)$, $\beta_j \sim N(0, \sigma_\beta^2)$, $(\alpha\beta)_{ij} \sim N(0, \sigma_{\alpha\beta}^2)$; and that the α_i's, β_j's, $(\alpha\beta)_{ij}$'s, and e_{ijk}'s are mutually and completely independent.

The following computations lead to the analysis of variance table.

(i) The cell totals:

$$y_{11.} = 71.5, \quad y_{12.} = 72.8, \quad y_{13.} = 70.3, \quad y_{14.} = 72.0,$$
$$y_{21.} = 71.5, \quad y_{22.} = 71.5, \quad y_{23.} = 72.9, \quad y_{24.} = 69.8,$$
$$y_{31.} = 70.7, \quad y_{32.} = 72.1, \quad y_{33.} = 72.0, \quad y_{34.} = 72.4.$$

TABLE 4.13
Screen Lengths (in inches) from a Quality Control Experiment

	Machine			
Operator	1	2	3	4
1	36.3	36.7	35.1	35.2
	35.2	36.1	35.2	36.8
2	35.2	35.3	36.8	34.9
	36.3	36.2	36.1	34.9
3	35.8	36.0	35.9	36.3
	34.9	36.1	36.1	36.1

Source: Burdick and Graybill (1992, p. 118). Used with permission.

(ii) The row totals:

$$y_{1..} = 286.6, \quad y_{2..} = 285.7, \quad y_{3..} = 287.2.$$

(iii) The column totals:

$$y_{.1.} = 213.7, \quad y_{.2.} = 216.4, \quad y_{.3.} = 215.2, \quad y_{.4.} = 214.2.$$

(iv) The grand total:

$$y_{...} = 859.5.$$

(v)

$$\sum_{i=1}^{3}\sum_{j=1}^{4}\sum_{k=1}^{2} y_{ijk}^2 = (36.3)^2 + (35.2)^2 + \cdots + (36.1)^2 = 30{,}789.67.$$

(vi)

$$\frac{y_{...}^2}{3 \times 4 \times 2} = \frac{(859.5)^2}{24} = 30{,}780.8438.$$

(vii)

$$\frac{1}{4 \times 2}\sum_{i=1}^{3} y_{i..}^2 = \frac{(286.6)^2 + (285.7)^2 + (287.2)^2}{8} = 30{,}780.9863.$$

TABLE 4.14
Analysis of Variance for the Screen Lengths Data of Table 4.13

Source of Variation	Degrees of Freedom	Sum of Squares	Mean Square	Expected Mean Square	F Value	p-Value
Operator	2	0.1425	0.071	$\sigma_e^2 + 2\sigma_{\alpha\beta}^2 + 4 \times 2\sigma_\alpha^2$	0.101	0.905
Machine	3	0.7112	0.237	$\sigma_e^2 + 2\sigma_{\alpha\beta}^2 + 3 \times 2\sigma_\beta^2$	0.339	0.798
Interaction	6	4.1975	0.700	$\sigma_e^2 + 2\sigma_{\alpha\beta}^2$	2.222	0.113
Error	12	3.7750	0.315	σ_e^2		
Total	23	8.8262				

(viii)

$$\frac{1}{3 \times 2} \sum_{j=1}^{4} y_{\cdot j \cdot}^2 = \frac{(213.7)^2 + (216.4)^2 + \cdots + (214.2)^2}{6} = 30{,}781.5550.$$

(ix)

$$\frac{1}{2} \sum_{i=1}^{3} \sum_{j=1}^{4} y_{ij\cdot}^2 = \frac{(71.5)^2 + (72.8)^2 + \cdots + (72.4)^2}{2} = 30{,}785.8950.$$

(x) $SS_T = 30{,}789.67 - 30{,}780.8438 = 8.8262.$
(xi) $SS_{TC} = 30{,}785.8950 - 30{,}780.8438 = 5.0512.$
(xii) $SS_A = 30{,}780.9863 - 30{,}780.8438 = 0.1425.$
(xiii) $SS_B = 30{,}781.5550 - 30{,}780.8438 = 0.7112.$
(xiv) $SS_{AB} = 5.0512 - 0.1425 - 0.7112 = 4.1975.$
(xv) $SS_E = 8.8262 - 5.0512 = 3.7750.$

These results along with the remaining computations are summarized in Table 4.14.

We can test the hypotheses of interest using the results shown in Table 4.14. The presence of the interaction is tested by comparing the ratio 2.222 with the theoretical F distribution with (6, 12) degrees of freedom which is not significant ($p = 0.113$). Hence, there is no evidence of the existence of any interaction effects. The existence of a main effect due to operators is tested by comparing the ratio 0.101 with the theoretical F distribution with (2, 6) degrees of freedom which is also not significant ($p = 0.905$). Similarly, the other main effect due to machines is tested by comparing the ratio 0.339 with the F distribution with (3, 6) degrees of freedom and this again is not significant ($p = 0.798$). Thus, we may conclude that there are no significant differences between the operators as well as between the machines, and also there is no evidence of any interaction between the two factors.

Furthermore, to assess the relative contribution of the variance components, we may obtain their estimates using formulae (4.7.20) through (4.7.23). Thus, we find that

$$\hat{\sigma}_e^2 = 0.315,$$
$$\hat{\sigma}_{\alpha\beta}^2 = \frac{1}{2}(0.700 - 0.315) = 0.193,$$
$$\hat{\sigma}_\beta^2 = \frac{1}{6}(0.237 - 0.700) = -0.077,$$

and

$$\hat{\sigma}_\alpha^2 = \frac{1}{8}(0.071 - 0.700) = -0.079.$$

The negative estimates are an indication that the corresponding variance components may be zero. The results are consistent with the tests of hypotheses performed earlier. It is further evident that the larger part of the variability arises in the replication of measurements.

4.16 WORKED EXAMPLE FOR MODEL III

The following example is taken from an experiment described in Youden (1951, pp. 64–65). An experiment was performed to determine the effect of time aging on the strength of cement. Three mixes of cement were prepared and six specimens were made from each mix. Three specimens from each mix were tested after two days and later after seven days. The test specimens were two-inch cubes that yielded under the given load and were measured in units of 10 pounds. The data are presented in Table 4.15.

The experiment just described constitutes a mixed effects model. The mixes are random components, a sample of three drawn from a large number of mixes. The results of the experiment should be valid for the entire distribution of mixes. On the other hand, effects of aging are fixed effects. The conclusions of the experiment will reveal whether the yield loads differ after two or seven days, these periods being fixed. Hence, the data of Table 4.15 should be analyzed using Model III. Since there are three observations for each mix and aging combination, this will enable the experimenter to test for the presence of interaction. Interaction terms cannot be ignored since it is quite possible that the three mixes differ after a long period of time without differing after a short period. In other words, the effect of an additional period of time is different for three mixes; that is, interaction is present.

TABLE 4.15
Yield Loads for Cement Specimens

Aging	Mix 1	Mix 2	Mix 3
2-Day Test	574	524	576
	564	573	540
	550	551	592
7-Day Test	1,092	1,028	1,066
	1,086	1,073	1,045
	1,065	998	1,055

Source: Youden (1951, p. 65). Used with permission.

The mathematical model for this experiment would be

$$y_{ijk} = \mu + \alpha_i + \beta_j + (\alpha\beta)_{ij} + e_{ijk} \quad \begin{cases} i = 1, 2 \\ j = 1, 2, 3 \\ k = 1, 2, 3, \end{cases}$$

where μ is general mean, α_i is the effect of the i-th "aging" ($\sum_{i=1}^{2} \alpha_i = 0$); β_j is the effect of the j-th "mix" and is a random variable assumed to be normally distributed with mean zero and variance σ_β^2; $(\alpha\beta)_{ij}$ is the interaction of the i-th "aging" with the j-th mix and is a random variable assumed to be normally distributed with mean zero and variance $\frac{2-1}{2}\sigma_{\alpha\beta}^2$ ($\sum_{i=1}^{2}(\alpha\beta)_{ij} = 0$, for $j = 1, 2, 3$); and e_{jjk}'s are experimental errors assumed to be independently and normally distributed with mean zero and variance σ_e^2.

The following computations lead to the analysis of variance table.

(i) The cell totals:

$$y_{11.} = 1,688, \quad y_{12.} = 1,648, \quad y_{13.} = 1,708,$$
$$y_{21.} = 3,243, \quad y_{22.} = 3,099, \quad y_{23.} = 3,166.$$

(ii) The row totals:

$$y_{1..} = 5,044, \quad y_{2..} = 9,508.$$

(iii) The column totals:

$$y_{.1.} = 4,931, \quad y_{.2.} = 4,747, \quad y_{.3.} = 4,874.$$

Two-Way Crossed Classification with Interaction

(iv) The grand total:
$$y_{...} = 5{,}044 + 9{,}508 = 14{,}552.$$

(v)
$$\sum_{i=1}^{2}\sum_{j=1}^{3}\sum_{k=1}^{3} y_{ijk}^2 = (574)^2 + (564)^2 + \cdots + (1{,}055)^2 = 12{,}882{,}026.$$

(vi)
$$\frac{y_{...}^2}{2 \times 3 \times 3} = \frac{(14{,}552)^2}{18} = 11{,}764{,}483.5556.$$

(vii)
$$\frac{1}{3 \times 3}\sum_{i=1}^{2} y_{i..}^2 = \frac{(5{,}044)^2 + (9{,}508)^2}{4} = 12{,}871{,}555.5556.$$

(viii)
$$\frac{1}{2 \times 3}\sum_{j=1}^{3} y_{.j.}^2 = \frac{(4{,}931)^2 + (4{,}747)^2 + (4{,}874)^2}{6} = 11{,}767{,}441.0000.$$

(ix)
$$\frac{1}{3}\sum_{i=1}^{2}\sum_{j=1}^{3} y_{ij.}^2 = \frac{(1{,}688)^2 + (1{,}648)^2 + \cdots + (3{,}166)^2}{3} = 12{,}875{,}639.3333.$$

(x) $SS_T = 12{,}882{,}026 - 11{,}764{,}483.5556 = 1{,}117{,}542.4444.$
(xi) $SS_{TC} = 12{,}875{,}639.3333 - 11{,}764{,}483.5556 = 1{,}111{,}155.7777.$
(xii) $SS_A = 12{,}871{,}555.5556 - 11{,}764{,}483.5556 = 1{,}107{,}072.0000.$
(xiii) $SS_B = 11{,}767{,}441.0000 - 11{,}764{,}483.5556 = 2{,}957.4444.$
(xiv) $SS_{AB} = 1{,}111{,}155.7777 - 1{,}107{,}072.0000 - 2{,}957.4444 = 1{,}126.3333.$
(xv) $SS_E = 1{,}117{,}542.4444 - 1{,}111{,}155.7777 = 6{,}386.6667.$

These results along with the remaining analysis of variance computations are summarized in Table 4.16. Note that the numerical values of F tests are calculated differently here than in the case of Model I or Model II. The test for interaction is the same, but the test for random effects involves MS_B/MS_E and the test for fixed effects involves MS_A/MS_{AB}. If we choose the level of significance $\alpha = 0.05$, we find from Appendix Table V that

$$F[1, 2; 0.95] = 18.51$$

TABLE 4.16
Analysis of Variance for the Yield Loads Data of Table 4.15

Source of Variation	Degrees of Freedom	Sum of Squares	Mean Square	Expected Mean Square	F Value	p-Value
Aging	1	1,107,072.0000	1,107,072.000	$\sigma_e^2 + 3\sigma_{\alpha\beta}^2 + \frac{3 \times 3}{2-1}\sum_{i=1}^{2}\alpha_i^2$	1,965.80	<0.001
Mix	2	2,957.4444	1,478.722	$\sigma_e^2 + 2 \times 3\sigma_\beta^2$	2.78	0.102
Interaction	2	1,126.3333	563.167	$\sigma_e^2 + 3\sigma_{\alpha\beta}^2$	1.06	0.377
Error	12	6,386.6667	532.222	σ_e^2		
Total	17	1,117,542.444				

and

$$F[2, 12; 0.95] = 3.89.$$

Comparing these values with the computed F values given in Table 4.16, we may reach the following conclusions:

(a) Do not reject the hypothesis of no interaction effects and conclude that the data do not give sufficient evidence of the existence of interaction between the "aging" and the "mixes" ($p = 0.377$).
(b) Do not reject the hypothesis of no "mixes" effects and conclude that the mean strength of cement does not vary in the population of mixes.
(c) Reject the hypothesis of no "aging" effects and conclude that there is a significant effect from the additional five days of aging.

Furthermore, we can make a comparison of two- and seven-day aging effects by using Tukey's and Scheffé's methods of simultaneous confidence interval. For the Tukey's procedure, we find from Appendix Table X that

$$q[a, (a-1)(b-1); 1-\alpha] = q[2, 2; 0.95] = 6.09.$$

So that

$$q[a, (a-1)(b-1); 1-\alpha]\sqrt{\frac{MS_{AB}}{bn}} = 6.09\sqrt{\frac{563.167}{3 \times 3}} = 48.17$$

and, from (4.12.13), a 95 percent simultaneous confidence interval for $\alpha_1 - \alpha_2$ is given as

$$(1,056.44 - 560.44) - 48.17 < \alpha_2 - \alpha_1 < (1,056.44 - 560.44) + 48.17$$

or

$$447.83 < \alpha_2 - \alpha_1 < 544.17.$$

For the Scheffé's procedure, we find from Appendix Table V that

$$S^2 = F[a-1, (a-1)(b-1); 1-\alpha] = F[1, 2; 0.95] = 18.51,$$

so that

$$\sqrt{S^2(a-1)(bn)^{-1} \text{MS}_{AB} \sum_{i=1}^{a} \ell_i^2} = \sqrt{18.51(1)(3 \times 3)^{-1}(563.167)(2)} = 48.13$$

and from (4.12.4) a 95 percent simultaneous confidence interval for $\alpha_2 - \alpha_1$ is given as

$$(1{,}056.44 - 560.44) - 48.13 < \alpha_2 - \alpha_1 < (1{,}056.44 - 560.44) + 48.13$$

or

$$447.87 < \alpha_2 - \alpha_1 < 544.13.$$

Notice that with $a = 2$, there is only one contrast and both Tukey's and Scheffé's procedures are equivalent to the usual t test.

Finally, suppose it is desired to determine the power of the test when the difference in time effect is as large as 400 psi. Since the test specimens are in units of 10 lbs, 400 psi corresponds to 40 units. Furthermore, since $\alpha_1 + \alpha_2 = 0$, this gives $\alpha_1 = -20.0$ and $\alpha_2 = 20.0$. Now, from Section 4.11, the normalized noncentrality parameter is

$$\phi = \sqrt{\frac{bn \sum_{i=1}^{a} \alpha_i^2}{a(\sigma_e^2 + n\sigma_\beta^2)}}$$

$$= \sqrt{\frac{3 \times 3\{(-20.0)^2 + (20.0)^2\}}{2(563.167)}}$$

$$= 2.53,$$

where an estimate of $\sigma_e^2 + n\sigma_\beta^2$ is obtained from $\text{MS}_{AB} = 563.167$. Since the Pearson-Hartley charts do not contain a power curve for $v_1 = 1$ and $v_2 = 2$, we calculate the power using the noncentral t distribution. The noncentrality parameter (δ) for the noncentral t distribution is determined as $\delta = \sqrt{a\phi} = \sqrt{2}(2.53) = 3.58$, Now, entering the Appendix Chart I with $\alpha = 0.05$, $df = 2$, and $\delta = 3.58$, the power is found to be about 0.48. The use of Appendix Tables VI and VII with appropriate interpolation gives essentially the same result. Notice that a very small number of degrees of freedom for the t test makes it quite insensitive.

4.17 USE OF STATISTICAL COMPUTING PACKAGES

As in Section 3.16, for a two-way fixed effects analysis of variance with an equal number of observations and no missing values, the recommended procedure is SAS ANOVA. If n_{ij}'s are unequal because of only few missing values, they could be replaced by their respective cell means and the data could be analyzed as in the preceding. However, if there is a wide disparity between n_{ij}'s, the GLM procedure should be used. The GLM produces Type I, Type II, Type III, and Type IV sums of squares. When some cells are empty, caution should be used in choosing an appropriate sum of squares. For a random or mixed model analysis, GLM with RANDOM or TEST option should be used. For equal n_{ij}'s in each cell, estimates of variance components can be readily obtained from the entries of the analysis of variance table. For unequal n_{ij}'s, PROC MIXED or VARCOMP must be used for estimating variance components. For the details of SAS commands, see Section 11.1.

Among SPSS procedures, the ANOVA would be a better choice for fixed effects analysis involving a balanced layout. For the design involving an unequal number of observations per cell and a random and mixed model analysis, GLM or MANOVA must be used. For the estimation of variance components, VARCOMP (available in Release 7.0 and 8.0) is the procedure of choice. For instructions regarding SPSS commands, see Section 11.2.

In using BMDP programs, as indicated in Section 3.15, the two programs suited for this model are 7D and 2V if the analysis involves only fixed effects in the model. However, when the number of observations in each cell is rather large, 7D would be a better choice since it could provide comparative histograms and descriptive statistics for data in each cell. Similar to GLM, 2V is a general purpose program for performing fixed effects analysis of variance for both balanced and unbalanced data sets. For the analysis involving random and mixed effects models, 3V and 8V can be used. For designs with equal n_{ij}'s in each cell, 8V is recommended since it is simpler to use. For unequal n_{ij}'s, 3V would be the preferred choice. This program also provides estimates of variance components using maximum likelihood and restricted maximum likelihood procedures.

Utmost care should be exercised in using packaged programs for unequal sample sizes and when some cells are empty. It is important to find out how the individual program or procedure handles the empty cells and the assumptions it makes about the interaction terms. The user should make sure that the program outputs the appropriate sums of squares for the tests of hypotheses of interest. For some further discussion and details in this regard, see Milliken and Johnson (1992, Chapter 14).

4.18 WORKED EXAMPLES USING STATISTICAL PACKAGES

In this section, we illustrate the applications of statistical packages to perform two-way analysis of variance with interaction for the data sets employed in examples presented in Sections 4.13 through 4.16. Figures 4.1 through 4.4 illustrate the program instructions and the output results for analyzing data given in Tables 4.9, 4.11, 4.13, and 4.15, using SAS ANOVA/GLM, SPSS MANOVA/GLM, and BMDP 7D/2V/8V procedures. The typical output provides the data format listed at the top, cell means, and the entries of the analysis of variance table. It should be noticed that in each case, the results are the same as those provided using manual computations in Sections 4.13 through 4.16. However, note that certain tests of significance in a mixed model may differ from one program to the other since they make different model assumptions.

4.19 THE MEANING AND INTERPRETATION OF INTERACTION

In the discussion of the two-way model (4.1.1), we have assumed the existence of interaction to take into account the fact that the two factors may not be independent; that is, the effects of one factor may vary with the levels of the other factor. Thus, for example, suppose that the yield of a chemical process depends on two factors: the concentration of the chemical and the operating temperature. Now, if the yield at different concentration levels varies with the level of the operating temperature, we would say that interaction is present. The lack or presence of interaction is marked by parallelism or nonparallelism in the plots of average treatment responses. For example, consider two levels for each factor A and B, denoted by (A_1, A_2) and (B_1, B_2), respectively. Some possible patterns for observed cell means and presence or lack of interactions are illustrated in Figure 4.5. The graphical illustrations allow a visual inspection of factor effects and their interactions. Any nonparallel change in the average response is an indication of the presence of an interaction.

The existence or nonexistence of interaction effects, as inferred from the F test of interaction, can have very important bearing on how one interprets and uses the results of an experiment. When two factors A and B interact, an important question arises as to whether the main effects of A and B are meaningful measures to interpret. Thus, if the hypothesis of interaction is rejected, we may conclude that the effects of A and B are not additive; that is, factors A and B interact. If this happens, testing the significance of A and B factor effects becomes meaningless under the present formulation of the model. Note that accepting the hypothesis about the A factor effects means that there are no differences in the various levels of A when averaged over the levels of B. However, in the presence of interaction this interpretation is meaningless. The presence of interaction means that the effect of one factor is dependent on the levels of the other. Similarly, rejecting the hypothesis about the A factor effects when interaction is present is also meaningless. The same argument holds true

254   The Analysis of Variance

```
DATA FITNESS;                          The SAS System
INPUT AGE FITNESS RUNNING;       Analysis of Variance Procedure
DATALINES;
1 1 669                    Dependent Variable: RUNNING
1 1 671                                    Sum of        Mean
1 2 602                    Source     DF   Squares      Square    F Value   Pr > F
. . .
3 3 713                    Model       8   378352.00   47294.00   147.38    0.0001
;                          Error       9     2888.00     320.89
PROC ANOVA;                Corrected  17   381240.00
CLASSES AGE FITNESS;       Total
MODEL RUNNING=AGE FITNESS           R-Square      C.V.      Root MSE   RUNNING Mean
AGE*FITNESS;                        0.992425   2.508876     17.913         714.00
RUN;
CLASS   LEVELS   VALUES    Source     DF   Anova SS  Mean Square  F Value   Pr > F
AGE        3     1 2 3
FITNESS    3     1 2 3     AGE         2   195733.00   97866.50   304.99    0.0001
NUMBER OF OBS. IN DATA     FITNESS     2   160867.00   80433.50   250.66    0.0001
SET=18                     AGE*FITNESS 4    21752.00    5438.00    16.95    0.0003
```

(i) SAS application: SAS ANOVA instructions and output for the two-way fixed effects analysis of variance with two observations per cell.

```
DATA LIST                         Analysis of Variance-Design 1
/AGE 1 FITNESS 3
 RUNNING 5-8.            Tests of Significance for RUNNING using UNIQUE sums of squares
BEGIN DATA.
1 1 669                  Source of Variation    SS       DF      MS        F      Sig of F
1 1 671
. . .                    WITHIN+RESIDUAL      2888.00     9     320.89
3 3 713                  AGE                195733.00     2   97866.50   304.99    .000
END DATA.                FITNESS            160867.00     2   80433.50   250.66    .000
MANOVA RUNNING BY        AGE BY FITNESS      21752.00     4    5438.00    16.95    .000
AGE(1,3) FITNESS
(1,3)                    (Model)            378352.00     8   47294.00   147.38    .000
/DESIGN=AGE              (Total)            381240.00    17   22425.88
FITNESS AGE BY
FITNESS.                 R-Squared =  .992             Adjusted R-Squared =  .986
```

(ii) SPSS application: SPSS MANOVA instructions and output for the two-way fixed effects analysis of variance with two observations per cell.

```
/INPUT      FILE='C:\SAHAI    BMDP7D - ONE- AND TWO-WAY ANALYSIS OF VARIANCE WITH
            \TEXTO\EJE8.TXT'.          DATA SCREENING Release: 7.0 (BMDP/DYNAMIC)
            FORMAT=FREE.      ---------------------------------------------------------
            VARIABLES=3.      | ANALYSIS OF VARIANCE                               TAIL |
/VARIABLE   NAMES=AGE,FIT,RUN.|SOURCE    SUM OF SQUARES  DF  MEAN SQUARE  F VALUE  PROB.
/GROUP      VARIABLE=AGE,FIT. |-------   --------------  --  -----------  -------  -----|
            CODES(AGE)=1,2,3. |AGE          195733.0000   2   97866.5000   304.99  0.0000|
            NAMES(AGE)=A40,A50|FITNESS      160867.0000   2   80433.5000   250.66  0.0000|
                     ,A60.    |INTERACTION   21752.0000   4    5438.0000    16.95  0.0003|
            CODES(FIT)=1,2,3. |ERROR          2888.0000   9     320.8889
            NAMES(FIT)=L,M,H. |---------------------------------------------------------
/HISTOGRAM  GROUPING=AGE,FIT.             ANALYSIS OF VARIANCE;
            VARIABLE=RUNNING.          VARIANCES ARE NOT ASSUMED TO BE EQUAL
/END                          |WELCH                    8, 4              1031.90  0.0000|
1 1 669                       |BROWN-FORSYTHE
1 1 671                       |AGE                      2, 2               304.99  0.0033|
. . .                         |FITNESS                  2, 2               250.66  0.0040|
3 3 713                       |INTERACTION              4, 2                16.95  0.0565|
                              |---------------------------------------------------------
```

(iii) BMDP application: BMDP 7D instructions and output for the two-way fixed effects analysis of variance with two observations per cell.

FIGURE 4.1 Program Instructions and Output for the Two-Way Fixed Effects Analysis of Variance with Two Observations per Cell: Data on Running Time (in seconds) to Complete a 1.5 Mile Course (Table 4.9).

Two-Way Crossed Classification with Interaction

```
DATA RATWEIGT;
INPUT SOURCE QUANTITY GAINS;
DATALINES;
1 1 81
1 1 100
1 1 102
1 1 104
1 1 107
. . .
3 2 107;
PROC GLM;
CLASSES SOURCE QUANTITY;
MODEL GAINS=SOURCE QUANTITY
SOURCE*QUANTITY;
RUN;
CLASS    LEVELS    VALUES
SOURCE     3       1 2 3
QUANTITY   2       1 2
NUMBER OF OBS. IN DATA
SET=52
```

```
                    The SAS System
              General Linear Models Procedure
Dependent Variable: GAINS
                         Sum of       Mean
Source           DF      Squares      Square    F Value   Pr > F
Model             5     4807.2934    961.4587    4.86     0.0012
Error            46     9105.0143    197.9351
Corrected
Total            51    13912.3077

              R-Square        C.V.      Root MSE    GAINS Mean
              0.345542     16.05761     14.069       87.615

Source           DF    Type I SS   Mean Square   F Value   Pr > F
SOURCE            2     302.1077     151.0538     0.76     0.4720
QUANTITY          1    2771.9325    2771.9325    14.00     0.0005
SOURCE*QUANTITY   2    1733.2532     866.6266     4.38     0.0182

Source           DF   Type III SS  Mean Square   F Value   Pr > F
SOURCE            2     228.7266     114.3633     0.58     0.5652
QUANTITY          1    3274.4985    3274.4985    16.54     0.0002
SOURCE*QUANTITY   2    1733.2532     866.6266     4.38     0.0182
```

(i) SAS application: SAS GLM instructions and output for the two-way fixed effects analysis of variance with unequal numbers of observations per cell.

```
DATA LIST
/SOURCE 1 QUANTITY 3
 GAINS 5-7.
BEGIN DATA.
1 1 81
1 1 100
1 1 102
1 1 104
. . .
3 2 107
END DATA.
MANOVA GAINS BY
SOURCE(1,3) QUANTITY(1,2)
/DESIGN=SOURCE QUANTITY
 SOURCE BY QUANTITY.
```

```
              Analysis of Variance-Design 1
Tests of Significance for GAINS using UNIQUE sums of squares

Source of Variation      SS       DF      MS         F      Sig of F
WITHIN+RESIDUAL       9105.01     46    197.94
SOURCE                 228.73      2    114.36      .58      .565
QUANTITY              3274.50      1   3274.50    16.54      .000
SOURCE BY QUANTITY    1733.25      2    866.63     4.38      .018

(Model)               4807.29      5    961.46     4.86      .001
(Total)              13912.31     51    272.79

R-Squared =  .346              Adjusted R-Squared =  .274
```

(ii) SPSS application: SPSS MANOVA instructions and output for the two-way fixed effects analysis of variance with unequal numbers of observations per cell.

```
/INPUT    FILE='C:\SAHAI
          \TEXTO\EJE9.TXT'.
          FORMAT=FREE.
          VARIABLES=3.
/VARIABLE NAMES=SOURCE,
          QUANTITY,GAINS
/GROUP    VARIABLE=S,Q.
          CODES(SOURCE)=1,2,3.
          NAMES(SOURCE)=B,P,C.
          CODES(QUANTITY)=1,2.
          NAMES(QUANTITY)=H,L.
/DESIGN   DEPENDENT=GAINS.
/END
1 1 81
. . .
3 2 107
```

```
BMDP2V - ANALYSIS OF VARIANCE AND COVARIANCE WITH
         REPEATED MEASURES. Release: 7.0 (BMDP/DYNAMIC)

ANALYSIS OF VARIANCE FOR THE 1-ST DEPENDENT VARIABLE

THE TRIALS ARE REPRESENTED BY THE VARIABLES:GAINS

SOURCE       SUM OF       D.F.    MEAN           F         TAIL
             SQUARES              SQUARE                   PROB.

MEAN       393454.04800    1    393454.04800   1987.79    0.0000
SOURCE        228.72657    2       114.36328      0.58    0.5652
QUANTITY     3274.49851    1      3274.49851     16.54    0.0002
SQ           1733.25321    2       866.62660      4.38    0.0182
ERROR        9105.01429   46       197.93509
```

(iii) BMDP application: BMDP 2V instructions and output for the two-way fixed effects analysis of variance with unequal numbers of observations per cell.

FIGURE 4.2 Program Instructions and Output for the Two-Way Fixed Effects Analysis of Variance with Unequal Numbers of Observations per Cell: Data on Weight Gains of Rats under Different Diets (Table 4.11).

```
DATA SCREENLT;
INPUT OPERATOR MACHINE
LENGTHS;
DATALINES;
1 1 36.3
1 1 35.2
1 2 36.7
1 2 36.1
. . . .
3 4 36.1
;
PROC GLM;
CLASSES OPERATOR MACHINE;
MODEL LENGTHS=OPERATOR
MACHINE. OPERATOR*MACHINE;
RANDOM OPERATOR MACHINE
OPERATOR*MACHINE;
TEST H=OPERATOR
E=OPERATOR*MACHINE;
TEST H=MACHINE
E=OPERATOR*MACHINE;
RUN;
CLASS    LEVELS    VALUES
OPERATOR   3       1 2 3
MACHINE    4       1 2 3 4
NUMBER OF OBS. IN DATA
SET=24
```

```
                        The SAS System
                 General Linear Models Procedure
Dependent Variable: LENGTHS
                     Sum of        Mean
Source           DF  Squares       Square     F Value   Pr > F
Model            11  5.0512500     0.4592045  1.46      0.2626
Error            12  3.7750000     0.3145833
Corrected        23  8.8262500
Total
                R-Square    C.V.    Root MSE  LENGTHS Mean
                0.572299   1.566149  0.5609    35.813
Source           DF  Type III SS  Mean Square  F Value Pr > F
OPERATOR          2  0.1425000    0.0712500    0.23    0.8007
MACHINE           3  0.7112500    0.2370833    0.75    0.5411
OPERATOR*MACHINE  6  4.1975000    0.6995833    2.22    0.1125
Source              Type III Expected Mean Square
OPERATOR            Var(Error) + 2 Var(OPERATOR*MACHINE)
                    + 8 Var(OPERATOR)
MACHINE             Var(Error) + 2 Var(OPERATOR*MACHINE)
                    + 6 Var(MACHINE)
OPERATOR*MACHINE    Var(Error) + 2 Var(OPERATOR*MACHINE)
Tests of Hypotheses using the Type III MS for
OPERATOR*MACHINE as an error term
Source           DF  Type III SS  Mean Square  F Value Pr > F
OPERATOR          2  0.1425000    0.0712500    0.10    0.9047
Source           DF  Type III SS  Mean Square  F Value Pr > F
MACHINE           3  0.7112500    0.23708333   0.34    0.7985
```

(i) SAS application: SAS GLM instructions and output for the two-way random effects analysis of variance with two observations per cell.

```
DATA LIST
/OPERATOR 1
 MACHINE 3
 LENGTHS 5-8(1)
BEGIN DATA.
1 1 36.3
1 1 35.2
1 2 36.7
1 2 36.1
1 3 35.1
1 3 35.2
1 4 35.2
. . .
3 4 36.1
END DATA.
GLM LENGTHS BY
 OPERATOR MACHINE
/DESIGN OPERATOR
 MACHINE
 OPERATOR*MACHINE
/RANDOM OPERATOR
 MACHINE.
```

```
Tests of Between-Subjects Effects  Dependent Variable: LENGTHS
Source              Type III SS   df  Mean Square    F       Sig.
OPERATOR  Hypothesis    .142       2   7.125E-02    .102    .905
          Error        4.197       6    .700(a)
MACHINE   Hypothesis    .711       3    .237        .339    .798
          Error        4.197       6    .700(a)
OPERATOR* Hypothesis   4.197       6    .700       2.224    .112
MACHINE   Error        3.775      12    .315(b)
a MS(OPERATOR*MACHINE)  b MS(Error)

                        Expected Mean Squares(a,b)
                          Variance Component
Source            Var(O)     Var(M)    Var(O*M)    Var(Error)
OPERATOR          8.000      .000       2.000       1.000
MACHINE           .000       6.000      2.000       1.000
OPERATOR*MACHINE  .000       .000       2.000       1.000
Error             .000       .000       .000        1.000
a For each source, the expected mean square equals the sum of the
coefficients in the cells times the variance components, plus a
quadratic term involving effects in the Quadratic Term cell.
b Expected Mean Squares are based on the Type III Sums of Squares.
```

(ii) SPSS application: SPSS GLM instructions and output for the two-way random effects analysis of variance with two observations per cell.

FIGURE 4.3 Program Instructions and Output for the Two-Way Random Effects Analysis of Variance with Two Observations per Cell: Data on Screen Lengths from a Quality Control Experiment (Table 4.13).

about the effects of factor B. Thus, when stating the effects of one factor it is necessary to specify the level of the other. This is the most important meaning of the interaction, namely, when interactions are present, the factors themselves cannot be evaluated individually. The presence of interactions requires that the factors be evaluated jointly rather than individually.

Two-Way Crossed Classification with Interaction

```
/INPUT     FILE=C:\SAHAI        BMDP8V  -  GENERAL MIXED MODEL ANALYSIS OF VARIANCE
           \TEXTO\EJE10.TXT'.           -  EQUAL CELL SIZES Release: 7.0 (BMDP/DYNAMIC)
           FORMAT=FREE.
           VARIABLES=2.         ANALYSIS OF VARIANCE FOR DEPENDENT VARIABLE   1
/VARIABLE  NAMES=L1,L2.          SOURCE       ERROR     SUM OF     D.F.   MEAN            F     PROB.
/DESIGN    NAMES=O,M,L.                       TERM      SQUARES           SQUARE
           LEVELS=3,4,2.        1 MEAN                  3.0780843E+4  1   30780.84348
           RANDOM=O,M,L.        2 OPERATOR    OM        0.1424993E+0  2       0.07125  0.10  0.9047
           MODEL='O,M,L(OM)'.   3 MACHINE     OM        0.7112480E+0  3       0.23708  0.34  0.7985
/END                            4 OM          L(OM)     4.1974911E+0  6       0.69958  2.22  0.1125
36.3 35.2                       5 L(OM)                 3.7749950E+0 12       0.31458
36.7 36.1
  .    .                          SOURCE           EXPECTED MEAN             ESTIMATES OF
36.3 36.1                                              SQUARE              VARIANCE COMPONENTS
ANALYSIS OF VARIANCE DESIGN     1 MEAN        24(1)+8(2)+6(3)+2(4)+(5)       1282.55145
INDEX              O   M   L    2 OPERATOR    8(2)+2(4)+(5)                    -0.07854
NUMBER OF LEVELS   3   4   2    3 MACHINE     6(3)+2(4)+(5)                    -0.07708
POPULATION SIZE  INF INF INF    4 OM          2(4)+(5)                          0.19250
MODEL     O, M, L(OM)           5 L(OM)       (5)                               0.31458
```

(iii) BMDP application: BMDP 8V instructions and output for the two-way random effects analysis of variance with two observations per cell.

FIGURE 4.3 (*continued*)

Significant interactions serve as a warning: treatment differences possibly do exist, but to specify exactly how the treatments differ, one must look within the levels of the other factor. The presence of the interaction effects is a signal that in any predictive use of the results, effects ascribed to a particular treatment representing one factor are best qualified by specifying the level of the other factor. This is especially important if one is going to try to use estimated effects in forecasting the result of a treatment to an experimental unit. If interaction effects are present, the best forecast can be made only if the particular levels of both factors are known.

When the observations suggest the presence of significant interactions, it is important to determine whether large interactions really do exist or whether there may be some other reasons for the presence of the interactions. Often large interactions may exist as a result of the dependent variable being measured on an inappropriate scale, and the use of a simple transformation may remove most of the interaction effects. Some simple transformations that are helpful in reducing the importance of interactions include the logarithmic, reciprocal, square, and square-root transformations (see Section 2.22 for a discussion of these transformations).

Sometimes, the investigator may think that there are no interactions; however, the data obtained may indicate a considerable amount of interactions. This could possibly happen purely by chance variation. On the other hand, such unexpectedly large interactions may simply occur due to the presence of outliers (observations much different from the rest of the data). The entire interaction may depend upon just one observation that may be wrong or an outlier. One should look at the data more carefully for the presence of an outlier before discarding them. If after further examination, the data look normal, there is the possibility of some complicated and unsuspected phenomenon that may require investigation. If the observations were not made using some random device, considerable time effects may be embedded in the data obtained. Thus,

```
DATA YIELDLOD;
INPUT AGING MIX YIELD;
DATALINES;
1 1 574
1 1 564
1 1 550
1 2 524
1 2 573
  . . .
2 3 1055
;
PROC GLM;
CLASSES AGING MIX;
MODEL YIELD=AGING MIX
 AGING*MIX;
RANDOM MIX AGING*MIX;
TEST H=AGING E=AGING*MIX;
RUN;
CLASS   `LEVELS   VALUES
AGING     2        1 2
MIX       3        1 2 3
NUMBER OF OBS. IN DATA
SET=18
```

```
                The SAS System
         General Linear Models Procedure
Dependent Variable: YIELD
              Sum of        Mean
Source    DF  Squares       Square      F Value   Pr > F
Model     5   1111155.7778  222231.1556 417.55    0.0001
Error     12  6386.6667     532.2222
Corrected 17  1117542.4444
Total

          R-Square    C.V.       Root MSE   YIELD Mean
          0.994285    2.8536212  23.069942  808.44444444

Source    DF  Type III SS   Mean Square  F Value   Pr > F
AGING     1   1107072.000   1107072.000  2080.09   0.0001
MIX       2   2957.444      1478.722     2.78      0.1020
AGING*MIX 2   1126.333      563.167      1.06      0.3774

Source    Type III Expected Mean Square
AGING     Var(Error) + 3 Var(AGING*MIX) + Q(AGING)
MIX       Var(Error) + 3 Var(AGING*MIX) + 6 Var(MIX)
AGING*MIX Var(Error) + 3 Var(AGING*MIX)
Tests of Hypotheses using the Type III MS for AGING*MIX as
an error term
Source    DF  Type III SS   Mean Square  F Value   Pr > F
AGING     1   1107072.000   1107072.000  1965.80   0.0005
```

(i) SAS application: SAS GLM instructions and output for the two-way mixed effects analysis of variance with three observations per cell.

```
DATA LIST
 /AGING 1 MIX 3
  YIELD 5-8.
 BEGIN DATA.
1 1  574
1 1  564
1 1  550
1 2  524
1 2  573
1 2  551
1 3  576
1 3  540
1 3  592
 . . .
2 3 1055
END DATA.
GLM YIELD BY
 AGING MIX
 /DESIGN AGING
  MIX AGING*MIX
 /RANDOM MIX.
```

```
         Tests of Between-Subjects Effects   Dependent Variable: YIELD

Source                   Type III SS    df   Mean Square     F         Sig.
AGING     Hypothesis     1107072.000    1    1107072.000     1965.798  .001
          Error          1126.333       2    563.167(a)
MIX       Hypothesis     2957.444       2    1478.722        2.626     .276
          Error          1126.333       2    563.167(a)
AGING*MIX Hypothesis     1126.333       2    563.167         1.058     .377
          Error          6386.667       12   532.222(b)
a MS(AGING*MIX)  b MS(Error)

                Expected Mean Squares(a,b)
                     Variance Component
Source     Var(MIX)   Var(AGING*MIX)   Var(Error)   Quadratic Term
AGING      .000       3.000            1.000        Aging
MIX        6.000      3.000            1.000
AGING*MIX  .000       3.000            1.000
Error      .000       .000             1.000
a For each source, the expected mean square equals the sum of the coeff-
icients in the cells times the variance components, plus a quadratic term
involving effects in the Quadratic Term cell. b Expected Mean Squares are
based on the Type III Sums of Squares.
```

(ii) SPSS application: SPSS GLM instructions and output for the two-way mixed effects analysis of variance with three observations per cell.

FIGURE 4.4 Program Instructions and Output for the Two-Way Mixed Effects Analysis of Variance with Three Observations per Cell: Data on Yield Loads for Cement Specimens (Table 4.15).

the error terms can no longer be assumed to be uncorrelated. In some cases, an uncontrolled variable may affect the results of the observations showing the presence of interactions. For example, in a laboratory experiment involving mice, the location of the cage may have an effect on the outcome and if this factor was left uncontrolled or the mice were not randomly assigned, we might observe an apparent interaction when there was none.

It has also been found that interactions frequently occur when the main effects are large. Interactions usually become less important by reducing the differences among the levels of treatment, and thus moderating the size of the main effects.

Two-Way Crossed Classification with Interaction 259

```
/INPUT     FILE='C:\SAHAI         BMDP8V - GENERAL MIXED MODEL ANALYSIS OF VARIANCE
           \TEXTO\EJE11.TXT'.            - EQUAL CELL SIZES Release: 7.0 (BMDP/DYNAMIC)
           FORMAT=FREE.           ANALYSIS OF VARIANCE FOR DEPENDENT VARIABLE 1
           VARIABLES=3.
/VARIABLE  NAMES=Y1,Y2,Y3.        SOURCE  ERROR   SUM OF        D.F.  MEAN           F       PROB.
/DESIGN    NAMES=AGING,MIX,               TERM    SQUARES             SQUARE
           YIELD.                 1 MEAN  MIX     11764483.56   1     11764483.6  7955.84  0.0001
           LEVELS=2,3,3.          2 AGING AM       1107072.00   1      1107072.0  1965.80  0.0005
           RANDOM=MIX.            3 MIX   Y(AM)       2957.44   2         1478.7     2.78  0.1020
           FIXED=AGING.           4 AM    Y(AM)       1126.33   2          563.2     1.06  0.3774
           MODEL='A,M,Y(AM)'.     5 Y(AM)             6386.67  12          532.2
/END                              SOURCE  EXPECTED MEAN               ESTIMATES OF
 574   564   550                          SQUARE                      VARIANCE COMPONENTS
 524   573   551                  1 MEAN  18(1)+6(3)+(5)              653500.26852
 576   540   592                  2 AGING  9(2)+3(4)+(5)              122945.42593
1092  1086  1065                  3 MIX    6(3)+(5)                      157.75000
1028  1073   998                  4 AM     3(4)+(5)                       10.31481
1066  1045  1055                  5 Y(AM)   (5)                          532.22222
```

(iii) BMDP application: BMDP 8V instructions and output for the two-way mixed effects analysis of variance with three observations per cell.

FIGURE 4.4 (*continued*)

For this reason, the presence of interaction effects can be most important to the interpretation of the experiment. Although it is necessary to consider possible interaction effects even in fairly simple experiments, the subject of interaction and of the interpretation that should be given to significant tests for interaction is neither simple nor fully explored. For a broad review of various aspects of interactions, see Cox (1984).

4.20 INTERACTION WITH ONE OBSERVATION PER CELL

In the discussion of model (3.1.1) in Chapter 3, we had assumed that there are no interaction terms. If the existence of interaction is assumed, model (3.1.1) becomes

$$y_{ij} = \mu + \alpha_i + \beta_j + (\alpha\beta)_{ij} + e_{ij}, \qquad (4.20.1)$$

where μ, α_i's, β_j's, and e_{ij}'s are defined as in model (3.1.1), and $(\alpha\beta)_{ij}$'s are the interaction effects between factors A and B, which are assumed to be constant under Model I and are randomly distributed with mean zero and variance $\sigma_{\alpha\beta}^2$ under Models II and III. Proceeding as before, the pertinent analysis of variance can be derived and is shown in Table 4.17.

On comparing Tables 3.2 and 4.17, it is seen that they are same except for the differences in the expressions of expected mean square column. In Table 4.17 if we let $(\alpha\beta)_{ij} = 0$ or $\sigma_{\alpha\beta}^2 = 0$ and change the word interaction to error, we have exactly the same analysis of variance table as in Table 3.2. However, if the assumption of no interaction is not tenable, we have the following inferential problems. Under Model I, no direct tests are possible since the hypothesis that either of the effects α_i's and β_j's or the interaction $(\alpha\beta)_{ij}$'s are zero gives us no suitable mean squares to compare. When the interactions are present, SS_{AB} has a noncentral chi-square distribution and the F ratios MS_A/MS_{AB} and

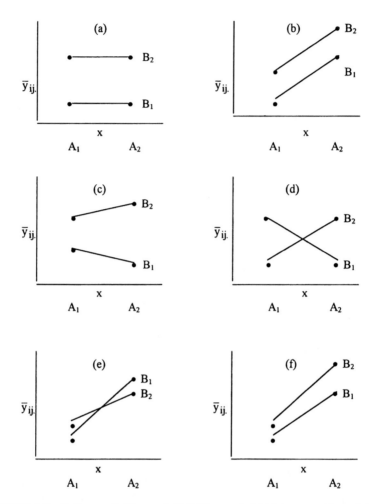

FIGURE 4.5 Patterns of Observed Cell Means and Existence or Nonexistence of Interaction Effects: (a) No effect of factor A, large effect of factor B, and no AB interaction; (b) Large effect of factor A, moderate effect of factor B, and no AB interaction; (c) No effect of factor A, large effect of factor B, and large AB interaction; (d) No effect of factor A, no effect of factor B, but large AB interaction; (e) Large effect of factor A, no effect of factor B, with small AB interaction; (f) Large effect of factor A, small effect of factor B, with small AB interaction. (The graphs (a) to (f) are obtained by representing the levels of factor A as values on the x-axis and plotting the cell means at those levels as values on the y-axis. Separate curves are drawn for each level of factor B. Alternatively, one could represent the levels of B on the x-axis and separate curves drawn for each level of factor A.)

TABLE 4.17
Analysis of Variance for Model (4.20.1)

Source of Variation	Degrees of Freedom	Sum of Squares	Mean Square	Expected Mean Square		
				Model I	Model II	Model III
Due to A	$a-1$	SS_A	MS_A	$\sigma_e^2 + \dfrac{b}{a-1}\sum_{i=1}^{a}\alpha_i^2$	$\sigma_e^2 + \sigma_{\alpha\beta}^2 + b\sigma_\alpha^2$	$\sigma_e^2 + \sigma_{\alpha\beta}^2 + \dfrac{b}{a-1}\sum_{i=1}^{a}\alpha_i^2$
Due to B	$b-1$	SS_B	MS_B	$\sigma_e^2 + \dfrac{a}{b-1}\sum_{j=1}^{b}\beta_j^2$	$\sigma_e^2 + \sigma_{\alpha\beta}^2 + a\sigma_\beta^2$	$\sigma_e^2 + \sigma_{\alpha\beta}^2 + a\sigma_\beta^2$
Interaction $A \times B$	$(a-1)(b-1)$	SS_{AB}	MS_{AB}	$\sigma_e^2 + \dfrac{1}{(a-1)(b-1)}\sum_{i=1}^{a}\sum_{j=1}^{b}(\alpha\beta)_{ij}^2$	$\sigma_e^2 + \sigma_{\alpha\beta}^2$	$\sigma_e^2 + \sigma_{\alpha\beta}^2$

MS_B/MS_{AB} have doubly noncentral F distributions.[8] The usual F tests for main effects α_i's and β_j's may be inefficient if there is appreciable interaction so that $\sum_{i=1}^{a}\sum_{j=1}^{b}(\alpha\beta)_{ij}^2 \neq 0$, since the denominator mean square will be inflated by the extra component. On the other hand, if either variance ratio is significant, it may be taken that the corresponding effect is real.

Even though there are no direct tests for interaction effects, Tukey (1949b) has devised a test that may be used for testing the existence of interaction terms. The null hypothesis and the alternate are:[9]

$$H_0 : (\alpha\beta)_{ij} = 0, \quad i = 1, 2, \ldots, a; j = 1, 2, \ldots, b$$

versus (4.20.2)

$$H_1 : \text{not all } (\alpha\beta)_{ij}\text{'s are zero.}$$

The procedure requires the computation of the sum of squares for nonadditivity defined by

$$SS_N = \frac{\left[\sum_{i=1}^{a}\sum_{j=1}^{b} y_{ij}(\bar{y}_{i.} - \bar{y}_{..})(\bar{y}_{.j} - \bar{y}_{..})\right]^2}{\sum_{i=1}^{a}(\bar{y}_{i.} - \bar{y}_{..})^2 \sum_{j=1}^{b}(\bar{y}_{.j} - \bar{y}_{..})^2}. \quad (4.20.3)$$

It can be shown that under H_0, the statistic

$$F^* = \frac{SS_N}{1} \bigg/ \frac{SS_{AB} - SS_N}{(a-1)(b-1) - 1} \quad (4.20.4)$$

is distributed as $F[1, (a-1)(b-1) - 1]$ variable. Note that there is one degree of freedom associated with SS_N and $(a-1)(b-1) - 1 = ab - a - b$ degrees of freedom are associated with $SS_{AB} - SS_N$. Thus, a large value of F^* leads to the rejection of H_0.

For computational purposes, the SS_N term can be further simplified by expanding the numerator in (4.20.3) into four terms and then rearranging them as follows

$$SS_N = \frac{\left[\sum_{i=1}^{a}\sum_{j=1}^{b} y_{ij} y_{i.} y_{.j} - y_{..}\left(\sum_{i=1}^{a}\frac{y_{i.}^2}{a} + \sum_{j=1}^{b}\frac{y_{.j}^2}{b} - \frac{y_{..}^2}{ab}\right)\right]^2}{ab(SS_A)(SS_B)}. \quad (4.20.5)$$

The first term in the numerator of (4.20.5), that is, $\sum_{i=1}^{a}\sum_{j=1}^{b} y_{ij} y_{i.} y_{.j}$, can

---

[8] For a definition of the doubly noncentral F distribution, see Appendix H.
[9] Technically speaking the interaction hypothesis in (4.20.2) is incorrect. Tukey (1949b) considered inreractions of the form $(\alpha\beta)_{ij} = G\alpha_i\beta_j$ with G fixed but unknown, leading to one-degree-of-freedom test of the interaction hypothesis $H_0 : G = 0$.

be more easily calculated by rewriting it as $\sum_{i=1}^{a} y_{i.}[\sum_{j=1}^{b} y_{ij} y_{.j}]$. The second term within parentheses in the numerator of (4.20.5) is equivalent to $SS_A + SS_B + y_{..}^2/ab$.

The test is commonly known as Tukey's one degree of freedom test for nonadditivity. It is discussed in detail by Scheffé (1959, pp. 129–134) and Rao (1973, pp. 249–255) and a numerical example appears in Ostle and Mensing (1975, Section 11.3).

Example 1. We illustrate Tukey's test for the data of Table 3.3. To calculate the test statistic (4.20.4), we obtain

$$\sum_{i=1}^{a}\sum_{j=1}^{b} y_{ij}(\bar{y}_{i.} - \bar{y}_{..})(\bar{y}_{.j} - \bar{y}_{..})$$
$$= 241(261.67 - 239.5)(226.25 - 239.5)$$
$$+ \cdots + 227(232.33 - 239.5)(237.25 - 239.5)$$
$$= -2{,}332.1405,$$

and

$$\sum_{i=1}^{a} (\bar{y}_{i.} - \bar{y}_{..})^2 = \frac{SS_A}{3} = \frac{3{,}141.667}{3} = 1{,}047.2223,$$

$$\sum_{j=1}^{b} (\bar{y}_{.j} - \bar{y}_{..})^2 = \frac{SS_B}{4} = \frac{1{,}683.500}{4} = 420.8750.$$

Hence, from (4.20.3), we have $SS_N = \dfrac{(-2{,}332.1405)^2}{(1{,}047.2223)(420.87750)} = 12.340$.
Finally, we obtain the test statistic (4.20.4) as

$$F* = \frac{12.340}{1} \bigg/ \frac{1{,}173.833 - 12.340}{6 - 1} = 0.05.$$

Assuming the level of significance at $\alpha = 0.05$, we obtain $F[1, 5; 0.95] = 6.61$. Since $F* < 6.61$, we may conclude that material and position do not interact ($p = 0.832$). The use of the no-interaction model for the data in Table 3.3, therefore, seems to be reasonable.

Remarks: (i) The power function of Tukey's test for nonadditivity has been studied by Ghosh and Sharma (1963), and Hegemann and Johnson (1976). Milliken and Graybill (1971) developed tests for interaction in the two-way model with missing data. In addition, a variety of tests that are sensitive to particular non-additivity structures have also been proposed in the literature (see, e.g., Mandel (1971), Hirotsu (1983), Miyakawa (1993)). Krishnaiah and Yochmowitz (1980), Johnson and Graybill (1972a), and Bolk

(1993) provide reviews of additivity tests. For a generalization of Tukey's test for a two-way classification to any general analysis of variance or experimental design model, see Milliken and Graybill (1970).

(ii) If Tukey's test shows the presence of interaction effects, some simple transformations such as a square-root or logarithmic transformation may be employed to see if the interaction can be removed or made negligible. Johnson and Graybill (1972b) discuss an approximate method of analysis in the presence of interaction effects.

(iii) Tukey's test of nonadditivity can be performed using SAS GLM procedure by first fitting a two-way model with factors A and B as sources. The predicted values from the fitted model are squared and then the procedure is run again with the MODEL statement that includes A, B and square predictions as sources. The square predictions do not appear in the class statement and appear as the last term of the MODEL statement (as a covariate). The following SAS codes illustrate the procedure:

```
data tukey;              data t;
input a b y;             set t;
datalines;               p2 = pred * pred;
...                      proc glm;
proc glm;                class a b;
class a b;               model y = a b p2;
model y = a b;           run;
output out = t t = pred;
```

Under Model II, it is possible to test for the main effects (i. e., $\sigma_\alpha^2 = 0$ or $\sigma_\beta^2 = 0$) by dividing their respective mean squares by the interaction mean square.[10] However, we have lost our F test for the hypothesis $\sigma_{\alpha\beta}^2 = 0$, although Tukey's one degree of freedom test for nonadditivity described in the foregoing can also be used here. In regard to the point estimation of variance components, the estimators (3.7.17) and (3.7.18) are still unbiased for σ_α^2 and σ_β^2, but our estimate (3.7.16) of the error variance is biased since $E(\mathrm{MS}_E) = \sigma_e^2 + \sigma_{\alpha\beta}^2$. There is not much one can do about the MS_E being a biased estimator of σ_e^2 except to assume that $\sigma_{\alpha\beta}^2 = 0$; if the assumption were incorrect, one would be erring on the conservative side because MS_E would tend to overestimate σ_e^2.

Similar remarks apply for Model III.

4.21 ALTERNATE MIXED MODELS

As remarked in Section 4.2, several different types of mixed models have been proposed in the statistical literature. Among other models proposed are those by Tukey (1949a), Wilk and Kempthorne (1955, 1956), Scheffé (1956b), and Smith and Murray (1984).[11] These models differ from the "standard" mixed model, discussed earlier in this chapter, in terms of the assumptions about the random effects β_j's and $(\alpha\beta)_{ij}$'s. In this section, we briefly describe one of these

---

[10] For a discussion of the robustness of the tests for σ_α^2 and σ_β^2, see Tan (1981).

[11] Smith and Murray (1984) proposed a model that employs covariance components to allow negative correlations among observations within the same cell.

TABLE 4.18
Analysis of Variance for an Alternate Mixed Model

Source of Variation	Degrees of Freedom	Sum of Squares	Mean Square	Expected Mean Square
Due to A	$a - 1$	SS_A	MS_A	$\sigma_e^2 + n\sigma_{\alpha\beta}^2 + \dfrac{bn}{a-1}\sum_{i=1}^{a}\alpha_i^2$
Due to B	$b - 1$	SS_B	MS_B	$\sigma_e^2 + n\sigma_{\alpha\beta}^2 + an\sigma_\beta^2$
Interaction $A \times B$	$(a-1)(b-1)$	SS_{AB}	MS_{AB}	$\sigma_e^2 + n\sigma_{\alpha\beta}^2$
Error	$ab(n-1)$	SS_E	MS_E	σ_e^2

alternate models. Suppose that the α_i's are fixed effects such that $\sum_{i=1}^{a}\alpha_i = 0$ and β_j's are independently distributed normal random variables with mean zero and variance σ_β^2. The interaction effects $(\alpha\beta)_{ij}$'s are also independently distributed normal random variables with mean zero and variance $\sigma_{\alpha\beta}^2$; and $(\alpha\beta)_{ij}$'s are independent of the β_j's. Note that the main difference between this mixed model and the "standard" mixed model discussed earlier is the assumption about the independence of the interaction effects.[12] The analysis of variance and expected mean squares for this model are shown in Table 4.18. On comparing this table to Table 4.2, we note that the only noticeable difference is the inclusion of the variance component $\sigma_{\alpha\beta}^2$ in the expected mean square for the random effects which does not appear in the "standard" model. (In fact, there are some other minor differences due to different definitions of the variance of the interaction effects in the two models, but these do not affect the analysis.)

Under this model, the hypothesis

$$H_0^B : \sigma_\beta^2 = 0 \quad \text{versus} \quad H_1^B : \sigma_\beta^2 > 0$$

would be tested by the statistic

$$F'_B = \frac{MS_B}{MS_{AB}}$$

in contrast with the statistic $F_B = MS_B/MS_E$ used in the "standard" model. The test is usually more conservative than the one based on the "standard" model, since MS_{AB} will in general be larger than MS_E. Again, the analysis of variance procedure may be used to estimate the variance components. From the mean square column of Table 4.18, we find that the only variance component which would have different estimates from the ones obtained in the "standard"

---

[12] The major criticism of the mixed model described here concerns the assumption of independence of the $(\alpha\beta)_{ij}$'s since it is felt that these random terms within a given level of the factor B will often be correlated.

model is σ_β^2, which is now estimated by

$$\hat{\sigma}_\beta^2 = \frac{MS_B - MS_{AB}}{an}.$$

The hypothesis concerning the fixed effects factor is

$$H_0^A : \alpha_i = 0 \text{ for all } i \quad \text{versus} \quad H_1^A : \alpha_i \neq 0 \text{ for at least one } i.$$

An exact α-level test of H_0^A is to reject H_0^A if $MS_A/MS_{AB} > F[a-1, (a-1)(b-1); 1-\alpha]$. Notice that this is the same test as the one under the 'standard' model discussed in Section 4.6. The fixed effects μ, $\mu + \alpha_i$'s, α_i's, $\alpha_i - \alpha_{i'}$ are, of course, estimated by

$$\hat{\mu} = \bar{y}_{...},$$
$$\widehat{\mu + \alpha_i} = \bar{y}_{i..}, \quad i = 1, 2, \ldots, a,$$
$$\hat{\alpha}_i = \bar{y}_{i..} - \bar{y}_{...}, \quad i = 1, 2, \ldots, a,$$

and

$$\widehat{\alpha_i - \alpha_{i'}} = \bar{y}_{i..} - \bar{y}_{i'..}, \quad i \neq i',$$

which are the same estimates as in the "standard" model. However, the variances of the estimates will differ from those in the "standard" model. Thus,

$$\text{Var}(\bar{y}_{...}) = \frac{\sigma_e^2 + n\sigma_{\alpha\beta}^2 + an\sigma_\beta^2}{abn},$$

$$\text{Var}(\bar{y}_{i..}) = \frac{\sigma_e^2 + n\sigma_{\alpha\beta}^2 + n\sigma_\beta^2}{bn},$$

$$\text{Var}(\bar{y}_{ij.}) = \frac{\sigma_e^2 + n\sigma_{\alpha\beta}^2 + n\sigma_\beta^2}{n},$$

and

$$\text{Var}(\bar{y}_{i..} - \bar{y}_{i'..}) = \frac{2(\sigma_e^2 + n\sigma_{\alpha\beta}^2)}{bn}.$$

An unbiased estimate of $\text{Var}(\bar{y}_{i..})$ can be obtained by using an appropriate linear combination of the mean squares. An approximate $1 - \alpha$ level confidence interval for $\mu + \alpha_i$ can be constructed using

$$\bar{y}_{i..} \pm t[\nu, 1 - \alpha/2]\sqrt{\frac{\hat{\sigma}_e^2 + n\hat{\sigma}_{\alpha\beta}^2 + n\hat{\sigma}_\beta^2}{bn}},$$

where the degrees of freedom ν will be estimated using the Satterthwaite procedure. Similarly, an exact confidence interval for $\alpha_i - \alpha_{i'}$ is given by

$$\bar{y}_{i..} - \bar{y}_{i'..} \pm t[(a-1)(b-1), 1-\alpha/2]\sqrt{\frac{2MS_{AB}}{bn}},$$

and an exact confidence interval for a general contrast of the form $\sum_{i=1}^{a} \ell_i \alpha_i$ ($\sum_{i=1}^{a} \ell_i = 0$) is $\sum_{i=1}^{a} \ell_i \bar{y}_{i..} \pm t[(a-1)(b-1), 1-\alpha/2]\sqrt{MS_{AB} \sum_{i=1}^{a} \ell_i^2 / bn}$, which is exactly the same as that given by (4.8.9).

The "standard" model as well the new mixed model described here are special cases of the mixed model discussed by Scheffé (1956b; 1959, pp. 261–274). According to the Scheffé's model, the observation y_{ijk} is represented by

$$y_{ijk} = m_{ij} + e_{ijk} \quad \begin{cases} i = 1, 2, \ldots, a \\ j = 1, 2, \ldots, b \\ k = 1, 2, \ldots, n, \end{cases}$$

where m_{ij} and e_{ijk} are mutually and completely independent random variables. Furthermore, m_{ij} is given by the linear structure:

$$m_{ij} = \mu + \alpha_i + \beta_j + (\alpha\beta)_{ij},$$

where

$$E(m_{ij}) = \mu + \alpha_i,$$

with

$$\sum_{i=1}^{a} \alpha_i = 0, \quad \beta_j = \frac{1}{b}\sum_{i=1}^{a} m_{ij} - \mu$$

and

$$\sum_{i=1}^{a} (\alpha\beta)_{ij} = 0, \quad j = 1, 2, \ldots, b.$$

Thus, for the Scheffé's model the restrictions on $(\alpha\beta)_{ij}$'s are the same as in the "standard" mixed model discussed in Section 4.2. The main difference between them arises in specifying the variance-covariance structure of the random components β_j and $(\alpha\beta)_{ij}$.

Remark: Scheffé assumes that the vectors $(\beta_j, (\alpha\beta)_{1j}, (\alpha\beta)_{2j}, \ldots, (\alpha\beta)_{aj}), j = 1, \ldots, b$ are independent multivariate normal vectors that satisfy the constraint $\sum_{i=1}^{a}(\alpha\beta)_{ij} = 0$ for each j. This implies that $(\alpha\beta)_{1j}, (\alpha\beta)_{2j}, \ldots, (\alpha\beta)_{aj}$ are dependent on β_j. He further defines a covariance structure for the $\{m_{ij}\}$ and it is possible to express the variances and covariances of the β_j's and $(\alpha\beta)_{ij}$'s indirectly by stating the elements of this variance-covariance matrix. Thus, the two mixed models discussed in this chapter are rather special cases of Scheffé's model. The analysis of the Scheffé's model is similar to the "standard model" and the F tests for testing hypotheses $H_o^B : \sigma_\beta^2 = 0$ and $H_o^{AB} : \sigma_{\alpha\beta}^2 = 0$ are exactly the same as for the "standard" model. However, the distribution theory of MS_A and MS_{AB} is much more complicated; and, in general, the statistic MS_A/MS_{AB} is not always distributed as an F variable when $H_o : \alpha_i = 0$ is true. The only way to obtain an exact test for this problem is to consider it in a multivariate framework, which leads to Hotelling's T^2 test (Scheffé, 1959, pp. 270–274). Scheffé avoids this procedure and instead suggests the use of the ratio MS_A/MS_{AB} which can be approximated as an F variable with $a - 1$ and $(a - 1)(b - 1)$ degrees of freedom. Another difference is that even though the hypothesis $H_o^{AB} : \sigma_{\alpha\beta}^2 = 0$ may be tested by the statistic MS_{AB}/MS_E, which under H_o^{AB} has an F distribution with $(a - 1)(b - 1)$ and $ab(n - 1)$ degrees of freedom, the power is not expressible in terms of the central or noncentral F distribution, since SS_{AB} is not distributed as constant times a chi-square variable when H_o^{AB} is false. The power of this test has been studied by Imhof (1958). For a discussion of multiple comparison methods for Scheffé's mixed model, see Hochberg and Tamhane (1983).

In view of numerous versions of mixed models, a natural question arises as to which model should be employed. Most people tend to favor the "standard model" and it is most often discussed in the literature. Furthermore, the results on expected mean squares under sampling from a finite population agree in form with those of the standard model (see Chapter IX). If the correlation values of the random components are not high, then either mixed model can be used and there are only minor differences between them. However, if the correlation values tend to be large, then Scheffé's model should be preferred. The choice between different mixed models should always be guided by the correlation structure of the observed data and to what extent the correlations between the random components affect the characteristics of tests and estimation procedures of different mixed models.

4.22 EFFECTS OF VIOLATIONS OF ASSUMPTIONS OF THE MODEL

The list of assumptions for the model (4.1.1) is almost an exact parallel to the list of assumptions for models (2.1.1) and (3.1.1). Similar assumptions are made for more complex experiments entailing higher-order classifications. Thus, as one may anticipate, the same violations of assumptions are possible in the two- or multi-way crossed classifications as in the one-way classification. In this section, we briefly summarize some known results concerning the effects of violations of assumptions on the inference of the model (4.1.1). Further discussions on this topic can be found in Scheffé (1959, Chapter X) and Miller (1986, Chapter IV).

MODEL I (FIXED EFFECTS)

For experiments involving balanced or nearly balanced designs, with relatively large numbers of observations per cell, the assumption of normality for error terms seems to be rather unimportant. However, for severely unbalanced experiments, the heavy-tailed or contaminated distributions may produce outliers and thereby distort the results on estimates and tests of significance. Thus, in an experiment, if the observations are suspected to depart from normality, then perhaps a balanced design with a correspondingly large number of observations per cell should be used. Furthermore, if the data yield an equal number of observations in each cell, then the requirement of equal error variance in each cell, if violated, may not involve any serious risk.

Krutchkoff (1989) carried out a simulation study to compare the performance of the usual F test along with a new procedure called the K test. The results indicated that the F test had larger type I error and decreased power. In designs involving unequal numbers of observations per cell, "... the size of the F test was inflated when the larger errors were on the cells with the smaller number of observations and deflated when the larger errors were on the cells with the larger number of observations." In both situations, there was a decrease in power; but the drop was much more serious for the latter case. However, the K test was generally insensitive to the heterogeneity of variances. Consequently, there are two good reasons for planning an experiment with an equal number of observations per cell: the experimental design will be balanced leading to simple exact tests and the possible consequences of heterogeneous variances will be minimized.

The assumption of independence seems to have major importance and its violation may lead to erroneous conclusions. (For a discussion of the problem of serial correlation created by observations taken in time sequence, see Section 3.17.) For this reason great care should be taken in the planning and analysis of experiments involving repeated observations to ensure the independence of error terms. Thus, random assignment of experimental units to the treatment combinations is especially important.

MODEL II (RANDOM EFFECTS)

The lack of nonnormality in any of the random effects can seriously affect the distribution theory of the sum of squares involving them. The point estimates of variance components are still unbiased but the effects of nonnormality on tests and confidence intervals can lead to erroneous results. In particular, the tests and confidence intervals on σ_e^2 are very sensitive to nonnormality. However, the statistics MS_A/MS_{AB} and MS_B/MS_{AB} for testing the effects of variance components are somewhat robust.

Very little is known about the effects of unequal variances and the lack of independence on the inferences for the two-way random model.

MODEL III (MIXED EFFECTS)

There are very few studies dealing with the effects of violation of assumptions on the inferences for the two-way mixed model. For balanced or nearly balanced designs and moderate departures from normality, the effect of nonnormal random effects $\{e_{ijk}\}$, $(\alpha\beta)_{ij}$, (β_j) on tests about the fixed effects $\{\alpha_i\}$ is expected to be rather small. In the presence of appreciable nonnormality, however, the consequences could be more serious. In regard to tests and confidence intervals for the variance components, the effects of nonnormality may be extremely misleading.

The problem of unequal variances could occur in terms of the variances of any of the random effects $\{e_{ijk}\}$, $\{(\alpha\beta)_{ij}\}$, and $\{\beta_j\}$. For tests on $\{\alpha_i\}$, the effect of varying σ_e^2 and $\sigma_{\alpha\beta}^2$ should be somewhat similar to the two-way fixed effects model (3.1.1) (with one observation per cell) since MS_{AB} is used in the denominator of the F test. The effect of varying σ_e^2, on testing the hypotheses concerning the variance components $\sigma_{\alpha\beta}^2$ and σ_β^2, should be similar to the one-way model (2.1.1) since MS_E is used in the denominators of the associated F statistics.

Not much is known concerning the effect of lack of independence of the random effects $\{e_{ijk}\}$, $\{(\alpha\beta)_{ij}\}$, and $\{\beta_j\}$ on inferences for the two-way mixed model.

EXERCISES

1. An industrial engineer wishes to determine whether four different makes of automobiles would yield the same mileage. An experiment is designed wherein a random sample of three cars of each make is selected from each of three cities, and each car given a test run with one gallon of gasoline. The results on the number of miles traveled are given as follows.

City	Make of Automobile			
	I	II	III	IV
Boston	24.4	23.6	27.1	22.6
	23.9	22.7	28.0	22.3
	25.5	22.9	27.4	23.6
Los Angeles	25.7	24.2	25.1	24.5
	26.5	23.9	26.8	24.2
	25.4	24.6	24.8	25.3
Dallas	23.9	23.7	27.3	24.4
	22.7	23.3	27.0	23.5
	25.1	24.8	26.6	24.1

(a) Why was it considered necessary to include three cities in the experiment rather than just one city?
(b) How would you obtain a random sample of three cars from a city?

(c) What assumptions are made about the populations, and what hypotheses can be tested?
(d) Describe the model and the assumptions for the experiment.
(e) Analyze the data and report the analysis of variance table.
(f) Test whether there are differences in mileage among the cities. Use $\alpha = 0.05$.
(g) Test whether there are differences in mileage between makes of automobiles. Use $\alpha = 0.05$.
(h) Test whether there are interaction effects between cities and makes of automobiles. Use $\alpha = 0.05$.

2. An experiment is designed to compare the corrosion effect on three leading metal products. Eighteen samples, six of each metal, were used in the experiment and they were assigned at random into six groups of three each. The first three groups had densities (kg/mm^3) taken after a test period of 30 hours and the next three groups were measured after a test period of 60 hours. The relevant data in certain standard units are given as follows.

Test Period (hrs)	Metal Product		
	Steel	Copper	Zinc
30	149	158	129
	126	129	124
	115	158	154
60	152	112	126
	142	152	151
	124	117	138

(a) Describe the model and the assumptions for the experiment.
(b) Analyze the data and report the analysis of variance table.
(c) Test whether there are differences in corrosion effect among the three metal products. Use $\alpha = 0.05$.
(d) Test whether there are differences in corrosion effect between the two test periods. Use $\alpha = 0.05$.
(e) Test whether there are interaction effects between metal products and test periods. Use $\alpha = 0.05$.
(f) Determine a 95 percent confidence interval for σ_e^2.
(g) Let α_i be the effect of the i-th test period and β_j be the effect of the j-th metal product. Determine simultaneous confidence intervals for $\alpha_1 - \alpha_2$ and $\beta_2 - \beta_3$ using an overall confidence level of 0.95.

3. A tool manufacturer wishes to study the effect of tool temperature and tool speed on a certain type of milling machine. An experiment was designed wherein two levels of tool temperature (300°F and 500°F) and four levels of tool speed (V_1, V_2, V_3, and V_4) were used, and three measurements were made for each combination of tool temperature

and tool speed. The relevant data on milling machine measurements in certain standard units are given as follows.

Tool Temperature (°F)	Tool Speed			
	V_1	V_2	V_3	V_4
300	4,783	5,720	5,185	5,530
	5,373	5,190	5,150	5,540
	5,383	5,523	5,397	5,155
500	5,225	5,837	5,131	5,493
	5,533	5,180	5,290	5,341
	5,145	5,190	5,235	5,390

(a) Describe the model and the assumptions for the experiment.
(b) Analyze the data and report the analysis of variance table.
(c) Test whether there are differences in milling machine measurements among the four tool speeds. Use $\alpha = 0.05$.
(d) Test whether there are differences in milling machine measurements between the two tool temperatures. Use $\alpha = 0.05$.
(e) Test whether there are interaction effects between tool temperatures and tool speeds. Use $\alpha = 0.05$.
(f) Determine a 95 percent confidence interval for σ_e^2.
(g) Let α_i be the effect of the i-th tool temperature and β_j be the effect of the j-th tool speed. Determine simultaneous confidence intervals for $\alpha_1 - \alpha_2$ and $\beta_2 - \beta_3$ using an overall confidence level of 0.95.

4. An experiment was performed to determine the "active life" for three specimens of punching dies P_1, P_2, and P_3 taken from seven punching machines, M_1, M_2, ..., M_7 in a certain factory. The relevant data on measurements in minutes are given as follows.

Punching Die	Punching Machine						
	M_1	M_2	M_3	M_4	M_5	M_6	M_7
P_1	35.8	37.7	36.8	36.9	39.3	37.4	41.3
	38.2	40.2	40.9	35.9	37.5	38.8	43.3
P_2	38.3	33.9	37.8	35.7	35.9	36.5	37.1
	36.1	37.2	38.5	33.1	37.3	38.3	36.4
P_3	38.7	39.7	38.9	37.3	36.9	38.1	40.0
	35.9	40.6	35.6	35.7	35.4	35.6	35.9

(a) Describe the model and the assumption for the experiment.
(b) Analyze the data and report the analysis of variance table.
(c) Test whether there are differences in "active life" among the seven punching machines. Use $\alpha = 0.05$.

(d) Test whether there are differences in "active life" among the three punching dies. Use $\alpha = 0.05$.
(e) Test whether there are interaction effects between punching dies and punching machines. Use $\alpha = 0.05$.
(f) Determine a 95 percent confidence interval for σ_e^2.
(g) Let α_i be the effect of the i-th punching die and β_j be the effect of the j-th punching machine. Determine simultaneous confidence intervals for $\alpha_1 - \alpha_2$ and $\beta_2 - \beta_3$ using an overall confidence level of 0.95.

5. A production engineer wishes to study the effect of cutting temperature and cutting pressure on the surface finish of the machined component. He designs an experiment wherein three levels of each factor are selected, and a factorial experiment with two replicates is run. The relevant data in certain standard units are given as follows.

		Temperature	
Pressure	Low	Medium	High
Low	51.5	51.8	51.3
	51.3	51.7	51.5
Medium	51.2	51.6	49.9
	51.4	51.7	51.2
High	51.6	51.9	51.5
	51.8	51.8	51.2

(a) Describe the model and the assumptions for the experiment.
(b) Analyze the data and report the analysis of variance table.
(c) Test whether there are differences in the surface finish among the three levels of temperature. Use $\alpha = 0.05$.
(d) Test whether there are differences in the surface finish between the three levels of pressure. Use $\alpha = 0.05$.
(e) Test whether there are interaction effects between cutting temperatures and cutting pressures. Use $\alpha = 0.05$.
(f) Determine a 95 percent confidence interval for σ_e^2.
(g) Let α_i be the effect of the i-th pressure and β_j be the effect of the j-th temperature. Determine simultaneous confidence intervals for $\alpha_1 - \alpha_2$ and $\beta_1 - \beta_3$ using an overall confidence level of 0.95.
(h) Evaluate the power of the test for detecting a true difference in pressure such that $\sum_{i=1}^{3} \alpha_i^2 = 0.10$, where α_i is the i-th level pressure effect.

6. The following table gives the partial results of the analysis of variance computations performed on the data of the life of five brands of plastic products used under five different process temperatures. Three plastics of each brand were used for each process temperature. Complete the analysis of variance table and perform the relevant tests

of hypotheses of interest to the experimenter. Why was it thought necessary to include different process temperatures in the experiment? Explain.

Source of Variation	Sum of Squares
Brand	311.23
Temperature	321.34
Interaction	....
Error	23.31
Total	915.7

7. It is suspected that the strength of a tensile specimen is affected by the strain rate and the temperature. A factorial experiment is designed wherein four temperatures are randomly selected for each of three strain rates. The relevant data in certain standard units are given as follows.

Strain Rate	Temperature (°F)			
(S^{-1})	100	200	300	400
0.10	81	86	89	106
	91	75	95	111
	67	79	99	103
0.20	109	105	106	111
	93	111	115	107
	95	95	102	106
0.30	106	111	115	111
	105	106	117	118
	109	102	106	114

(a) Describe the model and the assumptions for the experiment.
(b) Analyze the data and report the analysis of variance table.
(c) Test whether there are differences in the tensile strength among the levels of temperature. Use $\alpha = 0.05$.
(d) Test whether there are differences in the tensile strength between strain rates. Use $\alpha = 0.05$.
(e) Test whether there are interaction effects between levels of temperature and strain rates. Use $\alpha = 0.05$.
(f) Determine point and interval estimates of the variance components of the model.
(g) Determine a 95 percent confidence interval for the mean difference in response for strain rates of 0.10 and 0.30.
(h) Analyze the data using the alternate mixed model discussed in Section 4.21 and compare the results obtained from the two models.

8. A production control engineer wishes to study the factors that influence the breaking strength of metallic sheets. He designs an experiment wherein four machines and three robots are selected at random and a factorial experiment is performed using metallic sheets from the same production batch. The relevant data on breaking strength in certain standard units are given as follows.

	Machine			
Robot	1	2	3	4
1	112	113	111	113
	113	118	112	111
2	113	113	114	118
	115	114	112	117
3	119	115	117	123
	117	118	122	119

(a) Describe the model and the assumptions for the experiment.
(b) Analyze the data and report the analysis of variance table.
(c) Test whether there are differences in the breaking strength of the metallic sheets among machines. Use $\alpha = 0.05$.
(d) Test whether there are differences in the breaking strength of the metallic sheets between robots. Use $\alpha = 0.05$.
(e) Test whether there are interaction effects between machines and robots. Use $\alpha = 0.05$.
(f) Determine point and interval estimates of the variance components of the model.
(g) Determine the power of the test for detecting a machine effect such that $\sigma_\beta^2 = \sigma_e^2$, where σ_β^2 is the variance component for the machine factor and σ_e^2 is the error variance component.
(h) Suppose that the robots were selected at random, but only four machines were available for the test. Test for the main effects and interaction at the 5 percent level of significance. Does the new experimental situation affect either the analysis or the conclusions of your study?

9. A production control engineer wishes to study the thrust force generated by a lathe. He suspects that the cutting speed and the depth of cut of the material are the most important determining factors. He designs an experiment wherein four depths of cut are randomly selected and a high and low cutting speed chosen to represent the extreme operating conditions. The relevant data in certain standard units are given as follows.

	Depth of Cut		
Cutting Speed	0.01	0.03	0.05
Low	2.61	2.36	2.66
	2.69	2.39	2.77
High	2.74	2.77	2.85
	2.77	2.78	2.79

(a) Describe the model and the assumptions for the experiment.
(b) Analyze the data and report the analysis of variance table.
(c) Test whether there are differences in thrust force among depths of cut. Use $\alpha = 0.05$.
(d) Test whether there are differences in thrust force between cutting speeds. Use $\alpha = 0.05$.
(e) Test whether there are interaction effects between cutting speeds and depths of cut. Use $\alpha = 0.05$.
(f) Estimate the variance components of the model (point and interval estimates).

10. A quality control engineer wishes to study the influence of furnace temperature and type of material on the quality of a cast product. An experiment was designed to include three levels of furnace temperature (1200°F, 1250°F, and 1300°F) for each of three types of material. The relevant data in certain standard units are given as follows.

	Temperature		
Material	1200°F	1250°F	1300°F
1	791	2191	2493
	779	2198	2491
	781	2196	2497
2	761	2181	2439
	741	2145	2423
	789	2111	2399
3	757	2156	978
	786	2164	1115
	799	2177	999

(a) State the model and the assumptions for the experiment. Assume that both factors are fixed.
(b) Analyze the data and report the analysis of variance table.
(c) Does the material type affect the response? Use $\alpha = 0.05$.
(d) Does the temperature affect the response? Use $\alpha = 0.05$.
(e) Is there a significant interaction effect? Use $\alpha = 0.05$.

11. Crump (1946) reported the results of analysis of variance performed on the data from four successive genetic experiments on egg production with the same sample of 25 races of the common fruitfly (*Drosophila melanogaster*), 12 females being sampled from each race for each experiment. The observations were the total number of eggs produced by a female on the fourth day of laying. The mathematical model for this experiment would be

$$y_{ijk} = \mu + \alpha_i + \beta_j + (\alpha\beta)_{ij} + e_{ijk} \quad \begin{cases} i = 1, 2, 3, 4 \\ j = 1, 2, \ldots, 25 \\ k = 1, 2, \ldots, 12, \end{cases}$$

where μ is the general mean, α_i is the effect of the i-th experiment, β_j is the effect of the j-th race, $(\alpha\beta)_{ij}$ is the interaction of the i-th experiment with the j-th race, and the e_{ijk}'s are experimental errors. It is further assumed that $\alpha_i \sim N(0, \sigma_\alpha^2)$, $\beta_j \sim N(0, \sigma_\beta^2)$, $(\alpha\beta)_{ij} \sim N(0, \sigma_{\alpha\beta}^2)$; and that the α_i's, β_j's, $(\alpha\beta)_{ij}$'s, and e_{ijk}'s are mutually and completely independent. The analysis of variance computations of the data (not reported here) are carried out exactly as in Section 4.15 and the results on sums of squares are given as follows.

Analysis of Variance for Genetic Experiments Data

Source of Variation	Degrees of Freedom	Sum of Squares	Mean Square	Expected Mean Square	F Value	p-Value
Experiment		139,977				
Race		77,832				
Interaction		33,048				
Error		254,100				
Total		504,957				

Source: Crump (1946). Used with permission.

(a) Complete the remaining columns of the preceding analysis of variance table.
(b) Test whether there are differences in egg production among different experiments. Use $\alpha = 0.05$.
(c) Test whether there are differences in egg production among different races. Use $\alpha = 0.05$.

(d) Test whether there are interaction effects between experiments and races. Use $\alpha = 0.05$.
(e) Determine point and interval estimates for each of the variance components of the model.
(f) Suppose that 25 races used in the experiment are of particular interest to the experimenter; and thus, this factor is considered to have a fixed effect. Perform tests of hypotheses and obtain estimates of the variance components under the assumptions of the mixed model.

12. Box and Cox (1964) reported data from an experiment designed to investigate the effects of certain toxic agents. Groups of four animals were randomly allocated to three poisons and four treatments using a 3 × 4 replicate factorial design. The survival times (unit, 10 hrs) of animals were recorded and the data are given as follows.

	Treatment			
Poison	A	B	C	D
I	0.31	0.82	0.43	0.45
	0.45	1.10	0.45	0.71
	0.46	0.88	0.63	0.66
	0.43	0.72	0.76	0.62
II	0.36	0.92	0.44	0.56
	0.29	0.61	0.35	1.02
	0.40	0.49	0.31	0.71
	0.23	1.24	0.40	0.38
III	0.22	0.30	0.23	0.30
	0.21	0.37	0.25	0.36
	0.18	0.38	0.24	0.31
	0.23	0.29	0.22	0.33

Source: Box and Cox (1964). Used with permission.

(a) State the model and the assumptions for the experiment. Assume that both poison and treatment factors are fixed.
(b) Analyze the data and report the analysis of variance table.
(c) Does the poison type affect the survival time? Use $\alpha = 0.05$.
(d) Does the treatment affect the survival time? Use $\alpha = 0.05$.
(e) Is there a significant interaction effect? Use $\alpha = 0.05$.

13. Scheffé (1959, pp. 140–141) reported data from an experiment designed to study the variation in weight of hybrid female rats in a foster nursing. A two-factor factorial design was used with the factors in the two-way layout being the genotype of the foster mother and that of the litter. The weights in grams as litter averages at 28 days were recorded and the data are given as follows.

Two-Way Crossed Classification with Interaction

Genotype of Litter	Genotype of Foster Mother			
	A	F	I	J
A	61.5	55.0	52.5	42.0
	68.2	42.0	61.8	54.0
	64.0	60.2	49.5	61.0
	65.0		52.7	48.2
	59.7			39.6
F	60.3	50.8	56.5	51.3
	51.7	64.7	59.0	40.5
	49.3	61.7	47.2	
	48.0	64.0	53.0	
		62.0		
I	37.0	56.3	39.7	50.0
	36.3	69.8	46.0	43.8
	68.0	67.0	61.3	54.5
			55.3	
			55.7	
J	59.0	59.5	45.2	44.8
	57.4	52.8	57.0	51.5
	54.0	56.0	61.4	53.0
	47.0			42.0
				54.0

Source: Scheffé (1959, p. 140). Used with permission.

(a) State the model and the assumptions for the experiment.
(b) Analyze the data and report the analysis of variance table using the unweighted means analysis.
(c) Analyze the data and report the analysis of variance table using the weighted means analysis.
(d) Perform appropriate F tests, using the unweighted and weighted means analyses. Use $\alpha = 0.05$.
(e) Compare the results from the weighted and unweighted means analyses.

14. Davies and Goldsmith (1972, p. 154) reported data from an experiment designed to investigate sources of variability in testing strength of Portland cement. Several small samples of a sample of cement were mixed with water and worked for a fixed time, by three different persons (gaugers), and then were cast into cubes. The cubes were later tested for compressive strength by three other persons (breakers). Each gauger worked with 12 cubes which were then divided into three sets of four, and each breaker tested one set of four cubes from each gauger. All the testing was done on the same machine and the overall objective of the study was to investigate and quantify the relative magnitude of the variability in test results due to

individual differences between gaugers and between breakers. The data are shown below where measurements are given in the original units of pounds per square inch.

	Breaker					
Gauger	1		2		3	
1	5280	5520	4340	4400	4160	5180
	4760	5800	5020	6200	5320	4600
2	4420	5280	5340	4880	4180	4800
	5580	4900	4960	6200	4600	4480
3	5360	6160	5720	4760	4460	4930
	5680	5500	5620	5560	4680	5600

Source: Davies and Goldsmith (1972, p. 154). Used with permission.

(a) Describe the model and the assumptions for the experiment. Would you use Model I, Model II, or Model III. In the original experiment, the investigator's interest was in these particular gaugers and breakers.
(b) Analyze the data and report the analysis of variance table.
(c) Test whether there are differences in testing strength due to gaugers. Use $\alpha = 0.05$.
(d) Test whether there are differences in testing strength due to breakers. Use $\alpha = 0.05$.
(e) Test whether there are interaction effects between gaugers and breakers. Use $\alpha = 0.05$.
(f) Assuming that the gauger and breaker effects are random, estimate the variance components of the model (point and interval estimates) and determine their relative importance.

5 Three-Way and Higher-Order Crossed Classifications

5.0 PREVIEW

Many experiments and surveys involve three or more factors. Multifactor layouts entail data collection under conditions determined by several factors simultaneously. Such layouts usually provide more information and often can be even more economical than separate one-way or two-way designs. The models and analysis of variance for the case of three or more factors are straightforward extensions of the two-way crossed model. The methods of analysis of variance for the two-way crossed classification discussed in the preceding two chapters can thus be readily generalized to three-way and higher-order classifications. In this chapter, we study the three-way crossed classification in some detail because it serves as an illustration as to how the analysis can be extended when four or more factors are involved. Generalizations to four-way and higher-order classifications are briefly outlined.

5.1 MATHEMATICAL MODEL

Consider three factors A, B, and C having a, b, and c levels, respectively, and let there be n observations in each of the abc cells of the three-way layout. Let $y_{ijk\ell}$ be the ℓ-th observation corresponding to the i-th level of factor A, the j-th level of factor B, and the k-th level of factor C. Thus, there is a total of

$$N = abcn$$

observations in the study. The data involving a total of $N = abcn$ scores $y_{ijk\ell}$'s can then be schematically represented as in Table 5.1.

The notation employed here is a straightforward extension of the two-way crossed classification. As usual, a dot in the subscript indicates aggregation and a dot and a bar indicate averaging over the index represented by the dot. Thus,

TABLE 5.1
Data for a Three-Way Crossed Classification with n Replications per Cell

			A_1				A_2			B_b		A_a			B_b
Factor B		B_1	B_2	...	B_b	B_1	B_2	...	B_b		...	B_1	B_2	...	
Factor C	C_1	y_{1111}	y_{1211}	...	y_{1b11}	y_{2111}	y_{2211}	...	y_{2b11}	...	y_{a111}	y_{a211}	...	y_{ab11}
		y_{1112}	y_{1212}	...	y_{1b12}	y_{2112}	y_{2212}	...	y_{2b12}	...	y_{a112}	y_{a212}	...	y_{ab12}
		.	.		.	.	.		.			.	.		.
		y_{111n}	y_{121n}	...	y_{1b1n}	y_{211n}	y_{221n}	...	y_{2b1n}	...	y_{a11n}	y_{a21n}	...	y_{ab1n}
	C_2	y_{1121}	y_{1221}	...	y_{1b21}	y_{2121}	y_{2221}	...	y_{2b21}	...	y_{a121}	y_{a221}	...	y_{ab21}
		y_{1122}	y_{1222}	...	y_{1b22}	y_{2122}	y_{2222}	...	y_{2b22}	...	y_{a122}	y_{a222}	...	y_{ab22}
		.	.		.	.	.		.			.	.		.
		y_{112n}	y_{122n}	...	y_{1b2n}	y_{212n}	y_{222n}	...	y_{2b2n}	...	y_{a12n}	y_{a22n}	...	y_{ab2n}
	...														
	C_c	y_{11c1}	y_{12c1}	...	y_{1bc1}	y_{21c1}	y_{22c1}	...	y_{2bc1}	...	y_{a1c1}	y_{a2c1}	...	y_{abc1}
		y_{11c2}	y_{12c2}	...	y_{1bc2}	y_{21c2}	y_{22c2}	...	y_{2bc2}	...	y_{a1c2}	y_{a2c2}	...	y_{abc2}
		.	.		.	.	.		.			.	.		.
		y_{11cn}	y_{12cn}	...	y_{1bcn}	y_{21cn}	y_{22cn}	...	y_{2bcn}	...	y_{a1cn}	y_{a2cn}	...	y_{abcn}

Factor A

we employ the following notations for sample totals and means:

$$y_{ijk.} = \sum_{\ell=1}^{n} y_{ijk\ell}, \qquad \bar{y}_{ijk.} = y_{ijk.}/n;$$

$$y_{ij..} = \sum_{k=1}^{c}\sum_{\ell=1}^{n} y_{ijk\ell}, \qquad \bar{y}_{ij..} = y_{ij..}/cn;$$

$$y_{i.k.} = \sum_{j=1}^{b}\sum_{\ell=1}^{n} y_{ijk\ell}, \qquad \bar{y}_{i.k.} = y_{i.k.}/bn;$$

$$y_{.jk.} = \sum_{i=1}^{a}\sum_{\ell=1}^{n} y_{ijk\ell}, \qquad \bar{y}_{.jk.} = y_{.jk.}/an;$$

$$y_{i...} = \sum_{j=1}^{b}\sum_{k=1}^{c}\sum_{\ell=1}^{n} y_{ijk\ell}, \qquad \bar{y}_{i...} = y_{i...}/bcn;$$

$$y_{.j..} = \sum_{i=1}^{a}\sum_{k=1}^{c}\sum_{\ell=1}^{n} y_{ijk\ell}, \qquad \bar{y}_{.j..} = y_{.j..}/acn;$$

$$y_{..k.} = \sum_{i=1}^{a}\sum_{j=1}^{b}\sum_{\ell=1}^{n} y_{ijk\ell}, \qquad \bar{y}_{..k.} = y_{..k.}/abn;$$

and

$$y_{....} = \sum_{i=1}^{a}\sum_{j=1}^{b}\sum_{k=1}^{c}\sum_{\ell=1}^{n} y_{ijk\ell}, \qquad \bar{y}_{....} = y_{....}/abcn.$$

The analysis of variance model for this type of experimental layout is given as

$$y_{ijk\ell} = \mu + \alpha_i + \beta_j + \gamma_k + (\alpha\beta)_{ij} + (\alpha\gamma)_{ik} + (\beta\gamma)_{jk} + (\alpha\beta\gamma)_{ijk} + e_{ijk\ell} \quad \begin{cases} i = 1, \ldots, a \\ j = 1, \ldots, b \\ k = 1, \ldots, c \\ \ell = 1, \ldots, n, \end{cases} \quad (5.1.1)$$

where

μ is the general mean,
α_i is the effect of the i-th level of factor A,
β_j is the effect of the j-th level of factor B,
γ_k is the effect of the k-th level of factor C,
$(\alpha\beta)_{ij}, (\alpha\gamma)_{ik}, (\beta\gamma)_{jk}$ are the effects of the two-factor interactions $A \times B$, $A \times C$, and $B \times C$, respectively,
$(\alpha\beta\gamma)_{ijk}$ is the effect of the three-factor interaction $A \times B \times C$, and
e_{ijk} is the customary error term.

5.2 ASSUMPTIONS OF THE MODEL

The assumptions of the model (5.1.1) are as follows:

(i) $e_{ijk\ell}$'s are uncorrelated and randomly distributed with common mean zero and variance σ_e^2.

(ii) Under Model I, the α_i's, β_j's, γ_k's, $(\alpha\beta)_{ij}$'s, $(\alpha\gamma)_{ik}$'s, $(\beta\gamma)_{jk}$'s, and $(\alpha\beta\gamma)_{ijk}$'s are constants subject to the restrictions:

$$\sum_{i=1}^{a} \alpha_i = \sum_{j=1}^{b} \beta_j = \sum_{k=1}^{c} \gamma_k = 0,$$

$$\sum_{i=1}^{a} (\alpha\beta)_{ij} = \sum_{j=1}^{b} (\alpha\beta)_{ij} = \sum_{i=1}^{a} (\alpha\gamma)_{ik} = \sum_{k=1}^{c} (\alpha\gamma)_{ik}$$

$$= \sum_{j=1}^{b} (\beta\gamma)_{jk} = \sum_{k=1}^{c} (\beta\gamma)_{jk} = 0,$$

and

$$\sum_{i=1}^{a} (\alpha\beta\gamma)_{ijk} = \sum_{j=1}^{b} (\alpha\beta\gamma)_{ijk} = \sum_{k=1}^{c} (\alpha\beta\gamma)_{ijk} = 0.$$

(iii) Under Model II, α_i's, β_j's, γ_k's, $(\alpha\beta)_{ij}$'s, $(\alpha\gamma)_{ik}$'s, $(\beta\gamma)_{jk}$'s, $(\alpha\beta\gamma)_{ijk}$'s, and $e_{ijk\ell}$'s are mutually and completely uncorrelated random variables with mean zero and respective variances $\sigma_\alpha^2, \sigma_\beta^2, \sigma_\gamma^2, \sigma_{\alpha\beta}^2, \sigma_{\alpha\gamma}^2, \sigma_{\beta\gamma}^2, \sigma_{\alpha\beta\gamma}^2$, and σ_e^2.

(iv) Under Model III, several variations exist depending upon which factors are assumed fixed and which random. Suppose that factor A has fixed effects and factors B and C have random effects. In this case, α_i's are constants; β_j's, γ_k's, $(\alpha\beta)_{ij}$'s, $(\alpha\gamma)_{ik}$'s, $(\beta\gamma)_{jk}$'s, $(\alpha\beta\gamma)_{ijk}$'s, and $e_{ijk\ell}$'s are random variables with mean zero and respective variances $\sigma_\beta^2, \sigma_\gamma^2, \sigma_{\alpha\beta}^2, \sigma_{\alpha\gamma}^2, \sigma_{\beta\gamma}^2, \sigma_{\alpha\beta\gamma}^2$, and σ_e^2 subject to the restrictions:

$$\sum_{i=1}^{a} \alpha_i = 0 \qquad (5.2.1)$$

$$\sum_{i=1}^{a} (\alpha\beta)_{ij} = \sum_{i=1}^{a} (\alpha\gamma)_{ik} = \sum_{i=1}^{a} (\alpha\beta\gamma)_{ijk} = 0, \qquad (5.2.2)$$

for all j and k.

Note that all interaction terms in Model III are assumed to be random, since at least one of the factors involved is a random effects factor. Furthermore, the sums of effects involving the fixed factor are zero when summed over the fixed factor levels. The correlations between random effects resulting from restrictions (5.2.2) can be derived, but are not considered here. Other mixed effects models can be developed in a similar fashion. For example, a model analogous to the two-way mixed model discussed in Section 4.21 involves the restriction (5.2.1) but not (5.2.2). This implies that the random effects are mutually and completely uncorrelated random variables.[1]

5.3 PARTITION OF THE TOTAL SUM OF SQUARES

As before, the total sum of squares can be partitioned by starting with the identity:

$$
\begin{aligned}
y_{ijk\ell} - \bar{y}_{....} &= (\bar{y}_{i...} - \bar{y}_{....}) + (\bar{y}_{.j..} - \bar{y}_{....}) + (\bar{y}_{..k.} - \bar{y}_{....}) \\
&+ (\bar{y}_{ij..} - \bar{y}_{i...} - \bar{y}_{.j..} + \bar{y}_{....}) \\
&+ (\bar{y}_{i.k.} - \bar{y}_{i...} - \bar{y}_{..k.} + \bar{y}_{....}) \\
&+ (\bar{y}_{.jk.} - \bar{y}_{.j..} - \bar{y}_{..k.} + \bar{y}_{....}) \\
&+ (\bar{y}_{ijk.} - \bar{y}_{ij..} - \bar{y}_{i.k.} - \bar{y}_{.jk.} + \bar{y}_{i...} + \bar{y}_{.j..} + \bar{y}_{..k.} - \bar{y}_{....}) \\
&+ (y_{ijk\ell} - \bar{y}_{ijk.}).
\end{aligned}
\tag{5.3.1}
$$

Squaring each side and summing over i, j, k, and ℓ, and noting that the cross-product terms drop out, we obtain

$$SS_T = SS_A + SS_B + SS_C + SS_{AB} + SS_{AC} + SS_{BC} + SS_{ABC} + SS_E,$$

where

$$SS_T = \sum_{i=1}^{a} \sum_{j=1}^{b} \sum_{k=1}^{c} \sum_{\ell=1}^{n} (y_{ijk\ell} - \bar{y}_{....})^2,$$

$$SS_A = bcn \sum_{i=1}^{a} (\bar{y}_{i...} - \bar{y}_{....})^2,$$

$$SS_B = acn \sum_{j=1}^{b} (\bar{y}_{.j..} - \bar{y}_{....})^2,$$

$$SS_C = abn \sum_{k=1}^{c} (\bar{y}_{..k.} - \bar{y}_{....})^2,$$

$$SS_{AB} = cn \sum_{i=1}^{a} \sum_{j=1}^{b} (\bar{y}_{ij..} - \bar{y}_{i...} - \bar{y}_{.j..} + \bar{y}_{....})^2,$$

---

[1] For a more general formulation of the three-way mixed model as an extension of the two-way mixed model by Scheffé, see Imhof (1960).

$$SS_{AC} = bn \sum_{i=1}^{a} \sum_{k=1}^{c} (\bar{y}_{i.k.} - \bar{y}_{i...} - \bar{y}_{..k.} + \bar{y}_{....})^2,$$

$$SS_{BC} = an \sum_{j=1}^{b} \sum_{k=1}^{c} (\bar{y}_{.jk.} - \bar{y}_{.j..} - \bar{y}_{..k.} + \bar{y}_{....})^2,$$

$$SS_{ABC} = n \sum_{i=1}^{a} \sum_{j=1}^{b} \sum_{k=1}^{c} (\bar{y}_{ijk.} - \bar{y}_{ij..} - \bar{y}_{i.k.} - \bar{y}_{.jk.} + \bar{y}_{i...} + \bar{y}_{.j..} + \bar{y}_{..k.} - \bar{y}_{....})^2,$$

and

$$SS_E = \sum_{i=1}^{a} \sum_{j=1}^{b} \sum_{k=1}^{c} \sum_{\ell=1}^{n} (y_{ijk\ell} - \bar{y}_{ijk.})^2.$$

Here, SS_T is the total sum of squares; SS_A, SS_B, SS_C are the usual main effects sums of squares; SS_{AB}, SS_{AC}, SS_{BC} are the usual two-factor interaction sums of squares; SS_{ABC} is the three-factor interaction sum of squares; and SS_E is the error sum of squares.

Remark: In a three-way crossed classification, one can compute abc separate cell variances as $\sum_{\ell=1}^{n} (y_{ijk\ell} - \bar{y}_{ijk.})^2$ which can then be tested for homogeneity of variances (see Section 2.21).

5.4 MEAN SQUARES AND THEIR EXPECTATIONS

As usual the mean squares are obtained by dividing the sums of squares by the corresponding degrees of freedom. The degrees of freedom for main effects and two-factor interactions sums of squares correspond to those for the two-way classification. The number of degrees of freedom for the three-factor interaction is obtained by subtraction and corresponds to the number of independent linear relations among all the interaction terms $(\alpha\beta\gamma)_{ijk}$'s.

The expected mean squares are obtained in the same way as in the earlier derivations.[2] The results on the partition of the degrees of freedom and the sum of squares, and the expected mean squares are summarized in the form of an analysis of variance table as shown in Table 5.2.

5.5 TESTS OF HYPOTHESES: THE ANALYSIS OF VARIANCE F TESTS

By assuming the normality of the random components in model (5.1.1), the sampling distributions of mean squares can be derived in terms of central and

---

[2] One can use the algorithms formulated by Schultz (1955) and others to reproduce the results on expected mean squares rather quickly. See Appendix U for a discussion of the rules for finding expected mean squares.

TABLE 5.2
Analysis of Variance for Model (5.1.1)

Source of Variation	Degrees of Freedom	Sum of Squares	Mean Square	Model I A, B, and C Fixed	Model II A, B, and C Random	Model III A Fixed, B and C Random	Model III A Random, B and C Fixed
					Expected Mean Square		
Due to A	$a-1$	SS_A	MS_A	$\sigma_e^2 + \dfrac{bcn}{a-1}\sum_{i=1}^{a}\alpha_i^2$	$\sigma_e^2 + n\sigma_{\alpha\beta\gamma}^2 + cn\sigma_{\alpha\beta}^2 + bn\sigma_{\alpha\gamma}^2 + bcn\sigma_{\alpha}^2$	$\sigma_e^2 + n\sigma_{\alpha\beta\gamma}^2 + cn\sigma_{\alpha\beta}^2 + bn\sigma_{\alpha\gamma}^2 + \dfrac{bcn}{a-1}\sum_{i=1}^{a}\alpha_i^2$	$\sigma_e^2 + bcn\sigma_{\alpha}^2$
Due to B	$b-1$	SS_B	MS_B	$\sigma_e^2 + \dfrac{acn}{b-1}\sum_{j=1}^{b}\beta_j^2$	$\sigma_e^2 + n\sigma_{\alpha\beta\gamma}^2 + cn\sigma_{\alpha\beta}^2 + an\sigma_{\beta\gamma}^2 + acn\sigma_{\beta}^2$	$\sigma_e^2 + an\sigma_{\beta\gamma}^2 + acn\sigma_{\beta}^2$	$\sigma_e^2 + cn\sigma_{\alpha\beta}^2 + bn\sigma_{\alpha\gamma}^2 + \dfrac{acn}{b-1}\sum_{j=1}^{b}\beta_j^2$
Due to C	$c-1$	SS_C	MS_C	$\sigma_e^2 + \dfrac{abn}{c-1}\sum_{k=1}^{c}\gamma_k^2$	$\sigma_e^2 + n\sigma_{\alpha\beta\gamma}^2 + bn\sigma_{\alpha\gamma}^2 + an\sigma_{\beta\gamma}^2 + abn\sigma_{\gamma}^2$	$\sigma_e^2 + n\sigma_{\alpha\beta\gamma}^2 + bn\sigma_{\alpha\gamma}^2 + an\sigma_{\beta\gamma}^2 + abn\sigma_{\gamma}^2$	$\sigma_e^2 + cn\sigma_{\alpha\beta}^2$
Interaction $A\times B$	$(a-1)(b-1)$	SS_{AB}	MS_{AB}	$\sigma_e^2 + \dfrac{cn}{(a-1)(b-1)}\sum_{i=1}^{a}\sum_{j=1}^{b}(\alpha\beta)_{ij}^2$	$\sigma_e^2 + n\sigma_{\alpha\beta\gamma}^2 + cn\sigma_{\alpha\beta}^2$	$\sigma_e^2 + n\sigma_{\alpha\beta\gamma}^2 + cn\sigma_{\alpha\beta}^2$	$\sigma_e^2 + bn\sigma_{\alpha\gamma}^2$
Interaction $A\times C$	$(a-1)(c-1)$	SS_{AC}	MS_{AC}	$\sigma_e^2 + \dfrac{bn}{(a-1)(c-1)}\sum_{i=1}^{a}\sum_{k=1}^{c}(\alpha\gamma)_{ik}^2$	$\sigma_e^2 + n\sigma_{\alpha\beta\gamma}^2 + bn\sigma_{\alpha\gamma}^2$	$\sigma_e^2 + n\sigma_{\alpha\beta\gamma}^2 + bn\sigma_{\alpha\gamma}^2$	$\sigma_e^2 + n\sigma_{\alpha\beta\gamma}^2 + \dfrac{an}{(b-1)(c-1)}\sum_{j=1}^{b}\sum_{k=1}^{c}(\beta\gamma)_{jk}^2$
Interaction $B\times C$	$(b-1)(c-1)$	SS_{BC}	MS_{BC}	$\sigma_e^2 + \dfrac{an}{(b-1)(c-1)}\sum_{j=1}^{b}\sum_{k=1}^{c}(\beta\gamma)_{jk}^2$	$\sigma_e^2 + n\sigma_{\alpha\beta\gamma}^2 + an\sigma_{\beta\gamma}^2$	$\sigma_e^2 + an\sigma_{\beta\gamma}^2$	$\sigma_e^2 + n\sigma_{\alpha\beta\gamma}^2$
Interaction $A\times B\times C$	$(a-1)(b-1)(c-1)$	SS_{ABC}	MS_{ABC}	$\sigma_e^2 + \dfrac{n}{(a-1)(b-1)(c-1)}\sum_{i=1}^{a}\sum_{j=1}^{b}\sum_{k=1}^{c}(\alpha\beta\gamma)_{ijk}^2$	$\sigma_e^2 + n\sigma_{\alpha\beta\gamma}^2$	$\sigma_e^2 + n\sigma_{\alpha\beta\gamma}^2$	$\sigma_e^2 + n\sigma_{\alpha\beta\gamma}^2$
Error	$abc(n-1)$	SS_E	MS_E	σ_e^2	σ_e^2	σ_e^2	σ_e^2
Total	$abcn-1$	SS_T					

noncentral chi-square variables. The results are obvious extensions of the results given in Section 4.5 for the two-way classification. The tests for main and interaction effects can be readily obtained by the results of the sampling distributions of mean squares and their expectations. In the following we summarize the tests for fixed, random, and mixed effects models.

MODEL I (FIXED EFFECTS)

We note from the analysis of variance Table 5.2 that MS_A, MS_B, MS_C, MS_{AB}, MS_{AC}, MS_{BC}, and MS_{ABC} all have expectations equal to σ_e^2 if there are no factor effects of the type reflected by the corresponding mean squares. If there are such effects, each mean square has an expectation exceeding σ_e^2. Also, the expectation of MS_E is always σ_e^2 as was the case in the analyses of other models. Hence, the tests for factor effects and their interactions can be obtained by comparing the appropriate mean square against MS_E; the large value of the mean square ratio indicating the presence of the corresponding factor or interaction effect. The development of various test procedures follows the same pattern as in the case of two-way classification. In Table 5.3, we summarize the hypotheses of interests, corresponding test statistics, and the appropriate percentiles of the F distribution.

Remarks: (i) An examination of Table 5.2 reveals that if $n = 1$ there are no degrees of freedom associated with the error term. Thus, we must have at least 2 observations ($n \geq 2$) in order to determine a sum of squares due to error if all possible interactions are included in the model. For further discussion of this point, see Section 5.10.

(ii) If some of the interaction terms are zero, one may consider the possibility of pooling those terms with the error sum of squares. However, as discussed earlier in Section 4.6, the pooling of nonsignificant mean squares should be carried out with a great deal of discretion and not as a general rule. The pooling should probably be restricted to the mean squares corresponding to the effects that from prior experience are unlikely to yield significant results (not expected to be appreciable).

(iii) As in the case of two-way crossed classification, it may be of interest to consider the significance level associated with the experiment as a whole. Let $\alpha_1, \alpha_2, \ldots, \alpha_7$ be the significance levels of the seven F statistics, $F_A, F_B, \ldots, F_{ABC}$, respectively, and let α be the significance level comprising all seven tests. Then again it follows that $\alpha < 1 - \Pi_{i=1}^{7}(1 - \alpha_i)$. For example, if $\alpha_1 = \alpha_2 = \cdots = \alpha_7 = 0.05$, then $\alpha < 0.302$; and if $\alpha_1 = \alpha_2 = \cdots = \alpha_7 = 0.01$, then $\alpha < 0.068$.

MODEL II (RANDOM EFFECTS)

The appropriate test statistics for various hypotheses of interest can be determined by examining the expected mean squares in Table 5.2. However, for the first time, we encounter the difficulty that even under the normality assumption exact F tests may not be available for some of the hypotheses usually tested. There is no difficulty about testing the hypotheses on the three-factor or the two-factor interactions. Thus, from the expected mean square column of

TABLE 5.3
Tests of Hypotheses for Model (5.1.1) under Model I

Hypothesis	Test Statistic	Percentile
H_0^A : all $\alpha_i = 0$ versus H_1^A : not all $\alpha_i = 0$	$F_A = \dfrac{MS_A}{MS_E}$	$F[a-1, abc(n-1); 1-\alpha]$
H_0^B : all $\beta_j = 0$ versus H_1^B : not all $\beta_j = 0$	$F_B = \dfrac{MS_B}{MS_E}$	$F[b-1, abc(n-1); 1-\alpha]$
H_0^C : all $\gamma_k = 0$ versus H_1^C : not all $\gamma_k = 0$	$F_C = \dfrac{MS_C}{MS_E}$	$F[c-1, abc(n-1); 1-\alpha]$
H_0^{AB} : all $(\alpha\beta)_{ij} = 0$ versus H_1^{AB} : not all $(\alpha\beta)_{ij} = 0$	$F_{AB} = \dfrac{MS_{AB}}{MS_E}$	$F[(a-1)(b-1), abc(n-1); 1-\alpha]$
H_0^{AC} : all $(\alpha\gamma)_{ik} = 0$ versus H_1^{AC} : not all $(\alpha\gamma)_{ik} = 0$	$F_{AC} = \dfrac{MS_{AC}}{MS_E}$	$F[(a-1)(c-1), abc(n-1); 1-\alpha]$
H_0^{BC} : all $(\beta\gamma)_{jk} = 0$ versus H_1^{BC} : not all $(\beta\gamma)_{jk} = 0$	$F_{BC} = \dfrac{MS_{BC}}{MS_E}$	$F[(b-1)(c-1), abc(n-1); 1-\alpha]$
H_0^{ABC} : all $(\alpha\beta\gamma)_{ijk} = 0$ versus H_1^{ABC} : not all $(\alpha\beta\gamma)_{ijk} = 0$	$F_{ABC} = \dfrac{MS_{ABC}}{MS_E}$	$F[(a-1)(b-1)(c-1), abc(n-1); 1-\alpha]$

Table 5.2, we see that the hypothesis $H_0^{ABC} : \sigma_{\alpha\beta\gamma}^2 = 0$ versus $H_1^{ABC} : \sigma_{\alpha\beta\gamma}^2 > 0$ can be tested with the ratio MS_{ABC}/MS_E; and $H_0^{AB} : \sigma_{\alpha\beta}^2 = 0$ versus $H_1^{AB} : \sigma_{\alpha\beta}^2 > 0$ with MS_{AB}/MS_{ABC}, and so on. Now, suppose that we wish to test $H_0^A : \sigma_\alpha^2 = 0$ versus $H_1^A : \sigma_\alpha^2 > 0$ (cases H_0^B and H_0^C can, of course, be treated similarly). If we are willing to assume that $\sigma_{\alpha\beta}^2 = 0$, then an exact F test of H_0^A can be based on the statistic MS_A/MS_{AC}. In this case, SS_{AB} could be pooled with SS_{ABC} since they would have the same expected mean squares. Similarly, if we are willing to assume that $\sigma_{\alpha\gamma}^2 = 0$, we may test H_0^A with MS_A/MS_{AB} and pool SS_{AC} with SS_{ABC}. Furthermore, if we are willing to assume other variance components to be zero, there would be no difficulty in deducing exact tests, if any, of the standard hypotheses and pooling procedures obtained from the analysis of variance Table 5.2, by deleting in it the components assumed to be zero. However, if we are unwilling to assume that $\sigma_{\alpha\beta}^2 = 0$ or $\sigma_{\alpha\gamma}^2 = 0$, then no exact test of H_0^A can be found from Table 5.2.

An approximate F test of H_0^A can be obtained by using a procedure due to Satterthwaite (1946) and Welch (1936, 1956). (For a detailed discussion of the procedure, see Appendix K.) To illustrate the procedure for testing the hypothesis

$$H_0^A : \sigma_\alpha^2 = 0$$
$$\text{versus} \qquad\qquad\qquad\qquad (5.5.1)$$
$$H_1^A : \sigma_\alpha^2 > 0,$$

we note from Table 5.2 that

$$E(\text{MS}_{AB}) + E(\text{MS}_{AC}) - E(\text{MS}_{ABC}) = \sigma_e^2 + cn\sigma_{\alpha\beta}^2 + bn\sigma_{\alpha\gamma}^2 + n\sigma_{\alpha\beta\gamma}^2,$$

which is precisely equal to $E(\text{MS}_A)$ when $\sigma_\alpha^2 = 0$. Hence, the suggested F statistic is

$$F_A = \frac{\text{MS}_A}{\text{MS}_{AB} + \text{MS}_{AC} - \text{MS}_{ABC}}, \qquad (5.5.2)$$

which has an approximate F distribution with $a - 1$ and ν_a degrees of freedom, where ν_a is approximated by

$$\nu_a = \frac{(\text{MS}_{AB} + \text{MS}_{AC} - \text{MS}_{ABC})^2}{\dfrac{(\text{MS}_{AB})^2}{(a-1)(b-1)} + \dfrac{(\text{MS}_{AC})^2}{(a-1)(c-1)} + \dfrac{(\text{MS}_{ABC})^2}{(a-1)(b-1)(c-1)}}. \qquad (5.5.3)$$

Remarks: (i) Because of the lack of uniqueness of the approximate F ratio (different F ratios may result from the use of different linear combinations of mean squares) and because of the necessity of approximating the degrees of freedom, the procedure is of limited usefulness. However, if used with care, the test procedure can be of value. The reader is referred to Cochran (1951) for a detailed discussion of this problem.

(ii) Usually, the degrees of freedom given by (5.5.3) will not be an integer. One can then either interpolate in the F distribution table, or round to the nearest integer. In practice, the choice of the nearest interger will be more than adequate.

(iii) An alternative test statistic for testing the hypothesis (5.5.1) is

$$F_A' = \frac{\text{MS}_A + \text{MS}_{ABC}}{\text{MS}_{AB} + \text{MS}_{AC}}.$$

The approximate degrees of freedom for both the numerator and the denominator are obtained as in (5.5.3) using Satterthwaite's rule. Because of the need to estimate only the denominator degrees of freedom, the test criterion (5.5.2) might be expected to have better power but it suffers from the drawback that the approximation (5.5.2) is less accurate when the linear combination of mean squares contains a negative term. Moreover, the denominator of the test statistic (5.5.2) can assume a negative value. The problem, however, may be less important if the contribution of MS_{AB} is relatively small and the corresponding degrees of freedom are large. The reader is referred to Cochran and Cox (1957), Hudson and Krutchkoff (1968), and Gaylor and Hopper (1969) for some further discussions and treatment of this topic. The general consensus seems to be that the two statistics are comparable in terms of size and power performance under a wide range of parameter values (see, e.g., Davenport and Webester (1973); Lorenzen (1987)).

(iv) An alternative to the Satterthwaite approximation for estimating the degrees of freedom for F_A and F'_A has been proposed by Myers and Howe (1971), but the procedure has been found to provide a liberal test (Davenport (1975)).

(v) An alternative to an approximate F test of $H_0^A : \sigma_\alpha^2 = 0$ has been proposed by Jeyaratnam and Graybill (1980) which is tied to the lower confidence bound of σ_α^2. For a discussion of some other test procedures for this problem, see Naik (1974) and Seifert (1981). Birch et al. (1990) and Burdick (1994) provide results of a simulation study to compare several tests for the main effects variance components in model (5.1.1).

To obtain a procedure for testing the hypothesis

$$H_0^B : \sigma_\beta^2 = 0$$

versus

$$H_1^B : \sigma_\beta^2 > 0$$

interchange A and B (also a and b) in (5.5.2) and (5.5.3). Similarly, for the hypothesis

$$H_0^C : \sigma_\gamma^2 = 0$$

versus

$$H_1^C : \sigma_\gamma^2 > 0$$

interchange A and C (also a and c) in (5.5.2) and (5.5.3).

Finally, similar to Table 5.3 for Model I, the hypotheses of interests, corresponding test statistics, and the appropriate percentiles of the F distribution are summarized in Table 5.4.

MODEL III (MIXED EFFECTS)

Suppose that A is fixed and B and C are random. The approximate F test given by (5.5.2) is used to test

$$H_0 : \alpha_i = 0, \quad i = 1, 2, \ldots, a$$

versus

H_1 : not all α_i's are zero.

The other six F ratios test the following principal null hypotheses:

$$\sigma_\beta^2 = 0, \quad \sigma_\gamma^2 = 0, \quad \sigma_{\alpha\beta}^2 = 0, \quad \sigma_{\alpha\gamma}^2 = 0, \quad \sigma_{\beta\gamma}^2 = 0, \quad \text{and} \quad \sigma_{\alpha\beta\gamma}^2 = 0.$$

The results are summarized in Table 5.5. If B is the fixed factor and A and C are random, then interchange A and B (also a and b) in Table 5.5. Similarly, if

TABLE 5.4
Tests of Hypotheses for Model (5.1.1) under Model II

Hypothesis	Test Statistic*	Percentile
$H_0^A : \sigma_\alpha^2 = 0$ versus $H_1^A : \sigma_\alpha^2 > 0$	$F_A = \dfrac{MS_A}{MS_{AB} + MS_{AC} - MS_{ABC}}$	$F[a-1, v_a; 1-\alpha]$
$H_0^B : \sigma_\beta^2 = 0$ versus $H_1^B : \sigma_\beta^2 > 0$	$F_B = \dfrac{MS_B}{MS_{AB} + MS_{BC} - MS_{ABC}}$	$F[b-1, v_b; 1-\alpha]$
$H_0^C : \sigma_\gamma^2 = 0$ versus $H_1^C : \sigma_\gamma^2 > 0$	$F_C = \dfrac{MS_C}{MS_{AC} + MS_{BC} - MS_{ABC}}$	$F[c-1, v_c; 1-\alpha]$
$H_0^{AB} : \sigma_{\alpha\beta}^2 = 0$ versus $H_1^{AB} : \sigma_{\alpha\beta}^2 > 0$	$F_{AB} = \dfrac{MS_{AB}}{MS_{ABC}}$	$F[(a-1)(b-1), (a-1)(b-1)(c-1); 1-\alpha]$
$H_0^{AC} : \sigma_{\alpha\gamma}^2 = 0$ versus $H_1^{AC} : \sigma_{\alpha\gamma}^2 > 0$	$F_{AC} = \dfrac{MS_{AC}}{MS_{ABC}}$	$F[(a-1)(c-1), (a-1)(b-1)(c-1); 1-\alpha]$
$H_0^{BC} : \sigma_{\beta\gamma}^2 = 0$ versus $H_1^{BC} : \sigma_{\beta\gamma}^2 > 0$	$F_{BC} = \dfrac{MS_{BC}}{MS_{ABC}}$	$F[(b-1)(c-1), (a-1)(b-1)(c-1); 1-\alpha]$
$H_0^{ABC} : \sigma_{\alpha\beta\gamma}^2 = 0$ versus $H_1^{ABC} : \sigma_{\alpha\beta\gamma}^2 > 0$	$F_{ABC} = \dfrac{MS_{ABC}}{MS_E}$	$F[(a-1)(b-1)(c-1), abc(n-1); 1-\alpha]$

* For the test statistics F_A, F_B, and F_C, the denominator degrees of freedom v_a, v_b, and v_c are obtained using the formula (5.5.3) and its obvious analogues for v_b and v_c.

C is fixed and A and B are random, then interchange A and C (also a and c) in Table 5.5.

Next, suppose A is random and B and C are fixed. The results of all the principal hypotheses of interest are summarized in Table 5.6. Note that all the F tests are exact and no approximate tests are necessary. If the random factor is B, interchange the role of A and B (also a and b) in Table 5.6. If the random factor is C, interchange the role of A and C (also a and c) in Table 5.6.

5.6 POINT AND INTERVAL ESTIMATION

No new problems arise in obtaining unbiased estimators of variance components for random effects factors or in the estimation of contrasts for fixed effects factors in Models I or III. Confidence limits for contrasts for fixed effects factors are constructed by using the mean square employed in the denominator of the test statistic while testing for the effects of that factor. The degrees of freedom correspond to the mean square used in the denominator. The results on point estimation for Model I are summarized in Table 5.7.

Three-Way and Higher-Order Crossed Classifications

TABLE 5.5
Tests of Hypotheses for Model (5.1.1) under Model III
(A Fixed, B and C Random)

Hypothesis	Test Statistic*	Percentile
H_0^A : all $\alpha_i = 0$ versus H_1^A : not all $\alpha_i = 0$	$F_A = \dfrac{MS_A}{MS_{AB} + MS_{AC} - MS_{ABC}}$	$F[a-1, \nu_a; 1-\alpha]$
$H_0^B : \sigma_\beta^2 = 0$ versus $H_1^B : \sigma_\beta^2 > 0$	$F_B = \dfrac{MS_B}{MS_{BC}}$	$F[b-1, (b-1)(c-1); 1-\alpha]$
$H_0^C : \sigma_\gamma^2 = 0$ versus $H_1^C : \sigma_\gamma^2 > 0$	$F_C = \dfrac{MS_C}{MS_{BC}}$	$F[(c-1), (b-1)(c-1); 1-\alpha]$
$H_0^{AB} : \sigma_{\alpha\beta}^2 = 0$ versus $H_1^{AB} : \sigma_{\alpha\beta}^2 > 0$	$F_{AB} = \dfrac{MS_{AB}}{MS_{ABC}}$	$F[(a-1)(b-1), (a-1)(b-1)(c-1); 1-\alpha]$
$H_0^{AC} : \sigma_{\alpha\gamma}^2 = 0$ versus $H_1^{AC} : \sigma_{\alpha\gamma}^2 > 0$	$F_{AC} = \dfrac{MS_{AC}}{MS_{ABC}}$	$F[(a-1)(c-1), (a-1)(b-1)(c-1); 1-\alpha]$
$H_0^{BC} : \sigma_{\beta\gamma}^2 = 0$ versus $H_1^{BC} : \sigma_{\beta\gamma}^2 > 0$	$F_{BC} = \dfrac{MS_{BC}}{MS_E}$	$F[(b-1)(c-1), abc(n-1); 1-\alpha]$
$H_0^{ABC} : \sigma_{\alpha\beta\gamma}^2 = 0$ versus $H_1^{ABC} : \sigma_{\alpha\beta\gamma}^2 > 0$	$F_{ABC} = \dfrac{MS_{ABC}}{MS_E}$	$F[(a-1)(b-1)(c-1), abc(n-1); 1-\alpha]$

* For the test statistic F_A, the denominator degrees of freedom ν_a is determined using the formula (5.5.3).

An exact $100(1 - \alpha)$ percent confidence interval for σ_e^2 is

$$\frac{abc(n-1)MS_E}{\chi^2[abc(n-1), 1 - \alpha/2]} < \sigma_e^2 < \frac{abc(n-1)MS_E}{\chi^2[abc(n-1), \alpha/2]}. \qquad (5.6.1)$$

Confidence intervals for other parameters under Model I are obtained from the corresponding items in columns (2) and (3) of Table 5.7. For example, for making pairwise comparisons, Bonferroni intervals are given by

$$P\left[\bar{y}_{i\ldots} - \bar{y}_{i'\ldots} - \theta\sqrt{\frac{2MS_E}{bcn}} < \alpha_i - \alpha_{i'} < \bar{y}_{i\ldots} - \bar{y}_{i'\ldots} + \theta\sqrt{\frac{2MS_E}{bcn}} \right] \doteq 1 - \alpha, \qquad (5.6.2)$$

where

$$\theta = t[abc(n-1), 1 - \alpha/2m]$$

TABLE 5.6
Tests of Hypotheses for Model (5.1.1) under Model III
(*A* Random, *B* and *C* Fixed)

Hypothesis	Test Statistic	Percentile
$H_0^A : \sigma_\alpha^2 = 0$ versus $H_1^A : \sigma_\alpha^2 > 0$	$F_A = \dfrac{MS_A}{MS_E}$	$F[a-1, abc(n-1); 1-\alpha]$
$H_0^B :$ all $\beta_j = 0$ versus $H_1^B :$ not all $\beta_j = 0$	$F_B = \dfrac{MS_B}{MS_{AB}}$	$F[b-1, (a-1)(b-1); 1-\alpha]$
$H_0^C :$ all $\gamma_k = 0$ versus $H_1^C :$ not all $\gamma_k = 0$	$F_C = \dfrac{MS_C}{MS_{AC}}$	$F[c-1, (a-1)(c-1); 1-\alpha]$
$H_0^{AB} : \sigma_{\alpha\beta}^2 = 0$ versus $H_1^{AB} : \sigma_{\alpha\beta}^2 > 0$	$F_{AB} = \dfrac{MS_{AB}}{MS_E}$	$F[(a-1)(b-1), abc(n-1); 1-\alpha]$
$H_0^{AC} : \sigma_{\alpha\gamma}^2 = 0$ versus $H_1^{AC} : \sigma_{\alpha\gamma}^2 > 0$	$F_{AC} = \dfrac{MS_{AC}}{MS_E}$	$F[(a-1)(c-1), abc(n-1); 1-\alpha]$
$H_0^{BC} :$ all $(\beta\gamma)_{jk} = 0$ versus $H_1^{BC} :$ not all $(\beta\gamma)_{jk} = 0$	$F_{BC} = \dfrac{MS_{BC}}{MS_{ABC}}$	$F[(b-1)(c-1), (a-1)(b-1)(c-1); 1-\alpha]$
$H_0^{ABC} : \sigma_{\alpha\beta\gamma}^2 = 0$ versus $H_1^{ABC} : \sigma_{\alpha\beta\gamma}^2 > 0$	$F_{ABC} = \dfrac{MS_{ABC}}{MS_E}$	$F[(a-1)(b-1)(c-1), abc(n-1); 1-\alpha]$

is the t value with m being the number of intervals constructed. Similarly, $100(1-\alpha)$ percent Bonferroni intervals for the contrast

$$L = \sum_{i=1}^{a} \ell_i \alpha_i \left(\sum_{i=1}^{a} \ell_i = 0 \right)$$

are determined by

$$\hat{L} - \theta \sqrt{\frac{MS_E}{bcn} \sum_{i=1}^{a} \ell_i^2} < L < \hat{L} + \theta \sqrt{\frac{MS_E}{bcn} \sum_{i=1}^{a} \ell_i^2}. \quad (5.6.3)$$

Under Model III, with A fixed and B and C random, one is typically interested in contrasts of the form $\sum_{i=1}^{a} \ell_i \alpha_i (\sum_{i=1}^{a} \ell_i = 0)$; however, no exact interval for a linear contrast is available. To see this note that $\sum_{i=1}^{a} \ell_i \hat{\alpha}_i = \sum_{i=1}^{a} \ell_i \bar{y}_{i...}$ with

TABLE 5.7
Estimates of Parameters and Their Variances under Model I*

Parameter	Point Estimate	Variance of Estimate
μ	$\bar{y}_{....}$	$\sigma_e^2/abcn$
α_i	$\bar{y}_{i...} - \bar{y}_{....}$	$(a-1)\sigma_e^2/abcn$
$\mu + \alpha_i$	$\bar{y}_{i...}$	σ_e^2/bcn
$\mu + \alpha_i + \beta_j + (\alpha\beta)_{ij}$	$\bar{y}_{ij..}$	σ_e^2/cn
$\mu + \alpha_i + \beta_j + \gamma_k + (\alpha\beta)_{ij}$ $+ (\alpha\gamma)_{ik} + (\beta\gamma)_{jk}$ $+ (\alpha\beta\gamma)_{ijk}$	$\bar{y}_{ijk.}$	σ_e^2/n
$\alpha_i - \alpha_{i'}$	$\bar{y}_{i...} - \bar{y}_{i'...}$	$2\sigma_e^2/bcn$
$\sum_{i=1}^{a} \ell_i \alpha_i \left(\sum_{i=1}^{a} \ell_i = 0 \right)$	$\sum_{i=1}^{a} \ell_i \bar{y}_{i...}$	$\left(\sum_{i=1}^{a} \ell_i^2 \right) \sigma_e^2/bcn$
$(\alpha\beta)_{ij}$	$\bar{y}_{ij..} - \bar{y}_{i...} - \bar{y}_{.j..} + \bar{y}_{....}$	$(a-1)(b-1)\sigma_e^2/abcn$
$(\alpha\beta\gamma)_{ijk}$	$\bar{y}_{ijk.} - \bar{y}_{ij..} - \bar{y}_{i.k.} - \bar{y}_{.jk.}$ $+ \bar{y}_{i...} + \bar{y}_{.j..} + \bar{y}_{..k.} - \bar{y}_{....}$	$(a-1)(b-1)(c-1)\sigma_e^2/abcn$
σ_e^2	MS_E	$2\sigma_e^4/abc(n-1)$

* Estimates for β_j, $\mu + \beta_j$, $(\alpha\gamma)_{ik}$, and other parameters not included in Table 5.6 can be obtained by interchanging appropriate subscripts in the estimates shown in the table.

$\text{Var}(\sum_{i=1}^{a} \ell_i \bar{y}_{i...}) = (\sum_{i=1}^{a} \ell_i^2)(\sigma_e^2 + n\sigma_{\alpha\beta\gamma}^2 + cn\sigma_{\alpha\beta}^2 + bn\sigma_{\alpha\gamma}^2)/(bcn)$. Thus, Var $(\sum_{i=1}^{a} \ell_i \bar{y}_{i...})$ cannot be estimated using only a single mean square in the analysis of variance Table 5.2. Approximate intervals can be based on Satterthwaite procedure and a method due to Naik (1974). For further discussion of these and other related procedures including a numerical example, see Burdick and Graybill (1992, pp. 156–160). If two of the effects are fixed and one random, we have seen that exact tests exist for all the hypotheses of interest. Thus, exact intervals can be constructed for all estimable functions of fixed effect parameters. For example, with A and C fixed and B random, it can be shown that

$$\text{Var}(\bar{y}_{i...}) = \frac{\sigma_\beta^2}{b} + \frac{(a-1)\sigma_{\alpha\beta}^2}{ab} + \frac{\sigma_e^2}{bcn} \quad (5.6.4)$$

and

$$\text{Var}(\bar{y}_{i...} - \bar{y}_{i'...}) = \frac{2(\sigma_e^2 + cn\sigma_{\alpha\beta}^2)}{bcn}. \quad (5.6.5)$$

Remark: The results (5.6.4) and (5.6.5) can be derived as follows. First, since all the random and mixed terms are independent of each other so their cross-products will have expected value equal to zero. Further, it follows that $\text{Var}\{(\alpha\beta)_{ij}\} = E[(\alpha\beta)_{ij}^2] = [(a-1)/a]\sigma_{\alpha\beta}^2$ and $E = [(\alpha\beta)_{ij}(\alpha\beta)_{i'j'}] = 0$ for $j \neq j'$. Similar results hold for $(\beta\gamma)_{jk}$ and $(\alpha\beta\gamma)_{ijk}$. Now, $E(\bar{y}_{i...}) = \mu + \alpha_i$ and $\text{Var}(\bar{y}_{i...})$ is given by

$$\text{Var}(\bar{y}_{i...}) = E[\bar{y}_{i...} - \mu - \alpha_i]^2$$

$$= E\left[\frac{1}{bcn}\sum_{j=1}^{b}\sum_{k=1}^{c}\sum_{\ell=1}^{n}(\beta_j + (\alpha\beta)_{ij} + \gamma_k + (\alpha\gamma)_{ik}\right.$$

$$\left. + (\beta\gamma)_{jk} + (\alpha\beta\gamma)_{ijk} + e_{ijk\ell})\right]^2$$

$$= E\left[\frac{1}{b}\sum_{j=1}^{b}\beta_j + \frac{1}{b}\sum_{j=1}^{b}(\alpha\beta)_{ij} + \frac{1}{bcn}\sum_{j=1}^{b}\sum_{k=1}^{c}\sum_{\ell=1}^{n}e_{ijk\ell}\right]^2$$

$$= \frac{b}{b^2}\sigma_\beta^2 + \frac{(a-1)b}{ab^2}\sigma_{\alpha\beta}^2 + \frac{bcn}{b^2c^2n^2}\sigma_e^2$$

$$= \frac{\sigma_\beta^2}{b} + \frac{(a-1)\sigma_{\alpha\beta}^2}{ab} + \frac{\sigma_e^2}{bcn}.$$

Similarly, since $E[(\alpha\beta)_{ij}(\alpha\beta)_{i'j}] = -\sigma_{\alpha\beta}^2/a$ and $E[\bar{y}_{i...} - \bar{y}_{i'...}] = \alpha_i - \alpha_{i'}$, $\text{Var}(\bar{y}_{i...} - \bar{y}_{i'...})$ is given by

$$\text{Var}(\bar{y}_{i...} - \bar{y}_{i'...}) = E[\bar{y}_{i...} - \alpha_i - \bar{y}_{i'...} + \alpha_{i'}]^2$$

$$= E\left[\frac{1}{b}\sum_{j=1}^{b}(\alpha\beta)_{ij} - \frac{1}{b}\sum_{j'=1}^{b}(\alpha\beta)_{i'j'} + \frac{1}{bcn}\sum_{j=1}^{b}\sum_{k=1}^{c}\sum_{\ell=1}^{n}e_{ijk\ell}\right.$$

$$\left. - \frac{1}{bcn}\sum_{j'=1}^{b}\sum_{k'=1}^{c}\sum_{\ell'=1}^{n}e_{i'j'k'\ell'}\right]^2$$

$$= \frac{2(a-1)b}{ab^2}\sigma_{\alpha\beta}^2 - \frac{2E\left[\left(\sum_{j=1}^{b}(\alpha\beta)_{ij}\right)\left(\sum_{j'=1}^{b}(\alpha\beta)_{i'j'}\right)\right]}{b^2} + \frac{2\sigma_e^2}{bcn}$$

$$= \frac{2(a-1)\sigma_{\alpha\beta}^2}{ab} + \frac{2b\sigma_{\alpha\beta}^2}{ab^2} + \frac{2\sigma_e^2}{bcn}$$

$$= \frac{2(\sigma_e^2 + cn\sigma_{\alpha\beta}^2)}{bcn}.$$

The analysis of variance estimates of the variance components are readily obtained from Table 5.2 by setting the mean squares equal to the expected mean squares and solving for the desired variance components. For example, under Model III with A fixed and B and C random, the estimates of variance components are:

$$\hat{\sigma}_e^2 = \mathrm{MS}_E$$
$$\hat{\sigma}_{\alpha\beta\gamma}^2 = (\mathrm{MS}_{ABC} - \mathrm{MS}_E)/n,$$
$$\hat{\sigma}_{\beta\gamma}^2 = (\mathrm{MS}_{BC} - \mathrm{MS}_E)/an,$$
$$\hat{\sigma}_{\alpha\gamma}^2 = (\mathrm{MS}_{AC} - \mathrm{MS}_{ABC})/bn,$$
$$\hat{\sigma}_{\alpha\beta}^2 = (\mathrm{MS}_{AB} - \mathrm{MS}_{ABC})/cn,$$
$$\hat{\sigma}_\gamma^2 = (\mathrm{MS}_C - \mathrm{MS}_{BC})/abn,$$

and

$$\hat{\sigma}_\beta^2 = (\mathrm{MS}_B - \mathrm{MS}_{BC})/acn.$$

For some results on confidence intervals for individual variance components and certain sums and ratios of variance components under Models II and III, including numerical examples, see Burdick and Graybill (1992, pp. 131–136, 156–160).

5.7 COMPUTATIONAL FORMULAE AND PROCEDURE

Ordinarily, computer programs will be employed to perform the analysis of variance calculations involving three or more factors. For completeness, however, we present the necessary computational formulae:

$$\mathrm{SS}_T = \sum_{i=1}^{a}\sum_{j=1}^{b}\sum_{k=1}^{c}\sum_{\ell=1}^{n} y_{ijk\ell}^2 - \frac{y_{....}^2}{abcn},$$

$$\mathrm{SS}_A = \frac{1}{bcn}\sum_{i=1}^{a} y_{i...}^2 - \frac{y_{....}^2}{abcn},$$

$$\mathrm{SS}_B = \frac{1}{acn}\sum_{j=1}^{b} y_{.j..}^2 - \frac{y_{....}^2}{abcn},$$

$$\mathrm{SS}_C = \frac{1}{abn}\sum_{k=1}^{c} y_{..k.}^2 - \frac{y_{....}^2}{abcn},$$

$$\mathrm{SS}_{AB} = \frac{1}{cn}\sum_{i=1}^{a}\sum_{j=1}^{b} y_{ij..}^2 - \frac{1}{bcn}\sum_{i=1}^{a} y_{i...}^2 - \frac{1}{acn}\sum_{j=1}^{b} y_{.j..}^2 + \frac{y_{....}^2}{abcn},$$

$$\mathrm{SS}_{AC} = \frac{1}{bn}\sum_{i=1}^{a}\sum_{k=1}^{c} y_{i.k.}^2 - \frac{1}{bcn}\sum_{i=1}^{a} y_{i...}^2 - \frac{1}{abn}\sum_{k=1}^{c} y_{..k.}^2 + \frac{y_{....}^2}{abcn},$$

$$SS_{BC} = \frac{1}{an} \sum_{j=1}^{b} \sum_{k=1}^{c} y_{.jk.}^2 - \frac{1}{acn} \sum_{j=1}^{b} y_{.j..}^2 - \frac{1}{abn} \sum_{k=1}^{c} y_{..k.}^2 + \frac{y_{....}^2}{abcn},$$

$$SS_{ABC} = \frac{1}{n} \sum_{i=1}^{a} \sum_{j=1}^{b} \sum_{k=1}^{c} y_{ijk.}^2 - \frac{1}{cn} \sum_{i=1}^{a} \sum_{j=1}^{b} y_{ij..}^2 - \frac{1}{bn} \sum_{i=1}^{a} \sum_{k=1}^{c} y_{i.k.}^2,$$

$$- \frac{1}{an} \sum_{j=1}^{b} \sum_{k=1}^{c} y_{.jk.}^2 + \frac{1}{bcn} \sum_{i=1}^{a} y_{i...}^2 + \frac{1}{acn} \sum_{j=1}^{b} y_{.j..}^2$$

$$+ \frac{1}{abn} \sum_{k=1}^{c} y_{..k.}^2 - \frac{y_{....}^2}{abcn},$$

and

$$SS_E = \sum_{i=1}^{a} \sum_{j=1}^{b} \sum_{k=1}^{c} \sum_{\ell=1}^{n} y_{ijk\ell}^2 - \frac{1}{n} \sum_{i=1}^{a} \sum_{j=1}^{b} \sum_{k=1}^{c} y_{ijk.}^2.$$

Alternatively, SS_E can be obtained from the relation:

$$SS_E = SS_T - SS_A - SS_B - SS_C - SS_{AB} - SS_{AC} - SS_{BC} - SS_{ABC}.$$

Remark: The preceding computational formulae can be readily extended if four or more factors are involved. In Section 5.11, we illustrate some of these computational formulae for the case of the four-factor crossed classification model.

5.8 POWER OF THE ANALYSIS OF VARIANCE F TESTS

Under Model I, the power of each of the F tests summarized in Table 5.3 can be obtained in the manner described for the one-way and two-way classification models. The normalized noncentrality parameter ϕ needed for calculating the power of each F test can be obtained as follows:

$$\phi = \frac{1}{\sigma_e} \left[\frac{\text{Numerator of Second Term in the Expected Mean Squares Column in Table 5.2}}{\text{Corresponding Degrees of Freedom} + 1} \right]^{1/2}.$$

For example, for testing the null hypothesis that the three-factor interaction ABC is zero, we have

$$\phi_{ABC} = \frac{1}{\sigma_e} \left[\frac{n \sum_{i=1}^{a} \sum_{j=1}^{b} \sum_{k=1}^{c} (\alpha\beta\gamma)_{ijk}^2}{(a-1)(b-1)(c-1)+1} \right]^{1/2}.$$

For the two-factor interaction AB,

$$\phi_{AB} = \frac{1}{\sigma_e} \left[\frac{cn \sum_{i=1}^{a} \sum_{j=1}^{b} (\alpha\beta)_{ij}^2}{(a-1)(b-1)+1} \right]^{1/2},$$

and for the main effect A,

$$\phi_A = \frac{1}{\sigma_e} \left[\frac{bcn \sum_{i=1}^{a} \alpha_i^2}{a} \right]^{1/2},$$

and so on.

In an analogous manner, under Model II, for testing the null hypothesis that the three-factor interaction ABC is zero, we have

$$\lambda_{ABC} = \left[1 + \frac{n\sigma_{\alpha\beta\gamma}^2}{\sigma_e^2} \right]^{1/2}.$$

For the two-factor interaction AB,

$$\lambda_{AB} = \left[1 + \frac{cn\sigma_{\alpha\beta}^2}{\sigma_e^2 + n\sigma_{\alpha\beta\gamma}^2} \right]^{1/2}.$$

For testing the main effects, we have seen that no exact F tests are available. An approximate power of the pseudo-F tests discussed in Section 5.5 may, however, be computed (see, e.g., Scheffé (1959, p. 248)).

5.9 MULTIPLE COMPARISON METHODS

As before, multiple comparison methods can be utilized for the fixed as well as mixed effects models to test contrasts among cell means or factor level means. We briefly indicate the procedure for the fixed effects case.

When the null hypothesis about a certain main effect or interaction is rejected, Tukey, Scheffé, or other multiple comparison methods may be used to investigate specific contrasts of interest. For example, if H_0^{ABC} is rejected, we may be interested in comparing the contrasts of cell means

$$\mu_{ijk} = \mu + \alpha_i + \beta_j + \gamma_k + (\alpha\beta)_{ij} + (\alpha\gamma)_{ik} + (\beta\gamma)_{jk} + (\alpha\beta\gamma)_{ijk}.$$

Then multiple comparison methods may be used to investigate the general contrasts of the type

$$L = \sum_{i=1}^{a}\sum_{j=1}^{b}\sum_{k=1}^{c} \ell_{ijk}\mu_{ijk},$$

where

$$\sum_{i=1}^{a}\sum_{j=1}^{b}\sum_{k=1}^{c} \ell_{ijk} = 0.$$

An unbiased estimator of L is

$$\hat{L} = \sum_{i=1}^{a}\sum_{j=1}^{b}\sum_{k=1}^{c} \ell_{ijk}\bar{y}_{ijk.},$$

for which the estimated variance is

$$\widehat{\text{Var}}(\hat{L}) = \frac{\text{MS}_E}{n}\sum_{i=1}^{a}\sum_{j=1}^{b}\sum_{k=1}^{c} \ell_{ijk}^2.$$

For the Tukey's method involving pairwise comparisons, we have

$$T = q[abc, abc(n-1); 1-\alpha].$$

For the Scheffé's method involving general contrasts, we would have

$$S^2 = F[abc-1, abc(n-1); 1-\alpha].$$

Furthermore, if H_0^A is rejected, one may proceed to investigate contrasts involving α_i's of the form

$$L = \sum_{i=1}^{a} \ell_i \alpha_i,$$

where

$$\sum_{i=1}^{a} \ell_i = 0.$$

Again, an unbiased estimator of L is

$$\hat{L} = \sum_{i=1}^{a} \ell_i \bar{y}_{i...}$$

with estimated variance

$$\widehat{\mathrm{Var}}(\hat{L}) = \frac{\mathrm{MS}_E}{bcn} \sum_{i=1}^{a} \ell_i^2.$$

For the Tukey's procedure involving pairwise differences, we have

$$T = q[a, abc(n-1); 1-\alpha].$$

For the Scheffé's procedure involving general contrasts, we would have

$$S^2 = F[a-1, abc(n-1); 1-\alpha].$$

Contrasts based on β_j's and γ_k's can be investigated in an analogous manner.

5.10 THREE-WAY CLASSIFICATION WITH ONE OBSERVATION PER CELL

If there is only one observation per cell in model (5.1.1) (i.e., $n=1$), we cannot estimate the error variance σ_e^2 from within-cell replications. In this case, analysis of variance tests can be conducted only if it is possible to make an additional assumption that certain interactions are zero. Usually, we would assume that there is no three-factor interaction $A \times B \times C$. If it is possible to assume that the $A \times B \times C$ interaction is zero, then the corresponding mean square MS_{ABC} has expectation σ_e^2 and can be used as the error mean square MS_E to estimate the error variance σ_e^2. However, this layout does not allow separation of the three-factor interaction term from the within-cell variation or the error term.

The analysis of variance model in this case is written as

$$y_{ijk} = \mu + \alpha_i + \beta_j + \gamma_k + (\alpha\beta)_{ij} + (\alpha\gamma)_{ik} + (\beta\gamma)_{jk} + e_{ijk} \quad \begin{cases} i = 1, 2, \ldots, a \\ j = 1, 2, \ldots, b \\ k = 1, 2, \ldots, c. \end{cases} \quad (5.10.1)$$

All sums of squares and mean squares are calculated in the usual manner except that now $n = 1$. The definitional and computational formulae for the sums of squares are:

$$\mathrm{SS}_A = bc \sum_{i=1}^{a} (\bar{y}_{i..} - \bar{y}_{...})^2 = \frac{1}{bc} \sum_{i=1}^{a} y_{i..}^2 - \frac{1}{abc} y_{...}^2,$$

$$\mathrm{SS}_B = ac \sum_{j=1}^{b} (\bar{y}_{.j.} - \bar{y}_{...})^2 = \frac{1}{ac} \sum_{j=1}^{b} y_{.j.}^2 - \frac{1}{abc} y_{...}^2,$$

$$\mathrm{SS}_C = ab \sum_{k=1}^{c} (\bar{y}_{..k} - \bar{y}_{...})^2 = \frac{1}{ab} \sum_{k=1}^{c} y_{..k}^2 - \frac{1}{abc} y_{...}^2,$$

$$SS_{AB} = c \sum_{i=1}^{a} \sum_{j=1}^{b} (\bar{y}_{ij.} - \bar{y}_{i..} - \bar{y}_{.j.} + \bar{y}_{...})^2,$$

$$= \frac{1}{c} \sum_{i=1}^{a} \sum_{j=1}^{b} y_{ij.}^2 - \frac{1}{bc} \sum_{i=1}^{a} y_{i..}^2 - \frac{1}{ac} \sum_{j=1}^{b} y_{.j.}^2 + \frac{1}{abc} y_{...}^2,$$

$$SS_{AC} = b \sum_{i=1}^{a} \sum_{k=1}^{c} (\bar{y}_{i.k} - \bar{y}_{i..} - \bar{y}_{..k} + \bar{y}_{...})^2,$$

$$= \frac{1}{b} \sum_{i=1}^{a} \sum_{k=1}^{c} y_{i.k}^2 - \frac{1}{bc} \sum_{i=1}^{a} y_{i..}^2 - \frac{1}{ab} \sum_{k=1}^{c} y_{..k}^2 + \frac{1}{abc} y_{...}^2,$$

$$SS_{BC} = a \sum_{j=1}^{b} \sum_{k=1}^{c} (\bar{y}_{.jk} - \bar{y}_{.j.} - \bar{y}_{..k} + \bar{y}_{...})^2,$$

$$= \frac{1}{a} \sum_{j=1}^{b} \sum_{k=1}^{c} y_{.jk}^2 - \frac{1}{ac} \sum_{j=1}^{b} y_{.j.}^2 - \frac{1}{ab} \sum_{k=1}^{c} y_{..k}^2 + \frac{1}{abc} y_{...}^2,$$

and

$$SS_E = \sum_{i=1}^{a} \sum_{j=1}^{b} \sum_{k=1}^{c} (y_{ijk} - \bar{y}_{ij.} - \bar{y}_{i.k} - \bar{y}_{.jk} + \bar{y}_{i..} + \bar{y}_{.j.} + \bar{y}_{..k} - \bar{y}_{...})^2$$

$$= \sum_{i=1}^{a} \sum_{j=1}^{b} \sum_{k=1}^{c} y_{ijk}^2 - \frac{1}{c} \sum_{i=1}^{a} \sum_{j=1}^{b} y_{ij.}^2 - \frac{1}{b} \sum_{i=1}^{a} \sum_{k=1}^{c} y_{i.k}^2 - \frac{1}{a} \sum_{j=1}^{b} \sum_{k=1}^{c} y_{.jk}^2$$

$$+ \frac{1}{bc} \sum_{i=1}^{a} y_{i..}^2 + \frac{1}{ac} \sum_{j=1}^{b} y_{.j.}^2 + \frac{1}{ab} \sum_{k=1}^{c} y_{..k}^2 - \frac{1}{abc} y_{...}^2.$$

The resulting analysis of variance table is shown in Table 5.8. In the case of Model I, all mean squares are tested against the mean square for error (three-factor interaction). When the null hypothesis about a certain main effect or a two-factor interaction is rejected, Tukey, Scheffé or other multiple comparison methods may be used to investigate contrasts of interest. Under Models II and III, the appropriate test statistics for various hypotheses of interest can be determined by examining expected mean squares in Table 5.8. However, again, there are no exact F tests for testing the hypotheses about main effects under Model II. Pseudo-F tests discussed earlier in Section 5.5 can similarly be developed.

5.11 FOUR-WAY CROSSED CLASSIFICATION

The analysis of variance in a four-way classification is obtained as a straightforward generalization of the three-way classification and we discuss it only

TABLE 5.8
Analysis of Variance for Model (5.10.1)

Source of Variation	Degrees of Freedom	Sum of Squares	Mean Square	Expected Mean Square		
				Model I A Fixed, B and C Fixed	Model II A, B, and C Random	Model III A Random, B and C Fixed
Due to A	$a-1$	SS_A	MS_A	$\sigma_e^2 + \dfrac{bc}{a-1}\sum_{i=1}^{a}\alpha_i^2$	$\sigma_e^2 + c\sigma_{\alpha\beta}^2 + b\sigma_{\alpha\gamma}^2 + bc\sigma_\alpha^2$	$\sigma_e^2 + c\sigma_{\alpha\beta}^2 + b\sigma_{\alpha\gamma}^2$ $+ \dfrac{bc}{a-1}\sum_{i=1}^{a}\alpha_i^2$	$\sigma_e^2 + bc\sigma_\alpha^2$
Due to B	$b-1$	SS_B	MS_B	$\sigma_e^2 + \dfrac{ac}{b-1}\sum_{j=1}^{b}\beta_j^2$	$\sigma_e^2 + c\sigma_{\alpha\beta}^2 + a\sigma_{\beta\gamma}^2 + ac\sigma_\beta^2$	$\sigma_e^2 + c\sigma_{\alpha\beta}^2 + \dfrac{ac}{b-1}\sum_{j=1}^{b}\beta_j^2$
Due to C	$c-1$	SS_C	MS_C	$\sigma_e^2 + \dfrac{ab}{c-1}\sum_{k=1}^{c}\gamma_k^2$	$\sigma_e^2 + b\sigma_{\alpha\gamma}^2 + a\sigma_{\beta\gamma}^2 + ab\sigma_\gamma^2$	$\sigma_e^2 + b\sigma_{\alpha\gamma}^2 + \dfrac{ab}{c-1}\sum_{k=1}^{c}\gamma_k^2$
Interaction $A \times B$	$(a-1)(b-1)$	SS_{AB}	MS_{AB}	$\sigma_e^2 + \dfrac{c}{(a-1)(b-1)}\sum_{i=1}^{a}\sum_{j=1}^{b}(\alpha\beta)_{ij}^2$	$\sigma_e^2 + c\sigma_{\alpha\beta}^2$	$\sigma_e^2 + c\sigma_{\alpha\beta}^2$
Interaction $A \times C$	$(a-1)(c-1)$	SS_{AC}	MS_{AC}	$\sigma_e^2 + \dfrac{b}{(a-1)(c-1)}\sum_{i=1}^{a}\sum_{k=1}^{c}(\alpha\gamma)_{ik}^2$	$\sigma_e^2 + b\sigma_{\alpha\gamma}^2$	$\sigma_e^2 + b\sigma_{\alpha\gamma}^2$
Interaction $B \times C$	$(b-1)(c-1)$	SS_{BC}	MS_{BC}	$\sigma_e^2 + \dfrac{a}{(b-1)(c-1)}\sum_{j=1}^{b}\sum_{k=1}^{c}(\beta\gamma)_{jk}^2$	$\sigma_e^2 + a\sigma_{\beta\gamma}^2$	$\sigma_e^2 + a\sigma_{\beta\gamma}^2$
Interaction $A \times B \times C$ (Error)	$(a-1)(b-1)\times(c-1)$	SS_E	MS_{ABC}	σ_e^2	σ_e^2	$\sigma_e^2 + \dfrac{a}{(b-1)(c-1)}$ $\times\sum_{j=1}^{b}\sum_{k=1}^{c}(\beta\gamma)_{jk}^2$	σ_e^2
Total	$abc-1$	SS_T				

briefly. The model is given by

$$y_{ijk\ell m} = \mu + \alpha_i + \beta_j + \gamma_k + \delta_\ell + (\alpha\beta)_{ij} + (\alpha\gamma)_{ik}$$
$$+ (\alpha\delta)_{i\ell} + (\beta\gamma)_{jk} + (\beta\delta)_{j\ell} + (\gamma\delta)_{k\ell}$$
$$+ (\alpha\beta\gamma)_{ijk} + (\alpha\beta\delta)_{ij\ell} + (\alpha\gamma\delta)_{ik\ell}$$
$$+ (\beta\gamma\delta)_{jk\ell} + (\alpha\beta\gamma\delta)_{ijk\ell} + e_{ijk\ell m}$$

$$\begin{cases} i = 1, \ldots, a \\ j = 1, \ldots, b \\ k = 1, \ldots, c \\ \ell = 1, \ldots, d \\ m = 1, \ldots, n, \end{cases}$$

(5.11.1)

where $e_{ijk\ell m}$'s are independently and normally distributed with zero mean and variance σ_e^2. The assumptions on other effects can analogously be stated depending upon whether a factor is fixed or random. Note that the model equation (5.11.1) has 17 terms: a general mean, one main effect for each of the four factors, six two-factor interactions, four three-factor interactions, one four-factor interaction, and a residual or error term.

The usual identity $y_{ijk\ell m} - \bar{y}_{.....} =$ etc., contains the following groups of terms on its right-hand-side:

(i) Estimates of the four main effects, for example, $\bar{y}_{i....} - \bar{y}_{.....}$, which gives an estimate of α_i.
(ii) Estimates of the six two-way interactions, for example, $\bar{y}_{ij...} - \bar{y}_{i....} - \bar{y}_{.j...} + \bar{y}_{.....}$, which gives an estimate of $(\alpha\beta)_{ij}$.
(iii) Estimates of the four three-way interactions, for example, $\bar{y}_{ijk..} - \bar{y}_{ij...} - \bar{y}_{i.k..} - \bar{y}_{.jk..} + \bar{y}_{i....} + \bar{y}_{.j...} + \bar{y}_{..k..} - \bar{y}_{.....}$, which gives an estimate of $(\alpha\beta\gamma)_{ijk}$.
(iv) Estimate of the single four-way interaction, which will have the form $\bar{y}_{ijk\ell.} - [\bar{y}_{.....} + \text{four main effects} + \text{six two-way interactions} + \text{four three-way interactions}]$.
(v) The deviations of the individual observations from the cell means, for example, $y_{ijk\ell m} - \bar{y}_{ijk\ell.}$.

The partition of the total sum of squares is effected by squaring and summing over all indices on both sides of the identity $y_{ijk\ell m} - \bar{y}_{.....} =$ and so on. The typical sums of squares and corresponding computational formulae are:

$$SS_A = bcdn \sum_{i=1}^{a} (\bar{y}_{i....} - \bar{y}_{.....})^2$$

$$= \frac{1}{bcdn} \sum_{i=1}^{a} y_{i....}^2 - \frac{1}{abcdn} y_{.....}^2,$$

$$SS_{AB} = cdn \sum_{i=1}^{a} \sum_{j=1}^{b} (\bar{y}_{ij...} - \bar{y}_{i....} - \bar{y}_{.j...} + \bar{y}_{.....})^2$$

$$= \frac{1}{cdn} \sum_{i=1}^{a} \sum_{j=1}^{b} y_{ij...}^2 - \frac{1}{abcdn} y_{.....}^2 - (SS_A + SS_B),$$

$$SS_{ABC} = dn \sum_{i=1}^{a} \sum_{j=1}^{b} \sum_{k=1}^{c}$$
$$\times (\bar{y}_{ijk..} - \bar{y}_{ij...} - \bar{y}_{i.k..} - \bar{y}_{.jk..} + \bar{y}_{i....} + \bar{y}_{.j...} + \bar{y}_{..k..} - \bar{y}_{.....})^2$$
$$= \frac{1}{dn} \sum_{i=1}^{a} \sum_{j=1}^{b} \sum_{k=1}^{c} y_{ijk..}^2 - \frac{1}{abcdn} y_{.....}^2$$
$$- (SS_A + SS_B + SS_C + SS_{AB} + SS_{AC} + SS_{BC}),$$

$$SS_{ABCD} = \frac{1}{n} \sum_{i=1}^{a} \sum_{j=1}^{b} \sum_{k=1}^{c} \sum_{\ell=1}^{d} y_{ijk\ell.}^2 - \frac{1}{abcdn} y_{.....}^2 - (SS_A + SS_B + SS_C$$
$$+ SS_D + SS_{AB} + SS_{AC} + SS_{AD} + SS_{BC} + SS_{BD} + SS_{CD}$$
$$+ SS_{ABC} + SS_{ACD} + SS_{ABD} + SS_{BCD}),$$

$$SS_E = \sum_{i=1}^{a} \sum_{j=1}^{b} \sum_{k=1}^{c} \sum_{\ell=1}^{d} \sum_{m=1}^{n} (y_{ijk\ell m} - \bar{y}_{ijk\ell.})^2$$
$$= \sum_{i=1}^{a} \sum_{j=1}^{b} \sum_{k=1}^{c} \sum_{\ell=1}^{d} \sum_{m=1}^{n} y_{ijk\ell m}^2 - \frac{1}{n} \sum_{i=1}^{a} \sum_{j=1}^{b} \sum_{k=1}^{c} \sum_{\ell=1}^{d} y_{ijk\ell.}^2.$$

The degrees of freedom for the preceding sums of squares are $a-1$, $(a-1)(b-1)$, $(a-1)(b-1)(c-1)$, $(a-1)(b-1)(c-1)(d-1)$ and $abcd(n-1)$, respectively. The expected mean squares can be derived as before.[3] For example, under Model II,

$$E(MS_A) = \sigma_e^2 + n\sigma_{\alpha\beta\gamma\delta}^2 + dn\sigma_{\alpha\beta\gamma}^2 + cn\sigma_{\alpha\beta\delta}^2 + bn\sigma_{\alpha\gamma\delta}^2 + cdn\sigma_{\alpha\beta}^2$$
$$+ bdn\sigma_{\alpha\gamma}^2 + bcn\sigma_{\alpha\delta}^2 + bcdn\sigma_{\alpha}^2,$$
$$E(MS_{AB}) = \sigma_e^2 + n\sigma_{\alpha\beta\gamma\delta}^2 + dn\sigma_{\alpha\beta\gamma}^2 + cn\sigma_{\alpha\beta\delta}^2 + cdn\sigma_{\alpha\beta}^2,$$
$$E(MS_{ABC}) = \sigma_e^2 + n\sigma_{\alpha\beta\gamma\delta}^2 + dn\sigma_{\alpha\beta\gamma}^2,$$
$$E(MS_{ABCD}) = \sigma_e^2 + n\sigma_{\alpha\beta\gamma\delta}^2,$$

and

$$E(MS_E) = \sigma_e^2.$$

Expected mean squares under Model III depend on the particular combination of fixed and random factors. For example, for an experiment with A and C fixed, and B and D random, $(\alpha\beta\gamma\delta)_{ijk\ell}$'s are assumed to be distributed with mean

---

[3] One can use the rules formulated by Schultz (1955) and others to reproduce the results on expected mean squares rather quickly. See Appendix U for a discussion of rules for finding expected mean squares.

zero and variance $\sigma^2_{\alpha\beta\gamma\delta}$, subject to the restrictions that $\sum_{i=1}^{a}(\alpha\beta\gamma\delta)_{ijk\ell}=0=\sum_{k=1}^{c}(\alpha\beta\gamma\delta)_{ijk\ell}$. The assumptions imply that

$$E\left[\sum_{i=1}^{a}\sum_{k=1}^{c}(\alpha\beta\gamma\delta)^2_{ijk\ell}\right]=(a-1)(c-1)\sigma^2_{\alpha\beta\gamma\delta},$$

$$E\left[(\alpha\beta\gamma\delta)^2_{ijk\ell}\right]=\frac{(a-1)(c-1)}{ac}\sigma^2_{\alpha\beta\gamma\delta},$$

$E[(\alpha\beta\gamma\delta)_{ijk\ell}(\alpha\beta\gamma\delta)_{i'j'k'\ell'}]=0,\quad$ for $j=j', \ell\neq\ell'$ or both $j\neq j'$ and $\ell\neq\ell'$,

$$E[(\alpha\beta\gamma\delta)_{ijk\ell}(\alpha\beta\gamma\delta)_{i'jk\ell}]=-\left[\frac{c-1}{ac}\right]\sigma^2_{\alpha\beta\gamma\delta},\quad i\neq i',$$

$$E[(\alpha\beta\gamma\delta)_{ijk\ell}(\alpha\beta\gamma\delta)_{ijk'\ell}]=-\left[\frac{c-1}{ac}\right]\sigma^2_{\alpha\beta\gamma\delta},\quad k\neq k',$$

and

$$E[(\alpha\beta\gamma\delta)_{ijk\ell}(\alpha\beta\gamma\delta)_{i'jk'\ell}]=\frac{\sigma^2_{\alpha\beta\gamma\delta}}{ac},\quad i\neq i' \text{ and } k\neq k'.$$

Now, the results on expected mean squares follow readily. Finally, for a given model — fixed, random, or mixed — the point and interval estimates and tests of hypotheses corresponding to parameters of interest can be developed analogous to results for the three-way classification.

Remark: For a balanced crossed classification model involving only random effects, there are some simple rules to calculate the coefficients of the variance components in the expected mean square. The rules are stated as follows for a model containing four factors.

(a) All expected mean squares contain σ^2_e with coefficient 1.
(b) The coefficient of a variance component is zero or $abcdn$ divided by the product of the levels of the factors contained in the variance component. For example, the coefficient of $\sigma^2_{\alpha\beta\gamma\delta}$ is equal to $abcdn/abcd = n$.
(c) The coefficient of the variance component in the expected mean square of a main factor or interaction between factors is zero if the product of the levels of the factors contained in the variance component cannot be divided by the level of the factor or the product of the levels of the factors. For example, the coefficient of $\sigma^2_{\beta\gamma}$ in $E(MS_A)$ is zero since bc cannot be divided by a. Similarly the coefficient of $\sigma^2_{\beta\gamma\delta}$ in $E(MS_{AB})$ is zero since bcd cannot be divided by ab.
(d) A quick check on the correctness of the coefficients of variance components can be made by noting that for a given variance component, the weighted sum of

coefficients corresponding to all the mean squares, including that of the mean, is $abcdn$, where the weights are taken as the degrees of freedom.

5.12 HIGHER-ORDER CROSSED CLASSIFICATIONS

The reader should now be able to see how the four-way crossed classification analysis can be further generalized to five- and higher-order classifications. The formal symmetry in the sums of squares and degrees of freedom for the balanced case makes direct generalizations to the higher-order crossed classification models quite straightforward. For example, a full p-way crossed classification involving p crossed factors contains $2^p + 1$ terms in the model equation: a general mean, one main effect for each of the p-factors, $\binom{p}{2}$ two-factor interactions, $\binom{p}{3}$ three-factor interactions, and so on; and the total number of main effects and interactions to be tested is equal to $2^p - 1$. Computational formulae given in the preceding section can be readily extended if more than four factors are studied simultaneously. However, when the number of factors is large, the algebra becomes extremely tedious and the amount of computational work increases rapidly. Most of the algebra can be simplified by the use of "operators" as discussed by Bankier (1960a,b). The details on mechanization of the computational procedure on a digital computer can be found in the papers of Hartley (1956), Hemmerle (1964), and Bock (1963), and in the books by Peng (1967, pp. 47–50), Cooley and Lohnes (1962), and Dixon (1992). Hartley (1962) has suggested a simple and ingenious device of using a factorial analysis of variance without replication (with as many factors as necessary) to analyze many other designs on a digital computer, where data from any design are presented and analyzed as though they were a factorial experiment.

There are several procedures for deriving expected mean squares in an analysis of variance involving higher-order crossed classification models. They are, however, more readily written down following an easy set of rules. The interested reader is referred to papers by Schultz (1955), Cornfield and Tukey (1956), Millman and Glass (1967), Henderson (1959, 1969), Lorenzen (1977), and Blackwell et al. (1991), including books by Bennett and Franklin (1954), Scheffé (1959), and Lorenzen and Anderson (1993) for detailed discussions of these rules. A brief description of these rules is given in Appendix U. Finally, it should be stressed that in a higher-order crossed classification involving many factors, the complexity of the experiment as well as the analysis of data increases as the number of factors becomes large. In addition to providing a large number of experimental units, there are many interaction terms that must be evaluated and interpreted. Moreover, the tasks of evaluating the expected mean squares and performing the tests of significance for each source of variation also become increasingly complex. One common source of difficulty encountered in analyzing higher-order random and mixed factorials is that there is often no appropriate error term against which to test a given mean square. Frequently, the appropriate tests are carried out using an approximate procedure due to Satterthwaite (1946). It should, however, be mentioned that although the number of interaction terms in a higher-order classification increases rather rapidly, in many cases these interactions are so remote and difficult to interpret that they are frequently ignored and their sums of squares and degrees of freedom pooled with the residual.

We outline below an analysis of variance for the r-way classification involving factors, $A_1, A_2, \ldots, A_r$. Let a_i be the number of levels associated with the factor A_i ($i = 1, 2, \ldots, r$), and suppose there are n observations to be taken at every combination of the levels of $A_1, A_2, \ldots, A_r$. The model for a r-way classification can be written as

$$
\begin{aligned}
y_{i_1 i_2 \ldots i_r s} = \mu &+ (\alpha_1)_{i_1} + \cdots + (\alpha_r)_{i_r} + (\alpha_1 \alpha_2)_{i_1 i_2} + \cdots \\
&+ (\alpha_{r-1} \alpha_r)_{i_{r-1} i_r} + (\alpha_1 \alpha_2 \alpha_3)_{i_1 i_2 i_3} + \cdots \\
&+ (\alpha_{r-2} \alpha_{r-1} \alpha_r)_{i_{r-2} i_{r-1} i_r} + \cdots \\
&+ (\alpha_1 \alpha_2 \ldots \alpha_r)_{i_1 i_2 \ldots i_r} + e_{i_1 i_2 \ldots i_r s}
\end{aligned}
\quad
\begin{cases}
i_1 = 1, 2, \ldots, a_1 \\
i_2 = 1, 2, \ldots, a_2 \\
\vdots \\
i_r = 1, 2, \ldots, a_r \\
s = 1, 2, \ldots, n,
\end{cases}
$$
(5.12.1)

where $y_{i_1 i_2 \ldots i_r s}$ is the s-th observation corresponding to the i_1-th level of A_1, i_2-th level of $A_2, \ldots$, and i_r-th level of A_r; $-\infty < \mu < \infty$ is a constant; $(\alpha_j)_{i_j}$ is the effect of the i_j-th level of A_j ($j = 1, 2, \ldots, r$); $(\alpha_j \alpha_k)_{i_j i_k}$ is the effect of the interaction between the i_j-th level of A_j and the i_k-th level of A_k ($j < k = 1, 2, \ldots, r$); $(\alpha_j \alpha_k \alpha_\ell)_{i_j i_k i_\ell}$ is the interaction between the i_j-th level A_j, the i_k-th level of A_k, and the i_ℓ-th the level of A_ℓ ($j < k < \ell = 1, 2, \ldots, r$); $\ldots$; $(\alpha_1 \alpha_2 \ldots \alpha_r)_{i_1 i_2 \ldots i_r}$ is the interaction between the i_1-th level of A_1, the i_2-th level of $A_2, \ldots$, and the i_r-th level of A_r; and finally $e_{i_1 i_2 \ldots i_r s}$ is the customary error term.

The usual identity $y_{i_1 i_2 \ldots i_r s} - \bar{y}_{\ldots} =$ etc., contains the following groups of terms on its right-hand side:

- Estimates of r main effects; e.g., $\bar{y}_{i_j \ldots} - \bar{y}_{\ldots}$ which gives an estimate of $(\alpha_j)_{i_j}$ ($j = 1, 2, \ldots, r$).
- Estimates of $\binom{r}{2}$ two-way interactions; e.g., $\bar{y}_{i_j i_k \ldots} - \bar{y}_{i_j \ldots} - \bar{y}_{\cdot i_k \ldots} + \bar{y}_{\ldots}$, which gives an estimate of $(\alpha_j \alpha_k)_{i_j i_k}$ ($j < k = 1, 2, \ldots, r$).

$$
\begin{array}{cccccc}
\cdot & \cdot & \cdot & \cdot & \cdot & \cdot \\
\cdot & \cdot & \cdot & \cdot & \cdot & \cdot \\
\cdot & \cdot & \cdot & \cdot & \cdot & \cdot \\
\end{array}
$$

- Estimate of the single r-way interaction of the form $\bar{y}_{i_1 i_2 \ldots i_r \cdot} - [\bar{y}_{\ldots} + r$ main effects $+ \binom{r}{2}$ two-way interactions $+ \cdots + \binom{r}{r-1}(r-1)$-way interactions].
- The deviations of the individual observations from the cell means; e.g., $y_{i_1 i_2 \ldots i_r s} - \bar{y}_{i_1 i_2 \ldots i_r \cdot}$.

Three-Way and Higher-Order Crossed Classifications

The partition of the total sum of squares is effected by squaring and summing over all indices of the identity $y_{i_1 i_2 \ldots i_r s} - \bar{y}_{\ldots\ldots} = $ etc. The typical sums of squares can be expressed as follows:

$$SS_{A_1} = a_2 a_3 \ldots a_r n \sum_{i_1=1}^{a_1} (\bar{y}_{i_1 \ldots\ldots} - \bar{y}_{\ldots\ldots})^2, \text{ etc.,}$$

$$SS_{A_1 A_2} = a_3 a_4 \ldots a_r n \sum_{i_1=1}^{a_1} \sum_{i_2=1}^{a_2} (\bar{y}_{i_1 i_2 \ldots\ldots} - \bar{y}_{i_1 \ldots\ldots} - \bar{y}_{. i_2 \ldots\ldots} + \bar{y}_{\ldots\ldots})^2, \text{ etc.,}$$

. . . .
. . . .
. . . .

$$SS_{A_1 A_2 \ldots A_r} = n \sum_{i_1=1}^{a_1} \sum_{i_2=1}^{a_2} \ldots \sum_{i_r=1}^{a_r} [(\bar{y}_{i_1 i_2 \ldots i_r .} - (\bar{y}_{i_1 i_2 \ldots i_{r-1} ..} + \cdots) + \cdots)$$
$$+ (-1)^{r-2} (\bar{y}_{i_1 i_2 \ldots\ldots} + \cdots) + (-1)^{r-1} (\bar{y}_{i_1 \ldots\ldots} + \cdots)$$
$$+ (-1)^r \bar{y}_{\ldots\ldots})]^2,$$

and

$$SS_E = \sum_{i_1=1}^{a_1} \sum_{i_2=1}^{a_2} \ldots \sum_{i_r=1}^{a_r} \sum_{s=1}^{n} (y_{i_1 i_2 \ldots i_r s} - \bar{y}_{i_1 i_2 \ldots i_r .})^2.$$

The degrees of freedom for the above sums of square are $a_1 - 1$, $(a_1 - 1)(a_2 - 1), \ldots, (a_1 - 1)(a_2 - 1) \ldots (a_r - 1)$ and $a_1 a_2 \ldots a_r (n - 1)$, respectively.

Under Model I, $(\alpha_j)_{i_j} (j = 1, 2, \ldots, r)$, $(\alpha_j \alpha_k)_{i_j i_k}$ $(j < k = 1, 2, \ldots, r)$, $(\alpha_j \alpha_k \alpha_\ell)_{i_j i_k i_\ell}$ $(j < k < \ell = 1, 2, \ldots, r), \ldots$, and $(\alpha_1 \alpha_2 \ldots \alpha_r)_{i_1 i_2 \ldots i_r}$ are assumed to be constants subject to the following restrictions:

$$\sum_{i_j=1}^{a_j} (\alpha_j)_{i_j} = 0, \quad j = 1, 2, \ldots, r;$$

$$\sum_{i_j=1}^{a_j} (\alpha_j \alpha_k)_{i_j i_k} = \sum_{i_k=1}^{a_k} (\alpha_j \alpha_k)_{i_j i_k} = 0, \quad j < k = 1, 2, \ldots r;$$

$$\sum_{i_j=1}^{a_j} (\alpha_j \alpha_k \alpha_\ell)_{i_j i_k i_\ell} = \sum_{i_k=1}^{a_k} (\alpha_j \alpha_k \alpha_\ell)_{i_j i_k i_\ell}$$
$$= \sum_{i_\ell=1}^{a_\ell} (\alpha_j \alpha_k \alpha_\ell)_{i_j i_k i_\ell} = 0, \quad j < k < \ell = 1, 2, \ldots, r;$$

. . . . . .
. . . . . .

and

$$\sum_{i_1=1}^{a_1}(\alpha_1\alpha_2\ldots\alpha_r)_{i_1i_2\ldots i_r} = \sum_{i_2=1}^{a_2}(\alpha_1\alpha_2\ldots\alpha_r)_{i_1i_2\ldots i_r} = \cdots$$
$$= \sum_{i_r=1}^{a_r}(\alpha_1\alpha_2\ldots\alpha_r)_{i_1i_2\ldots i_r} = 0.$$

Furthermore, $e_{i_1i_2\ldots i_r s}$'s are uncorrelated and randomly distributed with zero means and variance σ_e^2. The expected mean squares are obtained as follows:

$$E(MS_{A_1}) = \sigma_e^2 + \frac{a_2 a_3 \ldots a_r n}{a_1 - 1}\sum_{i_1=1}^{a_1}(\alpha_1)_{i_1}^2, \text{ etc.,}$$

$$E(MS_{A_1 A_2}) = \sigma_e^2 + \frac{a_3 a_4 \ldots a_r n}{(a_1-1)(a_2-1)}\sum_{i_1=1}^{a_1}\sum_{i_2=1}^{a_2}(\alpha_1\alpha_2)_{i_1 i_2}^2, \text{ etc.,}$$

$$\cdot \quad \cdot \quad \cdot \quad \cdot \quad \cdot$$
$$\cdot \quad \cdot \quad \cdot \quad \cdot \quad \cdot$$
$$\cdot \quad \cdot \quad \cdot \quad \cdot \quad \cdot$$

$$E(MS_{A_1 A_2 \ldots A_r}) = \sigma_e^2 + \frac{n}{(a_1-1)\ldots(a_r-1)}$$
$$\times \sum_{i_1=1}^{a_1}\sum_{i_2=1}^{a_2}\cdots\sum_{i_r=1}^{a_r}(\alpha_1\alpha_2\ldots\alpha_r)_{i_1i_2\ldots i_r}^2,$$

and

$$E(MS_E) = \sigma_e^2.$$

Assuming normality for the error terms, all tests of hypotheses are carried out by an appropriate F statistic obtained as the ratio of the mean square of the effect being tested to the error mean square.

Under Model II, $(\alpha_j)_{i_j}$'s, $(\alpha_j\alpha_k)_{i_j i_k}$'s, $(\alpha_j\alpha_k\alpha_\ell)_{i_j i_k i_\ell}$'s, ..., $(\alpha_1\alpha_2\ldots\alpha_r)_{i_1 i_2\ldots i_r}$'s, and $e_{i_1 i_2\ldots i_r s}$'s are assumed to be mutually and completely uncorrelated random variables with zero means and variances $\sigma_{\alpha_j}^2, \sigma_{\alpha_j\alpha_k}^2, \sigma_{\alpha_j\alpha_k\alpha_\ell}^2, \ldots, \sigma_{\alpha_1\alpha_2\ldots\alpha_r}^2$, and σ_e^2 respectively. From (5.12.1), the variance of any observation is

$$\text{Var}(y_{i_1 i_2\ldots i_r s}) = \sigma_{\alpha_1}^2 + \cdots + \sigma_{\alpha_r}^2 + \sigma_{\alpha_1\alpha_2}^2 + \cdots + \sigma_{\alpha_{r-1}\alpha_r}^2 + \cdots + \sigma_{\alpha_1\alpha_2\ldots\alpha_r}^2 + \sigma_e^2,$$

and, thus, $\sigma_{\alpha_1}^2, \ldots, \sigma_{\alpha_r}^2; \sigma_{\alpha_1\alpha_2}^2, \ldots, \sigma_{\alpha_{r-1}\alpha_r}^2; \ldots; \sigma_{\alpha_1\alpha_2\ldots\alpha_r}^2;$ and σ_e^2, are the variance components of model (5.12.1). The expected value of the mean square corresponding to any source; for example, the interaction $A_{j_1} \times A_{j_2} \times \cdots \times A_{j_m}$ is obtained as follows:

$$\sigma_e^2 + n \sum q_{k_1 k_2 \ldots k_p} \sigma_{\alpha_{k_1} \alpha_{k_2} \ldots \alpha_{k_p}}^2,$$

where the summation is carried over all the variance components except σ_e^2. The coefficients $q_{k_1 k_2 \ldots k_p}$ of the variance components are given by

$$q_{k_1 k_2 \ldots k_p} = \begin{cases} \dfrac{a_1 a_2 \ldots a_r}{a_{k_1} a_{k_2} \ldots a_{k_p}}, & \text{if } (j_1, j_2, \ldots, j_m) \text{ is a subset of } (k_1, k_2, \ldots, k_p) \\ 0, & \text{otherwise.} \end{cases}$$

Remark: In many factorial experiments involving a r-way classification where the levels of the factor correspond to a fixed measure quantity, such as levels of temperature, quantity of fertilizer, etc., the investigator is often interested in studying the nature of the response surface. For instance, she may want to determine the value at which the response surface is maximum or minimum. For discussions of the response surface methodology, the reader is referred to Myers (1976), Box and Draper (1987), Myers and Montgomery (1995), and Khuri and Cornell (1996).

5.13 UNEQUAL SAMPLE SIZES IN THREE- AND HIGHER-ORDER CLASSIFICATIONS

When the sample sizes in three- or higher-order crossed classifications are not all equal, the procedures described in Section 4.10 can be used with the customary modifications. The formulae for the two-way model need simply be extended for experiments involving three and more factors. However, the computation of the analysis of variance in the general case of disproportionate frequencies tends to be extremely involved. Various aspects of the analysis of nonorthogonal three- and higher-order classifications have been considered by a number of authors. For further discussions and details the interested reader is referred to Kendall et al. (1983, Sections 35.43 and 35.44) and references cited therein.

In the following, we outline an analysis of variance for the unbalanced three-way crossed classification and indicate its extension to higher-order classifications. The model for the three-way crossed classification remains the same as in (5.1.1), except that the sample size corresponding to the (i, j, k)-th cell will now be denoted by n_{ijk}. We consider the analysis when the unequal sample sizes follow a proportional pattern; that is,

$$n_{ijk} = \frac{n_{i..} n_{.j.} n_{..k}}{N}.$$

For this case, the analysis of variance discussed earlier in this chapter can be employed with suitable modifications. For example, the definitional and computational formulae for the sums of squares are:

$$SS_T = \sum_{i=1}^{a} \sum_{j=1}^{b} \sum_{k=1}^{c} \sum_{\ell=1}^{n_{ijk}} (y_{ijk\ell} - \bar{y}_{....})^2 = \sum_{i=1}^{a} \sum_{j=1}^{b} \sum_{k=1}^{c} \sum_{\ell=1}^{n_{ijk}} y_{ijk\ell}^2 - \frac{y_{....}^2}{N}$$

$$SS_A = \sum_{i=1}^{a} n_{i..}(\bar{y}_{i...} - \bar{y}_{....})^2 = \sum_{i=1}^{a} \frac{y_{i...}^2}{n_{i..}} - \frac{y_{....}^2}{N},$$

$$SS_B = \sum_{j=1}^{b} n_{.j.}(\bar{y}_{.j..} - \bar{y}_{....})^2 = \sum_{j=1}^{b} \frac{y_{.j..}^2}{n_{.j.}} - \frac{y_{....}^2}{N},$$

$$SS_C = \sum_{k=1}^{c} n_{..k}(\bar{y}_{..k.} - \bar{y}_{....})^2 = \sum_{k=1}^{c} \frac{y_{..k.}^2}{n_{..k}} - \frac{y_{....}^2}{N},$$

$$SS_{AB} = \sum_{i=1}^{a} \sum_{j=1}^{b} n_{ij.}(\bar{y}_{ij..} - \bar{y}_{i...} - \bar{y}_{.j..} + \bar{y}_{....})^2$$

$$= \sum_{i=1}^{a} \sum_{j=1}^{b} \frac{y_{ij..}^2}{n_{ij.}} - \sum_{i=1}^{a} \frac{y_{i...}^2}{n_{i..}} - \sum_{j=1}^{b} \frac{y_{.j..}^2}{n_{.j.}} + \frac{y_{....}^2}{N},$$

$$SS_{AC} = \sum_{i=1}^{a} \sum_{k=1}^{c} n_{i.k}(\bar{y}_{i.k.} - \bar{y}_{i...} - \bar{y}_{..k.} + \bar{y}_{....})^2$$

$$= \sum_{i=1}^{a} \sum_{k=1}^{c} \frac{y_{i.k.}^2}{n_{i.k}} - \sum_{i=1}^{a} \frac{y_{i...}^2}{n_{i..}} - \sum_{k=1}^{c} \frac{y_{..k.}^2}{n_{..k}} + \frac{y_{....}^2}{N},$$

$$SS_{BC} = \sum_{j=1}^{b} \sum_{k=1}^{c} n_{.jk}(\bar{y}_{.jk.} - \bar{y}_{.j..} - \bar{y}_{..k.} + \bar{y}_{....})^2$$

$$= \sum_{j=1}^{b} \sum_{k=1}^{c} \frac{y_{.jk.}^2}{n_{.jk}} - \sum_{j=1}^{b} \frac{y_{.j..}^2}{n_{.j.}} - \sum_{k=1}^{c} \frac{y_{..k.}^2}{n_{..k}} + \frac{y_{....}^2}{N},$$

$$SS_{ABC} = \sum_{i=1}^{a} \sum_{j=1}^{b} \sum_{k=1}^{c} n_{ijk}(\bar{y}_{ijk.} - \bar{y}_{ij..} - \bar{y}_{i.k.} - \bar{y}_{.jk.} + \bar{y}_{i...} + \bar{y}_{.j..} + \bar{y}_{..k.} - \bar{y}_{....})^2$$

$$= \sum_{i=1}^{a} \sum_{j=1}^{b} \sum_{k=1}^{c} \frac{y_{ijk.}^2}{n_{ijk}} - \sum_{i=1}^{a} \sum_{j=1}^{b} \frac{y_{ij..}^2}{n_{ij.}} - \sum_{i=1}^{a} \sum_{k=1}^{c} \frac{y_{i.k.}^2}{n_{i.k}} - \sum_{j=1}^{b} \sum_{k=1}^{c} \frac{y_{.jk.}^2}{n_{.jk}}$$

$$+ \sum_{i=1}^{a} \frac{y_{i...}^2}{n_{i..}} + \sum_{j=1}^{b} \frac{y_{.j..}^2}{n_{.j.}} + \sum_{k=1}^{c} \frac{y_{..k.}^2}{n_{..k}} - \frac{y_{....}^2}{N},$$

and

$$SS_E = \sum_{i=1}^{a}\sum_{j=1}^{b}\sum_{k=1}^{c}\sum_{\ell=1}^{n_{ijk}}(y_{ijk\ell} - \bar{y}_{ijk.})^2$$

$$= \sum_{i=1}^{a}\sum_{j=1}^{b}\sum_{k=1}^{c}\sum_{\ell=1}^{n_{ijk}}y_{ijk\ell}^2 - \sum_{i=1}^{a}\sum_{j=1}^{b}\sum_{k=1}^{c}\frac{y_{ijk.}^2}{n_{ijk}},$$

where

$$y_{ijk.} = \sum_{\ell=1}^{n_{ijk}}y_{ijk\ell}, \qquad \bar{y}_{ijk.} = y_{ijk.}/n_{ijk},$$

$$y_{ij..} = \sum_{k=1}^{c}y_{ijk.}, \qquad \bar{y}_{ij..} = y_{ij..}/n_{ij.},$$

$$y_{i.k.} = \sum_{j=1}^{b}y_{ijk.}, \qquad \bar{y}_{i.k.} = y_{i.k.}/n_{i.k},$$

$$y_{.jk.} = \sum_{i=1}^{a}y_{ijk.}, \qquad \bar{y}_{.jk.} = y_{.jk.}/n_{.jk},$$

$$y_{i...} = \sum_{j=1}^{b}y_{ij..}, \qquad \bar{y}_{i...} = y_{i...}/n_{i..},$$

$$y_{.j..} = \sum_{i=1}^{a}y_{ij..}, \qquad \bar{y}_{.j..} = y_{.j..}/n_{.j.},$$

$$y_{..k.} = \sum_{i=1}^{a}y_{i.k.}, \qquad \bar{y}_{..k.} = y_{..k.}/n_{..k},$$

$$y_{....} = \sum_{i=1}^{a}y_{i...} = \sum_{j=1}^{b}y_{.j..} = \sum_{k=1}^{c}y_{..k.}, \quad \bar{y}_{....} = y_{....}/N,$$

$$n_{ij.} = \sum_{k=1}^{c}n_{ijk}, \quad n_{i.k} = \sum_{j=1}^{b}n_{ijk}, \quad n_{.jk} = \sum_{i=1}^{a}n_{ijk},$$

$$n_{i..} = \sum_{j=1}^{b}n_{ij.}, \quad n_{.j.} = \sum_{i=1}^{a}n_{ij.}, \quad n_{..k} = \sum_{i=1}^{a}n_{i.k},$$

and

$$N = \sum_{i=1}^{a}n_{i..} = \sum_{j=1}^{b}n_{.j.} = \sum_{k=1}^{c}n_{..k}.$$

The analysis of variance with expected mean squares for the fixed effects model

TABLE 5.9
Analysis of Variance for the Unbalanced Fixed Effects Model in (5.1.1) with Proportional Frequencies

Source of Variation	Degrees of Freedom	Sum of Squares	Mean Square	Expected Mean Square
Due to A	$a-1$	SS_A	MS_A	$\sigma_e^2 + \dfrac{1}{a-1} \sum_{i=1}^{a} n_{i..} \alpha_i^2$
Due to B	$b-1$	SS_B	MS_B	$\sigma_e^2 + \dfrac{1}{b-1} \sum_{j=1}^{b} n_{.j.} \beta_j^2$
Due to C	$c-1$	SS_C	MS_C	$\sigma_e^2 + \dfrac{1}{c-1} \sum_{k=1}^{c} n_{..k} \gamma_k^2$
Interaction $A \times B$	$(a-1)(b-1)$	SS_{AB}	MS_{AB}	$\sigma_e^2 + \dfrac{1}{(a-1)(b-1)} \times \sum_{i=1}^{a} \sum_{j=1}^{b} n_{ij.} (\alpha\beta)_{ij}^2$
Interaction $A \times C$	$(a-1)(c-1)$	SS_{AC}	MS_{AC}	$\sigma_e^2 + \dfrac{1}{(a-1)(c-1)} \times \sum_{i=1}^{a} \sum_{k=1}^{c} n_{i.k} (\alpha\gamma)_{ik}^2$
Interaction $B \times C$	$(b-1)(c-1)$	SS_{BC}	MS_{BC}	$\sigma_e^2 + \dfrac{1}{(b-1)(c-1)} \times \sum_{j=1}^{b} \sum_{k=1}^{c} n_{.jk} (\beta\gamma)_{jk}^2$
Interaction $A \times B \times C$	$(a-1)(b-1)(c-1)$	SS_{ABC}	MS_{ABC}	$\sigma_e^2 + \dfrac{1}{(a-1)(b-1)(c-1)} \times \sum_{i=1}^{a} \sum_{j=1}^{b} \sum_{k=1}^{c} n_{ijk} (\alpha\beta\gamma)_{ijk}^2$
Error	$N - abc$	SS_E	MS_E	σ_e^2

is given in Table 5.9. Under the assumption of normality, the test procedures are performed as in the case of corresponding balanced analysis. The expected mean squares for the random and mixed effects in the general case of disproportional frequencies are extremely involved and the interested reader is referred to Blischke (1966) for further information and details.

The model for the r-way crossed classification remains the same as in (5.12.1), except that the sample size corresponding to $(i_1, i_2, \ldots, i_r)$-th cell will now be denoted by $n_{i_1 i_2 \ldots i_r}$. The analysis when the unequal sample sizes follow a proportional pattern follows readily on the lines of three-way crossed classification outlined above. For further information and details, the reader is referred to Blischke (1968).

5.14 WORKED EXAMPLE FOR MODEL I

Anderson and Bancroft (1952, p. 291) reported data from an experiment designed to study the effect of electrolytic chromium plate as a source for the chromium impregnation of low-carbon steel wire. The experiment involved 18 treatments obtained as a combination of three diffusion temperatures (2200°F, 2350°F, 2500°F), three diffusion times (4, 8, and 12 hours), and two degrasing treatments (yes and no). Each treatment was applied on four wires giving a total of 72 determinations on average resistivities (in m-ohms/cm^3) which was the variable being studied. The data are given in Table 5.10.

The data in Table 5.10 can be regarded as a three-way classification with four observations per cell. Note that all three factors, temperature, time, and degrasing, should be regarded as fixed effects since the interest is directed only to the levels of the factors included in the experiment. The mathematical model for the experiment would be

$$y_{ijk\ell} = \mu + \alpha_i + \beta_j + \gamma_k + (\alpha\beta)_{ij} + (\alpha\gamma)_{ik} + (\beta\gamma)_{jk} + (\alpha\beta\gamma)_{ijk} + e_{ijk\ell} \quad \begin{cases} i = 1,2,3 \\ j = 1,2,3 \\ k = 1,2 \\ \ell = 1,2,3,4, \end{cases}$$

where μ is the general mean, α_i is the effect of the i-th level of temperature, β_j is the effect of the j-th level of time, γ_k is the effect of the k-th level of degrasing, $(\alpha\beta)_{ij}$ is the interaction between the i-th temperature and the j-th time, $(\alpha\gamma)_{ik}$ is the interaction between the i-th temperature and k-th degrasing, $(\beta\gamma)_{jk}$ is the interaction between the i-th time and the k-th degrasing, $(\alpha\beta\gamma)_{ijk}$ is the interaction between the i-th temperature, the j-th time and the k-th degrasing, and $e_{ijk\ell}$ is the customary error term. Furthermore, it is assumed that the α_i's, β_j's, γ_k's, $(\alpha\beta)_{ij}$'s, $(\alpha\gamma)_{ik}$'s, $(\beta\gamma)_{jk}$'s, and $(\alpha\beta\gamma)_{ijk}$'s are constants subject to the restrictions:

$$\sum_{i=1}^{3} \alpha_i = \sum_{j=1}^{3} \beta_j = \sum_{k=1}^{2} \gamma_k = 0,$$

$$\sum_{i=1}^{3} (\alpha\beta)_{ij} = \sum_{j=1}^{3} (\alpha\beta)_{ij} = \sum_{i=1}^{3} (\alpha\gamma)_{ik} = \sum_{k=1}^{2} (\alpha\gamma)_{ik}$$

$$= \sum_{j=1}^{3} (\beta\gamma)_{jk} = \sum_{k=1}^{2} (\beta\gamma)_{jk} = 0,$$

$$\sum_{i=1}^{3} (\alpha\beta\gamma)_{ijk} = \sum_{j=1}^{3} (\alpha\beta\gamma)_{ijk} = \sum_{k=1}^{2} (\alpha\beta\gamma)_{ijk} = 0,$$

and the $e_{ijk\ell}$'s are independently and normally distributed with mean zero and variance σ_e^2.

The first step in the analysis of variance computations is to form a three-way table of cell totals containing $y_{ijk.} = \sum_{\ell=1}^{4} y_{ijk\ell}$ (see Table 5.11). The next step

TABLE 5.10
Data on Average Resistivities (in m-ohms/cm³) for Electrolytic Chromium Plate Example

Time	2200°F						2350°F						2500°F					
	4 hrs		8 hrs		12 hrs		4 hrs		8 hrs		12 hrs		4 hrs		8 hrs		12 hrs	
Degrasing	Yes	No	Yes	No	Yes	No	Yes	No	Yes	No	Yes	No	Yes	No	Yes	No	Yes	No
	17.9	18.1	19.2	19.2	19.9	20.0	21.2	22.1	22.7	23.2	23.3	23.9	22.8	22.9	26.9	25.5	26.5	27.0
	18.0	18.9	19.0	19.3	20.1	20.2	20.4	20.2	22.7	21.8	23.5	23.6	22.3	24.0	26.9	26.6	26.8	26.2
	18.7	18.6	20.4	20.7	20.0	20.1	21.2	21.3	22.5	22.9	23.5	23.2	22.7	23.0	26.3	25.9	25.4	25.9
	19.0	19.1	19.2	20.4	20.8	20.5	21.2	22.6	22.5	22.3	22.9	23.7	23.3	23.0	26.9	26.8	27.2	26.9

Source: Anderson and Bancroft (1952, p. 291). Used with permission.

TABLE 5.11
Cell Totals y_{ijk}.

		Temperature								
		2200°F			2350°F			2500°F		
Time		4 hrs	8 hrs	12 hrs	4 hrs	8 hrs	12 hrs	4 hrs	8 hrs	12 hrs
Degrasing	Yes	73.6	77.8	80.8	84.0	90.4	93.2	91.1	107.0	105.9
	No	74.7	79.6	80.8	86.2	90.2	94.4	92.9	104.8	106.0

TABLE 5.12
Sums over Levels of Degrasing $y_{ij\cdot}$.

	Time			
Temperature (°F)	4 hrs	8 hrs	12 hrs	$y_{i\cdot\cdot}$
2200	148.3	157.4	161.6	467.3
2350	170.2	180.6	187.6	538.4
2500	184.0	211.8	211.9	607.7
$y_{\cdot j\cdot}$	502.5	549.8	561.1	$y_{\cdot\cdot\cdot\cdot}$ 1,613.4

TABLE 5.13
Sums over Levels of Time $y_{i\cdot k}$.

	Degrasing		
Temperature (°F)	Yes	No	$y_{i\cdot\cdot}$
2200	232.2	235.1	467.3
2350	267.6	270.8	538.4
2500	304.0	303.7	607.7
$y_{\cdot\cdot k}$	803.8	809.6	$y_{\cdot\cdot\cdot\cdot}$ 1,613.4

consists of forming sums over every index and every combination of indices. Thus, we sum over the levels of degrasing to get a temperature $(i) \times$ time (j) table containing $y_{ij\cdot} = \sum_{k=1}^{2} y_{ijk}$. (see Table 5.12). This table is then summed over the levels of time to obtain $y_{i\cdot\cdot}$ and over the levels of temperature to give $y_{\cdot j\cdot}$. The sum of $y_{i\cdot\cdot}$ is equal to the sum of $y_{\cdot j\cdot}$ which is the grand total $y_{\cdot\cdot\cdot}$. Tables 5.13 and 5.14 are obtained similarly.

TABLE 5.14
Sums over Levels of Temperature $y_{.jk.}$

	Degrasing		
Time (hrs)	Yes	No	$y_{.j..}$
4	248.7	253.8	502.5
8	275.2	274.6	549.8
12	279.9	281.2	561.1
$y_{..k.}$	803.8	809.6	$y_{....}$
			1,613.4

With these preliminary results, the subsequent analysis of variance calculations are fairly straightforward. Thus, from the computational formulae of Section 5.7, we have

$$SS_T = (17.9)^2 + (18.0)^2 + \cdots + (26.9)^2 - \frac{(1,613.4)^2}{3 \times 3 \times 2 \times 4}$$
$$= 36,677.86 - 36,153.6050$$
$$= 524.2550,$$

$$SS_A = \frac{1}{3 \times 2 \times 4}\{(467.3)^2 + (538.4)^2 + (607.7)^2\} - \frac{(1,613.4)^2}{3 \times 3 \times 2 \times 4}$$
$$= 36,564.2975 - 36,153.6050$$
$$= 410.6925,$$

$$SS_B = \frac{1}{3 \times 2 \times 4}\{(502.5)^2 + (549.8)^2 + (561.1)^2\} - \frac{(1,613.4)^2}{3 \times 3 \times 2 \times 4}$$
$$= 36,234.1458 - 36,153.6050$$
$$= 80.5408,$$

$$SS_C = \frac{1}{3 \times 3 \times 4}\{(803.8)^2 + (809.6)^2\} - \frac{(1,613.4)^2}{3 \times 3 \times 2 \times 4}$$
$$= 36,154.0722 - 36,153.6050$$
$$= 0.4672,$$

$$SS_{AB} = \frac{1}{2 \times 4}\{(148.3)^2 + (157.4)^2 + \cdots + (211.9)^2\}$$
$$- \frac{(1,613.4)^2}{3 \times 3 \times 2 \times 4} - SS_A - SS_B$$
$$= 36,659.6525 - 36,153.6050 - 410.6925 - 80.5408$$
$$= 14.8142,$$

Three-Way and Higher-Order Crossed Classifications 319

$$SS_{AC} = \frac{1}{3 \times 4}\{(232.2)^2 + (235.1)^2 + \cdots + (303.7)^2\}$$
$$- \frac{(1{,}613.4)^2}{3 \times 3 \times 2 \times 4} - SS_A - SS_C$$
$$= 36{,}565.0783 - 36{,}153.6050 - 410.6925 - 0.4672$$
$$= 0.3136,$$

$$SS_{BC} = \frac{1}{3 \times 4}\{(248.7)^2 + (253.8)^2 + \cdots + (281.2)^2\}$$
$$- \frac{(1{,}613.4)^2}{3 \times 3 \times 2 \times 4} - SS_B - SS_C$$
$$= 36{,}235.3150 - 36{,}153.6050 - 80.5408 - 0.4672$$
$$= 0.7020,$$

and

$$SS_{ABC} = \frac{1}{4}\{(73.6)^2 + (74.7)^2 + \cdots + (106.0)^2\} - \frac{(1{,}613.4)^2}{3 \times 3 \times 2 \times 4}$$
$$- SS_{AB} - SS_{AC} - SS_{BC} - SS_A - SS_B - SS_C$$
$$= 36{,}662.01 - 36{,}153.6050 - 14.8142 - 0.3136 - 0.7020$$
$$- 410.6925 - 80.5408 - 0.4672$$
$$= 0.8747.$$

Finally, by subtraction, the error sum of squares is given by

$$SS_E = 524.2550 - 410.6925 - 80.5408 - 0.4672 - 14.8142$$
$$- 0.3136 - 0.7020 - 0.8747$$
$$= 15.8500.$$

These results along with the remaining calculations are entered in Table 5.15.

From Table 5.15 it is evident that the variance ratio for the three-factor interaction (i.e., temperature × time × degrasing) is less than one indicating a nonsignificant effect ($p = 0.566$). The variance ratios for temperature × degrasing as well as time × degrasing interactions are also too low to achieve any significance. However, the variance ratio for temperature × time interactions is quite large and highly significant ($p < 0.001$). In terms of the main effects, the variance ratio for the degrasing effect is relatively small and nonsignificant ($p = 0.213$). However, the variance ratios for temperature and time main effects are extremely large and highly significant ($p < 0.001$). Hence, we may

TABLE 5.15
Analysis of Variance for the Resistivity Data of Table 5.10

Source of Variation	Degrees of Freedom	Sum of Squares	Mean Square	Expected Mean Square	F Value	p-Value
Temperature	2	410.6925	205.3463	$\sigma_e^2 + \dfrac{3 \times 2 \times 4}{3-1} \sum_{i=1}^{3} \alpha_i^2$	699.60	<0.001
Time	2	80.5408	40.2704	$\sigma_e^2 + \dfrac{3 \times 2 \times 4}{3-1} \sum_{j=1}^{3} \beta_j^2$	137.20	<0.001
Degrasing	1	0.4672	0.4672	$\sigma_e^2 + \dfrac{3 \times 3 \times 4}{2-1} \sum_{k=1}^{2} \gamma_k^2$	1.59	0.213
Temperature × Time	4	14.8142	3.7035	$\sigma_e^2 + \dfrac{2 \times 4}{(3-1)(3-1)} \sum_{i=1}^{3}\sum_{j=1}^{3} (\alpha\beta)_{ij}^2$	12.62	<0.001
Temperature × Degrasing	2	0.3136	0.1568	$\sigma_e^2 + \dfrac{3 \times 4}{(3-1)(2-1)} \sum_{i=1}^{3}\sum_{k=1}^{2} (\alpha\gamma)_{ik}^2$	0.53	0.589
Time × Degrasing	2	0.7020	0.3510	$\sigma_e^2 + \dfrac{3 \times 4}{(3-1)(2-1)} \sum_{j=1}^{3}\sum_{k=1}^{2} (\beta\gamma)_{jk}^2$	1.20	0.310
Temperature × Time × Degrasing	4	0.8747	0.2187	$\sigma_e^2 + \dfrac{4}{(3-1)(3-1)(2-1)} \sum_{i=1}^{3}\sum_{j=1}^{3}\sum_{k=1}^{2} (\alpha\beta\gamma)_{ijk}^2$	0.75	0.566
Error	54	15.8500	0.2935	σ^2		
Total	71	524.2550				

Three-Way and Higher-Order Crossed Classifications

reach the following conclusions:

(i) There are no three-factor (i.e., temperature × time × degrasing) interactions.
(ii) There are no two-factor interactions between degrasing and either of the other two factors — temperature and time. However, there are significant interactions between temperature and time.
(iii) There are two main effects, that is, due to temperature and time. There are no main effects for degrasing.

In view of the presence of significant temperature × time interactions, the tests for temperature and time main effects are not particularly meaningful. The researcher should first investigate the nature of temperature × time interactions before determining whether main effects are of any practical interest.

To study the nature of the temperature × time interactions, suppose the researcher wished to estimate separately, the differences in average resistivities for three diffusion temperatures for three diffusion times. The contrasts of interest are:

$$L_1 = \mu_{21.} - \mu_{11.}, \quad L_2 = \mu_{31.} - \mu_{21.}, \quad L_3 = \mu_{31.} - \mu_{11.},$$
$$L_4 = \mu_{22.} - \mu_{12.}, \quad L_5 = \mu_{32.} - \mu_{22.}, \quad L_6 = \mu_{32.} - \mu_{12.},$$
$$L_7 = \mu_{23.} - \mu_{13.}, \quad L_8 = \mu_{33.} - \mu_{23.}, \quad L_9 = \mu_{33.} - \mu_{13.}.$$

The preceding contrasts are estimated as

$$\hat{L}_1 = \frac{170.2}{8} - \frac{148.3}{8} = 2.74, \quad \hat{L}_2 = \frac{184.0}{8} - \frac{170.2}{8} = 1.73,$$
$$\hat{L}_3 = \frac{184.0}{8} - \frac{148.3}{8} = 4.46, \quad \hat{L}_4 = \frac{180.6}{8} - \frac{157.4}{8} = 2.90,$$
$$\hat{L}_5 = \frac{211.8}{8} - \frac{180.6}{8} = 3.90, \quad \hat{L}_6 = \frac{211.8}{8} - \frac{157.4}{8} = 6.80,$$
$$\hat{L}_7 = \frac{187.6}{8} - \frac{161.6}{8} = 3.25, \quad \hat{L}_8 = \frac{211.9}{8} - \frac{187.6}{8} = 3.04,$$
$$\hat{L}_9 = \frac{211.9}{8} - \frac{161.6}{8} = 6.29.$$

The estimated variances are obtained as

$$\widehat{\text{Var}}(\hat{L}_1) = \widehat{\text{Var}}(\hat{L}_2) = \cdots = \widehat{\text{Var}}(\hat{L}_9) = \frac{\text{MS}_E}{nc}[(1)^2 + (-1)^2]$$
$$= \frac{0.2935}{4 \times 2}(2) = 0.073.$$

The desired 95 percent Bonferroni intervals for the contrasts of interest are determined as

$$\hat{L}_i \pm t[54, 1 - 0.05/18]\sqrt{\widehat{\text{Var}}(\hat{L}_i)} = \hat{L}_i \pm 2.89 \times 0.270, \quad i = 1, \ldots, 9;$$

that is,

$1.96 = 2.74 - 2.89 \times 0.270 \leq \mu_{21.} - \mu_{11.} \leq 2.74 + 2.89 \times 0.270 = 3.52$
$0.95 = 1.73 - 2.89 \times 0.270 \leq \mu_{31.} - \mu_{21.} \leq 1.73 + 2.89 \times 0.270 = 2.51$
$3.68 = 4.46 - 2.89 \times 0.270 \leq \mu_{31.} - \mu_{11.} \leq 4.46 + 2.89 \times 0.270 = 5.24$
$2.12 = 2.90 - 2.89 \times 0.270 \leq \mu_{22.} - \mu_{12.} \leq 2.90 + 2.89 \times 0.270 = 3.68$
$3.12 = 3.90 - 2.89 \times 0.270 \leq \mu_{32.} - \mu_{22.} \leq 3.90 + 2.89 \times 0.270 = 4.68$
$6.02 = 6.80 - 2.89 \times 0.270 \leq \mu_{32.} - \mu_{12.} \leq 6.80 + 2.89 \times 0.270 = 7.58$
$2.47 = 3.25 - 2.89 \times 0.270 \leq \mu_{23.} - \mu_{13.} \leq 3.25 + 2.89 \times 0.270 = 4.03$
$2.26 = 3.04 - 2.89 \times 0.270 \leq \mu_{33.} - \mu_{23.} \leq 3.04 + 2.89 \times 0.270 = 3.82$
$5.51 = 6.29 - 2.89 \times 0.270 \leq \mu_{33.} - \mu_{13.} \leq 6.29 + 2.89 \times 0.270 = 7.07.$

The average resistivities for different combinations of diffusion temperatures and times indicate that average resistivity increases when going from lower to higher temperature levels. However, the increases are greater as one moves from lower to higher levels of diffusion time. The different influence of diffusion temperature, which depends on the diffusion time, implies that the temperature and time factors interact in their effect on resistivity. In view of the important interaction effects between temperature and time on average resistivities in the study findings, the researcher may decide that main effects due to temperature and time are not meaningful or of practical importance.

5.15 WORKED EXAMPLE FOR MODEL II

Johnson and Leone (1977, p. 861) reported data from an experiment designed to study the melting point of a homogeneous sample of hydroquinone. The experiment was performed with three analysts using three uncalibrated thermometers and working in three separate weeks. The data are given in Table 5.16.

The data in Table 5.16 can be regarded as a three-way classification with one observation per cell and all three factors can be regarded as random effects. The mathematical model for this experiment would be

$$y_{ijk} = \mu + \alpha_i + \beta_j + \gamma_k + (\alpha\beta)_{ij} + (\alpha\gamma)_{ik} + (\beta\gamma)_{jk} + e_{ijk} \quad \begin{cases} i = 1, 2, 3 \\ j = 1, 2, 3 \\ k = 1, 2, 3, \end{cases}$$

where μ is the general mean, α_i is the effect of the i-th thermometer, β_j is the effect of the j-th week, γ_k is the effect of the k-th analyst, $(\alpha\beta)_{ij}$ is the

TABLE 5.16
Data on the Melting Points of a Homogeneous Sample of Hydroquinone

				Thermometer					
		1			2			3	
Week	1	2	3	1	2	3	1	2	3
Analyst 1	174.0	173.5	174.5	173.0	173.5	173.0	171.5	172.5	173.0
2	173.0	173.0	173.5	172.0	173.0	173.5	171.0	172.0	171.5
3	173.5	173.0	173.0	173.0	173.5	172.5	173.0	173.0	172.5

Source: Johnson and Leone (1977, p. 861). Used with permission.

TABLE 5.17
Sums over Analysts $y_{ij.}$

		Week (j)		
Thermometer (i)	1	2	3	$y_{i..}$
1	520.5	519.5	521.0	1,561.0
2	518.0	520.0	519.0	1,557.0
3	515.5	517.5	517.0	1,550.0
$y_{.j.}$	1,554.0	1,557.0	1,557.0	$y_{...}$
				4,668.0

interaction of the i-th thermometer with the j-th week, $(\alpha\gamma)_{ik}$ is the interaction of the i-th thermometer with the k-th analyst, $(\beta\gamma)_{jk}$ is the interaction of the j-th week with the k-th analyst, and e_{ijk} is the customary error term. In order to estimate the error variance, it is assumed that thermometer × week × analyst interaction is zero. Furthermore, it is assumed that the α_i's, β_j's, γ_k's, $(\alpha\beta)_{ij}$'s, $(\alpha\gamma)_{ik}$'s, $(\beta\gamma)_{jk}$'s, and e_{ijk}'s are independently and normally distributed with zero means and variances σ_α^2, σ_β^2, σ_γ^2, $\sigma_{\alpha\beta}^2$, $\sigma_{\alpha\gamma}^2$, $\sigma_{\beta\gamma}^2$, and σ_e^2, respectively.

The first step in the analysis of variance computations consists of forming sums over every index and every combination of indices. Thus, we sum over analysts (k) to obtain a thermometer (i) × week (j) table containing $y_{ij.} = \sum_{k=1}^{3} y_{ijk}$ (see Table 5.17). This table is then summed over the levels of week (j) to obtain the thermometer totals ($y_{i..}$) and again over thermometers (i) to obtain week totals ($y_{.j.}$). Now, the sum of the thermometer totals is equal to the sum of the week totals, which gives the grand total ($y_{...}$). Finally, Tables 5.18 and 5.19 are obtained in a similar way.

With these preliminary results, the subsequent analysis of variance calculations are rather straightforward. Thus, from the computational formulae of

TABLE 5.18
Sums Over Weeks $y_{i.k}$

| Analyst (k) | Thermometer (i) | | | $y_{..k}$ |
	1	2	3	
1	522.0	519.5	517.0	1,558.5
2	519.5	518.5	514.5	1,552.5
3	519.5	519.0	518.5	1,557.0
$y_{i..}$	1,561.0	1,557.0	1,550.0	$y_{...}$ 4,668.0

TABLE 5.19
Sums Over Thermometer $y_{.jk}$

| Analyst (k) | Week (j) | | | $y_{..k}$ |
	1	2	3	
1	518.5	519.5	520.5	1,558.5
2	516.0	518.0	518.5	1,552.5
3	519.5	519.5	518.0	1,557.0
$y_{.j.}$	1,554.0	1,557.0	1,557.0	$y_{...}$ 4,668.0

Section 5.10, we have

$$SS_T = (174.0)^2 + (173.0)^2 + \cdots + (172.5)^2 - \frac{(4,668.0)^2}{3 \times 3 \times 3}$$
$$= 807,061 - 807,045.333 = 15.667,$$

$$SS_A = \frac{(1,561.0)^2 + (1,557.0)^2 + (1,550.0)^2}{3 \times 3} - \frac{(4,668.0)^2}{3 \times 3 \times 3}$$
$$= 807,052.222 - 807,045.333 = 6.889,$$

$$SS_B = \frac{(1,554.0)^2 + (1,557.0)^2 + (1,557.0)^2}{3 \times 3} - \frac{(4,668.0)^2}{3 \times 3 \times 3}$$
$$= 807,046.000 - 807,045.333 = 0.667,$$

$$SS_C = \frac{(1,558.5)^2 + (1,552.5)^2 + (1,557.0)^2}{3 \times 3} - \frac{(4,668.0)^2}{3 \times 3 \times 3}$$
$$= 807,047.500 - 807,045.333 = 2.167,$$

$$SS_{AB} = \frac{(520.5)^2 + (519.5)^2 + \cdots + (517.0)^2}{3} - \frac{(4,668.0)^2}{3 \times 3 \times 3} - SS_A - SS_B$$
$$= 807,054.000 - 807,045.333 - 6.889 - 0.667$$
$$= 1.111,$$

Three-Way and Higher-Order Crossed Classifications

$$SS_{AC} = \frac{(522.0)^2 + (519.5)^2 + \cdots + (518.5)^2}{3} - \frac{(4{,}668.0)^2}{3 \times 3 \times 3} - SS_A - SS_C$$
$$= 807{,}056.500 - 807{,}045.333 - 6.889 - 2.167$$
$$= 2.111,$$

and

$$SS_{BC} = \frac{(518.5)^2 + (519.5)^2 + \cdots + (518.0)^2}{3} - \frac{(4{,}668.0)^2}{3 \times 3 \times 3} - SS_B - SS_C$$
$$= 807{,}049.830 - 807{,}045.333 - 0.667 - 2.167$$
$$= 1.667.$$

Finally, by subtraction, the error (three-way interaction) sum of squares is given by

$$SS_E = 15.667 - 6.889 - 0.667 - 2.167 - 1.111 - 2.111 - 1.667$$
$$= 1.055.$$

These results along with the remaining calculations are shown in Table 5.20.

From Table 5.20 it is clear that the hypotheses on the two-factor interactions, namely,

$$H_0^{AB} : \sigma_{\alpha\beta}^2 = 0 \quad \text{versus} \quad H_1^{AB} : \sigma_{\alpha\beta}^2 > 0,$$
$$H_0^{AC} : \sigma_{\alpha\gamma}^2 = 0 \quad \text{versus} \quad H_1^{AC} : \sigma_{\alpha\gamma}^2 > 0,$$

and

$$H_0^{BC} : \sigma_{\beta\gamma}^2 = 0 \quad \text{versus} \quad H_1^{BC} : \sigma_{\beta\gamma}^2 > 0,$$

are all tested against the error (thermometer × week × analyst interaction) term. On examining the last two columns of Table 5.20, it is seen that only the thermometer × analyst interaction is significant ($p = 0.045$).

Now, we note that there are no exact F tests for testing the main effects hypotheses, namely,

$$H_0^A : \sigma_\alpha^2 = 0 \quad \text{versus} \quad H_1^A : \sigma_\alpha^2 > 0,$$
$$H_0^B : \sigma_\beta^2 = 0 \quad \text{versus} \quad H_1^B : \sigma_\beta^2 > 0,$$

and

$$H_0^C : \sigma_\gamma^2 = 0 \quad \text{versus} \quad H_1^C : \sigma_\gamma^2 > 0,$$

unless we are willing to assume that certain two-factor interactions are zero. Since we have just concluded that some two-factor interactions are insignificant, we can obtain exact tests for certain hypotheses by assuming the corresponding

TABLE 5.20
Analysis of Variance for the Melting Point Data of Table 5.16

Source of Variation	Degrees of Freedom	Sum of Squares	Mean Square	Expected Mean Square	F Value	p-Value
Thermometer	2	6.889	3.444	$\sigma_e^2 + 3\sigma_{\alpha\beta}^2 + 3\sigma_{\alpha\gamma}^2 + 3 \times 3\sigma_\alpha^2$	5.11	0.062
Week	2	0.667	0.333	$\sigma_e^2 + 3\sigma_{\alpha\beta}^2 + 3\sigma_{\beta\gamma}^2 + 3 \times 3\sigma_\beta^2$	0.59	0.589
Analyst	2	2.167	1.083	$\sigma_e^2 + 3\sigma_{\alpha\gamma}^2 + 3\sigma_{\beta\gamma}^2 + 3 \times 3\sigma_\gamma^2$	1.33	0.333
Thermometer × Week	4	1.111	0.278	$\sigma_e^2 + 3\sigma_{\alpha\beta}^2$	2.11	0.171
Thermometer × Analyst	4	2.111	0.528	$\sigma_e^2 + 3\sigma_{\alpha\gamma}^2$	4.00	0.045
Week × Analyst	4	1.667	0.417	$\sigma_e^2 + 3\sigma_{\beta\gamma}^2$	3.16	0.078
Error (Thermometer × Week × Analyst)	8	1.055	0.132	σ_e^2		
Total	26	15.667				

two-factor interactions to be zero. But, if we are unwilling to assume that any two-factor interactions are zero, then as we know there are no exact F tests available for this problem. We can, however, obtain pseudo-F tests as discussed in Section 5.5. In the following we illustrate the pseudo-F test for testing the hypothesis $H_0^A : \sigma_\alpha^2 = 0$ versus $H_1^A : \sigma_\alpha^2 > 0$.

From Table 5.20, we obtain

$$E(\text{MS}_{AB}) + E(\text{MS}_{AC}) - E(\text{MS}_E) = \sigma_e^2 + 3\sigma_{\alpha\beta}^2 + 3\sigma_{\alpha\gamma}^2,$$

which is precisely equal to $E(\text{MS}_A)$ when $\sigma_\alpha^2 = 0$. Hence, the desired test statistic is

$$F_A^* = \frac{\text{MS}_A}{\text{MS}_{AB} + \text{MS}_{AC} - \text{MS}_E} = \frac{3.444}{0.278 + 0.528 - 0.132} = 5.11,$$

which has an approximate F distribution with 2 and v_a degrees of freedom, where v_a (rounded to the nearest digit) is approximated by

$$v_a = \frac{(0.278 + 0.528 - 0.132)^2}{\frac{(0.278)^2}{4} + \frac{(0.528)^2}{4} + \frac{(-0.132)^2}{8}} \doteq 5.$$

For a level of significance of 0.05, we find from Appendix Table V that $F[2, 5; 0.95] = 5.79$. Since $F_A^* = 5.11 < 5.79$, we do not reject H_0^A and conclude that thermometers do not have a significant effect on melting point ($p = 0.062$). Similarly, approximate F tests can be readily performed for H_0^B and H_0^C yielding $F_B^* = 0.59$, $F_C^* = 1.33$, $v_b \doteq 5$, and $v_c \doteq 6$. The resulting p-values for F_B^* and F_C^* are 0.589 and 0.333, respectively, and we may conclude that weeks and analysts also do not have any significant effect on the melting point.

Finally, we can also obtain the unbiased estimates of the variance components, giving

$$\hat{\sigma}_e^2 = 0.132,$$

$$\hat{\sigma}_{\alpha\beta}^2 = \frac{0.278 - 0.132}{3} = 0.049,$$

$$\hat{\sigma}_{\alpha\gamma}^2 = \frac{0.528 - 0.132}{3} = 0.132,$$

$$\hat{\sigma}_{\beta\gamma}^2 = \frac{0.417 - 0.132}{3} = 0.095,$$

$$\hat{\sigma}_\alpha^2 = \frac{3.444 - 0.278 - 0.528 + 0.132}{3 \times 3} = 0.308,$$

$$\hat{\sigma}_\beta^2 = \frac{0.333 - 0.278 - 0.417 + 0.132}{3 \times 3} = -0.026,$$

TABLE 5.21
Data on Diamond Pyramid Hardness Number of Dental Fillings Made from Two Alloys of Gold

		Alloy					
		Gold Foil			Goldent		
Method		1	2	3	1	2	3
Dentist	1	792	772	782	824	772	803
	2	803	752	715	803	772	707
	3	715	792	762	724	715	606
	4	673	657	690	946	743	245
	5	634	649	724	715	724	627

Source: Halvorsen (1991, p. 145). Used with permission.

and

$$\hat{\sigma}_\gamma^2 = \frac{1.083 - 0.528 - 0.417 + 0.132}{3 \times 3} = 0.030.$$

The results on variance components estimates are consistent with those on tests of hypotheses. It should, however, be noted that although certain variance components seem to be relatively large, a rather small value for the error degrees of freedom makes the corresponding F test quite insensitive.

5.16 WORKED EXAMPLE FOR MODEL III

The following example is based on an experiment in dentistry designed to study the hardness of gold filling material. The data are taken from Halvorsen (1991, p. 145). Five dentists were asked to prepare the six types of gold filling material, sintered at the three temperatures, and using each of the three methods of condensation. The data in Table 5.21 represent a subset of the original data involving only two alloys of gold filling material and sintered at a single temperature.

The data in Table 5.21 can be regarded as a three-way classification with one observation per cell. Note that the alloy and method are fixed effects and the dentist is a random effect. The mathematical model for this experiment would be

$$y_{ijk} = \mu + \alpha_i + \beta_j + \gamma_k + (\alpha\beta)_{ij} + (\alpha\gamma)_{ik} + (\beta\gamma)_{jk} + e_{ijk} \quad \begin{cases} i = 1, 2 \\ j = 1, 2, 3 \\ k = 1, 2, 3, 4, 5, \end{cases}$$

TABLE 5.22
Sums over Dentists $y_{ij.}$

	Method (j)			
Alloy (i)	1	2	3	$y_{i..}$
Gold Foil	3,617	3,622	3,673	10,912
Goldent	4,012	3,726	2,988	10,726
$y_{.j.}$	7,629	7,348	6,661	$y_{...}$
				21,638

where μ is the general mean, α_i is the effect of the i-th alloy, β_j is the effect of the j-th method, γ_k is the effect of the k-th dentist, $(\alpha\beta)_{ij}$ is the interaction of the i-th alloy with the j-th method, $(\alpha\gamma)_{ik}$ is the interaction of the i-th alloy with the k-th dentist, $(\beta\gamma)_{jk}$ is the interaction of the j-th method with the k-th dentist, and e_{ijk} is the customary error term. In order to estimate the error variance, it is assumed that alloy × method × dentist interaction is zero. Furthermore, it is assumed that the α_i's, β_j's, and $(\alpha\beta)_{ij}$'s are constants subject to the restrictions:

$$\sum_{i=1}^{2}\alpha_i = \sum_{j=1}^{3}\beta_j = 0,$$

$$\sum_{i=1}^{2}(\alpha\beta)_{ij} = \sum_{j=1}^{3}(\alpha\beta)_{ij} = 0;$$

and γ_k's, $(\alpha\gamma)_{ik}$'s, $(\beta\gamma)_{jk}$'s, and e_{ijk}'s are independently and normally distributed with mean zero and variances $\sigma_\gamma^2, \sigma_{\alpha\gamma}^2, \sigma_{\beta\gamma}^2$, and σ_e^2, respectively; and subject to the restrictions

$$\sum_{i=1}^{2}(\alpha\gamma)_{ik} = \sum_{j=1}^{3}(\beta\gamma)_{jk} = 0 \quad (k = 1, 2, 3, 4, 5).$$

The first step in the analysis of variance computations consists of forming sums over every index and every combination of indices. Thus, we sum over dentists (k) to obtain an alloy (i) × method (j) table containing $y_{ij.} = \sum_{k=1}^{5} y_{ijk}$ (see Table 5.22). We then sum this table over methods (j) to obtain the alloy totals ($y_{i..}$) and again over alloys (i) to obtain method totals ($y_{.j.}$). Now, the sum of the alloy totals is equal to the sum of the method totals, which gives the grand total ($y_{...}$). Finally, Tables 5.23 and 5.24 are obtained in a similar way.

With these preliminary results, the subsequent analysis of variance calculations are rather straightforward. Thus, from the computational formulae of

TABLE 5.23
Sums over Methods $y_{i.k}$

Dentist (k)	Alloy (i)		$y_{..k}$
	Gold Foil	Goldent	
1	2,346	2,399	4,745
2	2,270	2,282	4,552
3	2,269	2,045	4,314
4	2,020	1,934	3,954
5	2,007	2,066	4,073
$y_{i..}$	10,912	10,776	$y_{...}$ 21,638

TABLE 5.24
Sums over Alloys $y_{.jk}$

Dentist (k)	Method (j)			$y_{..k}$
	1	2	3	
1	1,616	1,544	1,585	4,745
2	1,606	1,524	1,422	4,552
3	1,439	1,507	1,368	4,314
4	1,619	1,400	935	3,954
5	1,349	1,373	1,351	4,073
$y_{.j.}$	7,629	7,348	6,661	$y_{...}$ 21,638

Section 5.10, we have

$$SS_T = (792)^2 + (803)^2 + \cdots + (627)^2 - \frac{(21,638)^2}{2 \times 3 \times 5}$$
$$= 15,982,022 - 15,606,768.133$$
$$= 375,253.867,$$

$$SS_A = \frac{(10,912)^2 + (10,726)^2}{3 \times 5} - \frac{(21,638)^2}{2 \times 3 \times 5}$$
$$= 15,607,921.333 - 15,606,768.133$$
$$= 1,153.200,$$

$$SS_B = \frac{(7,629)^2 + (7,348)^2 + (6,661)^2}{2 \times 5} - \frac{(21,638)^2}{2 \times 3 \times 5}$$
$$= 15,656,366.600 - 15,606,768.133$$
$$= 49,598.467,$$

Three-Way and Higher-Order Crossed Classifications

$$SS_C = \frac{(4,745)^2 + (4,552)^2 + \cdots + (4,073)^2}{2 \times 3} - \frac{(21,638)^2}{2 \times 3 \times 5}$$
$$= 15,678,295.000 - 15,606,768.133$$
$$= 71,526.867,$$

$$SS_{AB} = \frac{(3,617)^2 + (3,622)^2 + \cdots + (2,988)^2}{5} - \frac{(21,638)^2}{2 \times 3 \times 5} - SS_A - SS_B$$
$$= 15,719,973.200 - 15,606,768.133 - 1,153.200 - 49,598.467$$
$$= 62,453.400,$$

$$SS_{AC} = \frac{(2,346)^2 + (2,270)^2 + \cdots + (2,066)^2}{3} - \frac{(21,638)^2}{2 \times 3 \times 5} - SS_A - SS_C$$
$$= 15,688,962.667 - 15,606,768.133 - 1,153.200 - 71,526.867$$
$$= 9,514.467,$$

$$SS_{BC} = \frac{(1,616)^2 + (1,606)^2 + \cdots + (1,351)^2}{2} - \frac{(21,638)^2}{2 \times 3 \times 5} - SS_B - SS_C$$
$$= 15,815,112.000 - 15,606,768.133 - 49,598.467 - 71,526.867$$
$$= 87,218.533.$$

Finally, by subtraction, the error (three-way interaction) sum of squares is given by

$$SS_E = 375,253.867 - 1,153.200 - 49,598.467 - 71,526.867 - 62,453.400$$
$$- 9,514.467 - 87,218.533$$
$$= 93,789.933.$$

These results along with the remaining calculations are shown in Table 5.25.

From Table 5.25 it is clear that all the two-factor interactions should be tested against the error term. It is immediately evident that the alloy × dentist and method × dentist interactions give F ratio values less than one and are clearly nonsignificant. Also, the alloy × method has an F value of 2.66, which again fails to reach the 5 percent level of significance ($p = 0.130$). The method main effect, when tested against the method × dentist interaction, has an F value of 2.27 which is also nonsignificant ($p = 0.165$). The alloy main effect, tested against the alloy × dentist interaction, gives an F ratio of less than 1 and is clearly nonsignificant ($p = 0.523$). The dentist effect, tested against error, has an F ratio of 1.53 which is again nonsignificant ($p = 0.282$).

One can also obtain unbiased estimates of the variance components $\sigma_e^2, \sigma_{\beta\gamma}^2, \sigma_{\alpha\gamma}^2$, and σ_γ^2, yielding

$$\hat{\sigma}_e^2 = 11,723.617,$$
$$\hat{\sigma}_{\beta\gamma}^2 = \frac{1}{2}(10,902.317 - 11,723.617) = -410.650,$$

TABLE 5.25
Analysis of Variance for the Gold Hardness Data of Table 5.21

Source of Variation	Degrees of Freedom	Sum of Squares	Mean Square	Expected Mean Square	F Value	p-Value
Alloy	1	1,153.200	1,153.200	$\sigma_e^2 + 3\sigma_{\alpha\gamma}^2 + \dfrac{3 \times 5}{2-1}\sum_{i=1}^{2}\alpha_i^2$	0.49	0.523
Method	2	49,598.467	24,799.234	$\sigma_e^2 + 2\sigma_{\beta\gamma}^2 + \dfrac{2 \times 5}{3-1}\sum_{j=1}^{3}\beta_j^2$	2.27	0.165
Dentist	4	71,526.867	17,881.717	$\sigma_e^2 + 3 \times 2\sigma_\gamma^2$	1.53	0.282
Alloy × Method	2	62,453.400	31,226.700	$\sigma_e^2 + \dfrac{5}{(2-1)(3-1)}\sum_{i=1}^{2}\sum_{j=1}^{3}(\alpha\beta)_{ij}^2$	2.66	0.130
Alloy × Dentist	4	9,514.467	2,378.617	$\sigma_e^2 + 3\sigma_{\alpha\gamma}^2$	0.20	0.930
Method × Dentist	8	87,218.533	10,902.317	$\sigma_e^2 + 2\sigma_{\beta\gamma}^2$	0.93	0.540
Error (Alloy × Method × Dentist)	8	93,788.933	11,723.617	σ_e^2		
Total	29	375,253.867				

Three-Way and Higher-Order Crossed Classifications

$$\hat{\sigma}^2_{\alpha\gamma} = \frac{1}{3}(2{,}378.617 - 11{,}723.617) = -3{,}115.000,$$

and

$$\hat{\sigma}^2_{\gamma} = \frac{1}{6}(17{,}881.717 - 11{,}723.617) = 1{,}026.350.$$

The negative estimates are probably an indication that the corresponding variance components may be zero. The point estimates of variance components are consistent with the results on tests of hypotheses. Finally, the confidence limits on contrasts for the fixed effects can be constructed in the usual way. For example, to obtain 95 percent confidence limits for the difference between the two alloy effects, we have

$$\bar{y}_{1..} - \bar{y}_{2..} = \frac{10{,}912}{3 \times 5} - \frac{10{,}726}{3 \times 5} = 12.400,$$

$$\text{Var}(\bar{y}_{1..} - \bar{y}_{2..}) = \frac{2}{3 \times 5}\text{MS}_{AC} = \frac{2}{3 \times 5}(2{,}378.617) = 317.149,$$

and

$$t[4, 0.975] = 2.776.$$

So the desired confidence limits are

$$12.400 \pm 2.776\sqrt{317.149} = (-37.037, 61.837).$$

One can similarly obtain confidence limits for the difference between any two method effects based on the t test. However, since there may be more than one comparison of interest, the multiple comparison techniques should generally be preferred.

5.17 USE OF STATISTICAL COMPUTING PACKAGES

Three-way and higher-order factorial models can be analyzed using either SAS ANOVA or GLM procedures. For a balanced design, the recommended procedure is ANOVA and for the unbalanced design, GLM must be used. The random and mixed model analyses can be handled by the use of RANDOM and TEST options. Approximate or pseudo-F tests can be carried out via GLM using the Satterthwaite procedure. For the estimation of variance components, PROC MIXED or VARCOMP must be used. For instructions regarding SAS commands see Section 11.1.

Among the SPSS procedures, either ANOVA or MANOVA could be used for a fixed effects analysis although ANOVA would be simpler. For the analyses involving random and mixed effects models, MANOVA or GLM must be used. For the estimation of variance components, VARCOMP is the procedure of choice. For instructions regarding SPSS commands see Section 11.2.

The BMDP programs described in Section 4.15 are also adequate for analyzing three-way and higher-order factorial models. No new problems arise for analyses involving higher-order factorials.

5.18 WORKED EXAMPLES USING STATISTICAL PACKAGES

In this section, we illustrate the applications of statistical packages to perform three-way analysis of variance for the data sets of examples presented in Sections 5.14 through 5.16. Figures 5.1 through 5.3 illustrate the program instructions and the output results for analyzing data in Tables 5.10, 5.16, and 5.21 using SAS ANOVA/GLM, SPSS MANOVA/GLM, and BMDP 2V/8V

```
DATA ELECCHRM;                      The SAS System
INPUT TEMP TIME DEGR          Analysis of Variance Procedure
RESIST;                    Dependent Variable: RESIST
DATALINES;                                Sum of         Mean
1 1 1 17.9                 Source          DF     Squares     Square   F Value   Pr > F
. . . .                    Model           17    508.40500   29.90618   101.89   0.0001
3 3 2 26.9;                Error           54     15.85000    0.29352
PROC ANOVA;                Corrected Total 71    524.25500
CLASSES TEMP TIME DEGR;
MODEL RESIST=TEMP TIME                     R-Square      C.V.    Root MSE   RESIST Mean
DEGR TEMP*TIME TEMP*DEGR                    0.969767   2.417732   0.5418      22.408
TIME*DEGR
TEMP*TIME*DEGR;            Source          DF   Anova SS   Mean Square   F Value   Pr > F
RUN;                       TEMP             2   410.69250    205.34625   699.60   0.0001
CLASS    LEVELS  VALUES    TIME             2    80.54083     40.27042   137.20   0.0001
TEMP        3    1 2 3     DEGR             1     0.46722      0.46722     1.59   0.2125
TIME        3    1 2 3     TEMP*TIME        4    14.81417      3.70354    12.62   0.0001
DEGR        2    1 2       TEMP*DEGR        2     0.31361      0.15681     0.53   0.5892
NUMBER OF OBS. IN DATA     TIME*DEGR        2     0.70194      0.35097     1.20   0.3104
SET=72                     TEMP*TIME*DEGR   4     0.87472      0.21868     0.75   0.5656
```

(i) SAS application: SAS ANOVA instructions and output for the three-way fixed effects analysis of variance.

```
DATA LIST                            Analysis of Variance-Design 1
/TEMP 1 TIME 3 DEGR 5
RESIST 7-10(1).            Tests of Significance for RESIST using UNIQUE sums of squares
BEGIN DATA.
1 1 1 17.9                 Source of Variation    SS     DF     MS        F      Sig of F
1 1 1 18.0
1 1 1 18.7                 WITHIN CELLS          15.85    54    .29
1 1 1 19.0                 TEMP                 410.69     2   205.35   699.60    .000
1 1 2 18.1                 TIME                  80.54     2    40.27   137.20    .000
1 1 2 18.9                 DEGR                    .47     1     .47     1.59     .212
1 1 2 18.6                 TEMP BY TIME          14.81     4    3.70    12.62     .000
1 1 2 19.1                 TEMP BY DEGR            .31     2     .16      .53     .589
. . .                      TIME BY DEGR            .70     2     .35     1.20     .310
3 3 2 26.9                 TEMP BY TIME BY DEGR    .87     4     .22      .75     .566
END DATA.                  (Model)              508.41    17   29.91   101.89     .000
MANOVA RESIST BY           (Total)              524.25    71    7.38
TEMP(1,3) TIME(1,3)
DEGR(1,2).                 R-Squared =   .970          Adjusted R-Squared =  .960
```

(ii) SPSS application: SPSS MANOVA instructions and output for the three-way fixed effects analysis of variance.

FIGURE 5.1 Program Instructions and Output for the Three-Way Fixed Effects Analysis of Variance: Data on Average Resistivities (in m-ohms/cm^3) for Electrolytic Chromium Plate Example (Table 5.10).

Three-Way and Higher-Order Crossed Classifications

```
/INPUT      FILE='C:\SAHAI          BMDP2V - ANALYSIS OF VARIANCE AND COVARIANCE WITH
            \TEXTO\EJE12.TXT'.               REPEATED MEASURES Release: 7.0 (BMDP/DYNAMIC)
            FORMAT=FREE.
            VARIABLES=4.            ANALYSIS OF VARIANCE FOR THE 1-ST DEPENDENT VARIABLE
/VARIABLE   NAMES=TE,TI,DEG,RES.    THE TRIALS ARE REPRESENTED BY THE VARIABLES:RESIST
/GROUP      VARIABLE=TE,TI,DEG.
            CODES(TE)=1,2,3.        SOURCE   SUM OF       D.F.    MEAN             F       TAIL
            NAMES(TE)=F1,F2,F3.              SQUARES              SQUARE                   PROB.
            CODES(TI)=1,2,3.        MEAN     36153.60507   1      36153.60507   123173.14  0.0000
            NAMES(TI)=H4,H8,H12.    G          410.69245   2        205.34623      699.60  0.0000
            CODES(DEGR)=1,2.        H           80.54083   2         40.27041      137.20  0.0000
            NAMES(DEGR)=YES,NO.     I            0.46722   1          0.46722        1.59  0.2125
/DESIGN     DEPENDENT=RESIST.       GH          14.81417   4          3.70354       12.62  0.0000
/END                                GI           0.31361   2          0.15681        0.53  0.5892
1 1 1 17.9                          HI           0.70194   2          0.35097        1.20  0.3104
. . . .                             GHI          0.87472   4          0.21868        0.75  0.5656
3 3 2 26.9                          ERROR       15.85000  54          0.29352
```

(iii) BMDP application: BMDP 2V instructions and output for the three-way fixed effects analysis of variance.

FIGURE 5.1 *(continued)*

```
DATA MELTPOIN;                                          The SAS System
INPUT THERMOM WEEK                                General Linear Models Procedure
ANALYST MELTINGP;            Dependent Variable: MELTINGP
DATALINES;                                           Sum of         Mean
1 1 1 174.0                  Source            DF    Squares        Square    F Value   Pr > F
1 1 2 173.0                  Model             18    14.611111      0.811728    6.15    0.0066
1 1 3 173.5                  Error              8     1.055556      0.131944
1 2 1 173.5                  Corrected Total   26    15.666667
1 2 2 173.0                                     R-Square     C.V.       Root MSE    MELTINGP Mean
1 2 3 173.0                                     0.932624   0.210101      0.3632       172.89
1 3 1 174.5                  Source            DF    Type III SS  Mean Square F Value  Pr > F
1 3 2 173.5                  THERMOM            2     6.8888889    3.4444444   26.11   0.0003
1 3 3 173.0                  WEEK               2     0.6666667    0.3333333    2.53   0.1411
2 1 1 173.0                  ANALYST            2     2.1666667    1.0833333    8.21   0.0115
2 1 2 172.0                  THERMOM*WEEK       4     1.1111111    0.2777778    2.11   0.1719
2 1 3 173.0                  THERMOM*ANALYST    4     2.1111111    0.5277778    4.00   0.0453
2 2 1 173.5                  WEEK*ANALYST       4     1.6666667    0.4166667    3.16   0.0780
2 2 2 173.0                  Source                   Type III Expected Mean Square
. . . .                      THERMOM                  Var(Error) + 3 Var(THERMOM*ANALYST)
3 3 3 172.5                                          + 3 Var(THERMOM*WEEK) + 9 Var(THERMOM)
;                            WEEK                     Var(Error) + 3 Var(WEEK*ANALYST)
PROC GLM;                                            + 3 Var(THERMOM*WEEK) + 9 Var(WEEK)
CLASSES THERMOM WEEK         ANALYST                  Var(Error) + 3 Var(WEEK*ANALYST)
ANALYST;                                             + 3 Var(THERMOM*ANALYST) + 9 Var(ANALYST)
MODEL MELTINGP=THERMOM       THERMOM*WEEK             Var(Error) + 3 Var(THERMOM*WEEK)
WEEK ANALYST THERMOM*        THERMOM*ANALYST          Var(Error) + 3 Var(THERMOM*ANALYST)
WEEK THERMOM*ANALYST         WEEK*ANALYST             Var(Error) + 3 Var(WEEK*ANALYST)
WEEK*ANALYST;                Source:THERMOM  Error:MS(THERMOM*WEEK)+MS(THERMOM*ANALYST)-MS(Err)
RANDOM THERMOM WEEK                                Denominator    Denominator
ANALYST THERMOM*WEEK         DF   Type III MS    DF       MS       F Value   Pr > F
THERMOM*ANALYST               2   3.4444444444    4.98   0.6736111111  5.1134  0.0621
WEEK*ANALYST /TEST;          Source: WEEK  Error: MS(THERMOM*WEEK) + MS(WEEK*ANALYST)-MS(Err)
RUN;                                               Denominator    Denominator
CLASS   LEVELS  VALUES       DF   Type III MS    DF       MS       F Value   Pr > F
THERMON    3     1 2 3        2   0.3333333333    4.88   0.5625     0.5926   0.5883
WEEK       3     1 2 3       Source:ANALYST Error:MS(THERMOM*ANALYST)+MS(WEEK*ANALYST)-MS(Err)
ANALYST    3     1 2 3                             Denominator    Denominator
NUMBER OF OBS. IN DATA       DF   Type III MS    DF       MS       F Value   Pr > F
SET=27                        2   1.0833333333    5.73   0.8125     1.3333   0.3346
```

(i) SAS application: SAS GLM instructions and output for the three-way random effects analysis of variance.

FIGURE 5.2 Program Instructions and Output for the Three-Way Random Effects Analysis of Variance: Data on the Melting Points of a Homogeneous Sample of Hydroquinone (Table 5.16).

```
DATA LIST          Tests of Between-Subjects Effects          Dependent Variable: MELTINGP
/THERMOM 1         Source                      Type III SS   df   Mean Square    F      Sig.
 WEEK 3            THERMOM    Hypothesis        6.889         2    3.444       5.113   .062
 ANALYST 5                    Error             3.355     4.981     .674(a)
 MELTINGP 7-11(1)  WEEK       Hypothesis         .667         2     .333        .593   .588
BEGIN DATA.                   Error             2.744     4.878     .562(b)
1 1 1 174.0        ANALYST    Hypothesis        2.167         2    1.083       1.333   .335
1 1 2 173.0                   Error             4.655     5.730     .812(c)
1 1 3 173.5        THERMOM*   Hypothesis        1.111         4     .278       2.105   .172
1 2 1 173.5         WEEK      Error             1.056         8     .132(d)
1 2 2 173.0        THERMOM*   Hypothesis        2.111         4     .528       4.000   .045
1 2 3 173.0         ANALYST   Error             1.056         8     .132(d)
1 3 1 174.5        WEEK*      Hypothesis        1.667         4     .417       3.158   .078
1 3 2 173.5         ANALYST   Error             1.056         8     .132(d)
1 3 3 173.0        a  MS(T*W)+MS(T*A)-MS(E)   b  MS(T*W)+MS(W*A)-1.000 MS(E)   c  MS(T*A)+MS(W*A)
2 1 1 173.0        -MS(E)   d  MS(E)
2 1 2 172.0
2 1 3 173.0                                   Expected Mean Squares(a,b)
2 2 1 173.5                                         Variance Component
 . . . .           Source        Var(T)  Var(W)  Var(A)  Var(T*W)  Var(T*A)  Var(W*A)  Var(E)
3 3 3 172.5        Intercept     9.000   9.000   9.000    3.000     3.000     3.000     1.000
END DATA.          THERMOM       9.000    .000    .000    3.000     3.000      .000     1.000
GLM MELTINGP BY    WEEK           .000   9.000    .000    3.000      .000     3.000     1.000
 THERMOM WEEK      ANALYST        .000    .000   9.000     .000     3.000     3.000     1.000
 ANALYST           THERMOM*WEEK   .000    .000    .000    3.000      .000      .000     1.000
/DESIGN THERMOM    THERMOM*ANALYST .000   .000    .000     .000     3.000      .000     1.000
 WEEK ANALYST      WEEK*ANALYST   .000    .000    .000     .000      .000     3.000     1.000
 THERMOM*WEEK      Error          .000    .000    .000     .000      .000      .000     1.000
 THERMOM*ANALYST   a  For each source, the expected mean square equals the sum of the
 WEEK*ANALYST      coefficients in the cells times the variance components, plus a
/RANDOM THERMOM    quadratic term involving effects in the Quadratic Term cell.  b  Expected
 WEEK ANALYST.     Mean Squares are based on the Type III Sums of Squares.
```

(ii) SPSS application: SPSS GLM instructions and output for the three-way random effects analysis of variance.

```
/INPUT      FILE='C:\SAHAI      BMDP8V - GENERAL MIXED MODEL ANALYSIS OF VARIANCE
            \TEXTO\EJE13.TXT'.         - EQUAL CELL SIZES Release: 7.0 (BMDP/DYNAMIC)
            FORMAT=FREE.        ANALYSIS OF VARIANCE FOR DEPENDENT VARIABLE   1
            VARIABLES=3.         SOURCE    ERROR    SUM OF       D.F.      MEAN          F      PROB.
/VARIABLE NAMES=A1,A2,A3.                  TERM    SQUARES                SQUARE
/DESIGN   NAMES=T,W,A.          1 MEAN             8.0704533E+5   1    8.0704533E+5
          LEVELS=3,3,3.         2 THERMOM          6.8888869E+0   2    3.4444444E+0
          RANDOM=T,W,A.         3 WEEK             0.6666667E+0   2    0.3333333E+0
          MODEL='T, W, A'.      4 ANALYST          2.1666667E+0   2    1.0833333E+0
/END                            5 TW         TWA   1.1111111E+0   4    0.2777778E+0   2.11   0.1719
174.0  173.0  173.5             6 TA         TWA   2.1111111E+0   4    0.5277778E+0   4.00   0.0453
173.5  173.0  173.0             7 WA         TWA   1.6666667E+0   4    0.4166667E+0   3.16   0.0780
174.5  173.5  173.0             8 TWA              1.0555556E+0   8    0.1319444E+0
173.0  172.0  173.0
173.5  173.0  173.5              SOURCE         EXPECTED MEAN                  ESTIMATES OF
173.0  173.5  172.5                                SQUARE                  VARIANCE COMPONENTS
171.5  171.0  173.0             1 MEAN       27(1)+9(2)+9(3)+9(4)+3(5)+3(6)+3(7)+8   29890.42824
172.5  172.0  173.0             2 THERMOM       9(2)+3(5)+3(6)+(8)                       0.30787
173.0  171.5  172.5             3 WEEK          9(3)+3(5)+3(7)+(8)                      -0.02546
ANALYSIS OF VARIANCE DESIGN     4 ANALYST       9(4)+3(6)+3(7)+(8)                       0.03009
 INDEX           T    W    A    5 TW            3(5)+(8)                                 0.04861
NUMBER OF LEVELS 3    3    3    6 TA            3(6)+(8)                                 0.13194
POPULATION SIZE INF  INF  INF   7 WA            3(7)+(8)                                 0.09491
MODEL     T, W, A               8 TWA           (8)                                      0.13194
```

(iii) BMDP application: BMDP 8V instructions and output for the three-way random effects analysis of variance.

FIGURE 5.2 *(continued)*

procedures. The typical output provides the data format listed at the top, all cell means, and the entries of the analysis of variance table. It should be noticed that in each case the results are the same as those provided using manual computations in Sections 5.14 through 5.16. However, note that certain tests of significance in a mixed model may differ from one program to the other since they make different model assumptions.

Three-Way and Higher-Order Crossed Classifications

```
DATA DIAMOND;                        The SAS System
INPUT ALLOY METHOD         General Linear Models Procedure
DENTIST NUMBER;         Dependent Variable: NUMBER
DATALINES;                              Sum of         Mean
1 1 1 792               Source      DF  Squares      Square    F Value  Pr > F
1 1 2 803               Model       21  281464.93333 13403.09206  1.14  0.4473
. . . .                 Error        8   93788.93333 11723.61667
2 3 5 627               Corrected Total 29 375253.86667
;                                       R-Square    C.V.    Root MSE    NUMBER Mean
PROC GLM;                               0.750065  15.011875  108.27565  721.26666667
CLASSES ALLOY METHOD    Source       DF  Type III SS  Mean Square  F Value  Pr > F
DENTIST;                ALLOY         1    1153.200    1153.200    0.10    0.7618
MODEL NUMBER=ALLOY      METHOD        2   49598.467   24799.233    2.12    0.1830
METHOD DENTIST          DENTIST       4   71526.867   17881.717    1.53    0.2829
ALLOY*METHOD            ALLOY*METHOD  2   62453.400   31226.700    2.66    0.1298
ALLOY*DENTIST           ALLOY*DENTIST 4    9514.467    2378.617    0.20    0.9297
METHOD*DENTIST;         METHOD*DENTIST 8  87218.533   10902.317    0.93    0.5396
RANDOM DENTIST          Source    Type III Expected Mean Square
ALLOY*DENTIST           ALLOY       Var(Error)+3 Var(ALLOY*DENTIST)+Q(ALLOY,ALLOY*METHOD)
METHOD*DENTIST;         METHOD      Var(Error)+2 Var(METHOD*DENTIST)+Q(METHOD,ALLOY*METHOD)
TEST H=ALLOY            DENTIST     Var(Error)+2 Var(METHOD*DENTIST)+3 Var(ALLOY*DENTIST)
E=ALLOY*DENTIST;                    + 6 Var(DENTIST)
TEST H=METHOD           ALLOY*METHOD   Var(Error) + Q(ALLOY*METHOD)
E=METHOD*DENTIST;       ALLOY*DENTIST  Var(Error) + 3 Var(ALLOY*DENTIST)
RUN;                    METHOD*DENTIST Var(Error) + 2 Var(METHOD*DENTIST)
CLASS   LEVELS VALUES   Tests of Hypotheses using the Type III MS for ALLOY*DENTIST as
ALLOY    2     1 2      an error term
METHOD   3     1 2 3    Source      DF  Type III SS Mean Square F Value Pr > F
DENTIST  5     1 2 3 4 5 ALLOY       1    1153.200    1153.200    0.48   0.5246
NUMBER OF OBS. IN DATA  Source      DF  Type III SS Mean Square F Value Pr > F
SET=30                  METHOD       2   49598.467   24799.233    2.27   0.1651
```

(i) SAS application: SAS GLM instructions and output for the three-way mixed effects analysis of variance.

```
DATA LIST              Tests of Between-Subjects Effects    Dependent Variable: NUMBER
/ALLOY 1
METHOD 3              Source             Type III SS   df     Mean Square       F    Sig
DENTIST 5             ALLOY   Hypothesis   1153.200     1       1153.200      .485   .525
NUMBER 7-9.                   Error        9514.467     4       2378.617(a)
BEGIN DATA.           METHOD  Hypothesis  49598.467     2      24799.233      2.275  .165
1 1 1 792.0                   Error       87218.533     8      10902.317(b)
1 1 2 803.0           DENTIST Hypothesis  71526.867     4      17881.717     11.482  .820
1 1 3 715.0                   Error         112.902    7.250E-02 1557.317(c)
1 1 4 673.0           ALLOY*  Hypothesis  62453.400     2      31226.700      2.664  .130
1 1 5 634.0           METHOD  Error       93788.933     8      11723.617(d)
1 2 1 772.0           ALLOY*  Hypothesis   9514.467     4       2378.617       .203  .930
1 2 2 752.0           DENTIST Error       93788.933     8      11723.617(d)
1 2 3 792.0           METHOD* Hypothesis  87218.533     8      10902.317       .930  .540
1 2 4 657.0           DENTIST Error       93788.933     8      11723.617(d)
1 2 5 649.0           a MS(A*D)  b MS(M*D)  c MS(A*D)+MS(M*D)-MS(E)  d MS(E)
1 3 1 782.0
1 3 2 715.0                              Expected Mean Squares(a,b)
2 3 5 762.0                                 Variance Component
. . . .               Source         Var(D)  Var(A*D)  Var(M*D)  Var(Error)  Quadratic Term
2 3 5 627.0           ALLOY           .000    3.000     .000      1.000       Alloy
END DATA.             METHOD          .000     .000    2.000      1.000       Method
GLM NUMBER BY         DENTIST        6.000    3.000    2.000      1.000
ALLOY METHOD          ALLOY*METHOD    .000     .000     .000      1.000       Alloy*Method
DENTIST               ALLOY*DENTIST   .000    3.000     .000      1.000
/DESIGN ALLOY         METHOD*DENTIST  .000     .000    2.000      1.000
METHOD DENTIST        Error           .000     .000     .000      1.000
ALLOY*METHOD          a For each source, the expected mean square equals the sum of the
ALLOY*DENTIST         coefficients in the cells times the variance components, plus a quadratic
METHOD*DENTIST        term involving effects in the Quadratic Term cell. b Expected Mean
/RANDOM DENTIST.      Squares are based on the Type III Sums of Squares.
```

(ii) SPSS application: SPSS GLM instructions and output for the three-way mixed effects analysis of variance.

FIGURE 5.3 Program Instructions and Output for the Three-Way Mixed Effects Analysis of Variance: Data on Diamond Pyramid Hardness Number of Dental Fillings Made from Two Alloys of Gold (Table 5.21).

```
/INPUT       FILE='C:\SAHAI           BMDP8V - GENERAL MIXED MODEL ANALYSIS OF VARIANCE
             \TEXTO\EJE14.TXT'.              - EQUAL CELL SIZES Release: 7.0 (BMDP/DYNAMIC)
             FORMAT=FREE.             ANALYSIS OF VARIANCE FOR DEPENDENT VARIABLE  1
             VARIABLES=5.               SOURCE    ERROR     SUM OF     D.F.    MEAN       F       PROB.
/VARIABLE    NAMES=D1,D2,D3,                      TERM      SQUARES            SQUARE
             D4,D5.                   1 MEAN      DENTIST   15606768.1  1   15606768.1  872.78   0.0000
/DESIGN      NAMES=ALLOY,METHOD,      2 ALLOY     AD            1153.2  1        1153.2   0.48   0.5246
             DENTIST.                 3 METHOD    MD           49598.5  2       24799.2   2.27   0.1651
             LEVELS=2,3,5.            4 DENTIST                71526.9  4       17881.7
             RANDOM=DENTIST.          5 AM        AMD          62453.4  2       31226.7   2.66   0.1298
             FIXED=ALLOY,METHOD.      6 AD                      9514.5  4        2378.6
             MODEL='A, M, D'.         7 MD                     87218.5  8       10902.3
/END                                  8 AMD                    93788.9  8       11723.6
792 803 715 673 634
772 752 792 657 649                     SOURCE   EXPECTED MEAN          ESTIMATES OF
782 715 762 690 724                              SQUARE                VARIANCE COMPONENTS
824 803 724 946 715                   1 MEAN     30(1)+6(4)             519629.54722
772 772 715 743 724                   2 ALLOY    15(2)+3(6)              -81.69444
803 707 606 245 627                   3 METHOD   10(3)+2(7)             1389.69167
ANALYSIS OF VARIANCE DESIGN           4 DENTIST  6(4)                   2980.28611
INDEX           A   M   D             5 AM       5(5)+(8)               3900.61667
NUMBER OF LEVELS 2   3   5            6 AD       3(6)                    792.87222
POPULATION SIZE  2   3  INF           7 MD       2(7)                   5451.15833
MODEL       A,  M,  D                 8 AMD      (8)                   11723.61667
```

(iii) BMDP application: BMDP 8V instructions and output for the three-way mixed effects analysis of variance.

FIGURE 5.3 *(continued)*

EXERCISES

1. A study was performed on 18 oxides and 18 hydroxides to compare the effect of corrosion on three metal products. Six samples from each oxide and each hydroxide of each metal product were assigned at random into six groups of three each. Three groups from each oxide and hydroxide had density measurements taken after a test period of 30 hours and the other three groups were measured after a test period of 60 hours. The relevant data are given as follows.

	Metal Products					
	Steel		Copper		Zinc	
	Corrosive Element		Corrosive Element		Corrosive Element	
Test Period (hrs)	O_2	OH	O_2	OH	O_2	OH
30	159	134	168	143	139	122
	135	154	139	145	134	127
	125	148	167	169	165	117
60	152	120	112	152	126	113
	142	163	152	127	151	146
	124	150	117	151	138	126

(a) Describe the mathematical model and the assumptions for the experiment.
(b) Analyze the data and report the analysis of variance table.

(c) Test whether there are differences in the effect of corrosion among the three metal products. Use $\alpha = 0.05$.
(d) Test whether there are differences in the effect of corrosion between the two test periods. Use $\alpha = 0.05$.
(e) Test whether there are differences in the effect of corrosion between oxides and hydroxides. Use $\alpha = 0.05$.
(f) Test the significance of different interaction effects. Use $\alpha = 0.05$.

2. The percentage of silicon carbide (SiC) concentration in an aluminum-silicon carbide (Al-SiC) composite, the fusion temperature, and the casting time of Al-SiC are being investigated for their effects on the strength of Al-SiC. Three levels of SiC concentration, three levels of fusion temperature, and two casting times are selected. A factorial experiment with two replicates was conducted and the following data in certain standard units (Mpa) were obtained.

	Casting Time					
	1			2		
Silicon Carbide	Temperature (°C)			Temperature (°C)		
Concentration (%)	1300	1400	1500	1300	1400	1500
10	186.6	187.7	189.8	188.4	189.6	190.6
	186.0	186.0	189.4	188.6	190.4	190.9
15	188.5	186.0	188.4	187.5	188.7	189.6
	187.2	186.9	187.6	188.1	188.0	189.0
20	187.5	185.6	187.4	187.6	187.0	188.5
	186.6	186.2	188.1	189.4	187.8	189.8

(a) State the model and the assumptions for the experiment. All factors may be regarded as fixed effects.
(b) Analyze the data and report the analysis of variance table.
(c) Test whether there are differences in the strength of the Al-SiC composite due to the casting time. Use $\alpha = 0.05$.
(d) Test whether there are differences in the strength of the Al-SiC composite due to the temperature. Use $a = 0.05$.
(e) Test whether there are differences in the strength of the Al-SiC composite due to the percentage of silicon carbide concentration. Use $\alpha = 0.05$.
(f) Test the significance of different interaction effects. Use $\alpha = 0.05$.

3. The production management department of a textile factory is studying the effect of several factors on the color of garments used to manufacture ladies' dresses. Three machinists, three production circuit times, and two relative humidities were selected and three small samples of garments were colored under each set of conditions. The completed garment was compared to a standard and a range of scale

was assigned. The data in certain standard units are given as follows.

Production Circuit Time	Relative Humidity					
	Low			High		
	Machinist			Machinist		
	1	2	3	1	2	3
40	28	32	36	29	43	39
	29	33	37	28	41	41
	30	31	34	33	40	44
50	41	39	38	42	39	39
	40	43	39	44	43	41
	41	44	40	40	41	36
60	33	40	31	31	41	33
	29	40	32	34	42	31
	32	39	30	30	39	29

(a) State the model and the assumptions for the experiment.
(b) Analyze the data and report the analysis of variance table.
(c) Test whether there are differences in coloring due to relative humidity. Use $\alpha = 0.05$.
(d) Test whether there are differences in coloring due to machinist. Use $\alpha = 0.05$.
(e) Test whether there are differences in coloring due to production circuit time. Use $\alpha = 0.05$.
(f) Test the significance of different interaction effects. Use $\alpha = 0.05$.
(g) Do exact tests exist for all effects? If not, use the pseudo-F tests discussed in this chapter.

4. Consider the following data from a factorial experiment involving three factors A, B, and C, where all factors are considered as having fixed effects.

	C_1			C_2			C_3		
	B_1	B_2	B_3	B_1	B_2	B_3	B_1	B_2	B_3
A_1	20.1	19.7	20.8	21.7	19.3	18.3	20.7	20.4	24.1
	33.4	18.5	14.7	20.3	17.8	16.7	19.2	18.6	18.6
	27.2	17.3	18.5	19.4	18.1	15.2	18.1	17.5	16.2
A_2	26.4	22.3	21.2	23.8	20.3	17.3	17.6	22.4	12.7
	19.5	20.4	19.6	22.4	22.1	18.5	19.1	20.7	16.3
	23.3	19.3	18.3	21.2	23.5	20.3	20.8	19.4	17.1

(a) State the model and the assumptions for the experiment.
(b) Analyze the data and report the analysis of variance table.
(c) Carry out tests of significance on all interactions. Use $\alpha = 0.05$.
(d) Carry out tests of significance on the main effects. Use $\alpha = 0.05$.
(e) Give an illustration of how a significant interaction has masked the effect of factor C.

5. Consider a factorial experiment involving three factors A, B, and C, and assume a three-way fixed effects model of the form

$$y_{ijk\ell} = \mu + \alpha_i + \beta_j + \gamma_k + (\beta\gamma)_{jk} + e_{ijk\ell}$$
$$(i = 1, 2, 3, 4; j = 1, 2; k = 1, 2, 3).$$

It is assumed that all other interactions are either nonexistent or negligible. The relevant data are given as follows.

	B_1			B_2		
	C_1	C_2	C_3	C_1	C_2	C_3
A_1	5.0	4.5	4.8	5.5	4.2	4.2
	5.8	5.2	5.4	4.3	4.4	4.8
A_2	4.7	3.8	4.2	3.8	3.8	4.8
	4.8	4.2	4.3	4.1	4.2	5.3
A_3	5.7	4.5	4.7	4.5	3.9	3.9
	5.8	4.7	5.5	4.7	4.5	4.6
A_4	4.5	4.3	4.2	2.3	3.8	4.5
	4.3	3.7	4.4	4.4	4.3	5.4

(a) State the model and the assumptions for the experiment.
(b) Analyze the data and report the analysis of variance table.
(c) Perform a test of significance on the $B \times C$ interaction. Use $\alpha = 0.05$.
(d) Perform tests of significance on the main effects, A, B, and C, using a pooled error mean square. Use $\alpha = 0.05$.
(e) Are two observations for each treatment combination sufficient if the power of the test for detecting differences among the levels of a factor C at the 0.05 level of significance is to be at least 0.8 when $\gamma_1 = -0.2$, $\gamma_2 = 0.4$, and $\gamma_3 = -0.2$? Use the same pooled estimate of σ_e^2 as obtained in the analysis of variance.

6. Ostle (1952) discussed the results of an analysis of variance of a three-factor factorial design. The experiment consisted of determining soluble matter in four extract solutions by pippetting in duplicate 25, 50, and 100 ml volumes of solution into dishes. The solution was evaporated and weighed for residues. The experiment was replicated by repeating it on three days. The researcher is interested in only the four extracts and only the three volumes used in the experiment. However, the days are considered to be a random sample of days. The mathematical model for this experiment would be:

$$y_{ijk\ell} = \mu + \alpha_i + \beta_j + \gamma_k + (\alpha\beta)_{ij} + (\alpha\gamma)_{ik} + (\beta\gamma)_{jk} + (\alpha\beta\gamma)_{ijk} + e_{ijk\ell} \quad \begin{cases} i = 1, 2, 3, 4 \\ j = 1, 2, 3 \\ k = 1, 2, 3 \\ \ell = 1, 2, \end{cases}$$

where μ is the general mean, α_j is the effect of the i-th extract, β_j is the effect of the j-th volume, γ_k is the effect of the k-th day, $(\alpha\beta)_{ij}$ is the interaction of the i-th extract with the j-th volume, $(\alpha\gamma)_{ik}$ is the interaction of the i-th extract with the k-th day, $(\beta\gamma)_{jk}$ is the interaction of the j-th volume with the k-th day, and $(\alpha\beta\gamma)_{ijk}$ is the interaction of the i-th extract with the j-th volume and the k-th day, and $e_{ijk\ell}$ is the customary error term. Furthermore, it is assumed that the α_i's, β_j's, and $(\alpha\beta)_{ij}$'s are constants subject to the restrictions:

$$\sum_{i=1}^{4} \alpha_i = \sum_{j=1}^{3} \beta_j = 0,$$

$$\sum_{i=1}^{4} (\alpha\beta)_{ij} = \sum_{j=1}^{3} (\alpha\beta)_{ij} = 0;$$

and the γ_k's, $(\alpha\gamma)_{ik}$'s, $(\beta\gamma)_{jk}$'s, and $e_{ijk\ell}$'s are independently and normally distributed with mean zero and variances σ_γ^2, $\sigma_{\alpha\gamma}^2$, $\sigma_{\beta\gamma}^2$, and σ_e^2, respectively; and subject to the restrictions:

$$\sum_{i=1}^{4} (\alpha\gamma)_{ik} = \sum_{j=1}^{3} (\beta\gamma)_{jk} = 0 \quad (k = 1, 2, 3).$$

The analysis of variance computations of the data (not reported here) are carried out exactly as in Section 5.16 and the results are given as follows.

Source of Variation	Degrees of Freedom	Mean Square	Expected Mean Square
Extract	3	161.5964	$\sigma_e^2 + 3 \times 2\sigma_{\alpha\gamma}^2 + \frac{3 \times 3 \times 2}{(4-1)} \sum_{i=1}^{4} \alpha_i^2$
Volume	2	0.02443	$\sigma_e^2 + 4 \times 2\sigma_{\beta\gamma}^2 + \frac{4 \times 3 \times 2}{(3-1)} \sum_{j=1}^{3} \beta_j^2$
Day	2	0.07535	$\sigma_e^2 + 4 \times 3 \times 2\sigma_\gamma^2$
Extract × Volume	6	0.00772	$\sigma_e^2 + 2\sigma_{\alpha\beta\gamma}^2 + \frac{3 \times 2}{(4-1)(3-1)} \sum_{i=1}^{4}\sum_{j=1}^{3} (\alpha\beta)_{ij}^2$
Extract × Day	6	0.03959	$\sigma_e^2 + 3 \times 2\sigma_{\alpha\gamma}^2$
Volume × Day	4	0.01501	$\sigma_e^2 + 4 \times 2\sigma_{\beta\gamma}^2$
Extract × Volume × Day	12	0.00654	$\sigma_e^2 + 2\sigma_{\alpha\beta\gamma}^2$
Error	36	0.00565	σ_e^2
Total	71		

Source: Ostle (1952). Used with permission.

(a) Test whether there are differences in soluble matter among different extracts. Use $\alpha = 0.05$.
(b) Test whether there are differences in soluble matter among different volumes. Use $\alpha = 0.05$.
(c) Test whether there are differences in soluble matter among different days. Use $\alpha = 0.05$.
(d) Test for the following interaction effects: extract × volume, extract × day, volume × day, and extract × volume × day. Use $\alpha = 0.05$.
(e) Determine appropriate point and interval estimates for each of the variance components of the model.
(f) It is found that the effects due to day, extract × day, and volume × day are all significant; that is, the method is unreliable in that the differences among volumes will not be the same on different days. Give one possible explanation of the excessive variation among days. How might this difficulty be overcome?

7. Damon and Harvey (1987, p. 316) reported data from an experiment with radishes involving a three-factor factorial design. There were two sources of nitrogen — ammonium sulfate and potassium nitrate, three levels of nitrogen, and two levels of treatments — nitrapyrin and no nitrapyrin. The following data are given where four observations are the fresh weights of the plants in grams/pot.

	Source of Nitrogen			
	Ammonium Sulfate		Potassium Nitrate	
Level	Nitrapyrin	No Nitrapyrin	Nitrapyrin	No Nitrapyrin
1	14.3	17.5	17.6	13.9
	15.9	16.7	24.0	16.8
	14.8	15.7	13.5	17.3
	20.8	15.1	17.9	12.6
2	37.5	39.6	43.3	45.8
	29.4	33.0	53.5	46.9
	33.8	52.8	49.3	48.0
	33.1	36.2	49.9	47.0
3	41.4	52.5	59.8	77.8
	49.9	53.4	98.4	87.6
	43.2	51.7	79.4	83.4
	40.1	52.2	80.0	84.7

Source: Damon and Harvey (1987, p. 316). Used with permission.

(a) State the model and the assumptions for the experiment. Consider it to be a fixed effects model.
(b) Analyze the data and report the analysis of variance table.
(c) Test whether there are differences in weights due to source of nitrogen. Use $\alpha = 0.05$.

(d) Test whether there are differences in weights due to levels of nitrogen. Use $\alpha = 0.05$.
(e) Test whether there are differences in weights due to levels of treatment. Use $\alpha = 0.05$.
(f) Test the significance of different interaction effects. Use $\alpha = 0.05$.

8. Scheffé (1959, pp. 145–146) reported data from an experiment conducted by Otto Dykstra, Jr., at the Research Center of the General Food Corporation, to study variation in moisture content of a certain food product. A four-factor factorial design, involving three kinds of salt, three amounts of salt, two amounts of acid, and two types of additives, was used. The moisture content (in grams) of samples in the experimental stage were recorded and the data given as follows.

		Amount of Acid			
		1		2	
Kind of Salt	Amount of Salt	Type of Additive		Type of Additive4	
		1	2	1	2
1	1	8	5	8	4
	2	17	11	13	10
	3	22	16	20	15
2	1	7	3	10	5
	2	26	17	24	19
	3	34	32	34	29
3	1	10	5	9	4
	2	24	14	24	16
	3	39	33	36	34

Source: Scheffé, (1959, p. 145). Used with permission.

(a) State the model and the assumptions for the experiment. Consider it to be a fixed effects model. Since there is no replication, one must think what error term to use.
(b) Analyze the data as the full factorial model and report the analysis of variance table.
(c) Test whether there are differences in moisture content due to kind of salt. Use $\alpha = 0.05$.
(d) Test whether there are differences in moisture content due to amount of salt. Use $\alpha = 0.05$.
(e) Test whether there are differences in moisture content due to amount of acid. Use $\alpha = 0.05$.
(f) Test whether there are differences in moisture content due to type of additives. Use $\alpha = 0.05$.
(g) Test the significance of different interaction effects. Use $\alpha = 0.05$.

9. Davies (1956, p. 275) reported data from a 2^4 factorial experiment designed to investigate the effect of acid strength, time of reaction, amount of acid, and temperature of reaction on the yield of an isatin derivative. The following data are given where observations are the yield of an isatin derivative measured in gms/100 gm of base material.

Acid Strength	Reaction Time	Temperature of Reaction			
		60°C		70°C	
		Amount of Acid		Amount of Acid	
		35 ml	45 ml	35 ml	45 ml
87%	15 min	6.08	6.31	6.79	6.77
	30 min	6.53	6.12	6.73	6.49
93%	15 min	6.04	6.09	6.68	6.38
	30 min	6.43	6.36	6.08	6.23

Source: Davies (1956, p. 275). Used with permission.

(a) State the model and the assumptions for the experiment. Consider it to be a fixed effects model.
(b) Analyze the data assuming that three- and four-factor interactions are negligible and report the analysis of variance table. (Davies stated that on technical grounds, the existence of three- and four-factor intereactions is unlikely.)
(c) Test whether there are differences in yield due to acid strength. Use $\alpha = 0.05$.
(d) Test whether there are differences in yield due to reaction time. Use $\alpha = 0.05$.
(e) Test whether there are differences in yield due to amount of acid. Use $\alpha = 0.05$.
(f) Test whether there are differences in yield due to temperature of reaction. Use $\alpha = 0.05$.
(g) Test the significance of two-factor interaction effects. Use $\alpha = 0.05$.

6 Two-Way Nested (Hierarchical) Classification

6.0 PREVIEW

In Chapters 3 through 5 we considered analysis of variance for experiments commonly referred to as crossed classifications. In a crossed-classification, data cells are formed by combining of each level of one factor with each level of every other factor. We now consider experiments involving two factors such that the levels of one factor occur only within the levels of another factor. Here, the levels of a given factor are all different across the levels of the other factor. More specifically, given two factors A and B, the levels of B are said to be nested within the levels of A, or more briefly B is nested within A, if every level of B appears with only a single level of A in the observations. This means that if the factor A has a levels, then the levels of B fall into a sets of $b_1, b_2, \ldots, b_a$ levels, respectively, such that the i-th set appears with the i-th level of A. These designs are commonly known as nested or hierarchical designs where the levels of factor B are nested within the levels of factor A.

For example, suppose an industrial firm procures a certain liquid chemical from three different locations. The firm wishes to investigate if the strength of the chemical is the same from each location. There are four barrels of chemicals available from each location and three measurements of strength are to be made from each barrel. The physical layout can be schematically represented as in Figure 6.1. This is a two-way nested or hierarchical design, with barrels nested within locations. In the first instance, one may ask why the two factors, locations and barrels, are not crossed. If the factors were crossed, then barrel 1 would always refer to the same barrel, barrel 2 would always refer to the same barrel, and so on. In this example, this is clearly not the situation since the barrels from each location are unique for that particular location. Thus, barrel 1 from location I has no relation to barrel 2 from any other location, and so on. To emphasize the point that barrels from each location are different barrels, we may recode the barrels as 1, 2, 3, and 4 from location I; 5, 6, 7, and 8 from location II; and 9, 10, 11, and 12 from location III. For another example, suppose that in order to study a certain characteristic of a product, samples of size 3 are taken from each of four spindles within each of three machines. Here, each

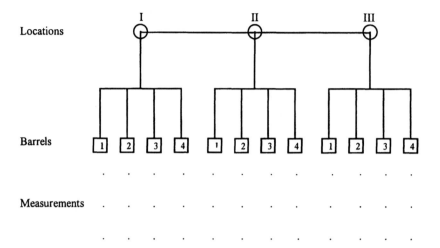

FIGURE 6.1 A Layout for the Two-Way Nested Design Where Barrels Are Nested within Locations.

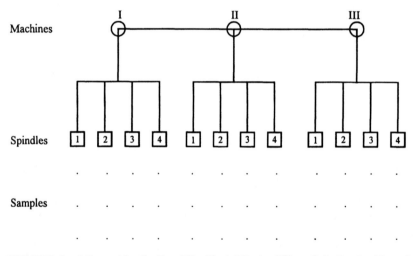

FIGURE 6.2 A Layout for the Two-Way Nested Design Where Spindles Are Nested within Machines.

separate spindle appears within a single machine and thus spindles are nested within machines. Again, the layout may be depicted as shown in Figure 6.2.

Both nested and crossed factors can occur in an experimental design. When each of the factors in an experiment is progressively nested within the preceding factor, it is called a completely or hierarchically nested design. Such nested designs are common in many fields of study and are particularly popular in surveys and industrial experiments similar to the ones previously described. In this

TABLE 6.1
Data for a Two-Way Nested Classification

A_1			A_2			...	A_a		
B_{11}	B_{12} ...	B_{1b}	B_{21}	B_{22} ...	B_{2b}	...	B_{a1}	B_{a2} ...	B_{ab}
y_{111}	y_{121} ...	y_{1b1}	y_{211}	y_{221} ...	y_{2b1}	...	y_{a11}	y_{a21} ...	y_{ab1}
y_{112}	y_{122} ...	y_{1b2}	y_{212}	y_{222} ...	y_{2b2}	...	y_{a12}	y_{a22} ...	y_{ab2}
.	.	.	.	.	.		.	.	.
.	.	.	.	.	.	...	.	.	.
y_{11k}	y_{12k} ...	y_{1bk}	y_{21k}	y_{22k} ...	y_{2bk}	...	y_{a1k}	y_{a2k} ...	y_{abk}
.	.	.	.	.	.		.	.	.
.	.	.	.	.	.	...	.	.	.
y_{11n}	y_{12n} ...	y_{1bn}	y_{21n}	y_{22n} ...	y_{2bn}	...	y_{a1n}	y_{a2n} ...	y_{abn}

and the following chapter, we consider only the hierarchically nested designs. The experiments involving both nested and crossed factors are considered in Chapter 8.

Remarks: (i) To distinguish the crossed-classification from the nested-classification, we can say that factors are crossed if neither is nested within the other.

(ii) When uncertain as to whether a factor is crossed or nested, try to renumber the levels of each factor. If the levels of the factor can be renumbered arbitrarily, then the factor is considered nested.

(iii) The nesting of factors can also occur in an experiment when the procedure restricts the randomization of factor-level combinations.

(iv) The experiments involving one-way classification can be thought of as one-way nested classification, where a factor corresponding to "replications" is nested within the main treatment factor of the experiment.

6.1 MATHEMATICAL MODEL

Consider two factors A and B having a and b levels, respectively. Let the b levels of factor B be nested under each level of factor A and let there be n replicates at each level of factor B. Let y_{ijk} be the observed score corresponding to the i-th level of factor A, j-th level of factor B nested within the i-th level of factor A, and the k-th replicate within the j-th level of B. The data involving the total of $N = a \times b \times n$ scores y_{ijk}'s can then be schematically represented as in Table 6.1.

The analysis of variance model for this type of experimental layout is taken as

$$y_{ijk} = \mu + \alpha_i + \beta_{j(i)} + e_{k(ij)} \quad \begin{cases} i = 1, 2, \ldots, a \\ j = 1, 2, \ldots, b \\ k = 1, 2, \ldots, n, \end{cases} \quad (6.1.1)$$

where $-\infty < \mu < \infty$ is the overall mean, α_i is the effect due to the i-th level of factor A, $\beta_{j(i)}$ is the effect due to the j-th level of factor B nested within the i-th level of factor A, and $e_{k(ij)}$ is the error term that takes into account the random variation within a particular cell. The subscript $j(i)$ means that the j-th level of B is nested within the i-th level of A. Note that in the nested model (6.1.1) the main effects for factor B are missing since the different levels of factor B are not the same for different levels of factor A. Furthermore, note that the model has no interaction term between A and B since every level of factor B does not appear with every level of factor A.

6.2 ASSUMPTIONS OF THE MODEL

For all models given by (6.1.1), it is assumed that the errors $e_{k(ij)}$'s are uncorrelated and randomly distributed with mean zero and variance σ_e^2. Other assumptions depend on whether the levels of A and B are fixed or random. If both factors A and B are fixed, we assume that

$$\sum_{i=1}^{a} \alpha_i = \sum_{j=1}^{b} \beta_{j(i)} = 0 \quad (i = 1, 2, \ldots, a).$$

That is, the A factor effects sum to zero and the B factor effects sum to zero within each level of A. Alternatively, if both A and B are random, then we assume that the α_i's and $\beta_{j(i)}$'s are mutually and completely uncorrelated and are randomly distributed with mean zero and variances σ_α^2 and σ_β^2, respectively. Mixed models with A fixed and B random or A random and B fixed are also widely used and will have analogous assumptions. For example, if we assume that A is fixed and B is random, then $\beta_{j(i)}$'s are uncorrelated and randomly distributed with mean zero and variance σ_β^2 subject to the restriction that $\sum_{i=1}^{a} \alpha_i = 0$.

6.3 ANALYSIS OF VARIANCE

Starting with the identity

$$y_{ijk} - \bar{y}_{...} = (\bar{y}_{i..} - \bar{y}_{...}) + (\bar{y}_{ij.} - \bar{y}_{i..}) + (y_{ijk} - \bar{y}_{ij.}), \tag{6.3.1}$$

squaring both sides, and summing over i, j, and k, the total sum of squares can be partitioned as

$$\sum_{i=1}^{a}\sum_{j=1}^{b}\sum_{k=1}^{n}(y_{ijk} - \bar{y}_{...})^2 = bn\sum_{i=1}^{a}(\bar{y}_{i..} - \bar{y}_{...})^2 + n\sum_{i=1}^{a}\sum_{j=1}^{b}(\bar{y}_{ij.} - \bar{y}_{i..})^2$$

$$+ \sum_{i=1}^{a}\sum_{j=1}^{b}\sum_{k=1}^{n}(y_{ijk} - \bar{y}_{ij.})^2, \tag{6.3.2}$$

with usual notations of dots and bars. The relation (6.3.2) is valid since the cross-product terms are equal to zero. The identity (6.3.2) states that the total sum of squares can be partitioned into the following components:

(i) a sum of squares due to factor A,
(ii) a sum of squares due to factor B within the levels of A, and
(iii) a sum of squares due to the residual or error.

As before, the equation (6.3.2) may be written symbolically as

$$SS_T = SS_A + SS_{B(A)} + SS_E. \qquad (6.3.3)$$

There are $abn - 1$ degrees of freedom for SS_T, $a - 1$ degrees of freedom for SS_A, $a(b - 1)$ degrees of freedom for $SS_{B(A)}$, and $ab(n - 1)$ degrees of freedom for the error. Note that $(a - 1) + a(b - 1) + ab(n - 1) = abn - 1$. The mean squares obtained by dividing each sum of squares on the right side of (6.3.3) are denoted by MS_A, $MS_{B(A)}$, and MS_E, respectively. The expected mean squares can be derived as before. The traditional analysis of variance summarizing the results of partitioning of the total sum of squares and degrees of freedom and that of the expected mean squares is shown in Table 6.2.

6.4 TESTS OF HYPOTHESES: THE ANALYSIS OF VARIANCE F TESTS

Under the assumption of normality, the mean squares, MS_A, $MS_{B(A)}$, and MS_E are independently distributed chi-square variables such that the ratio of any two mean squares is distributed as the variance ratio F. Table 6.2 suggests that if the levels of A and B are fixed, then the hypotheses

$$H_0^A : \text{all } \alpha_i = 0 \quad \text{versus} \quad H_1^A : \text{all } \alpha_i \neq 0$$

and

$$H_0^B : \text{all } \beta_{j(i)} = 0 \quad \text{versus} \quad H_1^B : \text{all } \beta_{j(i)} \neq 0$$

can be tested by the ratios

$$F_A = MS_A/MS_E$$

and

$$F_B = MS_{B(A)}/MS_E,$$

respectively. It can be readily shown that under the null hypotheses H_0^A and H_0^B, we have

$$F_A \sim F[a - 1, ab(n - 1)]$$

TABLE 6.2
Analysis of Variance for Model (6.1.1)

Source of Variation	Degrees of Freedom	Sum of Squares	Mean Square	Expected Mean Square			
				Model I	Model II	Model III	
				A Fixed, B Fixed	A Random, B Random	A Fixed, B Random	A Random, B Fixed
Due to A	$a-1$	SS_A	MS_A	$\sigma_e^2 + \dfrac{bn}{a-1}\sum_{i=1}^{a}\alpha_i^2$	$\sigma_e^2 + n\sigma_\beta^2 + bn\sigma_\alpha^2$	$\sigma_e^2 + n\sigma_\beta^2 + \dfrac{bn}{a-1}\sum_{i=1}^{a}\alpha_i^2$	$\sigma_e^2 + bn\sigma_\alpha^2$
B within A	$a(b-1)$	$SS_{B(A)}$	$MS_{B(A)}$	$\sigma_e^2 + \dfrac{n}{a(b-1)}\sum_{i=1}^{a}\sum_{j=1}^{b}\beta_{j(i)}^2$	$\sigma_e^2 + n\sigma_\beta^2$	$\sigma_e^2 + n\sigma_\beta^2$	$\sigma_e^2 + \dfrac{n}{a(b-1)}\sum_{i=1}^{a}\sum_{j=1}^{b}\beta_{j(i)}^2$
Error	$ab(n-1)$	SS_E	MS_E	σ_e^2	σ_e^2	σ_e^2	σ_e^2
Total	$abn-1$	SS_T					

and
$$F_B \sim F[a(b-1), ab(n-1)].$$

Thus, H_0^A and H_0^B are tested using the test statistics F_A and F_B respectively. Similarly, if both A and B are random factors, we test[1]

$$H_0^A : \sigma_\alpha^2 = 0 \quad \text{versus} \quad H_1^A : \sigma_\alpha^2 > 0$$

by

$$F_A = \text{MS}_A/\text{MS}_{B(A)}$$

and

$$H_0^B : \sigma_\beta^2 = 0 \quad \text{versus} \quad H_1^B : \sigma_\beta^2 > 0$$

by

$$F_B = \text{MS}_{B(A)}/\text{MS}_E.$$

Finally, if A is a fixed factor and B is random, then

$$H_0^A : \text{all } \alpha_i = 0 \quad \text{versus} \quad H_1^A : \text{all } \alpha_i \neq 0$$

is tested by

$$F_A = \text{MS}_A/\text{MS}_{B(A)}$$

and

$$H_0^B : \sigma_\beta^2 = 0 \quad \text{versus} \quad H_1^B : \sigma_\beta^2 > 0$$

is tested by

$$F_B = \text{MS}_{B(A)}/\text{MS}_E.$$

Remark: If one does not reject the null hypothesis $H_0^B : \sigma_\beta^2 = 0$, then $\text{MS}_{B(A)}$ might be considered to estimate the same population variance as does MS_E. Thus, as remarked earlier in Section 4.6, some authors recommend computing a pooled mean square by

---

[1] For some results on approximate tests for other hypotheses concerning σ_α^2 and σ_β^2, involving a non zero value and one or two-sided alternatives, see Hartung and Voet (1987) and Burdick and Graybill (1992, pp. 82–83).

pooling the sums of squares $SS_{B(A)}$ and SS_E and the corresponding degrees of freedom $a(b-1)$ and $ab(n-1)$ in Table 6.2. This will theoretically provide a more powerful test for differences due to levels of factor A. However, as indicated before, there is no widespread agreement on this matter and care should be exercised in resorting to pooling procedure.

6.5 POINT ESTIMATION

In this section, we present results on point estimation for parameters of interest under fixed, random, and mixed effects models.

MODEL I (FIXED EFFECTS)

If both factors A and B are fixed, the parameters μ, α_i's, and $\beta_{j(i)}$'s may be estimated by the least squares procedure by minimizing the quantity

$$Q = \sum_{i=1}^{a}\sum_{j=1}^{b}\sum_{k=1}^{n}(y_{ijk} - \mu - \alpha_i - \beta_{j(i)})^2 \qquad (6.5.1)$$

with respect to μ, α_i, and $\beta_{j(i)}$; and subject to the restrictions:

$$\sum_{i=1}^{a}\alpha_i = \sum_{j=1}^{b}\beta_{j(i)} = 0 \quad (i = 1, 2, \ldots, a). \qquad (6.5.2)$$

It can be readily shown by the methods of elementary calculus that the resulting solutions are:

$$\hat{\mu} = \bar{y}_{\ldots} \qquad (6.5.3)$$

$$\hat{\alpha}_i = \bar{y}_{i..} - \bar{y}_{\ldots}, \quad i = 1, 2, \ldots, a \qquad (6.5.4)$$

and

$$\hat{\beta}_{j(i)} = \bar{y}_{ij.} - \bar{y}_{i..}, \quad i = 1, 2, \ldots, a; \quad j = 1, 2, \ldots, b. \qquad (6.5.5)$$

It should be observed that the estimators (6.5.3) through (6.5.5) have considerable intuitive appeal; the A treatment effects are estimated by the average of all observations under each level of A minus the grand mean, and the B treatment effects within each level of A are estimated by the corresponding cell average minus the average under the level of A.

The estimators (6.5.3) through (6.5.5) have variances given by

$$\text{Var}(\hat{\mu}) = \sigma_e^2/abn, \qquad (6.5.6)$$

$$\text{Var}(\hat{\alpha}_i) = (a-1)\sigma_e^2/abn, \qquad (6.5.7)$$

and
$$\operatorname{Var}(\hat{\beta}_{j(i)}) = (b-1)\sigma_e^2/bn. \qquad (6.5.8)$$

The other parameters of interest may include: $\mu + \alpha_i$ (means of factor level A), pairwise differences $\alpha_i - \alpha_{i'}$, $\beta_{j(i)} - \beta_{j'(i)}$, and the contrasts of the type

$$\sum_{i=1}^{a} \ell_i \alpha_i \left(\sum_{i=1}^{a} \ell_i = 0 \right) \quad \text{and} \quad \sum_{j=1}^{b} \ell'_j \beta_{j(i)} \left(\sum_{j=1}^{b} \ell'_j = 0 \right).$$

Their respective estimates along with the variances are:

$$\widehat{\mu + \alpha_i} = \bar{y}_{i..}, \qquad \operatorname{Var}(\widehat{\mu + \alpha_i}) = \sigma_e^2/bn; \qquad (6.5.9)$$

$$\widehat{\alpha_i - \alpha_{i'}} = \bar{y}_{i..} - \bar{y}_{i'..}, \qquad \operatorname{Var}(\widehat{\alpha_i - \alpha_{i'}}) = 2\sigma_e^2/bn; \qquad (6.5.10)$$

$$\widehat{\beta_{j(i)} - \beta_{j'(i)}} = \bar{y}_{ij.} - \bar{y}_{ij'.}, \qquad \operatorname{Var}(\widehat{\beta_{j(i)} - \beta_{j'(i)}}) = 2\sigma_e^2/n; \qquad (6.5.11)$$

$$\sum_{i=1}^{a} \widehat{\ell_i \alpha_i} = \sum_{i=1}^{a} \ell_i \bar{y}_{i..}, \qquad \operatorname{Var}\left(\sum_{i=1}^{a} \widehat{\ell_i \alpha_i} \right) = \left(\sum_{i=1}^{a} \ell_i^2 \right) \sigma_e^2/bn; \qquad (6.5.12)$$

and

$$\sum_{j=1}^{b} \widehat{\ell'_j \beta_{j(i)}} = \sum_{j=1}^{b} \ell'_j \bar{y}_{ij.}, \qquad \operatorname{Var}\left(\sum_{j=1}^{b} \widehat{\ell'_j \beta_{j(i)}} \right) = \left(\sum_{j=1}^{b} \ell_j'^2 \right) \sigma_e^2/n. \qquad (6.5.13)$$

The best quadratic unbiased estimator of σ_e^2 is, of course, provided by the error mean square.

MODEL II (RANDOM EFFECTS)

When both factors A and B are random, the analysis of variance method can be used to estimate the variance components $\sigma_e^2, \sigma_\beta^2$, and σ_α^2. From the expected mean squares column under Model II of Table 6.2, we obtain

$$\hat{\sigma}_e^2 = \text{MS}_E, \qquad (6.5.14)$$

$$\hat{\sigma}_\beta^2 = (\text{MS}_{B(A)} - \text{MS}_E)/n, \qquad (6.5.15)$$

and

$$\hat{\sigma}_\alpha^2 = (\text{MS}_A - \text{MS}_{B(A)})/bn. \qquad (6.5.16)$$

The optimal properties of the analysis of variance estimators discussed in Section 2.9 also apply here. However, again $\hat{\sigma}_\beta^2$ and $\hat{\sigma}_\alpha^2$ can produce negative estimates.[2] A negative estimate may provide an indication that the corresponding variance component may be zero. One then might want to replace any negative estimates by zero, pool the adjacent mean squares, and subtract the pooled mean square from the next higher mean square for estimating the corresponding variance component. Finally, we may note that a minimum variance unbiased estimator for $\sigma_\beta^2/\sigma_e^2$ is given by

$$\frac{1}{n}\left[\frac{\text{MS}_{B(A)}}{\text{MS}_E}\left(1 - \frac{2}{ab(n-1)}\right) - 1\right]. \qquad (6.5.17)$$

MODEL III (MIXED EFFECTS)

Many applications of this design involve a mixed model with the main factor A fixed and the nested factor B random. For a mixed model situation, the fixed effects α_i's are estimated by

$$\hat{\alpha}_i = \bar{y}_{i..} - \bar{y}_{...}, \quad i = 1, 2, \ldots, a;$$

and the variance components σ_e^2 and σ_β^2 can be estimated by eliminating the line corresponding to A from the mean square column of Table 6.2 and applying the analysis of variance method to the next two lines. Table 6.3 summarizes the results on point estimates of some common parameters of interest.

6.6 INTERVAL ESTIMATION

In this section, we summarize results on confidence intervals for parameters of interest under fixed, random, and mixed effects models.

MODEL I (FIXED EFFECTS)

An exact confidence interval for σ_e^2 can be based on the chi-square distribution of $ab(n-1)\text{MS}_E/\sigma_e^2$. Thus, a $1 - \alpha$ level confidence interval for σ_e^2 is

$$\frac{ab(n-1)\text{MS}_E}{\chi^2[ab(n-1), 1-\alpha/2]} < \sigma_e^2 < \frac{ab(n-1)\text{MS}_E}{\chi^2[ab(n-1), \alpha/2]}. \qquad (6.6.1)$$

Furthermore, confidence intervals based on the t distribution for a particular treatment or factor level mean $\mu + \alpha_i$ or α_i can be readily obtained. For example,

$$\bar{y}_{i..} - t[ab(n-1), 1-\alpha/2]\sqrt{\frac{\text{MS}_E}{bn}} < \mu + \alpha_i$$

$$< \bar{y}_{i..} + t[ab(n-1), 1-\alpha/2]\sqrt{\frac{\text{MS}_E}{bn}}$$

---

[2] For a discussion of the maximum likelihood and other nonnegative estimation procedures and their properties, the reader is referred to Sahai (1974b, 1976).

TABLE 6.3
Point Estimates and Their Variances under Model III

Parameter	Point Estimate	Variance of Estimate
μ	$\bar{y}_{...}$	$(\sigma_e^2 + n\sigma_\beta^2)/abn$
$\mu + \alpha_i$	$\bar{y}_{i..}$	$(\sigma_e^2 + n\sigma_\beta^2)/bn$
α_i	$\bar{y}_{i..} - \bar{y}_{...}$	$(a-1)(\sigma_e^2 + n\sigma_\beta^2)/abn$
$\alpha_i - \alpha_{i'}$	$\bar{y}_{i..} - \bar{y}_{i'..}$	$2(\sigma_e^2 + n\sigma_\beta^2)/bn$
$\sum_{i=1}^{a}\ell_i\alpha_i \; \left(\sum_{i=1}^{a}\ell_i = 0\right)$	$\sum_{i=1}^{a}\ell_i\bar{y}_{i..}$	$\left(\sum_{i=1}^{a}\ell_i^2\right)(\sigma_e^2 + n\sigma_\beta^2)/bn$
σ_e^2	MS_E	$2\sigma_e^4/ab(n-1)$
$\sigma_e^2 + n\sigma_\beta^2$	MS_B	$2(\sigma_e^2 + n\sigma_\beta^2)^2/a(b-1)$
σ_β^2	$(MS_{B(A)} - MS_E)/n$	$\dfrac{2}{n^2}\left\{\dfrac{(\sigma_e^2 + n\sigma_\beta^2)^2}{a(b-1)} + \dfrac{\sigma_e^4}{ab(n-1)}\right\}$

and

$$(\bar{y}_{i..} - \bar{y}_{...}) - t[ab(n-1), 1-\alpha/2]\sqrt{\frac{(a-1)MS_E}{abn}} < \alpha_i$$
$$< (\bar{y}_{i..} - \bar{y}_{...}) + t[ab(n-1), 1-\alpha/2]\sqrt{\frac{(a-1)MS_E}{abn}}$$

give exact $1 - \alpha$ level confidence intervals for $\mu + \alpha_i$ and α_i, respectively. Similarly, to obtain confidence limits for $\alpha_i - \alpha_{i'}$, we note from (6.5.10) that

$$E(\bar{y}_{i..} - \bar{y}_{i'..}) = \alpha_i - \alpha_{i'}$$

and

$$\text{Var}(\bar{y}_{i..} - \bar{y}_{i'..}) = 2\sigma_e^2/bn.$$

The confidence limits can, therefore, be derived from the relation

$$\frac{(\bar{y}_{i..} - \bar{y}_{i'..}) - (\alpha_i - \alpha_{i'})}{\sqrt{2MS_E/bn}} \sim t[ab(n-1)]. \tag{6.6.2}$$

Similar results on confidence intervals for $\beta_{j(i)}$'s and any pairwise differences on them, using the t distribution, can also be obtained. However, multiple comparison methods, discussed in Section 6.9, are usually preferable.

MODEL II (RANDOM EFFECTS)

An exact confidence interval for the error variance component σ_e^2 is of course obtained as in (6.6.1). However, exact confidence intervals for the variance components σ_α^2 and σ_β^2 do not exist. One can, nevertheless, construct exact intervals for $\sigma_\beta^2/\sigma_e^2$, $\sigma_e^2/(\sigma_e^2 + \sigma_\beta^2)$, and $\sigma_\beta^2/(\sigma_e^2 + \sigma_\beta^2)$ by using results on the sampling distribution of the ratio of two mean squares. In particular, the probability is $1 - \alpha$ that the interval

$$\left[\frac{1}{n}\left(\frac{\text{MS}_{B(A)}}{\text{MS}_E} \cdot \frac{1}{F[a(b-1), ab(n-1); 1-\alpha/2]} - 1\right),\right.$$
$$\left.\frac{1}{n}\left(\frac{\text{MS}_{B(A)}}{\text{MS}_E} \cdot \frac{1}{F[a(b-1), ab(n-1); \alpha/2]} - 1\right)\right] \quad (6.6.1)$$

captures $\sigma_\beta^2/\sigma_e^2$. Exact intervals on $\sigma_e^2/(\sigma_e^2+\sigma_\beta^2)$ and $\sigma_\beta^2/(\sigma_e^2+\sigma_\beta^2)$ are obtained from (6.6.3) using appropriate transformations.[3] Similarly, from a combination of two confidence intervals, one for σ_e^2 and the other for $\sigma_e^2 + n\sigma_\beta^2$, one can obtain a conservative $100(1 - \alpha)$ percent confidence interval for σ_β^2 as

$$0 < \sigma_\beta^2 < \frac{1}{n}\left[\frac{a(b-1)\text{MS}_{B(A)}}{\chi^2[a(b-1), \alpha/2]} - \frac{ab(n-1)\text{MS}_E}{\chi^2[ab(n-1), 1-\alpha/2]}\right].$$

For the random effects model, one sometimes may also want to determine a confidence interval for μ. To obtain confidence limits for μ, we note that

$$E(\bar{y}_{...}) = \mu$$

and

$$\text{Var}(\bar{y}_{...}) = \frac{\sigma_e^2 + n\sigma_\beta^2 + bn\sigma_\alpha^2}{abn}.$$

Now, $\text{Var}(\bar{y}_{...})$ can be estimated by MS_A/abn, and it follows that

$$\frac{\bar{y}_{...} - \mu}{\sqrt{\text{MS}_A/abn}} \sim t[a-1].$$

The confidence limits for μ can now be determined using the standard normal theory.

---

[3] For some results on confidence intervals for the variance components σ_α^2 and σ_β^2, total variance $\sigma_e^2 + \sigma_\beta^2 + \sigma_\alpha^2$, the ratio of variance components $\sigma_\alpha^2/\sigma_e^2$, proportions of variability $\sigma_e^2/(\sigma_e^2 + \sigma_\alpha^2)$, $\sigma_\alpha^2/(\sigma_e^2+\sigma_\alpha^2)$, $\sigma_e^2/(\sigma_e^2+\sigma_\beta^2+\sigma_\alpha^2)$, $\sigma_\beta^2/(\sigma_e^2+\sigma_\beta^2+\sigma_\alpha^2)$, and $\sigma_\alpha^2/(\sigma_e^2+\sigma_\beta^2+\sigma_\alpha^2)$, see Burdick and Graybill (1992, pp. 80–90).

MODEL III (MIXED EFFECTS)

An exact confidence interval for σ_e^2 is of course given as in (6.6.1). An exact confidence interval for σ_β^2, however, does not exist. One can obtain an exact interval for $\sigma_\beta^2/\sigma_e^2$ by using a procedure based on the statistic $\text{MS}_{B(A)}/\text{MS}_E$. Also, as in the case of Model I, it is possible to obtain confidence intervals for $\mu, \alpha_i, \mu+\alpha_i, \alpha_i - \alpha_{i'}$, the contrast $\sum_{i=1}^{a} \ell_i \alpha_i (\sum_{i=1}^{a} \ell_i = 0)$, or any linear combination of the means $\sum_{i=1}^{a} \ell_i (\mu + \alpha_i)$, where the ℓ_i's are any set of constants. Thus, for example, exact $100(1-\alpha)$ percent confidence intervals for $\mu + \alpha_i$ and $\sum_{i=1}^{a} \ell_i (\mu + \alpha_i)$ are given by

$$\bar{y}_{i..} \pm t[a(b-1), 1-\alpha/2]\sqrt{\frac{\text{MS}_{B(A)}}{bn}}$$

and

$$\sum_{i=1}^{a} \ell_i \bar{y}_{i..} \pm t[a(b-1), 1-\alpha/2]\sqrt{\frac{\left(\sum_{i=1}^{a} \ell_i^2\right) \text{MS}_{B(A)}}{bn}},$$

respectively. However, again, multiple comparison methods discussed in Section 6.9 are to be preferred.

6.7 COMPUTATIONAL FORMULAE AND PROCEDURE

The computational formulae for the sums of squares may be obtained by expanding the corresponding definitional formulae and simplifying the algebra. They are:

$$\text{SS}_T = \sum_{i=1}^{a}\sum_{j=1}^{b}\sum_{k=1}^{n} y_{ijk}^2 - \frac{y_{...}^2}{abn},$$

$$\text{SS}_A = \frac{1}{bn}\sum_{i=1}^{a} y_{i..}^2 - \frac{y_{...}^2}{abn},$$

$$\text{SS}_{B(A)} = \frac{1}{n}\sum_{i=1}^{a}\sum_{j=1}^{b} y_{ij.}^2 - \frac{1}{bn}\sum_{i=1}^{a} y_{i..}^2,$$

and

$$\text{SS}_E = \sum_{i=1}^{a}\sum_{j=1}^{b}\sum_{k=1}^{n} y_{ijk}^2 - \frac{1}{n}\sum_{i=1}^{a}\sum_{j=1}^{b} y_{ij.}^2.$$

Note that $\text{SS}_{B(A)}$ can be written as

$$\text{SS}_{B(A)} = \sum_{i=1}^{a}\left[\frac{1}{n}\sum_{j=1}^{b} y_{ij.}^2 - \frac{1}{bn}y_{i..}^2\right].$$

This shows the idea that $SS_{B(A)}$ is the sum of squares between the levels of B for each level of A which is summed over all levels of A.

6.8 POWER OF THE ANALYSIS OF VARIANCE F TESTS

For the fixed effects or Model I, we can calculate the power of each of the F tests given in Section 6.4 in the usual way. Thus, when B effects are investigated, we have

$$\phi_{B(A)} = \frac{1}{\sigma_e} \left[\frac{n \sum_{i=1}^{a} \sum_{j=1}^{b} \beta_{j(i)}^2}{a(b-1)+1} \right]^{1/2}.$$

Similarly, for A effects, we have

$$\phi_{(A)} = \frac{1}{\sigma_e} \left[\frac{bn \sum_{i=1}^{a} \alpha_i^2}{a} \right]^{1/2}$$

and so on. Except for this change, the power calculations remain unchanged. Power formulae under Models II and III can similarly be obtained.

6.9 MULTIPLE COMPARISON METHODS

When the factor A effect is fixed and the null hypothesis concerning A effects is rejected, we may want to use Tukey, Scheffé, or other methods to investigate contrasts of interest. For example, if H_0^A is rejected, we may want to investigate contrasts involving α_i's of the form

$$L = \sum_{i=1}^{a} \ell_i \alpha_i \quad \left(\sum_{i=1}^{a} \ell_i = 0 \right),$$

which is estimated by

$$\hat{L} = \sum_{i=1}^{a} \ell_i \bar{y}_{i..}.$$

Now, if B is also fixed (i.e., we have a Model I situation), then

$$\widehat{\text{Var}}(\hat{L}) = \frac{\text{MS}_E}{bn} \sum_{i=1}^{a} \ell_i^2.$$

Two-Way Nested (Hierarchical) Classification 361

Further, for the Tukey's procedure involving pairwise differences, we have
$$T = q[a, ab(n-1); 1-\alpha].$$
For the Scheffé's procedure involving general contrasts, we would have
$$S^2 = F[a-1, ab(n-1); 1-\alpha].$$
In particular, for any m pairwise comparisons of the type $\alpha_i - \alpha_{i'}$, its $(1-\alpha)$-level Bonferroni intervals can be obtained as
$$(\bar{y}_{i..} - \bar{y}_{i'..}) \pm t[ab(n-1), 1-\alpha/2m]\sqrt{\frac{2\text{MS}_E}{bn}}.$$
Furthermore, since there are only $a-1$ independent comparisons of this form, the $(1-\alpha)$-level Scheffé's confidence intervals are given by
$$(\bar{y}_{i..} - \bar{y}_{i'..}) \pm \left\{(a-1)F[a-1, ab(n-1); 1-\alpha]\frac{2\text{MS}_E}{bn}\right\}^{\frac{1}{2}}.$$

Similarly, when H_0^B is rejected, similar simultaneous intervals for any contrast on $\beta_{j(i)}$'s or any pairwise differences on them can also be obtained. Thus, for any fixed i, the $(1-\alpha)$-level Bonferroni simultaneous confidence intervals for $\beta_{j(i)} - \beta_{j'(i)}$ are given by
$$(\bar{y}_{ij.} - \bar{y}_{ij'.}) \pm t[ab(n-1), \alpha/2m]\sqrt{\frac{2\text{MS}_E}{n}}.$$
The $(1-\alpha)$-level Scheffé's simultaneous confidence intervals are given by
$$(\bar{y}_{ij.} - \bar{y}_{ij'.}) \pm \left\{(b-1)F[b-1, ab(n-1); 1-\alpha]\frac{2\text{MS}_E}{n}\right\}^{\frac{1}{2}}.$$

For the case when B is random (i.e., we have a mixed effects or Model III situation),
$$\widehat{\text{Var}}(\hat{L}) = \frac{\text{MS}_{B(A)}}{bn}\sum_{i=1}^{a}\ell_i^2,$$
$$T = q[a, a(b-1); 1-\alpha],$$
and
$$S^2 = F[a-1, a(b-1); 1-\alpha].$$
For making pairwise comparisons, Bonferroni intervals are given by
$$(\bar{y}_{i..} - \bar{y}_{i'..}) \pm t[a(b-1), 1-\alpha/2m]\sqrt{\frac{2\text{MS}_{B(A)}}{bn}},$$

which gives an overall level of at least $1 - \alpha$ where m is the number of intervals constructed.

6.10 UNEQUAL NUMBERS IN THE SUBCLASSES

In experiments involving a nested classification, it is important to try to keep the sizes of nested factors (subsamples) equal. When the sizes of the subsamples are unequal, the analysis of variance and the expressions for the expected mean squares in the analysis of variance table become quite complicated. In this section, we indicate briefly an analysis of variance when there are unequal numbers in the subclasses.[4]

Define the following notations:

a = number of levels of factor A;
b_i = number of levels of factor B within the i-th level of factor A;
n_{ij} = number of replications from the j-th level of factor B within the i-th level of factor A;
$n_{i.} = \sum_{j=1}^{b_i} n_{ij}$ = number of observations at the i-th level of factor A; and
$N = \sum_{i=1}^{a} n_{i.} = \sum_{i=1}^{a} \sum_{j=1}^{b_i} n_{ij}$ = total number of observations in the experiment.

Now, the sums of squares can be defined by an analogy from the corresponding balanced case. They are:

$$SS_T = \sum_{i=1}^{a}\sum_{j=1}^{b_i}\sum_{k=1}^{n_{ij}} (y_{ijk} - \bar{y}_{...})^2 = \sum_{i=1}^{a}\sum_{j=1}^{b_i}\sum_{k=1}^{n_{ij}} y_{ijk}^2 - \frac{y_{...}^2}{N},$$

$$SS_A = \sum_{i=1}^{a} n_{i.}(\bar{y}_{i..} - \bar{y}_{...})^2 = \sum_{i=1}^{a} \frac{y_{i..}^2}{n_{i.}} - \frac{y_{...}^2}{N},$$

$$SS_{B(A)} = \sum_{i=1}^{a}\sum_{j=1}^{b_i} n_{ij}(\bar{y}_{ij.} - \bar{y}_{i..})^2 = \sum_{i=1}^{a}\sum_{j=1}^{b_i} \frac{y_{ij.}^2}{n_{ij}} - \sum_{i=1}^{a} \frac{y_{i..}^2}{n_{i.}},$$

and

$$SS_E = \sum_{i=1}^{a}\sum_{j=1}^{b_i}\sum_{k=1}^{n_{ij}} (y_{ijk} - \bar{y}_{ij.})^2 = \sum_{i=1}^{a}\sum_{j=1}^{b_i}\sum_{k=1}^{n_{ij}} y_{ijk}^2 - \sum_{i=1}^{a}\sum_{j=1}^{b_i} \frac{y_{ij.}^2}{n_{ij}},$$

where

$$y_{ij.} = \sum_{k=1}^{n_{ij}} y_{ijk}, \quad \bar{y}_{ij.} = y_{ij.}/n_{ij},$$

---

[4] For the design with unequal numbers in the subclasses there is no unique analysis of variance. The conventional analysis of variance being presented here is based on quadratics commonly known as Type I sums of squares.

Two-Way Nested (Hierarchical) Classification

$$y_{i..} = \sum_{j=1}^{b_i}\sum_{k=1}^{n_{ij}} y_{ijk}, \quad \bar{y}_{i..} = y_{i..}/n_{i.},$$

and

$$y_{...} = \sum_{i=1}^{a}\sum_{j=1}^{b_i}\sum_{k=1}^{n_{ij}} y_{ijk}, \quad \bar{y}_{...} = y_{...}/N.$$

The derivations for expected mean squares can be found in Scheffé (1959, pp. 255–258) and Graybill (1961, pp. 354–357). The resultant analysis of variance table is shown in Table 6.4. The coefficients of variance components in the expected mean square column are determined as follows:

$$\bar{n}_1 = \frac{N - \sum_{i=1}^{a}\frac{\sum_{j=1}^{b_i} n_{ij}^2}{n_{i.}}}{\sum_{i=1}^{a}(b_i - 1)}, \quad \bar{n}_2 = \frac{\sum_{i=1}^{a}\frac{\sum_{j=1}^{b_i} n_{ij}^2}{n_{i.}} - \frac{\sum_{i=1}^{a}\sum_{j=1}^{b_i} n_{ij}^2}{N}}{a - 1},$$

and

$$\bar{n}_3 = \frac{N - \frac{\sum_{i=1}^{a} n_{i.}^2}{N}}{a - 1}.$$

TESTS OF HYPOTHESES

Under Model I, the null hypothesis of interest, $H_0^B : \beta_{1(i)} = \beta_{2(i)} = \cdots = \beta_{b_i}(i) = 0$, subject to the constraint that $\sum_{j=1}^{b_i} n_{ij}\beta_{j(i)} = 0, i = 1, 2, \ldots a$, can be tested by the statistic

$$F_B = \frac{\text{MS}_{B(A)}}{\text{MS}_E},$$

which, when H_0^B is true, has an F distribution with $\sum_{i=1}^{a} b_i - a$ and $N - \sum_{i=1}^{a} b_i$ degrees of freedom. Similarly, one may be interested in testing whether the effects at each level of the factor A are the same, i.e., $H_0^A : \alpha_1 = \alpha_2 = \cdots = \alpha_a$. The hypothesis H_0^A, however, cannot be tested. It can be shown that the statistic

$$F_A = \frac{\text{MS}_A}{\text{MS}_E},$$

where F_A has an F distribution with $a - 1$ and $N - \sum_{i=1}^{a} b_i$ degrees of freedom, tests the hypothesis $H_0 : \alpha_i = 0, i = 1, 2, \ldots, a$, subject to the constraints that $\sum_{i=1}^{a} n_i \alpha_i = 0$ and $\sum_{j=1}^{b_i} n_{ij}\beta_{j(i)} = 0, i = 1, 2, \ldots, a$. For some further discussion and derivation of the results, see Searle (1987, Chapter III).

TABLE 6.4
Analysis of Variance for Model (6.1.1) Involving Unequal Numbers in the Subclasses

Source of Variation	Degrees of Freedom	Sum of Squares	Mean Square	Expected Mean Square		
				Model I* A Fixed, B Fixed	Model II A Random, B Random	Model III** A Fixed, B Random
Due to A	$a-1$	SS_A	MS_A	$\sigma_e^2 + \dfrac{\sum_{i=1}^{a} n_{i.}\alpha_i^2}{a-1}$	$\sigma_e^2 + \bar{n}_2\sigma_\beta^2 + \bar{n}_3\sigma_\alpha^2$	$\sigma_e^2 + \bar{n}_2\sigma_\beta^2 + \dfrac{\sum_{i=1}^{a} n_{i.}\alpha_i^2}{a-1}$
B within A	$\sum_{i=1}^{a} b_i - a$	$SS_{B(A)}$	$MS_{B(A)}$	$\sigma_e^2 + \dfrac{\sum_{i=1}^{a}\sum_{j=1}^{b_i} n_{ij}\beta_{j(i)}^2}{\sum_{i=1}^{a} b_i - a}$	$\sigma_e^2 + \bar{n}_1\sigma_\beta^2$	$\sigma_e^2 + \bar{n}_1\sigma_\beta^2$
Error	$N - \sum_{i=1}^{a} b_i$	SS_E	MS_E	σ_e^2	σ_e^2	σ_e^2
Total	$N-1$	SS_T				

* The side constraints under Model I are: $\sum_{i=1}^{a} n_{i.}\alpha_i = 0$ and $\sum_{j=1}^{b_i} n_{ij}\beta_{j(i)} = 0, i = 1, 2, \ldots, a$.

** The side constraint under Model III is $\sum_{i=1}^{a} n_{i.}\alpha_i = 0$.

Under Models II and III, it is evident that there are no simple F tests for the hypotheses relating factor A effects. This is because under the null hypothesis we do not have two mean squares in the analysis of variance table that estimate the same quantity. All three mean squares happen to estimate different quantities. In addition, although MS_E is distributed as constant times a chi-square random variable, $MS_{B(A)}$ and MS_A do not in general have a scaled chi-square distribution. Furthermore, $MS_{B(A)}$ and MS_A are not statistically independent. Approximate F tests can be developed using the Satterthwaite procedure discussed in Appendix K.[5] However, a test of the hypothesis:

$$H_0^B : \sigma_\beta^2 = 0 \quad \text{versus} \quad H_1^B : \sigma_\beta^2 > 0$$

can be carried out by the ratio

$$MS_{B(A)}/MS_E,$$

which has an F distribution with $\sum_{i=1}^{a} b_i - a$ and $N - \sum_{i=1}^{a} b_i$ degrees of freedom.

POINT AND INTERVAL ESTIMATION

Under Model I, when H_0^B is rejected, it is often of interest to construct a simultaneous confidence interval for $\beta_{j(i)} - \beta_{j'(i)}$, $i = 1, 2, \ldots, a$. For a fixed i, the $1 - \alpha$ level Bonferroni simultaneous confidence intervals for m independent comparisons of the form $\beta_{j(i)} - \beta_{j'(i)}$ are given by

$$(\bar{y}_{ij.} - \bar{y}_{ij'.}) \pm t\left[N - \sum_{i=1}^{a} b_i, 1 - \alpha/2m\right]\sqrt{\left(\frac{1}{n_{ij}} + \frac{1}{n_{ij'}}\right)MS_E}.$$

Furthermore, for any given i, since there are only $b_i - 1$ independent comparisons of this form, the $1 - \alpha$ level Scheffé's simultaneous confidence intervals are given by

$$(\bar{y}_{ij.} - \bar{y}_{ij'.}) \pm \left\{(b_i - 1)F\left[b_i - 1, N - \sum_{i=1}^{a} b_i; 1 - \alpha\right]\left(\frac{1}{n_{ij}} + \frac{1}{n_{ij'}}\right)MS_E\right\}^{\frac{1}{2}}.$$

---

[5] Some authors have ignored the unbalanced structure of the design and have used the conventional F test based on the statistic $MS_A/MS_{B(A)}$ with $a - 1$ and $\sum_{i=1}^{a} b_i - a$ degrees of freedom (see, e.g., Bliss (1976, p. 353)). A common procedure is to ignore the assumption of independence and chi-squaredness and construct an approximate F test using synthesis of mean squares based on the Satterthwaite procedure. For some further discussions and results on tests of hypotheses concerning variance components involving unequal sample sizes, see Cummings and Gaylor (1974), Tietjen (1974), Hussein and Milliken (1978b), Tan and Cheng (1984), Khuri (1987), and Hernandez et al. (1992).

Under Models II and III, if expected mean squares are equated to the corresponding mean squares in Table 6.4 and the resulting equations are solved for the variance components, these are the so-called analysis of variance estimates and are unbiased (Searle (1961)). For example, under Model II, the estimates of the variance components are[6]

$$\hat{\sigma}_e^2 = \text{MS}_E,$$

$$\hat{\sigma}_\beta^2 = \frac{1}{\bar{n}_1}(\text{MS}_{B(A)} - \text{MS}_E),$$

and

$$\hat{\sigma}_\alpha^2 = \frac{1}{\bar{n}_3}\left\{(\text{MS}_A - \text{MS}_{B(A)}) + \frac{\bar{n}_1 - \bar{n}_2}{\bar{n}_1}(\text{MS}_{B(A)} - \text{MS}_E)\right\}.$$

The expression $(\bar{n}_1 - \bar{n}_2)/\bar{n}_1$ is usually negligible, in which case the expression for $\hat{\sigma}_\alpha^2$ reduces to $(\text{MS}_A - \text{MS}_{B(A)})/\bar{n}_3$. The quantities $\bar{n}_1$ and $\bar{n}_2$ can be thought of as kinds of averages of the numbers of observations in the subgroups (n_{ij}) and they both reduce to n when $b_i = b$ and $n_{ij} = n$. Similarly, $\bar{n}_3$ can be thought of as an average of the numbers of observations corresponding to the levels of factor A and it reduces to bn when $b_i = b$ and $n_{ij} = n$.

Under Model II, in terms of the results on confidence intervals for the variance components, an exact confidence interval for σ_e^2 can be obtained by noting that the statistic $(N - \sum_{i=1}^a b_i)\text{MS}_E/\sigma_e^2$ has a chi-square distribution with $N - \sum_{i=1}^a b_i$ degrees of freedom. However, exact confidence intervals for σ_α^2 and σ_β^2 do not exist. A conservative $1 - \alpha$ level confidence interval for σ_β^2 can be obtained as

$$\frac{\left(\sum_{i=1}^a b_i - a\right)\text{MS}_B}{\bar{n}_1 \chi^2\left[\sum_{i=1}^a b_i - a, 1 - \alpha/2\right]} < \sigma_\beta^2 < \frac{\left(\sum_{i=1}^a b_i - a\right)\text{MS}_B}{\bar{n}_1 \chi^2\left[\sum_{i=1}^a b_i - a, \alpha/2\right]}.$$

Remark: Approximate confidence intervals for σ_α^2 and σ_β^2 have been proposed by Hernandez et al. (1992). Similarly, the problem of constructing exact confidence intervals on the ratios of variance components $\sigma_\alpha^2/\sigma_e^2$ and $\sigma_\beta^2/\sigma_e^2$ has been discussed by Seely and El-Bassiouni (1983) and Verdooren (1988). In addition, Burdick and Graybill (1985) and Hernandez and Burdick (1993) have considered confidence intervals for the total variance $\sigma_e^2 + \sigma_\beta^2 + \sigma_\alpha^2$, and Burdick et al. (1986a) and Sen et al. (1992) have developed confidence intervals on the proportions of variability $\sigma_e^2/(\sigma_e^2 + \sigma_\beta^2 + \sigma_\alpha^2)$, $\sigma_\beta^2/(\sigma_e^2 + \sigma_\beta^2 + \sigma_\alpha^2)$, and $\sigma_\alpha^2/(\sigma_e^2 + \sigma_\beta^2 + \sigma_\alpha^2)$. For a concise discussion of these and other results on

---

[6] The analysis of variance estimators for the unbalanced classification do not lead to the same estimates as the maximum likelihood estimators. The maximum likelihood equations are difficult to solve in unbalanced classifications. For some results on other estimation procedures, see Searle (1971b, pp. 475–477).

confidence intervals for variance components including numerical examples, see Burdick and Graybill (1992, pp. 98–109).

Under Model III, we are also in a difficult situation when we want to determine confidence intervals for the general mean μ, factor A level means $\mu + \alpha_i$, fixed effects α_i, pairwise differences $\alpha_i - \alpha_{i'}$, or the contrast $\sum_{i=1}^{a} \ell_i \alpha_i (\sum_{i=1}^{a} \ell_i = 0)$. For example, under Model III, the variance of a factor A level mean is

$$\text{Var}(\bar{y}_{i..}) = \frac{n_{i.} \sigma_e^2 + \sum_{j=1}^{b_i} n_{ij}^2 \sigma_\beta^2}{n_{i.}^2}$$

and the variance of the overall mean $\bar{y}_{...}$ is

$$\text{Var}(\bar{y}_{...}) = \frac{N \sigma_e^2 + \sum_{i=1}^{a} \sum_{j=1}^{b_i} n_{ij}^2 \sigma_\beta^2}{N^2}.$$

Furthermore,

$$\text{Var}(\bar{y}_{i..} - \bar{y}_{i'..}) = \frac{n_{i.} \sigma_e^2 + \sum_{j=1}^{b_i} n_{ij} \sigma_\beta^2}{n_{i.}^2} + \frac{n_{i'.} \sigma_e^2 + \sum_{j'=1}^{b_{i'}} n_{i'j'}^2 \sigma_\beta^2}{n_{i'.}^2}$$

and

$$\text{Var}(\bar{y}_{i..} - \bar{y}_{...}) = \frac{N - n_{i.}}{N n_{i.}} \sigma_e^2 + \left(\frac{N - 2n_{i.}}{N n_{i.}^2} \sum_{j=1}^{b_i} n_{ij}^2 + \frac{1}{N^2} \sum_{i=1}^{a} \sum_{j=1}^{b_i} n_{ij}^2 \right) \sigma_\beta^2.$$

Comparing these variances with the expected mean square expressions in Table 6.4, we notice that there are no simple estimates of the variances. Approximate methods involving Satterthwaite procedure can, however, be used to obtain the required confidence intervals. Burdick and Graybill (1992, pp. 170–171) give a numerical example illustrating methods of constructing confidence intervals for σ_e^2, σ_β^2, and $\sigma_\beta^2 / \sigma_e^2$.

Remark: In designing an experiment involving subsamples, it is suggested that subsamples of equal size preferably should be used. If during the course of the study, certain data are missing and the numbers in the various subsamples are unequal, then after some considerations of why the data are missing, the results can be analyzed by using the means of the subsamples. Such an analysis violates the assumptions of equal variance, but the variances will be approximately equal. A drawback in analyzing the cell means is that we no longer have the unbiased estimates of the variance components. Since F tests are relatively robust against heterogeneous variances, small differences in n_{ij} would not affect the conclusions. An advantage of cell means procedure is that

TABLE 6.5
Weight Gains of Chickens Placed on Four Feeding Treatments

	Treatments							
	LoCaLoL		LoCaHiL		HiCaLoL		HiCaHiL	
Pens	1	2	1	2	1	2	1	2
Weight gains	573	1,041	618	943	731	416	518	416
	636	814	926	640	845	729	782	729
	883	498	717	373	866	590	938	590
	550	890	677	907	729	552	755	552
	613	636	659	734	770	776	672	776
	901	685	817	1,050	787	657	576	657
Pen totals ($y_{ij.}$)	4,156	4,564	4,414	4,647	4,728	3,720	4,241	3,720
Treatment totals ($y_{i..}$)	8,720		9,061		8,448		7,961	

Source: Damon and Harvey (1987, p. 26). Used with permission.

the values tend to be normally distributed in view of the central limit theorem. If, however, the assumption of normality is not in question and we wish to estimate variance components, an alternative procedure is as follows. Missing values are replaced by the corresponding cell means and an analysis is carried out assuming an equal number of observations. The degrees of freedom corresponding to the residual or error mean square are decreased by one for each missing value. The consideration of why the values are missing should always be taken into account. It is appropriate to analyze the remaining data only if the loss of the missing data can be ascribed to have occurred by chance. The reader is referred to Yates (1934) for further discussion on this point.

6.11 WORKED EXAMPLE FOR MODEL I

Damon and Harvey (1987, p. 26) reported data on weight gains of chickens placed on four feeding treatments. The original data were supplied by Dr. Donald L. Anderson of the Department of Veterinary and Animal Sciences at the University of Massachusetts. The experiment involved the determination of weight gains (in grams) from 10 to 20 weeks of chickens placed on four feeding treatments obtained from combinations of high and low calcium and lysine. Weight determinations were made using six chickens in two pens from each of the four feeding treatments. The data are given in Table 6.5.

The data in Table 6.5 can be regarded as forming a two-way nested classification with pens nested within treatments and weight determinations made using six chickens from each pen. Here, treatments are fixed, and although pens are usually selected randomly, we analyze the data under the assumptions of a fixed effects model where both treatments and pens are considered to have

Two-Way Nested (Hierarchical) Classification

systematic effects. The mathematical model would be

$$y_{ijk} = \mu + \alpha_i + \beta_{j(i)} + e_{k(ij)} \quad \begin{cases} i = 1, 2, 3, 4 \\ j = 1, 2 \\ k = 1, 2, \ldots, 6, \end{cases}$$

where y_{ijk} is the k-th observation in the j-th pen and the i-th treatment, μ is the general mean, α_i is the effect of the i-th treatment, $\beta_{j(i)}$ is the effect of the j-th pen nested within the i-th treatment, and $e_{k(ij)}$ is the customary error term associated with the k-th observation, nested within the j-th pen within the i-th treatment. Also, the α_i's and $\beta_{j(i)}$'s are assumed to be fixed effects with $\sum_{i=1}^{4} \alpha_i = 0$, $\sum_{j=1}^{2} \beta_{j(i)} = 0$, $i = 1, 2, 3, 4$, and $e_{k(ij)}$'s are assumed to be independently and normally distributed with mean zero and variance σ_e^2.

Using the computational formulae for the sums of squares given in Section 6.6, we have

$$SS_T = (573)^2 + (636)^2 + \cdots + (657)^2 - \frac{(34,190)^2}{4 \times 2 \times 6}$$
$$= 25,515,538 - 24,353,252.083$$
$$= 1,162,286.917,$$

$$SS_A = \frac{(8,720)^2 + (9,601)^2 + (8,448)^2 + (7,961)^2}{2 \times 6} - \frac{(34,190)^2}{4 \times 2 \times 6}$$
$$= 24,407,195.500 - 24,353,252.083$$
$$= 53,943.417,$$

$$SS_{B(A)} = \frac{(4,156)^2 + (4,564)^2 + \cdots + (3,720)^2}{6}$$
$$- \frac{(8,720)^2 + (9,061)^2 + (8,448)^2 + (7,961)^2}{2 \times 6}$$
$$= 24,532,883.667 - 24,407,195.500$$
$$= 125,688.167,$$

and

$$SS_E = (573)^2 + (636)^2 + \cdots + (657)^2 - \frac{(4,156)^2 + (4,564)^2 + \cdots + (3,720)^2}{6}$$
$$= 25,515,538 - 24,532,883.667$$
$$= 982,654.333.$$

These results along with the remaining calculations are summarized in Table 6.6.

TABLE 6.6
Analysis of Variance for the Weight Gains Data of Table 6.5

Source of Variation	Degrees of Freedom	Sum of Squares	Mean Square	Expected Mean Square	F Value	p-Value
Treatments	3	53,943.417	17,981.139	$\sigma_e^2 + \dfrac{2 \times 6}{4-1}\sum_{i=1}^{4}\alpha_i^2$	0.732	0.539
Pens (within treatments)	4	125,688.167	31,422.042	$\sigma_e^2 + \dfrac{6}{4(2-1)}\sum_{i=1}^{4}\sum_{j=1}^{2}\beta_{j(i)}^2$	1.279	0.294
Error	40	982,654.333	24,566.358	σ_e^2		
Total	47	1,162,285.917				

Two-Way Nested (Hierarchical) Classification

The test of the hypothesis $H_0^{B(A)}$: all $\beta_{j(i)} = 0$ versus $H_1^{B(A)}$: all $\beta_{j(i)} \neq 0$ gives the variance ratio of 1.279 which is not significant ($p = 0.294$). Similarly, the test of the hypothesis H_0^A: all $\alpha_i = 0$ versus H_1^A: all $\alpha_i \neq 0$ gives the variance ratio of 0.732 which is also not significant ($p = 0.539$). Thus, there do not seem to be any significant differences in both treatment and pen effects.

6.12 WORKED EXAMPLE FOR MODEL II

Box et al. (1978, pp. 574–575) reported data from an experiment designed to estimate moisture content of the pigment paste. For this purpose, 15 batches of pigment paste used in the manufacture were randomly selected, each batch was independently sampled twice, and for each sample, two analyses were performed. The data are given in Table 6.7.

The model for this experiment is a two-way nested classification with samples nested within batches and two analyses (subsamples) made from each sample. We will analyze the data under the assumptions of a random effects model since both batches and samples are considered to have variable effects. The mathematical model would be

$$y_{ijk} = \mu + \alpha_i + \beta_{j(i)} + e_{k(ij)} \quad \begin{cases} i = 1, 2, \ldots, 15 \\ j = 1, 2 \\ k = 1, 2, \end{cases}$$

where y_{ijk} is the k-th observation (analysis) in the j-th sample in the i-th batch, μ is the general mean, α_i is the effect of the i-th batch, $\beta_{j(i)}$ is the effect of the j-th sample nested within the i-th batch, and $e_{k(ij)}$ is the effect of the k-th subsample nested within the j-th sample within the i-th batch (error term). Furthermore, the α_i's, $\beta_{j(i)}$'s, and $e_{k(ij)}$'s are assumed to be independently and normally distributed with mean zero and variances σ_α^2, σ_β^2, and σ_e^2, respectively.

Using the computational formulae for the sums of squares given in Section 6.7, we have

$$\begin{aligned} \text{SS}_T &= (40)^2 + (39)^2 + \cdots + (28)^2 - \frac{(1,607)^2}{15 \times 2 \times 2} \\ &= 45,149 - 43,040.817 \\ &= 2,108.183, \end{aligned}$$

$$\begin{aligned} \text{SS}_A &= \frac{(139)^2 + (105)^2 + \cdots + (130)^2}{2 \times 2} - \frac{(1,607)^2}{15 \times 2 \times 2} \\ &= 44,251.750 - 43,040.817 \\ &= 1,210.933, \end{aligned}$$

$$\text{SS}_{B(A)} = \frac{(79)^2 + (60)^2 + \cdots + (54)^2}{2} - \frac{(139)^2 + (105)^2 + \cdots + (130)^2}{2 \times 2}$$

TABLE 6.7
Moisture Content from Two Analyses on Two Samples of 15 Batches of Pigment Paste

	Batches									
	1		**2**		**3**		**4**		**5**	
Samples	1	2	1	2	1	2	1	2	1	2
Analyses	40	30	26	25	29	14	30	24	19	17
	39	30	28	26	28	15	31	24	20	17
Analysis totals ($y_{ij.}$)	79	60	54	51	57	29	61	48	39	34
Sample totals ($y_{i..}$)	139		105		86		109		73	

	Batches									
	6		**7**		**8**		**9**		**10**	
Samples	1	2	1	2	1	2	1	2	1	2
Analyses	33	26	23	32	34	29	27	31	13	27
	32	24	24	33	34	29	27	31	16	24
Analysis totals ($y_{ij.}$)	65	50	47	65	68	58	54	62	29	51
Sample totals ($y_{i..}$)	115		112		126		116		80	

	Batches									
	11		**12**		**13**		**14**		**15**	
Samples	1	2	1	2	1	2	1	2	1	2
Analyses	25	25	29	31	19	29	23	25	39	26
	23	27	29	32	20	30	24	25	37	28
Analysis totals ($y_{ij.}$)	48	52	58	63	39	59	47	50	76	54
Sample totals ($y_{i..}$)	100		121		98		97		130	

Source: Box et al. (1978, pp. 574–575). Used with permission.

$$= 45{,}121.500 - 45{,}251.750$$
$$= 869.750,$$

and

$$SS_E = (40)^2 + (39)^2 + \cdots + (28)^2 - \frac{(79)^2 + (60)^2 + \cdots + (54)^2}{2}$$
$$= 45{,}149 - 45{,}121.500$$
$$= 27.500.$$

TABLE 6.8
Analysis of Variance for the Moisture Content of Pigment Paste Data of Table 6.7

Source of Variation	Degrees of Freedom	Sum of Squares	Mean Square	Expected Mean Square	F Value	p-Value
Batches	14	1,210.933	86.495	$\sigma_e^2 + 2\sigma_\beta^2 + 4\sigma_\alpha^2$	1.492	0.226
Samples (within batches)	15	869.750	57.983	$\sigma_e^2 + 2\sigma_\beta^2$	63.231	<0.001
Error	30	27.500	0.917	σ_e^2		
Total	59	2,108.183				

These results along with the remaining calculations are summarized in Table 6.8. The test of the hypothesis $H_0^{B(A)} : \sigma_\beta^2 = 0$ versus $H_1^{B(A)} : \sigma_\beta^2 > 0$ gives the variance ratio of 63.231 which is highly significant ($p < 0.001$). However, the test of the hypothesis $H_0^A : \sigma_\alpha^2 = 0$ versus $H_1^A : \sigma_\alpha^2 > 0$ gives the F ratio of 1.49 which is not significant ($p = 0.226$). Thus, we reject the first null hypothesis but not the latter. The point estimates of the variance components are obtained as

$$\hat{\sigma}_e^2 = 0.917,$$

$$\hat{\sigma}_\beta^2 = \frac{1}{2}(57.983 - 0.917) = 28.533,$$

and

$$\hat{\sigma}_\alpha^2 = \frac{1}{4}(86.495 - 57.983) = 7.128.$$

These variance components account for 2.5, 78.0 and 19.5 percent of the total variation in the experimental data. The findings suggest that perhaps the largest single source of variability is the error arising in chemical sampling from the batches. The batch-to-batch variability also seems to be quite large although the results are not statistically significant. In order to estimate the mean of a batch, the estimated variance based on one subsample from one sample would be

$$\hat{\sigma}_e^2 + \hat{\sigma}_\beta^2 = 0.917 + 28.533 = 29.450.$$

The estimated variance based on two subsamples from one sample would be $\hat{\sigma}_e^2/2 + \hat{\sigma}_\beta^2 = 0.917/2 + 28.533 = 28.992$. Thus, there is very little gain in the precision of the estimation of a batch mean by using two samples rather than one. The use of two samples may, however, be useful as a check against any major errors.

Finally, we can obtain confidence limits for the overall mean μ of the process as follows. We have

$$\hat{\mu} = \bar{y}_{...} = 26.783,$$

$$\text{Var}(\bar{y}_{...}) = \frac{\sigma_e^2 + n\sigma_\beta^2 + bn\sigma_\alpha^2}{abn},$$

and

$$\widehat{\text{Var}}(\bar{y}_{...}) = \frac{\text{MS}_A}{abn} = \frac{86.495}{15 \times 2 \times 2} = 1.442.$$

Now, since

$$\frac{\bar{y}_{...} - \mu}{\sqrt{\widehat{\text{Var}}(\bar{y}_{...})}} \sim t[a-1] \quad \text{and} \quad t[14, 0.975] = 2.145,$$

the 95 percent confidence limits for μ are obtained as

$$26.783 \pm 2.145\sqrt{1.442} = (24.207, 29.359).$$

6.13 WORKED EXAMPLE FOR MODEL II: UNEQUAL NUMBERS IN THE SUBCLASSES

To give an example of Model II involving unequal sample sizes, consider the data in Table 6.9 taken from Graybill (1961, pp. 357–358). The data are artificial but the experiment is supposed to represent a breeding experiment where factor A is supposed to designate sires ($a = 4$) and factor B is supposed to designate dams ($b_1 = 3, b_2 = 4, b_3 = 2, b_4 = 3$) nested within sires. There are a total of $N = 52$ observations. The data in Table 6.9 can be regarded as forming a two-way nested classification with unequal sample sizes. Here, dams are nested within sires and sample determinations are made from each dam. Since both dams and sires are randomly selected, the data should be analyzed using a random effects model. The mathematical model would be

$$y_{ijk} = \mu + \alpha_i + \beta_{j(i)} + e_{k(ij)} \quad \begin{cases} i = 1, 2, \ldots, 4 \\ j = 1, 2, \ldots, b_i \\ k = 1, 2, \ldots, n_{ij}, \end{cases}$$

where y_{ijk} is the k-th observation for the j-th dam in the i-th sire, μ is the general mean, α_i is the effect of the i-th sire, $\beta_{j(i)}$ is the effect of the j-th dam nested within the i-th sire, and $e_{k(ij)}$ is the effect of the k-th observation nested within the j-th dam within the i-th sire (error term). Furthermore, the α_i's, $\beta_{j(i)}$'s, and $e_{k(ij)}$'s are assumed to be independently and normally distributed with mean zero and variances $\sigma_\alpha^2, \sigma_\beta^2$, and σ_e^2, respectively.

All the quantities needed for the analysis of variance computations outlined in Section 6.10 can be readily computed on an electronic calculator. The

TABLE 6.9
Data on Weight Gains from a Breeding Experiment

	Sires											
	1			2				3		4		
Dams	1	2	3	1	2	3	4	1	2	1	2	3
Weight gains	32	30	34	26	22	23	21	16	14	31	42	26
	31	26	30	20	31	21	21	20	18	34	43	25
	23	29	26	18	20	24	30	32	16	41	40	29
	26	28	34		21	26			17	40	35	40
		18	32			18					29	37
			31									
			26									
Dam totals ($y_{ij.}$)	112	131	213	64	94	112	72	68	65	146	189	157
No. in j-th dam (n_{ij}) (within i-th sire)	4	5	7	3	4	5	3	3	4	4	5	5
Sire totals ($y_{i..}$)	456			342				133		492		
No. in i-th sire ($n_{i.}$)	16			15				7		14		

Source: Graybill (1961, p. 358). Used with permission.

results are:

$$y_{...}^2/N = (1{,}423)^2/52 = 38{,}940.942,$$

$$\sum_{i=1}^{a}\sum_{j=1}^{b_i}\sum_{k=1}^{n_{ij}} y_{ijk}^2 = 41{,}811,$$

$$\sum_{i=1}^{a}\sum_{j=1}^{b_i} \frac{y_{ij.}^2}{n_{ij}} = 40{,}861.201,$$

and

$$\sum_{i=1}^{a} \frac{y_{i..}^2}{n_{i.}} = 40{,}610.885.$$

Thus,

$$SS_T = 41{,}811 - 38{,}940.942 = 2{,}870.058,$$

$$SS_A = 40{,}610.885 - 38{,}940.942 = 1{,}669.943,$$

$$SS_{B(A)} = 40{,}861.201 - 40{,}610.885 = 250.316,$$

and

$$SS_E = 41{,}811 - 40{,}861.201 = 949.799.$$

The corresponding degrees of freedom are determined as

Total: $N - 1 = 52 - 1 = 51$,

Sires: $a - 1 = 4 - 1 = 3$,

Dams (within sires): $\sum_{i=1}^{a} b_i - a = 12 - 4 = 8$,

Error: $N - \sum_{i=1}^{a} b_i = 52 - 12 = 40$.

For approximate tests of significance, we need to evaluate the coefficients of the variance components in the expected mean squares and then determine the linear combination of mean squares to be used as the denominator of an approximate F statistic using Satterthwaite procedure. The basic quantities needed to determine the coefficients of the variance components are:

$$N = \sum_{i=1}^{a} \sum_{j=1}^{b_i} n_{ij} = 52,$$

$$\sum_{i=1}^{a} \frac{n_{i.}^2}{N} = \frac{(16)^2 + (15)^2 + (7)^2 + (14)^2}{52} = 13.9615,$$

$$\sum_{i=1}^{a} \sum_{j=1}^{b_i} \frac{n_{ij}^2}{N} = \frac{(4)^2 + (5)^2 + \cdots + (5)^2}{52} = 4.6158,$$

and

$$\sum_{i=1}^{a} \sum_{j=1}^{b_i} \frac{n_{ij}^2}{n_{i.}} = \frac{(4)^2 + (5)^2 + (7)^2}{16} + \cdots + \frac{(4)^2 + (5)^2 + (5)^2}{14} = 17.8441.$$

Now, the coefficients of the variance components in the expected mean square column are given by

$$\bar{n}_1 = \frac{N - \sum_{i=1}^{a} \sum_{j=1}^{b_i} \frac{n_{ij}^2}{n_{i.}}}{\sum_{i=1}^{a} (b_i - 1)} = \frac{52 - 17.8441}{12 - 4} = 4.2695,$$

TABLE 6.10
Analysis of Variance for the Weight Gains Data of Table 6.9

Source of Variation	Degrees of Freedom	Sum of Squares	Mean Square	Expected Mean Square
Sires	3	1,669.943	556.648	$\sigma_e^2 + 4.4094\sigma_\beta^2 + 12.6795\sigma_\alpha^2$
Dams (within sires)	8	250.316	31.290	$\sigma_e^2 + 4.2695\sigma_\beta^2$
Error	40	949.799	23.745	σ_e^2
Total	51	2,870.058		

$$\bar{n}_2 = \frac{\sum_{i=1}^{a}\sum_{j=1}^{b_i}\frac{n_{ij}^2}{n_{i.}} - \sum_{i=1}^{a}\sum_{j=1}^{b_i}\frac{n_{ij}^2}{N}}{a-1} = \frac{17.8441 - 4.6158}{4-1} = 4.4094,$$

and

$$\bar{n}_3 = \frac{N - \sum_{i=1}^{a}\frac{n_{i.}^2}{N}}{a-1} = \frac{52 - 13.9615}{4-1} = 12.6795.$$

The resulting sums of squares, mean squares, and expected mean squares are summarized in Table 6.10.

The dams within sires can be tested directly against the error mean square, giving $F = 31.290/23.745 = 1.318$ ($p = 0.262$). The results are clearly nonsignificant. An approximate F test for sire effects can be carried out using the dams within sires mean square, giving $F = 556.648/31.290 = 17.790$ ($p < 0.001$), which is highly significant. However, to use Satterthwaite procedure, we first compute the coefficients as

$$\ell_2 = \bar{n}_2/\bar{n}_1 = 4.4094/4.2695 = 1.0328, \quad \ell_1 = 1 - \ell_2 = -0.0328$$

and the synthesized mean square is

$$-0.0328(23.745) + 1.0328(31.290) = 31.537.$$

The degrees of freedom for the synthesized mean square are

$$\nu' = \frac{(31.537)^2}{\frac{[-0.0328(23.745)]^2}{40} + \frac{[1.0328(31.290)]^2}{8}} \doteq 8.$$

The F ratio based on the synthesized mean square is $F = 556.648/31.537 = $

17.651 ($p < 0.001$), which gives essentially the same result as the earlier approximate test.

The estimates of the variance components σ_α^2, σ_β^2, and σ_e^2 are obtained as the solution to the following simultaneous equations:

$$556.648 = \sigma_e^2 + 4.4094\sigma_\beta^2 + 12.6795\sigma_\alpha^2,$$
$$31.290 = \sigma_e^2 + 4.2695\sigma_\beta^2,$$

and

$$23.745 = \sigma_e^2.$$

Therefore, the desired estimates are given by

$$\hat{\sigma}_e^2 = 23.745,$$
$$\hat{\sigma}_\beta^2 = \frac{31.290 - 23.745}{4.2695} = 1.767,$$

and

$$\hat{\sigma}_\alpha^2 = \frac{556.648 - 23.745 - 4.4094(1.767)}{12.6795} = 41.414.$$

These variance components account for 35.5, 2.6, and 61.9 percent of the total variation in the experimental data. The results on variance components estimates are consistent with those on tests of hypothesis. It is further evident from this analysis that the larger part of the variability in weight gains is attributable to sires. The variability between repeated measurements on a given dam is also quite large.

6.14 WORKED EXAMPLE FOR MODEL III

Snedecor and Cochran (1989, p. 250) reported data from an experiment designed to evaluate the breeding value of a set of five sires in raising pigs. Each sire was mated to two dams randomly selected from a group of dams and average daily weight gains of two pigs from each litter were recorded. The data are given in Table 6.11.

The data in Table 6.11 can be regarded as a two-way nested classification with dams nested within sires and average daily gains made from two pigs of each litter. Here, sires are fixed and dams are random, so we have a Model III situation. The mathematical model for the experiment would be

$$y_{ijk} = \mu + \alpha_i + \beta_{j(i)} + e_{k(ij)} \quad \begin{cases} i = 1, 2, 3, 4, 5 \\ j = 1, 2 \\ k = 1, 2, \end{cases} \quad (6.14.1)$$

TABLE 6.11
Average Daily Weight Gains of Two Pigs of Each Litter

	Sires									
	1		2		3		4		5	
Dams	1	2	1	2	1	2	1	2	1	2
Weight gains	2.77	2.58	2.28	3.01	2.36	2.72	2.87	2.31	2.74	2.50
	2.38	2.94	2.22	2.61	2.71	2.74	2.46	2.24	2.56	2.48
Dam totals ($y_{ij.}$)	5.15	5.52	4.50	5.62	5.07	5.46	5.33	4.55	5.30	4.98
Sire totals ($y_{i..}$)	10.67		10.12		10.53		9.88		10.28	

Source: Snedecor and Cochran (1989, p. 250). Used with permission.

where μ is the general mean, α_i is the effect of the i-th sire, $\beta_{j(i)}$ is the effect of the j-th dam nested within the i-th sire, and $e_{k(ij)}$ is the effect of the k-th observation nested within the j-th dam within the i-th sire (error term). Furthermore, the α_i's are fixed with $\sum_{i=1}^{5} \alpha_i = 0$, and $\beta_{j(i)}$'s and $e_{k(ij)}$'s are assumed to be independently and normally distributed with mean zero and variances σ_β^2 and σ_e^2, respectively.

To analyze the data of Table 6.11 according to model (6.1.1), the sums of squares using the computational formulae of Section 6.7 are obtained as

$$SS_T = (2.77)^2 + (2.38)^2 + \cdots + (2.48)^2 - \frac{(51.48)^2}{5 \times 2 \times 2}$$

$$= 133.5598 - 132.5095$$

$$= 1.0503,$$

$$SS_A = \frac{(10.67)^2 + (10.12)^2 + \cdots + (10.28)^2}{2 \times 2} - \frac{(51.48)^2}{5 \times 2 \times 2}$$

$$= 132.6092 - 132.5095$$

$$= 0.0997,$$

$$SS_{B(A)} = \frac{(5.15)^2 + (5.52)^2 + \cdots + (4.98)^2}{2}$$

$$- \frac{(10.67)^2 + (10.12)^2 + \cdots + (10.28)^2}{2 \times 2}$$

$$= 133.1728 - 132.6092$$

$$= 0.5636,$$

TABLE 6.12
Analysis of Variance for the Weight Gains Data of Table 6.11

Source of Variation	Degrees of Freedom	Sum of Squares	Mean Square	Expected Mean Square	F Value	p-Value
Sires	4	0.0997	0.0249	$\sigma_e^2 + 2\sigma_\beta^2 + 4\sum_{i=1}^{5}\alpha_i^2$	0.221	0.916
Dams (within sires)	5	0.5636	0.1127	$\sigma_e^2 + 2\sigma_\beta^2$	2.912	0.071
Error	10	0.3870	0.0387	σ_e^2		
Total	19	1.0503				

and

$$SS_E = (2.77)^2 + (2.38)^2 + \cdots + (2.48)^2$$
$$- \frac{(5.15)^2 + (5.52)^2 + \cdots + (4.98)^2}{2}$$
$$= 133.5598 - 133.1728$$
$$= 0.3870.$$

These results together with the remaining calculations are shown in Table 6.12. The test of the hypothesis $H_0^B: \sigma_\beta^2 = 0$ versus $H_1^B: \sigma_\beta^2 > 0$ gives the variance ratio of 2.912 which is less than its 5 percent critical value of 3.33 ($p = 0.071$). Similarly, the test of the hypothesis H_0^A: all $\alpha_i = 0$ versus $H_1^A: \alpha_i \neq 0$ for at least one $i = 1, 2, \ldots, 5$ gives the variance ratio of 0.221 which again falls substantially below its critical value of 5.19 at the 5 percent level ($p = 0.916$). Thus, we may conclude that there is probably no significant effect of either sires or dams within sires on average daily weight gains in these data. The estimates of variance components σ_e^2 and σ_β^2 are given by

$$\hat{\sigma}_e^2 = 0.0387$$

and

$$\hat{\sigma}_\beta^2 = \frac{1}{2}(0.1127 - 0.0387) = 0.037.$$

The results on variance components estimates are consistent with those on tests of hypotheses given previously.

6.15 USE OF STATISTICAL COMPUTING PACKAGES

For balanced nested designs involving only random factors, SAS NESTED is the procedure of choice. Although the NESTED procedure performs the F tests assuming a completely random model, the computations for sums of squares and mean squares remain equally valid under fixed and mixed model analysis. If some of the factors are crossed or any factor is fixed, PROC ANOVA is more appropriate for a fixed effects factorial model with balanced structure while GLM is more suited for a random or mixed effects model involving balanced or unbalanced data sets. In GLM, random and mixed model analyses can be handled via RANDOM and TEST options. For balanced designs, analysis of variance estimates of variance components are readily obtained from the output produced by either the NESTED or GLM procedure. For other methods of estimation of variance components, PROC MIXED or VARCOMP must be used. For instructions regarding SAS commands, see Section 11.1.

Among the SPSS procedures either MANOVA or GLM could be used for nested designs involving fixed, random, or mixed effects models. In SPSS GLM, the random or mixed effects of analysis of variance is performed by a RANDOM subcommand and the hypothesis testing for each effect is automatically carried out against the appropriate error term. In addition, GLM displays expected values of all the mean squares which can be used to estimate variance components. Furthermore, SPSS Release 7.5 incorporates a new procedure, VARCOMP, especially designed to estimate variance components. For instructions regarding SPSS commands, see Section 11.2.

Among the BMDP programs, 3V or 8V can be used for nested designs. 8V is especially designed for balanced data sets while 3V analyzes a general mixed model including balanced or unbalanced designs. In 3V, the procedures for estimating variance components include the restricted maximum likelihood and the maximum likelihood estimators. If the estimates obtained via the analysis of variance approach are nonnegative, they agree with those obtained using the restricted maximum likelihood procedure. The program 2V does not directly give the sums of squares for nested factors. However, the cross-factor sums of squares could be combined to produce desired sums of squares in a nested design.

6.16 WORKED EXAMPLES USING STATISTICAL PACKAGES

In this section, we illustrate the application of statistical packages to perform two-way nested analysis of variance for the data sets of examples presented in Sections 6.11 through 6.14. Figures 6.3 through 6.6 illustrate the program instructions and the output results for analyzing data in Tables 6.5, 6.7, 6.9, and 6.11 using SAS GLM/NESTED, SPSS MANOVA/GLM, and BMDP 3V/8V. The typical output provides the data format listed at the top, all cell means, and

```
DATA CHICKENS;                              The SAS System
INPUT TREATMEN PEN WEIGHT;            General Linear Models Procedure
DATALINES;
1 1 573                       Dependent Variable: WEIGHT
1 1 636                                      Sum of        Mean
1 1 883                       Source      DF  Squares     Square    F Value  Pr > F
  . . .                       Model        7  179631.58   25661.65   1.04    0.4163
4 2 657                       Error       40  982654.33   24566.36
;                             Corrected
PROC GLM;                     Total       47 1162285.92
CLASSES TREATMEN PEN;
MODEL WEIGHT=TREATMEN                R-Square    C.V.    Root MSE  WEIGHT Mean
  PEN(TREATMEN);                     0.154550  22.00455   156.74      712.29
RUN;
CLASS   LEVELS   VALUES       Source         DF  Type I SS  Mean Square  F Value  Pr > F
TREATMEN   4     1 2 3 4      TREATMEN        3  53943.42    17981.14    0.73    0.5391
PEN        2     1 2          PEN(TREATMEN)   4  125688.17   31422.04    1.28    0.2943
NUMBER OF OBS. IN DATA        Source         DF Type III SS Mean Square  F Value  Pr > F
SET=48                        TREATMEN        3  53943.42    17981.14    0.73    0.5391
                              PEN(TREATMEN)   4  125688.17   31422.04    1.28    0.2943
```

(i) SAS application: SAS GLM instructions and output for the two-way fixed effects nested analysis of variance.

```
DATA LIST                           Analysis of Variance-Design 1
/TREATMEN 1 PEN 3
 WEIGHT 5-8.               Tests of Significance for WEIGHT using UNIQUE sums of squares
BEGIN DATA.
1 1 573                    Source of Variation      SS       DF      MS       F   Sig of F
1 1 663
  . . .                    WITHIN CELLS          982654.33   40   24566.36
4 2 657                    PEN WITHIN TREATMEN   125688.17    4   31422.04  1.28   .294
END DATA.                  TREATMEN               53943.42    3   17981.14   .73   .539
MANOVA WEIGHT BY
TREATMEN(1,4)              (Model)               179631.58    7   25661.65  1.04   .416
PEN(1,2)                   (Total)              1162285.92   47   24729.49
/DESIGN=PEN WITHIN
  TREATMEN VS WITHIN       R-Squared       =  .155
  TREATMENT VS WITHIN.     Adjusted R-Squared = .007
```

(ii) SPSS application: SPSS MANOVA instructions and output for the two-way fixed effects nested analysis of variance.

```
/INPUT    FILE='C:\SAHAI         BMDP8V - GENERAL MIXED MODEL ANALYSIS OF VARIANCE
          \TEXTO\EJE15.TXT'.            - EQUAL CELL SIZES Release: 7.0 (BMDP/DYNAMIC)
          FORMAT=FREE.
          VARIABLES=6.           ANALYSIS OF VARIANCE FOR DEPENDENT VARIABLE  1
/VARIABLE NAMES=C1,C2,C3,
          C4,C5,C6.              SOURCE      ERROR  SUM OF    D.F.    MEAN      F     PROB.
/DESIGN   NAMES=T,P,C.                       TERM   SQUARES           SQUARE
          LEVELS=4,2,6.          1 MEAN      C(TP) 24353252.    1   24353252. 991.33  0.0000
          RANDOM=C.              2 TREATMNT  C(TP)    53943.    3      17981.   0.73  0.5391
          FIXED=T, P.            3 P(T)      C(TP)   125688.    4      31422.   1.28  0.2943
          MODEL='T,P(T),C(P)'.   4 C(TP)              982654.   40      24566.
/END
1 1 573                          SOURCE      EXPECTED MEAN           ESTIMATES OF
  . . .                                         SQUARE            VARIANCE COMPONENTS
4 2 657
ANALYSIS OF VARIANCE DESIGN      1 MEAN        48(1)+(4)              506847.61927
INDEX            T   P   C       2 TREATMNT    12(2)+(4)                -548.76829
NUMBER OF LEVELS 4   2   6       3 P(T)         6(3)+(4)                1142.61389
POPULATION SIZE  4   2  INF      4 C(TP)           (4)                 24566.35833
MODEL    T, P(T), C(P)
```

(iii) BMDP application: BMDP 8V instructions and output for the two-way fixed effects nested analysis of variance.

FIGURE 6.3 Program Instructions and Output for the Two-Way Fixed Effects Nested Analysis of Variance: Weight Gains Data for Example of Section 6.11 (Table 6.5).

Two-Way Nested (Hierarchical) Classification

```
DATA MOISTURE;                        The SAS System
INPUT BATCHES SAMPLE           Coefficients of Expected Mean Squares
MOISTURE;                    Source      BATCHES    SAMPLE      ERROR
DATALINES;                   BATCHES        4          2          1
1 1 40                       SAMPLE         0          2          1
1 1 39                       ERROR          0          0          1
1 2 30
1 2 30                       Variance Degrees of   Sum of                           Error
2 1 26                       Source   Freedom     Squares      F Value   Pr > F     Term
2 1 28                       TOTAL      59       2108.183333
2 2 25                       BATCHES    14       1210.933333    1.492    0.2256    SAMPLE
2 2 26                       SAMPLE     15        869.750000   63.255    0.0000    ERROR
3 1 29                       ERROR      30         27.500000
                             Variance                          Variance           Percent
. . .                        Source            Mean Square    Component          of Total
15 2 28                      TOTAL              35.731921      36.577976         100.0000
;                            BATCHES            86.495238       7.127976          19.4871
PROC NESTED;                 SAMPLE             57.983333      28.533333          78.0069
CLASSES BATCHES SAMPLE;      ERROR               0.916667       0.916667           2.5061
VAR MOISTURE;                Mean                              26.78333333
RUN;                         Standard error of mean             1.20066119
```

(i) SAS application: SAS NESTED instructions and output for the two-way random effects nested analysis of variance.

```
DATA LIST                Tests of Between-Subjects Effects   Dependent Variable: MOISTURE
/BATCHES 1-2
  SAMPLE 4               Source                  Type III SS   df   Mean Square     F      Sig.
  MOISTURE 6-7.          BATCHES   Hypothesis      1210.933    14      86.495      1.492   .226
BEGIN DATA.                        Error            869.750    15      57.983(a)
1 1 40                   SAMPLE    Hypothesis       869.750    15      57.983     63.255   .000
1 1 39                   (BACHES)  Error             27.500    30        .917(b)
1 2 30                   a  MS(SAMPLE(BATCHES))   b MS(Error)
1 2 30
2 1 26                                               Expected Mean Squares(a,b)
. . .                                                    Variance Component
15 2 28                  Source              Var(BATCHES)    Var(SAMPLE(BATCHES))    Var(Error)
END DATA.                BATCHES                4.000                2.000              1.000
GLM MOISTURE BY          SAMPLE(BATCHES)         .000                2.000              1.000
BATCHES SAMPLE           Error                   .000                 .000              1.000
/DESIGN BATCHES          a  For each source, the expected mean square equals the sum of the
  SAMPLE(BATCHES)        coefficients in the cells times the variance components, plus a
/RANDOM BATCHES          quadratic term involving effects in the Quadratic Term cell. b Expected
  SAMPLE.                Mean Squares are based on the Type III Sums of Squares.
```

(ii) SPSS application: SPSS GLM instructions and output for the two-way random effects nested analysis of variance.

```
/INPUT    FILE='C:\SAHAI        BMDP8V - GENERAL MIXED MODEL ANALYSIS OF VARIANCE
          \TEXTO\EJE16.TXT'.             - EQUAL CELL SIZES Release: 7.0 (BMDP/DYNAMIC)
          FORMAT=FREE.
          VARIABLES=2.          ANALYSIS OF VARIANCE FOR DEPENDENT VARIABLE   1
/VARIABLE NAMES=A1,A2.
/DESIGN   NAMES=B,S,A.          SOURCE    ERROR   SUM OF     D.F.   MEAN         F       PROB.
          LEVELS=15,2,2.                  TERM    SQUARES           SQUARE
          RANDOM=B,S,A.         1 MEAN    BATCH  43040.81667   1   43040.817   497.61   0.0000
          MODEL='B,S(B),A(S)'.  2 BATCH   S(B)    1210.93333  14      86.495     1.49   0.2256
/END                            3 S(B)    A(BS)    869.75000  15      57.983    63.25   0.0000
40 39                           4 A(BS)             27.50000  30       0.917
. .
26 28                           SOURCE         EXPECTED MEAN              ESTIMATES OF
ANALYSIS OF VARIANCE DESIGN                       SQUARE              VARIANCE COMPONENTS
INDEX          B   S   A        1 MEAN       60(1)+4(2)+2(3)+(4)           715.90536
NUMBER OF LEVELS 15  2   2      2 BATCH        4(2)+2(3)+(4)                 7.12798
POPULATION SIZE INF INF INF     3 S(B)           2(3)+(4)                   28.53333
MODEL   B, S(B), A(S)           4 A(BS)             (4)                      0.91667
```

(iii) BMDP application: BMDP 8V instructions and output for the two-way random effects nested analysis of variance.

FIGURE 6.4 Program Instructions and Output for the Two-Way Random Effects Nested Analysis of Variance: Moisture Content of Pigment Paste Data for Example of Section 6.12 (Table 6.7).

```
DATA BREEDING;                          The SAS System
INPUT SIRE DAM WEIGHT;            General Linear Models Procedure
DATALINES;                    Dependent Variable: WEIGHT
1 1 32                                    Sum of        Mean
1 1 31                        Source      DF   Squares      Square    F Value   Pr > F
1 1 23                        Model       11   1920.26007   174.56910   7.35    0.0001
1 1 26                        Error       40    949.79762    23.74494
1 2 30                        Corrected   51   2870.05769
1 2 26                        Total
1 2 29                                    R-Square    C.V.    Root MSE  WEIGHT Mean
1 2 28                                    0.669067   17.80672  4.87288     27.3654
1 2 18                        Source      DF   Type I SS   Mean Square  F Value  Pr > F
1 3 34                        SIRE         3   1669.94341   556.64780   23.44   0.0001
1 3 30                        DAM(SIRE)    8    250.31667    31.28958    1.32   0.2628
1 3 26                        Source      DF   Type III SS Mean Square  F Value  Pr > F
1 3 34                        SIRE         3   1594.12974   531.37658   22.38   0.0001
 . . .                        DAM(SIRE)    8    250.31667    31.28958    1.32   0.2628
4 3 37                        Source      Type III Expected Mean Square
;                             SIRE        Var(Error)+4.1311 Var(DAM(SIRE))+12.26 Var(SIRE)
PROC GLM;                     DAM(SIRE)   Var(Error)+4.2695 Var(DAM(SIRE))
CLASSES SIRE DAM;
MODEL WEIGHT = SIRE           Tests of Hypotheses for Random Model Analysis of Variance
DAM(SIRE);                    Source: SIRE Error: 0.9676*MS(DAM(SIRE)) + 0.0324*MS(Error)
RANDOM SIRE DAM(SIRE)/TEST;                            Denominator    Denominator
RUN;                          DF   Type III MS    DF       MS         F Value   Pr > F
CLASS   LEVELS   VALUES        3   531.37658135  8.41   31.045049505  17.1163   0.0006
SIRE       4     1 2 3 4      Source: DAM(SIRE)Error: MS(Error)
DAM        3     1 2 3                                 Denominator    Denominator
NUMBER OF OBS. IN DATA        DF   Type III MS    DF       MS         F Value   Pr > F
SET=52                         8   31.289583333   40   23.744940476    1.3177   0.2628
```

(i) SAS application: SAS GLM instructions and output for the two-way random effects nested analysis of variance with unequal numbers in the subclasses.

```
DATA LIST              Tests of Between-Subjects Effects       Dependent Variable: WEIGHT
/SIRE 1 DAM 3
 WEIGHT 5-6.           Source              Type III SS   df    Mean Square     F      Sig.
 BEGIN DATA.           SIRE    Hypothesis    1594.130     3     531.377      17.116   .001
1 1 32                         Error          261.114   8.411    31.045(a)
1 1 31                 DAM(SIRE) Hypothesis   250.317     8      31.290       1.318   .263
1 1 23                         Error          949.798    40      23.745(b)
1 1 26                 a .968 MS(D(S))+3.241E-02 MS(E)  b MS(Error)
1 2 30
1 2 26                                     Expected Mean Squares(a,b)
1 2 29                                         Variance Component
 . . .                 Source         Var(SIRE)    Var(DAM(SIRE))    Var(Error)
4 3 37                 SIRE            12.260          4.131           1.000
END DATA.              DAM(SIRE)         .000          4.269           1.000
GLM WEIGHT BY          Error             .000           .000           1.000
SIRE DAM               a For each source, the expected mean square equals the sum of the
/DESIGN SIRE           coefficients in the cells times the variance components, plus a
DAM(SIRE)              quadratic term involving effects in the Quadratic Term cell. b Expected
/RANDOM SIRE DAM.      Mean Squares are based on the Type III Sums of Squares.
```

(ii) SPSS application: SPSS GLM instructions and output for the two-way random effects nested analysis of variance with unequal numbers in the subclasses.

FIGURE 6.5 Program Instructions and Output for the Two-Way Random Effects Nested Analysis of Variance with Unequal Numbers in the Subclasses: Breeding Data for Example of Section 6.13 (Table 6.9).

Two-Way Nested (Hierarchical) Classification

```
/INPUT     FILE='C:\SAHAI
           \TEXTO\EJE17.TXT'.
           FORMAT=FREE.
           VARIABLES=3.
/VARIABLE  NAMES=SIRE,DAM,WEIGHT.
/GROUP     CODES(SIRE)=1,2,3,4.
           NAMES(SIRE)=S1,S2,S3,S4.
           CODES(DAM)=1,2,3,4.
           NAMES(DAM)=D1,D2,D3,D4.
/DESIGN    DEPENDENT=WEIGHT.
           RANDOM=SIRE.
           RANDOM=DAM, SIRE.
           RNAMES=S,'D(S)'.
           METHOD=REML.
/END
1 1 32
. . .
4 3 37
```

```
BMDP3V - GENERAL MIXED MODEL ANALYSIS OF VARIANCE
            Release: 7.0      (BMDP/DYNAMIC)

         DEPENDENT VARIABLE WEIGHT

PARAMETER  ESTIMATE  STANDARD  EST/.    TWO-TAIL PROB.
                     ERROR     ST.DEV   (ASYM. THEORY)
ERR.VAR.   23.480    5.184
CONSTANT   26.441    3.477     7.603    0.000
SIRE       45.601   39.793
DAM(SIRE    2.135    3.779

TESTS OF FIXED EFFECTS BASED ON ASYMPTOTIC VARIANCE
-COVARIANCE MATRIX

SOURCE     F-STATISTIC  DEGREES OF   PROBABILITY
                        FREEDOM
CONSTANT      57.81      1   51        0.00000
```

(iii) BMDP application: BMDP 3V instructions and output for the two-way random effects nested analysis of variance with unequal numbers in the subclasses.

FIGURE 6.5 (continued)

```
DATA LITTER;
INPUT SIRE DAM WEIGHT;
DATALINES;
1 1 2.77
1 1 2.38
1 2 2.58
1 2 2.94
. . . .
5 2 2.48
;
PROC GLM;
CLASSES SIRE DAM;
MODEL WEIGHT=SIRE DAM(SIRE);
RANDOM DAM(SIRE);
TEST H=SIRE E=DAM(SIRE);
RUN;
CLASS    LEVELS    VALUES
SIRE       5       1 2 3 4 5
DAM        2       1 2
NUMBER OF OBS. IN DATA
SET=20
```

```
                    The SAS System
              General Linear Models Procedure
Dependent Variable: WEIGHT
             Sum of        Mean
Source   DF  Squares       Square     F Value  Pr > F
Model     9  0.66328000    0.07369778   1.90   0.1649
Error    10  0.38700000    0.03870000
Corrected
Total    19  1.05028000

         R-Square    C.V.      Root MSE    WEIGHT Mean
         0.631527   7.6427022  0.19672316  2.574000

Source   DF  Type III SS   Mean Square  F Value  Pr > F
SIRE      4  0.09973000    0.02493250    0.64   0.6433
DAM(SIRE) 5  0.56355000    0.11271000    2.91   0.0707
Source       Type III Expected Mean Square
SIRE         Var(Error) + 2 Var(DAM(SIRE)) + Q(SIRE)
DAM(SIRE)    Var(Error) + 2 Var(DAM(SIRE))
Tests of Hypotheses using the Type III MS for DAM(SIRE) as
an error term
Source   DF  Type III SS   Mean Square  F Value  Pr > F
SIRE      4  0.09973000    0.02493250    0.22   0.9155
```

(i) SAS application: SAS GLM instructions and output for the two-way mixed effects nested analysis of variance.

FIGURE 6.6 Program Instructions and Output for the Two-Way Mixed Effects Nested Analysis of Variance: Average Daily Weight Gains Data for Example of Section 6.14 (Table 6.11).

```
DATA LIST              Tests of Between-Subjects Effects           Dependent Variable: WEIGHT
/SIRE 1 DAM 3
 WEIGHT 5-8(2)   Source             Type III SS   df   Mean Square     F      Sig.
 BEGIN DATA.     SIRE     Hypothesis   9.973E-02    4   2.493E-02     .221    .916
 1 1 2.77                 Error            .564    5    .113(a)
 1 1 2.38       DAM(SIRE) Hypothesis       .564    5    .113         2.912    .071
 1 2 2.58                 Error            .387   10   3.870E-02(b)
 1 2 2.94       a MS(DAM(SIRE))  b  MS(Error)
 2 1 2.28
 2 1 2.22                                 Expected Mean Squares(a,b)
 2 2 3.01                                     Variance Component
  . . .         Source         Var(DAM(SIRE))        Var(Error)       Quadratic Term
 5 2 2.48       SIRE               2.000               1.000               Sire
 END DATA.      DAM(SIRE)          2.000               1.000
 GLM WEIGHT BY  Error               .000               1.000
 SIRE DAM       a For each source, the expected mean square equals the sum of the
 /DESIGN SIRE   coefficients in the cells times the variance components, plus a quadratic
  DAM(SIRE)     term involving effects in the Quadratic Term cell.  b Expected Mean Squares
 /RANDOM SIRE.  are based on the Type III Sums of Squares.
```

(ii) SPSS application: SPSS GLM instructions and output for the two-way mixed effects nested analysis of variance.

```
/INPUT    FILE='C:\SAHAI      BMDP8V - GENERAL MIXED MODEL ANALYSIS OF VARIANCE
          \TEXTO\EJE18.TXT'.          - EQUAL CELL SIZES Release: 7.0 (BMDP/DYNAMIC)
          FORMAT=FREE.
          VARIABLES=2.        ANALYSIS OF VARIANCE FOR DEPENDENT VARIABLE  1
/VARIABLE NAMES=PIG1, PIG2.
/DESIGN   NAMES=SIRE,DAM,PIG.  SOURCE    ERROR     SUM OF    D.F.    MEAN          F        PROB.
          LEVELS=5, 2, 2.                TERM      SQUARES           SQUARE
          RANDOM=DAM, PIG.
          FIXED=SIRE.          1 MEAN    D(S)     132.509519  1    132.509519   1175.67    0.0000
          MODEL='S,D(S),P(D)'. 2 SIRE    D(S)       0.099730  4      0.024933      0.22    0.9155
/END                           3 D(S)    P(SD)      0.563550  5      0.112710      2.91    0.0707
2.77  2.38                     4 P(SD)              0.387000 10      0.038700
  .     .
2.50  2.48                     SOURCE        EXPECTED MEAN            ESTIMATES OF VARIANCE
ANALYSIS OF VARIANCE DESIGN                    SQUARE                    COMPONENTS
INDEX         S  D   P         1 MEAN       20(1)+2(3)+(4)                 6.61984
NUMBER OF LEVELS 5 2  2        2 SIRE        4(2)+2(3)+(4)                -0.02194
POPULATION SIZE  5 INF INF     3 D(S)           2(3)+(4)                   0.03700
MODEL     S, D(S), P(D)        4 P(SD)            (4)                      0.03870
```

(iii) BMDP application: BMDP 8V instructions and output for the two-way mixed effects nested analysis of variance.

FIGURE 6.6 *(continued)*

the entries of the analysis of variance table. It should be noticed that in each case the results are the same as those provided using manual computations in Sections 6.11 through 6.14. However, note that in an unbalanced design, certain tests of significance may differ from one program to the other since they use different types of sums of squares.

EXERCISES

1. An experiment was designed to study the ignition rate of dynamite from three different explosive-forming processes. Four types of dynamite were randomly selected from each explosive-forming process

and three measurements of ignition rate were made on each type. The data in certain standard units are given as follows.

Explosive Process	1				2				3			
Dynamite Type	1	2	3	4	1	2	3	4	1	2	3	4
	28.1	23.0	18.3	18.1	23.1	26.3	21.2	38.0	17.1	38.1	41.1	28.0
	32.3	31.1	20.6	19.0	20.2	27.7	24.6	30.0	18.0	24.5	37.5	32.3
	29.5	23.5	17.5	16.6	17.6	24.8	20.3	28.0	23.7	27.3	53.6	36.5

(a) Describe the model and the assumptions for the experiment.
(b) Analyze the data and report the analysis of variance table.
(c) Test whether there are differences in ignition rates among the three explosive-forming processes. Use $\alpha = 0.05$.
(d) Test whether there are differences in ignition rates among dynamite types within the explosive processes. Use $\alpha = 0.05$.
(e) Estimate the variance components of the model and determine 95 percent confidence intervals for them.

2. A manufacturing company wishes to study the tensile strength of yarns produced on four different looms. An experiment was designed wherein 12 machinists were selected at random and each loom was run by three different machinists and two specimens from each machinist were obtained and tested. The data in certain standard units are given as follows.

Loom	1			2			3			4		
Machinist	1	2	3	1	2	3	1	2	3	1	2	3
	38.2	53.5	15.3	61.3	41.5	35.3	47.1	22.5	14.7	15.5	19.3	21.6
	21.6	51.5	26.7	58.3	38.5	27.3	34.3	25.7	26.3	32.3	35.7	26.5

(a) Describe the model and the assumptions for the experiment.
(b) Analyze the data and report the analysis of variance table.
(c) Test whether there are differences in the tensile strength among the four looms. Use $\alpha = 0.05$.
(d) Test whether there are differences in the tensile strength among machinists within looms. Use $\alpha = 0.05$.
(e) Estimate the variance components of the model.

3. A manufacturing company wishes to study the material variability of a particular product being manufactured on three different machines. Each machine operates in two shifts and four samples are randomly

chosen from each shift. The data in certain standard units are given as follows.

Machine	1		2		3	
Shift	1	2	1	2	1	2
	23.5	19.3	25.1	23.5	25.0	27.3
	20.7	20.5	26.5	21.3	19.5	26.5
	22.9	21.3	24.7	22.6	23.4	26.3
	23.3	19.7	25.3	24.7	22.3	25.8

(a) Describe the model and the assumptions for the experiment.
(b) Analyze the data and report the analysis of variance table.
(c) Test whether there are differences in the material variability among the machines. Use $\alpha = 0.05$.
(d) Test whether there are differences in the material variability between the shifts within the machines. Use $\alpha = 0.05$.

4. An industrial firm wishes to streamline production scheduling by assigning one time standard to a particular class of machines. An experiment was designed wherein three machines are randomly selected and each machine is assigned to a different group of three operators selected at random. Each operator uses the machine three times at different periods during a given week. The data in certain standard units are given as follows.

Machine	1			2			3		
Operator	1	2	3	1	2	3	1	2	3
	103.2	104.1	103.8	99.5	102.6	99.7	107.4	106.0	105.4
	104.3	104.6	102.7	99.8	101.7	101.2	107.6	103.0	104.4
	105.1	103.7	101.5	98.7	103.5	101.7	108.1	104.2	103.7

(a) Describe the model and the assumptions for the experiment.
(b) Analyze the data and report the analysis of variance table.
(c) Test whether there are differences in completion time among machines. Use $\alpha = 0.05$.
(d) Test whether there are differences in completion time betweem operators within machines. Use $\alpha = 0.05$.
(e) Estimate the variance components of the model and determine 95 percent confidence intervals for them.

5. A public health official wishes to test the difference in mean fluoride concentration of water in a community. An experiment was designed

Two-Way Nested (Hierarchical) Classification

wherein three water samples were taken from each of three sources of water supply and three determinations of fluoride content were performed on each of the nine samples. The data in milligrams fluoride per liter of water are given as follows.

Supply	1			2			3		
Sample	1	2	3	1	2	3	1	2	3
	1.7	1.9	1.8	1.9	1.9	2.0	2.4	2.7	2.8
	1.8	1.7	1.6	2.0	2.1	1.8	2.6	2.6	2.5
	1.9	1.8	1.7	2.1	2.2	1.9	2.5	2.8	2.6

(a) Describe the model and the assumptions for the experiment.
(b) Analyze the data and report the analysis of variance table.
(c) Test the hypothesis that there is no difference in mean fluoride content between samples at a source of water supply. Use $\alpha = 0.05$.
(d) Test the hypothesis that there is no difference in mean fluoride concentration between the sources of water supply. Use $\alpha = 0.05$.

6. A nutritional scientist wishes to test the difference in protein levels in animals fed on different dietary regimens. An experiment is designed wherein four animals of a certain species are subjected to each of two dietary regimens and three samples of blood are drawn from each animal to determine the protein content (in mg/100 ml blood). The data are given as follows.

Dietary Regimen	1				2			
Animal	1	2	3	4	1	2	3	4
	3.44	3.51	3.54	3.52	4.88	4.91	4.79	4.95
	3.46	3.50	3.52	3.53	4.84	4.89	4.81	4.93
	3.47	3.47	3.59	3.57	4.85	4.87	4.82	4.92

(a) Describe the model and the assumptions for the experiment.
(b) Analyze the data and report the analysis of variance table.
(c) Test the hypothesis that there are no differences in mean protein levels between animals fed on a given dietary regimen. Use $\alpha = 0.05$.
(d) Test the hypothesis that there are no differences in mean protein levels between the two dietary regimens. Use $\alpha = 0.05$.

7. An experiment is designed involving a two-stage nested design with the levels of factor B nested within the levels of factor A. The relevant data are given as follows.

Factor A	1		2			3	
Factor B	1	2	1	2	3	1	2
	12.2	3.5	11.2	8.3	7.0	13.2	5.7
	10.1	7.3	13.3	10.2	6.3	9.3	6.2
	14.2		15.1	9.1	3.5	10.1	
			12.5				

(a) Describe the model and the assumptions for the experiments. It is assumed that both factors A and B are fixed.
(b) Analyze the data and report the analysis of variance table.
(c) Test whether there are differences in the levels of factor A. Use $\alpha = 0.05$.
(d) Test whether there are differences in the levels of factor B within A. Use $\alpha = 0.05$.

8. A study is conducted to investigate whether a batch of material is homogenous by randomly sampling the material in five different vats from the large number of vats produced. Three bags are chosen at random from each vat. Finally, two independent analyses are made on each sample to determine the percentage of a particular substance. The data in certain standard units are given as follows.

Vat	1			2			3			4			5		
Bag	1	2	3	1	2	3	1	2	3	1	2	3	1	2	3
	44.3	41.0	44.3	41.0	38.7	42.1	39.9	45.4	46.5	38.7	38.7	46.5	36.5	36.5	38.9
	44.3	43.2	44.3	42.1	38.7	42.1	38.7	44.3	45.4	37.6	39.9	46.5	36.5	35.4	39.8

(a) Describe the model and the assumptions for the experiment.
(b) Analyze the data and report the analysis of variance table.
(c) Test the hypothesis that there is no difference in mean percentage values between bags within a vat. Use $\alpha = 0.05$.
(d) Test the hypothesis that there is no difference in mean percentage values between vats. Use $\alpha = 0.05$.
(e) Estimate the variance components of the model and determine 95 percent confidence intervals on them.

9. Hicks (1956) reported the results of an experiment involving strain measurements on each of four seals made on each of the four heads of each of the five sealing machines. The coded raw data are given as follows.

Two-Way Nested (Hierarchical) Classification

Machine	1				2				3				4				5			
Head	1	2	3	4	1	2	3	4	1	2	3	4	1	2	3	4	1	2	3	4
	6	13	1	7	10	2	4	0	0	10	8	7	11	5	1	0	1	6	3	3
	2	3	10	4	9	1	1	3	0	11	5	2	0	10	8	8	4	7	0	7
	0	9	0	7	7	1	7	4	5	6	0	5	6	8	9	6	7	0	2	4
	8	8	6	9	12	10	9	1	5	7	7	4	4	3	4	5	9	3	2	0

Source: Hicks (1956, p. 14). Used with permission.

(a) Describe the model and the assumptions for the experiment.
(b) Analyze the data and report the analysis of variance table.
(c) Test the hypothesis that there is no difference in mean strain values between the heads within a machine. Use $\alpha = 0.05$.
(d) Test the hypothesis that there is no difference in mean strain values between the machines. Use $\alpha = 0.05$.

10. Sokal and Rohlf (1995, p. 294) reported data from an experiment designed to investigate variation in the blood pH of female mice. The experiment was carried out on 15 dams that were mated over a period of time with either two or three sires. Each sire was mated to different dams and measurements were made on the blood pH reading of a female offspring. The following data refer to a subset of 5 dams which have been randomly selected from 15 dams in the experiment.

Dam	1		2		3			4		5		
Sire	1	2	1	2	1	2	3	1	2	1	2	3
pH Reading	7.48	7.48	7.38	7.37	7.41	7.47	7.53	7.39	7.50	7.39	7.43	7.46
	7.48	7.53	7.48	7.31	7.42	7.36	7.40	7.31	7.44	7.37	7.38	7.44
	7.52	7.43	7.46	7.45	7.36	7.43	7.44	7.30	7.40	7.33	7.44	7.37
	7.54	7.39			7.41	7.47	7.38	7.40	7.41	7.45	7.43	7.54
						7.41	7.45	7.48		7.42		

Source: Sokal and Rohlf (1995, p. 294). Used with permission.

(a) Describe the model and the assumptions for the experiment.
(b) Analyze the data and report the analysis of variance table.
(c) Test whether there are differences in mean pH readings between dams. Use $\alpha = 0.05$.
(d) Test whether there are differences in mean pH readings between sires within dams. Use $\alpha = 0.05$.
(e) Estimate the variance components of the model.

11. Marcuse (1949) reported results from an experiment designed to investigate the moisture content of cheese. The experiment was

conducted by sampling three different lots from the large number of lots produced. Two samples of cheese were selected at random from each lot. Finally, two subsamples per sample were chosen and independent analyses made on each subsample to determine the percentage of moisture content. The data are given as follows.

Lot	1		2		3	
Sample	1	2	1	2	1	2
	39.02	38.96	35.74	35.58	37.02	35.70
	38.79	39.01	35.41	35.52	36.00	36.04

Source: Marcuse (1949). Used with permission.

(a) Describe the model and the assumptions for the experiment.
(b) Analyze the data and report the analysis of variance table.
(c) Test the hypothesis that there are no differences in mean percentage values of the moisture content between samples within lots. Use $\alpha = 0.05$.
(d) Test the hypothesis that there are no differences in mean percentage values of the moisture content between lots. Use $\alpha = 0.05$.
(e) Estimate the variance components of the model and determine 95 percent confidence intervals on them.

12. Sokal and Rohlf (1995, p. 276) reported data from a biological experiment involving 12 female mosquito pupae. The mosquitos were randomly assigned into three rearing cages with each cage receiving 4 pupae. The reported responses are independent measurements of left wings of the mosquito and the data are given as follows.

Cage	1				2				3			
Mosquito	1	2	3	4	1	2	3	4	1	2	3	4
	58.5	77.8	84.0	70.1	69.8	56.0	50.7	63.8	56.6	77.8	69.9	62.1
	59.5	80.9	83.6	68.3	69.8	54.5	49.3	65.8	57.5	79.2	69.2	64.5

Source: Sokal and Rohlf (1995, p. 276). Used with permission.

(a) Describe the model and assumptions for the experiment.
(b) Analyze the data and report the analysis of variance table.
(c) Test the hypothesis that there are no differences in mean measurement values between mosquitos within a cage. Use $\alpha = 0.05$.
(d) Test the hypothesis that there are no differences in mean measurement values between cages. Use $\alpha = 0.05$.
(e) Estimate the variance components of the model and determine 95 percent confidence intervals on them.

13. Steel and Torrie (1980, p. 154) reported data from a greenhouse experiment that examined the growth of mint plants. A large group of plants were assigned at random to pots with each pot receiving four plants. Treatments were randomly assigned to pots with each treatment receiving three pots. There were 6 fixed treatments representing combinations of cross factors with 3 levels of hours of daylight and 2 levels of temperatures. Observations were made on individual plants where the response variable was the one week stem growth of the mint plant. The data are given as follows.

Treatment*	1			2			3		
Pot	1	2	3	1	2	3	1	2	3
	3.5	2.5	3.0	5.0	3.5	4.5	5.0	5.5	5.5
	4.0	4.5	3.0	5.5	3.5	4.0	4.5	6.0	4.5
	3.0	5.5	2.5	4.0	3.0	4.0	5.0	5.0	6.5
	4.5	5.0	3.0	3.5	4.0	5.0	4.5	5.0	5.5

Treatment*	4			5			6		
Pot	1	2	3	1	2	3	1	2	3
	8.5	6.5	7.0	6.0	6.0	6.5	7.0	6.0	11.0
	6.0	7.0	7.0	5.5	8.5	6.5	9.0	7.0	7.0
	9.0	8.0	7.0	3.5	4.5	8.5	8.5	7.0	9.0
	8.5	6.5	7.0	7.0	7.5	7.5	8.5	7.0	8.0

Source: Steel and Torrie (1990, p. 154). Used with permission.

* Treatments representing combinations of hours of daylight and temperatures are defined as follows:

	Hours of Daylight		
Temperature	8	12	16
Low	1	2	3
High	4	5	6

(a) Describe the model and assumptions for the experiment.
(b) Analyze the data and report the analysis of variance table.
(c) Test the hypothesis that there are no differences in mean stem growths between pots within a treatment. Use $\alpha = 0.05$.
(d) Test the hypothesis that there are no differences in mean stem growths between the treatments. Use $\alpha = 0.05$.
(e) Estimate the variance components of the model and determine 95 percent confidence intervals on them.

(f) Estimate the contrast representing the difference between two treatments defined as the low temperature–8 hours and high temperature–8 hours and set a 95 percent confidence interval.

(g) Estimate the contrast defined as: Low–8 + Low–12 + Low–16 − High–8 − High–12 − High–16 and set a 95 percent confidence interval.

14. Sokal and Rohlf (1994, p. 364) reported data from an experiment designed to investigate the effects of breed and maturity of pure-bred cows on butterfat content. Five breeds of pure-bred dairy cattle were taken from Canadian records and random samples of 10 mature (≥ 5 years old) and 10 two-year-old cows were selected from each of five breeds. The following data give average butterfat percentages for each cow.

Breed	Ayrshire		Canadian		Guernsey		Holstein-Fresian		Jersey	
Cow	Mature	2-yr	Mature	2-yr	Mature	2-yr	Mature	2-yr	Mature	2-yr
	3.74	4.44	3.92	4.29	4.54	5.30	3.40	3.79	4.80	5.75
	4.01	4.37	4.95	5.24	5.18	4.50	3.55	3.66	6.45	5.14
	3.77	4.25	4.47	4.43	5.75	4.59	3.83	3.58	5.18	5.25
	3.78	3.71	4.28	4.00	5.04	5.04	3.95	3.38	4.49	4.76
	4.10	4.08	4.07	4.62	4.64	4.83	4.43	3.71	5.24	5.18
	4.06	3.90	4.10	4.29	4.79	4.55	3.70	3.94	5.70	4.22
	4.27	4.41	4.38	4.85	4.72	4.97	3.30	3.59	5.41	5.98
	3.94	4.11	3.98	4.66	3.88	5.38	3.93	3.55	4.77	4.85
	4.11	4.37	4.46	4.40	5.28	5.39	3.58	3.55	5.18	6.55
	4.25	3.53	5.05	4.33	4.66	5.97	3.54	3.43	5.23	5.72

Source: Sokal and Rohlf (1994, p. 364). Used with permission.

(a) Describe the model and the assumptions for the experiment. Would you use Model I, Model II, or Model III. Explain.
(b) Analyze the data and report the analysis of variance table.
(c) Test the hypothesis that there are no differences in average percentage values of the butterfat content between mature and two-year old cows. Use $\alpha = 0.05$.
(d) Test the hypothesis that there are no differences in average percentage values of the butterfat content between the breeds. Use $\alpha = 0.05$.
(e) If you assumed any of the factors to be random, estimate the variance components of the model and construct 95 percent confidence intervals on them.

7 Three-Way and Higher-Order Nested Classifications

7.0 PREVIEW

The results of the preceding chapter can be readily extended to the case of three-way and the general q-way nested or hierarchical classifications. As an example of a three-way nested classification, suppose a chemical company wishes to examine the strength of a certain liquid chemical. The chemical is made in large vats and then is barreled. To study the strength of the chemical, an analyst randomly selects three different vats of the product. Three barrels are selected at random from each vat and then three samples are taken from each barrel. Finally, two independent measurements are made on each sample. The physical layout can be depicted schematically as shown in Figure 7.1. In this experiment, barrels are nested within the levels of the factor vats and samples are nested within the levels of the factor barrels. This is the so-called three-way nested classification having two replicates or measurements. In this chapter, we consider the three-way nested classification and indicate its generalization to higher-order nested classifications.

7.1 MATHEMATICAL MODEL

Consider three factors A, B, and C having a, b, and c levels respectively. The b levels of factor B are nested under each level of A and c levels of factor C are nested under each level of factor B (within A), and there are n replicates within the combination of levels of A, B, and C. The analysis of variance model for this type of experimental layout is taken as

$$y_{ijk\ell} = \mu + \alpha_i + \beta_{j(i)} + \gamma_{k(ij)} + e_{\ell(ijk)} \quad \begin{cases} i = 1, 2, \ldots, a \\ j = 1, 2, \ldots, b \\ k = 1, 2, \ldots, c \\ \ell = 1, 2, \ldots, n, \end{cases} \quad (7.1.1)$$

where μ is the general mean, α_i is the effect due to the i-th level of factor A,

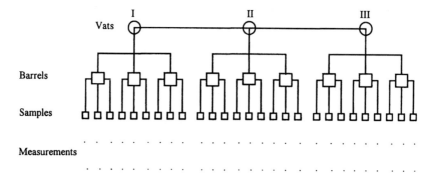

FIGURE 7.1 A Layout for the Three-Way Nested Design Where Barrels Are Nested within Vats and Samples Are Nested within Barrels.

$\beta_{j(i)}$ is the effect due to the j-th level of factor B within the i-th level of factor A, $\gamma_{k(ij)}$ is the effect due to the k-th level of factor C within the j-th level of factor B and the i-th level of factor A, and $e_{\ell(ijk)}$ is the error term that represents the variation within each cell.

When all the factors have systematic effects, Model I is applicable to the data in (7.1.1). When all the factors are random, Model II is appropriate; and when some factors are fixed and others are random, a mixed model or Model III is the appropriate one. The assumptions under Models I, II, and III are exact parallels to that of the model (6.1.1). For example, if we assume that A is fixed and B and C are random, then α_i's are unknown fixed constants with the restriction that $\sum_{i=1}^{a} \alpha_i = 0$, and $\beta_{j(i)}$'s, $\gamma_{k(ij)}$'s, and $e_{\ell(ijk)}$'s are mutually and completely uncorrelated random variables with zero means and variances σ_β^2, σ_γ^2, and σ_e^2 respectively.

7.2 ANALYSIS OF VARIANCE

The calculations of the sums of squares and the analysis of variance for the three-way nested design are similar to the analysis for the two-way nested design presented in Chapter 6. The formulae for the sums of squares together with their computational forms are simple extensions of the formulae for the two-way nested design given in Section 6.7. Thus, starting with the identity

$$y_{ijk\ell} - \bar{y}_{....} = (\bar{y}_{i...} - \bar{y}_{....}) + (\bar{y}_{ij..} - \bar{y}_{i...}) + (\bar{y}_{ijk.} - \bar{y}_{ij..}) + (y_{ijk\ell} - \bar{y}_{ijk.}),$$

the total sum of squares is partitioned as

$$SS_T = SS_A + SS_{B(A)} + SS_{C(B)} + SS_E,$$

where

$$SS_T = \sum_{i=1}^{a}\sum_{j=1}^{b}\sum_{k=1}^{c}\sum_{\ell=1}^{n}(y_{ijk\ell} - \bar{y}_{....})^2 = \sum_{i=1}^{a}\sum_{j=1}^{b}\sum_{k=1}^{c}\sum_{\ell=1}^{n}y_{ijk\ell}^2 - \frac{y_{....}^2}{abcn},$$

$$SS_A = bcn\sum_{i=1}^{a}(\bar{y}_{i...} - \bar{y}_{....})^2 = \frac{1}{bcn}\sum_{i=1}^{a}y_{i...}^2 - \frac{y_{....}^2}{abcn},$$

$$SS_{B(A)} = cn\sum_{i=1}^{a}\sum_{j=1}^{b}(\bar{y}_{ij..} - \bar{y}_{i...})^2 = \frac{1}{cn}\sum_{i=1}^{a}\sum_{j=1}^{b}y_{ij..}^2 - \frac{1}{bcn}\sum_{i=1}^{a}y_{i...}^2,$$

$$SS_{C(B)} = n\sum_{i=1}^{a}\sum_{j=1}^{b}\sum_{k=1}^{c}(\bar{y}_{ijk.} - \bar{y}_{ij..})^2 = \frac{1}{n}\sum_{i=1}^{a}\sum_{j=1}^{b}\sum_{k=1}^{c}y_{ijk.}^2 - \frac{1}{cn}\sum_{i=1}^{a}\sum_{j=1}^{b}y_{ij..}^2,$$

and

$$SS_E = \sum_{i=1}^{a}\sum_{j=1}^{b}\sum_{k=1}^{c}\sum_{\ell=1}^{n}(y_{ijk\ell} - \bar{y}_{ijk.})^2 = \sum_{i=1}^{a}\sum_{j=1}^{b}\sum_{k=1}^{c}\sum_{\ell=1}^{n}y_{ijk\ell}^2 - \frac{1}{n}\sum_{i=1}^{a}\sum_{j=1}^{b}\sum_{k=1}^{c}y_{ijk.}^2,$$

with

$$y_{ijk.} = \sum_{\ell=1}^{n}y_{ijk\ell}, \qquad \bar{y}_{ijk.} = y_{ijk.}/n,$$

$$y_{ij..} = \sum_{k=1}^{c}\sum_{\ell=1}^{n}y_{ijk\ell}, \qquad \bar{y}_{ij..} = y_{ij..}/cn,$$

$$y_{i...} = \sum_{j=1}^{b}\sum_{k=1}^{c}\sum_{\ell=1}^{n}y_{ijk\ell}, \qquad \bar{y}_{i...} = y_{i...}/bcn,$$

$$y_{....} = \sum_{i=1}^{a}\sum_{j=1}^{b}\sum_{k=1}^{c}\sum_{\ell=1}^{n}y_{ijk\ell}, \quad \text{and} \quad \bar{y}_{....} = y_{....}/abcn.$$

It should be noticed that the nested sums of squares are related to the sums of squares for main effects and interactions, considering all the factors being crossed, as follows:

$$SS_{B(A)} = SS_B + SS_{AB}, \quad SS_{C(B)} = SS_C + SS_{AC} + SS_{BC} + SS_{ABC}.$$

The expected mean squares can be obtained by proceeding directly as earlier or using the general rules for obtaining the expected mean squares. The resultant analysis of variance is summarized in Table 7.1. There are various types of mixed models that may arise. The analysis of variance table contains the expectations of mean squares for the case when A and B are fixed and C is random. If we

TABLE 7.1
Analysis of Variance for Model (7.1.1)

Source of Variation	Degrees of Freedom	Sum of Squares	Mean Square	Expected Mean Square		
				Model I A, B, and C Fixed	Model II A, B, and C Random	Model III A and B Fixed, C Random
Due to A	$a-1$	SS_A	MS_A	$\sigma_e^2 + \dfrac{bcn}{a-1}\sum_{i=1}^{a}\alpha_i^2$	$\sigma_e^2 + n\sigma_\gamma^2 + cn\sigma_\beta^2 + bcn\sigma_\alpha^2$	$\sigma_e^2 + n\sigma_\gamma^2 + \dfrac{bcn}{a-1}\sum_{i=1}^{a}\alpha_i^2$
B within A	$a(b-1)$	$SS_{B(A)}$	$MS_{B(A)}$	$\sigma_e^2 + \dfrac{cn}{a(b-1)}\sum_{i=1}^{a}\sum_{j=1}^{b}\beta_{j(i)}^2$	$\sigma_e^2 + n\sigma_\gamma^2 + cn\sigma_\beta^2$	$\sigma_e^2 + n\sigma_\gamma^2 + \dfrac{cn}{a(b-1)}\sum_{i=1}^{a}\sum_{j=1}^{b}\beta_{j(i)}^2$
C within B	$ab(c-1)$	$SS_{C(B)}$	$MS_{C(B)}$	$\sigma_e^2 + \dfrac{n}{ab(c-1)}\sum_{i=1}^{a}\sum_{j=1}^{b}\sum_{k=1}^{c}\gamma_{k(ij)}^2$	$\sigma_e^2 + n\sigma_\gamma^2$	$\sigma_e^2 + n\sigma_\gamma^2$
Error	$abc(n-1)$	SS_E	MS_E	σ_e^2	σ_e^2	σ_e^2
Total	$abcn-1$	SS_T				

assume a mixed model with factor A having a systematic effect and factors B and C having random effects, then analysis is the same as under Model II except that σ_α^2 is now replaced by $\sum_{i=1}^{a} \alpha_i^2/(a-1)$. The remaining four cases that fall under the mixed model (i.e., those in which factors B and C have opposite effects) are left as an exercise. It should, however, be pointed out that the term involving $\beta_{j(i)}$ disappears from the expectation of MS_A when factor B has a systematic effect, and the term involving $\gamma_{k(ij)}$ disappears from the expectations of MS_A and $MS_{B(A)}$ when factor C has a systematic effect. Thus, special care is needed when determining the appropriate tests to be made.

7.3 TESTS OF HYPOTHESES AND ESTIMATION

Under the assumption of normality, the four mean squares are independently distributed as multiples of a chi-square variable such that the ratio of any two mean squares is distributed as the variance ratio F.[1] The expected mean square column of Table 7.1 suggests the proper test statistics to be employed for testing the particular hypotheses of interest. Thus, under Model I, F tests for the effects of all three factors can be performed by dividing the corresponding mean squares by MS_E. Under Model II, tests for the existence of the main effects of factor A and the two nested factors B and C all exist. The factor A effect is tested by means of the ratio $MS_A/MS_{B(A)}$, factor B by the ratio $MS_{B(A)}/MS_{C(B)}$, and factor C by the ratio $MS_{C(B)}/MS_E$.[2] Under Model III, with factor A having a systematic effect and factors B and C having random effects, the tests for all the factor effects would be the same as indicated under Model II. The tests for other variations of mixed models are obtained similarly.

The variance components for various model factors are readily estimated by using the customary analysis of variance procedure. For example, under Model II, the desired estimators are:

$$\hat{\sigma}_e^2 = MS_E,$$
$$\hat{\sigma}_\gamma^2 = (MS_{C(B)} - MS_E)/n,$$
$$\hat{\sigma}_\beta^2 = (MS_{B(A)} - MS_{C(B)})/cn, \qquad (7.3.1)$$

and

$$\hat{\sigma}_\alpha^2 = (MS_A - MS_{B(A)})/bcn.$$

The estimators (7.3.1) are the so-called best unbiased estimators as discussed before; but the estimates for σ_α^2, σ_β^2, and σ_γ^2 can be negative. It should be

---

[1] For a proof of this result see Scheffé (1959, pp. 251–254).
[2] These are all exact F tests and their power is readily expressed in terms of the (central) F distribution.

noticed that the estimation of variance components is especially simple for hierarchically nested designs. One simply obtains the difference between the mean squares for the factor involving the variance component of interest and the one following it; and the resulting difference is divided by the coefficient of the variance component in the expected mean square. An exact confidence interval for σ_e^2 can be constructed by noting that $abc(n-1)\text{MS}_E/\sigma_e^2$ has a chi-square distribution with $abc(n-1)$ degrees of freedom.[3]

Under Model III, with A fixed and B and C random, exact confidence intervals on means, $\mu + \alpha_i$'s, and a linear combination of means, $\sum_{i=1}^{a} \ell_i(\mu + \alpha_i)$, can be obtained as in Section 6.6. Thus, exact $100(1-\alpha)$ percent confidence intervals for $\mu + \alpha_i$ and $\sum_{i=1}^{a} \ell_i(\mu + \alpha_i)$ are given by

$$\bar{y}_{i\ldots} \pm t[a(b-1), 1-\alpha/2]\sqrt{\frac{\text{MS}_{B(A)}}{bcn}}$$

and

$$\sum_{i=1}^{a} \ell_i \bar{y}_{i\ldots} \pm t[a(b-1), 1-\alpha/2]\sqrt{\frac{\left(\sum_{i=1}^{a} \ell_i^2\right)\text{MS}_{B(A)}}{bcn}},$$

respectively.[4]

7.4 UNEQUAL NUMBERS IN THE SUBCLASSES

Consider three factors A, B, and C where B is nested within A and C is nested within B. Suppose each A level has b_i B levels, each B level has c_{ij} C levels, and n_{ijk} samples are taken from each C level. Here, the model remains the same as (7.1.1), where $i = 1, 2, \ldots, a$; $j = 1, 2, \ldots, b_i$; $k = 1, 2, \ldots, c_{ij}$; and $\ell = 1, 2, \ldots, n_{ijk}$. The total number of observations is

$$N = \sum_{i=1}^{a} \sum_{j=1}^{b_i} \sum_{k=1}^{c_{ij}} n_{ijk}.$$

---

[3] For a discussion of methods for constructing confidence intervals for individual variance components σ_γ^2, σ_β^2, and σ_α^2, the total variance $\sigma_e^2 + \sigma_\gamma^2 + \sigma_\beta^2 + \sigma_\alpha^2$, the variance ratio $\sigma_\gamma^2/\sigma_e^2$, and the proportions of variability $\sigma_e^2/(\sigma_e^2 + \sigma_\gamma^2)$, $\sigma_\gamma^2/(\sigma_e^2 + \sigma_\alpha^2)$, $\sigma_e^2/(\sigma_e^2 + \sigma_\gamma^2 + \sigma_\beta^2 + \sigma_\alpha^2)$, $\sigma_\gamma^2/(\sigma_e^2 + \sigma_\gamma^2 + \sigma_\beta^2 + \sigma_\alpha^2)$, $\sigma_\beta^2/(\sigma_e^2 + \sigma_\gamma^2 + \sigma_\beta^2 + \sigma_\alpha^2)$, and $\sigma_\alpha^2/(\sigma_e^2 + \sigma_\gamma^2 + \sigma_\beta^2 + \sigma_\alpha^2)$, see Burdick and Graybill (1992, pp. 92–96).

[4] Formulae for selecting b, c, and n to minimize the cost of obtaining a sample have been derived by Marcuse (1949) and Vidmar and Brunden (1980).

Three-Way and Higher-Order Nested Classifications 401

Now, the sums of squares in the analysis of variance are computed as follows:

$$SS_A = \sum_{i=1}^{a} n_{i..}(\bar{y}_{i...} - \bar{y}_{....})^2 = \sum_{i=1}^{a} \frac{y_{i...}^2}{n_{i..}} - \frac{y_{....}^2}{N},$$

$$SS_{B(A)} = \sum_{i=1}^{a}\sum_{j=1}^{b_i} n_{ij.}(\bar{y}_{ij..} - \bar{y}_{i...})^2 = \sum_{i=1}^{a}\sum_{j=1}^{b_i} \frac{y_{ij..}^2}{n_{ij.}} - \sum_{i=1}^{a} \frac{y_{i...}^2}{n_{i..}},$$

$$SS_{C(B)} = \sum_{i=1}^{a}\sum_{j=1}^{b_i}\sum_{k=1}^{c_{ij}} n_{ijk}(\bar{y}_{ijk.} - \bar{y}_{ij..})^2 = \sum_{i=1}^{a}\sum_{j=1}^{b_i}\sum_{k=1}^{c_{ij}} \frac{y_{ijk.}^2}{n_{ijk}} - \sum_{i=1}^{a}\sum_{j=1}^{b_i} \frac{y_{ij..}^2}{n_{ij.}},$$

and

$$SS_E = \sum_{i=1}^{a}\sum_{j=1}^{b_i}\sum_{k=1}^{c_{ij}}\sum_{\ell=1}^{n_{ijk}} (y_{ijk\ell} - \bar{y}_{ijk.})^2$$
$$= \sum_{i=1}^{a}\sum_{j=1}^{b_i}\sum_{k=1}^{c_{ij}}\sum_{\ell=1}^{n_{ijk}} y_{ijk\ell}^2 - \sum_{i=1}^{a}\sum_{j=1}^{b_i}\sum_{k=1}^{c_{ij}} \frac{y_{ijk.}^2}{n_{ijk}},$$

where the customary notations for totals and means are employed. The resultant analysis of variance is summarized in Table 7.2. The derivations for expected mean squares can be found in Ganguli (1941) and Scheffé (1959, pp. 255–258). The coefficients of variance components under the expected mean square columns are determined as follows:

$$\bar{n}_1 = \frac{N - k_6}{c - b}, \qquad \bar{n}_2 = \frac{k_6 - k_5}{b - a},$$

$$\bar{n}_3 = \frac{N - k_4}{b - a}, \qquad \bar{n}_4 = \frac{k_5 - k_3}{a - 1},$$

$$\bar{n}_5 = \frac{k_4 - k_2}{a - 1}, \quad \text{and} \quad \bar{n}_6 = \frac{N - k_1}{a - 1},$$

where

$$b = \sum_{i=1}^{a} b_i, \qquad c = \sum_{i=1}^{a}\sum_{j=1}^{b_i} c_{ij},$$

$$k_1 = \sum_{i=1}^{a} n_{i..}^2 / N, \qquad k_2 = \sum_{i=1}^{a}\sum_{j=1}^{b_i} n_{ij.}^2 / N,$$

$$k_3 = \sum_{i=1}^{a}\sum_{j=1}^{b_i}\sum_{k=1}^{c_{ij}} n_{ijk}^2 / N, \qquad k_4 = \sum_{i=1}^{a}\sum_{j=1}^{b_i} n_{ij.}^2 / n_{i..},$$

$$k_5 = \sum_{i=1}^{a}\sum_{j=1}^{b_i}\sum_{k=1}^{c_{ij}} n_{ijk}^2 / n_{i..}, \quad \text{and} \quad k_6 = \sum_{i=1}^{a}\sum_{j=1}^{b_i}\sum_{k=1}^{c_{ij}} n_{ijk}^2 / n_{ij.}.$$

TABLE 7.2
Analysis of Variance for Model (7.1.1) Involving Unequal Numbers in the Subclasses

Source of Variation	Degrees of Freedom	Sum of Squares	Mean Square	Expected Mean Square		
				Model I*	Model II	Model III**
				A, B, and C Fixed	A, B, and C Random	A Fixed, B and C Random
Due to A	$a - 1$	SS_A	MS_A	$\sigma_e^2 + \dfrac{\sum_{i=1}^{a} n_{i..}\alpha_i^2}{a-1}$	$\sigma_e^2 + \bar{n}_4\sigma_\gamma^2 + \bar{n}_5\sigma_\beta^2 + \bar{n}_6\sigma_\alpha^2$	$\sigma_e^2 + \bar{n}_4\sigma_\gamma^2 + \bar{n}_5\sigma_\beta^2 + \dfrac{\sum_{i=1}^{a} n_{i..}\alpha_i^2}{a-1}$
B within A	$\sum_{i=1}^{a} b_i - a$	$SS_{B(A)}$	$MS_{B(A)}$	$\sigma_e^2 + \dfrac{\sum_{i=1}^{a}\sum_{j=1}^{b_i} n_{ij.}\beta_{j(i)}^2}{b-a}$	$\sigma_e^2 + \bar{n}_2\sigma_\gamma^2 + \bar{n}_3\sigma_\beta^2$	$\sigma_e^2 + \bar{n}_2\sigma_\gamma^2 + \bar{n}_3\sigma_\beta^2$
C within B	$\sum_{i=1}^{a}\sum_{j=1}^{b_i} c_{ij} - \sum_{i=1}^{a} b_i$	$SS_{C(B)}$	$MS_{C(B)}$	$\sigma_e^2 + \dfrac{\sum_{i=1}^{a}\sum_{j=1}^{b_i}\sum_{k=1}^{c_{ij}} n_{ijk}\gamma_{k(ij)}^2}{c-b}$	$\sigma_e^2 + \bar{n}_1\sigma_\gamma^2$	$\sigma_e^2 + \bar{n}_1\sigma_\gamma^2$
Error	$N - \sum_{i=1}^{a}\sum_{j=1}^{b_i} c_{ij}$	SS_E	MS_E	σ_e^2	σ_e^2	σ_e^2
Total	$N - 1$	SS_T				

* The side constraints under Model I are: $\sum_{i=1}^{a} n_{i..}\alpha_i = 0$; $\sum_{j=1}^{b_i} n_{ij.}\beta_{j(i)} = 0, i = 1, 2, \ldots, a$; and $\sum_{k=1}^{c_{ij}} n_{ijk}\gamma_{k(ij)} = 0, j = 1, 2, \ldots, b_i; i = 1, 2, \ldots, a$.

** The side constraint under Model III is $\sum_{i=1}^{a} n_{i..}\alpha_i = 0$.

From Table 7.2, it is evident that no simple F tests are possible except for testing factor C effects. Again, this is so because under the null hypothesis we do not have two mean squares in the analysis of variance table that estimate the same quantity. Furthermore, the mean squares other than MS_E are not distributed as constant times a chi-square random variable, and they are not statistically independent except that MS_E is independent of the other mean squares. For the approximate tests of significance, we determine the coefficients of the variance components, calculate estimates of the variance components, and then determine linear combinations of mean squares to use as the denominators of the F ratios in the Satterthwaite approximation. The variance components estimates as usual are obtained by solving the equations obtained by equating the mean squares to their respective expected values. These are the so-called analysis of variance estimates (Mahamunulu (1963)). The formulae for these estimators including the expressions for their sampling variances are also given in Searle (1971b, pp. 477–479) and Searle et al. (1992, pp. 431–433).

An exact confidence interval for σ_e^2 can be obtained as in Section 6.10. Burdick and Graybill (1992, pp. 109–116) indicate a method for constructing confidence intervals for σ_α^2, σ_β^2, and σ_γ^2 with a numerical example.

7.5 FOUR-WAY NESTED CLASSIFICATION

In this section, we briefly review the analysis of variance for the four-way nested classification which is the obvious extension of the three-way nested analysis of variance. The model is

$$y_{ijk\ell m} = \mu + \alpha_i + \beta_{j(i)} + \gamma_{k(ij)} + \delta_{\ell(ijk)} + e_{m(ijk\ell)} \quad \begin{cases} i = 1, \ldots, a \\ j = 1, \ldots, b \\ k = 1, \ldots, c \\ \ell = 1, \ldots, d \\ m = 1, \ldots, n, \end{cases} \quad (7.5.1)$$

where the meaning of each symbol and the assumptions of the model are readily stated. Starting with the identity

$$y_{ijk\ell m} - \bar{y}_{.....} = (\bar{y}_{i....} - \bar{y}_{.....}) + (\bar{y}_{ij...} - \bar{y}_{i....}) + (\bar{y}_{ijk..} - \bar{y}_{ij...}) + (\bar{y}_{ijk\ell.} - \bar{y}_{ijk..}) + (y_{ijk\ell m} - \bar{y}_{ijk\ell.}),$$

the total sum of squares is partitioned as

$$SS_T = SS_A + SS_{B(A)} + SS_{C(B)} + SS_{D(C)} + SS_E,$$

where

$$SS_T = \sum_{i=1}^{a}\sum_{j=1}^{b}\sum_{k=1}^{c}\sum_{\ell=1}^{d}\sum_{m=1}^{n}(y_{ijk\ell m} - \bar{y}_{.....})^2$$

$$= \sum_{i=1}^{a}\sum_{j=1}^{b}\sum_{k=1}^{c}\sum_{\ell=1}^{d}\sum_{m=1}^{n}y_{ijk\ell m}^2 - \frac{1}{abcdn}y_{.....}^2,$$

$$SS_A = bcdn\sum_{i=1}^{a}(\bar{y}_{i....} - \bar{y}_{.....})^2$$

$$= \frac{1}{bcdn}\sum_{i=1}^{a}y_{i....}^2 - \frac{1}{abcdn}y_{.....}^2,$$

$$SS_{B(A)} = cdn\sum_{i=1}^{a}\sum_{j=1}^{b}(\bar{y}_{ij...} - \bar{y}_{i....})^2$$

$$= \frac{1}{cdn}\sum_{i=1}^{a}\sum_{j=1}^{b}y_{ij...}^2 - \frac{1}{bcdn}\sum_{i=1}^{a}y_{i....}^2,$$

$$SS_{C(B)} = dn\sum_{i=1}^{a}\sum_{j=1}^{b}\sum_{k=1}^{c}(\bar{y}_{ijk..} - \bar{y}_{ij...})^2$$

$$= \frac{1}{dn}\sum_{i=1}^{a}\sum_{j=1}^{b}\sum_{k=1}^{c}y_{ijk..}^2 - \frac{1}{cdn}\sum_{i=1}^{a}\sum_{j=1}^{b}y_{ij...}^2,$$

$$SS_{D(C)} = n\sum_{i=1}^{a}\sum_{j=1}^{b}\sum_{k=1}^{c}\sum_{\ell=1}^{d}(\bar{y}_{ijk\ell.} - \bar{y}_{ijk..})^2$$

$$= \frac{1}{n}\sum_{i=1}^{a}\sum_{j=1}^{b}\sum_{k=1}^{c}\sum_{\ell=1}^{d}y_{ijk\ell.}^2 - \frac{1}{dn}\sum_{i=1}^{a}\sum_{j=1}^{b}\sum_{k=1}^{c}y_{ijk..}^2,$$

and

$$SS_E = \sum_{i=1}^{a}\sum_{j=1}^{b}\sum_{k=1}^{c}\sum_{\ell=1}^{d}\sum_{m=1}^{n}(y_{ijk\ell m} - \bar{y}_{ijk\ell.})^2$$

$$= \sum_{i=1}^{a}\sum_{j=1}^{b}\sum_{k=1}^{c}\sum_{\ell=1}^{d}\sum_{m=1}^{n}y_{ijk\ell m}^2 - \frac{1}{n}\sum_{i=1}^{a}\sum_{j=1}^{b}\sum_{k=1}^{c}\sum_{\ell=1}^{d}y_{ijk\ell.}^2,$$

with the usual notations of dots and bars. The corresponding mean squares denoted by MS_A, $MS_{B(A)}$, $MS_{C(B)}$, $MS_{D(C)}$, and MS_E are obtained by dividing the sums of squares by the respective degrees of freedom. The resultant analysis of variance is summarized in Table 7.3. The nested classifications having more than four factors have the analysis of variance tables with the same general pattern.

TABLE 7.3
Analysis of Variance for Model (7.5.1)

Source of Variation	Degrees of Freedom	Sum of Squares	Mean Square	Expected Mean Square		
				Model I $A, B, C,$ and D Fixed	Model II $A, B, C,$ and D Random	Model III A and B Fixed, C and D Random
Due to A	$a-1$	SS_A	MS_A	$\sigma_e^2 + \dfrac{bcdn}{a-1}\sum_{i=1}^{a}\alpha_i^2$	$\sigma_e^2 + n\sigma_\delta^2 + dn\sigma_\gamma^2 + cdn\sigma_\beta^2 + bcdn\sigma_\alpha^2$	$\sigma_e^2 + n\sigma_\delta^2 + dn\sigma_\gamma^2 + \dfrac{bcdn}{a-1}\sum_{i=1}^{a}\alpha_i^2$
B within A	$a(b-1)$	$SS_{B(A)}$	$MS_{B(A)}$	$\sigma_e^2 + \dfrac{cdn}{a(b-1)}\sum_{i=1}^{a}\sum_{j=1}^{b}\beta_{j(i)}^2$	$\sigma_e^2 + n\sigma_\delta^2 + dn\sigma_\gamma^2 + cdn\sigma_\beta^2$	$\sigma_e^2 + n\sigma_\delta^2 + dn\sigma_\gamma^2 + \dfrac{cdn}{a(b-1)}\sum_{i=1}^{a}\sum_{j=1}^{b}\beta_{j(i)}^2$
C within B	$ab(c-1)$	$SS_{C(B)}$	$MS_{C(B)}$	$\sigma_e^2 + \dfrac{dn}{ab(c-1)}\sum_{i=1}^{a}\sum_{j=1}^{b}\sum_{k=1}^{c}\gamma_{k(ij)}^2$	$\sigma_e^2 + n\sigma_\delta^2 + dn\sigma_\gamma^2$	$\sigma_e^2 + n\sigma_\delta^2 + dn\sigma_\gamma^2$
D within C	$abc(d-1)$	$SS_{D(C)}$	$MS_{D(C)}$	$\sigma_e^2 + \dfrac{n}{abc(d-1)}\sum_{i=1}^{a}\sum_{j=1}^{b}\sum_{k=1}^{c}\sum_{t=1}^{d}\delta_{t(ijk)}^2$	$\sigma_e^2 + n\sigma_\delta^2$	$\sigma_e^2 + n\sigma_\delta^2$
Error	$abcd(n-1)$	SS_E	MS_E	σ_e^2	σ_e^2	σ_e^2
Total	$abcdn-1$	SS_T				

The expected mean square column of Table 7.3 suggests the proper test statistics to be employed for testing the particular hypotheses of interest. For example, under Model II, note that each expected mean square contains all the terms of the expected mean square that follows it in the table. Thus, an appropriate F statistic to test the statistical significance of any factor effects is determined as the mean square of the factor of interest divided by the mean square immediately following it. The unbiased estimators of the variance components are also readily obtained using the customary analysis of variance procedure. In particular, the best unbiased estimators of the variance components under Model II are obtained simply by using the differences between the mean square of the factor of interest and the one immediately following it; that is,

$$\hat{\sigma}_e^2 = \text{MS}_E,$$
$$\hat{\sigma}_\delta^2 = (\text{MS}_{D(C)} - \text{MS}_E)/n,$$
$$\hat{\sigma}_\gamma^2 = (\text{MS}_{C(B)} - \text{MS}_{D(C)})/dn,$$
$$\hat{\sigma}_\beta^2 = (\text{MS}_{B(A)} - \text{MS}_{C(B)})/cdn,$$

and

$$\hat{\sigma}_\alpha^2 = (\text{MS}_A - \text{MS}_{B(A)})/bcdn.$$

As in Section 7.3, an exact confidence interval for σ_e^2 can be obtained by noting that $abcd(n-1)\text{MS}_E/\sigma_e^2$ has a chi-square distribution with $abcd(n-1)$ degrees of freedom. For a discussion of methods for constructing confidence intervals for other variance components, including certain sums and ratios of variance components, see Burdick and Graybill (1992, pp. 92–95).

7.6 GENERAL q-WAY NESTED CLASSIFICATION

The results of a nested classification can be readily generalized to the case of q completely nested factors. Such a design is also called a $(q+1)$-stage nested design. The general q-way nested classification model is a direct extension of the model (7.5.1) and can be written as

$$y_{ijk\ldots pqr} = \mu + \beta_{j(i)} + \gamma_{k(ij)} + \cdots + \delta_{q(ijk\ldots p)} + e_{r(ijk\ldots pq)} \quad \begin{cases} i = 1, 2, \ldots, a \\ j = 1, 2, \ldots, b \\ k = 1, 2, \ldots, c \\ \ldots\ldots\ldots \\ r = 1, 2, \ldots, n, \end{cases}$$
(7.6.1)

where μ is the general mean, $\alpha_i, \beta_{j(i)}, \gamma_{k(ij)}, \ldots, \delta_{q(ijk\ldots p)}$ are the effects due to the i-th level of factor A, the j-th level of factor B, the k-th level of factor $C, \ldots,$

the q-th level of factor Q, and $e_{r(ijk...pq)}$ is the customary error term that represents the variation within each cell. The assumptions of the model (7.6.1) are readily stated depending upon whether the levels of factors $A, B, C, \ldots, Q$ are fixed or random. For example, under Model II, the α_i's, $\beta_{j(i)}$'s, $\gamma_{k(ij)}$'s, $\ldots, \delta_{q(ijk...p)}$'s, and $e_{r(ijk...pq)}$'s are independent (normal) random variables with means of zero and variances $\sigma_\alpha^2, \sigma_\beta^2, \sigma_\gamma^2, \ldots$, and σ_e^2, respectively. For this balanced model, all the mean squares are distributed as constant times a chi-square random variable and they are statistically independent. Under the assumptions of the model of interest, the results on tests of hypotheses and estimation of variance components can be obtained in the manner described earlier. For a discussion of methods for constructing confidence intervals for the variance components, see Burdick and Graybill (1992, Section 5.3). For the analysis of a general q-way nested classification with unequal numbers in the subclasses, we use the same approach as given in Section 7.4. The details on analysis of variance, tests of hypotheses, and variance components estimation can be found in Gates and Shiue (1962) and Gower (1962). Khuri (1990) presents some exact tests for random models when all stages except the last one are balanced.

7.7 WORKED EXAMPLE FOR MODEL II

Brownlee (1953, p. 117) reported data from an experiment carried out to determine whether a batch of material was homogeneous. The material was sampled in six different vats. The matter from each vat was wrung in a centrifuge and bagged. Two bags were randomly selected from each vat and two samples were taken from each bag. Finally, for each sample, two independent determinations were made for the percentage of an ingredient. The data are given in Table 7.4.

The experimental structure follows a three-way nested or hierarchical classification and the mathematical model is

$$y_{ijk\ell} = \mu + \alpha_i + \beta_{j(i)} + \gamma_{k(ij)} + e_{\ell(ijk)} \quad \begin{cases} i = 1, 2, \ldots, 6 \\ j = 1, 2 \\ k = 1, 2 \\ \ell = 1, 2, \end{cases}$$

where $y_{ijk\ell}$ is the ℓ-th determination (analysis) of the k-th sample, of the j-th bag and for the i-th vat, μ is the general mean, α_i is the effect of the i-th vat, $\beta_{j(i)}$ is the effect of the j-th bag within the i-th vat, $\gamma_{k(ij)}$ is the effect of the k-th sample within the j-th bag within the i-th vat, and $e_{\ell(ijk)}$ is the customary error term. Furthermore, in this example, it is reasonable to assume that all factors are random and thus the α_i's, $\beta_{j(i)}$'s, $\gamma_{k(ij)}$'s, and $e_{\ell(ijk)}$'s are all independently and normally distributed with mean zero and variances $\sigma_\alpha^2, \sigma_\beta^2, \sigma_\gamma^2$, and σ_e^2, respectively.

TABLE 7.4
Percentage of Ingredient of a Batch of Material

Bags	1				2				3				4				5				6			
Samples	1		2		1		2		1		2		1		2		1		2		1		2	
Determinations	1	2	1	2	1	2	1	2	1	2	1	2	1	2	1	2	1	2	1	2	1	2	1	2
	29	28	29	27	29	27	26	24	32	29	25	30	29	30	28	30	30	27	25	26	29	31	29	29
	29	27	29	28	29	28	27	25	30	30	27	31	29	31	28	28	29	27	28	26	31	32	30	31
$Y_{ijk.}$	58	55	58	55	58	55	53	49	62	59	52	61	58	61	56	58	59	54	53	52	60	63	59	60
$Y_{ij..}$	113		113		113		102		121		113		119		114		113		105		123		119	
$Y_{i...}$	226				215				234				233				218				242			

Source: Brownlee (1953, p. 117). Used with permission.

Using the computational formulae for the sums of squares given in Section 7.2, we have

$$SS_T = (29)^2 + (29)^2 + \cdots + (31)^2 - \frac{(1{,}368)^2}{6 \times 2 \times 2 \times 2}$$

$$= 39{,}156 - 38{,}988$$

$$= 168.000,$$

$$SS_A = \frac{(226)^2 + (215)^2 + \cdots + (242)^2}{2 \times 2 \times 2} - \frac{(1{,}368)^2}{6 \times 2 \times 2 \times 2}$$

$$= 39{,}054.250 - 38{,}988$$

$$= 66.250,$$

$$SS_{B(A)} = \frac{(113)^2 + (113)^2 + \cdots + (119)^2}{2 \times 2} - \frac{(226)^2 + (215)^2 + \cdots + (242)^2}{2 \times 2 \times 2}$$

$$= 39{,}090.500 - 39{,}054.250$$

$$= 36.250,$$

$$SS_{C(B)} = \frac{(58)^2 + (55)^2 + \cdots + (60)^2}{2} - \frac{(113)^2 + (113)^2 + \cdots + (119)^2}{2 \times 2}$$

$$= 39{,}136 - 39{,}090.500$$

$$= 45.500,$$

and

$$SS_E = (29)^2 + (29)^2 + \cdots + (31)^2 - \frac{(58)^2 + (55)^2 + \cdots + (60)^2}{2}$$

$$= 39{,}156 - 39{,}136$$

$$= 20.000.$$

These results along with the remaining calculations are summarized in Table 7.5. The test of the hypothesis $H_0^{C(B)}: \sigma_\gamma^2 = 0$ versus $H_1^{C(B)}: \sigma_\gamma^2 > 0$ gives the variance ratio of 4.55 which is highly significant ($p < 0.001$). The test of the hypothesis $H_0^{B(A)}: \sigma_\beta^2 = 0$ versus $H_1^{B(A)}: \sigma_\beta^2 > 0$ gives the variance ratio of 1.59 which is not significant ($p = 0.232$). Finally, the test of the hypothesis $H_0^A: \sigma_\alpha^2 = 0$ versus $H_1^A: \sigma_\alpha^2 > 0$ gives the variance ratio of 2.19 which is again not significant ($p = 0.184$). However, note that the F test for vats has so few degrees of freedom that it may not be able to detect significant differences even if there really are important differences among them. Thus, we may conclude that although there seems to be significant variability among samples within bags and some variability between vats, there is no indication of any differences among bags within vats. The estimates of the variance components are

TABLE 7.5
Analysis of Variance for the Material Homogeneity Data of Table 7.4

Source of Variation	Degrees of Freedom	Sum of Squares	Mean Square	Expected Mean Square	F Value	p-Value
Vats	5	66.250	13.250	$\sigma_e^2 + 2\sigma_\gamma^2 + 2 \times 2\sigma_\beta^2 + 2 \times 2 \times 2\sigma_\alpha^2$	2.19	0.183
Bags (within vats)	6	36.250	6.042	$\sigma_e^2 + 2\sigma_\gamma^2 + 2 \times 2\sigma_\beta^2$	1.59	0.232
Samples (within bags)	12	45.500	3.792	$\sigma_e^2 + 2\sigma_\gamma^2$	4.55	<0.001
Error	24	20.000	0.833	σ_e^2		
Total	47	168.000				

given by

$$\hat{\sigma}_e^2 = 0.833,$$
$$\hat{\sigma}_\gamma^2 = \frac{1}{2}(3.792 - 0.833) = 1.480,$$
$$\hat{\sigma}_\beta^2 = \frac{1}{4}(6.042 - 3.792) = 0.563,$$

and

$$\hat{\sigma}_\alpha^2 = \frac{1}{8}(13.250 - 6.042) = 0.901.$$

These variance components account for 22.0, 39.2, 14.9, and 23.9 percent of the total variation in material content in this experiment. It is evident from this analysis that the batch of material under investigation is highly inhomogeneous and the larger part of this variability arises in bagging the material. The variability between repeated analyses on a given sample is also quite large, and there also seem to be appreciable differences between contents of each vat.

We further refine the preceding analysis by resorting to the method of pooling. In our earlier analysis we have seen that the variance component due to bags within vats (σ_β^2) is not statistically significant. We can thus pool its mean square with the samples within bags mean square to get a new estimate of $\sigma_e^2 + 2\sigma_\alpha^2$ equal to $(45.500 + 36.250)/18 = 4.542$ with 18 degrees of freedom. Now, the hypothesis on the variance component due to vats (σ_α^2) is tested by the between vats mean square against the pooled value of the between samples within bags mean square. The variance ratio for the test is $13.250/4.542 = 2.92$ with 5 and 18 degrees of freedom, respectively. Note that in contrast to the unpooled analysis, this value is significant at the 5 percent level ($p = 0.041$). Finally, the pooled estimates of variance components are now given by

$$\hat{\sigma}_e^2 = 0.833,$$
$$\hat{\sigma}_\gamma^2 = \frac{1}{2}(4.542 - 0.833) = 1.855,$$
$$\hat{\sigma}_\beta^2 = 0,$$

and

$$\hat{\sigma}_\alpha^2 = \frac{1}{8}(13.250 - 4.542) = 1.089.$$

These variance components account for 22.1, 49.1, 0, and 28.8 percent of the total variation in the material. The results of pooled analysis are similar to the earlier analysis. Thus, it is seen that there is an appreciable variability between vats. The variability between bags from a given vat is not large enough to be statistically significant. The variability between samples from a given bag is extremely large and, in fact, may account for nearly half of the total variation. The variability between duplicate analyses of a given sample is also quite large, probably the second most important component of variability in the process.

7.8 WORKED EXAMPLE FOR MODEL II: UNEQUAL NUMBERS IN THE SUBCLASSES

Damon and Harvey (1987, p. 29) reported data from an experiment to determine the number of diatoms at different locations on a river. (The original data were supplied by Dr. Richard Larsen of the Department of Fisheries and Wild Life of the University of Massachusetts.) The experiment entailed determination of the number of diatoms at two randomly selected locations, two or three bricks at each location, and one or two slides attached to each brick. Thus, there are two hierarchies of nesting, bricks nested within locations and slides nested within bricks. The number of diatoms per square centimeter colonizing each glass slide were determined and the data are given in Table 7.6.

The design structure follows a three-way nested or hierarchical classification and the mathematical model is

$$y_{ijk\ell} = \mu + \alpha_i + \beta_{j(i)} + \gamma_{k(ij)} + e_{\ell(ijk)} \quad \begin{cases} i = 1, 2 \\ j = 1, 2, \ldots, b_i \\ k = 1, 2, \ldots, c_{ij} \\ \ell = 1, 2, \ldots, n_{ijk}, \end{cases} \quad (7.8.1)$$

where $y_{ijk\ell}$ is the ℓ-th observation on the k-th slide, on the j-th brick and at the i-th location. Here, the number of bricks in the i-th location is designated as b_i ($b_1 = 3$, $b_2 = 2$), the number of slides in the $j(i)$-th brick (location) subclass

TABLE 7.6
Number of Diatoms per Square Centimeter Colonizing Glass Slides*

	Location 1					Location 2			
	Brick 1		Brick 2		Brick 3	Brick 1		Brick 2	
	Slide 1	Slide 2	Slide 1	Slide 2	Slide 1	Slide 1	Slide 2	Slide 1	Slide 2
	102	500	142	119	243	500	822	826	642
	111	480	125	114	189	165	743	750	710
	400	112	221	461	362	752	263	682	720
	380	103	464	382	264	142	321	522	584
	210	225		510		921	620	650	650
	245	361		380		792	584	621	
		842				871	841		
		657				900			
$y_{ijk.}$	1,448	3,280	952	1,966	1,058	5,043	4,194	4,051	3,306
n_{ijk}	6	8	4	6	4	8	7	6	5
$y_{ij..}$	4,728		2,918		1,058	9,237		7,357	
$n_{ij.}$	14		10		4	15		11	
$y_{i...}$	8,704					16,594			
$n_{i..}$	28					26			

Source: Damon and Harvey (1987, p. 29). Used with permission.

* Numbers have been coded to simplify computation.

is designated as c_{ij} ($c_{11}=2$, $c_{12}=2$, $c_{13}=1$, $c_{21}=2$, $c_{22}=2$), and the number of observations in the $k(ij)$-th slide (brick (location)) sub-subclass is designated as n_{ijk} ($n_{111}=6$, $n_{112}=8$, $n_{121}=4$, $n_{122}=6$, $n_{131}=4$, $n_{211}=8$, $n_{212}=7$, $n_{221}=6$, $n_{222}=5$). Furthermore, in the model equation (7.8.1), μ is the general mean, α_i is the effect of the i-th location, $\beta_{j(i)}$ is the effect of the j-th brick in the i-th location, $\gamma_{k(ij)}$ is the effect of the k-th slide on the j-th brick in the i-th location, and $e_{\ell(ijk)}$ is the customary error term. Finally, we will assume that all factors are random; that is, the α_i's, $\beta_{j(i)}$'s, $\gamma_{k(ij)}$'s, and $e_{\ell(ijk)}$'s are independently and normally distributed with mean zero and variances σ_α^2, σ_β^2, σ_γ^2, and σ_e^2, respectively.

All the quantities needed for the analysis of variance computations outlined in Section 7.4 can be readily computed on an electronic calculator. The results are:

$$y_{....}^2/N = (25{,}298)^2/54 = 11{,}851{,}644.52,$$

$$\sum_{i=1}^{a}\sum_{j=1}^{b_i}\sum_{k=1}^{c_{ij}}\sum_{\ell=1}^{n_{ijk}} y_{ijk\ell}^2 = 15{,}370{,}364,$$

Three-Way and Higher-Order Nested Classifications 413

$$\sum_{i=1}^{a}\sum_{j=1}^{b_i}\sum_{k=1}^{c_{ij}}\frac{y_{ijk.}^2}{n_{ijk}} = \frac{(1,448)^2}{6} + \frac{(3,280)^2}{8} + \cdots + \frac{(3,306)^2}{5}$$
$$= 13,457,673.97,$$
$$\sum_{i=1}^{a}\sum_{j=1}^{b_i}\frac{y_{ij..}^2}{n_{ij.}} = \frac{(4,728)^2}{14} + \frac{(2,918)^2}{10} + \cdots + \frac{(7,357)^2}{11}$$
$$= 13,336,666.51,$$

and

$$\sum_{i=1}^{a}\frac{y_{i..}^2}{n_{i..}} = \frac{(8,704)^2}{28} + \frac{(16,594)^2}{26}$$
$$= 13,296,501.96.$$

Thus,

$$SS_T = 15,370,364 - 11,851,644.52 = 3,518,719.48,$$
$$SS_A = 13,296,501.96 - 11,851,644.52 = 1,444,857.44,$$
$$SS_{B(A)} = 13,336,666.51 - 13,296,501.96 = 40,164.55,$$
$$SS_{C(B)} = 13,457,673.97 - 13,336,666.51 = 121,007.46,$$

and

$$SS_E = 15,370,364 - 13,457,673.97 = 1,912,690.03.$$

The corresponding degrees of freedom are computed as:

Total: $N - 1 = 54 - 1 = 53,$
Locations: $a - 1 = 2 - 1 = 1,$
Bricks (within locations): $\sum_{i=1}^{a} b_i - a = 5 - 2 = 3,$
Slides (within bricks): $\sum_{i=1}^{a}\sum_{j=1}^{b_i} c_{ij} - \sum_{i=1}^{a} b_i = 9 - 5 = 4,$
Error: $N - \sum_{i=1}^{a}\sum_{j=1}^{b} c_{ij} = 54 - 9 = 45.$

For appropriate tests of significance, we must evaluate the coefficients of the variance components in the expected mean squares, and determine the linear combinations of mean squares to be used as the denominator of an approximate F statistic using Satterthwaite procedure. The basic quantities needed to determine the coefficients of the variance components are computed as follows:

$$N = \sum_{i=1}^{a}\sum_{j=1}^{b_i}\sum_{k=1}^{c_{ij}} n_{ijk} = 6 + 8 + \cdots + 5 = 54,$$

$$k_1 = \frac{\sum_{i=1}^{a} n_{i..}^2}{N} = \frac{(28)^2 + (26)^2}{54} = 27.0370,$$

$$k_2 = \frac{\sum_{i=1}^{a}\sum_{j=1}^{b_i} n_{ij.}^2}{N} = \frac{(14)^2 + (10)^2 + \cdots + (11)^2}{54} = 12.1852,$$

$$k_3 = \frac{\sum_{i=1}^{a}\sum_{j=1}^{b_i}\sum_{k=1}^{c_{ij}} n_{ijk}^2}{N} = \frac{(6)^2 + (8)^2 + \cdots + (5)^2}{54} = 6.3333,$$

$$k_4 = \sum_{i=1}^{a}\sum_{j=1}^{b_i} \frac{n_{ij.}^2}{n_{i..}} = \frac{(14)^2 + (10)^2 + (4)^2}{28} + \frac{(15)^2 + (11)^2}{26} = 24.4506,$$

$$k_5 = \sum_{i=1}^{a}\sum_{j=1}^{b_i}\sum_{k=1}^{c_{ij}} \frac{n_{ijk}^2}{n_{i..}} = \frac{(6)^2 + (8)^2 + \cdots + (4)^2}{28} + \frac{(8)^2 + (7)^2 + \cdots + (5)^2}{26}$$
$$= 12.6923,$$

and

$$k_6 = \sum_{i=1}^{a}\sum_{j=1}^{b_i}\sum_{k=1}^{c_{ij}} \frac{n_{ijk}^2}{n_{ij.}} = \frac{(6)^2 + (8)^2}{14} + \frac{(4)^2 + (6)^2}{10} + \cdots + \frac{(6)^2 + (5)^2}{11}$$
$$= 29.4217.$$

Now, the coefficients of the variance components in the expected mean square column are given by

$$\bar{n}_1 = \frac{N - k_6}{c - b} = \frac{54 - 29.4217}{4} = 6.1446,$$

$$\bar{n}_2 = \frac{k_6 - k_5}{b - a} = \frac{29.4217 - 12.6923}{3} = 5.5765,$$

$$\bar{n}_3 = \frac{N - k_4}{b - a} = \frac{54 - 24.4506}{3} = 9.8498,$$

$$\bar{n}_4 = \frac{k_5 - k_3}{a - 1} = \frac{12.6923 - 6.3333}{1} = 6.3590,$$

$$\bar{n}_5 = \frac{k_4 - k_2}{a - 1} = \frac{24.4506 - 12.1852}{1} = 12.2654,$$

TABLE 7.7
Analysis of Variance for the Diatom Data of Table 7.6

Source of Variation	Degrees of Freedom	Sum of Squares	Mean Square	Expected Mean Square
Locations	1	1,444,857.44	1,444,857.440	$\sigma_e^2 + 6.3590\sigma_\gamma^2 + 12.2654\sigma_\beta^2 + 26.9630\sigma_\alpha^2$
Bricks (within locations)	3	40,164.55	13,388.183	$\sigma_e^2 + 5.5765\sigma_\gamma^2 + 9.8498\sigma_\beta^2$
Slides (within bricks)	4	121,007.46	30,251.865	$\sigma_e^2 + 6.1446\sigma_\gamma^2$
Error	45	1,912,690.03	42,504.223	σ_e^2
Total	53	3,518,719.48		

and

$$\bar{n}_6 = \frac{N - k_1}{a - 1} = \frac{54 - 27.0370}{1} = 26.9630.$$

The results on sums of squares, mean squares, and expected mean squares are summarized in Table 7.7.

The slides within bricks effects can be tested directly against the error mean square, giving $F = 30{,}251.865/42{,}504.223 = 0.712$ ($p = 0.588$). The results are clearly nonsignificant. An approximate F test for bricks within locations can be obtained using the slides within bricks mean square, giving $F = 13{,}388.183/30{,}251.865 = 0.443$ ($p = 0.735$). However, to use the Satterthwaite procedure, we first compute the coefficient as

$$\ell_2 = \bar{n}_2/\bar{n}_1 = 5.5765/6.1446 = 0.9075, \quad \ell_1 = 1 - \ell_2 = 0.0925;$$

and the synthesized mean square is

$$0.0925(42{,}504.223) + 0.9075(30{,}251.865) = 31{,}385.208.$$

The number of degrees of freedom for the synthesized mean square (rounded to the nearest digit) is

$$\nu' = \frac{(31{,}385.208)^2}{\frac{[0.0925(42{,}504.223)]^2}{45} + \frac{[0.9075(30{,}251.865)]^2}{4}} \doteq 5.$$

The F ratio based on the synthesized mean square is $F = 13{,}388.183/31{,}385.208 = 0.427$ ($p = 0.743$), which gives nearly the same result as before; that is, bricks within location effects are also not significant. Similar procedures are used to test for location effects. For example, an approximate F test for location effects can be obtained using the bricks within location mean square, giving

$F = 1{,}444{,}857.440/13{,}388.183 = 107.920\,(p = 0.002)$. To use the Satterthwaite procedure, the coefficients are

$$\ell_5 = \bar{n}_5/\bar{n}_3 = 12.2654/9.8498 = 1.2452,$$
$$\ell_4 = \bar{n}_4/\bar{n}_1 - \ell_5 \bar{n}_2/\bar{n}_1$$
$$= (6.3590/6.1446) - 1.2452(5.5765/6.1446) = -0.0952,$$
$$\ell_3 = 1 - \ell_4 - \ell_5 = 1 - (-0.0952) - 1.2452 = -0.1500$$

and the synthesized mean square is

$$-0.1500(42{,}504.223) + (-0.0952)(30{,}251.865) + 1.2452(13{,}388.183)$$
$$= 7{,}415.354.$$

The number of degrees of freedom for the synthesized mean square (rounded to the nearest digit) is

$$v'' = \frac{(7{,}415.354)^2}{\dfrac{[-0.1500(42{,}504.223)]^2}{45} + \dfrac{[-0.0952(30{,}251.865)]^2}{4} + \dfrac{[1.2452(13{,}388.183)]^2}{3}}$$
$$\doteq 2.$$

The F ratio based on the synthesized mean square is $1{,}444{,}857.440/7{,}415.354 = 194.847$. Again, the results are highly significant ($p < 0.001$).

The estimates of the variance components σ_α^2, σ_β^2, σ_γ^2, and σ_e^2 are obtained as the solution to the following simultaneous equations:

$$1{,}444{,}857.440 = \sigma_e^2 + 6.3590\sigma_\gamma^2 + 12.2654\sigma_\beta^2 + 26.9630\sigma_\alpha^2,$$
$$13{,}388.183 = \sigma_e^2 + 5.5765\sigma_\gamma^2 + 9.8498\sigma_\beta^2,$$
$$30{,}251.865 = \sigma_e^2 + 6.1446\sigma_\gamma^2,$$

and

$$42{,}502.223 = \sigma_e^2.$$

Therefore, the desired estimates are given by

$$\hat{\sigma}_e^2 = 42{,}504.223,$$
$$\hat{\sigma}_\gamma^2 = \frac{30{,}251.865 - 42{,}504.223}{6.1446} = -1{,}994.004,$$
$$\hat{\sigma}_\beta^2 = \frac{13{,}388.183 - 42{,}504.223 - 5.5765(-1{,}994.004)}{9.8498} = -1{,}827.091,$$

and

$$\hat{\sigma}_\alpha^2 = \frac{1{,}444{,}857.440 - 42{,}504.223 - 6.3590(-1{,}994.004) - 12.2654(-1{,}827.091)}{26.9630}$$
$$= 53{,}311.690.$$

Three-Way and Higher-Order Nested Classifications 417

The negative estimates are probably an indication that the corresponding variance components may be zero. The point estimates of variance components are consistent with the results on tests of hypotheses. It is further evident from the analysis that the most of the variation in the number of diatoms is due to different location on the river.

7.9 WORKED EXAMPLE FOR MODEL III

Sokal and Rohlf (1995, p. 289) reported data from an experiment designed to analyze glycogen content of rat livers. For each of the three treatments — control, compound 217, and compound 217 plus sugar — used in the experiment, three preparations of rat livers from each of the two rats were analyzed and duplicate readings were made for each preparation. The data are given in Table 7.8.

The design structure follows a three-way nested or hierarchical classification and the mathematical model is

$$y_{ijk\ell} = \mu + \alpha_i + \beta_j + \gamma_{k(ij)} + e_{\ell(ijk)} \quad \begin{cases} i = 1, 2, 3 \\ j = 1, 2 \\ k = 1, 2, 3 \\ \ell = 1, 2, \end{cases} \quad (7.9.1)$$

where $y_{ijk\ell}$ is the ℓ-th observation (reading) on the k-th preparation, on the j-th rat and for the i-th treatment, μ is the general mean, α_i is the effect of the i-th treatment, $\beta_{j(i)}$ is the effect of the j-th rat within the i-th treatment, and $\gamma_{k(ij)}$ is the effect of the k-th preparation within the j-th rat within the i-th treatment, and $e_{\ell(ijk)}$ is the customary error term. Furthermore, the α_i's are considered to be fixed effects with $\sum_{i=1}^{3} \alpha_i = 0$, and the $\beta_{j(i)}$'s, $\gamma_{k(ij)}$'s, and $e_{\ell(ijk)}$'s are assumed to be independently and normally distributed with mean zero and variances $\sigma_\beta^2, \sigma_\gamma^2$, and σ_e^2, respectively.

Using the computational formulae for the sums of squares given in Section 7.2, we have

$$SS_T = (131)^2 + (130)^2 + \cdots + (127)^2 - \frac{(5,120)^2}{3 \times 2 \times 3 \times 2}$$
$$= 731,508 - 728,177.778$$
$$= 3,330.222,$$

$$SS_A = \frac{(1,686)^2 + (1,812)^2 + (1,622)^2}{2 \times 3 \times 2} - \frac{(5,120)^2}{3 \times 2 \times 3 \times 2}$$
$$= 729,735.333 - 728,177.778$$
$$= 1,557.555,$$

$$SS_{B(A)} = \frac{(795)^2 + (891)^2 + \cdots + (816)^2}{3 \times 2} - \frac{(1,686)^2 + (1,812)^2 + (1,622)^2}{2 \times 3 \times 2}$$
$$= 730,533 - 729,735.333$$
$$= 797.667,$$

TABLE 7.8
Glycogen Content of Rat Livers in Arbitrary Units

	Treatments																	
	Control						Compound 217						Compound 217 plus Sugar					
Rats	1			2			1			2			1			2		
Preparations	1	2	3	1	2	3	1	2	3	1	2	3	1	2	3	1	2	3
Readings	131	131	136	150	140	160	157	154	147	151	147	162	134	138	135	138	139	134
	130	125	142	148	143	150	145	142	153	155	147	152	125	138	136	140	138	127
$Y_{ijk.}$	261	256	278	298	283	310	302	296	300	306	294	314	259	276	271	278	277	261
$Y_{ij..}$	795			891			898			914			806			816		
$Y_{i...}$	1,686						1,812						1,622					

Source: Sokal and Rohlf (1995, p. 289). Used with permission.

$$SS_{C(B)} = \frac{(261)^2 + (256)^2 + \cdots + (261)^2}{2} - \frac{(795)^2 + (891)^2 + \cdots + (816)^2}{3 \times 2}$$

$$= 731,127 - 730,533$$

$$= 594.000,$$

and

$$SS_E = (131)^2 + (130)^2 + \cdots + (127)^2 - \frac{(261)^2 + (256)^2 + \cdots + (261)^2}{2}$$

$$= 731,508 - 731,127$$

$$= 381.000.$$

These results along with the remaining computations are summarized in Table 7.9. The test of the hypothesis $H_0^{C(B)}: \sigma_\gamma^2 = 0$ versus $H_1^{C(B)}: \sigma_\gamma^2 > 0$ gives the variance ratio 2.34 which barely reaches its 5 percent critical value of 2.342 ($p = 0.050$). The test of the hypothesis $H_0^{B(A)}: \sigma_\beta^2 = 0$ versus $H_1^{B(A)}: \sigma_\beta^2 > 0$ gives the variance ratio of 5.37 which clearly exceeds its 5 percent critical value of 3.49 ($p = 0.014$). Finally, the test of the hypothesis H_0^A: all $\alpha_i = 0$ versus H_1^A: all $\alpha_i \neq 0$ gives the variance ratio of 2.93 which is too low to reach its 5 percent critical value of 9.55 ($p = 0.197$). Thus, we may conclude that although there seem to be significant differences among preparations within rats and among rats within treatments, there is no indication of any differences between the treatments. However, note that the F test for treatments has so few degrees of freedom that it may not be able to detect significant differences even if there are really important differences among them. Perhaps repetition of the experiment using more rats per treatment is indicated. The estimates of variance components $\sigma_e^2, \sigma_\gamma^2$, and σ_β^2 are given by

$$\hat{\sigma}_e^2 = 21.167,$$

$$\hat{\sigma}_\gamma^2 = \frac{1}{2}(49.500 - 21.167) = 14.167,$$

and

$$\hat{\sigma}_\beta^2 = \frac{1}{6}(265.889 - 49.500) = 36.065.$$

These variance components account for 29.6, 19.8, and 50.5 percent of the total variation in glycogen content in this experiment. It is evident from this analysis that the large part of the variability arises among rats within treatments. Readings within preparations and preparations within rats also seem to account for a significant portion of the total variability in the experiment. However, we cannot establish significant differences among treatments.

TABLE 7.9
Analysis of Variance for the Glycogen Data of Table 7.8

Source of Variation	Degrees of Freedom	Sum of Squares	Mean Square	Expected Mean Square	F Value	p-Value
Treatments	2	1,557.555	778.778	$\sigma_e^2 + 2\sigma_\gamma^2 + 6\sigma_\beta^2 + \dfrac{12\sum_{i=1}^{3}\alpha_i^2}{3-1}$	2.93	0.197
Rats (within treatments)	3	797.667	265.889	$\sigma_e^2 + 2\sigma_\gamma^2 + 6\sigma_\beta^2$	5.37	0.014
Preparations (within rats)	12	594.000	49.500	$\sigma_e^2 + 2\sigma_\gamma^2$	2.34	0.050
Error	18	381.000	21.167	σ_e^2		
Total	35	3,330.222				

7.10 USE OF STATISTICAL COMPUTING PACKAGES

The use of SAS, SPSS, and BMDP programs for analyzing three- and higher-order nested factors is the same as described in Section 6.15 for the case of two-way nested designs. No new problems arise for analysis involving higher-order nested designs.

7.11 WORKED EXAMPLES USING STATISTICAL PACKAGES

In this section, we illustrate the application of statistical packages to perform three-way nested analysis of variance for the data sets employed in examples presented in Sections 7.7 through 7.9. Figures 7.2 through 7.4 illustrate the program instructions and the output results for analyzing data in Tables 7.4, 7.6, and 7.8 using SAS GLM, SPSS GLM, and BMDP 3V/8V procedures. The typical output provides the data format listed at the top, all cell means, and the entries of the analysis of variance table. It should be noticed that in each case the results are the same as those provided using manual computations in Sections 7.7 through 7.9. However, note that in an unbalanced design, certain tests of significance may differ from one program to the other since they use different types of sums of squares.

```
DATA INGREDIENT;                              The SAS System
INPUT VAT BAG SAMPLE                    General Linear Models Procedure
PERCENT;                         Dependent Variable: PERCENT
DATALINES;                                        Sum of          Mean
1 1 1 29                         Source       DF  Squares         Square      F Value  Pr > F
1 1 1 29                         Model        23  148.00000000    6.43478261    7.72   0.0001
1 1 2 28                         Error        24   20.00000000    0.83333333
1 1 2 27                         Corrected    47  168.00000000
. . . .                          Total
6 2 2 31                                      R-Square      C.V.      Root MSE    PERCENT Mean
;                                             0.880952   3.2030559   0.91287093   28.50000000
PROC GLM;                        Source       DF   Type III SS   Mean Square  F Value  Pr > F
CLASSES VAT BAG SAMPLE;          VAT           5    66.250000    13.250000     15.90   0.0001
MODEL PERCENT=VAT BAG(VAT)       BAG(VAT)      6    36.250000     6.041667      7.25   0.0002
SAMPLE(BAG VAT);                 SAMPLE(VAT*BAG) 12  45.500000    3.791667      4.55   0.0008
RANDOM VAT BAG(VAT)              Source             Type III Expected Mean Square
SAMPLE(BAG VAT);                 VAT            Var(Error)+2 Var(SAMPLE(VAT*BAG))+4Var(BAG(VAT))
TEST H=VAT E=BAG(VAT);                         + 8 Var(VAT)
TEST H=BAG(VAT);                 BAG(VAT)       Var(Error)+2 Var(SAMPLE(VAT*BAG))+4Var(BAG(VAT))
E=SAMPLE(BAG VAT);               SAMPLE(VAT*BAG) Var(Error) + 2 Var(SAMPLE(VAT*BAG))
RUN;
CLASS  LEVELS  VALUES            Tests of Hypotheses using the Type III MS for BAG(VAT) as an
VAT     6      1 2 3 4 5 6       error term
BAG     2      1 2               Source       DF   Type III SS   Mean Square  F Value  Pr > F
SAMPLE  2      1 2               VAT           5    66.250000    13.250000     2.19   0.1834
NUMBER OF OBS. IN DATA           Source       DF   Type III SS   Mean Square  F Value  Pr > F
SET=48                           BAG(VAT)      6    36.25000000   6.04166667    1.59   0.2316
```

(i) SAS application: SAS GLM instructions and output for the three-way random effects nested analysis of variance.

FIGURE 7.2 Program Instructions and Output for the Three-Way Random Effects Nested Analysis of Variance: Material Homogeneity Data for Example of Section 7.7 (Table 7.4).

```
DATA LIST              Tests of Between-Subjects Effects          Dependent Variable: PERCENT
/VAT 1
 BAG 3                 Source              Type III SS    df    Mean Square       F      Sig.
 SAMPLE 5              VAT     Hypothesis    66.250       5       13.250        2.193    .183
 PERCENT 7-8.                  Error         36.250       6        6.042(a)
BEGIN DATA.            BAG(VAT) Hypothesis   36.250       6        6.042        1.593    .232
1 1 1 29                       Error         45.500      12        3.792(b)
1 1 1 29               SAMPLE(BAG Hypothesis 45.500      12        3.792        4.550    .001
1 1 2 28                (VAT))  Error        20.000      24        0.833(c)
1 1 2 27               a  MS(BAG(VAT))  b  MS(SAMPLE(BAG(VAT)))  c  MS(ERROR)
. . . .
6 2 2 31                                            Expected Mean Squares(a,b)
END DATA.                                               Variance Component
 GLM PERCENT           Source            Var(VAT)   Var(BAG(VAT))  Var(SAMPLE(BAG))  Var(Error)
  BY VAT BAG           VAT                 8.000       4.000            2.000          1.000
  SAMPLE.              BAG(VAT)             .000       4.000            2.000          1.000
 /DESIGN  VAT          SAMPLE(BAG(VAT))     .000        .000            2.000          1.000
  BAG(VAT)             Error                .000        .000             .000          1.000
  SAMPLE(BAG           a  For each source, the expected mean square equals the sum of the
  (VAT))               coefficients in the cells times the variance components, plus a quadratic
 /RANDOM VAT BAG       term involving effects in the Quadratic Term cell. b Expected Mean Squares
  SAMPLE.              are based on the Type III Sums of Squares.
```

(ii) SPSS application: SPSS GLM instructions and output for the three-way random effects nested analysis of variance.

```
/INPUT     FILE='C:\SAHAI\       BMDP8V - GENERAL MIXED MODEL ANALYSIS OF VARIANCE
           TEXTO\EJE19.TXT'.            - EQUAL CELL SIZES Release: 7.0 (BMDP/DYNAMIC)
           FORMAT=FREE.
           VARIABLES=2.          ANALYSIS OF VARIANCE FOR DEPENDENT VARIABLE   1
/VARIABLE  NAMES=D1,D2.
/DESIGN    NAMES=V,B,S,D.         SOURCE   ERROR     SUM OF       D.F.  MEAN         F      PROB.
           LEVELS=6,2,2,2.                 TERM      SQUARES            SQUARE
           RANDOM=V,B,S,D.        1 MEAN   VAT      38988.00000    1   38988.000   2942.49  0.0000
           MODEL='V,B(V),S(B),    2 VAT    B(V)        66.25000    5      13.250      2.19  0.1834
                  D(S)'.          3 B(V)   S(VB)       36.25000    6       6.042      1.59  0.2316
/END                              4 S(VB)  D(VBS)      45.50000   12       3.792      4.55  0.0008
29 29                             5 D(VBS)             20.00000   24       0.833
. .                               SOURCE    EXPECTED MEAN                ESTIMATES OF
29 31                                         SQUARE                  VARIANCE COMPONENTS
ANALYSIS OF VARIANCE DESIGN       1 MEAN   48(1)+8(2)+4(3)+2(4)+(5)      811.97396
INDEX       V  B  S  D            2 VAT    8(2)+4(3)+2(4)+(5)              0.90104
NUM LEVELS  6  2  2  2            3 B(V)   4(3)+2(4)+(5)                   0.56250
POPULATION INF INF INF INF        4 S(VB)  2(4)+(5)                        1.47917
MODEL     V, B(V), S(B), D(S)     5 D(VBS) (5)                             0.83333
```

(iii) BMDP application: BMDP 8V instructions and output for the three-way random effects nested analysis of variance.

FIGURE 7.2 (*continued*)

Three-Way and Higher-Order Nested Classifications

```
DATA DIATOMS;                          The SAS System
INPUT LOCATION BRICK             General Linear Models Procedure
SLIDE DIATOMS;              Dependent Variable: DIATOMS
DATALINES;                                    Sum of          Mean
1 1 1 102                   Source        DF  Squares       Square    F Value  Pr > F
1 1 1 111                   Model          8  1606029.4    200753.7     4.72   0.0003
1 1 1 400                   Error         45  1912690.0     42504.2
1 1 1 380                   Corrected
1 1 1 210                   Total         53  3518719.5
1 1 1 245                                 R-Square    C.V.    Root MSE  DIATOMS Mean
1 1 2 500                                 0.456424  44.00719   206.17      468.48
1 1 2 480                   Source        DF  Type I SS   Mean Square  F Value  Pr > F
1 1 2 112                   LOCATION       1  1444857.4    1444857.4    33.99   0.0001
1 1 2 103                   Source        DF  Type I SS   Mean Square  F Value  Pr > F
1 1 2 225                   BRICK(LOCATION) 3   40164.6     13388.2     0.31    0.8144
1 1 2 361                   SLIDE(LOCATION*BRICK) 4 121007.5  30251.9   0.71    0.5882
1 1 2 842                   Source        DF  Type III SS Mean Square  F Value  Pr > F
1 1 2 657                   LOCATION       1  1483396.9    1483396.9   34.90    0.0001
1 2 1 142                   BRICK(LOCATION) 3   34822.4     11607.5     0.27    0.8445
1 2 1 125                   SLIDE(LOCATION*BRICK) 4 121007.5  30251.9   0.71    0.5882
1 2 1 221                   Source         Type III Expected Mean Square
  . . .                     LOCATION       Var(Error) + 5.591 Var(SLIDE(LOCATION*BRICK))
2 2 2 650                                + 10.271 Var(BRICK(LOCATION)) + 24.489
;                           Var(LOCATION)
PROC GLM;                   BRICK(LOCATION)  Var(Error) + 5.4151 Var(SLIDE(LOCATION*BRICK))
CLASSES LOCATION BRICK                     + 9.6922 Var(BRICK(LOCATION))
SLIDE;                      SLIDE(LOCATION*BRICK)  Var(Error) + 6.1446 Var(SLIDE(LOCATION*BRICK))
MODEL DIATOMS=LOCATION      Source: LOCATION   Error: 1.0598*MS(BRICK(LOCATION)) -
BRICK(LOCATION)             0.024*MS(SLIDE(LOCATION*BRICK)) - 0.0357*MS(Error)
SLIDE(BRICK LOCATION);                      Denominator    Denominator
RANDOM LOCATION             DF  Type III MS   DF    MS     F Value   Pr > F
BRICK(LOCATION)              1  1483396.9269  2.00 10055.801769 147.5165  0.0067
SLIDE(BRICK LOCATION)/      Source: BRICK(LOCATION) Error:0.8813*MS(SLIDE(LOCATION*BRICK))
TEST;                       +.1187*MS(Error)
RUN;                                         Denominator    Denominator
CLASS    LEVELS VALUES      DF  Type III MS   DF    MS     F Value   Pr > F
LOCATION    2    1 2         3  11607.477802  5.64 31706.429974  0.3661  0.7807
BRICK       3    1 2 3      Source: SLIDE(LOCATION*BRICK) Error: MS(Error)
SLIDE       2    1 2                         Denominator    Denominator
NUMBER OF OBS. IN DATA      DF  Type III MS   DF    MS     F Value   Pr > F
SET=54                       4  30251.865341  45  42504.222937  0.7117  0.5882
```

(i) SAS application: SAS GLM instructions and output for the three-way random effects nested analysis of variance with unequal numbers in the subclasses.

```
DATA LIST            Tests of Between-Subjects Effects      Dependent Variable: DIATOMS
/LOCATION 1
 BRICK  3                                   Type III SS  df  Mean Square    F      Sig
 SLIDE  5            Source    LOCATION  Hypothesis 1483396.927  1  1483396.927 147.517  .007
 DIATOMS 7-9.                  Error           20086.860 1.998   10055.802(a)
BEGIN DATA.                    BRICK(LOC) Hypothesis  34822.434  3   11607.478    .366   .781
1 1 1 102                      Error          178807.755 5.639  31706.430(b)
1 1 1 111                      SLIDE(BRICK Hypothesis 121007.461  4   30251.865    .712   .588
1 1 1 400                      (LOCATION)) Error     1912690.000  45  42504.229(c)
1 1 1 380            a 1.060 MS(B(L))-2.403E-02 MS(S(B(L)))-3.572E-02 MS(E) b .881 MS (S(B(L)))+
  . . .              .119 MS(E) c MS(E)
2 2 2 650.
END DATA.                                    Expected Mean Squares(a,b)
GLM DIATOMS                                    Variance Component
BY LOCATION          Source            Var(LOC)  Var(B(LOC))  Var(S(B)   Var(Error)
 BRICK SLIDE         LOCATION          24.489      10.271       5.591      1.000
/RANDOM LOCATION     BRICK(LOCATION)     .000       9.692       5.415      1.000
 BRICK SLIDE         SLIDE(BRICK(LOCATION)) .000    .000        6.145      1.000
/DESIGN              Error                .000      .000         .000      1.000
 LOCATION            a For each source, the expected mean square equals the sum of the
 BRICK(LOCATION)     coefficients in the cells times the variance components, plus a quadratic
 SLIDE(BRICK         term involving effects in the Quadratic Term cell. b Expected Mean Squares
 (LOCATION)).        are based on the Type III Sums of Squares.
```

(ii) SPSS application: SPSS GLM instructions and output for the three-way random effects nested analysis of variance with unequal numbers in the subclasses.

FIGURE 7.3 Program Instructions and Output for the Three-Way Random Effects Nested Analysis of Variance with Unequal Numbers in the Subclasses: Diatom Data for Example of Section 7.8 (Table 7.6).

```
/INPUT     FILE='C:\SAHAI              BMDP3V - GENERAL MIXED MODEL ANALYSIS OF VARIANCE
           \TEXTO\EJE20.TXT'.                   Release: 7.0         (BMDP/DYNAMIC)
           FORMAT=FREE.
           VARIABLES=4.                        DEPENDENT VARIABLE DIATOM
/VARIABLE  NAMES=LOCATION,BRICK,
           SLIDE,DIATOM.                PARAMETER  ESTIMATE   STANDARD    EST/     TWO-TAIL PROB.
/GROUP     CODES(LOCATION)=1,2.                               ERROR       ST.DEV.  (ASYM. THEORY)
           NAMES(LOCATION)=L1,L2.
           CODES(BRICK)=1,2,3.          ERR.VAR.   39881.962  7821.496
           NAMES(BRICK)=B1,B2,B3.       CONSTANT     474.377   163.687    2.898       0.004
           CODES(SLIDE)=1,2.            LOCATION   52107.607 75783.658
           NAMES(SLIDE)=S1,S2.          BRK(LOC)       0.000     0.000
/DESIGN    DEPENDENT=DIATOM.            SLD(BRK)       0.000     0.000
           RANDOM=LOCATION.
           RANDOM=LOCATION, BRICK.
           RANDOM=BRICK, SLIDE.         TESTS OF FIXED EFFECTS BASED ON ASYMPTOTIC VARIANCE
           RNAMES=L,'B(L)','S(B)'.      -COVARIANCE MATRIX
           METHOD=REML.
/END                                    SOURCE        F-STATISTIC    DEGREES OF      PROBABILITY
1 1 1 102                                                            FREEDOM
. . . .
2 2 2 650                               CONSTANT         8.40         1    53         0.00545
```

(iii) BMDP application: BMDP 3V instructions and output for the three-way random effects nested analysis of variance with unequal numbers in the subclasses.

FIGURE 7.3 (*continued*)

```
DATA GLYCOGEN;                                       The SAS System
INPUT TREATMNT $ RAT                         General Linear Models Procedure
PREPARAT GLYCOGEN;           Dependent Variable: GLYCOGEN
DATALINES;                                            Sum of        Mean
CONTROL 1 1 131              Source           DF     Squares       Square     F Value   Pr > F
CONTROL 1 1 130              Model            17    2949.22222   173.48366      8.20    0.0001
CONTROL 1 2 131              Error            18     381.00000    21.16667
CONTROL 1 2 125              Corrected        35    3330.22222
CONTROL 1 3 136              Total
CONTROL 1 3 142                              R-Square      C.V.     Root MSE    GLYCOGEN Mean
CONTROL 2 1 150                              0.885593    3.234884   4.60072      142.222
CONTROL 2 1 148              Source           DF   Type III SS  Mean Square  F Value  Pr > F
CONTROL 2 2 140              TREATMNT          2    1557.55556   778.77778    36.79   0.0001
CONTROL 2 2 143              RAT(TREATMNT)     3     797.66667   265.88889    12.56   0.0001
  .                          PREPARAT(TREATMNT*RAT)12 594.00000   49.50000     2.34   0.0503
C217SUG 2 3 127              Source                Type III Expected Mean Square
;                            TREATMNT             Var(Error) + 2 Var(PREPARAT(TREATMNT*RAT))
PROC GLM;                                         + 6 Var(RAT(TREATMNT)) + Q(TREATMNT)
CLASSES TREATMNT RAT         RAT(TREATMNT)        Var(Error) + 2 Var(PREPARAT(TREATMNT*RAT))
PREPARAT;                                         + 6 Var(RAT(TREATMNT))
MODEL GLYCOGEN=TREATMNT      PREPARAT(TREATMNT*RAT) Var(Error) + 2 Var(PREPARAT(TREATMNT*RAT))
RAT(TREATMNT)                Tests of Hypotheses for Mixed Model Analysis of Variance
PREPARAT(RAT TREATMNT);      Source: TREATMNT   Error: MS(RAT(TREATMNT))
RANDOM RAT(TREATMNT)                             Denominator   Denominator
PREPARAT(RAT TREATMNT)/          DF   Type III MS    DF        MS       F Value  Pr > F
TEST;                             2   778.77777778    3   265.88888889   2.9290   0.1971
RUN;                         Source: RAT(TREATMNT)   Error: MS(PREPARAT(TREATMNT*RAT))
CLASS   LEVELS   VALUES                          Denominator   Denominator
TREATMNT   3      C217           DF   Type III MS    DF        MS       F Value  Pr > F
           C217SUG  CONTROL       3   265.88888889   12      49.5        5.3715   0.0141
RAT        2     1 2             Source: PREPARAT(TREATMNT*RAT)  Error: MS(Error)
PREPARAT   3     1 2 3                           Denominator   Denominator
NUMBER OF OBS. IN DATA           DF   Type III MS    DF        MS       F Value  Pr > F
SET=36                           12      49.5        18    21.16666666   2.3386   0.0503
```

(i) SAS application: SAS GLM instructions and output for the three-way mixed effects nested analysis of variance.

FIGURE 7.4 Program Instructions and Output for the Three-Way Mixed Effects Nested Analysis of Variance: Glycogen Data for Example of Section 7.9 (Table 7.8).

Three-Way and Higher-Order Nested Classifications 425

```
DATA LIST              Tests of Between-Subjects Effects        Dependent Variable: GLYCOGEN
/TREATMNT  1
  RAT      3           Source                    Type III SS   df   Mean Square     F      Sig.
  PREPARAT 5           TREATMNT   Hypothesis       1557.556    2     778.778      2.929    .197
  GLYCOGEN 7-9.                   Error            797.667     3     265.889(a)
BEGIN DATA.            RAT(TREATMN) Hypothesis     797.667     3     265.889      5.371    .014
1 1 1 131                          Error            594.000   12      49.500(b)
1 1 1 130              PREPARAT(RAT Hypothesis      594.000   12      49.500      2.339    .050
1 1 2 131              (TREATMNT))  Error            381.000   18      21.167(c)
1 1 2 125              a  MS(RAT(TREATMNT))  b  MS(PREPARAT(RAT(TREATMNT))
 . . .                 c  MS(E)
3 2 3 127
END DATA.                                       Expected Mean Squares(a,b)
GLM GLYCOGEN                                        Variance Component
 BY TREATMNT RAT       Source                  Var(R(T)) Var(P(R(T))) Var(Error) Quadratic Term
 PREPARAT              TREATMNT                  6.000     2.000        1.000      TREATMNT
 /DESIGN               RAT(TREATMNT)             6.000     2.000        1.000
  TREAMNT              PREPARAT(RAT(TREATMNT))    .000     2.000        1.000
  RAT(TREATMNT)        Error                      .000      .000        1.000
  PREPARAT(RAT         a  For  each source, the expected mean square equals the sum of the
  (TREATMNT)).         coefficients in the cells times the variance components, plus a quadratic
 /RANDOM RAT           term involving effects in the Quadratic Term cell. b Expected Mean Squares
  PREPARAT.            are based on the Type III Sums of Squares.
```

(ii) SPSS application: SPSS GLM instructions and output for the three-way mixed effects nested analysis of variance.

```
/INPUT    FILE='C:\SAHAI       BMDP8V - GENERAL MIXED MODEL ANALYSIS OF VARIANCE
          \TEXTO\EJE21.TXT'.           - EQUAL CELL SIZES Release: 7.0 (BMDP/DYNAMIC)
          FORMAT=FREE.
          VARIABLES=2.          ANALYSIS OF VARIANCE FOR DEPENDENT VARIABLE  1
/VARIABLE NAMES=R1,R2.
/DESIGN   NAMES=T,R,P,D.         SOURCE    ERROR    SUM OF    D.F.   MEAN        F      PROB.
          LEVELS=3,2,3,2.                  TERM    SQUARES            SQUARE
          RANDOM=R,P,D.         1 MEAN      R(T)  728177.778   1   728177.78  2738.65  0.0000
          FIXED=T.              2 TREATMNT  R(T)    1557.556   2      778.78     2.93  0.1971
          MODEL='T,R(T),P(R),   3 R(T)     P(TR)     797.667   3      265.89     5.37  0.0141
          D(P)'.                4 P(TR)    D(TRP)    594.000  12       49.50     2.34  0.0503
/END                            5 D(TRP)             381.000  18       21.17
131 130
 . .                              SOURCE      EXPECTED MEAN             ESTIMATES OF
134 127                                          SQUERE               VARIANCE COMPONENTS
ANALYSIS OF VARIANCE DESIGN     1 MEAN      36(1)+6(3)+2(4)+(5)         20219.77469
INDEX          T  R  P  D       2 TREATMNT  12(2)+6(3)+2(4)+(5)            42.74074
NUM LEVELS     3  2  3  2       3 R(T)       6(3)+2(4)+(5)                 36.06481
POPULATION SIZE 3 INF INF INF   4 P(TR)      2(4)+(5)                      14.16667
MODEL          T, R(T),P(R),D(P) 5 D(TRP)     (5)                          21.16667
```

(iii) BMDP application: BMDP 8V instructions and output for the three-way mixed effects nested analysis of variance.

FIGURE 7.4 (*continued*)

EXERCISES

1. An experiment is performed to investigate alloy hardness using a three-way nested design having two fixed alloys with different chemistries, three heats within each alloy, two ingots within each heat, and two determinations are made on each ingot. The data in certain standard units are given as follows.

Alloys	1						2					
Heat	1		2		3		1		2		3	
Ingot	1	2	1	2	1	2	1	2	1	2	1	2
	31	18	86	60	56	69	33	14	74	66	52	26
	54	21	58	58	45	36	23	30	53	55	68	33

(a) Describe the model and the assumptions for the experiment. It is assumed that alloys and heats are fixed and ingots are random.
(b) Analyze the data and report the analysis of variance table.
(c) Test whether there are differences in mean hardness levels between alloy chemistries. Use $\alpha = 0.05$.
(d) Test whether there are differences in mean hardness levels between heats within alloys. Use $\alpha = 0.05$.
(e) Test whether there are differences in mean hardness levels between ingots within heats. Use $\alpha = 0.05$.
(f) Estimate the variance components of the model and determine 95 percent confidence intervals on them.

2. A chemical company wishes to examine the strength of a certain liquid chemical. The chemical is made in large vats and is then barreled. A random sample of three different vats is selected, three barrels are selected at random from each vat, and then two samples are taken for each barrel. Finally, two independent measurements are made on each sample. The data in certain standard units are given as follows.

Vat	1						2						3					
Barrel	1		2		3		1		2		3		1		2		3	
Sample	1	2	1	2	1	2	1	2	1	2	1	2	1	2	1	2	1	2
	4.3	4.0	4.3	4.6	4.7	4.9	4.8	4.6	4.7	4.5	4.3	4.5	5.0	5.3	5.1	5.0	5.0	5.1
	4.1	4.5	4.5	4.4	4.4	4.3	4.7	4.5	4.5	4.7	4.7	5.1	4.8	5.2	4.8	5.2	4.7	4.9

(a) Describe the model and the assumptions for the experiment.
(b) Analyze the data and report the analysis of variance table.
(c) Test whether there are differences in mean strength levels between vats. Use $\alpha = 0.05$.
(d) Test whether there are differences in mean strength levels between barrels within vats. Use $\alpha = 0.05$.
(e) Test whether there are differences in mean strength levels between samples within barrels. Use $\alpha = 0.05$.

(f) Estimate the variance components of the model and determine 95 percent confidence interval on them.
3. Consider an experiment designed to study heat transfer in the molds utilized in manufacturing household plastics. A company has two plants that manufacture household plastics. Two furnaces are randomly selected from each plant and two molds are drawn from each furnace. The response variable of interest is the mold temperature, and five temperatures are recorded from each mold. The data from test results of the experiment are given as follows.

Plant	1				2			
Furnace	1		2		1		2	
Mold	1	2	1	2	1	2	1	2
Temperature (°C)	468	473	474	475	481	481	480	480
	476	477	475	473	480	477	478	477
	474	470	478	472	481	477	482	481
	481	473	472	475	480	482	480	474
	473	478	474	475	480	480	480	481

(a) Describe the model and the assumptions for the experiment. It is assumed that the effect due to plant is a fixed effect whereas furnace and mold are random factors.
(b) Analyze the data and report the analysis of variance table.
(c) Test whether there are differences in mean temperature levels between the plants. Use $\alpha = 0.05$.
(d) Test whether there are differences in mean temperature levels between furnaces within plants. Use $\alpha = 0.05$.
(e) Test whether there are differences in mean temperature levels between molds within furnaces. Use $\alpha = 0.05$.
(f) Estimate the variance components of the model and determine 95 percent confidence intervals on them.
4. Bliss (1967, p. 354) reported data from an experiment designed to investigate variation in insecticide residue on celery. The experiment was carried out on 11 randomly selected plots of celery which were sprayed with insecticide and residue was measured from plants selected in three stages. Three samples of plants were selected from each plot and one or two subsamples were selected from each sample. Finally one or two independent measurements on residue were made on each subsample. The following data refer to a subset of 6 plots that have been randomly selected from 11 plots in the experiment.

Plot	1						2						3					
Sample	1		2		3		1		2		3		1		2		3	
Subsample	1	2	1	2	1	1	2	1	2	1	1	2	1	2	1			
Residue	0.52	040	0.26	0.54	0.52	0.18	0.31	0.13	0.25	0.10	0.52	0.55	0.33	0.26	0.41			
	0.43	0.52				0.24	0.29				0.66	0.40						

Plot	4						5						6					
Sample	1		2		3		1		2		3		1		2		3	
Subsample	1	2	1	2	1	1	2	1	2	1	1	2	1	2	1			
Residue	0.77	0.51	0.44	0.50	0.44	0.50	0.60	0.60	0.71	0.92	0.24	0.48	0.53	0.50	0.39			
	0.56	0.60				0.67	0.53				0.36	0.30						

Source: Bliss (1967, p. 354). Used with permission.

(a) Describe the model and the assumption for the experiment.
(b) Analyze the data and report the analysis of variance table.
(c) Test whether there are differences in mean residue levels between plots. Use $\alpha = 0.05$.
(d) Test whether there are differences in mean residue levels between samples within plots. Use $\alpha = 0.05$.
(e) Test whether there are differences in mean residue levels between subsamples within samples. Use $\alpha = 0.05$.
(f) Estimate the variance components of the model and determine 95 percent confidence intervals on them.

5. Anderson and Bancroft (1952, p. 333) reported the results of an experiment designed to study some of the factors affecting the variability of estimates of various soil properties. The experiment was conducted on 20 fields by sampling two sections from each field. Two samples consisting of a composite of 20 borings were taken from each section and finally two subsamples were drawn from each sample. The data were analyzed for several soil properties and the following table gives an analysis of variance for the magnesium data.

Analysis of Variance for the Magnesium Data

Source of Variation	Degrees of Freedom	Mean Square	Expected Mean Square
Field		0.1809	
Section (within fields)		0.0545	
Sample (within sections)		0.0080	
Subsample (within samples)		0.0005	

Source: Anderson and Bancroft (1952, p. 333). Used with permission.

(a) State the model and the assumptions for the experiment.
(b) Complete the missing columns of the preceding analysis of variance table.

(c) Test whether there are differences in mean levels of magnesium between samples within sections. Use $\alpha = 0.05$.
(d) Test whether there are differences in mean levels of magnesium between sections within fields. Use $\alpha = 0.05$.
(e) Test whether there are differences in mean levels of magnesium between fields. Use $\alpha = 0.05$.
(f) Estimate the variance components of the model and determine 95 percent confidence intervals on them.

6. Anderson* and Bancroft (1952, pp. 334–335) described an experiment designed to test various molds for their efficacy in the manufacturing of *streptomycin*. A trial experiment to assess variability at various stages of production process is to be run. There are five stages in the production process: The initial *incubation stage* in a test tube, *a primary inoculation period* in a petridish, *a secondary inoculation period*, *a fermentation period* in a bath, and the *final assay* of the quantity of streptomycin produced. The number of test tubes to be used at different stages are as follows: $a = 5, b = 2, c = 2, d = 2$, and $n = 2$, giving a total of 80 assays for the final analysis. Let σ_α^2, $\sigma_{\beta(\alpha)}^2$, $\sigma_{\gamma(\beta)}^2$, $\sigma_{\delta(\gamma)}^2$, and σ_e^2 be the variance components associated with the five stages of the production process; and consider the following analysis of variance table.

Analysis of Variance for the Streptomycin Production Data

Source of Variation	Degrees of Freedom	Mean Square	Expected Mean Square
Incubation stage		MS_A	
Primary inoculation (within incubation stage)		$MS_{B(A)}$	
Secondary inoculation (within primary inoculation)		$MS_{C(B)}$	
Fermentation (within secondary inoculation)		$MS_{D(C)}$	
Final assay (within fermentation)		$MS_{E(D)}$	
Error		MS_E	

(a) State the model and the assumptions for the experiment considering all effects are random.
(b) Complete the missing columns of the preceding analysis of variance table.
(c) Determine algebraic expressions for the estimates of the variance components as functions of the mean squares.

---

* Dr. R.L. Anderson first proposed this design for an experiment conducted at the Purdue University in 1950. Used with permission.

8 Partially Nested Classifications

8.0 PREVIEW

In the preceding chapters, we discussed classification models involving several factors that are either all crossed or all nested. Occasionally, in a multifactor experiment, some factors will be crossed and others nested. Such designs are called partially nested (hierarchical), crossed-nested, nested-factorial, or mixed-classification designs. For example, suppose that in a study involving an industrial experiment it is desired to test three different methods of a production process. For each method, five operators are employed. The experiment is carried out over a period of four days and three observations are obtained for each combination of method, operator, and day. Because of the nature of the experiment, the five operators employed under Method I are really individuals different from the five operators under Method II or Method III, and the five operators under Method II are different from those under Method III. The physical layout of such an experiment can be depicted schematically as shown in Figure 8.1. In this experiment, the days are crossed with the methods and operators, and operators are nested within methods.

8.1 MATHEMATICAL MODEL

Consider three factors A, B, and C having a, b, and c levels, respectively. Let b levels of factor B be nested under each level of A and let the c levels of factor C be crossed with a levels of factor A and b levels of factor B. The model for this type of experimental layout can be written as

$$y_{ijk\ell} = \mu + \alpha_i + \beta_{j(i)} + \gamma_k + (\alpha\gamma)_{ik} + (\beta\gamma)_{jk(i)} + e_{\ell(ijk)} \quad \begin{cases} i = 1, 2, \ldots, a \\ j = 1, 2, \ldots, b \\ k = 1, 2, \ldots, c \\ \ell = 1, 2, \ldots, n, \end{cases} \quad (8.1.1)$$

where μ is the general mean, α_i is the effect due to the i-th level of factor A, $\beta_{j(i)}$ is the effect due to the j-th level of factor B within the i-th level of factor A, γ_k is the effect due to the k-th level of factor C, $(\alpha\gamma)_{ik}$ is the interaction of the

		Days			
Method	Operator	1	2	3	4
I	1(I)	· · ·	· · ·	· · ·	· · · ·
	2(I)	· · ·	· · ·	· · ·	· · · ·
	3(I)	· · ·	· · ·	· · ·	· · · ·
	4(I)	· · ·	· · ·	· · ·	· · · ·
	5(I)	· · ·	· · ·	· · ·	· · · ·
II	1(II)	· · ·	· · ·	· · ·	· · · ·
	2(II)	· · ·	· · ·	· · ·	· · · ·
	3(II)	· · ·	· · ·	· · ·	· · · ·
	4(II)	· · ·	· · ·	· · ·	· · · ·
	5(II)	· · ·	· · ·	· · ·	· · · ·
III	1(III)	· · ·	· · ·	· · ·	· · · ·
	2(III)	· · ·	· · ·	· · ·	· · · ·
	3(III)	· · ·	· · ·	· · ·	· · · ·
	4(III)	· · ·	· · ·	· · ·	· · · ·
	5(III)	· · ·	· · ·	· · ·	· · · ·

FIGURE 8.1 A Layout for the Partially Nested Design Where Days Are Crossed with Methods and Operators Are Nested within Methods.

i-th level of factor A with the k-th level of factor C, $(\beta\gamma)_{jk(i)}$ is the interaction of the j-th level of factor B with the k-th level of factor C within the i-th level of factor A, and $e_{\ell(ijk)}$ is the usual error term. Notice that no $A \times B$ interaction can exist, because the levels of factor B occur within different levels of factor A. Similarly, there can be no three-way interaction $A \times B \times C$.

Under Model I, the α_i's, $\beta_{j(i)}$'s, $(\alpha\gamma)_{ik}$'s, and $(\beta\gamma)_{jk(i)}$'s are constants subject to the restrictions:

$$\sum_{i=1}^{a}\alpha_i = \sum_{k=1}^{c}\gamma_k = 0,$$

$$\sum_{i=1}^{a}(\alpha\gamma)_{ik} = \sum_{k=1}^{c}(\alpha\gamma)_{ik} = 0,$$

$$\sum_{j=1}^{b}\beta_{j(i)} = 0 \quad \text{for each } i,$$

$$\sum_{j=1}^{b}(\beta\gamma)_{jk(i)} = 0 \quad \text{for each } (i,k),$$

$$\sum_{k=1}^{c}(\beta\gamma)_{jk(i)} = 0 \quad \text{for each } j(i),$$

Partially Nested Classifications

and the $e_{\ell(ijk)}$'s are uncorrelated and randomly distributed with mean zero and variance σ_e^2. However, for a fixed k, the $(\beta\gamma)_{jk(i)}$'s do not sum to zero over i for a fixed j.

Under Model II, we assume that the α_i's, $\beta_{j(i)}$'s, γ_k's, $(\alpha\gamma)_{ik}$'s, $(\beta\gamma)_{jk(i)}$'s, and $e_{\ell(ijk)}$'s are uncorrelated and randomly distributed with zero means and variances $\sigma_\alpha^2, \sigma_{\beta(\alpha)}^2, \sigma_\gamma^2, \sigma_{\alpha\gamma}^2, \sigma_{\beta\gamma(\alpha)}^2$, and σ_e^2, respectively. Thus, $\sigma_\alpha^2, \sigma_{\beta(\alpha)}^2, \sigma_\gamma^2, \sigma_{\alpha\gamma}^2$, $\sigma_{\beta\gamma(\alpha)}^2$, and σ_e^2 are the variance components of the model (8.1.1).

Various types of mixed models are possible and their assumptions can analogously be stated. For example, with A and C fixed and B random, we assume that α_i's, γ_k's, and $(\alpha\gamma)_{ik}$'s are constants subject to the restrictions:

$$\sum_{i=1}^{a} \alpha_i = \sum_{k=1}^{c} \gamma_k = 0,$$

$$\sum_{i=1}^{a} (\alpha\gamma)_{ik} = \sum_{k=1}^{c} (\alpha\gamma)_{ik} = 0.$$

Furthermore, the $\beta_{j(i)}$'s, $(\beta\gamma)_{jk(i)}$'s, and $e_{\ell(ijk)}$'s are randomly distributed with zero means and variances $\sigma_{\beta(\alpha)}^2$, $\sigma_{\beta\gamma(\alpha)}^2$, and σ_e^2, respectively; and the three groups of random variables are pairwise uncorrelated. The random effects $(\beta\gamma)_{jk(i)}$'s, however, are correlated due to the restrictions:

$$\sum_{k=1}^{c} (\beta\gamma)_{jk(i)} = 0, \quad \text{for each } j(i).$$

8.2 ANALYSIS OF VARIANCE

The identity corresponding to the model (8.1.1) is

$$\begin{aligned}
y_{ijk\ell} - \bar{y}_{....} &= (\bar{y}_{i...} - \bar{y}_{....}) + (\bar{y}_{ij..} - \bar{y}_{i...}) + (\bar{y}_{..k.} - \bar{y}_{....}) \\
&\quad + (\bar{y}_{i.k.} - \bar{y}_{i...} - \bar{y}_{..k.} + \bar{y}_{....}) + (\bar{y}_{ijk.} - \bar{y}_{ij..} - \bar{y}_{i.k.} + \bar{y}_{i...}) \\
&\quad + (y_{ijk\ell} - \bar{y}_{ijk.}).
\end{aligned} \quad (8.2.1)$$

Note that the terms on the right-hand side of (8.2.1) are the sample estimates of the terms on the right-hand side of the model (8.1.1) excluding the grand mean. The first and third terms are similar to main effects in a crossed classification model (5.1.1). The second term is analogous to an ordinary nested term such as the second term in (6.3.1). The fourth term is an ordinary two-way interaction similar to the fifth term of (5.3.1). The fifth term can be obtained by considering it as the difference between $\bar{y}_{ijk.}$ and the term obtained as the general mean $\bar{y}_{....}$ + the factor A effect (i.e., $\bar{y}_{i...} - \bar{y}_{....}$) + the factor B within A effect (i.e.,

$\bar{y}_{ij..} - \bar{y}_{i...}) + $ the factor C effect (i.e., $\bar{y}_{..k.} - \bar{y}_{....}$) + the $A \times C$ interaction (i.e., $\bar{y}_{i.k.} - \bar{y}_{i...} - \bar{y}_{..k.} + \bar{y}_{....}$); that is,

$$\bar{y}_{ijk.} - [\bar{y}_{....} + (\bar{y}_{i...} - \bar{y}_{....}) + (\bar{y}_{ij..} - \bar{y}_{i...}) + (\bar{y}_{..k.} - \bar{y}_{....})$$
$$+ (\bar{y}_{i.k.} - \bar{y}_{i...} - \bar{y}_{..k.} + \bar{y}_{....})]$$
$$= \bar{y}_{ijk.} - \bar{y}_{ij..} - \bar{y}_{i.k.} + \bar{y}_{i...}. \qquad (8.2.2)$$

Alternatively, partially hierarchical models can be looked upon as degenerate cases of completely crossed models. For example, suppose that the B effect is fully crossed with A, so that there will be a B main effect $\bar{y}_{.j..} - \bar{y}_{....}$ and an $A \times B$ interaction $\bar{y}_{ij..} - \bar{y}_{i...} - \bar{y}_{.j..} + \bar{y}_{....}$. Now, noting that the B effect is not really a main effect and combining it with its interaction with A, we obtain

$$(\bar{y}_{.j..} - \bar{y}_{....}) + (\bar{y}_{ij..} - \bar{y}_{i...} - \bar{y}_{.j..} + \bar{y}_{....}) = \bar{y}_{ij..} - \bar{y}_{i...}, \qquad (8.2.3)$$

which is precisely the second term on the right-hand side of (8.2.1). Similarly, if B were a crossed effect, then it would have an interaction with C, and its interaction with A would also have an interaction with C. But since B is not really a crossed effect, these two interactions are combined to obtain

$$(\bar{y}_{.jk.} - \bar{y}_{.j..} - \bar{y}_{..k.} + \bar{y}_{....})$$
$$+ (\bar{y}_{ijk.} - \bar{y}_{ij..} - \bar{y}_{i.k.} - \bar{y}_{.jk.} + \bar{y}_{i...} + \bar{y}_{.j..} + \bar{y}_{..k.} - \bar{y}_{....})$$
$$= \bar{y}_{ijk.} - \bar{y}_{ij..} - \bar{y}_{i.k.} + \bar{y}_{i...}, \qquad (8.2.4)$$

which is equivalent to (8.2.2).

The same reasoning also holds in the determination of the degrees of freedom. For the B within A effect, each level of A contributes $b-1$ degrees of freedom and since there are a levels of A, the total number of degrees of freedom are $a(b-1)$. However, using the argument of (8.2.3), the degrees of freedom would be

$$(b-1) + (a-1)(b-1) = a(b-1), \qquad (8.2.5)$$

which gives exactly the same value. For the $B \times C$ within A interaction, since C has $c-1$ degrees of freedom and B within A has $a(b-1)$ degrees of freedom, their interaction will have $a(b-1)(c-1)$ degrees of freedom. From the argument of (8.2.4), the degrees of freedom will be

$$(b-1)(c-1) + (a-1)(b-1)(c-1) = a(b-1)(c-1), \qquad (8.2.6)$$

which again gives the same result.

Partially Nested Classifications

Now, performing the operations of squaring and summing over all indices of (8.2.1), we obtain the following partition of the total sum of squares:

$$SS_T = SS_A + SS_{B(A)} + SS_C + SS_{AC} + SS_{BC(A)} + SS_E,$$

where

$$SS_T = \sum_{i=1}^{a}\sum_{j=1}^{b}\sum_{k=1}^{c}\sum_{\ell=1}^{n}(y_{ijk\ell} - \bar{y}_{....})^2,$$

$$SS_A = bcn\sum_{i=1}^{a}(\bar{y}_{i...} - \bar{y}_{....})^2,$$

$$SS_{B(A)} = cn\sum_{i=1}^{a}\sum_{j=1}^{b}(\bar{y}_{ij..} - \bar{y}_{i...})^2,$$

$$SS_C = abn\sum_{k=1}^{c}(\bar{y}_{..k.} - \bar{y}_{....})^2,$$

$$SS_{AC} = bn\sum_{i=1}^{a}\sum_{k=1}^{c}(\bar{y}_{i.k.} - \bar{y}_{i...} - \bar{y}_{..k.} + \bar{y}_{....})^2,$$

$$SS_{BC(A)} = n\sum_{i=1}^{a}\sum_{j=1}^{b}\sum_{k=1}^{c}(\bar{y}_{ijk.} - \bar{y}_{ij..} - \bar{y}_{i.k.} + \bar{y}_{i...})^2,$$

and

$$SS_E = \sum_{i=1}^{a}\sum_{j=1}^{b}\sum_{k=1}^{c}\sum_{\ell=1}^{n}(y_{ijk\ell} - \bar{y}_{ijk.})^2.$$

The corresponding mean squares are denoted by MS_A, $MS_{B(A)}$, MS_C, MS_{AC}, $MS_{BC(A)}$, and MS_E respectively. The expected values of mean squares can be derived as before. Bennett and Franklin (1954, pp. 410–427) give a general procedure for obtaining the expected values in partially nested classifications. The resultant analysis of variance is shown in Table 8.1. The proper test statistic for any main effect or interaction of interest can be obtained from an examination of the analysis of variance table. The variance components estimates are obtained by equating mean squares to their respective expected values and solving the resultant equations for the corresponding variance components.

Remark: In a partially nested situation, it is useful to remember the following rule of thumb for calculating the degrees of freedom. The number of degrees of freedom for a crossed-factor is one less than the number of levels of the factor; for a nested factor the number of degrees of freedom is equal to the product of the quantity above multiplied by the number of levels of all the factors within which it is nested.

TABLE 8.1
Analysis of Variance for Model (8.1.1)

Source of Variation	Degrees of Freedom	Sum of Squares	Mean Square	Expected Mean Square		
				Model I A, B, and C Fixed	Model II A, B, and C Random	Model III A and C Fixed, B Random
Due to A	$a-1$	SS_A	MS_A	$\sigma_e^2 + \dfrac{bcn}{a-1}\sum_{i=1}^{a}\alpha_i^2$	$\sigma_e^2 + n\sigma_{\beta\gamma(\alpha)}^2 + bn\sigma_{\alpha\gamma}^2 + cn\sigma_{\beta(\alpha)}^2 + bcn\sigma_\alpha^2$	$\sigma_e^2 + cn\sigma_{\beta(\alpha)}^2 + \dfrac{bcn}{a-1}\sum_{i=1}^{a}\alpha_i^2$
Due to B (within A)	$a(b-1)$	$SS_{B(A)}$	$MS_{B(A)}$	$\sigma_e^2 + \dfrac{cn}{a(b-1)}\sum_{i=1}^{a}\sum_{j=1}^{b}\beta_{j(i)}^2$	$\sigma_e^2 + n\sigma_{\beta\gamma(\alpha)}^2 + cn\sigma_{\beta(\alpha)}^2$	$\sigma_e^2 + cn\sigma_{\beta(\alpha)}^2$
Due to C	$c-1$	SS_C	MS_C	$\sigma_e^2 + \dfrac{abn}{c-1}\sum_{k=1}^{c}\gamma_k^2$	$\sigma_e^2 + n\sigma_{\beta\gamma(\alpha)}^2 + bn\sigma_{\alpha\gamma}^2 + abn\sigma_\gamma^2$	$\sigma_e^2 + n\sigma_{\beta\gamma(\alpha)}^2 + \dfrac{abn}{c-1}\sum_{k=1}^{c}\gamma_k^2$
Due to A × C	$(a-1)(c-1)$	SS_{AC}	MS_{AC}	$\sigma_e^2 + \dfrac{bn}{(a-1)(c-1)}\sum_{i=1}^{a}\sum_{k=1}^{c}(\alpha\gamma)_{ik}^2$	$\sigma_e^2 + n\sigma_{\beta\gamma(\alpha)}^2 + bn\sigma_{\alpha\gamma}^2$	$\sigma_e^2 + n\sigma_{\beta\gamma(\alpha)}^2 + \dfrac{bn}{(a-1)(c-1)}\sum_{i=1}^{a}\sum_{k=1}^{c}(\alpha\gamma)_{ik}^2$
Due to B × C (within A)	$a(b-1)(c-1)$	$SS_{BC(A)}$	$MS_{BC(A)}$	$\sigma_e^2 + \dfrac{n}{a(b-1)(c-1)}\sum_{i=1}^{a}\sum_{j=1}^{b}\sum_{k=1}^{c}(\beta\gamma)_{jk(i)}^2$	$\sigma_e^2 + n\sigma_{\beta\gamma(\alpha)}^2$	$\sigma_e^2 + n\sigma_{\beta\gamma(\alpha)}^2$
Error	$abc(n-1)$	SS_E	MS_E	σ_e^2	σ_e^2	σ_e^2
Total	$abcn$	SS_T				

8.3 COMPUTATIONAL FORMULAE AND PROCEDURE

The following formulae may be used for calculating the sums of squares:

$$SS_T = \sum_{i=1}^{a}\sum_{j=1}^{b}\sum_{k=1}^{c}\sum_{\ell=1}^{n} y_{ijk\ell}^2 - \frac{y_{....}^2}{abcn},$$

$$SS_A = \frac{1}{bcn}\sum_{i=1}^{a} y_{i...}^2 - \frac{y_{....}^2}{abcn},$$

$$SS_{B(A)} = \frac{1}{cn}\sum_{i=1}^{a}\sum_{j=1}^{b} y_{ij..}^2 - \frac{1}{bcn}\sum_{i=1}^{a} y_{i...}^2,$$

$$SS_C = \frac{1}{abn}\sum_{k=1}^{c} y_{..k.}^2 - \frac{y_{....}^2}{abcn},$$

$$SS_{AC} = \frac{1}{bn}\sum_{i=1}^{a}\sum_{k=1}^{c} y_{i.k.}^2 - \frac{1}{bcn}\sum_{i=1}^{a} y_{i...}^2 - \frac{1}{abn}\sum_{k=1}^{c} y_{..k.}^2 + \frac{y_{....}^2}{abcn},$$

$$SS_{BC(A)} = \frac{1}{n}\sum_{i=1}^{a}\sum_{j=1}^{b}\sum_{k=1}^{c} y_{ijk.}^2 - \frac{1}{cn}\sum_{i=1}^{a}\sum_{j=1}^{b} y_{ij..}^2 - \frac{1}{bn}\sum_{i=1}^{a}\sum_{k=1}^{c} y_{i.k.}^2$$
$$+ \frac{1}{bcn}\sum_{i=1}^{a} y_{i...}^2,$$

and

$$SS_E = \sum_{i=1}^{a}\sum_{j=1}^{b}\sum_{k=1}^{c}\sum_{\ell=1}^{n} y_{ijk\ell}^2 - \frac{1}{n}\sum_{i=1}^{a}\sum_{j=1}^{b}\sum_{k=1}^{c} y_{ijk.}^2.$$

Examining the forms of $SS_{B(A)}$ and $SS_{BC(A)}$, we notice that these formulae can be written as

$$SS_{B(A)} = \sum_{i=1}^{a}\left[\frac{1}{cn}\sum_{j=1}^{b} y_{ij..}^2 - \frac{1}{bcn} y_{i...}^2\right]$$

and

$$SS_{BC(A)} = \sum_{i=1}^{a}\left[\frac{1}{n}\sum_{j=1}^{b}\sum_{k=1}^{c} y_{ijk.}^2 - \frac{1}{cn}\sum_{j=1}^{b} y_{ij..}^2 - \frac{1}{bn}\sum_{k=1}^{c} y_{i.k.}^2 + \frac{1}{bcn} y_{i...}^2\right].$$

Thus, $SS_{B(A)}$ can be obtained by first calculating the sums of squares among levels of B for each level of A, then pooling over all levels of A; and $SS_{BC(A)}$ can be obtained by first computing the sums of squares among levels of B and C for each level of A, then pooling over all levels of A. Their degrees of freedom can also be determined in a similar manner.

8.4 A FOUR-FACTOR PARTIALLY NESTED CLASSIFICATION

In this section, we briefly outline the analysis of variance for a four-factor partially nested classification. Consider four factors A, B, C, and D having a, b, c, and d levels, respectively. Let b levels of factor B be nested under each level of A, let c levels of factor C be nested under each level of B, and let d levels of factor D be crossed with a levels of A, b levels of B, and c levels of C. The model for this type of experimental layout can be written as

$$y_{ijk\ell m} = \mu + \alpha_i + \beta_{j(i)} + \gamma_{k(ij)} + \delta_\ell + (\alpha\delta)_{i\ell} \\ + (\beta\delta)_{j\ell(i)} + (\gamma\delta)_{k\ell(ij)} + e_{m(ijk\ell)} \quad \begin{cases} i = 1, \ldots, a \\ j = 1, \ldots, b \\ k = 1, \ldots, c \\ \ell = 1, \ldots, d \\ m = 1, \ldots, n, \end{cases} \quad (8.4.1)$$

where the meaning of each symbol and the assumptions of the model are readily stated. Starting with the identity

$$\begin{aligned} y_{ijk\ell m} - \bar{y}_{\ldots\ldots} &= (\bar{y}_{i\ldots\ldots} - \bar{y}_{\ldots\ldots}) + (\bar{y}_{ij\ldots\ldots} - \bar{y}_{i\ldots\ldots}) + (\bar{y}_{ijk\ldots} - \bar{y}_{ij\ldots\ldots}) \\ &+ (\bar{y}_{\ldots\ell\ldots} - \bar{y}_{\ldots\ldots}) + (\bar{y}_{i\ldots\ell\ldots} - \bar{y}_{i\ldots\ldots} - \bar{y}_{\ldots\ell\ldots} + \bar{y}_{\ldots\ldots}) \\ &+ (\bar{y}_{ij.\ell\ldots} - \bar{y}_{ij\ldots\ldots} - \bar{y}_{i\ldots\ell\ldots} + \bar{y}_{i\ldots\ldots}) \\ &+ (\bar{y}_{ijk\ell\ldots} - \bar{y}_{ijk\ldots} - \bar{y}_{ij.\ell\ldots} + \bar{y}_{ij\ldots\ldots}) \\ &+ (y_{ijk\ell m} - \bar{y}_{ijk\ell.}), \end{aligned}$$

the total sum of squares is partitioned as

$$SS_T = SS_A + SS_{B(A)} + SS_{C(B)} + SS_D + SS_{AD} + SS_{BD(A)} + SS_{CD(B)} + SS_E,$$

where

$$SS_T = \sum_{i=1}^{a} \sum_{j=1}^{b} \sum_{k=1}^{c} \sum_{\ell=1}^{d} \sum_{m=1}^{n} (y_{ijk\ell m} - \bar{y}_{\ldots\ldots})^2,$$

$$SS_A = bcdn \sum_{i=1}^{a} (\bar{y}_{i\ldots\ldots} - \bar{y}_{\ldots\ldots})^2,$$

$$SS_{B(A)} = cdn \sum_{i=1}^{a} \sum_{j=1}^{b} (\bar{y}_{ij\ldots\ldots} - \bar{y}_{i\ldots\ldots})^2,$$

$$SS_{C(B)} = dn \sum_{i=1}^{a} \sum_{j=1}^{b} \sum_{k=1}^{c} (\bar{y}_{ijk\ldots} - \bar{y}_{ij\ldots\ldots})^2,$$

Partially Nested Classifications

$$SS_D = abcn \sum_{\ell=1}^{d} (\bar{y}_{...\ell.} - \bar{y}_{.....})^2,$$

$$SS_{AD} = bcn \sum_{i=1}^{a} \sum_{\ell=1}^{d} (\bar{y}_{i..\ell.} - \bar{y}_{i....} - \bar{y}_{...\ell.} + \bar{y}_{.....})^2,$$

$$SS_{BD(A)} = cn \sum_{i=1}^{a} \sum_{j=1}^{b} \sum_{\ell=1}^{d} (\bar{y}_{ij.\ell.} - \bar{y}_{ij...} - \bar{y}_{i..\ell.} + \bar{y}_{i....})^2,$$

$$SS_{CD(B)} = n \sum_{i=1}^{a} \sum_{j=1}^{b} \sum_{k=1}^{c} \sum_{\ell=1}^{d} (\bar{y}_{ijk\ell.} - \bar{y}_{ijk..} - \bar{y}_{ij.\ell.} + \bar{y}_{ij...})^2,$$

and

$$SS_E = \sum_{i=1}^{a} \sum_{j=1}^{b} \sum_{k=1}^{c} \sum_{\ell=1}^{d} \sum_{m=1}^{n} (y_{ijk\ell m} - \bar{y}_{ijk\ell.})^2,$$

with the usual notations of dots and bars. The corresponding mean squares denoted by MS_A, $MS_{B(A)}$, $MS_{C(B)}$, MS_D, MS_{AD}, $MS_{BD(A)}$, $MS_{CD(B)}$, and MS_E are obtained by dividing the sums of squares by the respective degrees of freedom. The resultant analysis of variance is summarized in Table 8.2. The proper test statistic for any main effect or interaction of interest can be obtained from an examination of the analysis of variance table. The variance components estimates as usual are obtained by solving the equations obtained by equating the mean squares to their respective expected values.

8.5 WORKED EXAMPLE FOR MODEL II

Schultz (1954) discussed the results of an analysis of variance performed on data on the calcium, phosphorous, and magnesium content of turnip leaves. The data were obtained as follows. "Duplicate [microchemical] analyses were made on each of four randomly-selected leaves from each of four turnip plants picked at random .... Duplicate determinations were made on each ash solution from a particular leaf ..... The analyses of the two sets of ash solutions were made at different times." The analysis of variance for the calcium data are given in Table 8.3.

It is evident from the structure of the experiment that the plants are crossed with ashings and leaves are nested within plants. Since both plants and leaves within plants were randomly selected, both these factors should be regarded as random. In addition, the factor ashing should also be assumed as random inasmuch as two ashings might be regarded as coming from repeated experiments on the same leaves (a new random sample from each leaf might be taken at future periods, ashed, and then analyzed in duplicate).

TABLE 8.2
Analysis of Variance for Model (8.4.1)

Source of Variation	Degrees of Freedom	Sum of Squares	Mean Square	Expected Mean Square* Model I	Expected Mean Square* Model II
Due to A	$a-1$	SS_A	MS_A	$\sigma_e^2 + \dfrac{bcdn}{a-1}\sum_{i=1}^{a}\alpha_i^2$	$\sigma_e^2 + n\sigma_{\gamma\delta}^2 + cn\sigma_{\beta\delta}^2 + bcn\sigma_{\alpha\delta}^2 + dn\sigma_{\gamma}^2$ $+ cdn\sigma_{\beta}^2 + bcdn\sigma_{\alpha}^2$
Due to B (within A)	$a(b-1)$	$SS_{B(A)}$	$MS_{B(A)}$	$\sigma_e^2 + \dfrac{cdn}{a(b-1)}\sum_{i=1}^{a}\sum_{j=1}^{b}\beta_{j(i)}^2$	$\sigma_e^2 + n\sigma_{\gamma\delta}^2 + cn\sigma_{\beta\delta}^2 + dn\sigma_{\gamma}^2 + cdn\sigma_{\beta}^2$
Due to C (within B)	$ab(c-1)$	$SS_{C(B)}$	$MS_{C(B)}$	$\sigma_e^2 + \dfrac{dn}{ab(c-1)}\sum_{i=1}^{a}\sum_{j=1}^{b}\sum_{k=1}^{c}\gamma_{k(ij)}^2$	$\sigma_e^2 + n\sigma_{\gamma\delta}^2 + dn\sigma_{\gamma}^2$
Due to D	$d-1$	SS_D	MS_D	$\sigma_e^2 + \dfrac{abcn}{d-1}\sum_{\ell=1}^{d}\delta_\ell^2$	$\sigma_e^2 + n\sigma_{\gamma\delta}^2 + cn\sigma_{\beta\delta}^2 + bcn\sigma_{\alpha\delta}^2$
Due to $A \times D$	$(a-1)(d-1)$	SS_{AD}	MS_{AD}	$\sigma_e^2 + \dfrac{bcn}{(a-1)(d-1)}\sum_{i=1}^{a}\sum_{\ell=1}^{d}(\alpha\delta)_{i\ell}^2$	$\sigma_e^2 + n\sigma_{\gamma\delta}^2 + cn\sigma_{\beta\delta}^2 + bcn\sigma_{\alpha\delta}^2$
Due to $B \times D$ (within A)	$a(b-1)(d-1)$	$SS_{BD(A)}$	$MS_{BD(A)}$	$\sigma_e^2 + \dfrac{cn}{a(b-1)(d-1)}\sum_{i=1}^{a}\sum_{j=1}^{b}\sum_{\ell=1}^{d}(\beta\delta)_{j\ell(i)}^2$	$\sigma_e^2 + n\sigma_{\gamma\delta}^2 + cn\sigma_{\beta\delta}^2$
Due to $C \times D$ (within B)	$ab(c-1)(d-1)$	$SS_{CD(B)}$	$MS_{CD(B)}$	$\sigma_e^2 + \dfrac{n}{ab(c-1)(d-1)}\sum_{i=1}^{a}\sum_{j=1}^{b}\sum_{k=1}^{c}\sum_{\ell=1}^{d}(\gamma\delta)_{k\ell(ij)}^2$	$\sigma_e^2 + n\sigma_{\gamma\delta}^2$
Error	$abcd(n-1)$	SS_E	MS_E	σ_e^2	σ_e^2
Total	$abcdn-1$	SS_T			

* Various types of fixed models are possible and corresponding expected mean squares can similarly be obtained.

TABLE 8.3
Analysis of Variance for the Calcium Content of Turnip Leaves Data

Source of Variation	Degrees of Freedom	Mean Square	Expected Mean Square	F Value	p-Value
Plants	3	6.202154	$\sigma_e^2 + 2\sigma_{\beta\gamma(\alpha)}^2 + 4 \times 2\sigma_{\alpha\gamma}^2 + 2 \times 2\sigma_{\beta(\alpha)}^2 + 4 \times 2 \times 2\sigma_\alpha^2$		
Leaves (within plants)	12	0.605917	$\sigma_e^2 + 2\sigma_{\beta\gamma(\alpha)}^2 + 2 \times 2\sigma_{\beta(\alpha)}^2$	43.38	<0.001
Ashings	1	0.02945	$\sigma_e^2 + 2\sigma_{\beta\gamma(\alpha)}^2 + 4 \times 2\sigma_{\alpha\gamma}^2 + 4 \times 4 \times 2\sigma_\gamma^2$	0.88	0.417
Plants × Ashings	3	0.033569	$\sigma_e^2 + 2\sigma_{\beta\gamma(\alpha)}^2 + 4 \times 2\sigma_{\alpha\gamma}^2$	2.40	0.119
Ashings × Leaves (within plants)	12	0.013968	$\sigma_e^2 + 2\sigma_{\beta\gamma(\alpha)}^2$	3.92	<0.001
Error	32	0.003560	σ_e^2		
Total	63				

Source: Schultz (1954). Used with permission.

The mathematical model for the experimental design would be

$$y_{ijk\ell} = \mu + \alpha_i + \beta_{j(i)} + \gamma_k + (\alpha\gamma)_{ik} + (\beta\gamma)_{jk(i)} + e_{\ell(ijk)} \quad \begin{cases} i = 1,2,3,4 \\ j = 1,2,3,4 \\ k = 1,2 \\ \ell = 1,2, \end{cases}$$

where μ is the general mean, α_i is the effect of the i-th plant, $\beta_{j(i)}$ is the effect of the j-th leaf within the i-th plant, γ_k is the effect of the k-th ashing, $(\alpha\gamma)_{ik}$ is the interaction of the i-th plant with the k-th ashing, $(\beta\gamma)_{jk(i)}$ is the interaction of the j-th leaf with the k-th ashing within the i-th plant, and $e_{\ell(ijk)}$ is the customary error term (analysis in duplicates). Under the assumption that all the effects are random, the α_i's, $\beta_{j(i)}$'s, γ_k's, $(\alpha\gamma)_{ik}$'s, $(\beta\gamma)_{jk(i)}$'s, and $e_{\ell(ijk)}$'s are normally distributed with zero means and variances σ_α^2, $\sigma_{\beta(\alpha)}^2$, σ_γ^2, $\sigma_{\alpha\gamma}^2$, $\sigma_{\beta\gamma(\alpha)}^2$, and σ_e^2, respectively.

The test of the null hypothesis that the variance component due to a particular source is zero can be based on the ratio of the mean square of the source to that mean square whose expectation is the same as the expectation of the mean square being tested except for the component due to the source of variation being tested which is equal to zero under the null hypothesis of no effect. Thus, the ashings × leaves (within plants) interaction is tested against the error mean square and the difference is highly significant ($p < 0.001$). Similarly, the plants × ashings interaction is tested against ashings × leaves (within plants) and has probability of occurrence greater than 10 percent due to chance alone ($p = 0.119$). Among main effects, ashings is tested against plants × ashings and is not found to be significant ($p = 0.417$); leaves (within plants) is tested against ashings × leaves (within plants) and is found to be highly significant ($p < 0.001$). The test of significance for plants, however, does not have an exact test. As discussed in Section 5.5, an approximate test may be constructed using the test statistic

$$F' = \frac{MS_A}{MS_{AC} + MS_{B(A)} - MS_{BC(A)}},$$

which has an approximate F distribution with df_A and v' degrees of freedom where

$$v' = \frac{(MS_{AC} + MS_{B(A)} - MS_{BC(A)})^2}{\frac{(MS_{AC})^2}{df_{AC}} + \frac{(MS_{B(A)})^2}{df_{B(A)}} + \frac{(MS_{BC(A)})^2}{df_{BC(A)}}}.$$

In the example, the values of F' and v' (rounded to the nearest digit) are found to be:

$$F' = \frac{6.202154}{0.033569 + 0.605917 - 0.013968} = 9.92$$

and

$$v' = \frac{(0.033569 + 0.605917 - 0.013968)^2}{\frac{(0.033569)^2}{3} + \frac{(0.605917)^2}{12} + \frac{(0.013968)^2}{12}} \doteq 13.$$

From Appendix Table V, it is found that the values as large as 9.92 at 3 and 13 degrees of freedom occur in less than 1 percent of trials due to chance alone ($p = 0.001$). Thus, there is strong evidence that plants are to be regarded as differing significantly in calcium content.

As pointed out in Section 5.5, an alternate test for plants may be based on the statistic

$$F'' = \frac{MS_A + MS_{BC(A)}}{MS_{AC} + MS_{B(A)}},$$

which has an approximate F distribution with v_1'' and v_2'' degrees of freedom where

$$v_1'' = \frac{(MS_A + MS_{BC(A)})^2}{\frac{(MS_A)^2}{df_A} + \frac{(MS_{BC(A)})^2}{df_{BC(A)}}}$$

and

$$v_2'' = \frac{(MS_{AC} + MS_{B(A)})^2}{\frac{(MS_{AC})^2}{df_{AC}} + \frac{(MS_{B(A)})^2}{df_{B(A)}}}.$$

Again, in the example at-hand, the values of F'', v_1'', and v_2'' (rounded to the nearest digit) are found to be:

$$F'' = \frac{6.202154 + 0.013968}{0.033569 + 0.605917} = 9.72,$$

$$v_1'' = \frac{(6.202154 + 0.013968)^2}{\frac{(6.202154)^2}{3} + \frac{(0.013968)^2}{12}} \doteq 3,$$

and

$$v_2'' = \frac{(0.033569 + 0.605917)^2}{\frac{(0.033569)^2}{3} + \frac{(0.605917)^2}{12}} \doteq 13.$$

These values are essentially the same as for the test statistic F' and we reach exactly the same conclusion as before.

Finally, the estimates of the variance components are give by

$$\hat{\sigma}_e^2 = 0.00356,$$

$$\hat{\sigma}_{\beta\gamma(\alpha)}^2 = \frac{1}{2}(0.013968 - 0.003560) = 0.00520,$$

$$\hat{\sigma}_{\alpha\gamma}^2 = \frac{1}{8}(0.033569 - 0.013968) = 0.00245,$$

$$\hat{\sigma}_\gamma^2 = \frac{1}{32}(0.02945 - 0.033569) = -0.00013,$$

$$\hat{\sigma}_{\beta(\alpha)}^2 = \frac{1}{4}(0.605917 - 0.013968) = 0.14799,$$

and

$$\hat{\sigma}_\alpha^2 = \frac{1}{16}(6.202154 - 0.033569 - 0.605917 + 0.013968) = 0.34854.$$

Assuming that $\sigma_\gamma^2 = 0$, these variance components account for 0.7, 1.0, 0.5, 0.0, 29.2 and 68.6 percent of the total variation. Thus, the largest single source of variability is attributable to variation between plants and may account for nearly 70 percent of the total variation. Leaves within plants are also quite variable and may account for most of the remaining variation. Although the effect due to ashings × leaves (within plants) interactions is statistically significant, it accounts for only 1 percent of the total variation.

TABLE 8.4
Measured Strengths of Tire Cords from Two Plants Using Different Production Processes

		\multicolumn{12}{c}{Distance (yds)}											
		\multicolumn{2}{c}{0}	\multicolumn{2}{c}{500}	\multicolumn{2}{c}{1,000}	\multicolumn{2}{c}{1,500}	\multicolumn{2}{c}{2,000}	\multicolumn{2}{c}{2,500}
		Duplicate		Duplicate		Duplicate		Duplicate		Duplicate		Duplicate	
Plant	Bobbin	1	2	1	2	1	2	1	2	1	2	1	2
1	1(1)	−1	−5	−2	−8	−2	3	−3	−4	0	−1	−12	4
	2(1)	1	10	1	2	2	2	10	−4	−4	3	4	8
	3(1)	2	−3	5	−5	1	−1	−6	1	2	5	7	5
	4(1)	6	10	1	5	0	5	−2	−2	1	1	5	9
	5(1)	−1	−8	5	−10	1	−5	1	−4	5	−5	3	6
	6(1)	−1	−10	−8	−8	−2	2	0	−3	−8	−1	−2	−4
	7(1)	−9	−2	5	−2	7	−2	−2	−2	−1	2	10	5
	8(1)	0	2	−5	−2	5	3	10	−1	4	1	7	−1
2	1(2)	10	8	−5	6	2	13	7	15	17	14	18	11
	2(2)	9	12	6	15	15	12	18	16	13	10	9	11
	3(2)	0	8	12	6	2	0	5	4	18	8	6	8
	4(2)	5	9	2	16	15	5	21	18	15	11	18	15
	5(2)	−1	−1	11	19	12	10	1	20	13	9	4	6
	6(2)	7	16	15	11	12	12	8	12	22	11	12	21
	7(2)	−5	1	−2	10	12	15	2	13	10	10	7	5
	8(2)	10	9	10	15	9	16	12	11	18	20	11	15

Source: Akutowicz and Traux (1956, Table 1, p. 4). Used with permission.

8.6 WORKED EXAMPLE FOR MODEL III

Akutowicz and Traux (1956) described an experiment designed to investigate the variability of the strength of tire cord. Prior to the establishment of control of cord testing laboratories, data were obtained from two plants that used different production processes to make nominally the same kind of tire cord. A random sample of eight bobbins of cord was selected from each plant and six 500-yard intervals over the length of each bobbin were determined. In order to give as nearly as possible "duplicate" measurements, adjacent pairs of breaks were made at each interval measuring the recorded strength in 0.1 lb deviations from 21.5 lb. The coded raw data are given in Table 8.4.

It is evident that the structure of this experiment is somewhat different from the crossed and nested classification models discussed in the earlier chapters. If the bobbins were crossed with the plants, so that the first bobbin with plant 1 had some correspondence with the first bobbin with plant 2, and the second bobbin in like manner, and so on, then one would have a three-way crossed classification with replication in the cells. However, clearly, this is not the situation. The bobbins are not crossed with plants; rather they are nested within

Partially Nested Classifications

the plants since eight bobbins of cords were selected at random from each plant. Similarly, if the distances were obtained as random samples from each bobbin, so that they were nested within the bobbins, with no crossing between distance 1 and bobbins or plants, then one would have a completely nested or hierarchical classification. However, again, this is not so. In this experiment, the distances were chosen at 500-yard intervals over the length of each bobbin and thus constitute a fixed effect which is crossed with bobbins and plants. Thus, the experimental structure conforms to the partially nested or hierarchical design, described earlier in this chapter, where the distances are crossed with the plants and bobbins and bobbins are nested within plants.

The mathematical model for the experimental design would be

$$y_{ijk\ell} = \mu + \alpha_i + \beta_{j(i)} + \gamma_k + (\alpha\gamma)_{ik} + (\beta\gamma)_{jk(i)} + e_{\ell(ijk)} \quad \begin{cases} i = 1, 2 \\ j = 1, 2, \ldots, 8 \\ k = 1, 2, \ldots, 6 \\ \ell = 1, 2, \end{cases}$$

where μ is the general mean, α_i is the effect of the i-th plant, $\beta_{j(i)}$ is the effect of the j-th bobbin nested within the i-th plant, γ_k is the effect of the k-th distance, $(\alpha\gamma)_{ik}$ is the interaction of the i-th plant with the k-th distance, $(\beta\gamma)_{jk(i)}$ is the interaction of the k-th distance with the j-th bobbin within the i-th plant, and $e_{\ell(ijk)}$ is the customary error term. Furthermore, the α_i's, γ_k's, and $(\alpha\gamma)_{ik}$'s are fixed effects with the constraints:

$$\sum_{i=1}^{2} \alpha_i = 0, \quad \sum_{k=1}^{6} \gamma_k = 0, \quad \sum_{i=1}^{2} (\alpha\gamma)_{ik} = 0, \quad \sum_{k=1}^{6} (\alpha\gamma)_{ik} = 0;$$

and the $\beta_{j(i)}$'s, $(\beta\gamma)_{jk(i)}$'s, and $e_{\ell(ijk)}$'s are random effects that are independently and normally distributed with mean zero and variances $\sigma_{\beta(\alpha)}^2$, $\sigma_{\beta\gamma(\alpha)}^2$, and σ_e^2, respectively.

To calculate the sums of squares, we first form the bobbins within plants totals ($y_{ij..}$), plant × distance totals ($y_{i.k.}$), plant totals ($y_{i...}$), distance totals ($y_{..k.}$), and grand total ($y_{....}$), as shown in Table 8.5. The other quantities needed in the calculations of the sums of squares are:

$$\frac{y_{....}^2}{2 \times 8 \times 6 \times 2} = \frac{(1{,}016)^2}{192} = 5{,}376.333,$$

$$\sum_{i=1}^{2}\sum_{j=1}^{8}\sum_{k=1}^{6}\sum_{\ell=1}^{2} y_{ijk\ell}^2 = (-1)^2 + (1)^2 + \cdots + (15)^2 = 15{,}788,$$

$$\frac{1}{2}\sum_{i=1}^{2}\sum_{j=1}^{8}\sum_{k=1}^{6} y_{ijk.}^2 = \frac{(-6)^2 + (-10)^2 + \cdots + (26)^2}{2} = 13{,}697,$$

$$\frac{1}{6 \times 2}\sum_{i=1}^{2}\sum_{j=1}^{8} y_{ij..}^2 = \frac{(-31)^2 + (35)^2 + \cdots + (156)^2}{12} = 11{,}347.167,$$

TABLE 8.5
Calculation of Cell and Marginal Totals

Plant	Bobbin	\multicolumn{6}{c}{$y_{ijk.}$ Distance (yds)}	$y_{ij..}$	$y_{i...}$
		0	500	1,000	1,500	2,000	2,500		
1	1(1)	−6	−10	1	−7	−1	−8	−31	
	2(1)	11	3	4	6	−1	12	35	
	3(1)	−1	0	0	−5	7	12	13	
	4(1)	16	6	5	−4	2	14	39	31
	5(1)	−9	−5	−4	−3	0	9	−12	
	6(1)	−11	−16	0	−3	−9	−6	−45	
	7(1)	−11	3	5	−4	1	15	9	
	8(1)	2	−7	8	9	5	6	23	
	$y_{1.k.}$	−9	−26	19	−11	4	54		
2	1(2)	18	1	15	22	31	29	116	
	2(2)	21	21	27	34	23	20	146	
	3(2)	8	18	2	9	26	14	77	
	4(2)	14	18	20	39	26	33	150	985
	5(2)	−2	30	22	21	22	10	103	
	6(2)	23	26	24	20	33	33	159	
	7(2)	−4	8	27	15	20	12	78	
	8(2)	19	25	25	23	38	26	156	
	$y_{2.k.}$	97	147	162	183	219	177	$y_{....}$	
	$y_{..k.}$	88	121	181	172	223	231		1,016

$$\frac{1}{8 \times 2} \sum_{i=1}^{2} \sum_{k=1}^{6} y_{i.k.}^2 = \frac{(-9)^2 + (-26)^2 + \cdots + (177)^2}{16} = 10,888.250,$$

$$\frac{1}{8 \times 6 \times 2} \sum_{i=1}^{2} y_{i...}^2 = \frac{(31)^2 + (985)^2}{96} = 10,116.521,$$

$$\frac{1}{2 \times 8 \times 2} \sum_{k=1}^{6} y_{..k.}^2 = \frac{(88)^2 + (121)^2 + \cdots + (231)^2}{32} = 5,869.375.$$

Now, the sum of squares for plants is an ordinary main effect; that is,

$$SS_A = \frac{1}{8 \times 6 \times 2} \sum_{i=1}^{2} y_{i...}^2 - \frac{y_{....}^2}{2 \times 8 \times 6 \times 2}$$

$$= 10,116.521 - 5,376.333$$

$$= 4,740.188.$$

Partially Nested Classifications

The sum of squares for bobbins within plants is an ordinary nested effect; that is,

$$SS_{B(A)} = \frac{1}{6 \times 2} \sum_{i=1}^{2} \sum_{j=1}^{8} y_{ij..}^2 - \frac{1}{8 \times 6 \times 2} \sum_{i=1}^{2} y_{i...}^2$$
$$= 11{,}347.167 - 10{,}116.521$$
$$= 1{,}230.646.$$

The sum of squares for distance is again an ordinary main effect; that is,

$$SS_C = \frac{1}{2 \times 8 \times 2} \sum_{k=1}^{6} y_{..k.}^2 - \frac{y_{....}^2}{2 \times 8 \times 6 \times 2}$$
$$= 5{,}869.375 - 5{,}376.333$$
$$= 493.042.$$

The sum of squares for plant × distance is an ordinary two-way interaction; that is,

$$SS_{AC} = \frac{1}{8 \times 2} \sum_{i=1}^{2} \sum_{k=1}^{6} y_{i.k.}^2 - \frac{1}{8 \times 6 \times 2} \sum_{i=1}^{2} y_{i...}^2 - \frac{1}{2 \times 8 \times 2} \sum_{k=1}^{6} y_{..k.}^2$$
$$+ \frac{y_{....}^2}{2 \times 8 \times 6 \times 2}$$
$$= 10{,}888.250 - 10{,}116.521 - 5{,}869.375 + 5{,}376.333$$
$$= 278.687.$$

The sum of squares for distance × bobbin within plants is

$$SS_{BC(A)} = \frac{1}{2} \sum_{i=1}^{2} \sum_{j=1}^{8} \sum_{k=1}^{6} y_{ijk.}^2 - \frac{1}{6 \times 2} \sum_{i=1}^{2} \sum_{j=1}^{8} y_{ij..}^2 - \frac{1}{8 \times 2} \sum_{i=1}^{2} \sum_{k=1}^{6} y_{i.k.}^2$$
$$+ \frac{1}{8 \times 6 \times 2} \sum_{i=1}^{2} y_{i...}^2$$
$$= 13{,}697 - 11{,}347.167 - 10{,}888.250 + 10{,}116.521$$
$$= 1{,}578.104.$$

The total sum of squares is

$$SS_T = \sum_{i=1}^{2} \sum_{j=1}^{8} \sum_{k=1}^{6} \sum_{\ell=1}^{2} y_{ijk\ell}^2 - \frac{y_{....}^2}{2 \times 8 \times 6 \times 2}$$
$$= 15{,}788 - 5{,}376.333$$
$$= 10{,}411.667.$$

Finally, the error sum of squares is obtained by subtraction as

$$SS_E = SS_T - SS_A - SS_{B(A)} - SS_C - SS_{AC} - SS_{BC(A)}$$
$$= 10{,}411.667 - 4{,}740.188 - 1{,}230.646 - 493.042$$
$$\quad - 278.687 - 1{,}578.104$$
$$= 2{,}090.999.$$

The complete analysis of variance is shown in Table 8.6.

The plants, tested against bobbins within plants, is evidently highly significant ($p < 0.001$). Similarly, bobbins within plants, tested against the error term, also seem to differ quite significantly ($p < 0.001$). Distances also appear to differ significantly, that is, have considerable effect on cord strength ($p = 0.002$). There is some evidence of interaction between plant and distance ($p = 0.041$). However, there does not seem to be any interaction between distance and bobbin within plants, indicating that the effect of different distances is probably the same for all bobbins ($p = 0.426$). The variance components σ_e^2, $\sigma_{\beta\gamma(\alpha)}^2$, and $\sigma_{\beta(\alpha)}^2$ are estimated as

$$\hat{\sigma}_e^2 = 21.781,$$

$$\hat{\sigma}_{\beta\gamma(\alpha)}^2 = \frac{1}{2}(22.544 - 21.781) = 0.382,$$

and

$$\hat{\sigma}_{\beta(\alpha)}^2 = \frac{1}{12}(87.903 - 21.781) = 5.510.$$

It is evident from the preceding analysis that there is a great deal of variability in the strength of tire cord and the larger part of this variability arises due to differences in the manufacturing processes of the two plants. The distances also differ quite significantly. The variability between bobbins within a given plant is quite large and the duplicate measurements on adjacent pairs also differ considerably.

8.7 USE OF STATISTICAL COMPUTING PACKAGES

Among the SAS procedures, PROC ANOVA and PROC NESTED cannot be used to analyze a partially nested model since they are written for either completely crossed or completely nested designs. PROC GLM is the procedure of choice for analyzing this type of model. Again, the analysis involving a random or mixed effects model can be handled via RANDOM and TEST options. PROC MIXED or VARCOMP can be used for the estimation of variance components. For instructions regarding SAS commands, see Section 11.1.

Partially Nested Classifications

TABLE 8.6
Analysis of Variance for the Tire Cords Strength Data of Table 8.4

Source of Variation	Degrees of Freedom	Sum of Squares	Mean Square	Expected Mean Square	F Value	p-Value
Plants	1	4,740.188	4,740.188	$\sigma_e^2 + 6 \times 2\sigma_{\beta(\alpha)}^2 + \dfrac{8 \times 6 \times 2}{2-1}\sum_{i=1}^{2}\alpha_i^2$	53.93	<0.001
Bobbins (within plants)	14	1,230.646	87.903	$\sigma_e^2 + 6 \times 2\sigma_{\beta(\alpha)}^2$	4.04	<0.001
Distances	5	493.042	98.608	$\sigma_e^2 + 2\sigma_{\beta\gamma(\alpha)}^2 + \dfrac{2 \times 8 \times 2}{6-1}\sum_{k=1}^{6}\gamma_k^2$	4.37	0.002
Plants × Distances	5	278.687	55.737	$\sigma_e^2 + 2\sigma_{\beta\gamma(\alpha)}^2 + \dfrac{8 \times 2}{(2-1)(6-1)}\sum_{i=1}^{2}\sum_{k=1}^{6}(\alpha\gamma)_{ik}^2$	2.47	0.041
Distances × Bobbins (within plants)	70	1,578.104	22.544	$\sigma_e^2 + 2\sigma_{\beta\gamma(\alpha)}^2$	1.04	0.426
Error	96	2,090.999	21.781	σ_e^2		
Total	191	10,411.667				

```
DATA STRENGTHS;                         The SAS System
INPUT PLANT BOBBIN                General Linear Models Procedure
DISTANCE $ DUPLICA    Dependent Variable: STRENGTH
STRENGTH;                            Sum of        Mean
DATALINES;            Source     DF  Squares      Square    F Value  Pr > F
1 1 0YD 1 -1          Model      95  8320.6666    87.5859    4.02    0.0001
1 1 0YD 2 -5          Error      96  2091.0000    21.7812
1 2 0YD 1  1          Corrected
1 2 0YD 2 10          Total     191  10411.6666667
1 3 0YD 1  2
1 3 0YD 2 -3                    R-Square        C.V.      Root MSE    STRENGTH Mean
. . . . .                       0.799168      88.196006   4.6670387   5.29166667
2 8 2500YD 2 15       Source              DF  Type III SS  Mean Square  F Value  Pr > F
;                     PLANT                1   4740.1875    4740.1875  217.63    0.0001
PROC GLM;             BOBBIN(PLANT)       14   1230.6458      87.9033    4.04    0.0001
CLASSES PLANT BOBBIN  DISTANCE             5    493.0417      98.6083    4.53    0.0009
DISTANCE;             PLANT*DISTANCE       5    278.6875      55.7375    2.56    0.0322
MODEL STRENGTH = PLANT BOBBIN*DISTAN(PLANT) 70  1578.1042      22.5443    1.04    0.4340
BOBBIN(PLANT) DISTANCE Source            Type III Expected Mean Square
PLANT*DISTANCE        PLANT             Var(Error) + 2 Var(BOBBIN*DISTAN(PLANT))
BOBBIN*DISTANCE(PLANT);                +12 Var(BOBBIN(PLANT)) + Q(PLANT,PLANT*DISTANCE)
RANDOM BOBBIN(PLANT)  BOBBIN(PLANT)     Var(Error) + 2 Var(BOBBIN*DISTAN(PLANT))
BOBBIN*DISTANCE(PLANT);                   + 12 Var(BOBBIN(PLANT))
TEST H=PLANT          DISTANCE          Var(Error) + 2 Var(BOBBIN*DISTAN(PLANT))
E=BOBBIN(PLANT)                           + Q(DISTANCE,PLANT*DISTANCE)
TEST H=DISTANCE       PLANT*DISTANCE    Var(Error) + 2 Var(BOBBIN*DISTAN(PLANT))
E=BOBBIN*DISTANCE(PLANT)                  + Q(PLANT*DISTANCE)
TEST H=PLANT*DISTANCE BOBBIN*DISTAN(PLANT) Var(Error) + 2 Var(BOBBIN*DISTAN(PLANT))
E=BOBBIN*DISTANCE(PLANT); Tests of Hypotheses using the Type III MS for BOBBIN(PLANT) as an
RUN;                  error term
CLASS   LEVELS VALUES Source              DF  Type III SS  Mean Square  F Value  Pr > F
PLANT     2    1 2 3 4 PLANT               1   4740.1875    4740.1875   53.93    0.0001
BOBBIN    8    1 2 3 4 Tests of Hypotheses using the Type III MS for BOBBIN*DISTAN(PLANT)
               5 6 7 8 as an error term
DISTANCE  6           Source              DF  Type III SS  Mean Square  F Value  Pr > F
 0YD    500YD  1000YD DISTANCE             5   493.0416      98.6083    4.37    0.0016
1500YD  2000YD 2500YD Tests of Hypotheses using the Type III MS for BOBBIN*DISTAN(PLANT)
NUMBER OF OBS. IN DATA as an error term
SET=192               Source              DF  Type III SS  Mean Square  F Value  Pr > F
                      PLANT*DISTANCE       5   278.6875      55.7375    2.47    0.0403
```

(i) SAS application: SAS GLM instructions and output for the partially nested mixed effects analysis of variance.

```
DATA LIST          Tests of Between-Subjects Effects   Dependent Variable: STRENGTH
/PLANT 1
 BOBBIN 3         Source                    Type III SS  df  Mean Square    F       Sig.
 DISTANCE 5       PLANT         Hypothesis   4740.188     1   4740.188    53.925   .000
 DUPLICA 7                      Error        1230.646    14     87.903(a)
 STRENGTH 9-11.   BOBBIN(PLANT) Hypothesis   1230.646    14     87.903     3.899   .000
BEGIN DATA.                     Error        1578.104    70     22.544(b)
1 1 1 1  -1       DISTANCE      Hypothesis    493.042     5     98.608     4.374   .002
1 1 1 2  -5                     Error        1578.104    70     22.544(b)
1 2 1 1   1       PLANT*DISTANCE Hypothesis   278.688     5     55.738     2.472   .040
1 2 1 2  10                     Error        1578.104    70     22.544(b)
1 3 1 1   2       BOBBIN*DISTANCE Hypothesis 1578.104    70     22.544     1.035   .434
1 3 1 2  -3       (PLANT)       Error        2091.000    96     21.781(c)
. . . .           a  MS(BOBBIN(PLANT))  b MS(BOBBIN*DISTANCE (PLANT))  c  MS(Error)
2 8 6 2  15
END DATA.                                    Expected Mean Squares(a,b)
GLM STRENGTH BY                              Variance Component
PLANT             Source          Var(B(P))   Var(B*D(P))  Var(Error)   Quadratic Term
BOBBIN            PLANT            12.000      2.000        1.000       Plant
DISTANCE          BOBBIN(PLANT)    12.000      2.000        1.000
DUPLICA           DISTANCE           .000      2.000        1.000       Distance
/DESIGN PLANT     PLANT*DISTANCE     .000      2.000        1.000       Plant*Distance
 BOBBIN(PLANT)    BOBBIN*DISTANCE(PLANT) .000  2.000        1.000
 DISTANCE         Error              .000       .000        1.000
 PLANT*DISTANCE   a  For each source, the expected mean square equals the sum of the
 DISTANCE*BOBBIN  coefficients in the cells times the variance components, plus a quadratic
 (PLANT)          term involving effects in the Quadratic Term cell. b Expected Mean Squares
 /RANDOM BOBBIN.  are based on the Type III Sums of Squares.
```

(ii) SPSS application: SPSS GLM instructions and output for the partially nested mixed effects analysis of variance.

FIGURE 8.2 Program Instructions and Outputs for the Partially Nested Mixed Effects Analysis of Variance: Measured Strengths of Tire Cords from Two Plants Using Different Production Processes (Table 8.4).

Partially Nested Classifications

```
/INPUT      FILE='C:\SAHAI              BMDP8V  - GENERAL MIXED MODEL ANALYSIS OF VARIANCE
            \TEXTO\EJE22.TXT'.                  - EQUAL CELL SIZES Release: 7.0 (BMDP/DYNAMIC)
            FORMAT=FREE.
            VARIABLES=2.                ANALYSIS OF VARIANCE FOR DEPENDENT VARIABLE   1
/VARIABLE   NAMES=REP1,REP2.              SOURCE      ERROR   SUM OF   D.F.  MEAN         F      PROB.
/DESIGN     NAMES=PLANT,DEISTANCE,                    TERM    SQUARES        SQUARE
                  BOBBIN,REPLIC.        1   MEAN      B(P)    5376.333  1   5376.33    61.16    0.0000
            LEVELS=2,6,8,2.             2   PLANT     B(P)    4740.188  1   4740.19    53.93    0.0000
            RANDOM=BOBBIN,REPLIC.       3   DISTANCE  DB(P)    493.042  5     98.61     4.37    0.0016
            FIXED=PLANT,DISTANCE.       4   B(P)      R(PDB)  1230.646 14     87.90     4.04    0.0000
            MODEL='P,B(P),D,R(PBD)'.    5   PD        DB(P)    278.688  5     55.74     2.47    0.0403
/END                                    6   DB(P)     R(PDB)  1578.104 70     22.54     1.04    0.4343
1 1 1 1 -1                              7   R(PDB)            2091.000 96     21.78
1 1 1 2 -5
1 1 1 1  1                                SOURCE          EXPECTED MEAN            ESTIMATES OF
1 1 1 2 10                                                    SQUARE               VARIANCE COMPONENTS
. . . . .                               1   MEAN        192(1)+12(4)+(7)               27.54391
2 8 6 2 15                              2   PLANT        96(2)+12(4)+(7)               48.46129
ANALYSIS OF VARIANCE DESIGN             3   DISTANCE     32(3)+2(6)+(7)                 2.37700
INDEX          P    D    B    R         4   B(P)           12(4)+(7)                    5.51017
NUMBER OF LEVELS 2   6   8    2         5   PD            16(5)+2(6)+(7)                2.07457
POPULATION SIZE  2    6   INF  INF      6   DB(P)           2(6)+(7)                    0.38155
MODEL     F,  B(P),  D, R(PBD)          7   R(PDB)            (7)                      21.78125
```

(iii) BMDP application: BMDP 8V instructions and output for the partially nested mixed effects analysis of variance.

FIGURE 8.2 (*continued*)

Among the SPSS procedures, either MANOVA or GLM could be used for the analysis involving random and mixed effects models. For the estimation of variance components, SPSS VARCOMP will be the procedure of choice. For instructions regarding SPSS commands, see Section 11.2.

Among the BMDP programs, 3V or 8V can be used for partially nested designs. For designs with balanced structure, 8V is preferable; 2V can also be used but the special methods of combining crossed factor sums of squares must be used for obtaining sums of squares corresponding to nested factors.

8.8 WORKED EXAMPLE USING STATISTICAL PACKAGES

In this section, we illustrate the application of statistical packages to perform partially nested analysis of variance for the data set of the example presented in Section 8.6. Figure 8.2 illustrates the program instructions and the output results for analyzing data in Table 8.4 using SAS GLM, SPSS GLM, and BMDP 8V. The typical output provides the data format listed at the top, all cell means, and the entries of the analysis of variance table. Note that the results are the same as those provided using manual computations in Section 8.6.

EXERCISES

1. An experiment is designed to study the performance of three different lathes. Each lathe has three different speeds where the product is manufactured and each was operated at two different feed rates. The runs are made in random order and three observations are taken from each speed. The relevant data in certain standard units are as

follows.

Lathe	I			II			III		
Speed	1	2	3	1	2	3	1	2	3
Feed rates									
Low	41.2	40.8	43.3	39.2	40.2	39.9	39.9	40.9	40.7
	37.4	41.9	43.9	40.6	41.8	42.8	40.1	40.5	39.9
	38.7	42.1	44.2	41.1	40.9	41.4	40.2	39.8	38.8
High	31.4	35.2	32.8	31.2	31.2	33.1	31.3	30.3	31.8
	33.4	37.4	33.2	32.1	32.2	34.2	33.2	34.5	29.1
	34.2	36.7	31.9	33.4	34.9	30.9	32.4	33.1	31.9

(a) Describe the model and the assumptions for the experiment. It is assumed that all three factors are fixed.
(b) Analyze the data and report the analysis of variance table.
(c) Test whether there are differences in lathes. Use $\alpha = 0.05$.
(d) Test whether there are differences in feed rates. Use $\alpha = 0.05$.
(e) Test whether there are differences in speeds. Use $\alpha = 0.05$.
(f) Suppose that a large number of feed rates are available, and the two taken for the experiment are selected randomly. Modify the analysis of variance in part (b) to give the expected mean squares for this case and estimate the variance components of the model.

2. An experiment is designed to study the microhardness of high-strength steel purchased from three different foundries. Each foundry supplied the steel in three different lengths of bars: 3.0, 3.50, or 4.0 inches. Inasmuch as the production of different lengths of bar from a common ingot required different extrusion techniques, this factor may be important. Moreover, the bars were forged from ingots produced at different temperatures. Each foundry provided two test specimens of each bar from three different temperatures. The resulting data in certain standard units are as follows.

Foundry	I			II			III		
Temperature	°C			°C			°C		
	1100	1200	1300	1100	1200	1300	1100	1200	1300
Bar Length (in.)									
3.0	1.841	1.957	1.846	1.912	1.957	1.926	1.858	1.886	1.935
	1.869	1.911	1.817	1.874	1.993	1.931	1.897	1.879	1.926
3.5	1.927	1.919	1.861	1.885	1.995	1.957	1.884	1.871	1.993
	1.911	1.973	1.849	1.879	1.986	1.968	1.875	1.876	1.975
4.0	1.898	1.957	1.884	1.858	1.973	1.947	1.912	1.891	1.929
	1.893	1.993	1.826	1.826	1.939	1.953	1.873	1.882	1.934

(a) Describe the model and the assumptions for the experiment. It is assumed that all three factors are fixed.
(b) Analyze the data and report the analysis of variance table.
(c) Test whether there are differences in the microhardness of steel purchased from the three foundries. Use $\alpha = 0.05$.
(d) Test whether there are differences in the microhardness of steel of different lengths of bar. Use $\alpha = 0.05$.
(e) Suppose that bars may be acquired in many lengths and the three lengths being used in the experiment were selected randomly. Make the necessary modifications in the analysis of variance in part (b) to reflect the expected mean squares for this situation and estimate the variance components of the model.

3. Brownlee (1965, p. 544) reported data from an experiment involving five laboratories that participated in measuring the brightness of six lamps of each of two types. The brightness of each lamp type was measured in all five laboratories. The data on values of candle power measured at different laboratories are as follows.

Type	Lamps	Laboratory*				
		A	B	C	D	E
I	1	741	768	770	772	738
	2	731	763	755	742	724
	3	731	763	757	760	728
	4	759	779	775	774	752
	5	738	758	750	750	730
	6	770	795	800	800	768
II	1	625	650	655	651	615
	2	590	611	605	625	588
	3	602	630	640	630	605
	4	578	607	640	608	581
	5	578	604	605	608	573
	6	625	673	670	664	631

* All figures have been multiplied by 100 and then 1,000 subtracted from them.
Source: Brownlee (1965, p. 544). Used with permission.

(a) Describe the model and the assumptions for the experiment.
(b) Analyze the data and report the analysis of variance table.
(c) Test whether there are differences in laboratories. Use $\alpha = 0.05$.
(d) Test whether there are differences in lamp types. Use $\alpha = 0.05$.
(e) Test whether there are differences in lamps within types. Use $\alpha = 0.05$.
(f) Determine 95 percent confidence limits for the difference between lamp types. (It is assumed that the laboratory × lamp type interaction is negligible.)

(g) Determine 95 percent confidence limits for the difference between laboratory A and laboratory E. (Again it may be assumed that laboratory × lamp type interaction is negligible.)

(h) Determine 95 percent confidence limits for the difference between lamp type I in laboratory A and lamp type II in laboratory B.

(i) Estimate the component of variance for lamps within types.

4. Brownlee (1965, p. 545) reported data from an experiment involving five laboratories that took part in a test comparison of their measurement procedures for evaluating the impact strength of a type of fiberboard. Panels from two batches of board were tested by each of the five laboratories for each batch in duplicate on three days. The three days reported in the experiment were different for each laboratory. The data on impact strengths are as follows.

		Laboratory				
Day	Batch	A	B	C	D	E
1	1	1483	1449	1499	1428	1509
		1496	1400	1472	1401	1439
	2	1504	1465	1506	1407	1480
		1505	1423	1537	1416	1429
2	1	1441	1477	1483	1404	1416
		1416	1471	1509	1419	1441
	2	1477	1418	1578	1455	1364
		1457	1445	1486	1435	1441
3	1	1450	1446	1489	1414	1419
		1478	1398	1435	1446	1444
	2	1435	1424	1499	1423	1437
		1478	1426	1491	1442	1438

Source: Brownlee (1965, p. 545). Used with permission.

(a) Describe the model and the assumptions for the experiment.
(b) Analyze the data and report the analysis of variance table.
(c) Test whether there are differences in laboratories. Use $\alpha = 0.05$.
(d) Test whether there are differences in days. Use $\alpha = 0.05$.
(e) Test whether there are differences in batches. Use $\alpha = 0.05$.
(f) Determine 95 percent confidence limits for the mean difference between laboratories A and E.
(g) Estimate the within day component of variance averaged over laboratories.

(h) Estimate the between day component of variance averaged over laboratories.
5. Desmond (1954) reported the results of an experiment in testing the operation of voltage regulators involving four setting stations. From each of these stations, three regulators were randomly selected and each was tested at four test stations. The data are given as follows.

Setting Station	Regulator No.	Test Station			
		1	2	3	4
I	1	16.5	16.1	16.2	16.0
	2	15.9	15.4	15.8	15.5
	3	16.9	15.9	16.0	15.8
II	1	16.7	16.1	15.7	16.2
	2	17.0	16.4	16.4	16.4
	3	16.3	16.1	16.1	15.8
III	1	17.0	16.1	15.8	16.0
	2	16.6	16.3	15.9	15.7
	3	16.3	15.9	16.2	15.3
IV	1	16.8	16.7	16.3	16.2
	2	16.1	16.0	16.0	15.6
	3	16.2	16.1	16.1	16.0

Source: Desmond (1954). Used with permission.

(a) Describe the model and the assumptions for the experiment. Assume a fixed effect model for setting stations and test stations.
(b) Analyze the data and report the analysis of variance table.
(c) Test whether there are differences in test stations. Use $\alpha = 0.05$.
(d) Test whether there are differences in setting stations. Use $\alpha = 0.05$.
(e) Test whether there are differences in regulators within setting stations. Use $\alpha = 0.05$.
(f) Present a table of the means of each setting station with estimated standard error for each mean.
(g) Determine 95 percent confidence limits for the mean differences between test stations 1 and 4.
6. Consider the experiment described in the worked example in Section 8.5. The analysis of variance for the phosphorous data is performed in exactly the same manner as for the calcium data and the results are as follows.

Analysis of Variance for the Phosphorous Content of Turnip Leaves Data

Source of Variation	Degrees of Freedom	Mean Square	Expected Mean Square
Plants	3	0.056375	$\sigma_e^2 + 2\sigma_{\beta\gamma(\alpha)}^2 + 4 \times 2\sigma_{\alpha\gamma}^2 + 2 \times 2\sigma_{\beta(\alpha)}^2 + 4 \times 2 \times 2\sigma_\alpha^2$
Leaves (within plants)	12	0.035786	$\sigma_e^2 + 2\sigma_{\beta\gamma(\alpha)}^2 + 2 \times 2\sigma_{\beta(\alpha)}^2$
Ashings	1	0.000467	$\sigma_e^2 + 2\sigma_{\beta\gamma(\alpha)}^2 + 4 \times 2\sigma_{\alpha\gamma}^2 + 4 \times 4 \times 2\sigma_\gamma^2$
Plants × Ashings	3	0.000664	$\sigma_e^2 + 2\sigma_{\beta\gamma(\alpha)}^2 + 4 \times 2\sigma_{\alpha\gamma}^2$
Ashings × Leaves (within plants)	12	0.000935	$\sigma_e^2 + 2\sigma_{\beta\gamma(\alpha)}^2$
Error	32	0.000457	σ_e^2
Total	63		

Source: Schultz (1954). Used with permission.

(a) Test whether there are differences in effects due to plants. Use $\alpha = 0.05$.
(b) Test whether there are differences in effects due to leaves within plants. Use $\alpha = 0.05$.
(c) Test whether there are differences in effects due to ashings. Use $\alpha = 0.05$.
(d) Test whether there are differences in effects due to plants × ashings. Use $\alpha = 0.05$.
(e) Test whether there are differences in effects due to ashings × leaves within plants. Use $\alpha = 0.05$.
(f) Estimate the variance components of the model and determine 95 percent confidence intervals on them.

7. Anderson (1954) reported the results of an experiment designed to compare the absorption properties of ceramic compositions. There are 15 ceramic compositions and the experiment was performed under three different temperatures. Two batches of each composition were prepared and two firings were made at each temperature. Finally, one observation was made for each firing of a batch, giving a total of 180 observations. The mathematical model for this experiment would be:

$$y_{ijk\ell} = \mu + \alpha_i + \beta_{j(i)} + \gamma_k + \delta_{\ell(k)} + (\alpha\gamma)_{ik} + (\alpha\delta)_{i\ell(k)} + (\beta\gamma)_{jk(i)} + e_{\ell(ijk)} \quad \begin{cases} i = 1, 2, 3 \\ j = 1, 2 \\ k = 1, 2, .., 15 \\ \ell = 1, 2, \end{cases}$$

where μ is the general mean, α_i is the effect of the i-th temperature, $\beta_{j(i)}$ is the effect of the j-th firing within the i-th temperature, γ_k is the effect of the k-th composition, $\delta_{\ell(k)}$ is the effect of the ℓ-th batch within the k-th composition, $(\alpha\gamma)_{ik}$ is the interaction of the i-th temperature with the k-th composition, $(\alpha\delta)_{i\ell(k)}$ is the interaction of the i-th temperature with the ℓ-th batch within the k-th composition, $(\beta\gamma)_{jk(i)}$ is the interaction of the j-th firing with the k-th composition within the i-th temperature, and $e_{\ell(ijk)}$ is the customary error term. Note that it is assumed that there are no $(\beta\delta)_{j\ell(ik)}$ interactions. Under the assumption that all effects are random, the α_i's, $\beta_{j(i)}$'s, γ_k's, $\delta_{\ell(k)}$'s, $(\alpha\gamma)_{ik}$'s, $(\alpha\delta)_{i\ell(k)}$'s, $(\beta\gamma)_{jk(i)}$'s, and $e_{\ell(ijk)}$'s are normally distributed with mean zero and variances σ_α^2, $\sigma_{\beta(\alpha)}^2$, σ_γ^2, $\sigma_{\delta(\gamma)}^2$, $\sigma_{\alpha\gamma}^2$, $\sigma_{\alpha\delta(\gamma)}^2$, $\sigma_{\beta\gamma(\alpha)}^2$, and σ_e^2, respectively. The analysis of variance table is given as follows.

Analysis of Variance for the Ceramic Compositions Data

Source of Variation	Degrees of Freedom	Mean Square	Expected Mean Square
Temperatures	2	1,179.9900	$\sigma_e^2 + 2\sigma_{\beta\gamma(\alpha)}^2 + 2\sigma_{\alpha\delta(\gamma)}^2 + 4\sigma_{\alpha\gamma}^2 + 3\sigma_{\beta(\alpha)}^2 + 60\sigma_\alpha^2$
Firings (within temperatures)	3	0.1521	$\sigma_e^2 + 2\sigma_{\beta\gamma(\alpha)}^2 + 30\sigma_{\beta(\alpha)}^2$
Compositions	14	10.3400	$\sigma_e^2 + 2\sigma_{\beta\gamma(\alpha)}^2 + 2\sigma_{\alpha\delta(\gamma)}^2 + 4\sigma_{\alpha\gamma}^2 + 6\sigma_{\delta(\gamma)}^2 + 12\sigma_\gamma^2$
Batches (within compositions)	15	0.7405	$\sigma_e^2 + 2\sigma_{\alpha\delta(\gamma)}^2 + 6\sigma_{\delta(\gamma)}^2$
Temperatures × Compositions	28	1.1130	$\sigma_e^2 + 2\sigma_{\beta\gamma(\alpha)}^2 + 2\sigma_{\alpha\delta(\gamma)}^2 + 4\sigma_{\alpha\gamma}^2$
Temperatures × Batches (within compositions)	30	0.0857	$\sigma_e^2 + 2\sigma_{\alpha\delta(\gamma)}^2$
Firings × Compositions (within temperatures)	42	0.0818	$\sigma_e^2 + 2\sigma_{\beta\gamma(\alpha)}^2$
Error	45	0.0631	σ_e^2

Source: Anderson (1954). Used with permission.

(a) Test whether there are differences in absorption properties among different temperatures. Use $\alpha = 0.05$.
(b) Test whether there are differences in absorption properties among firings within temperatures. Use $\alpha = 0.05$.
(c) Test whether there are differences in absorption properties among compositions. Use $\alpha = 0.05$.

(d) Test whether there are differences in absorption properties among batches within compositions.
(e) Test for the following interaction effects: temperatures × compositions, temperatures × batchs (within compositions), and firings × compositions (within temperatures).
(f) Determine the point and interval estimates for each of the variance components of the model.

8. Consider a variation of the four-factor partially nested classification described in Section 8.4, where now b levels of factor B are nested under each level of A and d levels of factor D are nested under each level of C; that is, model (8.4.1) is now given by

$$y_{ijk\ell m} = \mu + \alpha_i + \beta_{j(i)} + \gamma_k + \delta_{\ell(k)} + (\alpha\gamma)_{ik} + (\alpha\delta)_{i\ell(k)} + (\beta\gamma)_{jk(i)} + e_{m(ijk\ell)} \quad \begin{cases} i = 1, \ldots, a \\ j = 1, \ldots, b \\ k = 1, \ldots, c \\ \ell = 1, \ldots, d \\ m = 1, \ldots, n, \end{cases}$$

where the meaning of each symbol and the assumptions of the model are readily stated. Note that it is assumed that certain second- and higher-order interactions are zero.

(a) Describe the assumptions of the model under Models I, II, and III. Under Model III assume that factors A and C are fixed and B and D are random.
(b) Develop the analysis of variance including expected mean squares under the assumptions of Models I, II, and III.
(c) Describe appropriate F tests for the effects of factors A, B, C, and D and their interactions for all the three models assuming normality for the random effects.

9. Consider a three-factor study where factors A and B are crossed and factor C is nested within factors A and B. Further, suppose that A, B, and C are all random having a, b, and c levels respectively and there are n replications in each cell. The mathematical model for this type of layout would be

$$y_{ijk\ell} = \mu + \alpha_i + \beta_j + (\alpha\beta)_{ij} + \gamma_{k(ij)} + e_{\ell(ijk)} \quad \begin{cases} i = 1, 2, \ldots, a \\ j = 1, 2, \ldots, b \\ k = 1, 2, \ldots, c \\ \ell = 1, 2, \ldots, n, \end{cases}$$

where α_i is the effect of the i-th level of factor A, β_j is the effect of the j-th level of factor B, $(\alpha\beta)_{ij}$ is the interaction between the i-th level of factor A and the j-th level of factor B, $\gamma_{k(ij)}$ is the effect of the k-th level of factor C within the combination of the i-th level of actor A and the j-th level of factor B, and $e_{\ell(ijk)}$ is the customary error term. It is

Partially Nested Classifications

assumed that the random effects α_i's, $\beta_{j(i)}$'s, $(\alpha\beta)_{ij}$'s, $\gamma_{k(ij)}$'s, and $e_{\ell(ijk)}$'s are all mutually and completely uncorrelated random variables with zero means and variances σ_α^2, σ_β^2, $\sigma_{\alpha\beta}^2$, $\sigma_{\gamma(\alpha\beta)}^2$, and σ_e^2 respectively. The overall sum of squares is partitioned as

$$SS_T = SS_A + SS_B + SS_{AB} + SS_{C(AB)} + SS_E,$$

where

$$SS_T = \sum_{i=1}^{a}\sum_{j=1}^{b}\sum_{k=1}^{c}\sum_{\ell=1}^{n}(y_{ijk\ell} - \bar{y}_{....})^2,$$

$$SS_A = bcn\sum_{i=1}^{a}(\bar{y}_{i...} - \bar{y}_{....})^2, \quad SS_B = acn\sum_{j=1}^{b}(\bar{y}_{.j..} - \bar{y}_{....})^2,$$

$$SS_{AB} = cn\sum_{i=1}^{a}\sum_{j=1}^{b}(\bar{y}_{ij..} - \bar{y}_{i...} - \bar{y}_{.j..} + \bar{y}_{....})^2,$$

$$SS_{C(AB)} = n\sum_{i=1}^{a}\sum_{j=1}^{b}\sum_{k=1}^{c}(\bar{y}_{ijk.} - \bar{y}_{ij..})^2,$$

and

$$SS_E = \sum_{i=1}^{a}\sum_{j=1}^{b}\sum_{k=1}^{c}\sum_{\ell=1}^{n}(y_{ijk\ell} - \bar{y}_{ijk.})^2.$$

Finally, the analysis of variance table for this model is shown as follows.

Source of Variation	Degrees of Freedom	Sum of Squares	Mean Square	Expected Mean Square
Factor A	$a - 1$	SS_A	MS_A	$\sigma_e^2 + n\sigma_{\gamma(\alpha\beta)}^2 + cn\sigma_{\alpha\beta}^2 + bcn\sigma_\alpha^2$
Factor B	$b - 1$	SS_B	MS_B	$\sigma_e^2 + n\sigma_{\gamma(\alpha\beta)}^2 + cn\sigma_{\alpha\beta}^2 + acn\sigma_\beta^2$
Interaction $A \times B$	$(a-1)(b-1)$	SS_{AB}	MS_{AB}	$\sigma_e^2 + n\sigma_{\gamma(\alpha\beta)}^2 + cn\sigma_{\alpha\beta}^2$
Factor C (within A and B)	$ab(c - 1)$	$SS_{C(AB)}$	$MS_{C(AB)}$	$\sigma_e^2 + n\sigma_{\gamma(\alpha\beta)}^2$
Error	$abc(n - 1)$	SS_E	MS_E	σ_e^2

(a) Develop the results on expected mean squares using the rules given in Appendix U.
(b) Assuming normality, determine the tests of hypotheses for testing the effects corresponding to factors A, B, C, and the interaction $A \times B$.

(c) Determine the estimators of the variance components based on the analysis of variance procedure.
(d) Repeat parts (a) through (c) for a mixed model analysis where A is fixed and B and C are random.
(e) Repeat parts (a) through (c) for a mixed model analysis where A and B are fixed and C is random.

10. Consider a four factor study where B is nested within A, D is nested within A, B, and C; and A and C are crossed. Further, suppose that A, B, C, and D are all random having a, b, c, and d levels respectively and there are n replications in each cell. The mathematical model for this type of layout would be

$$y_{ijk\ell m} = \mu + \alpha_i + \beta_{j(i)} + \gamma_k + (\alpha\gamma)_{ik} + (\beta\gamma)_{jk(i)} + \delta_{\ell(ijk)} + e_{m(ijk\ell)} \quad \begin{cases} i = 1, 2, \ldots, a \\ j = 1, 2, \ldots, b \\ k = 1, 2, \ldots, c \\ \ell = 1, 2, \ldots, d \\ m = 1, 2, \ldots, n, \end{cases}$$

where α_i is the effect of the i-th level of factor A, $\beta_{j(i)}$ is the effect of the j-th level of factor B within the i-th level of factor A, γ_k is the effect of the k-th level of factor C, $(\alpha\gamma)_{ik}$ is the interaction between the i-th level of factor A and the k-th level of factor C, $(\beta\gamma)_{jk(i)}$ is the interaction between the j-th level of factor B and the k-th level of factor C within the i-th level of factor A, $\delta_{\ell(ijk)}$ is the effect of the ℓ-th level of factor D within the combination of the i-th level of factor A, the j-th level of factor B, and the k-th level of factort C, and $e_{m(ijk\ell)}$ is the customary error term. It is assumed that the random effects α_i's, $\beta_{j(i)}$'s, γ_k's, $(\alpha\gamma)_{ik}$'s, $(\beta\gamma)_{jk(i)}$'s, $\delta_{\ell(ijk)}$'s, and $e_{m(ijk\ell)}$'s are all mutually and completely uncorrelated random variables with zero means and variances σ_α^2, $\sigma_{\beta(\alpha)}^2$, σ_γ^2, $\sigma_{\alpha\gamma}^2$, $\sigma_{\beta\gamma(\alpha)}^2$, $\sigma_{\delta(\alpha\beta\gamma)}^2$, and σ_e^2 respectively.

(a) Develop a partitioning of the total sum of squares corresponding to four main effects (including two nested), the two interactions, and a residual term.
(b) Report the analysis of variance table including expected mean squares.
(c) Assuming normality, determine the tests of hypotheses for testing the effects corresponding to factors A, B, C, and D, and the interactions $A \times D$ and $B \times D$ (within A).
(d) Determine the estimators of variance components based on the analysis of variance procedure.
(e) Repeat parts (b) through (d) for a mixed model analysis where factors A and C are fixed and B and D are random.

9 Finite Population and Other Models

9.0 PREVIEW

As discussed earlier, so far in this volume we have been primarily concerned with random effects models or Model II based on the infinite population theory, that is, when the treatments included in the experiment are assumed to be a random sample from a population of treatments having infinite size or when the experimenter selects the levels at random from a large number of possible levels of a factor usually considered as infinite. However, as described in Section 1.4, there are situations when the treatments selected may be a sample from a finite population and then the assumptions of an infinite population may be inappropriate. For example, in a large laboratory, there could be a total of 10 analysts and the data obtained on just three of them could be used to make inferences concerning a new method for the determination of arginine content as used by the entire group of 10 analysts.

The finite population model is also of interest because if we let the population sizes go to infinity, then we obtain Model II, and if we decrease the population size until it equals the sample size (so that the sample comprises the entire population), then we obtain Model I. If some population sizes are increased to infinity while others decreased to sample sizes, we are in a Model III situation. Under finite population models, the calculations for sums of squares, degrees of freedom, and mean squares remain the same. The difference lies in the derivation of expected mean squares and consequently in the estimation of the parameters and the testing of hypotheses. In this chapter, we briefly present the results for the finite population models. These models were first considered by Tukey (1949 a, c, 1950), Cornfield and Tukey (1956), and Bennett and Franklin (1954, Chapter VII). The interested reader is advised to go over these references for a more thorough treatment of the topic.

9.1 ONE-WAY FINITE POPULATION MODEL

For a one-way classification, with a groups or a levels of factor A and n observations per group, the mathematical model under finite population theory is

TABLE 9.1
Analysis of Variance for Model (9.1.1)

Source of Variation	Degrees of Freedom	Sum of Squares	Mean Square	Expected Mean Square
Between	$a - 1$	SS_B	MS_B	$\sigma_e^2 + n\sigma_\alpha^2$
Within	$a(n - 1)$	SS_W	MS_W	σ_e^2
Total	$an - 1$	SS_T		

the same as (2.1.1), namely,

$$y_{ij} = \mu + \alpha_i + e_{ij} \quad \begin{cases} i = 1, 2, \ldots, a \\ j = 1, 2, \ldots, n, \end{cases} \quad (9.1.1)$$

where y_{ij} is the value of the j-th observation in the i-th group. However, now the assumptions are as follows:

(i) As before, μ is the constant general effect.
(ii) There is a population of effects due to factor A of size A with mean zero and variance σ_α^2. The α_i's are assumed to be a random sample of size a from this population. We denote a particular level of A in the population by α_I, where $I = 1, 2, \ldots, A$. The α_I's in the population satisfy the condition $\sum_{I=1}^{A} \alpha_I = 0$.
(iii) Sampling is random in each group and independent among different groups. The e_{ij}'s are a random sample of size n from an infinite population with mean zero and variance σ_e^2.
(iv) We make the following definitions of population variances, that is, the variance components of the model (9.1.1),

$$\sigma_\alpha^2 = \frac{1}{A - 1} \sum_{I=1}^{A} \alpha_I^2$$

and

$$\sigma_e^2 = E(e_{ij}^2).$$

For the finite population model (9.1.1), the entire analysis of variance, including the sums of squares, mean squares, and expected mean squares, remains the same and is summarized in Table 9.1. Thus, in Table 9.1, if $A = a$, the definition of σ_α^2 corresponds to the Model I case of Table 2.1; and if $A = \infty$, it corresponds to the Model II case of Table 2.1.

9.2 TWO-WAY CROSSED FINITE POPULATION MODEL

For a two-way crossed classification, with factor A having a levels, factor B having b levels, and n replications per cell, the mathematical model under finite population is the same as (4.1.1); that is,

$$y_{ijk} = \mu + \alpha_i + \beta_j + (\alpha\beta)_{ij} + e_{ijk} \quad \begin{cases} i = 1, 2, \ldots, a \\ j = 1, 2, \ldots, b \\ k = 1, 2, \ldots, n, \end{cases} \quad (9.2.1)$$

where y_{ijk} is the score of the k-th observation at the i-th level of factor A and the j-th level of factor B. However, underlying model (9.2.1), we now have the following assumptions:

(i) As before, μ is the constant general effect.
(ii) There is a population of main effects due to factor A of size A with mean zero and variance σ_α^2. The α_i's are assumed to be a random sample of size a from this population. We denote a particular level of A in the population by α_I, where $I = 1, 2, \ldots, A$. The α_I's satisfy the condition $\sum_{I=1}^{A} \alpha_I = 0$.
(iii) There is a population of main effects due to factor B of size B with mean zero and variance σ_β^2. The β_j's are assumed to be a random sample of size b from this population. We denote a particular level of B in the population by β_J where $J = 1, 2, \ldots, B$. The β_J's satisfy the condition $\sum_{J=1}^{B} \beta_J = 0$.
(iv) For each combination of a potential level of A with a potential level of B, there is a population of interaction effects of size $A \times B$ with mean zero and variance $\sigma_{\alpha\beta}^2$. Selecting a particular I and a particular J determines the row and column and hence the cell that forms their interaction, and with this cell is associated the interaction $(\alpha\beta)_{IJ}$. The interaction terms satisfy the conditions:

$$\sum_{I=1}^{A} (\alpha\beta)_{IJ} = 0, \quad \text{for each } J$$

and

$$\sum_{J=1}^{B} (\alpha\beta)_{IJ} = 0, \quad \text{for each } I.$$

(v) Sampling is at random in each cell and independent between different cells. The e_{ijk}'s are a random sample of size n from an infinite population with mean zero and variance σ_e^2.
(vi) We make the following definitions of the population variances, that is, the variance components of model (9.2.1):

$$\sigma_\alpha^2 = \frac{1}{A-1} \sum_{I=1}^{A} \alpha_I^2,$$

$$\sigma_\beta^2 = \frac{1}{B-1} \sum_{J=1}^{B} \beta_J^2,$$

$$\sigma_{\alpha\beta}^2 = \frac{1}{(A-1)(B-1)} \sum_{I=1}^{A} \sum_{J=1}^{B} (\alpha\beta)_{IJ}^2,$$

TABLE 9.2
Analysis of Variance for Model (9.2.1)

Source of Variation	Degrees of Freedom	Sum of Squares	Mean Square	Expected Mean Square
Factor **A**	$a-1$	SS_A	MS_A	$\sigma_e^2 + n\left(1 - \dfrac{b}{B}\right)\sigma_{\alpha\beta}^2 + bn\sigma_\alpha^2$
Factor **B**	$b-1$	SS_B	MS_B	$\sigma_e^2 + n\left(1 - \dfrac{a}{A}\right)\sigma_{\alpha\beta}^2 + an\sigma_\beta^2$
Interaction **A** × **B**	$(a-1)(b-1)$	SS_{AB}	MS_{AB}	$\sigma_e^2 + n\sigma_{\alpha\beta}^2$
Error	$ab(n-1)$	SS_E	MS_E	σ_e^2
Total	$abn-1$	SS_T		

and

$$\sigma_e^2 = E(e_{ijk}^2).$$

For the finite population model (9.2.1), the sums of squares and mean squares are the same as those shown in Table 4.2. However, the expected mean squares are those shown in Table 9.2. The derivation of expected mean squares involves some tedious algebra and can be found in Cornfield and Tukey (1956), Bennett and Franklin (1954, pp. 368–373), and Brownlee (1965, pp. 489–498). We simply mention here the results on the covariance structure of the α_i's, β_j's, and $(\alpha\beta)_{ij}$'s which are employed in finding expected mean squares. Thus,

$$\mathrm{Cov}(\alpha_i, \alpha_{i'}) = -\frac{\sigma_\alpha^2}{A}, \quad i \neq i'$$

$$\mathrm{Cov}(\beta_j, \beta_{j'}) = -\frac{\sigma_\beta^2}{B}, \quad j \neq j'$$

$$\mathrm{Cov}\{(\alpha\beta)_{ij}, (\alpha\beta)_{i'j'}\} = \begin{cases} \dfrac{\sigma_{\alpha\beta}^2}{AB}, & i \neq i', j \neq j' \\ -\dfrac{(1 - 1/B)}{A}\sigma_{\alpha\beta}^2, & i \neq i', j = j' \\ -\dfrac{(1 - 1/A)}{B}\sigma_{\alpha\beta}^2, & i = i', j \neq j'. \end{cases}$$

Furthermore, because the α_i's and β_j's are selected independently, their covariances are zero. The values of $(\alpha\beta)_{ij}$'s, on the other hand, in general depend on the i-th level of **A** and the j-th level of **B** and thus $(\alpha\beta)_{ij}$'s are not independent of the α_i's and β_j's in the sample. However, it can be shown that the covariance between α_i and $(\alpha\beta)_{i'j}$ is zero irrespective of whether $i = i'$; similarly, the covariance between β_j and $(\alpha\beta)_{ij'}$ is zero. Thus, we obtain the following results:

$$\text{Cov}(\alpha_i, \beta_j) = 0,$$
$$\text{Cov}(\alpha_i, (\alpha\beta)_{ij}) = 0,$$
$$\text{Cov}(\alpha_i, (\alpha\beta)_{i'j}) = 0,$$
$$\text{Cov}(\beta_j, (\alpha\beta)_{ij}) = 0,$$

and

$$\text{Cov}(\beta_j, (\alpha\beta)_{ij'}) = 0.$$

TESTS OF HYPOTHESES

The results on expected mean squares provide a valuable guide to deciding which mean squares in the analysis of variance table are to be compared. For example, from Table 9.2, it is seen that a hypothesis test is only available to test for the existence of interaction in the general case. To develop tests for the main effects, we must limit ourselves to certain special cases described in the following:

(i) In Table 9.2, if the samples of factors A and B levels correspond to the entire population, so that $A = a$, $B = b$, then the expected values of the mean squares are those given in Table 4.2 for Model I and the finite population model becomes exactly Model I. Hence, the tests for the main effects are obtained by dividing the mean squares by the error mean square.

(ii) If the levels of factors A and B are infinitely large so that $1 - a/A$ and $1 - b/B$ both tend to 1, then all covariances approach to zero and the expected values of the mean squares are given exactly as in Table 4.2 for Model II and the finite population model becomes exactly Model II. Hence, the main effects are tested against the interaction mean square.

(iii) If the α_i's are samples from an infinite population and the β_j's are samples from the entire population (i.e., $A = \infty$ and $B = b$), then the expected values of the mean squares are given exactly as in Table 4.2 for Model III and the finite population model becomes exactly Model III. Thus, the factor A effect is tested against the error and the factor B effect is tested against the interaction mean square.

In the following we give the details of F tests of the hypotheses of interest for the general case of the finite population model including the point and interval estimation of the parameters.

F TESTS

An F test of the hypothesis

$$H_0^{AB} : \sigma_{\alpha\beta}^2 = 0 \quad \text{versus} \quad H_1^{AB} : \sigma_{\alpha\beta}^2 > 0$$

can be performed by using the statistic MS_{AB}/MS_E. Now, to test

$$H_0^B : \sigma_\beta^2 = 0 \quad \text{versus} \quad H_1^B : \sigma_\beta^2 > 0, \qquad (9.2.2)$$

we notice that under H_0^B,

$$E[MS_B] = \sigma_e^2 + n\left(1 - \frac{a}{A}\right)\sigma_{\alpha\beta}^2.$$

Since there is no other mean square with this expectation, an exact F test for the hypothesis (9.2.2) is not available. However, it is possible to determine a conservative or an approximate F test.

To obtain a conservative test, note that when $\sigma_\beta^2 = 0$, the quantity

$$\frac{MS_B \Big/ \left[\sigma_e^2 + n\left(1 - \frac{a}{A}\right)\sigma_{\alpha\beta}^2\right]}{MS_{AB}/(\sigma_e^2 + n\sigma_{\alpha\beta}^2)} \qquad (9.2.3)$$

has an F distribution with $b - 1$ and $(a - 1)(b - 1)$ degrees of freedom, respectively. It cannot, however, be used for testing the hypothesis (9.2.2), since the expressions involving unknown parameters do not cancel out and therefore the statistic (9.2.3) cannot be evaluated. However, if we change the coefficient of $\sigma_{\alpha\beta}^2$ from $n(1 - \frac{a}{A})$ to n by neglecting a/A, then $\sigma_e^2 + n\sigma_{\alpha\beta}^2$ cancels out from both the numerator and the denominator, and (9.2.3) reduces to the statistic

$$F_B = MS_B/MS_{AB},$$

which is readily evaluated. Note that in this way we have reduced the value of the statistic (9.2.3), and, thus, if we compare it with $F[b - 1, (a - 1)(b - 1); 1 - \alpha]$, we have made it harder to reject the null hypothesis. Such a test is conservative, since if we try to construct a test with significance level α, we actually have one with level $\leq \alpha$. When A is sufficiently large compared to a, this test is generally adequate. In addition, we can also develop a psuedo-F test for the hypothesis (9.2.2) by using a linear combination of the mean squares whose expected value is equal to $\sigma_e^2 + n(1 - \frac{a}{A})\sigma_{\alpha\beta}^2$. We note from Table 9.2

Finite Population and Other Models

that $E[(1 - \frac{a}{A})MS_{AB} + (\frac{a}{A})MS_E] = \sigma_e^2 + n(1 - \frac{a}{A})\sigma_{\alpha\beta}^2$. Hence, the suggested F statistic is

$$F'_B = \frac{MS_B}{\left(1 - \frac{a}{A}\right)MS_{AB} + \left(\frac{a}{A}\right)MS_E}, \quad (9.2.4)$$

which has an approximate F distribution with $b - 1$ and v_b degrees of freedom, where v_b is approximated by

$$v_b = \frac{\left(\left(1 - \frac{a}{A}\right)MS_{AB} + \left(\frac{a}{A}\right)MS_E\right)^2}{\frac{\left(1 - \frac{a}{A}\right)^2 (MS_{AB})^2}{(a-1)(b-1)} + \frac{\left(\frac{a}{A}\right)^2 (MS_E)^2}{ab(n-1)}}. \quad (9.2.5)$$

Analogous tests — conservative as well approximate — can be constructed for the hypothesis $H_0^A : \sigma_\alpha^2 = 0$ versus $H_1^A : \sigma_\alpha^2 > 0$.

POINT ESTIMATION

The parameter μ is clearly estimated by

$$\hat{\mu} = \bar{y}_{...},$$

where

$$\text{Var}(\bar{y}_{...}) = \frac{1}{abn}\left[\sigma_e^2 + n\left(1 - \frac{a}{A}\right)\left(1 - \frac{b}{B}\right)\sigma_{\alpha\beta}^2 + an\left(1 - \frac{b}{B}\right)\sigma_\beta^2 \right.$$
$$\left. + bn\left(1 - \frac{a}{A}\right)\sigma_\alpha^2\right], \quad (9.2.6)$$

which is estimated unbiasedly as

$$\widehat{\text{Var}}(\bar{y}_{...}) = \frac{1}{abn}\left[\frac{ab}{AB}MS_E - \left(1 - \frac{a}{A}\right)\left(1 - \frac{b}{B}\right)MS_{AB} + \left(1 - \frac{b}{B}\right)MS_B \right.$$
$$\left. + \left(1 - \frac{a}{A}\right)MS_A\right]. \quad (9.2.7)$$

The other parameters of interest are the variance components σ_e^2, $\sigma_{\alpha\beta}^2$, σ_β^2, and σ_α^2. From Table 9.2, on equating mean squares to their respective expected values and solving for the variance components, we obtain

$$\hat{\sigma}_e^2 = \mathrm{MS}_E, \qquad (9.2.8)$$

$$\hat{\sigma}_{\alpha\beta}^2 = \frac{1}{n}(\mathrm{MS}_{AB} - \mathrm{MS}_E), \qquad (9.2.9)$$

$$\hat{\sigma}_\beta^2 = \frac{1}{an}\left[\mathrm{MS}_B - \left(1 - \frac{a}{A}\right)\mathrm{MS}_{AB} - \frac{a}{A}\mathrm{MS}_E\right], \qquad (9.2.10)$$

and

$$\hat{\sigma}_\alpha^2 = \frac{1}{bn}\left[\mathrm{MS}_A - \left(1 - \frac{b}{B}\right)\mathrm{MS}_{AB} - \frac{b}{B}\mathrm{MS}_E\right]. \qquad (9.2.11)$$

INTERVAL ESTIMATION

To find a confidence interval for μ, we note that there is no entry in the mean square column of Table 9.2 which is a multiple of $\mathrm{Var}(\bar{y}_{...})$ given by (9.2.6). Thus, we cannot find an exact confidence interval for μ. However, as earlier, we can determine an approximate confidence interval using the Satterthwaite procedure. Thus, from (9.2.6) and (9.2.7), we have

$$\frac{\nu\mathrm{MS}}{\sigma_e^2 + n\left(1 - \frac{a}{A}\right)\left(1 - \frac{b}{B}\right)\sigma_{\alpha\beta}^2 + an\left(1 - \frac{b}{B}\right)\sigma_\beta^2 + bn\left(1 - \frac{a}{A}\right)\sigma_\alpha^2} \sim \chi^2[\nu],$$

where

$$\mathrm{MS} = \frac{ab}{AB}\mathrm{MS}_E - \left(1 - \frac{a}{A}\right)\left(1 - \frac{b}{B}\right)\mathrm{MS}_{AB} + \left(1 - \frac{b}{B}\right)\mathrm{MS}_B$$
$$+ \left(1 - \frac{a}{A}\right)\mathrm{MS}_A \qquad (9.2.12)$$

and ν is calculated from

$$\nu = \frac{(\mathrm{MS})^2}{\left(\frac{ab}{AB}\right)^2 \frac{(\mathrm{MS}_E)^2}{ab(n-1)} + \left(1 - \frac{a}{A}\right)^2\left(1 - \frac{b}{B}\right)^2 \frac{(\mathrm{MS}_{AB})^2}{(a-1)(b-1)} + \left(1 - \frac{b}{B}\right)^2 \frac{(\mathrm{MS}_B)^2}{b-1} + \left(1 - \frac{a}{A}\right)^2 \frac{(\mathrm{MS}_A)^2}{a-1}}.$$
$$(9.2.13)$$

Thus,

$$\frac{\bar{y}_{...} - \mu}{\sqrt{\frac{\mathrm{MS}}{abn}}} \text{ approx} \sim t[\nu], \qquad (9.2.14)$$

Finite Population and Other Models

and a $1 - \alpha$ level confidence interval for μ is given by

$$\bar{y}_{...} \pm t[v, 1 - \alpha/2]\sqrt{MS/abn}. \qquad (9.2.15)$$

These limits are approximate because of the approximation involved in (9.2.14).

Now, an examination of the mean square and expected mean square columns of Table 9.2 indicates which variance components or their functions can be easily estimated by a confidence interval. Thus, as in the case of the infinite population model, a $100(1 - \alpha)$ percent confidence interval for σ_e^2 is given by

$$\frac{ab(n-1)MS_E}{\chi^2[ab(n-1), 1-\alpha/2]} < \sigma_e^2 < \frac{ab(n-1)MS_E}{\chi^2[ab(n-1), \alpha/2]}. \qquad (9.2.16)$$

Furthermore,

$$\frac{(a-1)(b-1)MS_{AB}}{\sigma_e^2 + n\sigma_{\alpha\beta}^2} \sim \chi^2[(a-1)(b-1)]$$

and a $1 - \alpha$ level confidence interval for $\sigma_e^2 + n\sigma_{\alpha\beta}^2$ is given by

$$\frac{(a-1)(b-1)MS_{AB}}{\chi^2[(a-1)(b-1), 1-\alpha/2]} < \sigma_e^2 + n\sigma_{\alpha\beta}^2 < \frac{(a-1)(b-1)MS_{AB}}{\chi^2[(a-1)(b-1), \alpha/2]}. \qquad (9.2.17)$$

Likewise

$$\frac{(b-1)MS_B}{\sigma_e^2 + n\left(1 - \frac{a}{A}\right)\sigma_{\alpha\beta}^2 + an\sigma_\beta^2} \sim \chi^2[b-1]$$

and a $1 - \alpha$ level confidence interval for $\sigma_e^2 + n(1 - \frac{a}{A})\sigma_{\alpha\beta}^2 + an\sigma_\beta^2$ is given by

$$\frac{(b-1)MS_B}{\chi^2[b-1, 1-\alpha/2]} < \sigma_e^2 + n\left(1 - \frac{a}{A}\right)\sigma_{\alpha\beta}^2 + an\sigma_\beta^2 < \frac{(b-1)MS_B}{\chi^2[b-1, \alpha/2]}. \qquad (9.2.18)$$

Similarly, a $1 - \alpha$ level confidence interval for $\sigma_e^2 + n(1 - \frac{b}{B})\sigma_{\alpha\beta}^2 + bn\sigma_\alpha^2$ is given by

$$\frac{(a-1)MS_A}{\chi^2[a-1, 1-\alpha/2]} < \sigma_e^2 + n\left(1 - \frac{b}{B}\right)\sigma_{\alpha\beta}^2 + bn\sigma_\alpha^2 < \frac{(a-1)MS_A}{\chi^2[a-1, \alpha/2]}. \qquad (9.2.19)$$

Unfortunately, as noted in earlier chapters, the expressions in the expected mean square column are the only quantities for which we can find exact confidence intervals in this manner. Thus, there do not exist exact confidence intervals for $\sigma_{\alpha\beta}^2$, σ_{β}^2, and σ_{α}^2, and we have to resort to some approximate results. There are various methods for obtaining approximate intervals (see, e.g., Bennett and Franklin, 1954, Chapter VII). In the following, we give a method for obtaining a conservative confidence interval in the sense that the confidence level is at least $1 - \alpha$. Inasmuch as variance components are nonnegative, it is possible to obtain a conservative confidence interval by simply deleting the undesired terms (nuisance parameters) in the usual confidence intervals obtained by using the chi-square table. For example, from (9.2.17), we can delete the term containing σ_e^2 and it will yield a conservative $100(1 - \alpha)$ percent confidence interval for $\sigma_{\alpha\beta}^2$ as

$$\frac{(a-1)(b-1)\text{MS}_{AB}}{n\chi^2[(a-1)(b-1), 1-\alpha/2]} < \sigma_{\alpha\beta}^2 < \frac{(a-1)(b-1)\text{MS}_{AB}}{n\chi^2[(a-1)(b-1), \alpha/2]}. \tag{9.2.20}$$

Similarly, from (9.2.18) and (9.2.19), one can obtain conservative $100(1 - \alpha)$ percent confidence intervals for σ_{β}^2 and σ_{α}^2 given by

$$\frac{(b-1)\text{MS}_B}{an\chi^2[b-1, 1-\alpha/2]} < \sigma_{\beta}^2 < \frac{(b-1)\text{MS}_B}{an\chi^2[b-1, \alpha/2]} \tag{9.2.21}$$

and

$$\frac{(a-1)\text{MS}_A}{bn\chi^2[a-1, 1-\alpha/2]} < \sigma_{\alpha}^2 < \frac{(a-1)\text{MS}_A}{bn\chi^2[a-1, \alpha/2]}. \tag{9.2.22}$$

9.3 THREE-WAY CROSSED FINITE POPULATION MODEL

The finite population model for the replicated three-way crossed classification is a natural generalization of the two-way crossed finite population model (9.2.1). Thus, with factor A at A levels, factor B at B levels, factor C at C levels, and n replications per cell, the model equation may be written as

$$y_{ijk\ell} = \mu + \alpha_i + \beta_j + \gamma_k + (\alpha\beta)_{ij} + (\alpha\gamma)_{ik} + (\beta\gamma)_{jk} + (\alpha\beta\gamma)_{ijk} + e_{ijk\ell} \quad \begin{cases} i = 1, 2, \ldots, a \\ j = 1, 2, \ldots, b \\ k = 1, 2, \ldots, c \\ \ell = 1, 2, \ldots, n, \end{cases} \tag{9.3.1}$$

where the terms on the right-hand side of equation (9.3.1) have the familiar meanings and correspond to the general mean; main effects due to factors

A, B, C; interactions A × B, A × C, B × C, A × B × C; and the error term, respectively.

The assumptions of the finite population model (9.3.1) are stated in a way similar to those for the finite two-way model (9.2.1). Thus, for example, we suppose that the α_i's are a random sample of size a from a population of size A, and the β_j's and γ_k's are random samples of sizes b and c from populations of sizes B and C, respectively. In addition, in the population, the various parameters sum to zero over each index; that is,

$$\sum_{I=1}^{A} \alpha_I = \sum_{J=1}^{B} \beta_J = \sum_{K=1}^{C} \gamma_K = 0,$$

$$\sum_{I=1}^{A} (\alpha\beta)_{IJ} = \sum_{J=1}^{B} (\alpha\beta)_{IJ} = \sum_{I=1}^{A} (\alpha\gamma)_{IK} = \sum_{K=1}^{C} (\alpha\gamma)_{IK}$$

$$= \sum_{J=1}^{B} (\beta\gamma)_{JK} = \sum_{K=1}^{C} (\beta\gamma)_{JK} = 0,$$

(9.3.2)

$$\sum_{I=1}^{A} (\alpha\beta\gamma)_{IJK} = \sum_{J=1}^{B} (\alpha\beta\gamma)_{IJK} = \sum_{K=1}^{C} (\alpha\beta\gamma)_{IJK} = 0;$$

and we make the following definitions of the population variances, that is, the variance components of model (9.3.1):

$$\sigma_\alpha^2 = \frac{1}{A-1} \sum_{I=1}^{A} \alpha_I^2, \text{ etc.,}$$

$$\sigma_{\alpha\beta}^2 = \frac{1}{(A-1)(B-1)} \sum_{I=1}^{A} \sum_{J=1}^{B} (\alpha\beta)_{IJ}^2, \text{ etc.,}$$

(9.3.3)

and

$$\sigma_{\alpha\beta\gamma}^2 = \frac{1}{(A-1)(B-1)(C-1)} \sum_{I=1}^{A} \sum_{J=1}^{B} \sum_{K=1}^{C} (\alpha\beta\gamma)_{IJK}^2.$$

The sums of squares and mean squares are the same as those shown in Table 5.2. The expected mean squares, which can be derived by an extension of the method used for the two-way model (see, e.g., Cornfield and Tukey (1956)), are displayed in Table 9.3.

Again, it is readily seen that when $A = a$, $B = b$, and $C = c$, the expected values of the mean squares are those given in Table 5.2 for Model I and the finite population model becomes exactly Model I. When A, B, and C are all infinite, then the expected values of the mean squares are those given in Table 5.2 for Model II and the finite model becomes exactly Model II. When A is infinite, $B = b$, and $C = c$, we have a Model III situation where factor A is random and

TABLE 9.3
Expected Mean Squares for Model (9.3.1)

Source	Expected Mean Square
A	$\sigma_e^2 + n\left(1-\dfrac{b}{B}\right)\left(1-\dfrac{c}{C}\right)\sigma_{\alpha\beta\gamma}^2 + bn\left(1-\dfrac{c}{C}\right)\sigma_{\alpha\gamma}^2 + cn\left(1-\dfrac{b}{B}\right)\sigma_{\alpha\beta}^2 + bcn\sigma_\alpha^2$
B	$\sigma_e^2 + n\left(1-\dfrac{a}{A}\right)\left(1-\dfrac{c}{C}\right)\sigma_{\alpha\beta\gamma}^2 + an\left(1-\dfrac{c}{C}\right)\sigma_{\beta\gamma}^2 + cn\left(1-\dfrac{a}{A}\right)\sigma_{\alpha\beta}^2 + acn\sigma_\beta^2$
C	$\sigma_e^2 + n\left(1-\dfrac{a}{A}\right)\left(1-\dfrac{b}{B}\right)\sigma_{\alpha\beta\gamma}^2 + an\left(1-\dfrac{b}{B}\right)\sigma_{\beta\gamma}^2 + bn\left(1-\dfrac{a}{A}\right)\sigma_{\alpha\gamma}^2 + abn\sigma_\gamma^2$
A × B	$\sigma_e^2 + n\left(1-\dfrac{c}{C}\right)\sigma_{\alpha\beta\gamma}^2 + cn\sigma_{\alpha\beta}^2$
A × C	$\sigma_e^2 + n\left(1-\dfrac{b}{B}\right)\sigma_{\alpha\beta\gamma}^2 + bn\sigma_{\alpha\gamma}^2$
B × C	$\sigma_e^2 + n\left(1-\dfrac{a}{A}\right)\sigma_{\alpha\beta\gamma}^2 + an\sigma_{\beta\gamma}^2$
A × B × C	$\sigma_e^2 + n\sigma_{\alpha\beta\gamma}^2$
Error	σ_e^2

factors B and C are fixed; and the case when A and B are infinite and $C = c$, gives a Model III with factors A and B as random and factor C as fixed.

9.4 FOUR-WAY CROSSED FINITE POPULATION MODEL

The finite population model for the four-way classification is the obvious extension of two and three-way finite population models and we survey it only briefly. Thus, with factor A at A levels, factor B at B levels, factor C at C levels, factor D at D levels, and n replicates per cell, the model is

$$y_{ijk\ell m} = \mu + \alpha_i + \beta_j + \gamma_k + \delta_\ell + (\alpha\beta)_{ij} + (\alpha\gamma)_{ik} \\ + (\alpha\delta)_{i\ell} + (\beta\gamma)_{jk} + (\beta\delta)_{j\ell} + (\gamma\delta)_{k\ell} + (\alpha\beta\gamma)_{ijk} \\ + (\alpha\beta\delta)_{ij\ell} + (\alpha\gamma\delta)_{ik\ell} + (\beta\gamma\delta)_{jk\ell} + (\alpha\beta\gamma\delta)_{ijk\ell} \\ + e_{ijk\ell m}$$

$$\begin{cases} i = 1, \ldots, a \\ j = 1, \ldots, b \\ k = 1, \ldots, c \\ \ell = 1, \ldots, d \\ m = 1, \ldots, n, \end{cases}$$

(9.4.1)

where the terms on the right-hand side of equation (9.4.1) have familiar meanings. For example, the α_i's are a random sample of size a from a population of size A and sum to zero in the population; for example,

$$\sum_{I=1}^{A} \alpha_I = 0;$$

Finite Population and Other Models

with corresponding results for the other main effects. Similarly, the two-way interactions sum to zero in the population over each index; for example,

$$\sum_{I=1}^{A}(\alpha\beta)_{IJ} = \sum_{J=1}^{B}(\alpha\beta)_{IJ} = 0,$$

and so on. The three-way and four-way interactions also sum to zero in the population over each index; for example,

$$\sum_{I=1}^{A}(\alpha\beta\gamma)_{IJK} = \sum_{J=1}^{B}(\alpha\beta\gamma)_{IJK} = \sum_{K=1}^{C}(\alpha\beta\gamma)_{IJK} = 0,$$

and so on, and

$$\sum_{I=1}^{A}(\alpha\beta\gamma\delta)_{IJKL} = \sum_{J=1}^{B}(\alpha\beta\gamma\delta)_{IJKL} = \sum_{K=1}^{C}(\alpha\beta\gamma\delta)_{IJKL}$$
$$= \sum_{L=1}^{D}(\alpha\beta\gamma\delta)_{IJKL} = 0.$$

We define $\sigma_\alpha^2, \sigma_{\alpha\beta}^2, \sigma_{\alpha\beta\gamma}^2, \sigma_{\alpha\beta\gamma\delta}^2$, etc., in an analogous manner as in (9.3.3).

The sums of squares and mean squares are the same as defined in Section 5.11. The expected mean squares can again be derived by an extension of the method used for the two- and three-way finite population models. Thus, for example,

$$E(\text{MS}_A) = \sigma_e^2 + n\left(1 - \frac{b}{B}\right)\left(1 - \frac{c}{C}\right)\left(1 - \frac{d}{D}\right)\sigma_{\alpha\beta\gamma\delta}^2$$
$$+ nd\left(1 - \frac{b}{B}\right)\left(1 - \frac{c}{C}\right)\sigma_{\alpha\beta\gamma}^2 + nc\left(1 - \frac{b}{B}\right)\left(1 - \frac{d}{D}\right)\sigma_{\alpha\beta\delta}^2$$
$$+ nb\left(1 - \frac{c}{C}\right)\left(1 - \frac{d}{D}\right)\sigma_{\alpha\gamma\delta}^2 + ncd\left(1 - \frac{b}{B}\right)\sigma_{\alpha\beta}^2$$
$$+ nbd\left(1 - \frac{c}{C}\right)\sigma_{\alpha\gamma}^2 + nbc\left(1 - \frac{d}{D}\right)\sigma_{\alpha\delta}^2 + nbcd\sigma_\alpha^2, \text{ etc.,}$$

$$E(\text{MS}_{AB}) = \sigma_e^2 + n\left(1 - \frac{c}{C}\right)\left(1 - \frac{d}{D}\right)\sigma_{\alpha\beta\gamma\delta}^2 + nd\left(1 - \frac{c}{C}\right)\sigma_{\alpha\beta\gamma}^2$$
$$+ nc\left(1 - \frac{d}{D}\right)\sigma_{\alpha\beta\delta}^2 + ncd\sigma_{\alpha\beta}^2, \text{ etc.,}$$

$$E(\text{MS}_{ABC}) = \sigma_e^2 + n\left(1 - \frac{d}{D}\right)\sigma_{\alpha\beta\gamma\delta}^2 + nd\sigma_{\alpha\beta\gamma}^2, \text{ etc.,}$$

and

$$E(\text{MS}_{ABCD}) = \sigma_e^2 + n\sigma_{\alpha\beta\gamma\delta}^2.$$

9.5 NESTED FINITE POPULATION MODELS

Similar to the crossed classification models, we can develop finite population models for the nested classification. We briefly consider here the two-way nested finite population model. Thus, corresponding to the infinite population model (6.1.1), we have

$$y_{ijk} = \mu + \alpha_i + \beta_{j(i)} + e_{k(ij)} \quad \begin{cases} i = 1, 2, \ldots, a \\ j = 1, 2, \ldots, b \\ k = 1, 2, \ldots, n, \end{cases} \quad (9.5.1)$$

where the usual assumptions of the finite population model (9.5.1) are as follows:

(i) As before, μ is the constant general effect.
(ii) There is a population of α_I's of size A with $\sum_{I=1}^{A} \alpha_I = 0$ and α_i's are random samples of size a from this population.
(iii) Associated with each I is a population of $\beta_{J(I)}$'s of size B. For each of these populations of size B, we have the condition that $\sum_{J=1}^{B} \beta_{J(I)} = 0$ for each $I = 1, 2, \ldots, A$. The $\beta_{j(i)}$'s are random samples of size b from these populations. It should be noted, however, that for each value of I, the entire set of B $\beta_{J(I)}$'s sum to zero; but, in general, $\beta_{j(i)}$'s do not sum to zero for the sample b unless $b = B$. Also, the $\beta_{J(I)}$'s do not, in general, sum to zero within a row; that is, $\sum_{I=1}^{A} \beta_{J(I)} \ne 0$.
(iv) The $e_{k(ij)}$'s are a random sample of size n from an infinite population with mean zero and variance σ_e^2.
(v) We make the following definitions of the finite population variances or the variance components of model (9.5.1):

$$\sigma_\alpha^2 = \frac{1}{A-1} \sum_{I=1}^{A} \alpha_I^2,$$

$$\sigma_\beta^2 = \frac{1}{A(B-1)} \sum_{I=1}^{A} \sum_{J=1}^{B} \beta_{J(I)}^2,$$

and

$$\sigma_e^2 = E(e_{k(ij)}^2).$$

Again, the sums of squares and mean squares are the same as those given in Table 6.2. However, the expected mean squares are those shown in Table 9.4. The derivation of the expected mean squares follows the same general approach of the crossed situation and can be found in Bennett and Franklin (1954, pp. 358–363). Note that in Table 9.4 if both factors A and B constitute the entire population (i. e., $A = a$ and $B = b$), then the expected values of the mean squares become identical to those given in Table 6.2 for Model I. If both

TABLE 9.4
Analysis of Variance for Model (9.5.1)

Source of Variation	Degrees of Freedom	Sum of Squares	Mean Square	Expected Mean Square
Factor **A**	$a-1$	SS_A	MS_A	$\sigma_e^2 + n\left(1 - \dfrac{b}{B}\right)\sigma_\beta^2 + bn\sigma_\alpha^2$
Factor **B** within **A**	$a(b-1)$	$SS_{B(A)}$	$MS_{B(A)}$	$\sigma_e^2 + n\sigma_\beta^2$
Error	$ab(n-1)$	SS_E	MS_E	σ_e^2
Total	$abn-1$	SS_T		

$A = \infty$, $B = \infty$, we get Model II; and the case with $A = a$ and $B = \infty$ gives Model III.

The finite population model (9.5.1) can be extended similarly to higher-order nested classifications.

9.6 UNBALANCED FINITE POPULATION MODELS

In the preceding sections, we have dealt mainly with finite population models having balanced sampling. The details on the models involving unequal sampling can be found in the papers of Gaylor and Hartwell (1969) and Searle and Fawcett (1970).

9.7 WORKED EXAMPLE FOR A FINITE POPULATION MODEL

Consider an industrial experiment involving 3 machines and 4 operators. Machines were randomly selected from a set of 10 machines and operators were chosen at random from a group of 12 available operators. Three observations were made on each of the 12 machine-operator combinations and the data on production output are given in Table 9.5.

This is an example of a two-way crossed finite population model with replication where both machines and operators are randomly selected from populations involving only a finite number of elements. The mathematical model for this experiment would be

$$y_{ijk} = \mu + \alpha_i + \beta_j + (\alpha\beta)_{ij} + e_{ijk} \quad \begin{cases} i = 1, 2, 3 \\ j = 1, 2, 3, 4 \\ k = 1, 2, 3, \end{cases}$$

where μ is the constant general effect, α_i is the effect of the i-th machine, β_j is the effect of the j-th operator, $(\alpha\beta)_{ij}$ is the interaction effect of the i-th machine with the j-th operator, and e_{ijk} is the customary error term. Furthermore, it is

TABLE 9.5
Production Output from an Industrial Experiment

	Operator			
Machine	1	2	3	4
1	26.3	26.0	25.7	25.0
	26.9	25.2	26.0	25.3
	27.2	24.6	26.2	25.0
2	26.7	26.0	26.2	25.5
	27.0	26.4	26.1	24.7
	26.9	26.6	27.3	26.0
3	26.8	26.6	26.5	26.2
	27.0	27.0	27.5	25.5
	27.2	26.9	27.0	25.7

assumed that the α_i's are a random sample of size 3 from a population of α_I's that satisfy the condition $\sum_{I=1}^{10} \alpha_I = 0$; β_j's are a random sample of size 4 from a population of β_J's that satisfy the condition $\sum_{J=1}^{12} \beta_J = 0$; $(\alpha\beta)_{ij}$'s are a random sample from a population of $(\alpha\beta)_{IJ}$'s that satisfy the conditions $\sum_{I=1}^{10}(\alpha\beta)_{IJ} = 0$ for each J and $\sum_{J=1}^{12}(\alpha\beta)_{IJ} = 0$, for each I; and e_{ijk}'s are a random sample of size 3 from an infinite population with mean zero and variance σ_e^2. The population variances of the α_i's, β_j's, and $(\alpha\beta)_{ij}$'s are defined as follows:

$$\sigma_\alpha^2 = \frac{1}{10-1} \sum_{I=1}^{10} \alpha_I^2,$$

$$\sigma_\beta^2 = \frac{1}{12-1} \sum_{J=1}^{12} \beta_J^2,$$

and

$$\sigma_{\alpha\beta}^2 = \frac{1}{(10-1)(12-1)} \sum_{I=1}^{10} \sum_{J=1}^{12} (\alpha\beta)_{IJ}^2.$$

The analysis of variance computations for degrees of freedom, sums of squares, and mean squares are performed as in the case of an infinite population model and the results are shown in Table 9.6.

Assuming normality we can test the hypotheses of interest using the results shown in Table 9.6. The results on expected mean squares provide a valuable guide to deciding which mean squares are to be compared. The interaction hypothesis $H_0^{AB} : \sigma_{\alpha\beta}^2 = 0$ versus $H_1^{AB} : \sigma_{\alpha\beta}^2 > 0$ is tested by comparing the ratio

TABLE 9.6
Analysis of Variance for the Production Output Data of Table 9.5

Source of Variation	Degrees of Freedom	Sum of Squares	Mean Square	Expected Mean Square
Machine	2	4.6250	2.312	$\sigma_e^2 + 3\left(1 - \frac{4}{12}\right)\sigma_{\alpha\beta}^2 + 4 \times 3\sigma_\alpha^2$
Operator	3	10.3364	3.445	$\sigma_e^2 + 3\left(1 - \frac{4}{10}\right)\sigma_{\alpha\beta}^2 + 3 \times 3\sigma_\beta^2$
Interaction	6	1.6261	0.271	$\sigma_e^2 + 3\sigma_{\alpha\beta}^2$
Error	24	4.5000	0.188	σ_e^2
Total	35	21.0875		

$0.271/0.188 = 1.44$ with the percentile of the theoretical F distribution with (6, 24) degrees of freedom which is not significant ($p = 0.241$). To test the hypothesis regarding the presence of a main effect due to operator (i.e., $H_0^B : \sigma_\beta^2 = 0$ versus $H_1^B : \sigma_\beta^2 > 0$), we notice that there does not exist an exact F test. However, as noted in Section 9.2, a conservative test can be performed by comparing the ratio $3.445/0.271 = 12.71$ with the percentile of the theoretical F distribution with (3, 6) degrees of freedom which is highly significant ($p = 0.005$). Similarly, a conservative test of the hypothesis regarding the presence of a main effect due to machine (i.e., $H_0^A : \sigma_\alpha^2 = 0$ versus $H_1^A : \sigma_\alpha^2 > 0$), is performed by comparing the ratio $2.312/0.271 = 8.53$ with the percentile of the theoretical F distribution with (2, 6) degrees of freedom which is more significant than the 2 percent level of significance ($p = 0.017$). In addition, we can also perform psuedo-F tests for the hypotheses considered previously. As discussed in Section 9.2, a psuedo-F test for the hypothesis $H_0^B : \sigma_\beta^2 = 0$ versus $H_1^B : \sigma_\beta^2 > 0$ is performed using the statistic

$$F_B' = \frac{\mathrm{MS}_B}{\left(1 - \frac{a}{A}\right)\mathrm{MS}_{AB} + \left(\frac{a}{A}\right)\mathrm{MS}_E},$$

which has an approximate F distribution with 3 and v_b' degrees of freedom where

$$v_b' = \frac{\left(\left(1 - \frac{a}{A}\right)\mathrm{MS}_{AB} + \left(\frac{a}{A}\right)\mathrm{MS}_E\right)^2}{\dfrac{\left(1 - \frac{a}{A}\right)^2 (\mathrm{MS}_{AB})^2}{(a-1)(b-1)} + \dfrac{\left(\frac{a}{A}\right)^2 (\mathrm{MS}_E)^2}{ab(n-1)}}.$$

In the example at-hand, F'_B and v'_b (rounded to the nearest digit) are found to be

$$F'_B = \frac{3.445}{\left(1 - \frac{3}{10}\right)(0.271) + \left(\frac{3}{10}\right)(0.188)} = 14.00$$

and

$$v'_b = \frac{\left(\left(1 - \frac{3}{10}\right)(0.271) + \left(\frac{3}{10}\right)(0.188)\right)^2}{\dfrac{\left(1 - \frac{3}{10}\right)^2 (0.271)^2}{6} + \dfrac{\left(\frac{3}{10}\right)^2 (0.188)^2}{24}} \doteq 10.$$

These values lead to essentially the same result as the conservative test obtained earlier with even higher significance ($p < 0.001$). Similarly, a psuedo-F test for the hypothesis $H_0^A : \sigma_\alpha^2 = 0$ versus $\sigma_\alpha^2 > 0$ is determined using the statistic

$$F'_A = \frac{MS_A}{\left(1 - \frac{b}{B}\right) MS_{AB} + \left(\frac{b}{B}\right) MS_E},$$

which has an approximate F distribution with 2 and v'_a degrees of freedom where

$$v'_a = \frac{\left(\left(1 - \frac{b}{B}\right) MS_{AB} + \left(\frac{b}{B}\right) MS_E\right)^2}{\dfrac{\left(1 - \frac{b}{B}\right)^2 (MS_{AB})^2}{(a-1)(b-1)} + \dfrac{\left(\frac{b}{B}\right)^2 (MS_E)^2}{ab(n-1)}}.$$

Again, in the example at-hand, the values of F'_A and v'_a (rounded to the nearest digit) are found to be

$$F'_A = \frac{2.312}{\left(1 - \frac{4}{12}\right)(0.271) + \left(\frac{4}{12}\right)(0.188)} = 9.50$$

and

$$v'_a = \frac{\left(\left(1-\frac{4}{12}\right)(0.271) + \left(\frac{4}{12}\right)(0.188)\right)^2}{\dfrac{\left(1-\frac{4}{12}\right)^2 (0.271)^2}{6} + \dfrac{\left(\frac{4}{12}\right)^2 (0.188)^2}{24}} \doteq 11.$$

These values also lead to essentially the same result as the conservative test obtained earlier with even higher significance ($p = 0.004$).

Now, to assess the relative contribution of individual variance components, we may obtain their estimates using formulae (9.2.8) through (9.2.11). Thus, we find that

$$\hat{\sigma}_e^2 = 0.188,$$

$$\hat{\sigma}_{\alpha\beta}^2 = \frac{1}{13}(0.271 - 0.188) = 0.028,$$

$$\hat{\sigma}_\beta^2 = \frac{1}{3\times 3}\left[3.445 - \left(1-\frac{3}{10}\right)(0.271) - \left(\frac{3}{10}\right)(0.188)\right] = 0.355,$$

and

$$\hat{\sigma}_\alpha^2 = \frac{1}{4\times 3}\left[2.312 - \left(1-\frac{4}{12}\right)(0.271) - \left(\frac{4}{12}\right)(0.188)\right] = 0.172.$$

These components account for 25.3, 3.8, 47.8 and 23.1 percent of the total variation. The results are consistent with the tests of hypotheses performed earlier.

We can further proceed to obtain confidence intervals for the variance components. To determine a 95 percent confidence interval for σ_e^2, we have

$$MS_E = 0.188, \quad \chi^2[24, 0.025] = 12.397, \quad \text{and} \quad \chi^2[24, 0.975] = 39.980;$$

Substituting the values in (9.2.16), the desired 95 percent confidence interval for σ_e^2 is given by

$$P\left[\frac{24\times 0.188}{39.380} < \sigma_e^2 < \frac{24\times 0.188}{12.397}\right] = 0.95$$

or

$$P\left[0.114 < \sigma_e^2 < 0.363\right] = 0.95.$$

As noted in Section 9.2, there do not exist exact confidence intervals for $\sigma_{\alpha\beta}^2$, σ_β^2, and σ_α^2; however, we can obtain their conservative confidence intervals. Using formulae (9.2.20) through (9.2.22), it can be verified that the conservative confidence intervals for $\sigma_{\alpha\beta}^2$, σ_β^2, and σ_α^2 are given as follows:

$$P[0.038 < \sigma_{\alpha\beta}^2 < 0.437] \geq 0.95,$$
$$P[0.123 < \sigma_\beta^2 < 5.220] \geq 0.95,$$

and

$$P[0.052 < \sigma_\alpha^2 < 7.707] \geq 0.95.$$

It should be remarked that the variance components are in general highly variable and the lengths of their confidence intervals given previoulsy attest to the fact that there is great deal of uncertainty involved in their estimates.

Finally, we construct a confidence interval for the general constant μ. As noted earlier, there does not exist an exact confidence interval for this problem. However, we can obtain an approximate 95 percent confidence interval for μ as

$$\bar{y}_{...} \pm t[\nu, 0.975]\sqrt{\frac{MS}{abn}},$$

where MS and ν are defined as in (9.2.12) and (9.2.13), respectively. For the example at-hand, $\bar{y}_{...}$, MS, ν (rounded to the nearest digit), and $t[\nu, 0.975]$ are found to be

$$\bar{y}_{...} = 26.242,$$

$$MS = \left(\frac{3 \times 4}{10 \times 12}\right)(0.188) - \left(1 - \frac{3}{10}\right)\left(1 - \frac{4}{12}\right)(0.271)$$
$$+ \left(1 - \frac{4}{12}\right)(3.445) + \left(1 - \frac{3}{10}\right)(2.312) = 3.807,$$

$$\nu \doteq \frac{\left(\left(\frac{3\times 4}{10\times 12}\right)(0.188) - \left(1 - \frac{3}{10}\right)\left(1 - \frac{4}{12}\right)(0.271) + \left(1 - \frac{4}{12}\right)(3.445) + \left(1 - \frac{3}{10}\right)(2.312)\right)^2}{\frac{\left(\frac{3\times 4}{10\times 12}\right)^2 (0.188)^2}{24} + \frac{\left(1-\frac{3}{10}\right)^2\left(1-\frac{4}{12}\right)^2(0.271)^2}{6} + \frac{\left(1-\frac{4}{12}\right)^2(3.445)^2}{3} + \frac{\left(1-\frac{3}{10}\right)^2(2.312)^2}{2}}$$

$$\doteq 5,$$

and

$$t[5, 0.975] = 2.571.$$

Substituting these values in the preceding formula the desired 95 percent confidence limits for μ are determined as

$$26.242 \pm 2.571\sqrt{\frac{3.807}{36}} = (25.406, 27.078).$$

9.8 OTHER MODELS

Throughout this volume, we have been concerned mainly with such terms as Models I, II, and III, or fixed, random, and mixed models, depending on the nature of the factors in the experiment. For Model I, it was assumed that for all factors the levels employed in the experiment make up the population of levels of interest. When the levels used in the experiment constitute a sample from an infinite population of levels, Model II is appropriate. A case involving at least one factor fixed and others random was termed as Model III.

In the preceding sections of this chapter, we have considered the so-called finite population models, in which the error terms are assumed to be random variables from an infinite population; but the levels of the factors are assumed to be random samples from a finite population of levels, and use is made of the fact that the variance of the mean of a random sample of n from a finite population of size N with variance σ^2 is given by $(1 - \frac{n}{N})\frac{\sigma^2}{n}$. The extra factor $(1 - \frac{n}{N})$ is known as the finite population correction. If we let $f = n/N$, the finite population correction is $1 - f$. In this way, the tables of the expected values of the mean squares for various crossed and nested classifications were readily obtained.

Tukey (1949a) emphasized the restrictiveness of these models and proposed to extend the range by defining more complex models. These models have received very little attention in statistical literature, except in some theoretical works. It is not possible to provide any further discussion on this topic here. Plackett (1960) presents an excellent review of many of these models.

9.9 USE OF STATISTICAL COMPUTING PACKAGES

The use of SAS, SPSS, and BMDP programs for analyzing finite population models is the same as described in earlier chapters for crossed, nested, and partially nested factors. The computations of degrees of freedom, sums of squares, and mean squares as obtained earlier also remain valid for the finite population models. However, the expected values of mean squares must be provided using the results outlined in this chapter. The results on tests of hypotheses, point estimates, and confidence intervals can then be obtained using procedures developed in this chapter.

EXERCISES

1. Consider a two-way crossed finite population model involving three varieties of wheat and 3 different fertilizers. The three varieties of

wheat are selected randomly from a finite population of nine varieties of interest to the experimenter. Similarly, three fertilizers are taken at random from a finite population of twelve fertilizers available for the experiment. The data on yields in bushels/acre are given as follows.

		Fertilizer		
Variety		1	2	3
I		60	52	65
		61	50	66
		62	58	68
II		75	60	71
		79	61	72
		77	62	73
III		76	59	74
		77	60	75
		78	61	77

(a) State the mathematical model and the assumptions for the experiment.
(b) Analyze the data and report the analysis of variance table.
(c) Test whether there are significant interaction effects among varieties and fertilizers. Use $\alpha = 0.05$.
(d) Perform conservative and psuedo-F tests to determine whether there are differences in yields among the varieties of wheat. Use $\alpha = 0.05$.
(e) Perform conservative and psuedo-F tests to determine whether there are differences in yields among different fertilizers. Use $\alpha = 0.05$.
(f) Estimate the variance components of the model.
(g) Determine an exact 95 percent confidence interval for the error variance component.
(h) Determine approximate 95 percent confidence intervals for other variance components of the model.
(i) Determine an approximate 95 percent confidence interval for the general mean μ using the Satterthewaite procedure.

10 Some Simple Experimental Designs

10.0 PREVIEW

In the previous chapters we developed techniques suitable for analyzing experimental data. It is important at this point to consider the manner in which the experimental data were collected as this greatly influences the choice of the proper technique for data analysis. If an experiment has been properly designed or planned, the data will have been collected in the most efficient manner for the problem being considered. Experimental design is the sequence of steps initially taken to ensure that the data will be obtained in such a way that analysis will lead immediately to valid statistical inferences. The purpose of statistically designing an experiment is to collect the maximum amount of useful information with a minimum expenditure of time and resources. It is important to remember that the design of the experiment should be as simple as possible consistent with the objectives and requirements of the problem. The purpose of this chapter is to introduce some basic principles of experimental design and discuss some commonly employed experimental designs of general applications.

10.1 PRINCIPLES OF EXPERIMENTAL DESIGN

Three basic principles in designing an experiment are: replication, randomization, and control. The application of these principles ensures validity of the analysis and increases its sensitivity, and thus they are crucial to any scientific experiment. We briefly discuss each of these principles in the following.

REPLICATION

The first principle of a designed experiment is replication, which is merely a complete repetition of the basic experiment. It refers to running all the treatment combinations again, at a later time period, where each treatment is applied to several experimental units. It provides an estimate of the magnitude of the experimental error and also makes tests of significance of effects possible.

RANDOMIZATION

The second principle of a designed experiment is that of randomization, which helps to ensure against any unintentional bias in the experimental units and/or treatment combinations and can form a sound basis for statistical inference. Here, an experimental unit is a unit to which a single treatment combination is applied in a single replication of the experiment. The term treatment or treatment combinations means the experimental conditions that are imposed on an experimental unit in a particular experiment. If the data are random, it is safe to assume that the experimental errors are independently distributed. However, errors associated with the experimental units that are adjacent in time or space will tend to be correlated, thus violating the assumption of independence. Randomization helps to make this correlation as small as possible so that the analysis can be carried out as though the assumption of independence were true. Furthermore, it allows for unbiased estimates and valid tests of significance of the effects of treatments. In addition, although many extraneous variables affecting the response in a designed experiment do not vary in a completely random manner, it is reasonable to assume that their cumulative effect varies in a random manner. The randomization of treatments to experimental units has the effect of randomly assigning the error terms (associated with experimental units) to the treatments and thus satisfying the assumptions required for the validity of statistical inference. The idea was originally introduced by Fisher (1926) and has been further elaborated by Greenberg (1951), Kempthorne[1] (1955, 1977), and Lorenzen (1984). There are a number of randomization methods available for assigning treatments to experimental units (see, e.g., Cochran and Cox (1957); Cox (1958a)).

CONTROL

The third principle of a designed experiment is that of control, which refers to the way in which experimental units in a particular design are balanced, blocked, and grouped. Balancing means the assignment of the treatment combinations to the experimental units in such a way that a balanced or systematic configuration is obtained. Otherwise, it is unbalanced or we simply say that there are missing data. Blocking is the assignment of experimental units to blocks in such a manner that the units within a particular block are as homogeneous as possible. Grouping refers to the placement of homogeneous experimental units into different groups to which separate treatments may be assigned. Balancing, blocking, and grouping can be achieved in various ways and at various stages of the experiment and their choice is indicated by the availability of the experimental conditions. The application of control results in the reduction of experimental error, which in turn leads to a more sensitive analysis.

Detailed discussions of these and other principles involved in designing an experiment can be found in books on experimental design (see, e.g., Cochran and Cox (1957); Cox (1958a)). In the succeeding sections we discuss some simple experimental designs for general application. Complex designs employed

---

[1] Kempthorne (1977) stresses the necessity of randomization for the validity of error assumptions.

in many agricultural and biomedical experimentation are not considered here. The reader is referred to excellent books by Federer (1955), Cochran and Cox (1957), Gill (1978), Fleiss (1986), Hicks (1987), Winer et al. (1991), Hinkelman and Kempthorne (1994), Kirk (1995), Steel et al. (1997), among others, for a discussion of designs not included here. Also, complete details of mathematical models and statistical analysis are not given here since they readily follow from the same type of statistical models and the principle of partitioning of the sum of squares described in full detail in earlier chapters.

Remark: There is a voluminous literature on experimental design and many excellent sources of reference are currently available. Herzberg and Cox (1959) have given bibliographies on experimental designs. Federer and Balaam (1972) provided an exclusive bibliography on designs (the arrangement of treatment in an experiment) and treatment designs (the selection of treatments employed in an experiment) for the period prior to 1968. Federer and Federer (1973) presented a partial bibliography on statistical designs for the period 1968 through 1971. Federer (1980, 1981a,b) in a three-part article gave a bibliography on experimental designs from 1972 through 1978. For recent developments in design of experiemnts, covering the literature of 1975 through 1980, see Atkinson (1982). For an annotated bibliography of the books on design of experiments see Hahn (1982).

10.2 COMPLETELY RANDOMIZED DESIGN

In a completely randomized design, the treatments are allocated entirely by chance. In other words, all experimental units are considered the same and no division or grouping among them exists. The design is entirely flexible in that any number of treatments or replications may be used. The replications may vary from treatment to treatment and all available experimental material can be utilized. Among other advantages of this design include the simplicity of the statistical analysis even for the case of missing data. The relative loss of information due to missing data is less for the completely randomized design than for any other design.

In a completely randomized design all the variability among the experimental units goes into the experimental error. The completely randomized design should be used when the experimental material is homogeneous or missing values are expected to occur. The design is also appropriate in small experiments when an increase in accuracy from other designs does not outweigh the loss of degrees of freedom due to experimental error. The main disadvantage to the completely randomized design is that it is often inefficient.

MODEL AND ANALYSIS

If we take n_i replications for each treatment or treatment combination in a completely random manner, then the analysis of variance model for the experiment

TABLE 10.1
Analysis of Variance for the Completely Randomized Design with Equal Sample Sizes

Source of Variation	Degrees of Freedom	Sums of Squares	Mean Square	Expected Mean Square		F Value
				Model I	Model II	
Treatment	$a-1$	SS_τ	MS_τ	$\sigma_e^2 + \dfrac{n\sum_{i=1}^{a}\tau_i^2}{a-1}$	$\sigma_e^2 + n\sigma_\tau^2$	MS_τ/MS_E
Error	$a(n-1)$	SS_E	MS_E	σ_e^2		
Total	$an-1$	SS_T				

is given by

$$y_{ij} = \mu + \tau_i + e_{ij} \quad \begin{cases} i = 1, 2, \ldots, a \\ j = 1, 2, \ldots, n_i, \end{cases} \quad (10.2.1)$$

where y_{ij} is the j-th observation corresponding to the i-th treatment, $-\infty < \mu < \infty$ is the general mean, τ_i is the effect due to i-th treatment, and e_{ij} is the error associated with the i-th treatment and the j-th observation. As before, the assumptions inherent in the model are linearity, normality, additivity, independence, and homogeneity of variances. Clearly, model (10.2.1) is the same as the one-way classification model (2.1.1) and the appropriate analysis will be the one-way analysis of variance as described in Chapter 2.

There are two models associated with the model equation (10.2.1). Model I is concerned with only the treatments present in the experiment and under Model II treatments are assumed to be a random sample from an infinite population of treatments. Model I requires that $\sum_{i=1}^{a} n_i \tau_i = 0$ and under Model II the τ_i's are assumed to be normal random variables with mean zero and variance σ_τ^2. The steps in the analysis of this model are identical to that discussed in Chapter 2. The complete analysis of variance for the balanced case (i.e., when $n_1 = n_2 = \cdots = n_a = n$) is shown in Table 10.1 and that for the unbalanced case in Table 10.2.

The hypothesis of interest under fixed effects or Model I is

$$H_0: \tau_1 = \tau_2 = \cdots = \tau_a = 0$$
versus $\quad\quad\quad\quad\quad\quad\quad\quad\quad\quad\quad\quad (10.2.2)$
$$H_1: \text{at least one } \tau_i \neq 0.$$

In Model II, we are still interested in the hypothesis of no treatment effects; however, the τ_i's are random variables with mean zero and variance σ_τ^2. In this case the hypothesis of no treatment effects is

$$H_0: \sigma_\tau^2 = 0 \quad \text{versus} \quad H_1: \sigma_\tau^2 > 0. \quad (10.2.3)$$

TABLE 10.2
Analysis of Variance for the Completely Randomized Design with Unequal Sample Sizes

Source of Variation	Degrees of Freedom	Sums of Squares	Mean Square	Expected Mean Square		F Value
				Model I	Model II*	
Treatment	$a-1$	SS_τ	MS_τ	$\sigma_e^2 + \dfrac{\sum_{i=1}^{a} n_i \tau_i^2}{a-1}$	$\sigma_e^2 + n_o \sigma_\tau^2$	MS_τ/MS_E
Error	$\sum_{i=1}^{a} n_i - a$	SS_E	MS_E	σ_e^2	σ_e^2	
Total	$\sum_{i=1}^{a} n_i - 1$	SS_T				

* $n_o = \left(\left(\sum_{i=1}^{a} n_i \right)^2 - \sum_{i=1}^{a} n_i^2 \right) \bigg/ \left(\sum_{i=1}^{a} n_i \right) (a-1).$

The statistic $F = MS_\tau/MS_E$, which has an F distribution with $a-1$ and $a(n-1)$ ($\sum_{i=1}^{a} n_i - a$, for the unbalanced case) degrees of freedom, is used to test the hypothesis (10.2.2) or (10.2.3). A more general hypothesis on σ_τ^2 may be of the form

$$H'_0 : \sigma_\tau^2/\sigma_e^2 \leq \rho_o \quad \text{versus} \quad H'_1 : \sigma_\tau^2/\sigma_e^2 > \rho_o,$$

where ρ_o is a specified value of $\rho_o = \sigma_\tau^2/\sigma_e^2$. As in (2.7.11), this hypothesis is tested by the statistic $(1 + n\rho_o)^{-1}(MS_\tau/MS_E)$ which has an F distribution with $a-1$ and $a(n-1)$ degrees of freedom.

For the estimation of the variance components σ_e^2 and σ_τ^2, which are of interest under Model II, we can, as before, employ the analysis of variance procedure. The estimators thus obtained are given by

$$\hat{\sigma}_e^2 = MS_E$$
and
$$\hat{\sigma}_\tau^2 = \frac{1}{n}(MS_\tau - MS_E). \tag{10.2.4}$$

For all other details of the analysis of the model (10.2.1), refer to Chapter 2.

WORKED EXAMPLE

Fisher (1958, p. 262) reported data on the weights of mangold roots collected by Mercer and Hall in a uniformity trial with 20 strips of land using a completely randomized design to test five different treatments each in quadruplicate. The data are given in Table 10.3.

TABLE 10.3
Data on Weights of Mangold Roots in a Uniformity Trial

\<td colspan=5\> Treatment				
A	B	C	D	E
3376	3504	3430	3404	3253
3361	3416	3334	3210	3314
3366	3244	3291	3168	3287
3330	3195	3029	3118	3085

Source: Fisher (1958, p. 262). Used with permission.

TABLE 10.4
Analysis of Variance for the Data on Weights of Mangold Roots

Source of Variation	Degrees of Freedom	Sums of Squares	Mean Square	F value	p-value
Treatment	4	58,725.500	14,681.375	0.95	0.461
Error	15	231,040.250	15,402.683		
Total	19	289,765.750			

The analysis of variance calculations are readily performed and the results are summarized in Table 10.4. The outputs illustrating the applications of statistical packages to perform the analysis of variance are presented in Figure 10.1. Here, the ratio of mean squares is not significant ($p = 0.461$) and the conclusion would be that there are no significant differences among the treatments.

10.3 RANDOMIZED BLOCK DESIGN

If the experimental units are divided into a number of groups and a complete replication of all treatments is allocated to each group, we have the so-called randomized complete block design. The randomized block design was developed by Fisher (1926). The randomization is carried out separately in each group of experimental units, which is usually designated as a block. Here, an attempt is made to contain the major variations between blocks so that the experimental error in each group is relatively small. Thus, the blocks may be constructed so as to coincide with the degree of variability in experimental material. For example, in agricultural experimentation, each observation of, say, yield, comes from a plot of land, and we may group adjacent plots that are relatively homogeneous to form a block. In executing the experiment, we randomly allocate the treatments to the plots in the first block and then repeat the randomization for the second and other remaining blocks.

Some Simple Experimental Designs 489

```
DATA MANGOLD;
INPUT TRTMENT $ WEIGHT;
DATALINES;
A 3376
A 3361
 . .
 . .
E 3085
;
PROC ANOVA;
CLASSES TRTMENT;
MODEL WEIGHT=TRTMENT;
RUN;
CLASS    LEVELS    VALUES
TRTMENT    5       A B C D E
NUMBER OF OBS. IN DATA
SET=20
```

```
                    The SAS System
                 Analysis of Variance Procedure

Dependent Variable: WEIGHT
                       Sum of          Mean
Source        DF      Squares         Square      F Value    Pr > F
Model          4     58725.5000     14681.3750     0.95      0.4610
Error         15    231040.2500     15402.6833
Corrected     19    289765.7500
Total
              R-Square      C.V.       Root MSE    WEIGHT Mean
              0.202665   3.7771452    124.10755    3285.7500000

Source        DF    Anova SS     Mean Square    F Value    Pr > F
TRTMENT        4    58725.50      14681.37       0.95      0.4610
```

(i) SAS application: SAS ANOVA instructions and output for the completely randomized design.

```
DATA LIST
/TRTMENT 1
 WEIGHT 3-6
BEGIN DATA.
1 3376.00
1 3361.00
1 3366.00
1 3330.00
 . .
 . .
5 3085.00
END DATA.
ONEWAY WEIGHT BY
TRTMENT (1,5)
/STATISTICS=ALL.
```

```
               Test of Homogeneity of Variances

                      Levene
                      Statistic    df1    df2    Sig.
           WEIGHT      1.940        4     15    .156

                              ANOVA
                  Sum of Squares   df   Mean Square    F     Sig.
WEIGHT Between Groups  58725.500    4   14681.375   .953   .461
       Within Groups  231040.250   15   15402.683
       Total          289765.750   19
```

(ii) SPSS application: SPSS ONEWAY instructions and output for the completely randomized design.

```
/INPUT     FILE='C:\SAHAI
           \TEXTO\EJE23.TXT'.
           FORMAT=FREE.
           VARIABLES=2.
/VARIABLE  NAMES=TRT,WEIGHT.
/GROUP     CODES(TRT)=1,2,3,
                      4,5.
           NAMES(TRT)=A,B,C,
                      D,E.
/HISTOGRA  GROUPING=TRT.
           VARIABLES=WEIGHT.
/END
1 3376
1 3361  .
 . .
5 3085
```

```
BMDP7D - ONE- AND TWO-WAY ANALYSIS OF VARIANCE WITH
         DATA SCREENING Release: 7.0 (BMDP/DYNAMIC)
-----------------------------------------------------------
|ANALYSIS OF VARIANCE TABLE FOR MEANS                TAIL |
|SOURCE    SUM OF SQUARES  DF  MEAN SQUARE  F VALUE PROB.|
|------    --------------  --  -----------  ------- -----|
|TRTMENT     58725.5000     4   14681.3750    0.95  0.4610|
|ERROR      231040.2500    15   15402.6833                |
|---------------------------------------------------------|
                   EQUALITY OF MEANS TESTS;
             VARIANCES ARE NOT ASSUMED TO BE EQUAL
|WELCH                    4, 6              2.00   0.2133|
|BROWN-FORSYTHE           4,11              0.95   0.4701|
|---------------------------------------------------------|
|LEVENE'S TEST FOR VARIANCES 4, 15          1.94   0.1559|
-----------------------------------------------------------
```

(iii) BMDP application: BMDP 7D instructions and output for the completely randomized design.

FIGURE 10.1 Program Instructions and Output for the Completely Randomized Design: Data on Weights of Mangold Roots in a Uniformity Trial (Table 10.3).

To illustrate the layout of a randomized block design, let us consider eight treatments, say, $T_1, T_2, \ldots, T_8$, corresponding to eight levels of a factor to be included in each of five blocks. Figure 10.2 shows such an experimental layout. Note that the treatments are randomly allocated within each block. It is evident that this layout is quite different from the completely randomized experiment where there will be a single randomization of eight treatments repeated five times to 40 plots.

Block 1	T_2	T_1	T_5	T_6	T_4	T_3	T_8	T_7
Block 2	T_5	T_3	T_4	T_7	T_6	T_1	T_2	T_8
Block 3	T_7	T_6	T_8	T_1	T_2	T_4	T_3	T_5
Block 4	T_1	T_5	T_6	T_4	T_3	T_7	T_2	T_8
Block 5	T_4	T_6	T_1	T_3	T_8	T_2	T_5	T_7

FIGURE 10.2 A Layout of a Randomized Block Design.

Model and Analysis

The analysis of variance model for a randomized complete block design with one observation per experimental unit is given by

$$y_{ij} = \mu + \beta_i + \tau_j + e_{ij} \quad \begin{cases} i = 1, 2, \ldots, b \\ j = 1, 2, \ldots, t, \end{cases} \quad (10.3.1)$$

where y_{ij} denotes the observed value corresponding to the i-th block and the j-th treatment; $-\infty < \mu < \infty$ is the general mean, β_i is the effect of the i-th block, τ_j is the effect of the j-th treatment, and e_{ij} is the customary error term. Clearly, the model (10.3.1) is the same as the model equation (3.1.1) for the two-way crossed classification with one observation per cell. Thus, the analysis is identical to that discussed in Chapter 3 with the only difference that the factor A now designates "blocks" and the factor B denotes "treatments." There are three versions of model (10.3.1) (i.e., Models I, II, and III) depending on whether blocks or treatments or both are chosen at random.

Both Blocks and Treatments Fixed

In a randomized block experiment, both blocks and treatments may be fixed. In this case, the β_i's and τ_j's are fixed constants with the restrictions that

$$\sum_{i=1}^{b} \beta_i = 0 = \sum_{j=1}^{t} \tau_j,$$

and the e_{ij}'s are normal random variables with mean zero and variance σ_e^2. The analysis of variance in Table 3.2 can now be rewritten in the notation of model (10.3.1) as shown in Table 10.5.

TABLE 10.5
Analysis of Variance for the Randomized Block Design

Source of Variation	Degrees of Freedom	Sum of Squares	Mean Square	Expected Mean Square			F Value
				Model I	Model II	Model III	
Block	$b-1$	SS_B	MS_B	$\sigma_e^2 + \dfrac{t}{b-1}\sum_{i=1}^{b}\beta_i^2$	$\sigma_e^2 + t\sigma_\beta^2$	$\sigma_e^2 + t\sigma_\beta^2$	MS_B/MS_E
Treatment	$t-1$	SS_τ	MS_τ	$\sigma_e^2 + \dfrac{b}{t-1}\sum_{j=1}^{t}\tau_j^2$	$\sigma_e^2 + b\sigma_\tau^2$	$\sigma_e^2 + \dfrac{b}{t-1}\sum_{j=1}^{t}\tau_j^2$	MS_τ/MS_E
Error	$(b-1)(t-1)$	SS_E	MS_E	σ_e^2	σ_e^2	σ_e^2	
Total	$bt-1$	SS_T					

Here, the hypothesis

$$H_0^\tau : \tau_1 = \tau_2 = \cdots = \tau_t = 0$$
versus (10.3.2)
$$H_1^\tau : \tau_j \neq 0 \text{ for at least one } j, j = 1, 2, \ldots, t$$

is of primary interest and is tested by the statistic

$$F_\tau = \frac{\mathrm{MS}_\tau}{\mathrm{MS}_E}.$$

If $F_\tau > F[t-1, (b-1)(t-1); 1-\alpha]$, then H_0^τ will be rejected and the conclusion is that there are significant differences among the treatments.

The hypothesis

$$H_0^B : \beta_1 = \beta_2 = \cdots = \beta_b = 0$$
versus (10.3.3)
$$H_1^B : \beta_i \neq 0 \text{ for at least one } i, i = 1, 2, \ldots, b,$$

although of minor importance, may be tested in a similar manner by the statistic

$$F_B = \frac{\mathrm{MS}_B}{\mathrm{MS}_E}.$$

However, due to the manner in which the experiment is set up, the hypothesis (10.3.3) should not be tested except as a check on the blocking of the experiment. The whole purpose of a randomized block design is to reduce experimental error and get a more efficient test of (10.3.2). Therefore, if the statistic F_B is nonsignificant, there is strong evidence of improperly carried out blocking. In that case, the entire experiment should be repeated with more careful attention to the assignment of the treatments to the experimental units.[2]

Both Blocks and Treatments Random

In a randomized block experiment, both blocks and treatments may be randomly chosen, and then we will have a Model II or a random model. Here, all β_i's, τ_j's, and e_{ij}'s are mutually and completely independent normal random variables with mean zero and variances σ_β^2, σ_τ^2, and σ_e^2, respectively. The analysis of variance in this case is also given by Table 10.5. The hypotheses on σ_β^2 and σ_τ^2 can be tested by the same statistics as in the case when both blocks and treatments are fixed.

Blocks Random and Treatments Fixed

In a randomized block experiment, the blocks may be chosen randomly from a population of blocks, but the treatments may be fixed. In this case, we will have

---

[2] For further discussions of this issue, see Lentner et al. (1989) and Samuels et al. (1991).

a Model III or mixed model, where τ_j's are fixed constants with the restriction that

$$\sum_{j=1}^{t} \tau_j = 0;$$

and the β_i's and e_{ij}'s are mutually and completely independent normal random variables with mean zero and variance σ_β^2 and σ_e^2, respectively. The analysis of variance in this case is again given by Table 10.5. The hypotheses about β_i's and τ_j's can similarly be tested as in the case when both blocks and treatments are fixed..

Blocks Fixed and Treatments Random

In a randomized block experiment the treatments may be chosen at random from a population of treatments, but the blocks may be fixed. In this case, we again have a mixed model situation. The assumptions and tests of hypotheses are as given in the preceding case with the roles of the β_i's and τ_j's being reversed.

Remark: It should be noticed that if block effects had been ignored in the analysis, the analysis of variance would be the same as shown in Table 10.5, except that now the block and residual sum of squares would be pooled giving an error sum of squares equal to $SS_B + SS_E$ with $b(t-1)$ degrees of freedom. Thus, the test for the hypothesis (10.3.2) would be inefficient since all the variation between blocks has been lumped with the experimental error. Furthermore, note that the analysis of variance model (10.3.1) for a randomized complete block design looks identical to the two-way crossed classification model (3.2.1) with one observation per cell. However, the assignment of experimental units to treatments in these two layouts is quite different. In a randomized block design, the t treatments are randomized within a block whereas in a two-way crossed model, $a \times b$ treatment combinations are completely randomized to $a \times b$ experimental units. Thus, the interpretation of the two models is quite different. The randomized block design of course can be extended to problems involving two or more factors.

MISSING OBSERVATIONS

The problem of missing data is treated similarly to that discussed in Section 3.10 for the two-way classification with one observation per cell.

RELATIVE EFFICIENCY OF THE DESIGN

The relative efficiency (RE) of an experimental design in comparison to any other design can be evaluated in terms of the variance of the treatment.[3] In a

---

[3] In general, the relative efficiency is defined as the ratio of two variances. Thus, given two estimators T_1 and T_2 of the same parameter, the relative efficiency of T_1 compared to T_2 is defined as $\text{Var}(T_2)/\text{Var}(T_1)$. The preceding derivation depends essentially upon this type of comparison of variances. For another approach to relative efficiency of a design, see Cochran (1937).

TABLE 10.6
Data on Yields of Wheat Straw from a Randomized Block Experiment

Block	Treatment			
	1	2	3	4
1	332	412	542	730
2	260	384	472	590
3	202	362	516	294
4	210	348	458	560

Source: Anderson (1946). Used with permission.

randomized block design (RBD) with b blocks and t treatments, let MS_B and MS_E be the block and error mean squares, respectively. If a completely randomized design (CRD) were used with the same number $b \times t$ of experimental units as the RBD, then an estimate of the error variance would be obtained as

$$\frac{(b-1)\text{MS}_B + b(t-1)\text{MS}_E}{bt-1}.$$

However, the error mean square of the RBD with the same number of experimental units is actually MS_E. Hence, the RE of the RBD compared to CRD is given by

$$\text{RE} = \frac{(b-1)\text{MS}_B + b(t-1)\text{MS}_E}{(bt-1)\text{MS}_E}.$$

REPLICATIONS

In using a randomized block design (RBD), it is sometimes desirable to replicate each block-treatment combination on r experimental units. Such a design is commonly known as generalized randomized block design (GRBD). The principal advantage of the GRBD over the RBD lies in the fact that it allows the estimation of interaction effects between blocks and treatments. The analysis of this design proceeds in exactly the same manner as for the two-way crossed classification with interactions discussed in Chapter 4 with the only difference that the factor A now designates 'blocks' and the factor B denotes "treatments."

WORKED EXAMPLE

Anderson (1946) reported data on the yields of wheat straw from an experiment using a randomized block design with four blocks and four treatments. A portion of the data are given in Table 10.6.

TABLE 10.7
Analysis of Variance for the Data on Yields of Wheat Straw

Source of Variation	Degrees of Freedom	Sums of Squares	Mean Square	F value	p-value
Block	3	4,362	18,120.667	2.60	0.117
Treatment	3	206,394	68,798.000	9.88	0.003
Error	9	62,700	6,966.667		
Total	15	323,456			

The analysis of variance calculations are readily performed and the results are summarized in Table 10.7. The outputs illustrating the applications of statistical packages to perform the analysis of variance are presented in Figure 10.3. Here, the ratio of mean squares for treatments is highly significant ($p = 0.003$) and there is very strong evidence of real treatment differences. The block effects seem to be insignificant ($p = 0.117$) and there may be some question regarding the effectiveness of the blocking.

10.4 LATIN SQUARE DESIGN

The randomized block design was used to reduce experimental error by eliminating a source of variation in experimental units by utilizing the principle of blocking. The Latin square design eliminates two extraneous sources of variation in experimental units by using two-way or double blocking on the experimental units. The rows and columns are then used for two mutually orthogonal systems of blocks and the letters are used for treatments. In agricultural experiments, the rows and columns are usually strips of land, with row strips at right angles to the column strips, and the plots are the intersection of strips in different directions. In this sense we can say that the Latin square is an extension of the randomized block design.

In general, a Latin square for p treatments, or a $p \times p$ Latin square, is a square matrix with p rows and p columns. Each of the resulting p^2 cells contains one of the p letters. Each letter corresponds to one of the treatments and each letter occurs once and only once in each row and each column. A Latin square of any order can be obtained most easily by simply writing the n letters in their natural order in the first column and then completing each row by other letters cyclically, that is, with symbols again in the same order except that the last letter is followed by the first. Some of the examples of Latin squares are given in Figure 10.4. Appendix X contains some more representations of Latin squares from 3×3 to 12×12. Some more examples are given in Norton

```
DATA WHEATSTRAW;                         The SAS System
INPUT BLOCK TRTMENT YIELD;          Analysis of Variance Procedure
DATALINES;
1 1 332                          Dependent Variable: YIELD
1 2 412                                          Sum of       Mean
1 3 542                          Source    DF   Squares      Square    F Value   Pr > F
. . .
4 4 560                          Model      6  260756.0000  43459.3333   6.24    0.0079
;                                Error      9   62700.0000   6966.6666
PROC ANOVA;                      Corrected
CLASSES BLOCK TRTMENT;           Total     15  323456.0000
MODEL YIELD=BLOCK TRTMENT;
RUN;                                  R-Square      C.V.     Root MSE   YIELD Mean
CLASS    LEVELS    VALUES              0.806156   20.015962  83.466560    417.0000
BLOCK       4      1 2 3 4
TRTMENT     4      1 2 3 4       Source    DF   Anova SS   Mean Square  F Value   Pr > F
NUMBER OF OBS. IN DATA
SET=16                           BLOCK      3   54362.0000  18120.6667    2.60    0.1165
                                 TRTMENT    3  206394.0000  68798.0000    9.88    0.0033
```

(i) SAS application: SAS ANOVA instructions and output for the randomized block design.

```
DATA LIST                            Analysis of Variance--Design 1
/BLOCK 1 TRTMENT 3
 YIELD 5-7.                     Tests of Significance for YIELD using UNIQUE sums of squares
BEGIN DATA.
1 1 332                         Source of Variation        SS      DF     MS       F    Sig of F
1 2 412
1 3 542                         RESIDUAL               62700.00     9   6966.67
. . .                           BLOCK                  54362.00     3  18120.67   2.60    .117
4 4 560                         TRTMENT               206394.00     3  68798.00   9.88    .003
END DATA.
MANOVA YIELD BY                 (Model)               260756.00     6  43459.33   6.24    .008
BLOCK(1,4)                      (Total)               323456.00    15  21563.73
TRTMENT(1,4)
/DESIGN=BLOCK                   R-Squared       =      .806
  TRTMENT.                      Adjusted R-Squared =   .677
```

(ii) SPSS application: SPSS MANOVA instructions and output for the randomized block design.

```
/INPUT     FILE='C:\SAHAI       BMDP2V - ANALYSIS OF VARIANCE AND COVARIANCE WITH
         .  \TEXTO\EJE24.TXT'.           REPEATED MEASURES Release: 7.0 (BMDP/DYNAMIC)
           FORMAT=FREE.
           VARIABLES=3.         ANALYSIS OF VARIANCE FOR THE 1-ST DEPENDENT VARIABLE
/VARIABLE  NAMES=BL,TRE,YIELD.  THE TRIALS ARE REPRESENTED BY THE VARIABLES:YIELD
/GROUP     VARIABLE=BL, TRE.
           CODES(BL)=1,2,3,4.   THE HIGHEST ORDER INTERACTION IN EACH TABLE HAS BEEN
           NAMES(BL)=B1,B2,B3,  REMOVED FROM THE MODEL SINCE THERE IS ONE SUBJECT PER
              B4.               CELL
           CODES(TRE)=1,2,3,4
           NAMES(TRE)=T1,T2,T3, SOURCE      SUM OF        D.F   MEAN             F      TAIL
              T4.                           SQUARES             SQUARE                  PROB.
/DESIGN    DEPENDENT=YIELD.
/END                            MEAN      2782224.00000    1  2782224.00000  399.36   0.0000
1 1 332                         BLOCK       54362.00000    3    18120.66667    2.60   0.1165
. . .                           TREATM     206394.00000    3    68798.00000    9.88   0.0033
4 4 560                         ERROR       62700.00000    9     6966.66667
```

(iii) BMDP application: BMDP 2V instructions and output for the randomized block design.

FIGURE 10.3 Program Instructions and Output for the Randomized Block Design: Data on Yields of Wheat Straw from a Randomized Block Design (Table 10.6).

(1939), Cochran and Cox (1957, pp. 145–146), and Fisher and Yates (1963, pp. 86–89).

```
  3 x 3         4 x 4           5 x 5             6 x 6

  A B C       A B C D        A B C D E        A B C D E F

  B C A       B C D A        B C D E A        B C D E F A

  C A B       C D A B        C D E A B        C D E F A B

              D A B C        D E A B C        D E F A B C

                             E A B C D        E F A B C D

                                              F A B C D E
```

FIGURE 10.4 Some Selected Latin Squares.

For a given size p, there are many different $p \times p$ Latin squares that can be constructed. For example, there are 576 different possible 4×4 Latin squares, 161,280 different 5×5 squares, 812,851,200 different 6×6 squares, 61,428,210,278,400 different 7×7 squares, and the number of possible squares increases vastly as the size of p increases. The smallest Latin square that can be used is a 3×3 design. Latin squares larger than 9×9 are rarely used due to the difficulty of finding equal numbers of groups for the rows, columns, and treatments. The randomization procedures for Latin squares were initially given by Yates (1937a) and are also described by Fisher and Yates (1963). The proper randomization scheme consists of selecting at random one of the appropriate size Latin squares from those available. Randomization can also be carried out by randomly permuting first the rows and then the columns, and finally randomly assigning the treatments to the letters.

Latin squares were first employed in agricultural experiments where soil conditions often vary row-wise as well as column-wise. Treatments were applied in a field using a Latin square design in order to randomize for any differences in fertility in different directions of the field. However, the design was soon found to be useful in many other scientific and industrial experiments. Latin squares are often used to study the effects of three factors, where the factors corresponding to the rows and columns are of interest in themselves and not introduced for the main purpose of reducing experimental error. Note that in a Latin square there are only p^2 experimental units to be used in the experiment instead of the p^3 possible experimental units needed in a complete three-way layout. Thus, the use of the Latin square design results in the savings in observations by a factor of $1/p$ observations over the complete three-way layout. However, this reduction is gained at the cost of the assumption of additivity or the absence of interactions among the factors. Thus, in a Latin square, it is very difficult (often impossible) to detect interaction between factors. To study interactions, other layouts such as factorial designs are needed.

MODEL AND ANALYSIS

The analysis of variance model for a Latin square design is

$$y_{ijk} = \mu + \alpha_i + \beta_j + \tau_k + e_{ijk} \quad \begin{cases} i = 1, 2, \ldots, p \\ j = 1, 2, \ldots, p \\ k = 1, 2, \ldots, p, \end{cases} \quad (10.4.1)$$

where y_{jjk} denotes the observed value corresponding to the i-th row, the j-th column, and the k-th treatment; $-\infty < \mu < \infty$ is the overall mean, α_i is the effect of the i-th row, β_j is the effect of the j-th column, τ_k is the effect of the k-th treatment, and e_{ijk} is the random error. The model is completely additive; that is, there are no interactions between rows, columns, and treatments. Furthermore, since there is only one observation in each cell, only two of the three subscripts i, j, and k are needed to denote a particular observation. This is a consequence of each treatment appearing exactly once in each row and column.

The analysis of variance consists of partitioning the total sum of squares of the $N = p^2$ observations into components of rows, columns, treatments, and error by using the identity

$$y_{ijk} - \bar{y}_{...} = (\bar{y}_{i..} - \bar{y}_{...}) + (\bar{y}_{.j.} - \bar{y}_{...}) + (\bar{y}_{..k} - \bar{y}_{...}) \\ + (y_{ijk} - \bar{y}_{i..} - \bar{y}_{.j.} - \bar{y}_{..k} + 2\bar{y}_{...}).$$

Squaring each side and summing over i, j, k, and noting that (i, j, k) take on only p^2 values, we obtain

$$SS_T = SS_R + SS_C + SS_\tau + SS_E, \quad (10.4.2)$$

where

$$SS_T = \sum_{i=1}^{p}\sum_{j=1}^{p}\sum_{k=1}^{p}(y_{ijk} - \bar{y}_{...})^2,$$

$$SS_R = p\sum_{i=1}^{p}(\bar{y}_{i..} - \bar{y}_{...})^2,$$

$$SS_C = p\sum_{j=1}^{p}(\bar{y}_{.j.} - \bar{y}_{...})^2,$$

$$SS_\tau = p\sum_{k=1}^{p}(\bar{y}_{..k} - \bar{y}_{...})^2,$$

TABLE 10.8
Analysis of Variance for the Latin Square Design

Source of Variation	Degrees of Freedom	Sums of Squares	Mean Square	Expected Mean Square* Model I	Model II	F Value
Row	$p-1$	SS_R	MS_R	$\sigma_e^2 + \dfrac{p}{p-1}\sum_{i=1}^{p}\alpha_i^2$	$\sigma_e^2 + p\sigma_\alpha^2$	MS_R/MS_E
Column	$p-1$	SS_C	MS_C	$\sigma_e^2 + \dfrac{p}{p-1}\sum_{j=1}^{p}\beta_j^2$	$\sigma_e^2 + p\sigma_\beta^2$	MS_C/MS_E
Treatment	$p-1$	SS_τ	MS_τ	$\sigma_e^2 + \dfrac{p}{p-1}\sum_{k=1}^{p}\tau_k^2$	$\sigma_e^2 + p\tau_k^2$	MS_τ/MS_E
Error	$(p-1)(p-2)$	SS_E	MS_E	σ_e^2		
Total	p^2-1	SS_T				

* The expected mean squares for the mixed model are not shown, but they can be obtained by replacing the appropriate term by the corresponding term as one changes from fixed to random effect; for example, replacing $\sum_{i=1}^{p}\alpha_i^2/(p-1)$ by σ_α^2.

and

$$SS_E = \sum_{i=1}^{p}\sum_{j=1}^{p}\sum_{k=1}^{p}(y_{ijk} - \bar{y}_{i..} - \bar{y}_{.j.} - \bar{y}_{..k} + 2\bar{y}_{...})^2.$$

The corresponding degrees of freedom are partitioned as

$$\begin{array}{ccccc} \text{Total} & \text{Rows} & \text{Columns} & \text{Treatments} & \text{Error} \\ p^2-1 = (p-1) + & (p-1) + & (p-1) & +(p-1)(p-2) \end{array}$$

The usual assumptions of the fixed effects model are:

$$\sum_{i=1}^{p}\alpha_i = \sum_{j=1}^{p}\beta_j = \sum_{k=1}^{p}\tau_k = 0$$

and the e_{ijk}'s are normal random variables with mean zero and variance σ_e^2. Under the assumptions of the random effects model, the α_i's, β_j's, and τ_k's are also normal random variables with mean zero and variances σ_α^2, σ_β^2, and σ_τ^2, respectively. Other assumptions leading to a mixed model can also be made. Now, the expected mean squares are readily derived and the complete analysis of variance is shown in Table 10.8. Furthermore, it can be shown that under the fixed effects model, each sum of squares on the right-hand side of (10.4.2) divided by σ_e^2 is an independently distributed chi-square random variable.

Under Model I, the appropriate statistic for testing the hypothesis of no treatments effects, that is,

$$H_0^\tau: \tau_1 = \tau_2 = \cdots = \tau_p = 0$$

versus

$$H_1^\tau: \tau_k \neq 0 \text{ for at least one } k, k = 1, 2, \ldots, p,$$

is

$$F_\tau = \text{MS}_\tau/\text{MS}_E,$$

which is distributed as $F[p-1, (p-1)(p-2)]$ under the null hypothesis and as $F'[p-1, (p-1)(p-2); \lambda]$ under the alternative, where

$$\lambda = \frac{p}{2\sigma_e^2(p-1)} \sum_{k=1}^{p} \tau_k^2.$$

The only hypothesis generally of interest in a Latin square design is the one concerning the equality of treatments under Model I as given previously. However, one may also test for no row effects and no column effects by forming the ratio MS_R/MS_E or MS_C/MS_E. However, since the rows and columns represent restrictions on randomization, these tests may not be appropriate. If rows and columns represent factors and any real interactions are present, they will inflate the MS_E and will make the tests less sensitive. If $p \leq 4$, the design is considered to be inadequate for providing sufficient degrees of freedom for estimating experimental error.

POINT AND INTERVAL ESTIMATION

Estimates of various parameters of interest in a Latin square design are readily obtained along with their sample variances. For example, under Model I, we have

$$\hat{\mu} = \bar{y}_{...}, \qquad \text{Var}(\bar{y}_{...}) = \sigma_e^2/p^2;$$

$$\widehat{\mu + \alpha_i} = \bar{y}_{i..}, \qquad \text{Var}(\bar{y}_{i..}) = \sigma_e^2/p;$$

$$\hat{\alpha}_i = \bar{y}_{i..} - \bar{y}_{...}, \qquad \text{Var}(\bar{y}_{i..} - \bar{y}_{...}) = (p-1)\sigma_e^2/p^2;$$

$$\widehat{\alpha_i - \hat{\alpha}_{i'}} = \bar{y}_{i..} - \bar{y}_{i'..}, \qquad \text{Var}(\bar{y}_{i..} - \bar{y}_{i'..}) = 2\sigma_e^2/p;$$

$$\widehat{\sum_{i=1}^{p} \ell_i \alpha_i} = \sum_{i=1}^{p} \ell_i \bar{y}_{i..}, \qquad \text{Var}\left(\sum_{i=1}^{p} \ell_i \bar{y}_{i..}\right) = \left(\sum_{i=1}^{p} \ell_i^2\right) \sigma_e^2/p \quad \left(\sum_{i=1}^{p} \ell_i = 0\right);$$

$$\hat{\sigma}_e^2 = \text{MS}_E, \qquad \text{Var}(\text{MS}_E) = 2\sigma_e^4/(p-1)(p-2).$$

A $100(1-\alpha)$ percent confidence interval for σ_e^2 is given by

$$\frac{(p-1)(p-2)\text{MS}_E}{\chi^2[(p-1)(p-2), 1-\alpha/2]} < \sigma_e^2 < \frac{(p-1)(p-2)\text{MS}_E}{\chi^2[(p-1)(p-2), \alpha/2]}.$$

Some Simple Experimental Designs

Confidence intervals for fixed effects parameters considered previously can be constructed from the results of their sampling variances.

POWER OF THE F TEST

One can calculate the power of the test in the same manner as discussed in earlier chapters. For example, the noncentrality parameter ϕ with respect to the hypothesis H_0^τ is given by

$$\phi = \frac{1}{\sigma_e}\sqrt{\sum_{k=1}^{p}\tau_k^2}.$$

Here, $\nu_1 = p - 1$ and $\nu_2 = (p - 1)(p - 2)$. Except for this modification, the power calculations remain unchanged.

MULTIPLE COMPARISONS

For Models I and III, Tukey, Scheffé, and other procedures described in Section 2.19 may be readily adapted for use with the Latin square design. For example, consider a contrast of the form

$$L = \sum_{i=1}^{p}\ell_i\alpha_i \quad \left(\sum_{i=1}^{p}\ell_i = 0\right),$$

which is estimated by

$$\hat{L} = \sum_{i=1}^{p}\ell_i\bar{y}_{i..}.$$

Then, using the Tukey's method, $\hat{L}$ is significantly different from zero with confidence coefficient $1 - \alpha$ if

$$\frac{\hat{L}}{\sqrt{p^{-1}\mathrm{MS}_E}\left(\frac{1}{2}\sum_{i=1}^{p}|\ell_i|\right)} > q[p,(p-1)(p-2);1-\alpha].$$

If the Scheffé's method is applied to these comparisons, then $\hat{L}$ is significantly different from zero with confidence coefficient $1 - \alpha$ if

$$\frac{\hat{L}}{\sqrt{(p-1)\mathrm{MS}_E\left(\sum_{i=1}^{p}\ell_i^2/p\right)}} > \{F[p-1,(p-1)(p-2);1-\alpha]\}^{1/2}.$$

Similar modifications are made for other contrasts and procedures.

COMPUTATIONAL FORMULAE

The computation of sums of squares can be performed easily by using the following computational formulae:

$$SS_R = \frac{1}{p}\sum_{i=1}^{p} y_{i..}^2 - \frac{y_{...}^2}{p^2},$$

$$SS_C = \frac{1}{p}\sum_{j=1}^{p} y_{.j.}^2 - \frac{y_{...}^2}{p^2},$$

$$SS_\tau = \frac{1}{p}\sum_{k=1}^{p} y_{..k}^2 - \frac{y_{...}^2}{p^2},$$

$$SS_T = \sum_{i=1}^{p}\sum_{j=1}^{p}\sum_{k=1}^{p} y_{ijk}^2 - \frac{y_{...}^2}{p^2},$$

and

$$SS_E = SS_T - SS_R - SS_C - SS_\tau.$$

MISSING OBSERVATIONS

When a single observation y_{ijk} is missing, its value is estimated by

$$\hat{y}_{ijk} = \frac{p(y'_{i..} + y'_{.j.} + y'_{..k}) - 2y'_{...}}{(p-1)(p-2)}, \qquad (10.4.3)$$

where the primes indicate the previously defined totals with one observation missing. After substituting the estimate (10.4.3) for the missing value, the sums of squares are calculated in the usual way. To correct the treatment mean squares for possible bias, the quantity

$$\frac{[y'_{...} - y'_{i..} - y'_{.j.} - (p-1)y'_{..k}]^2}{(p-1)^3(p-2)^2}.$$

is subtracted from the treatment mean square. The variance of the mean of the treatment with a missing value is

$$\text{Var}(\bar{y}_{..k}) = \left[\frac{1}{p-1} + \frac{2}{p(p-1)(p-2)}\right]\sigma_e^2$$

and the variance of the difference between two treatment means (involving one with the missing value) is

$$\left[\frac{2}{p} + \frac{1}{(p-1)(p-2)}\right]\sigma_e^2$$

which is slightly larger than the usual expression $2\sigma_e^2/p$ for the case of no missing value.

For several missing values, more complicated methods are generally required. Formulae giving explicit expressions for several missing values can be found in Kramer and Glass (1960). However, for a few missing values, an iterative scheme may be used. The procedure is to make repeated use of the formula (10.4.3). When all missing values have been estimated, the analysis of variance is performed in the usual way with the degrees of freedom equal to the number of missing values subtracted from the total and error. Detailed discussions on handling cases with two or more missing values can be found in Steele and Torrie (1980, pp. 227–228) and Hinkelman and Kempthorne (1994, Chapter 10). The analysis of the design when a single row, column, or treatment is missing is given by Yates (1936b). The methods of analysis when more than one row, column, or treatment is missing are described by Yates and Hale (1939) and DeLury (1946).

TESTS FOR INTERACTION

Tukey (1955) and Abraham (1960) have generalized Tukey's one degree of freedom test for nonadditivity to Latin squares. Snedecor and Cochran (1989, pp. 291–294) and Neter et al. (1990, pp. 1096–1098) provide some additional details and numerical examples. For some further discussion of the topic, see Milliken and Graybill (1972). Effects of nonadditivity in Latin squares have been discussed by Wilk and Kempthorne (1957) and Cox (1958b).

RELATIVE EFFICIENCY OF THE DESIGN

Suppose instead of a Latin square design (LSD), a randomized block design (RBD) with p rows as blocks is used. An estimate of the error variance would then be given by

$$\frac{(p-1)\text{MS}_C + (p-1)^2 \text{MS}_E}{p(p-1)}.$$

The preceding formula comes from the fact that the column mean square would be pooled with the error mean square as there are no columns in the RBD. However, the LSD under the same experimental conditions actually has the error mean square MS_E. Hence, the relative efficiency (RE) of LSD relative to RBD with rows as blocks (called column efficiency) is given by

$$\begin{aligned}\text{RE}_{\text{column}} &= \frac{(p-1)\text{MS}_C + (p-1)^2 \text{MS}_E}{p(p-1)\text{MS}_E} \\ &= \frac{\text{MS}_C + (p-1)\text{MS}_E}{p\text{MS}_E}.\end{aligned}$$

Similarly, if the columns are treated as blocks, then the RE of LSD relative to RBD (called row efficiency) is given by

$$\text{RE}_{\text{row}} = \frac{\text{MS}_R + (p-1)\text{MS}_E}{p\text{MS}_E}.$$

REPLICATIONS

In using a small size Latin square, it is often desirable to replicate it. The usual model for a Latin square with r replications is

$$y_{ijk\ell} = \mu + \alpha_i + \beta_j + \tau_k + \rho_\ell + e_{ijk\ell} \quad \begin{cases} i = 1, 2, \ldots, p \\ j = 1, 2, \ldots, p \\ k = 1, 2, \ldots, p \\ \ell = 1, 2, \ldots, r, \end{cases}$$

where $y_{ijk\ell}$ denotes the observed value corresponding to the i-th row, the j-th column, the k-th treatment, and the ℓ-th replication; $-\infty < \mu < \infty$ is the overall mean, α_i is the effect of the i-th row, β_j is the effect of the j-th column, τ_k is the effect of the k-th treatment, ρ_ℓ is the effect of the r-th replication, and $e_{ijk\ell}$ is the customary error term. When a Latin square is replicated, it is important to know whether it is replicated using the same blocking variables or there are additional versions of one or both blocking variables. The analysis of variance for the general case in which a Latin square is replicated r times using the same blocking variables proceeds in the same manner as before. However, now, an additional source of variation due to replicates is introduced. The degrees of freedom for the rows, columns, and treatments are the same, i.e., $p - 1$, but the degrees of freedom for the total, replicates, and error are given by $rt^2 - 1, r - 1$, and $(p - 1)[r(p + 1) - 3]$ respectively. When a Latin square is replicated with additional versions of the row (column) blocking variable, the analysis remains the same except that now the degrees of freedom for the rows (columns) and the error are $r(p - 1)$ and $(p - 1)(rp - 2)$ respectively. When a Latin square is replicated with additional versions of both row and column blocking variables, the degrees of freedom for the rows, columns, and error are now given by $r(p - 1), r(p - 1)$, and $(p - 1)[r(p - 1) - 1]$ respectively.

Remark: Latin squares were proposed as experimental designs by R. A. Fisher (1925, 1926) and in 1924 he made some early applications of Latin squares in the design of an experiment in a forest nursery. A Latin square experiment for testing the differences among four treatments for warp breakage, where time periods and looms were used as rows and columns, has been described by Tippett (1931). Davies (1954) describes one of the earliest industrial applications of Latin squares related to wear-testing experiments of four materials where the runs and positions of a machine were represented as rows and columns. For a survey of Latin square designs in agricultural experiments, see Street and

TABLE 10.9
Data on Responses of Monkeys to Different Stimulus Conditions

Monkey	Week				
	1	2	3	4	5
1	194 (B)	369 (D)	344 (C)	380 (A)	693 (E)
2	202 (D)	142 (B)	200 (A)	356 (E)	473 (C)
3	335 (C)	301 (A)	493 (E)	338 (B)	528 (D)
4	515 (E)	590 (C)	552 (B)	677 (D)	546 (A)
5	184 (A)	421 (E)	355 (D)	284 (C)	366 (B)

Source: Snedecor (1955). Used with permission.

TABLE 10.10
Analysis of Variance for the Data on Responses of Monkeys to Different Stimulus Conditions

Source of Variation	Degrees of Freedom	Sums of Squares	Mean Square	F value	p-value
Monkey	4	262,961.040	65,740.260	18.51	<0.001
Week	4	144,515.440	36,128.860	10.17	<0.001
Stimulus	4	111,771.440	27,942.860	7.87	0.002
Error	12	42,628.320	3,552.360		
Total	24	561,876.240			

Street (1988). The principal reference book on Latin squares is by Dénes and Keedwell (1974). For a discussion of combinatorial problems in Latin squares, see Street and Street (1987).

WORKED EXAMPLE

Snedecor (1955) reported data from an experiment conducted to study responses of pairs of monkeys to a certain kind of stimulus under a variety of conditions. The responses were measured on five pairs of monkeys during five successive weeks under five different conditions using a Latin square design. The data are given in Table 10.9 where the letter within parentheses represents the stimulus condition used.

The analysis of variance calculations are readily performed and the results are summarized in Table 10.10. The outputs illustrating the applications of statistical packages to perform the analysis of variance are presented in Figure 10.5.

The Analysis of Variance

```
DATA MONKEYS;
INPUT MONKEY WEEK
STIMULUS $ RESPONSE;
DATALINES;
1 1 B 194
. . . ·.
5 5 B 366
;
PROC ANOVA;
CLASSES MONKEY WEEK
STIMULUS;
MODEL RESPONSE=MONKEY
WEEK STIMULUS;
RUN;
CLASS   LEVELS   VALUES
MONKEY     5     1 2 3 4 5
WEEK       5     1 2 3 4 5
STIMULUS   5     A B C D E
NUMBER OF OBS. IN DATA
SET=25
```

             The SAS System
       Analysis of Variance Procedure
Dependent Variable: RESPONSE

Source	DF	Sum of Squares	Mean Square	F Value	Pr > F
Model	12	519247.92000	43270.66000	12.18	0.0001
Error	12	42628.32000	3552.36000		
Corrected Total	24	561876.24000			

R-Square	C.V.	Root MSE	RESPONSE Mean
0.924132	15.145781	59.601678	393.52000000

Source	DF	Anova SS	Mean Square	F Value	Pr > F
MONKEY	4	262961.0400	65740.2600	18.51	0.0001
WEEK	4	144515.4400	36128.8600	10.17	0.0008
STIMULUS	4	111771.4400	27942.8600	7.87	0.0024

(i) SAS application: SAS ANOVA instructions and output for the Latin square design.

```
DATA LIST
/MONKEY 1 WEEK 3
 STIMULUS 5
 RESPONSE 7-9.
BEGIN DATA.
1 1 2 194
1 1 4 202
. . . ·.
5 5 2 366
END DATA.
MANOVA RESPONSE BY
 MONKEY(1,5)
 WEEK(1,5)
 STIMULUS(1,5)
/DESIGN=MONKEY
 WEEK STIMULUS.
```

Analysis of Variance--Design 1

Tests of Significance for RESPONSE using UNIQUE sums of squares

Source of Variation	SS	DF	MS	F	Sig of F
RESIDUAL	42628.32	12	3552.36		
MONKEY	262961.04	4	65740.26	18.51	.000
WEEK	144515.44	4	36128.86	10.17	.001
STIMULUS	111771.44	4	27942.86	7.87	.002
(Model)	519247.92	12	43270.66	12.18	.000
(Total)	561876.24	24	23411.51		

R-Squared = .924
Adjusted R-Squared = .848

(ii) SPSS application: SPSS MANOVA instructions and output for the Latin square design.

```
/INPUT     FILE='C:\SAHAI
           \TEXTO\EJE25.TXT'.
           FORMAT=FREE.
           VARIABLES=4.
/VARIABLE  NAMES=M,W,S,RESP.
/GROUP     VARIABLE=M,W,S.
           CODES(M)=1,2,3,4,5.
           NAMES(M)=M1,...,M5.
           CODES(W)=1,2,3,4,5.
           NAMES(W)=W1,...,W5.
           CODES(S)=1,2,3,4,5.
           NAMES(S)=A,B,C,D,E.
/DESIGN    DEPENDENT=RESP.
           INCLUDE=1,2,3.
/END
1 1 2 194
. . . ·.
5 5 2 366
```

BMDP2V - ANALYSIS OF VARIANCE AND COVARIANCE WITH
REPEATED MEASURES Release: 7.0 (BMDP/DYNAMIC)

ANALYSIS OF VARIANCE FOR THE 1-ST DEPENDENT VARIABLE

THE TRIALS ARE REPRESENTED BY THE VARIABLES:RESPONSE

SOURCE	SUM OF SQUARES	D.F.	MEAN SQUARE	F	TAIL PROB.
MEAN	3871449.76000	1	3871449.76000	1089.82	0.0000
MONKEY	262961.04000	4	65740.26000	18.51	0.0000
WEEK	144515.44000	4	36128.86000	10.17	0.0008
STIMULUS	111771.44000	4	27942.86000	7.87	0.0024
ERROR	42628.32000	12	3552.36000		

(iii) BMDP application: BMDP 2V instructions and output for the Latin square design.

FIGURE 10.5 Program Instructions and Output for the Latin Square Design: Data on Responses of Monkeys to Different Stimulus Conditions (Table 10.9).

Here, there is a very significant effect due to stimulus conditions. The effects due to monkeys and weeks are also highly significant. The use of the Latin square design seems to be highly effective.

10.5 GRAECO-LATIN SQUARE DESIGN

We have seen that the Latin square design is effective for controlling two sources of external variation. The principle can be further extended to control more sources of variation. The Graeco-Latin square is one such design that can be used to control three sources of variation. The design is also useful for investigating simultaneous effects of four factors: rows, columns, Latin letters, and Greek letters, in a single experiment. The Graeco-Latin square design is obtained by juxtaposing or superimposing two Latin squares, one with treatments denoted by Latin letters and the other with treatments denoted by Greek letters, such that each Latin letter appears once and only once with each Greek letter. The designs have been constructed for all numbers of treatments[4] from 3 to 12. Some selected Graeco-Latin squares are shown in Figure 10.6. Some more examples are given in Appendix Y, Cochran and Cox (1957, pp. 146–147), and Fisher and Yates (1963, pp. 86–89).

```
   3 × 3              4 × 4                  5 × 5
Aα  Bγ  Cβ       Aα  Bγ  Cδ  Dβ       Aα  Bγ  Cε  Dβ  Eδ
Bβ  Cα  Aγ       Bβ  Aδ  Dγ  Cα       Bβ  Cδ  Dα  Eγ  Aε
Cγ  Aβ  Bα       Cγ  Dα  Aβ  Bδ       Cγ  Dε  Eβ  Aδ  Bα
                 Dδ  Cβ  Bα  Aγ       Dδ  Eα  Aγ  Bε  Cβ
                                      Eε  Aβ  Bδ  Cα  Dγ
```

FIGURE 10.6 Some Selected Graeco-Latin Squares.

MODEL AND ANALYSIS

The analysis of variance model for a Graeco-Latin square design is

$$y_{ijk\ell} = \mu + \alpha_i + \beta_j + \tau_k + \delta_\ell + e_{ijk\ell} \quad \begin{cases} i = 1, 2, \ldots, p \\ j = 1, 2, \ldots, p \\ k = 1, 2, \ldots, p \\ \ell = 1, 2, \ldots, p, \end{cases}$$

---

[4] Graeco-Latin squares exist for all orders except 1, 2, and 6. The problem of nonexistence of Graeco-Latin squares for certain values of p goes back well over 200 years, when the Swiss mathematician Euler (1782) conjectured that no $p \times p$ Graeco-Latin square exists for $p = 4m + 2$ where m is a positive integer. In 1900, Euler's conjecture was shown to be true for $m = 1$; that is, there does not exist a 6×6 Graeco-Latin square. However, his conjecture was shown to be false for $m \geq 2$ by Bose and Shrikhande (1959) and Parker (1959).

where $y_{ijk\ell}$ is the observation corresponding to the i-th row, the j-th column, the k-th Greek letter, and the ℓ-th Latin letter; $-\infty < \mu < \infty$ is the overall mean, α_i is the effect of the i-th row, β_j is the effect of the j-th column, τ_k is the effect of the k-th Greek letter, δ_ℓ is the effect of the ℓ-th Latin letter, and $e_{ijk\ell}$ is the random error. The model assumes additivity of the effects of all four factors; that is, there are no interactions between rows, columns, Greek letters, and Latin letters. Furthermore, note that only two of the four subscripts i, j, k, and ℓ are needed to identify a particular observation. This is a consequence of each Greek letter appearing exactly once with each Latin letter and in each row and column.

The analysis of variance of the design is very similar to the Latin square design. The partitioning of the total sum of squares of the $N = p^2$ observations into components of rows, columns, Greek letters, Latin letters, and the error is given by

$$SS_T = SS_R + SS_C + SS_G + SS_L + SS_E,$$

where

$$SS_T = \sum_{i=1}^{p}\sum_{j=1}^{p}\sum_{k=1}^{p}\sum_{\ell=1}^{p}(y_{ijk\ell} - \bar{y}_{....})^2,$$

$$SS_R = p\sum_{i=1}^{p}(\bar{y}_{i...} - \bar{y}_{....})^2,$$

$$SS_C = p\sum_{j=1}^{p}(\bar{y}_{.j..} - \bar{y}_{....})^2,$$

$$SS_G = p\sum_{k=1}^{p}(\bar{y}_{..k.} - \bar{y}_{....})^2,$$

$$SS_L = p\sum_{\ell=1}^{p}(\bar{y}_{...\ell} - \bar{y}_{....})^2,$$

and

$$SS_E = \sum_{i=1}^{p}\sum_{j=1}^{p}\sum_{k=1}^{p}\sum_{\ell=1}^{p}(y_{ijk\ell} - \bar{y}_{i...} - \bar{y}_{.j..} - \bar{y}_{..k.} - \bar{y}_{...\ell} + 3\bar{y}_{....})^2.$$

The corresponding degrees of freedom are partitioned as

Total	Rows	Columns	Greek Letters	Latin Letters	Error
$p^2 - 1 =$	$(p-1) +$	$(p-1) +$	$(p-1) +$	$(p-1) +$	$(p-1)(p-3)$

TABLE 10.11
Analysis of Variance for the Graeco-Latin Square Design

Source of Variation	Degrees of Freedom	Sums of Squares	Mean Square	Expected Mean Square* Model I	Model II	F Value
Row	$p-1$	SS_R	MS_R	$\sigma_e^2 + \dfrac{p}{p-1}\sum_{i=1}^{p}\alpha_i^2$	$\sigma_e^2 + p\sigma_\alpha^2$	MS_R/MS_E
Column	$p-1$	SS_C	MS_C	$\sigma_e^2 + \dfrac{p}{p-1}\sum_{j=1}^{p}\beta_j^2$	$\sigma_e^2 + p\sigma_\beta^2$	MS_C/MS_E
Greek Letter	$p-1$	SS_G	MS_G	$\sigma_e^2 + \dfrac{p}{p-1}\sum_{k=1}^{p}\tau_k^2$	$\sigma_e^2 + p\sigma_\tau^2$	MS_G/MS_E
Latin Letter	$p-1$	SS_L	MS_L	$\sigma_e^2 + \dfrac{p}{p-1}\sum_{\ell=1}^{p}\delta_\ell^2$	$\sigma_e^2 + p\sigma_\delta^2$	MS_L/MS_E
Error	$(p-1)(p-3)$	SS_E	MS_E	σ_e^2	σ_e^2	
Total	p^2-1	SS_T				

* The expected mean squares for the mixed model are not shown, but they can be obtained by replacing the appropriate term by the corresponding term as one changes from fixed to random effect; for example, replacing $\sum_{i=1}^{p}\alpha_i^2/(p-1)$ by σ_α^2.

The usual assumptions of the fixed effects model are:

$$\sum_{i=1}^{p}\alpha_i = \sum_{j=1}^{p}\beta_j = \sum_{k=1}^{p}\tau_k = \sum_{\ell=1}^{p}\delta_\ell = 0$$

and the $e_{ijk\ell}$'s are normal random variables with mean zero and variance σ_e^2. Under the assumptions of the random effects model, the α_i's, β_j's, τ_k's, and δ_ℓ's are also normal random variables with mean zero and variances $\sigma_\alpha^2, \sigma_\beta^2, \sigma_\tau^2, \sigma_\delta^2$, and σ_e^2, respectively. Other assumptions leading to a mixed model can also be made. Now, the expected mean squares are readily derived and the complete analysis of variance is shown in Table 10.11. The null hypotheses of equal effects for rows, columns, Greek letters, and Latin letters are tested by dividing the corresponding mean squares by the error mean square.

Remark: The Graeco-Latin square design has not been used much because the experimental units cannot be easily balanced in all three groupings. Some early applications were described by Dunlop (1933) for testing 5 feeding treatments on pigs and by Tippett (1934) involving an industrial experiment. Perry et al. (1980) describe an application and advantages and disadvantages of the design in experiments for comparing different insect sex attractants. Discussions of the analysis of variance of the design when some observations are missing can be found in Yates (1933), Nair (1940), Davies (1960), and Dodge and Shah (1977). When $p < 6$, the number of error degrees of freedom is rather inadequate and the design is not practical (see Cochran and Cox (1957, p. 133)).

TABLE 10.12
Data on Photographic Density for Different Brands of Flash Bulbs*

	Camera				
Film	1	2	3	4	5
1	0.64 ($A\alpha$)	0.70 ($B\gamma$)	0.73 ($C\varepsilon$)	0.66 ($D\beta$)	0.66 ($E\delta$)
2	0.62 ($B\beta$)	0.63 ($C\delta$)	0.69 ($D\alpha$)	0.70 ($E\gamma$)	0.78 ($A\varepsilon$)
3	0.65 ($C\gamma$)	0.72 ($D\varepsilon$)	0.68 ($E\beta$)	0.64 ($A\delta$)	0.74 ($B\alpha$)
4	0.64 ($D\delta$)	0.73 ($E\alpha$)	0.68 ($A\gamma$)	0.74 ($B\varepsilon$)	0.72 ($C\beta$)
5	0.74 ($E\varepsilon$)	0.73 ($A\beta$)	0.67 ($B\delta$)	0.74 ($C\alpha$)	0.78 ($D\gamma$)

Source: Johnson and Leone (1964, p. 175). Used with permission.

* The original experiment reported duplicate measurements. Only the first set of readings are presented here.

TABLE 10.13
Analysis of Variance for the Data on Photographic Density for Different Brands of Flash Bulbs

Source of Variation	Degrees of Freedom	Sums of Squares	Mean Square	F value	p-value
Film	4	0.00950	0.00237	7.18	0.010
Camera	4	0.01558	0.00389	11.79	0.002
Brand	4	0.00026	0.00006	0.18	0.936
Filter	4	0.02398	0.00599	18.15	<0.001
Error	8	0.00267	0.00033		
Total	24	0.05198			

WORKED EXAMPLE

Johnson and Leone (1964, p. 175) presented data from an experiment conducted to study the effect of different brands of flash bulbs on photographic density. A 5 × 5 Graeco-Latin square design with 5 varieties of cameras, 5 film types, and 5 filter types was used. The data are given in Table 10.12 where the Roman letter within parentheses represents the brand and the Greek letter represents the filter type.

The analysis of variance calculations are readily performed and the results are shown in Table 10.13. The outputs illustrating the applications of statistical packages to perform analysis of variance are presented in Figure 10.7. There does not seem to be a significant effect of different brands of flash bulbs on

```
DATA PHOTOGRPH;                           The SAS System
INPUT FILM CAMERA BRAND                Analysis of Variance Procedure
$ FILTER $ DENSITY;
CARDS;                        Dependent Variable: DENSITY
1 1 A α 0.64
                                          Sum of        Mean
. . . .                       Source   DF Squares       Square      F Value    Pr > F
5 5 D γ 0.78;
PROC ANOVA;                   Model    16  0.04930400   0.00308150   9.23       0.0017
CLASSES FILM CAMERA           Error     8  0.00267200   0.00033400
BRAND FILTER;                 Corrected
MODEL DENSITY=FILM            Total    24  0.05197600
CAMERA BRAND FILTER;
RUN;                                  R-Square    C.V.       Root MSE    DENSITY Mean
CLASS  LEVELS  VALUES                 0.948592  2.6243060   0.01827567   0.69640000
FILM     5      1 2 3 4 5
CAMERA   5      1 2 3 4 5     Source   DF Anova SS    Mean Square  F Value   Pr > F
BRAND    5      A B C D E     FILM      4  0.00949600   0.00237400   7.11     0.0096
FILTER   5      α β γ δ ε     CAMERA    4  0.01557600   0.00389400  11.66     0.0020
NUMBER OF OBS. IN DATA        BRAND     4  0.00025600   0.00006400   0.19     0.9361
SET=25                        FILTER    4  0.02397600   0.00599400  17.95     0.0005
```

(i) SAS application: SAS ANOVA instructions and output for the Graeco-Latin square design.

```
DATA LIST                          Analysis of Variance--Design 1
/FILM 1 CAMERA 3
 BRAND 5                   Tests of Significance for DENSITY using UNIQUE sums of squares
 FILTER 7
 DENSITY 9-12(2).
BEGIN DATA.                Source of Variation      SS      DF     MS      F       Sig of F
1 1 1 1 0.64
. . . . .                  RESIDUAL                .00       8    .00
5 5 4 3 0.78               FILM                    .01       4    .00    7.11      .010
END DATA.                  CAMERA                  .02       4    .00   11.66      .002
MANOVA DENSITY BY          BRAND                   .00       4    .00     .19      .936
 FILM(1,5)                 FILTER                  .02       4    .01   17.95      .000
 CAMERA(1,5)
 BRAND(1,5)                (Model)                 .05      16    .00    9.23      .002
 FILTER(1,5)               (Total)                 .05      24    .00
/DESIGN=FILM
 CAMERA BRAND              R-Squared     =        .949
 FILTER.                   Adjusted R-Squared =   .846
```

(ii) SPSS application: SPSS MANOVA instructions and output for the Graeco-Latin square design.

```
/INPUT    FILE='C:\SAHAI              BMDP2V - ANALYSIS OF VARIANCE AND COVARIANCE WITH
          \TEXTO\EJE26.TXT'.                    REPEATED MEASURES Release: 7.0 (BMDP/DYNAMIC)
          FORMAT=FREE.
          VARIABLES=5.
/VARIABLE NAMES=F,C,B,FI,DENS.        ANALYSIS OF VARIANCE FOR THE 1-ST DEPENDENT VARIABLE
/GROUP    VARIABLE=F,C,B,FI.
          CODES(F)=1,2,3,4,5.
          NAMES(F)=F1,…,F5.           THE TRIALS ARE REPRESENTED BY THE VARIABLES:DENSITY
          CODES(C)=1,2,3,4,5.
          NAMES(C)=C1,…,C5.
          CODES(B)=1,2,3,4,5.         SOURCE    SUM OF    D.F.    MEAN        F        TAIL
          NAMES(B)=B1,…,B5.                     SQUARES           SQUARE               PROB.
          CODES(FI)=1,2,3,4,5.
          NAMES(FI)=FI1,…,FI5.
/DESIGN   DEPENDENT=DENSITY.          MEAN      12.12432    1    12.12432   36300.42  0.0000
          INCLUDE=1, 2, 3, 4.         G          0.00950    4     0.00237       7.11  0.0096
/END                                  H          0.01558    4     0.00389      11.66  0.0020
1 1 1 1 0.64                          I          0.00026    4     0.00006       0.19  0.9361
. . . . .                             J          0.02398    4     0.00599      17.95  0.0005
5 5 4 3 0.78                          ERROR      0.00267    8     0.00033
```

(iii) BMDP application: BMDP 2V instructions and output for the Graeco-Latin square design.

FIGURE 10.7 Program Instructions and Output for the Graeco-Latin Square Design: Data on Photographic Density for Different Brands of Flash Bulbs (Table 10.12).

photographic density. The use of a Graeco-Latin square design in reducing variability due to varieties of camera, film types, and filter types seems to be highly effective.

10.6 SPLIT-PLOT DESIGN

Split-plot design can be considered as a special case of the two-factor randomized block design where one wants to obtain more precise information about one factor and also about the interaction between the two factors, the second factor being of secondary importance to the experimenter. Thus, suppose there are two factors A and B having a and b levels, respectively. As described in the previous sections, one might use a completely randomized design by completely randomizing the $a \times b$ treatment combinations, or a randomized block design (in, say, r randomized blocks), each block containing $a \times b$ plots. Alternatively, suppose we wish to evaluate the effects of factor B and the interaction between the factors A and B with greater precision than the effects of factor A. In this situation, one could arrange the treatments of factor A in a randomized block design of r blocks as described earlier. Each of the $a \times r$ plots can then be divided into b subplots so that the treatments of factor B can now be allocated at random over each subplot. This design yields more precise information on the factor allocated to the split- or subplots at the expense of less precise information to the factor assigned to the whole-plots.

As explained previously, the principal advantage of this type of design lies in the fact that since no attempt is being made to obtain an accurate information of factor A, larger plots can be used to allocate the first a treatments of factor A without any consideration of the variability within the blocks. If $a = 3$, $b = 4$, and $r = 3$, a split-plot design may be laid out as shown in Figure 10.8.

Block 1			Block II			Block III		
A_2	A_1	A_3	A_1	A_3	A_2	A_3	A_1	A_2
B_3	B_4	B_1	B_3	B_2	B_4	B_2	B_4	B_1
B_2	B_3	B_3	B_2	B_4	B_3	B_1	B_1	B_3
B_1	B_1	B_2	B_1	B_3	B_2	B_3	B_3	B_4
B_4	B_2	B_4	B_4	B_1	B_1	B_4	B_2	B_2

FIGURE 10.8 A Layout of a Split-Plot Design.

Note that the essential feature of a split-plot design is that instead of $a \times b \times r$ experimental units obtained after random allocation over the entire $a \times b \times r$

units as in a completely randomized design, or obtained after r separate randomizations over $a \times b$ units, as in a simple randomized block design, they are obtained by first randomizing treatments of factor B on the b subplots (this randomization being performed $a \times r$ times) and then randomizing the treatments of factor A onto the a whole-plots (this randomization being performed r times, once for each of the r blocks). The name split-plot has its origin in agricultural experimentations where the terms whole-plots (large areas of land) and subplots (small areas of land) are in common use.

In a split-plot design when there is a choice, the more important treatments requiring a higher level of precision should be assigned to the subplots and the treatments of secondary importance should be assigned to the whole-plots. However, in many industrial and laboratory experiments, the treatments that cannot be administered in small scale are applied to whole-plots and the treatments that can be conveniently applied to small scale are assigned to the subplots. This choice of a split-plot design is dictated purely by administrative and logistic considerations rather than the precision of the desired information.

MODEL AND ANALYSIS

The model for the split-plot design described previously is

$$y_{ijk} = \mu + B_i + \alpha_j + e_{ij} + \beta_k + (B\beta)_{ik} + (\alpha\beta)_{jk} + \varepsilon_{ijk} \quad \begin{cases} i = 1, \ldots, r \\ j = 1, \ldots, a \\ k = 1, \ldots, b, \end{cases}$$

where μ is the general or overall mean, B_i is the effect of the i-th block, α_j is the effect of the j-th treatment of factor A, e_{ij} is the whole-plot error; β_k is the effect of the k-th treatment of factor B, $(B\beta)_{ik}$ is the interaction between the i-th block and the k-th treatment of factor B, $(\alpha\beta)_{jk}$ is the interaction between the j-th treatment of factor A and the k-th treatment of factor B, and ε_{ijk} is the subplot error. Note that e_{ij} is the same as the $(B\alpha)_{ij}$ interaction and ε_{ijk} is the same as the $(B\alpha\beta)_{ijk}$ interaction.

Usually, the blocks are considered as random and factors A and B are fixed. Thus, the B_i's, $(B\beta)_{ik}$'s, e_{ij}'s, and ε_{ijk}'s are normally distributed with mean zero and variances $\sigma_B^2, \sigma_{B\beta}^2, \sigma_e^2$, and σ_ε^2, respectively. If both A and B are random, the $\alpha_{j's}$, $\beta_{k's}$, and $(\alpha\beta)_{jk's}$ are assumed to be normally distributed with zero means and variances $\sigma_\alpha^2, \sigma_\beta^2$, and $\sigma_{\alpha\beta}^2$, respectively. Mixed models with A fixed and B random, or B fixed and A random can also arise and their assumptions are analogously stated. The analysis of variance is performed in exactly the same manner as before. Thus, the total sum of squares is partitioned by the identity

$$SS_T = SS_{B\ell} + SS_A + SS_E + SS_B + SS_{B\ell \times B} + SS_{A \times B} + SS_\epsilon,$$

where

$$SS_T = \sum_{i=1}^{r}\sum_{j=1}^{a}\sum_{k=1}^{b}(y_{ijk} - \bar{y}_{...})^2,$$

$$SS_{B\ell} = ab\sum_{i=1}^{r}(\bar{y}_{i..} - \bar{y}_{...})^2,$$

$$SS_A = rb\sum_{j=1}^{a}(\bar{y}_{.j.} - \bar{y}_{...})^2,$$

$$SS_E = b\sum_{i=1}^{r}\sum_{j=1}^{a}(\bar{y}_{ij.} - \bar{y}_{i..} - \bar{y}_{.j.} + \bar{y}_{...})^2,$$

$$SS_B = ra\sum_{k=1}^{b}(\bar{y}_{..k} - \bar{y}_{...})^2,$$

$$SS_{B\ell \times B} = a\sum_{i=1}^{r}\sum_{k=1}^{b}(\bar{y}_{i.k} - \bar{y}_{i..} - \bar{y}_{..k} + \bar{y}_{...})^2,$$

$$SS_{A \times B} = r\sum_{j=1}^{a}\sum_{k=1}^{b}(\bar{y}_{.jk} - \bar{y}_{.j.} - \bar{y}_{..k} + \bar{y}_{...})^2,$$

and

$$SS_\epsilon = \sum_{i=1}^{r}\sum_{j=1}^{a}\sum_{k=1}^{b}(y_{ijk} - \bar{y}_{ij.} - \bar{y}_{i.k} - \bar{y}_{.jk} + \bar{y}_{i..} + \bar{y}_{.j.} + \bar{y}_{..k} - \bar{y}_{...})^2.$$

The complete analysis of variance including the degrees of freedom and expected mean squares is shown in Table 10.14. When both A and B are fixed, we can test block and factor A effects against the whole-plot error. Similarly, $B\ell \times B$ and $A \times B$ interactions can be tested against the subplot error. The factor B effect can be tested against the $B\ell \times B$ interaction. Sometimes, the $B\ell \times B$ interaction is also considered to be negligible and not included in the model. Then it is pooled with the subplot error, and the B main effect as well as $A \times B$ interaction are tested against the subplot error. Under Models II and III, however, exact tests may not always exist and psuedo-F tests as discussed in Section 5.5 would have to be employed.

Remarks: (i) The split-plot technique may also be applied to a Latin square design. The $a \times a$ Latin square corresponds to the whole-plot treatments. Each whole-plot can be further subdivided into b subplots. Now, A treatments are applied randomly to whole-plots and B treatments are applied randomly to subplots within a whole-plot. Statistical analysis of such a design proceeds on lines similar to that of a randomized block. The first stage is an analysis of the a^2 whole-plots and the second stage is an analysis of the subplots within whole-plots.

(ii) The problem of estimating missing values in a split-plot design has been studied by Anderson (1946) and Khargonkar (1948). Formulae for estimating the standard errors

TABLE 10.14
Analysis of Variance for the Split-Plot Design

Source of Variation	Degrees of Freedom	Sum of Squares	Mean Square	Expected Mean Square			
						Model III	
				Model I	Model II	A Random, B Fixed	A Fixed, B Random
Block	$r-1$	$SS_{B\ell}$	$MS_{B\ell}$	$\sigma_\varepsilon^2 + b\sigma_e^2 + ab\sigma_B^2$	$\sigma_\varepsilon^2 + b\sigma_e^2 + a\sigma_{B\beta}^2 + ab\sigma_B^2$	$\sigma_\varepsilon^2 + b\sigma_e^2 + a\sigma_{B\beta}^2 + ab\sigma_B^2$	$\sigma_\varepsilon^2 + b\sigma_e^2 + a\sigma_{B\beta}^2 + \dfrac{rb}{a-1} \times \sum_{j=1}^{a}\alpha_j^2$
Factor A	$a-1$	SS_A	MS_A	$\sigma_\varepsilon^2 + b\sigma_e^2 + \dfrac{rb}{a-1} \times \sum_{j=1}^{a}\alpha_j^2$	$\sigma_\varepsilon^2 + b\sigma_e^2 + r\sigma_{\alpha\beta}^2 + rb\sigma_\alpha^2$	$\sigma_\varepsilon^2 + b\sigma_e^2 + r\sigma_{\alpha\beta}^2 + rb\sigma_\alpha^2$	$\sigma_\varepsilon^2 + b\sigma_e^2 + r\sigma_{\alpha\beta}^2 + ra\sigma_\beta^2$
Whole-Plot Error (Interaction $B\ell \times A$)	$(r-1)(a-1)$	SS_E	MS_E	$\sigma_\varepsilon^2 + b\sigma_e^2$	$\sigma_\varepsilon^2 + b\sigma_e^2$	$\sigma_\varepsilon^2 + b\sigma_e^2$	$\sigma_\varepsilon^2 + b\sigma_e^2$
Factor B	$b-1$	SS_B	MS_B	$\sigma_\varepsilon^2 + a\sigma_{B\beta}^2 + \dfrac{ra}{b-1} \times \sum_{k=1}^{h}\beta_k^2$	$\sigma_\varepsilon^2 + a\sigma_{B\beta}^2 + r\sigma_{\alpha\beta}^2 + ra\sigma_\beta^2$	$\sigma_\varepsilon^2 + a\sigma_{B\beta}^2 + r\sigma_{\alpha\beta}^2 + \dfrac{ra}{b-1} \times \sum_{k=1}^{h}\beta_k^2$	$\sigma_\varepsilon^2 + a\sigma_{B\beta}^2 + ra\sigma_\beta^2$
Interaction $B\ell \times B$	$(r-1)(b-1)$	$SS_{B\ell \times B}$	$MS_{B\ell \times B}$	$\sigma_\varepsilon^2 + a\sigma_{B\beta}^2$	$\sigma_\varepsilon^2 + a\sigma_{B\beta}^2$	$\sigma_\varepsilon^2 + a\sigma_{B\beta}^2$	$\sigma_\varepsilon^2 + a\sigma_{B\beta}^2$
Interaction $A \times B$	$(a-1)(b-1)$	$SS_{A \times B}$	$MS_{A \times B}$	$\sigma_\varepsilon^2 + \dfrac{r}{(a-1)(b-1)} \times \sum_{j=1}^{a}\sum_{k=1}^{h}(\alpha\beta)_{jk}^2$	$\sigma_\varepsilon^2 + r\sigma_{\alpha\beta}^2$	$\sigma_\varepsilon^2 + r\sigma_{\alpha\beta}^2$	$\sigma_\varepsilon^2 + r\sigma_{\alpha\beta}^2$
Subplot Error (Interaction $B\ell \times A \times B$)	$(r-1)(a-1)(b-1)$	SS_ε	MS_ε	σ_ε^2	σ_ε^2	σ_ε^2	σ_ε^2
Total	$abr-1$	SS_T					

of differences between two means involving missing values are given by Cochran and Cox (1957, pp. 302–303) and are also reported in Steel and Torrie (1980, pp. 388–390). A more complete description of this design can be found in the books by Cochran and Cox (1957, Chapter 7), Steel and Torrie (1980, Chapter 16), Fleiss (1986, Chapter 13), Damon and Harvey (1987, Chapter 7), Snedecor and Cochran (1989, pp. 324–329), and Hinkelman and Kempthorne (1994, Chapter 13).

WORKED EXAMPLE

Steel and Torrie (1980, p. 387) reported data from an experiment conducted by J. W. Lambert, at the University of Minnesota, to compare the effect of row spacing on the yields of two varieties of soybean. A split-plot design was used with a variety as a whole-plot, which was then divided into four subplots, and row spacing was applied to subplots. The varieties as whole-plot treatments were allocated in six blocks using a randomized complete block layout. The data on yields in bushels per acre for six blocks are given in Table 10.15.

The analysis of variance computations are readily performed and the results are summarized in Table 10.16. The outputs illustrating the applications of statistical packages to perform the analysis of variance are presented in Figure 10.9. In performing tests of significance, blocks and whole-plot error (block × variety interaction) are considered as random leading to expected mean squares shown in Table 10.16. We may conclude that there are highly significant differences due to both varieties and row spacings. No significant differences are found due to either blocks, or block × spacing and variety × spacing interactions.

10.7 OTHER DESIGNS

The designs described so far in this chapter are relatively simple, commonly used designs. There are a great number of other designs that differ mainly due to experimental conditions, such as limitations on resources, and the attempt to reduce the error variance. In this section, we briefly review some designs that are occasionally useful in scientific experimentation. Further details can be found in Kempthorne (1952), Federer (1955), Cochran and Cox (1957), and Das and Giri (1976).

INCOMPLETE BLOCK DESIGNS

In a randomized block design, each treatment must be present in every block. However, when there are too many treatments, it may not be possible to accommodate all factor levels or treatment combinations in each block because of limitations of the size of the block (amount of work or space) or lack of experimental resources. To overcome this problem, randomized block designs are used in which every treatment does not occur in every block. These designs are commonly known as incomplete block designs. There are several types of

TABLE 10.15
Data on Yields of Two Varieties of Soybean

	Block											
	1		2		3		4		5		6	
	Variety*		Variety		Variety		Variety		Variety		Variety	
Row Spacing (in.)	OM	B	OM	B	OM	B	OM	B	OM	B	OM	B
18	33.6	28.0	37.1	25.5	34.1	28.3	34.6	29.4	35.4	27.3	36.1	28.3
24	31.1	23.7	34.5	26.2	30.5	27.0	32.7	25.8	30.7	26.8	30.3	23.8
30	33.0	23.5	29.5	26.8	29.2	24.9	30.7	23.3	30.7	21.4	27.9	22.0
36	28.4	25.0	29.9	25.3	31.6	25.6	32.3	26.4	28.1	24.6	26.9	24.5
42	31.4	25.7	28.3	23.2	28.9	23.4	28.6	25.6	18.5	24.5	33.4	22.9

Source: Steel and Torrie (1980, p. 387). Used with permission.

*OM = Ottawa Mandarin, B = Blackhawk.

TABLE 10.16
Analysis of Variance for the Data on Yields of Two Varieties of Soybean

Source of Variation	Degrees of Freedom	Sums of Squares	Mean Square	Expected Mean Square	F value	p-value
Block	5	30.3588	6.0718	$\sigma_\varepsilon^2 + 5\sigma_e^2 + 2 \times 5\sigma_B^2$	2.019	0.230
Variety	1	477.7082	477.7082	$\sigma_\varepsilon^2 + 5\sigma_e^2 + \dfrac{6 \times 5}{2-1} \times \sum_{j=1}^{2} \alpha_j^2$	158.823	<0.001
Whole-plot error (Block × Variety)	5	15.0388	3.0078	$\sigma_\varepsilon^2 + 5\sigma_e^2$	0.559	0.730
Row spacing	4	206.1043	51.5261	$\sigma_\varepsilon^2 + 2\sigma_{B\beta}^2 + \dfrac{6 \times 2}{5-1} \times \sum_{k=1}^{5} \beta_k^2$	11.789	<0.001
Block × Spacing	20	87.4137	4.3707	$\sigma_\varepsilon^2 + 2\sigma_{B\beta}^2$	0.812	0.677
Variety × Spacing	4	25.4543	6.3636	$\sigma_\varepsilon^2 + \dfrac{6}{(2-1)(5-1)} \times \sum_{j=1}^{2}\sum_{k=1}^{5}(\alpha\beta)_{jk}^2$	1.183	0.348
Subplot error (Block × Variety × Spacing)	20	107.6237	5.3812	σ_ε^2		
Total	59	949.7018				

The Analysis of Variance

```
DATA SOYBEAN;
INPUT BLOCK SPACING $
VARIETY $ YIELD;
DATALINES;
1 18" OM 33.6
1 24" OM 31.1
1 30" OM 33.0
1 36" OM 28.4
1 42" OM 31.4
1 18" B  28.0
1 24" B  23.7
1 30" B  23.5
1 36" B  25.0
   .  .   .
6 42" B  22.9
;
PROC GLM;
CLASSES BLOCK
VARIETY SPACING;
MODEL YIELD=BLOCK
VARIETY SPACING
BLOCK*VARIETY
SPACING*BLOCK
VARIETY*SPACING;
RANDOM BLOCK BLOCK*
VARIETY BLOCK*SPACING;
TEST H=BLOCK
E=BLOCK*VARIETY;
TEST H=VARIETY
E=BLOCK*VARIETY;
TEST H=SPACING
E=BLOCK*SPACING
RUN;
CLASS LEVELS VALUES
BLOCK  6  1 2 3 4
          5 6
VARIETY 2 B OM
SPACING 5 18" 24"
          30" 36" 42"
NUMBER OF OBS. IN DATA
SET=60
```

```
                    The SAS System
              General Linear Models Procedure
Dependent Variable: YIELD
                  Sum of        Mean
Source       DF  Squares       Square     F Value  Pr > F
Model        39  842.07816667  21.59174786  4.01   0.0008
Error        20  107.62366667   5.38118333
Corrected    59  949.70183333
Total

             R-Square   C.V.     Root MSE   YIELD Mean
             0.886676   8.2518685  2.3197378  28.11166667

Source          DF  Type III SS  Mean Square  F Value  Pr > F
BLOCK            5   30.358833    6.071767    1.13    0.3776
VARIETY          1  477.708167  477.708167   88.77    0.0001
SPACING          4  206.104333   51.526083    9.58    0.0002
BLOCK*VARIETY    5   15.038833    3.007767    0.56    0.7301
BLOCK*SPACING   20   87.413667    4.370683    0.81    0.6768
VARIETY*SPACING  4   25.454333    6.363583    1.18    0.3486

Source          Type III Expected Mean Square
BLOCK           Var(Error) + 2 Var(BLOCK*SPACING) +5 Var(BLOCK*VARIETY)
                + 10 Var(BLOCK)
VARIETY         Var(Error) + 5 Var(BLOCK*VARIETY)
                + Q(VARIETY,VARIETY*SPACING)
SPACING         Var(Error) + 2 Var(BLOCK*SPACING)
                + Q(SPACING,VARIETY*SPACING)
BLOCK*VARIETY   Var(Error) + 5 Var(BLOCK*VARIETY)
BLOCK*SPACING   Var(Error) + 2 Var(BLOCK*SPACING)
VARIETY*SPACING Var(Error) + Q(VARIETY*SPACING)

Tests of Hypotheses using the Type III MS for BLOCK*VARIETY
as an error term
Source   DF  Type III SS  Mean Square  F Value  Pr > F
BLOCK     5   30.35883333  6.07176667   2.02    0.2296

Tests of Hypotheses using the Type III MS for BLOCK*VARIETY
as an error term
Source   DF  Type III SS   Mean Square   F Value  Pr > F
VARIETY   1  477.70816667  477.7081666  158.82    0.0001

Tests of Hypotheses using the Type III MS for BLOCK*SPACING
as an error term
Source   DF  Type III SS   Mean Square   F Value  Pr > F
SPACING   4  206.10433333   51.52608333   11.79   0.0001
```

(i) SAS application: SAS GLM instructions and output for the split-plot design.

```
DATA LIST
/BLOCK 1
 SPACING 3-4
 VARIETY 6
 YIELD 8-11(1).
BEGIN DATA.
1 18 1 33.6
1 24 1 31.1
1 30 1 33.0
1 36 1 28.4
1 42 1 31.4
1 18 2 28.0
1 24 2 23.7
1 30 2 23.5
1 36 2 25.0
1 42 2 25.7
2 18 1 37.1
2 24 1 34.5
2 30 1 29.5
2 36 1 29.9
2 42 1 28.3
 .  .  .
6 42 2 22.9
END DATA.
GLM YIELD BY BLOCK
SPACING VARIETY
/DESIGN=BLOCK
VARIETY SPACING
BLOCK*SPACING
BLOCK*VARIETY
SPACING*VARIETY
/RANDOM BLOCK.
```

```
Tests of Between-Subjects Effects    Dependent Variable: YIELD
Source              Type III SS   df   Mean       F       Sig.
                                       Square
BLOCK    Hypothesis   30.359       5   6.072    3.040   .421
         Error         1.891     .947  1.997(a)
VARIETY  Hypothesis  477.708       1  477.708  158.825  .000
         Error        15.039       5   3.008(b)
SPACING  Hypothesis  206.104       4   51.526   11.789  .000
         Error        87.414      20   4.371(c)
BLOCK*   Hypothesis   87.414      20   4.371             .812   .677
SPACING  Error       107.624      20   5.381(d)
BLOCK*   Hypothesis   15.039       5   3.008             .559   .730
VARIETY  Error       107.624      20   5.381(d)
SPACING* Hypothesis   25.454       4   6.364            1.183   .349
VARIETY  Error       107.624      20   5.381(d)

a MS(B*S)+MS(B*V)-MS(E)  b MS(B*V)  c MS(B*S)  d MS(Error)

              Expected Mean Squares(a,b)
                 Variance Component
Source          Var(B)  Var(B*S)  Var(B*V)  Var(Error)  Quadratic Term
BLOCK           10.000   2.000    5.000     1.000
VARIETY          .000     .000    5.000     1.000       Variety
SPACING          .000    2.000     .000     1.000       Spacing
BLOCK*SPACING    .000    2.000     .000     1.000
BLOCK*VARIETY    .000     .000    5.000     1.000
SPACING*VARIETY  .000     .000     .000     1.000       Variety*Spacing
Error            .000     .000     .000     1.000

a For each source, the expected mean square equals the sum of the
coefficients in the cells times the variance components, plus a
quadratic term involving effects in the Quadratic Term cell. b Expected
Mean Squares are based on the Type III Sums of Squares.
```

(ii) SPSS application: SPSS GLM instructions and output for the split-plot design.

FIGURE 10.9 Program Instructions and Output for the Split-Plot Design: Data on Yields of Two Varieties of Soybean (Table 10.15).

Some Simple Experimental Designs

```
/INPUT     FILE='C:\SAHAI
           \TEXTO\EJE27.TXT'.
           FORMAT=FREE.
           VARIABLES=5.
/VARIABLE  NAMES=S1,S2,S3,S4,
           S5.
/DESIGN    NAMES=BLOCK,
           VARIETY,
           SPACING.
           LEVELS=6, 2, 5.
           RANDOM=BLOCK.
           FIXED=VARIETY,
           SPACING.
           MODEL='B, V, S'.
/END
33.6 31.1 33.0 28.4 31.4
28.0 23.7 23.5 25.0 25.7
 .    .    .    .    .
36.1 30.3 27.9 26.9 33.4
28.3 23.8 22.0 24.5 22.9
ANALYSIS OF VARIANCE DESIGN
INDEX            B   V   S
NUMBER OF LEVELS 6   2   5
POPULATION SIZE  INF 2   5
MODEL            B,  V,  S
```

```
BMDP8V - GENERAL MIXED MODEL ANALYSIS OF VARIANCE
       - EQUAL CELL SIZES Release: 7.0 (BMDP/DYNAMIC)
ANALYSIS OF VARIANCE FOR DEPENDENT VARIABLE  1

  SOURCE   ERROR   SUM OF    D.F.  MEAN       F        PROB.
           TERM    SQUARES         SQUARE
1 MEAN     BLOCK   47415.9477 1    47415.948  7809.25  0.0000
2 BLOCK            30.3588    5    6.072
3 VARIETY  BV      477.7082   1    477.708    158.83   0.0001
4 SPACING  BS      206.1043   4    51.526     11.79    0.0000
5 BV               15.0388    5    3.008
6 BS               87.4137    20   4.371
7 VS       BVS     25.4543    4    6.364      1.18     0.3486
8 BVS              107.6237   20   5.381

  SOURCE      EXPECTED MEAN      ESTIMATES OF VARIANCE
              SQUARE             COMPONENTS
1 MEAN        60(1)+10(2)        790.16460
2 BLOCK       10(2)              0.60718
3 VARIETY     30(3)+5(5)         15.82335
4 SPACING     12(4)+2(6)         3.92962
5 BV          5(5)               0.60155
6 BS          2(6)               2.18534
7 VS          6(7)+(8)           0.16373
8 BVS         (8)                5.38118
```

(iii) BMDP application: BMDP 8V instructions and output for the split-plot design.

FIGURE 10.9 (*continued*)

incomplete block designs, the simplest of which involve blocks of equal size and all treatments equally replicated. If an incomplete block design has t treatments, b blocks with c experimental units within each block and there are r replications of each treatment, then the number of times any two treatments appear together in a block is $\lambda = r(c-1)/(t-1) = n(c-1)/t(t-1)$ where $n = tr$. When it is desired to make all treatment comparisons with equal precision, the incomplete block designs are formed such that every pair of treatments occurs together the same number of times. Such designs are called balanced incomplete block designs and were originally proposed by Yates (1936a). For a list of some useful balanced incomplete block designs, see Box et al. (1978, pp. 270–274). Balanced incomplete block designs do not always exist or may result in excessively large block sizes. To reduce the number of blocks required in an experiment, the experimenter can employ designs known as partially balanced incomplete block designs in which different pairs of treatments appear together a different number of times. For further discussions of incomplete block designs, see Cochran and Cox (1957, Chapters 9 and 13) and Cox (1958a, pp. 231–245).

Lattice Designs

Lattice designs are a class of incomplete block designs introduced by Yates (1937b) to increase the precision of treatment comparisons in agricultural crop cultivate trials. The designs are also sometimes called quasi-factorials because of their analogy to confounding in factorial experiments. For example, if k^2 treatments are to be compared, one can arrange them as the points of a two-dimensional lattice and regard the points as representing the treatments in a two-factor experiment. Suppose a balanced incomplete block layout with k^2 treatment is arranged in $b = k(k + 1)$ blocks with k units per block and $r = k + 1$ replicates for each treatment. Such a design is called a balanced lattice. In a balanced lattice, the number of treatments is always an exact square and the size of the block is the square root of this number. Incomplete lattice designs are grouped to form separate replications. In a balanced incomplete lattice, every pair of treatments occurs once in the same incomplete block. This allows the same degree of precision for all treatment pairs being compared. Lattice designs may involve a large number of treatments and in order to reduce the size of the design, partially balanced lattice designs are also used. Further details of the lattice designs are given in Kempthorne (1952), Federer (1955), and Cochran and Cox (1957). The SAS PROC LATTICE performs the analysis of variance and analysis of simple covariance using experimental data obtained from a lattice design. The procedure analyzes data from balanced square lattices, partially balanced square lattices, and some other rectangular lattices. For further information and applications of PROC LATTICE, see SAS Institute (1997, Chapter 14).

YOUDEN SQUARES

Youden squares are constructed by a rearrangement of certain of the balanced incomplete block designs and possess the property of "two-way control" of Latin squares. They are special types of incomplete Latin squares in which the number of columns, rows, and treatments are not all equal. If a column or row is deleted from a Latin square, the remaining layout is always a Youden square. However, omission of two or more rows or columns does not in general produce a Youden square. Youden squares can also be thought of as symmetrically balanced incomplete block designs by means of which two sources of variation can be controlled. These designs were developed by Youden (1937, 1940) in investigations involving greenhouse experiments. The name Youden square was given by Yates (1936b). The standard analysis of variance of a Youden square design is similar to that of a balanced incomplete randomized block design. A detailed treatment of planning and analysis of Youden squares is given in Natrella (1963, Section 13.6). A table of Youden squares is given in Davies (1960) and other types of incomplete Latin squares are discussed by Cochran and Cox (1957, Chapter 13).

CROSS-OVER DESIGNS

In most experimental designs, each subject is assigned only to a single treatment during the entire course of the experiment. In a cross-over design, the total duration of the experiment is divided into several periods and the treatment of each subject changes from each period to the next. In a cross-over study involving k treatments, each treatment is allocated to an equal number of subjects and is applied to each subject in k different time periods. Since the order of treatment assignment to experimental units may have some consequences regarding the effectiveness of different treatments, the order of treatment is chosen randomly so as to eliminate the order effects. This type of design is particularly suited for animal and human subjects. The intervening period between the assignment of different treatments depends on the objectives of the experiment and other experimental considerations. For example, suppose in an experiment involving human subjects, the effect of two treatments is investigated. In the first period, half of the subjects are randomly assigned to treatment 1 and the other half to treatment 2. At the end of the study period, the subjects are evaluated for the desired response and sufficient time is allowed so that the biological effect of each treatment is eliminated. In the second period, the subjects who were assigned treatment 1 are given treatment 2 and vice versa. The cross-over designs can be analyzed as a set of Latin squares with rows as time periods, columns as subjects, and treatments as letters. Cross- over designs have been used successfully in clinical trials, bioassay, and animal nutrition experiments. For further discussions of cross-over designs see Cochran and Cox (1957, Section 4.4), Cox (1958a, Chapter 13), John (1971, Chapter 6), John and Quenouille (1977, Chapter 11), Fleiss (1986, Chapter 10), Jones and Kenward (1989), Senn (1993), and Ratkowski et al. (1993).

Repeated Measures Designs

Any design involving k ($k \leq 2$) successive measurements on the same subject is called a repeated measures design. In a repeated measures design, subjects are crossed with the factor involving repeated measures. The k measurements may correspond to different times, trials, or experimental conditions. For example, blood pressures may be measured at successive time periods, say, once a week, for a group of patients attending a clinic, or animals injected with different drugs and measurements made after each injection. If possible, the order of assignment of k repeated measures should be selected randomly. Of course, when repeated measures are taken in different time sequences, it is not possible to include randomization. In repeated measures designs, each subject acts as his or her own control. This helps to control for variability between subjects since the same subject is measured repeatedly. Thus, repeated measures designs are used to control for the presence of many extraneous factors while at the same time limiting the total number of experimental units. A major concern in repeated measures designs are that no carry-over or residual effects are present from treatment at one time period to response at the next time period. Thus, as in the case of cross-over designs, sufficient time must be allowed to eliminate any carry-over effect from the previous treatment. When this cannot be achieved, cross-over designs are to be preferred. It is important to point out that it is incorrect to analyze the time dimension in repeated measures studies by the straightforward application of the analysis of variance. For a complete coverage of repeated measures designs, see Fleiss (1986, Chapter 8), Maxwell and Delaney (1990), Winer et al. (1991), and Kirk (1995). For a book-length treatment of the topic see Crowder and Hand (1990) and Lindsey (1993). The latter work also includes a fairly extensive and classified bibliography on repeated measures. Hedayat and Afsarinejad (1975, 1978) have given an extensive survey and bibliography on repeated measures designs. For analysis of repeated measures data using SAS procedures, PROC GLM and PROC MIXED, see Littell et al. (1996, Chapter 3).

HYPER-GRAECO-LATIN AND HYPER SQUARES

The principle of Latin and Graeco-Latin square designs can be further extended to control for four or more sources of variation. Hyper-Graeco-Latin square is a design which can be used to control four sources of variation. The design can also be used to investigate simultaneous effects of five factors: rows, columns, Latin letters, Greek letters, and Hebrew letters, in a single experiment. The hyper-Graeco-Latin square design is obtained by juxtaposing or superimposing three Latin squares, one with treatments denoted by Greek letters, the second with treatments denoted by Latin letters, and the third with treatments denoted by Hebrew letters, such that each Hebrew letters appears once and only once with each Greek and Latin letters. The number of Latin squares that can be combined in forming hyper-Graeco-Latin squares is limited. For example, no more than three orthogonal 4×4 Latin squares can be combined and no more than four orthogonal 5×5 Latin squares can be combined. The sum of squares formulae for rows, columns, Greek letters, Latin letters, and Hebrew letters follow the same general pattern as the corresponding formulae in Latin and Graeco-Latin square designs. The concept of superimposing two or more orthogonal Latin squares in forming Graeco-Latin and hyper-Graeco-Latin squares can be extended even further. A $p \times p$ hypersquare is a design in which three or more orthogonal $p \times p$ Latin squares are superimposed. In general, one can investigate a maximum of $p + 1$ factors if a complete set of $p - 1$ orthogonal Latin squares is available. In such a design, one would utilize all $(p + 1)(p - 1) = p^2 - 1$ degrees of freedom, so that an independent estimate of the error variance would be required. Of course, the researcher must assume that there would be no interactions between factors when using hypersquares. For a detailed discussion of hyper-Graeco-Latin squares, and other hypersquares, see Federer (1955).

MAGIC AND SUPER MAGIC LATIN SQUARES

These are Latin square designs with additional restrictions placed on the grouping of treatments within a Latin square in order to reduce the error term. For this purpose, additional smaller squares or rectangles are formed within a Latin square in order to remove additional variation from the error term. If the use of squares or rectangles to remove variation is done in only one direction, the design is called a magic Latin square. If the technique is used to control variation in both directions, the design is a called super magic Latin square. These designs were initially developed by Gertrude M. Cox and have been used in sugarcane research in Hawaii and at the Geneva Experimental Station in New York.

Split-Split-Plot Design

In a split-plot design, each subplot may be further subdivided into a number of sub-subplots to which a third set of treatments corresponding to c levels of a factor C may be applied. In a split-split-plot design, three factors are assigned to the various levels of experimental units, using three distinct stages of randomization. The a levels of factor A are randomly assigned to whole-plots; b levels of factor B are randomly assigned to subplots within a whole-plot; and c levels of factor C are randomly assigned to sub-subplots within a subplot. For such a design there will be three error variances: whole-plot error for the A treatments, subplot error for the B treatments, and sub-subplot error for the C treatments. The details of statistical analysis follow the same general pattern as that of the split-plot design. Finally, it should be noted that in a split-split-plot design, the three error sums of squares and their corresponding degrees of freedom would add up to the sum of squares and the degrees of freedom for the single error term if the experiment were conducted in a standard randomized block design of abc units. For further information about the split-split-plot design, see Anderson and McLean (1974, Section 7.2) and Koch et al. (1988).

2^p Design and Fractional Replications

In many experimental works involving a large number of factors, a very useful factorial design for preliminary exploration is a 2^p design. This design has p treatment factors, each having two levels, giving a total of 2^p treatment combinations. Thus, in any replication of this design, 2^p experimental units are required. In a 2^p design, there are p main effects, $\binom{p}{2}$ two-way interactions, $\binom{p}{3}$ three-way interactions, etc., and finally one p-way interaction. Note that all the main effects and each one of the interactions have only one degree of freedom. If the design is replicated in b blocks each containing 2^p experimental units, then there are $(2^p - 1)(b - 1)$ degrees of freedom available for the error term. If the number of experimental units available is limited, it may not be possible to replicate the design. In such a case there will ble no degrees of freedom available for the error term. However, if the higher-order interactions can be assumed to be negligible, which is often the case, one can pool the sums of squares for these interactions in order to obtain an estimate of the error mean square. If some of the higher-order interactions are not zero, the F test for the main effects and the lower-order interactions will tend to be conservative. If there are large number of treatment factors and the available resources are limited, it may be necessary to use a replication of only a fraction of the total number of treatment combinations. In a design involving a fractional replication, some of the effects cannot be estimated since they are confounded with one or more other effects. Usually, the choice of a fractional replication is made such that the effects considered to be of importance are confounded only with the effects that can be assumed to be negligible. For a complete discussion of 2^p and other factorial designs, and their fractional replications, see Kempthorne (1952), Cochran and Cox (1957), and Box et al. (1978).

10.8 USE OF STATISTICAL COMPUTING PACKAGES

Completely randomized designs (CRD) can be analyzed exactly as the one-way analysis of variance. The use of SAS, SPSS, and BMDP programs for this analysis is described in Section 2.15. Similarly, randomized block designs (RBD) can be analyzed exactly as the two-way analysis of variance with $n(n \geq 1)$ observations per cell. The use of appropriate programs for this type of analysis is described in Sections 3.17 and 4.17. Latin squares (LSD) and Graeco-Latin squares (GLS) can be analyzed similar to three-way and four-way crossed-classification models without interactions. For example, with SAS, one can use either PROC ANOVA or PROC GLM for all of them. The important instructions for both procedures are the CLASS and the MODEL statements. For the CRD, these are

CLASS TRT;
MODEL Y = TRT;

and for the RBD, these are

CLASS BLC TRT;
MODEL Y = BLC TRT;

where TRT, BLC, and Y designate treatment, block, and response, respectively. Similarly, for the LSD, we have

CLASS ROW COL TRT;
MODEL Y = ROW COL TRT;

and, for the GLS, we have

CLASS ROW COL GRG ROM;
Y = ROW COL GRG ROM;

where ROW, COL, GRG, and ROM designate row, column, Greek letter, and Roman letter factors, respectively. For the split-plot design, these statements are

CLASS BLC A B
MODEL Y = BLC A BLC*A B BLC*B A*B;

where BLC stands for blocks (replications) and A and B are whole-plot and subplot treatments, respectively.

Some Simple Experimental Designs

If some of the factors are to be treated as random and the researcher is interested in estimating variance components, one may employ appropriate procedures in SAS, SPSS, and BMDP for this purpose. For example, in a Latin square with rows and columns regarded as random factors, the following SAS codes may be used to analyze the design via PROC MIXED procedure:

> PROC MIXED;
> CLASS ROW COL TRT;
> MODEL Y = TRT;
> RANDOM ROW COL;
> RUN;

EXERCISES

1. An experiment was designed to compare four different feeds in regard to the gain in weight of cattle. Twenty cattle were divided at random into four groups of five each and each group was placed on a different feed. After a certain duration of time, the weight gains in kilograms for each of the cattle was recorded and the data given as follows.

Feed A	Feed B	Feed C	Feed D
34	64	111	96
45	49	34	85
35	52	122	91
49	47	27	88
44	58	29	94

    (a) Describe the mathematical model and the assumptions for the experiment.
    (b) Analyze the data and report the analysis of variance table.
    (c) Perform an appropriate F test for the hypothesis that the mean weight gains for all the feed are the same. Use $\alpha = 0.05$.
    (d) If the hypothesis in part (c) is rejected, find 95 percent simultaneous confidence intervals for the contrasts between each pair of feeds using Tukey's and Scheffé's methods.
    (e) Carry out the test for homoscedasticity at $\alpha = 0.01$ by employing

    (i) Bartlett's test,

    (ii) Hartley's test,

    (iii) Cochran's test.

2. Three methods of teaching were compared to determine their comparative value on student's learning ability. Thirty students of comparable ability were randomly divided into three groups of 10 each, and each group received instruction using a different method. After completion of instruction, the learning score for each student was determined and

the data given as follows.

Method A	Method B	Method C
161	179	134
131	261	176
186	311	153
281	176	186
213	196	131
155	163	131
221	221	157
167	232	164
191	264	175
216	259	133

(a) Describe the mathematical model and the assumptions for the experiment.
(b) Analyze the data and report the analysis of variance table.
(c) Perform an appropriate F test for the hypothesis that the mean learning scores for the three methods are the same. Use $\alpha = 0.05$.
(d) If the hypothesis in part (c) is rejected, find 95 percent simultaneous confidence intervals for the three single contrasts between each pair of teaching methods using Tukey's and Scheffé's procedures. Use $\alpha = 0.01$.
(e) Carry out the test for homoscedasticity at $\alpha = 0.01$ by employing
  (i) Bartlett's test,
  (ii) Hartley's test,
  (iii) Cochran's test.

3. Steel and Torrie (1980, p. 144) reported data from an experiment, conducted by F. R. Urey, Department of Zoology, University of Wisconsin, on estrogen assay of several solutions that had been subjected to an *in vitro* inactivation technique. Twenty-eight rats were randomly assigned to six different solutions and a control group and the uterine weight of the rat was used as a measure of the estrogen activity. The uterine weight in milligrams for each rat was recorded and the data given as follows.

1	2	3	4	5	6	Control
84.4	64.4	75.2	88.4	56.4	65.6	89.8
116.0	79.8	62.4	90.2	83.2	79.4	93.8
84.0	88.0	62.4	73.2	90.4	65.6	88.4
68.6	69.4	73.8	87.8	85.6	70.2	112.6

Source: Steel and Torrie (1980, p. 144). Used with permission.

(a) Describe the mathematical model and the assumptions for the experiment.
(b) Analyze the data and report the analysis of variance table.
(c) Perform an appropriate F test for the hypothesis that the mean uterine weights for all the groups are the same. Use $\alpha = 0.05$.
(d) If the hypothesis in part (c) is rejected, find 95 percent confidence intervals for the single contrast comparing control with the mean of all the other six treatments and interpret your results. Use $\alpha = 0.05$.
(e) Carry out the test for homoscedasticity at $\alpha = 0.01$ by employing

   (i) Bartlett's test,

   (ii) Hartley's test,

   (iii) Cochran's test.

4. Fisher and McDonald (1978, p. 45) reported data from an experiment designed to study the effect of experience on errors in the reading of chest x-rays. Ten radiologists participated in the study and were classified into one of three groups: senior staff, junior staff, and residents. Each radiologist was asked whether the left ventricle was normal and the response was compared to the results of ventriculography. The percentage of errors for each radiologist was determined and the data given as follows.

Senior Staff	Junior Staff	Residents
7.3	13.3	14.7
7.4	10.6	23.0
	15.0	22.7
	20.7	26.6

Source: Fisher and McDonald (1978, p. 46). Used with permission.

(a) Describe the mathematical model and the assumptions for the experiment.
(b) Analyze the data and report the analysis of variance table.
(c) Perform an appropriate F test for the hypothesis that the mean percentage errors for the three groups of radiologists are the same. Use $\alpha = 0.05$.
(d) Transform each percentage to its arcsine value and then perform a second analysis of variance on the transformed data, comparing the results with those obtained in part (c).

5. Lorenzen and Anderson (1993, p. 46) reported data from an experiment designed to study the effect of honey on haemoglobin in children. A completely randomized design was used with 12 children, 6 given a

tablespoon of honey added to a cup of milk, and 6 not given honey over a period of six straight weeks. The data are given as follows.

Honey	Control
19	14
12	8
9	4
17	4
24	11
22	15

Source: Lorenzen and Anderson (1993, p. 46). Used with permission.

(a) Describe the mathematical model and the assumptions for the experiment.
(b) Analyze the data and report the analysis of variance table.
(c) Perform an appropriate F test for the hypothesis that the mean haemoglobin levels in two groups of children are the same. Use $\alpha = 0.05$.
(d) Carry out the test for homoscedasticity using the Snedecor's F test. Use $\alpha = 0.05$.

6. A randomized block experiment was conducted with five treatments and five blocks. The following table gives partial results on analysis of variance.

Source of Variation	Degrees of Freedom	Sum of Squares	Mean Square	F Value	p-Value
Treatment		150.2			
Block		49.1			
Error		—			
Total		295.2			

(a) Describe the mathematical model and the assumptions for the experiment.
(b) Complete the analysis of variance table.
(c) Perform an appropriate F test for the hypothesis that all the treatment means are the same. Use $\alpha = 0.05$.

7. An experiment is designed to compare mileage of four brands of gasoline. Inasmuch as mileage will vary according to road and other driving conditions, five different categories of driving conditions are included in the experiment. A randomized block design is used and each brand of gasoline is randomly selected to fill five cars. Finally, each car is

randomly assigned a given driving condition. The data are given as follows.

Driving condition	Brand of Gasoline			
	B_1	B_2	B_3	B_4
D_1	48.1	39.1	51.6	49.1
D_2	34.6	46.1	45.6	35.1
D_3	47.6	48.6	41.6	39.6
D_4	30.6	43.6	39.6	31.6
D_5	39.6	45.1	44.1	21.1

(a) Describe the mathematical model and the assumptions for the experiment.
(b) Analyze the data and report the analysis of variance table.
(c) Do the brands have a significant effect on the mileage? Use $\alpha = 0.05$.
(d) Do the driving conditions have a significant effect on mileage? Use $\alpha = 0.05$.
(e) If there are significant differences in mileage due to brands, use a suitable multiple comparison procedure to determine which brands differ. Use $\alpha = 0.01$.

8. An agricultural experiment was designed to study the potential for grain yield of four different varieties of wheat. A randomized block design with five blocks was used and each variety was planted in each of the blocks. The data on yields are given as follows.

Block	Variety			
	I	II	III	IV
1	152.4	153.4	150.9	145.5
2	154.1	152.8	154.4	146.1
3	154.4	156.3	155.3	149.8
4	155.1	156.1	152.3	148.1
5	156.5	154.6	155.2	148.9

(a) Describe the mathematical model and the assumptions for the experiment.
(b) Analyze the data and report the analysis of variance table.
(c) Do the varieties have a significant effect on the yield? Use $\alpha = 0.05$.
(d) Do the blocks have a significant effect on the yield? Use $\alpha = 0.05$.
(e) If there are significant differences in yields due to varieties, use a suitable multiple comparison method to determine which varieties differ. Use $\alpha = 0.01$.

9. An experiment was designed to study the reaction time among rats under the influence of three different treatments. Four rats were chosen for the experiment and three treatments were administered on each rat on three different days, and the order in which each rat received a treatment was random. The data on reaction time in seconds are given as follows.

	Treatment		
Rat	A	B	C
1	6.8	11.5	8.4
2	5.6	9.7	7.3
3	2.2	7.4	3.9
4	3.7	8.3	5.7

(a) Describe the mathematical model and the assumptions for the experiment.
(b) Analyze the data and report the analysis of variance table.
(c) Do the treatments have a significant effect on the reaction time? Use $\alpha = 0.05$.
(d) Do the rats have a significant effect on the reaction time? Use $\alpha = 0.05$.
(e) If there are significant differences in reaction time due to treatments, use a suitable multiple comparison method to determine which treatments differ. Use $\alpha = 0.01$.

10. Anderson and Bancroft (1952, p. 245) reported data from an experiment conducted by Middleton and Chapman at Laurinburg, North Carolina, to compare eight varieties of oats. The experiment involved a randomized block design with five blocks and the yields of grain in grams for a 16-foot row were recorded. The data are given as follows.

	Variety							
Block	I	II	III	IV	V	VI	VII	VIII
1	296	402	437	303	469	345	324	488
2	357	390	334	319	405	342	339	374
3	340	431	426	310	442	358	357	401
4	331	340	320	260	487	300	352	338
5	348	320	296	242	394	308	220	320

Source: Anderson and Bancroft (1952, p. 245). Used with permission.

(a) Describe the mathematical model and the assumptions for the experiment.
(b) Analyze the data and report the analysis of variance table.
(c) Perform an appropriate F test for the hypothesis that the mean yields for all the varieties are the same. Use $\alpha = 0.05$.

(d) If there are significant differences in mean yields among the varieties, use a suitable multiple comparison procedure to determine which varieties differ. Use $\alpha = 0.01$.

(e) What is the efficiency of this design compared with a completely randomized design?

11. Fisher and McDonald (1978, p. 66) reported data from an experiment designed to study the effect of different heat treatments of the dietary protein of young rats on the sulfur-containing free amino acids in the plasma. The experiment involved a randomized block design with three blocks and six different treatments of heated soybean protein. The plasma-free crystine levels in rats fed on different treatments were recorded (μ moles/100 mℓ) and the data are given as follows (where each observation is the average for four rats).

Block	Heat Treatment					
	I	II	III	IV	V	VI
1	4.0	4.0	4.1	3.8	4.5	3.8
2	4.6	5.7	5.2	4.9	5.6	5.3
3	4.9	6.1	5.4	5.2	5.9	5.7

Source: Fisher and McDonald (1978, p. 66). Used with permission.

(a) Describe the mathematical model and the assumptions for the experiment.

(b) Analyze the data and report the analysis of variance table.

(c) Perform an appropriate F test for the hypothesis that the mean levels of plasma-free crystines in rats fed on different treatments are the same. Use $\alpha = 0.05$.

(d) If there are significant differences in mean levels of plasma-free crystines due to treatments, use a suitable multiple comparison method to determine which treatments differ. Use $\alpha = 0.01$.

(e) What is the efficiency of this design compared with a completely randomized design?

12. John (1971, p. 64) reported data from an experiment involving a randomized block design with three blocks and 12 treatments including a control. The yields, in ounces, of cured tobacco leaves were recorded and the data are given as follows.

Block	Treatment											
	I	II	III	IV	V	VI	VII	VIII	IX	X	XI	Control
1	76	82	76	70	76	70	82	88	81	74	67	79
2	70	70	73	74	73	83	74	65	67	67	67	78
3	80	73	77	62	86	84	80	80	81	76	79	63

Source: John (1971, p. 64). Used with permission.

(a) Describe the mathematical model and the assumptions for the experiment.
(b) Analyze the data and report the analysis of variance table.
(c) Perform an appropriate F test for the hypothesis that the mean yields for all the treatments are the same. Use $\alpha = 0.05$.
(d) If the hypothesis in part (c) is rejected, use Dunnett's procedure to test differences between the control and each of the other treatment means. Use $\alpha = 0.01$.
(e) As an alternative to Dunnett's procedure, one might wish to compare control versus the mean of the other 11 treatments. Set up the necessary contrast and test the implied null hypothesis. Use $\alpha = 0.01$.

13. John and Quenouille (1977) reported data from a randomized block experiment to test the efficacy of five levels of application of potash on the Pressley strength index of cotton. The levels of potash consisted of pounds of K_2O per unit area, expressed as units, and the experiment was carried out in three blocks. The data are as follows.

	Treatment				
Block	I	II	III	IV	V
1	7.62	8.14	7.76	7.17	7.46
2	8.00	8.15	7.73	7.57	7.68
3	7.93	7.87	7.74	7.80	7.21

Source: John and Quenouille (1977). Used with permission.

(a) Describe the mathematical model and the assumptions for the experiment.
(b) Analyze the data and report the analysis of variance table.
(c) Do the treatments have a significant effect on the strength of cotton? Use $\alpha = 0.05$.
(d) Do the blocks have a significant effect on the strength of cotton? Use $\alpha = 0.05$.
(e) If there are significant differences in mean levels of the Pressley index of cotton, use a suitable multiple comparison method to determine which treatments differ. Use $\alpha = 0.01$.

14. Snee (1985) reported data from an experiment designed to investigate the effect of a drug added to the feed of chicks on their growth. There were three treatments: standard feed (control group), standard feed and a low dose of drug, standard feed and a high dose of drug. The experimental units included a group of chicks fed and reared in the same bird house. Eight blocks of three experimental units each were laid out with physically adjacent units assigned to the same block. The

data are given in the following where each observation is the average weight (lbs) per bird at maturity.

	Treatment		
Block	Control	Low dose	High dose
1	3.93	3.99	3.96
2	3.78	3.96	3.94
3	3.88	3.96	4.02
4	3.93	4.03	4.06
5	3.84	4.10	3.94
6	3.75	4.02	4.09
7	3.98	4.06	4.17
8	3.84	3.92	4.12

Source: Snee (1985). Used with permission.

(a) Describe the mathematical model and the assumptions for the experiment.
(b) Analyze the data and report the analysis of variance table.
(c) Perform an appropriate F test for the hypothesis that the mean weights of chicks fed on different treatments are the same. Use $\alpha = 0.05$.
(d) If there are significant differences in mean weights due to treatments, use a suitable multiple comparison method to determine which treatments differ. Use $\alpha = 0.01$.
(e) What is the efficiency of this design compared with a completely randomized design?

15. Steel and Torrie (1980, p. 202) reported unpublished data, courtesy of R. A. Linthurst and E. D. Seneca, North Carolina State University, Raleigh, North Carolina (paper title, "Aeration, Nitrogen, and Salinity as Determinants of *Spartina alterniflora* Growth Response,), who conducted a greenhouse experiment on the growth of *Spartina alterniflora* in order to study the effects of salinity, nitrogen and aeration. The dried weight of all aerial plant material was recorded and the data are given as follows.

Block	Treatment*											
	I	II	III	IV	V	VI	VII	VIII	IX	X	XI	XII
1	11.8	18.8	21.3	83.3	8.8	26.2	20.4	50.2	2.2	8.8	1.4	25.8
2	8.1	15.8	22.3	25.3	8.1	19.5	8.5	47.7	3.3	7.6	15.3	22.6
3	22.6	37.1	19.8	55.1	2.1	17.8	8.2	16.4	11.1	6.0	10.2	17.9
4	4.1	22.1	49.0	47.6	10.0	20.3	4.8	25.8	2.7	7.4	0.0	14.0

* Treatment combinations are defined as follows.

	Treatment Code											
Number	I	II	III	IV	V	VI	VII	VIII	IX	X	XI	XII
Salinity parts/thousand	15	15	15	15	30	30	30	30	45	45	45	45
Nitrogen kg/hectare	0	0	168	168	0	0	168	168	0	0	168	168
Aeration (0 = none, 1 = saturation)	0	1	0	1	0	1	0	1	0	1	0	1

Source: Steel and Torrie (1980, p. 202). Used with permission.

(a) Describe the mathematical model and the assumptions for the experiment.
(b) Analyze the data and report the analysis of variance table.
(c) Perform an appropriate F test for the hypothesis that there are no differences in response due to treatments. Use $\alpha = 0.05$.
(d) Perform an appropriate F test for the hypothesis that there are no differences in response due to differences in salinity. Use $\alpha = 0.05$.
(e) Perform an appropriate F test for the hypothesis that there are no differences in response due to differences in nitrogen treatments. Use $\alpha = 0.05$.
(f) Perform an appropriate F test of the hypothesis that there are no differences in response due to differences in aeration treatments. Use $\alpha = 0.05$.
(g) Is the nitrogen contrast orthogonal to the aeration contrast?

16. A researcher wants to study two treatment factors A and C using a factorial arrangement along with a randomized block design. Assume that the factors A and C have a and c levels respectively giving a total of ac treatment combinations. There are b blocks and each treatment combination is randomly assigned to ac experimental units within each block. The mathematical model for this design is given by

$$y_{ijk} = \mu + \beta_i + \alpha_j + \gamma_k + (\alpha\gamma)_{jk} + e_{ijk} \quad \begin{cases} i = 1, 2, \ldots, b \\ j = 1, 2, \ldots, a \\ k = 1, 2, \ldots, c, \end{cases}$$

where y_{ijk} is the observed response corresponding to the i-th block, the j-th level of factor A, and the k-th level of factor C; $-\infty < \mu < \infty$ is the overall mean, β_i is the effect of the i-th block, α_j is the effect of the j-th level of factor A, γ_k is the effect of the k-th level of factor C, $(\alpha\gamma)_{jk}$ is the interaction between the j-th level of factor A and the k-th level of factor C, and e_{ijk} is the customary error term.

(a) State the assumptions of the model if all the effects are considered to be fixed.
(b) State the assumptions of the model if all the effects are considered to be random.

(c) State the assumptions of the model if the block effects are considered to be random and A and C effects are considered to be fixed.

(d) Report the analysis of variance table including expected mean squares under the assumptions of fixed, random, and mixed models as stated in parts (a) through (c).

(e) Assuming normality for the random effects in parts (a) through (c), develop tests of hypotheses for testing the effects corresponding to the block, factors A and C, and the $A \times C$ interaction.

(f) Determine the estimators of the variance components based on the analysis of variance procedure under the assumptions of the random and mixed model.

17. A psychological experiment was designed to study the effect of five learning devices. In order to control for directional bias on learning, a Latin square design was used with five subjects using five different orders. The test scores are given in the following where the letter within parentheses represents the learning device used.

Subject	Order of Test				
	1	2	3	4	5
1	105 (D)	195 (C)	185 (A)	135 (E)	170 (B)
2	165 (C)	185 (B)	150 (E)	190 (A)	150 (D)
3	155 (A)	150 (D)	185 (C)	155 (B)	85 (E)
4	165 (B)	195 (E)	135 (D)	110 (C)	105 (A)
5	245 (E)	240 (A)	170 (B)	175 (D)	135 (C)

(a) Describe the mathematical model and the assumptions for the experiment.

(b) Analyze the data and report the analysis of variance table.

(c) Perform an appropriate F test for the hypothesis that the mean test scores for all the devices are the same, and state which device you would recommend for use. Use $\alpha = 0.05$.

(d) Use Tukey's multiple comparison method to determine whether there are significant differences among the top three learning devices. Use $\alpha = 0.01$.

(e) Comment on the usefulness of the Latin square design in this case.

(f) What is the efficiency of the design compared to the randomized block design?

18. An experiment was designed to compare grain yields of five different varieties of corn. A 5×5 Latin square design was used to control for fertility gradients due to rows and columns. The data on yields were given as follows where the letter within parentheses represents the variety of the corn.

	Column				
Row	1	2	3	4	5
1	65.9 (A)	66.0 (D)	68.0 (B)	67.4 (E)	63.4 (C)
2	68.2 (B)	67.6 (E)	68.6 (A)	68.6 (C)	66.4 (D)
3	68.6 (C)	68.3 (B)	68.0 (E)	67.2 (D)	69.3 (A)
4	64.1 (D)	62.9 (A)	66.2 (C)	67.0 (B)	66.8 (E)
5	61.2 (E)	62.7 (C)	61.6 (D)	67.8 (A)	64.9 (B)

(a) Describe the mathematical model and the assumptions for the experiment.
(b) Analyze the data and report the analysis of variance table.
(c) Perform an appropriate F test for the hypothesis that the mean yields for all the varieties are the same, and state which variety you would recommend for planting. Use $\alpha = 0.05$.
(d) Use Tukey's multiple comparison method to determine whether there are significant differences among the top three varieties. Use $\alpha = 0.01$.
(e) Comment on the usefulness of the Latin square design in this case.
(f) What is the efficiency of the design compared to the randomized block design?

19. Anderson and Bancroft (1952, p. 247) reported data from an experiment conducted at the University of Hawaii to compare six different legume intercycle crops for pineapples. A Latin square design was used and the data on yields in 10-gram units were given as follows where the letter within parentheses represents the variety of the legume.

	Column					
Row	1	2	3	4	5	6
1	220 (B)	98 (F)	149 (D)	92 (A)	282 (E)	169 (C)
2	74 (A)	238 (E)	158 (B)	228 (C)	48 (F)	188 (D)
3	118 (D)	279 (C)	118 (F)	278 (E)	176 (B)	65 (A)
4	295 (E)	222 (B)	54 (A)	104 (D)	213 (C)	163 (F)
5	187 (C)	90 (D)	242 (E)	96 (F)	66 (A)	122 (B)
6	90 (F)	124 (A)	195 (C)	109 (B)	79 (D)	211 (E)

Source: Anderson and Bancroft (1952, p. 247). Used with permission.

(a) Describe the mathematical model and the assumptions for the experiment.
(b) Analyze the data and report the analysis of variance table.
(c) Perform an appropriate F test for the hypothesis that the mean yields for all the varieties are the same. Use $\alpha = 0.05$.

(d) Use Tukey's multiple comparison method to make pairwise comparisons among the three legumes with the highest yield. Use $\alpha = 0.01$.
(e) Comment on the usefulness of the Latin square design in this case.
(f) What is the efficiency of the design compared to the randomized block design?
20. Damon and Harvey (1987, p. 315) reported data from an experiment conducted by Scott Werme of the Department of Veterinary and Animal Sciences at the University of Massachusetts to study the effect of the level of added okara in a ration of total digestible nutrient (TDN). A 4×4 Latin square design involving four treatments (levels of added okara), four sheep, and four periods was used. The data on the TDN levels (in percents) in the total ration are given as follows where the letter within parentheses represents the treatment.

	Sheep			
Period	1	2	3	4
1	61.60 (A)	75.83 (D)	68.17 (C)	65.61 (B)
2	62.05 (B)	58.63 (A)	70.33 (D)	67.22 (C)
3	66.91 (C)	68.10 (B)	58.43 (A)	71.98 (D)
4	69.87 (D)	67.25 (C)	63.32 (B)	60.30 (A)

Source: Damon and Harvey (1987, p. 315). Used with permission.

(a) Describe the mathematical model and the assumptions for the experiment.
(b) Analyze the data and report the analysis of variance table.
(c) Perform an appropriate F test for the hypothesis that the mean TDN levels in the total ration for all the treatments are the same. Use $\alpha = 0.05$.
(d) Use Tukey's multiple comparison procedure to make pairwise comparisons among the three top treatments. Use $\alpha = 0.01$.
(e) Comment on the usefulness of the Latin square design in this case.
(f) What is the efficiency of the design compared to the randomized block design?
21. Fisher (1958, pp. 267–268) reported data on root weights for mangolds from five different treatments found by Mercer and Hall in 25 plots. The following table gives data in a Latin square layout where letters (A, B, C, D, E) representing five different treatments are distributed randomly in such a way that each appears once in each row and each column.

	Column				
Row	1	2	3	4	5
1	376 (D)	371 (E)	355 (C)	356 (B)	335 (A)
2	316 (B)	338 (D)	336 (E)	356 (A)	332 (C)
3	326 (C)	326 (A)	335 (B)	343 (D)	330 (E)
4	317 (E)	343 (B)	330 (A)	327 (C)	336 (D)
5	321 (A)	332 (C)	317 (D)	318 (E)	306 (B)

Source: Fisher (1958, pp. 267–268). Used with permission.

(a) Describe the mathematical model and the assumptions for the experiment.
(b) Analyze the data and report the analysis of variance table.
(c) Perform an appropriate F test for the hypothesis that the mean root weights for mangolds for all the treatments are the same. Use $\alpha = 0.05$.
(d) Use Tukey's multiple comparison procedure to make pairwise comparisons among the three treatments with the highest mean weight. Use $\alpha = 0.01$.
(e) Comment on the usefulness of the Latin square design in this case.
(f) What is the efficiency of the design compared to the randomized block design?

22. Steel and Torrie (1980, p. 225) reported data from an experiment designed to study moisture content of turnip greens. A Latin square design involving five plants, five leaf sizes, and five treatments was used. Treatments were times of weighing since moisture losses might be anticipated in a 70°F laboratory as the experiment progressed. The data on moisture content (in percent) are given as follows where the letter within parentheses represents a treatment.

	Leaf size (1 = smallest, 5 = largest)				
Plant	1	2	3	4	5
1	86.67 (E)	87.15 (D)	88.29 (A)	88.95 (C)	89.62 (B)
2	85.40 (B)	84.77 (E)	85.40 (D)	87.54 (A)	86.93 (C)
3	87.32 (C)	88.53 (B)	88.50 (E)	89.99 (D)	89.68 (A)
4	84.92 (A)	85.00 (C)	87.29 (B)	87.85 (E)	87.08 (D)
5	84.88 (D)	86.16 (A)	87.83 (C)	85.83 (B)	88.51 (E)

Source: Steel and Torrie (1980, p. 225). Used with permission.

(a) Describe the mathematical model and the assumptions for the experiment.

(b) Analyze the data and report the analysis of variance table.
(c) Perform an appropriate F test for the hypothesis that the mean moisture content for all the treatments are the same. Use $\alpha = 0.05$.
(d) Use Tukey's multiple comparison procedure to make pairwise comparisons among the three treatments with the highest moisture content. Use $\alpha = 0.01$.
(e) Comment on the usefulness of the Latin square design in this case.
(f) What is the efficiency of the design compared to the randomized block design?

23. An experiment was designed to study the effect of fertilizers on yields of wheat. A 4×4 Graeco-Latin square design involving four fertilizers, four varieties of wheat, four rows, and four columns was used. The data are given as follows where the Roman letter within parentheses represents the fertilizer and the Greek letter represents the wheat variety.

	Column			
Row	1	2	3	4
1	135.4 ($C\beta$)	124.4 ($B\gamma$)	114.6 ($D\delta$)	135.4 ($A\alpha$)
2	114.2 ($B\alpha$)	105.2 ($C\delta$)	113.6 ($A\gamma$)	124.4 ($D\beta$)
3	114.0 ($A\delta$)	116.1 ($D\alpha$)	134.2 ($B\beta$)	119.6 ($C\gamma$)
4	114.9 ($D\gamma$)	164.4 ($A\beta$)	134.5 ($C\alpha$)	118.9 ($B\delta$)

(a) Describe the mathematical model and the assumptions for the experiment.
(b) Analyze the data and report the analysis of variance table.
(c) Perform an appropriate F test for the hypothesis that the mean yields for all the fertilizers are the same, and state which fertilizer you would recommend for use. Use $\alpha = 0.05$.
(d) Perform an appropriate test for the hypothesis that the mean yields for all the varieties are the same, and state which variety you would recommend for use. Use $\alpha = 0.05$.
(e) Comment on the usefulness of the Graeco-Latin square design in this case.

24. An experiment was designed to study the effect of diet on cholesterol. A Graeco-Latin square design involving five diets, five time periods, five technicians, and five laboratories was used. Subjects were fed the diets for different time periods and cholesterol was measured. The data are given as follows where the Roman letter within parentheses represents the diet and the Greek letter represents the period.

	Laboratory				
Technician	1	2	3	4	5
1	175.5 (Aα)	165.5 (Bβ)	168.8 (Cγ)	155.5 (Dδ)	162.2 (Eε)
2	157.7 (Bγ)	170.0 (Cδ)	167.7 (Dε)	160.0 (Eα)	170.0 (Aβ)
3	170.0 (Cε)	161.5 (Dα)	165.5 (Eβ)	174.4 (Aγ)	162.2 (Bδ)
4	154.4 (Dβ)	164.4 (Eγ)	171.1 (Aδ)	163.3 (Bε)	166.6 (Cα)
5	160.0 (Eδ)	173.3 (Aγ)	166.6 (Bα)	166.6 (Cβ)	163.3 (Dγ)

(a) Describe the mathematical model and the assumptions for the experiment.
(b) Analyze the data and report the analysis of variance table.
(c) Perform an appropriate F test for the hypothesis that the mean cholesterol levels for all the diets are the same. Use $\alpha = 0.05$.
(d) Comment on the usefulness of the Graeco-Latin square design in this case.

25. An experiment was designed to compare the efficiencies of four different operators using three different machines. A split-plot design was used where the output of one machine constitutes a whole-plot, which was then divided into four subplots for the four operators. The experiment was repeated four times and the data given as follows.

Machine 1				Machine 2				Machine 3			
Operator				Operator				Operator			
1	2	3	4	1	2	3	4	1	2	3	4
170.7	135.1	167.5	166.3	163.0	138.0	161.0	155.0	157.3	143.4	161.0	160.5
161.1	140.5	160.8	162.5	166.1	142.7	160.7	154.4	157.1	136.5	165.6	164.3
161.8	142.5	158.7	160.2	160.0	132.6	156.5	155.1	156.0	135.3	160.6	155.1
163.5	146.0	158.5	161.2	156.7	147.7	150.5	150.5	154.5	142.2	154.5	158.1

(a) Describe the mathematical model and the assumptions for the experiment.
(b) Analyze the data and report the analysis of variance table.
(c) Test whether there are significant differences in output of the three machines. Use $\alpha = 0.05$.
(d) Test whether there are significant differences in output of the four operators. Use $\alpha = 0.05$.
(e) Is there a significant interaction effect between machines and operators? Use $\alpha = 0.05$.
(f) Comment on the usefulness of the split-plot design in this case.

26. Consider a split-plot design involving six treatments $A_1, A_2, \ldots, A_6$ that are assigned at random to six whole-plots. Each whole-plot is then divided into two subplots for testing treatments B_1 and B_2 involving three replications. The data on yields are given as follows.

A_1		A_2		A_3		A_4		A_5		A_6	
B_1	B_2	B_1	B_2	B_1	B_2	B_1	B_2	B_1	B_2	B_1	B_2
111	107	110	113	111	127	111	123	125	120	131	130
107	125	117	115	110	119	121	129	127	126	127	123
118	123	109	119	116	117	113	117	123	123	122	116

(a) Describe the mathematical model and the assumptions for the experiment.
(b) Analyze the data and report the analysis of variance table.
(c) Test whether there are significant differences in yields of the whole-plot treatments. Use $\alpha = 0.05$.
(d) Test whether there are significant differences in yields of the subplot treatments. Use $\alpha = 0.05$.
(e) Is there a significant interaction effect between whole-plot and subplot treatments? Use $\alpha = 0.05$.
(f) Comment on the usefulness of the split-plot design in this case.

27. John (1971, p. 98) reported data from an experiment described by Yates (1937b) carried out at the Rothamsted Experimental Station, Harpenden, England, with two factors, varieties of oats and quantity of manure. A split-plot design was used with the variety as the whole-plot treatment and the manure as the subplot treatment. There were six blocks of three plots each and each plot was divided into four subplots. One plot in each block was planted with each of the three varieties of oats and each subplot was assigned at random to one of the four levels of manure. The data on yields are given as follows.

	Block								
	1			2			3		
Manure	Variety			Variety			Variety		
Level (Tons/Acre)	A_1	A_2	A_3	A_1	A_2	A_3	A_1	A_2	A_3
No manure	111	117	105	74	64	70	61	70	96
0.01	130	114	140	89	103	89	91	108	124
0.02	157	161	118	81	132	104	97	126	121
0.03	174	141	156	122	133	117	100	149	144

	Block								
	4			5			6		
Manure	Variety			Variety			Variety		
Level (Tons/Acre)	A_1	A_2	A_3	A_1	A_2	A_3	A_1	A_2	A_3
No manure	62	80	63	68	60	89	53	89	97
0.01	90	82	70	64	102	129	74	82	99
0.02	100	94	109	112	89	132	118	86	119
0.03	116	126	99	86	96	124	113	104	121

Source: John (1971, p. 99). Used with permission.

(a) Describe the mathematical model and the assumptions for the experiment.
(b) Analyze the data and report the analysis of variance table.
(c) Test whether there are significant differences in yields of the six blocks. Use $\alpha = 0.05$.
(d) Test whether there are significant differences in yields of the three varieties. Use $\alpha = 0.05$.
(e) Test whether there are significant differences in yields of the four manures. Use $\alpha = 0.05$.
(f) Is there a significant interaction effect between varieties and manures? Use $\alpha = 0.05$.
(g) Comment on the usefulness of the split-plot design in this case.

11 Analysis of Variance Using Statistical Computing Packages

11.0 PREVIEW

The widespread availability of modern high speed mainframes and microcomputers and myriad accompanying software have made it much simpler to perform a wide range of statistical analyses. The use of statistical computing packages or software can make it possible even for a relatively inexperienced person to utilize computers to perform a statistical analysis. Although there are numerous statistical packages that can perform the analysis of variance, we have chosen to include for this volume three statistical packages that are most widely used by scientists and researchers throughout the world and that have become standards in the field.[1] The packages are the Statistical Analysis System (SAS), the Statistical Product and Service Solutions (SPSS),[2] and the Biomedical Programs (BMDP). In the following we provide a brief introduction to these packages and their use for performing an analysis of variance, and related statistical tests of significance.[3,4,5]

11.1 ANALYSIS OF VARIANCE USING SAS

SAS is an integrated system of software products for data management, report writing and graphics, business forecasting and decision support, applications

---

[1] A recent survey of faculty regarding their commercial software preference for advanced analysis of variance courses found that the most frequently used packages were SAS, SPSS, and BMDP (Tabachnick and Fidell (1991)).

[2] The acronym SPSS initially originated from Statistical Package for the Social Sciences. However, the term Statistical Product and Solutions really applies to the SPSS acronym as used as the company name. The package is simply known as SPSS.

[3] For a discussion of computational algorithms and the construction of computer programs for the analysis of variance as used in statistical packages for the analysis of designed experiments, see Heiberger (1989).

[4] For a listing of a series of analysis of variance and related programs that can be run on microcomputer systems, see Wolach (1983).

[5] For some further discussions of the use of SAS, SPSS, and BMDP software in performing analysis of variance including numerical examples illustrating various designs, see Colleyer and Enns (1987).

research, and project management. The statistical analysis procedures in the SAS system are among the finest available. They range from simple descriptive statistics to complex multivariate techniques. One of the advantages of SAS software is the variety of procedures — from elementary analysis to the most sophisticated statistical procedures, of which GLM (*General Linear Models*) is the flagship — that can be performed. SAS software provides tremendous flexibility and is currently available on thousands of computer facilities throughout the world.

For a discussion of instructions for creating SAS files and running SAS procedures, the reader is referred to the SAS manuals and other publications documenting SAS procedures. The SAS manuals, SAS Language and Procedures: Usage (SAS Institute, 1989, 1991), SAS Language: Reference (SAS Institute, 1990a), SAS Procedures Guide (SAS Institute, 1990b), and SAS/STAT User's Guide (SAS Institute, 1990c), together provide exhaustive coverage of the SAS software. The reference and procedure manuals provide an introduction to data handling and data management procedures including some descriptive statistics procedures, and the STAT manual covers inferential statistical procedures. The SAS Introductory Guide for PCs (SAS Institute, 1992) is a very useful publication for the beginner which provides an elementary discussion of some basic and commonly used data management and statistical procedures, including several simple ANOVA designs. Some other related publications providing easy-to-use instructions for running SAS procedures together with a broad coverage of statistical techniques and interpretation of SAS data analysis include DiIorio (1991), Friendly (1991), Freund and Littell (1991), Littell et al. (1991), Miron (1993), Spector (1993), Aster (1994), Burch and King (1994), Hatcher and Stepanski (1994), Herzberg (1994), Jaffe (1994), Elliott (1995), Everitt and Derr (1996), DiIorio and Hardy (1996), Cody and Smith (1997), and Schlotzhauer and Littell (1997).

There are several SAS procedures for performing an analysis of variance. PROC ANOVA is a very useful procedure that can be used for analyzing a wide variety of *anova* designs known as balanced designs that contain an equal number of observations in each submost subcell. However, this procedure is somewhat limited in scope and cannot be used for the unbalanced anova designs that contain an unequal number of observations in each submost subcell. Moreover, in a multifactorial experiment, PROC ANOVA is appropriate when all the effects are fixed. PROC GLM, which stands for *General Linear Models*, is more general and can accommodate both balanced and unbalanced designs. However, the procedure is more complicated, requiring more memory and execution time; it runs much slower and is considerably more expensive to use than PROC ANOVA. The other two SAS procedures which are more appropriate for anova models involving random effects include NESTED and VARCOMP. PROC NESTED is specially configured for anova designs where all factors are hierarchically nested and involve only random effects. The NESTED procedure is computationally more efficient than GLM for nested designs. Although PROC

NESTED is written for a completely random effects model, the computations of the sums of squares and mean squares are the same for all the models. If some of the factors are crossed or any factor is fixed, PROC ANOVA is more appropriate for balanced data involving only fixed effects and GLM for balanced or unbalanced data involving random effects. PROC VARCOMP is especially designed for estimating variance components and currently implements four methods of variance components estimation. In addition, SAS has recently introduced a new procedure, PROC MIXED, which fits a variety of mixed linear models and produces appropriate statistics to enable one to make statistical inference about the data. Traditional mixed linear models contain both fixed and random effects parameters, and PROC MIXED fits not only the traditional variance components models but models containing other covariance structures as well. PROC MIXED can be considered as a generalization of the GLM procedure in the sense that although PROC GLM fits standard linear models, PROC MIXED fits a wider class of mixed linear models. PROC GLM produces all Types I to IV tests of fixed effects, but PROC MIXED computes only Type I and Type III. Instead of producing traditional analysis of variance estimates, PROC MIXED computes REML and ML estimates and optionally computes MIVQUE (0) estimates which are similar to analysis of variance estimates. PROC MIXED subsumes the VARCOMP procedure except that it does not include the Type I method of estimating variance components. For further information and applications of PROC MIXED to random and mixed linear models, see Littel et al. (1996) and SAS Institute (1997).

Remark: The output from GLM produces four different kinds of sums of squares, labelled as Types I, II, III, and IV, which can be thought of, respectively, as "sequential," "each-after-all-others," "Σ-restrictions" and "hypotheses." The Type I sum of squares of an effect is calculated in a hierarchical or sequential manner by adjusting each term only for the terms that precede it in the model. The Type II sum of squares of an effect is calculated by adjusting each term by all other terms that do not contain the effect in question. If the model contains only the main effects, then each effect is adjusted for every other terms in the model. The Type III sum of squares of an effect is calculated by adjusting for any other terms that do not contain it and are orthogonal to any effects that contain it. The Type IV sum of squares of an effect is designed for situations in which there are empty cells and for any effect in the model, if it is not contained in any other term, Type IV = Type III = Type II. For balanced designs, all types of sums of squares are identical; they add up to the total sum of squares and form an unique and orthogonal decomposition. For unbalanced designs with no interaction models, Types II, III, and IV sums of squares are the same. Finally, for unbalanced designs with no empty cells, Types III and IV are equivalent. Furthermore, output from GLM includes a special form of estimable functions for each of the sums of squares listed under Types I, II, III, and IV. The importance of the estimable functions is that each provides a basis for formulating the associated hypothesis for the corresponding sum of squares. For further information on sums of squares from GLM, see Searle (1987, Section 12.2).

The following is an example of SAS commands necessary to run the ANOVA procedure:

PROC ANOVA;
CLASS {list of factors};
MODEL {dependent variable(s)} = {list of effects};

The class statement contains the keyword CLASS followed by a list of factors or variables of classification. The model statement contains the keyword MODEL followed by the list of dependent variable(s), followed by the equality symbol (=) which in turn is followed by the list of factors representing main effects as well as interaction effects. Interaction effects between different factors are designated by the factor names connected by an asterisk (*). The same commands are required for executing other analysis of variance procedures as well. Let Y designate the dependent variable and consider a factorial experiment involving three factors A, B, and C. The following is a typical model statement containing all the main and two factor interaction effects

MODEL $Y = A \ \ B \ \ C \ \ A*B \ \ A*C \ \ B*C$;

To analyze a factorial experiment with the full factorial model, a simpler way to specify a model statement is to list all the factors separated by a vertical slash. For example, the full factorial model involving factors A, B, C, and D can be written as

MODEL $Y = A \mid B \mid C \mid D$;

In terms of the choice of a fixed, random, or mixed effects model, the procedures require that the user supply information about which factors are to be considered fixed and which random. In some cases, if appropriate specifications for fixed and random effects are not provided, the analysis of variance tests provided by the procedure may differ from what the user wants them to be. In GLM, one can designate a factor as random by an additional statement containing the keyword RANDOM followed by the names of all effects, including interactions, which are to be treated as random. When some of the factors are designated as RANDOM, the GLM procedure will perform a mixed effects analysis including a default analysis by treating all factors as fixed. For example, given a mixed effects model with A fixed and B random, the following commands may be used to execute a GLM program

PROC GLM;
CLASS $A \ \ B$;
MODEL $Y = A \ \ B \ \ A*B$;
RANDOM $B \ \ A*B$;

The program will produce expected mean squares using "alternate" mixed model analysis, but will use the error mean square for testing the hypotheses for both the fixed factor A and the random factor B. If the analyst wants to use an alternate test, he has an option to designate the hypothesis to be tested and the appropriate error term by employing the following command:

TEST H = {effect to be tested} E = {error term};

For example, if the analyst wants to use the interaction mean square for the error term for testing the hypothesis for the random factor B, the following commands may be used:

PROC GLM;
CLASS A B;
MODEL Y = A B A*B;
TEST H = B E = A*B;

If several different hypotheses are to be tested, multiple TEST statements can be used. It should be noted that the TEST statement does not override the default option involving the test of a hypothesis against the error term which is always performed in addition to the hypothesis test implied by the TEST statement.

Remarks: (i) Given the specifications of the random and fixed effects via the RANDOM option, the GLM will print out the coefficients of the expected mean squares. The researcher can then determine how to make any test using expected mean squares. When exact tests do not exist, PROC GLM can still be used followed by the Satterthwaite's procedure.

(ii) The expected means squares and tests of significance are computed based on 'alternate' mixed model theory without the assumption that fixed-by-random interactions sum to zero across levels of the fixed factor. Thus, the tests based on RANDOM option will differ from those based on 'standard' mixed model theory.

(iii) The RANDOM statement provides for the following two options which can be placed at the end of the statement following a slash (/):
Q: This option provides a complete listing of all quadratic forms in the fixed effects that appear in the expected mean square.
TEST: This option provides that an F test be performed for each effect specified in the model using an appropriate error term as determined by the expected mean squares under the assumptions of the random or mixed model. When more than one mean square must be combined to determine the appropriate error term, a pseudo-F test based on Satterthwaite's approximation is performed.

(iv) For balanced designs, the ANOVA procedure can be used instead of GLM. However, the ANOVA performs only the fixed effects analysis and does not allow the option of the RANDOM statement.

For using the NESTED procedure, no specifications for MODEL or RANDOM statements are required since the procedure is designed for hierarchically nested designs involving only random factors. The order of nesting of the factors is indicated by the CLASS statement. In addition, a VAR statement is needed to specify the dependent variable. The following commands will execute the NESTED procedure where the factor B is nested within factor A:

> PROC NESTED;
> CLASS A B;
> VAR Y;

The program involving the preceding statements will perform two tests of hypotheses; one for the factor A effects against the mean square for B and the other for the factor B effects against the error mean square. The GLM procedure can also be used for analyzing designs involving nested factors. The nesting is indicated in the MODEL statement by the use of parentheses containing the factor within which the factor preceding the left parenthesis is being nested. For example, to indicate that the factor B is nested within the factor A, the following MODEL statement could be used:

> MODEL $Y = A$ $B(A)$;

If the factor B is random, one could perform the appropriate F test by using the RANDOM statement in the following SAS commands:

> PROC GLM;
> CLASS A B;
> MODEL $Y = A$ $B(A)$;
> RANDOM $B(A)$ / TEST;

Alternatively, one could replace the preceding RANDOM statement by the following TEST option:

> TEST $H = A$ $E = B(A)$;

Multiple nesting involving several factors, say, A, B, C, and D, where D is nested within C, C within B, and B within A, is indicated by the following MODEL statement:

> MODEL $Y = A$ $B(A)$ $C(BA)$ $D(CBA)$;

If factors C and D are crossed rather than nested, the above MODEL statement will be modified as:

> MODEL $Y = A B(A)$ $C*D(BA)$;

For the problems involving estimation of variance components, PROC VARCOMP is a better choice over other analysis of variance procedures. It produces estimates of variance components assuming all factors are random. For performing a mixed model analysis, the user must indicate the factors that are to be treated as random. All the SAS commands involving the CLASS and MODEL statements are used the same way as in other procedures. For example, the following commands can be used to estimate the variance components associated with the random factors A, B, and C and their interaction effects.

PROC VARCOMP;
CLASS A B C;
MODEL $Y = A \mid B \mid C$;

It should be pointed out that the VARCOMP does not perform F tests. However, it does produce sums of squares and mean squares that can be readily used to perform the required F tests manually.

PROC MIXED employs the same specifications for the CLASS statement as GLM but differs in the specifications of the MODEL statement. The right side of the MODEL statement now contains only the fixed-effect factors. The random-effect factors do not appear in the MODEL statement instead they are listed under the RANDOM statement. Thus, the MODEL and RANDOM statements are core essential statements in the application of the PROC MIXED procedure. Given two factors A and B with A fixed and B random, the following commands can be used to run the procedure:

PROC MIXED;
CLASS A B;
MODEL $Y = A$;
RANDOM B $A*B$;

The program will perform significance tests for fixed effects (in this case for factor A) and will compute REML estimates of the variance components including the error variance component.

Remark: PROC MIXED computes variance components estimates for random effect factors listed in the RANDOM statement including the error variance component; and performs significance tests for fixed effects specified in the MODEL statement. By default, the significance tests are based on likelihood principle and are equivalent to the conventional F tests for balanced data (default option). Variance components estimates are computed using the restricted maximum likelihood (REML) procedure (default option). For balanced data sets with nonnegative estimates, the REML estimates are identical to the traditional anova estimates. However, for designs with unbalanced structure, REML estimates generally differ from the anova estimates. Furthermore, unlike the ANOVA and GLM procedures, PROC MIXED does not directly compute or print sums of squares. Instead, it shows REML estimates of variance components and prints a separate table of tests of fixed effects that contains results of significance tests for the fixed-effect factors specified in the MODEL statement.

11.2 ANALYSIS OF VARIANCE USING SPSS

Statistical Package for the Social Sciences was initially developed by N. H. Nie and his coworkers at the National Opinion Research Center at the University of Chicago. It was officially shortened to SPSS when the first version of SPSS^X was released in 1983. In Release 4.0 of the new product it was simply known as SPSS (the X was dropped). Current versions are SPSS for Windows, the most recent of which is now Release 9.0. This package is an integrated system of computer programs originally developed for the analysis of social sciences data. The package provides great flexibility in data formation, data transformation, and manipulation of files. Some of the procedures currently available include descriptive analysis, simple and partial correlations, one-way and multiway analysis of variance, linear and nonlinear regression, loglinear analysis, reliability and life tables, and a variety of multivariate methods.

For a discussion of instructions for preparing SPSS files and running a SPSS procedure, the reader is referred to the SPSS manuals and other publications documenting SPSS procedures. For applications involving SPSS for Windows, SPSS Base 7.5 for Windows User's Guide (SPSS Inc., 1997a) provides the most comprehensive and complete coverage of data management, graphics, and basic statistical procedures. The other two manuals: SPSS Professional Statistics 7.5 (SPSS Inc., 1997b) and SPSS Advanced Statistics (SPSS Inc., 1997c) are also very useful publications for a broad coverage and documentation of intermediate and advanced level statistical procedures and interpretation of SPSS data analysis. Some other related publications providing easy-to-use instructions for running SPSS programs include Crisler (1991), Hedderson (1991), Frude (1993), Hedderson and Fisher (1993), Lurigio et al. (1995), and Coakes and Steed (1997). The monograph by Levine (1991) is a very useful guide to SPSS for performing analysis of variance and other related procedures.

There are several SPSS procedures for performing analysis of variance. The simplest procedure is ONEWAY for performing one-way analysis of variance.[6] The other programs will also do the one-way analysis of variance, but ONEWAY has a number of attractive features which make it a very useful procedure. In addition to providing the standard analysis of variance table, the ONEWAY will produce the following statistics for each group: number of cases, minimum, maximum, mean, standard deviation, and standard error, and 95 percent confidence interval for the mean. The ONEWAY also provides for testing linear, quadratic and polynomial relations across means of ordered groups as well as user specified a prior contrasts (coefficient for means) to test for specific linear relations among the means (e.g., comparing means of two treatment groups

---

[6] The MEANS procedure is even more basic than ONEWAY, although it does require few extra mouse clicks or a subcommand keyword to obtain analysis of variance results.

against that of a control group). Additional features include a number of multiple comparison tests including more than a dozen methods for testing pairwise differences in means and 10 multiple range tests for identifying subsets of means that are not different from each other.

Three other SPSS procedures that are of common use for analyzing higher level designs include ANOVA, MANOVA, and GLM. The ANOVA is used for performing analysis of variance for a factorial design involving two or more factors. For a model with five or fewer factors, the default ANOVA option provides for a full factorial analysis including all the interaction terms up to order five. The user has the option to control the number of interaction terms to be included in the model and any interaction effects that are not computed are pooled into the residual or error sum of squares. However, the user control over interactions in ANOVA is limited to specifying a maximum order of interactions to include. This means that all interactions at and below that level are included. For example, in a design with factors A, B, and C, the model that ANOVA will fit (assuming no empty cells) with all three factors used are: main effects only, main effects and all three two-way interactions and the full factorial model. The user does not have an option of fitting a model such as A, B, C, $A*B$ where some but not all of the interactions of a particular model are included.

The MANOVA and GLM are probably the most versatile and complex of all the SPSS procedures and can accommodate both balanced and unbalanced designs including nested or nonfactorial designs, multivariate data and analyses involving random and mixed effects models. Both procedures are based on a general linear model program and can allow multiple design subcommands. However, the GLM only honors the last DESIGN subcommand it encounters. The main difference between the two procedures in terms of statistical design is that while the MANOVA uses the full-rank reparametrization, the GLM uses the generalized inverse approach to accommodate a non-full rank overparametrized model. In SPSS 7.5 version, the MANOVA is available only through syntax commands, while the GLM is available both in syntax and via dialog boxes. In addition, the GLM offers a variety of features unavailable in MANOVA (see, e.g., SPSS Inc., 1997c, pp. 345–346). For example, the GLM tests for univariate homogeneity of variance assumption using the Levene test and provides for a number of different multiple comparison tests for unadjusted one-way factor means while these options are unavailable in MANOVA.

Remark: In Release 8.0, ANOVA is available via command syntax only; and in its place a new procedure UNIANOVA for performing univariate analysis of variance has been introduced. UNIANOVA is simply a univariate version of the GLM procedure restricted primarily for handling designs with one dependent variable.

In a SPSS analysis of variance procedure, all the dependent and independent variables are specified by using the keyword denoting the name of the procedure followed by the listing of the dependent variable first which is separated from the independent variables or factors using the keyword BY. The levels of a factor are specified by the use of parentheses that give the minimum and maximum numeric value of the levels of the given factor. For example, if the factor A has three levels coded as 1, 2, and 3 and B has four levels coded as 1, 2, 3, and 4, and Y denotes the dependent variable then a one-factor analysis of variance using ONEWAY and a full factorial analysis using ANOVA and MANOVA procedures can be performed using the following commands:

$$\text{ONEWAY } Y \text{ BY } A(1, 3)$$
$$\text{ANOVA } Y \text{ BY } A(1, 3) \, B(1, 4)$$
$$\text{MANOVA } Y \text{ BY } A(1, 3) \, B(1, 4).$$

Analysis of variance involving more than two factors can similarly be performed using either ANOVA or MANOVA procedures. For example, with three factors A, B, and C, each with three levels, a full factorial analysis can be performed using the following statements:

$$\text{ANOVA } Y \text{ BY } A(1, 3) \, B(1, 3) \, C(1, 3)$$
$$\text{MANOVA } Y \text{ BY } A(1, 3) \, B(1, 3) \, C(1, 3).$$

For four factors A, B, C, and D, with A having 2 levels, B 3 levels, C 4 levels, and D 5 levels, the statements are:

$$\text{ANOVA } Y \text{ BY } A(1, 2) \, B(1, 3) \, C(1, 4) \, D(1, 5)$$
$$\text{MANOVA } Y \text{ BY } A(1, 2) \, B(1, 3) \, C(1, 4) \, D(1, 5).$$

The syntax for GLM does not require the use of code levels for the factors appearing in the model and is simply written as:

$$\text{GLM } Y \text{ BY } A \, B$$
$$\text{GLM } Y \text{ BY } A \, B \, C$$
$$\text{GLM } Y \text{ BY } A \, B \, C \, D,$$
etc.

Remark: In Release 7.5 (and 8.0), ONEWAY does not require range specifications and does not honor them if they are included.

The preceding commands for ANOVA, MANOVA, and GLM procedures without any further qualifications will assume a full factorial model. In MANOVA and GLM, a factorial model could be made more explicit by an additional statement separated from the MANOVA/GLM statement by a slash (/), and consisting of the keyword DESIGN followed by the symbol "=" which in turn is followed by a listing of all the factors including interactions. The interaction $A \times B$ between the two factors A and B is indicated by the use of the keyword BY connecting the two factors (i.e., A BY B). In GLM, one can either use the keyword BY or the asterisk (*) to join the factors involved in the interaction term. For example, in the MANOVA/GLM command given previously, the user can explicitly indicate the full factorial model by the following commands:

> MANOVA Y BY A(1, 2) B(1, 3)
> /DESIGN = A, B, A BY B.
> GLM Y BY A B
> /DESIGN = A, B, A*B.

If the model is not a full factorial, the DESIGN statement simply lists the effects to be estimated and tested. For example, with three factors A, B, and C, a model containing main effects and the interactions $A \times B$ and $B \times C$ can be specified by the following MANOVA commands:

> MANOVA Y BY A(1, 2) B(1, 3) C(1, 4)
> /DESIGN = A, B, C, A BY B, B BY C.

ANOVA does not have an option for a design subcommand, and models other than the full factorial are specified by using the MAXORDERS subcommand to suppress interactions below a certain level.

MANOVA and GLM procedures also allow the use of nested models for nesting one effect within another. In MANOVA, nested models are indicated by connecting the two factors being nested by the keyword WITHIN (or just W). For example, for a two-factor nested design with factor B nested within factor A and the response units nested within factor B, the following commands are used for the MANOVA procedure:

> MANOVA Y BY A(1, 3) B(1, 3)
> /DESIGN A, B WITHIN A.

In GLM, one can either use the keyword WITHIN or a pair of parentheses to indicate the desired nesting. Thus, the nesting in the above example is indicated by

> GLM Y BY A B
> /DESIGN A, B(A).

Multiple nesting involving several factors is specified by the repeated use of the keyword WITHIN. In GLM, one can also use more than one pair of parentheses where each pair must be enclosed or nested within another pair. For example, to specify a three-factor completely nested design where B is nested within A and C is nested within B, the following commands are used for the MANOVA/GLM procedures:

> MANOVA Y BY $A(1, 3)$ $B(1, 4)$ $C(1, 4)$
> /DESIGN A, B WITHIN A, C WITHIN B WITHIN A.
> GLM Y BY $A B C$
> /DESIGN A, $B(A)$, $C(B(A))$.

Remarks: (i) The DESIGN subcommand in MANOVA cannot be simply written as /DESIGN B WITHIN A; otherwise it would produce nonsensical sums of squares with unbalanced data and an incorrect term and degrees of freedom in balanced or unbalanced designs. MANOVA's reparameterization method in general requires that hierarchies be maintained when listing effects on the DESIGN subcommand. That is, /DESIGN B WITHIN A without A generally makes no sense, and /DESIGN $A*B$ without both A and B in the model also makes no sense. Neither term estimates what the user wants it to estimate unless hierarchy is maintained.

(ii) In GLM, the subcommand /DESIGN $B(A)$ will fit the same model as /DESIGN A, B, $A*B$, or /DESIGN A, $B(A)$, or /DESIGN B, $A(B)$, or /DESIGN A, $A*B$, or /DESIGN B, $A*B$, or /DESIGN $A*B$, but the interpretation of the parameter estimates will differ in various cases. GLM doesn't reparametrize, so if one leaves out contained effects while including containing ones, some of the parameters for the containing effects that would normally be redundant no longer are, and the degrees of freedom and interpretation of the fitted effects are altered. In order to fit the standard nested model, the subcommand should be specified as /DESIGN A, $B(A)$.

For the analysis of variance models containing random factors, the MANOVA or GLM procedures must be used. In MANOVA, special F tests involving a random and mixed model analysis are performed by specifying denominator mean squares via the use of the keyword VS within the design statement. All the denominator mean squares, other than the error mean square (which is referred to as WITHIN), must be named by a numeric code from 1 to 10. One can then test against the number assigned to the error term. It is generally convenient to test against the error term before defining it as long as it is defined on the same design subcommand. For example, in a factorial analysis involving the random factors A and B where the main effects for the factors A and B are to be tested against the $A \times B$ mean square and the $A \times B$ effects are to be tested against the error mean square, the following commands are required:

> MANOVA Y BY $A(1, 2)$ $B(1, 3)$
> /DESIGN $=$ A VS 1
> B VS 1
> A BY B $=$ 1 VS WITHIN.

In the design statement given above, the first two lines specify that factors A and B are to be tested against the error term 1. The third line specifies that the error term 1 is defined as the $A \times B$ interaction, which in turn is to be tested against the error mean square (defined by the keyword WITHIN). Suppose, in the preceding, example, A is fixed, B is a random factor, and A is to be tested against $A \times B$ interaction as error term 1 and B is to be tested against the usual error mean square. To perform such a mixed model analysis, we would use the following command sequence:

> MANOVA Y BY $A(1, 2)$ $B(1, 3)$
> /DESIGN = A VS 1
>   A BY B = 1 VS WITHIN
>   B VS WITHIN.

Similarly, consider a two-factor nested design where a random factor B is nested within a fixed factor A. Now, the following commands will execute appropriate tests of the effects due to A and $B(A)$ factors:

> MANOVA Y BY $A(1, 2)$ $B(1, 3)$
> /DESIGN = A VS 1
>   B WITHIN A = 1 VS WITHIN.

The first line in the design statement specifies that the factor A effect is to be tested against the error term 1. The second line specifies the error term 1 as the mean square due to the $B(A)$ effect, and further specifies that this effect is to be tested against the usual error mean square (designated by the keyword WITHIN). For a three-factor completely nested design where a random factor C is nested within a random factor B which in turn is nested within a random factor A, the following commands are required to perform appropriate tests of the effects due to A, $B(A)$, and $C(B)$:

> MANOVA Y BY $A(1, 2)$ $B(1, 3)$ $C(1, 4)$
> /DESIGN = A VS 1
>   B WITHIN A = 1 VS 2
>   C WITHIN B WITHIN A = 2 VS WITHIN.

In the above example, A is tested against the error term number 1 which is defined as B WITHIN A mean square; B is tested against the error term number 2 which is defined as the C WITHIN B WITHIN A mean square; and C is tested against the usual error mean square.

Remarks: (i) The default analysis without VS specification assumes that all factors are fixed. The use of the keyword VS within the design statement is the main device available for tailoring of the analysis involving random and mixed effects models.

(ii) The assignment of error terms via VS specification are appropriate only for balanced designs. For unbalanced designs, psuedo-F tests based on Satterthwaite procedure will generally be required.

In GLM, the random or mixed effects analysis of variance is performed by a subcommand containing the keyword RANDOM followed by names of all the factors which are to be treated as random. If a factor A is specified as a random effect, then all the two-factor and higher order interaction effects containing the specified effect are automatically treated as random effects. When the RANDOM subcommand is used, the appropriate error terms for testing hypotheses concerning all the effects in the model are determined automatically. When more than one mean squares must be combined to determine the appropriate error term, a pseudo-F test based on the Satterthwaite procedure is performed. Several random effects can be specified on a single RANDOM subcommand; or one may use more than one RANDOM subcommand which have an accumulated effect. For example, to perform a two-factor random effects analysis of variance involving factors A and B, the following statements are required:

GLM Y BY A B
/DESIGN A, B, $A*B$
/RANDOM A, B.

In the above example, the hypothesis testing for each effect will be automatically carried out against the appropriate error term. Thus, the main effects A and B are tested against the $A \times B$ interaction which in turn is tested against the usual error term. Suppose, in the example above, that A is fixed and B is random. To perform a mixed model analysis, we would use the following sequence of statements:

GLM Y BY A B
/DESIGN A, B, $A*B$
/RANDOM B.

In the above example, the effects B and $A \times B$ are treated as random effects. A and B are tested against $A \times B$ interaction while $A*B$ is tested against the usual error term. In addition, GLM also allows the option of a user specified error term to test a hypothesis via the use of a subcommand TEST. To use this subcommand, the user must specify both the hypothesis term and the error term separated by the keyword VS. The hypothesis term must be a valid effect specified or implied in the DESIGN subcommand and must precede the keyword VS. The error term can be a numerical value or a linear combination of valid effects. A coefficient in a linear combination can be a real number or a fraction. If a value is specified for the error term, one must specify the number of degrees of freedom following the keyword DF. The degrees of freedom must be a positive real number. Multiple TEST subcommands are allowed and are executed independently. Thus, in the two-factor mixed model example considered above, suppose an alternate mixed model analysis is performed where both A and B main effects are to be tested against the $A \times B$ interaction. Now, the following statements are required to perform the appropriate test via the TEST subcommand:

GLM Y BY A B
/DESIGN A, B, A*B
/RANDOM B
/TEST A VS A*B
/TEST B VS A*B.

Further, suppose that B effect is to be pooled with $A \times B$ term and A is to be tested against the pooled mean square. To achieve this, the following syntax is required:

GLM Y BY A B
/DESIGN A, B, A*B
/TEST A VS B + A*B.

In addition to ONEWAY, ANOVA, MANOVA, and GLM procedures, SPSS 7.5 (and 8.0) incorporates a new procedure, VARCOMP, especially designed to estimate variance components in a random or mixed effects analysis of variance model. It can be used through syntax commands or via dialog boxes. Similar to GLM, the random factors are specified by the use of a RANDOM subcommand. There must be at least one RANDOM subcommand with one random factor; several random factors can be specified on a single RANDOM statement or one can use multiple RANDOM subcommands which have a cumulative effect. Four methods of estimation are available in the VARCOMP procedure which can be specified via the use of a METHOD subcommand. For example, in a two-factor mixed model with A fixed and B random, the following syntax will estimate the variance components using the maximum likelihood method:

VARCOMP Y BY A B
/DESIGN A, B, A*B
/RANDOM B
/METHOD = ML.

Remarks: (i) Four methods of estimation in the VARCOMP procedure are: analysis of variance (ANOVA), maximum likelihood (ML), restricted maximum likelihood (REML), and minimum norm quadratic unbiased estimator (MINQUE). The ANOVA, ML, REML, and MINQUE methods of estimation are specified by the keywords SSTYPE (n) $(n = 1$ or $3)$, REML, ML, and MINQUE (n) $(n = 0$ or $1)$ respectively on the METHOD subcommand. The default method of estimation is MINQUE(1). MINQUE(1) assigns unit weight to both the random effects and the error term while MINQUE (0) assigns zero weight to the random effects and unit weight to the error term. The ANOVA method uses Type I and Type III sums of squares designated by the keywords SSTYPE(1) and SSTYPE(3) respectively; the latter being the default option for this method.

(ii) When using ML and REML methods of estimation, the user can specify the numerical tolerance for checking singularity, convergence criterion for checking relative change in the objective function, and the maximum number of iterations by use of the following keywords on the CRITERIA subcommand: (a) EPS(n) — epsilon value used in checking singularity, $n > 0$ and the default is 1. 0E-8; (b) CONVERGE(n) — convergence value, $n > 0$ and the default is 1.0 E-8; (c) ITERATE(n) — value of the number of iterations to be performed, n must be a positive interger and the default is 50.

(iii) The user can control the display of optional output by the following keywords on the PRINT subcommand: (a) EMS — When using SSTYPE(n) on the METHOD subcommand, this option prints expected mean squares of all the effects; (b) HISTORY(n) — When using ML or REML on the METHOD subcommand, this option prints a table containing the value of the objective function and variance component estimates at every n-th iteration. The value of n is a positive integer and the default is 1; (c) SS — When using SSTYPE(n) on the METHOD subcommand, this option prints a table containing sums of squares, degrees of freedom, and mean squares for each source of variation.

11.3 ANALYSIS OF VARIANCE USING BMDP[7]

BMDP Programs are successors to BMD (biomedical) computer programs developed under the direction of W. J. Dixon at the Health Service Computing Facility of the Medical Center of the University of California during early 1960s. Since 1975 the BMDP series has virtually replaced the BMD package. The BMDP series provides the user with more flexible descriptive language, newly available statistical procedures, powerful computing algorithms, and the capability of performing repeated analyses from the same data file. The BMDP programs are arranged in six categories: data description, contingency table, multivariate methods, regression, analysis of variance, and special programs. Some of the procedures covered in the BMDP series are: contingency tables, regression analysis, nonparametric methods, robust estimators, the analysis of repeated measures, and graphical output which includes histograms, bivariate plots, normal probability plots, residual plots, and factor loading plots.

For a discussion of instructions for creating BMDP files and running BMDP programs, the reader is referred to BMDP manuals (Dixon 1992). In addition, a comprehensive volume, *Applied Statistics: A Handbook of BMDP Analyses* by Snell (1987), provides easy instructions to enable the student to use BMDP programs to analyze statistical data. The main BMDP program for performing analysis of variance is the BMDP 2V. It is a flexible general purpose program for analysis of variance, and can handle both balanced and unbalanced designs. It performs analysis of variance or covariance for a wide variety of fixed effects models. It can accommodate any number of grouping factors including repeated

---

[7] BMDP software is now owned and distributed by SPSS, Inc., Chicago, Illinois.

measures, which, however, must be crossed (not nested). The program can also distinguish between group factors from repeated measures factors. In addition, there are several other programs including 7D, 3V, and 8V that can be used to perform an analysis of variance. 7D performs one- and two-way fixed effects analysis of variance; and its output includes descriptive statistics and side-by-side histograms for all the groups, and other diagnostics for a thorough data screening including a separate tally of missing or out-of-range values. In addition, it performs Welch and Brown-Forsythe tests for homogeneity of variances and an analysis of variance based on trimmed means including confidence intervals for each group. 3V performs an analysis of variance for a general mixed model including balanced or unbalanced designs. It uses maximum and restricted maximum likelihood approaches to estimate the fixed effects and the variance components of the model. The output includes descriptive statistics for the dependent variable and for any covariate(s). It also provides for a number of other optional outputs including parameter estimates for specified hypotheses and the likelihood ratio tests with degrees of freedom and p-values. The program 8V can perform an analysis of variance for a general mixed model having balanced data. It can handle crossed, nested, and partially nested designs involving either fixed, random, or mixed effects models. The output from 8V includes an analysis of variance table with columns for expected mean squares defined in terms of variance components, the F values, including overall mean, cell means, and corresponding standard deviations, and estimates of variance components.

Generally, using the simplest possible program adequate for analyzing a given *anova* design is recommended. The programs 7D and 2V are commonly used for the fixed effects model whereas 3V and 8V can be used for the random and mixed effects models. For designs involving balanced data, 8V is recommended since it is simpler to use. For designs with unbalanced data, 3V should be used. For random and mixed effects models, in addition to performing standard analysis of variance, the program 3V also provides variance components estimates using maximum likelihood and restricted maximum likelihood procedures. In addition, BMDP has a number of other programs, namely, 3D, 1V, 4V, and 5V, which can be employed for certain special types of anova designs. 3D performs two-group tests (with equal or unequal group variances, paired or independent groups, Levene's test for the equality of group variances, trimmed t test, and the nonparametric forms of the tests including Spearman correlation and Wilcoxon's signed-rank and rank-sum tests) and 1V performs one-way analysis of covariance to test the equality of adjusted group means. 4V performs univariate and multivariate analysis of variance and covariance for a wide variety of models including repeated measures, split-plot, and cross-over designs. The output includes, among other things, a summary of descriptive statistics, multivariate statistics, and analysis of covariance. Finally, 5V analyzes repeated measures data for a wide variety of models including those with unequal variances, covariances with specified patterns, and missing data.

11.4 USE OF STATISTICAL PACKAGES FOR COMPUTING POWER

Consider an analysis of variance F test of a fixed factor effect to be performed against the critical value $F[\nu_1, \nu_2; 1 - \alpha]$. Let λ be the noncentrality parameter under the alternative hypothesis. The following SAS command can be used to calculate the required power:

$$PR = 1 - PROBF (FC, NU1, NU2, LAMBDA),$$

where PR, FC, NU1, NU2, LAMBDA are simply SAS names chosen to denote power, $F[\nu_1, \nu_2; 1 - \alpha]$, ν_1, ν_2, and λ, respectively. For an F test involving a random factor, where the distribution of the test statistic under the alternative hypothesis depends only on the (central) F distribution, the power can be calculated using the following command,

$$PR = 1 - PROBF (FC, KAPA, NU1, NU2),$$

where KAPA is the proportionality factor determined as a function of the design parameters and the values of the variance components under the alternative hypothesis.

The power calculations should generally be performed for different values of ν_1, ν_2, λ, and KAPA. This would be a useful exercise for investigating the range of power values possible for different values of the design parameters and factor size effect to be detected under the alternative. These results could then be used to set the parameters of the given analysis of variance design so as to provide adequate power in order to detect a factor effect of a given size.

In SPSS MANOVA, exact and approximate power values can be comptued via the POWER subcommand. If the POWER is specified by itself without any keyword, MANOVA calculates approximate power values of all F tests at 0.05 significant level. The following keywords are available on the POWER subcommand:

APPROXIMATE — This option calculates approximate power values which are generally accurate to three decimal places and are much more economical to calculate than the exact values. This is the default option if no keyword is used.

EXACT — This option calculates exact power values using the noncentral incomplete beta distribution.

F(a) — This option permits the specification of the alpha value at which the power is to be calculated. The value of alpha must be a number between 0 and 1, exclusive. The default alpha value is 0.05.

In SPSS GLM, the observed power for each F test at 0.05 significance level is displayed by default. An alpha value other than 0.05 can be specified by the use of the keyword ALPHA(n) in the CRITERIA subcommand. The value of n must be between 0 and 1, exclusive.

Remarks: (i) In SPSS Release 8.0, the observed power for each F test is no longer printed by default. Instead checkboxes in the Options dialog box are available for effect size estimates and observed power, and corresponding keywords ETASQ and OPOWER in the PRINT subcommand.

(ii) SPSS also has inverse distribution functions and noncentral t, χ^2, and F distributions which can be used for power analysis in the same manner as in SAS.

11.5 USE OF STATISTICAL PACKAGES FOR MULTIPLE COMPARISON PROCEDURES

Most multiple comparison tests are more readily performed manually using the results on means, sums of squares, and mean squares provided by a computer analysis. The hand computations are quite simple and the use of a computing program does not necessarily save any time. However, some of the more recent methods are more suited to a computer analysis. For all three computing packages considered here, the programs designed to perform t tests do not allow the use of a custom-made error mean square (EMS). Some programs in some packages that allow the use of a t test on a contrast also do not permit the use of a custom-made EMS. SAS provides the user the option of specifying a custom-made EMS whereas the error term employed in SPSS ONEWAY and BMDP is fixed. Most statistical packages will perform all possible pairwise comparisons as the default when the option to perform multiple comparisons is requested. More general comparisons are also available, but may require some restrictions in terms of which procedures may be used.

SAS can perform pairwise comparisons using either the PROC ANOVA or the GLM. There are 16 different multiple comparison procedures, including Tukey, Scheffé, LSD, Bonferroni, Newman-Keuls, and Duncan, which can be readily implemented using the following statement for both the procedures:

MEANS {name of independent variable}/{SAS name of the multiple comparison to be used}

For example, in order to perform Duncan's multiple comparison on a balanced one-way experimental layout using PROC ANOVA, the following commands may be used:

```
PROC ANOVA;
CLASS A;
MODEL Y = A;
MEANS A / DUNCAN;
```

Both ANOVA and GLM procedures allow the use of as many multiple comparisons as the user may want simply by listing their SAS names and separating the names with blanks. The name of the independent variable in the MEANS statement must be the name chosen for the independent variable when the data file is created. For testing more general contrasts than pairwise comparisons, the sum of squares for a contrast can be obtained by a CONTRAST command. For example, the contrast $C1$ with four coefficients can be tested by the statements:

> CLASS A;
> MODEL $Y = A$;
> CONTRAST '$C1$'
> A 1 1 $-$1 $-$1;

Any number of orthogonal contrasts can be tested by including more lines in the CONTRAST command, for example,

> CONTRAST '$C1$'
> A 1 1 $-$1 $-$1;
> CONTRAST '$C2$'
> A 1 $-$1 1 $-$1;
> CONTRAST '$C3$'
> A 1 $-$1 $-$1 1;

Remark: In Release 6.12, the LSMEANS statements in PROC GLM and PROC MIXED with ADJUST options offer several different multiple comparison procedures. In addition, PROC MULTTEST can input a set of p-values and adjust them for multiplicity.

SPSS can perform multiple comparison tests by using the ONEWAY and GLM procedures. In ONEWAY, multiple comparisons are implemented by the use of RANGES and POSTHOC subcommands. The RANGES subcommand allows for the following seven tests specified by the respective keywords: Least Significant Difference (LSD), Bonferroni (BONFERRONI), Dunn-Šidák (SIDAK), Tukey (TUKEY), Scheffé (SCHEFFE), Newman-Keuls (SNK) and Duncan (DUNCAN). (LSDMOD and MODLSD are acceptable for the Bonferroni procedure.) The RANGES subcommand provides an option for several tests in the same run by including more lines in the RANGES subcommand, e.g.,

> /RANGES = LSD
> /RANGES = TUKEY
> $\vdots$
> etc.

Remark: The LSD test does not maintain a single overall α-level. However, Bonferroni and Dunn-Šidák procedures can be performed by adjusting the overall α-level according to the number of desired comparisons and then using the LSD for the modified α-level. The default α-level for the LSD is 0.05. If some other α-level is desired, it is specified within parentheses immediately following the keyword LSD in the RANGE statement, for example,

$$/\text{RANGES} = \text{LSD } (0.0167).$$

The POSTHOC subcommand offers options for twenty different multiple comparisons including those that test the pairwise differences among means and those that identify homogenous subsets of means that are not different from each other. The latter tests are commonly known as multiple range tests. Among different types of multiple comparisons available via POSTHOC subcommand are: Bonferroni, Šidák, Tukey's honestly significant difference, Hochberg's GT2, Gabriel, Dunnett, Ryan-Einot-Gabriel-Welsch F test (R-E-G-W F), Ryan-Einot-Gabriel-Welsch range test (R-E-G-W Q), Tamhane's T2, Dunnett's T3, Games-Howell, Dunnett's C, Duncan's multiple range test, Student-Newman-Keuls (S-N-K), Tukey's b, Waller-Duncan, Scheffé, and least-significant difference. One can use as many of these tests in one run as one wants, either using one POSTHOC subcommand, or a stack of them.

Remark: Tukey's honestly significant difference test, Hochberg's GT2, Gabriel's test, and Scheffé's test are multiple comparison tests and range tests. Other available range tests are Tukey's b, S-N-K (Student-Newman-Keuls), Duncan, R-E-G-W F (Ryan-Einot-Gabriel-Welsch F test), R-E-G-W Q (Ryan-Einot-Gabriel-Welsch range test), and Waller-Duncan. Among available multiple comparison tests are Bonferroni, Tukey's honestly significant difference test, Šidák, Gabriel, Hochberg, Dunnett, Scheffé, and LSD (least significant difference). Multiple comparison tests that do not assume equal variances are Tamhane's T2, Dunnett's T3, Games-Howell, and Dunnett's C.

For performing more general contrasts than the pairwise comparisons between the means, the subcommand RANGES is replaced by CONTRAST. The coefficients defining the contrast are specified immediately after the keyword CONTRAST and separated from it by an equality sign (=) in between. The coefficients can be separated by spaces or commas. For example, the contrast with four coefficients $1, 1, -1, -1$ is tested by the statement

$$/\text{CONTRAST} = 1\ 1\ -1\ -1$$

or

$$/\text{CONTRAST} = 1, 1, -1, -1.$$

Similar to pairwise multiple comparison procedures, a number of contrasts can be selected by including more lines in the CONTRAST subcommand; for example,

$$/\text{CONTRAST} = 1\ 1\ -1\ -1$$
$$/\text{CONTRAST} = 1\ -1\ 1\ -1$$
$$/\text{CONTRAST} = 1\ -1\ -1\ 1$$
$$\vdots$$

etc.

Also, one can use both RANGES and CONTRAST subcommands following the same ONEWAY command. However, it is necessary that all the RANGES subcommands be placed consecutively followed by all the CONTRAST subcommands. The CONTRAST subcommand previously outlined is equivalent to performing a t test and is appropriate for a single a priori comparison of interest. If the researcher is interested in many simultaneous comparisons, which include complex contrasts as well as pairwise, a multiple comparison procedure based on general contrasts needs to be employed. Although Scheffé's method can be accessed by the SPSS, it can be used only for pairwise comparisons. For more complex contrasts, however, one can use the CONTRAST statement to get all the quantities needed for the Scheffé's comparison and then perform the computations manually.

Remarks: (i) One can list fractional coefficients in defining a contrast and thereby indicating that averages are in fact being tested; however, they are not required.

(ii) Comparisons of means that do not define a contrast (do not add up to zero) can be tested using the CONTRAST subcommand. SPSS will analyze a "noncontrast" but will flag a warning message that a noncontrast is being tested.

All the tests available in ONEWAY are also available in GLM (and UNIANOVA in Release 8.0) for performing multiple comparisons between the means of a factor for the dependent variable which are produced via the use of POSTHOC subcommand. The user can specify one or more effects to be tested which, however, must be a fixed main effect appearing or implied in the design subcommand. The value of type I error can be specified using the keyword ALPHA on the CRITERIA subcommand. The default alpha value is 0.05 and the default confidence level is 0.95. GLM also allows the use of an optional error term which can be defined following an effect to be tested by using the keyword VS after the test specification. The error term can be any single effect that is not the intercept or a factor tested on POSTHOC subcommand. Thus, it can be an interaction effect, and it can be fixed or random. Furthermore, GLM allows the use of multiple POSTHOC subcommands which are executed independently. This way the user can test different effects against different error terms. The output for test used for pairwise comparisons includes the difference between each pair of compared means, the confidence interval for the difference, and the significance.

GLM also allows tests for contrasts via the CONTRAST subcommand. The name of the factor is specified within a parenthesis following the subcommand CONTRAST. After enclosing the factor name within the parenthesis, one must enter an equal sign followed by one of the CONTRAST keywords. In addition to the user defined contrasts available via the keyword SPECIAL, several different types of contrasts including Helmert and polynomial contrasts are also available. Although one can specify only one factor per CONTRAST subcommand, multiple contrast subcommands within the same design are allowed. Values specified after the keyword SPECIAL are stored in a matrix in row order. For example, if factor A has three levels, then CONTRAST (A) = SPECIAL (1 1 1 1 −1 0 0 1 −1) produces the following contrast matrix:

$$\begin{bmatrix} 1 & 1 & 1 \\ 1 & -1 & 0 \\ 0 & 1 & -1 \end{bmatrix}.$$

Suppose the factor A has three levels and Y designates the dependent variable being analyzed. The following example illustrates the use of a polynomial contrast:

GLM Y BY A
/CONTRAST (A) = POLYNOMIAL (1, 2, 4).

The specified contrast indicates that the three levels of A are actually in the proportion 1:2:4. For illustration and examples of other contrast types, see SPSS Advanced Statistics 7.5 (SPSS Inc., 1997c, APPENDIX A, pp. 535–540).

Similar to the SPSS, BMDP performs multiple comparisons by using the one-way analysis of variance program. BMDP 7D contains a number of multiple comparison procedures including Tukey, Scheffé, LSD, Bonferroni, Newman-Keuls, Duncan, and Dunnett. However, the reported level of significance for comparisons are for one-sided tests and should be multiplied by two to obtain the overall significance level associated with a family of two-sided tests. BMDP 7D can perform all pairwise comparisons by using the following statement:

/COMPARISON {BMDP name of multiple comparison to be used}.

The program provides the option of several multiple comparison procedures that may be selected by including more procedure names consecutively; for example,

/COMPARISON TUKEY
SCHEFFE
BONFERRONI
NK
DUNCAN.

11.6 USE OF STATISTICAL PACKAGES FOR TESTS OF HOMOSCEDASTICITY

All three packages (SAS, SPSS, and BMDP) provide several tests of homoscedasticity or homogeneity of variances. The MEANS statement in SAS

GLM procedure now includes various options for testing homogeneity of variances in a one-way model and for computing Welch's variance-weighted one-way analysis of variance test for differences between groups when the group variances are unequal. The user can choose between Bartlett, Levene, Brown-Forsythe, and O'Brien's tests by specifying the following options in the MEANS statement:

> HOVTEST = BARTLETT
> HOVTEST = LEVENE < (TYPE = ABS/SQUARE) >
> HOVTEST = BF
> HOVTEST = OBRIEN < (W = number) >

If no test is specified in the HOVTEST option, by default Levene's test (type = square) is performed. Welch's analysis of variance is requested via option WELCH in the MEANS statement. For a simple one-way model with dependent variable Y and single factor classification A, the following code illustrates the use of HOVTEST (default) and WELCH options in the MEANS statement for the GLM procedure:

> PROC GLM;
> CLASS = A;
> MODEL $Y = A$;
> MEANS A / HOVTEST WELCH;
> RUN;

For further information and applications of homogeneity of variance tests in SAS GLM, see SAS Institute (1997, pp. 356–359).

SPSS ONEWAY provides an option to calculate Levene statistic for homogeneity of group variances. In MANOVA procedure the user can request the option HOMOGENEITY as a keyword in its PRINT subcommand. HOMOGENEITY performs tests for the homogeneity of variance of the dependent varaible across the cells of the design. One or more of the following specifications can be included in the parentheses after the keyword HOMOGENEITY:

BARTLETT — This option performs and displays Bartlett-Box F test.
COCHRAN — This option performs and displays Cochran's C test.
ALL — This option performs and displays both Bartlett-Box F test and Cochran's C test. This is the default option if HOMOGENEITY is requested without further specifications.

SPSS GLM procedure also performs Levene's test for equality of variances for the dependent variable across all cells formed by combinations of between subject factors. However, the Levene test in GLM (and UNIANOVA in Release 8.0) uses the residuals from whatever model is fitted and will differ from the description given above if the model is not a full factorial, or if there are covariates. The test can be requested using the HOMOGENEITY subcommand. BMDP 7D performs Levene's test of homogeneity of variances among the cell when performing one- and two-way analysis of variance for fixed effects.

11.7 USE OF STATISTICAL PACKAGES FOR TESTS OF NORMALITY

Tests of skewness and kurtosis can be performed using BMDP and SPSS since they provide standard errors of both statistics. SAS UNIVARIATE implements the Shapiro-Wilk W test when the sample is 2000 or less and a modified Kolmogorov-Smirnov (K-S) test when the sample size is greater than 2000. Also, SPSS EXAMINE procedure offers the K-S Lilliefors and Shapiro-Wilk tests, and NPAR TESTS offers a one-sample K-S test without the Lilliefors correction. The K-S tests in both cases are performed for sample sizes which are sufficiently large, while the Shapiro-Wilk test is given when sample size is 50 or less.

Appendices

A STUDENT'S t DISTRIBUTION

If the random variable X has a normal distribution with mean μ and variance σ^2, we denote this by $X \sim N(\mu, \sigma^2)$. Let $X_1, X_2, \ldots, X_n$ be a random sample from the $N(\mu, \sigma^2)$ distribution. Then it is known from the central limit theorem that $\bar{X} = \sum_{i=1}^{n} X_i/n \sim N(\mu, \sigma^2/n)$. Applying the technique of standardization to $\bar{X}$, we have

$$Z = \frac{\bar{X} - \mu}{\sigma/\sqrt{n}} \sim N(0, 1).$$

If σ is unknown, then it is usually replaced by $S = \sqrt{\sum_{i=1}^{n}(X_i - \bar{X})^2/(n-1)}$ and the statistic Z becomes $(\bar{X} - \mu)/(S/\sqrt{n})$. Now, we no longer have a standard normal variate but a new variable whose distribution is known as the Student's t distribution.

Before 1908, this statistic was treated as an approximate Z variate in large sample experiments. William S. Gosset, a statistician at the Guinness Brewery in Dublin, Ireland, empirically studied the distribution of the statistic $(\bar{X} - \mu)/(S/\sqrt{n})$ for small samples, when the sampling was from a normal distribution with mean μ and variance σ^2. Since the company did not allow the publication of research by its scientists, he published his findings under the pseudonym "Student." In 1923, Ronald A. Fisher theoretically derived the distribution of the statistic $(\bar{X} - \mu)/(S/\sqrt{n})$ and since then it has come to be known as the Student's t distribution. The distribution is completely determined by a single parameter $v = n - 1$, known as the degrees of freedom. The distribution has the same general form as the distribution of Z, in that they both are symmetric about a mean of zero. Both distributions are bell-shaped, but the t distribution is more variable. In large samples, the t distribution is well approximated by the standard normal distribution. It is only for small samples that the distinction between the two distributions becomes important.

We use the symbol $t[v]$ to denote a t random variable with v degrees of freedom. A $100(1 - \alpha)$th percentile of the $t[v]$ random variable is denoted by $t[v, 1 - \alpha]$ and is the point on the $t[v]$ curve for which

$$P\{t[v] \leq t[v, 1 - \alpha]\} = 1 - \alpha.$$

Useful tables of percentiles of the t distribution are given by Hald (1952), Owen (1962), Fisher and Yates (1963), and Pearson and Hartley (1970). Appendix

Table III gives certain selected values of percentiles of $t[\nu]$. The mean and variance of a $t[\nu]$ variable are:

$$E(t[\mu]) = 0$$

and

$$\text{Var}(t[\mu]) = \frac{\nu}{\nu - 2}, \quad \nu > 2.$$

B CHI-SQUARE DISTRIBUTION

If $Z_1, Z_2, \ldots, Z_\nu$ are independently distributed random variables and each Z_i has the $N(0, 1)$ distribution, then the random variable

$$V = \sum_{i=1}^{\nu} Z_i^2$$

is said to have a chi-square distribution with ν degrees of freedom. Note that if $X_1, X_2, \ldots, X_\nu$ are independently distributed and X_i has the $N(\mu_i, \sigma_i^2)$ distribution, then the random variable

$$\sum_{i=1}^{\nu} \frac{(X_i - \mu_i)^2}{\sigma_i^2}$$

also has the chi-square distribution with ν degrees of freedom.

We use the symbol $\chi^2[\nu]$ to denote a chi-square random variable with ν degrees of freedom. The chi-square random variable has the reproductive property; that is, the sum of two chi-square random variables is again a chi-square random variable. More specifically, if V_1 and V_2 are independent random variables, where

$$V_1 \sim \chi^2[\nu_1]$$

and

$$V_2 \sim \chi^2[\nu_2],$$

then

$$V_1 + V_2 \sim \chi^2[\nu_1 + \nu_2].$$

The chi-square distribution was developed by Karl Pearson in 1900. A $100(1 - \alpha)$th percentile of the $\chi^2[\nu]$ random variable is denoted by $\chi^2[\nu, 1 - \alpha]$ and is the point on the $\chi^2[\nu]$ curve for which

$$P\{\chi^2[\nu] \leq \chi^2[\nu, 1 - \alpha]\} = 1 - \alpha.$$

Useful tables of percentiles of chi-square distributions are given by Hald and Sinkbaek (1950), Vanderbeck and Cook (1961), Harter (1964a,b), and Pearson and Hartley (1970). Appendix Table IV gives certain selected values of percentiles of $\chi^2[\nu]$. The mean and variance of a $\chi^2[\nu]$ variable are:

$$E(\chi^2[\nu]) = \nu$$

and

$$\text{Var}(\chi^2[\nu]) = 2\nu.$$

C SAMPLING DISTRIBUTION OF $(n-1)S^2/\sigma^2$

Let $X_1, X_2, \ldots, X_n$ be a random sample from the $N(\mu, \sigma^2)$ distribution. Define the sample mean $\bar{X}$ and variance S^2 as

$$\bar{X} = \frac{\sum_{i=1}^{n} X_i}{n}$$

and

$$S^2 = \frac{\sum_{i=1}^{n}(X_i - \bar{X})^2}{n-1}.$$

The sampling distribution of the statistic $(n-1)S^2/\sigma^2$ is of particular interest in many statistical problems and we consider its distribution here.

By the addition and subtraction of the sample mean $\bar{X}$, it is easy to see that

$$\sum_{i=1}^{n}(X_i - \mu)^2 = \sum_{i=1}^{n}[(X_i - \bar{X}) + (\bar{X} - \mu)]^2$$

$$= \sum_{i=1}^{n}(X_i - \bar{X})^2 + \sum_{i=1}^{n}(\bar{X} - \mu)^2 + 2(\bar{X} - \mu)\sum_{i=1}^{n}(X_i - \bar{X})$$

$$= \sum_{i=1}^{n}(X_i - \bar{X})^2 + n(\bar{X} - \mu)^2.$$

Dividing each term on both sides of the equality by σ^2 and substituting $(n-1)S^2$ for $\sum_{i=1}^{n}(X_i - \bar{X})^2$, we obtain

$$\frac{1}{\sigma^2}\sum_{i=1}^{n}(X_i - \mu)^2 = \frac{(n-1)S^2}{\sigma^2} + \frac{(\bar{X} - \mu)^2}{\sigma^2/n}.$$

Now, from Appendix B, it follows that $\sum_{i=1}^{n}(X_i - \mu)^2/\sigma^2$ is a chi-square random variable with n degrees of freedom. The second term on the right-hand side of the equality is the square of a standard normal variate, since $\bar{X}$ is a normal random variable with mean μ and variance σ^2/n. Therefore, the quantity $(\bar{X} - \mu)^2/(\sigma^2/n)$ has a chi-square distribution with one degree of freedom. Using advanced statistical techniques, one can also show that the two chi-square variables, $(n-1)S^2/\sigma^2$ and $(\bar{X} - \mu)^2/(\sigma^2/n)$, are independent (see, e.g., Hogg and Craig, (1995, pp. 214–217)). Thus, from the reproductive property of the chi-square variable, it follows that $(n-1)S^2/\sigma^2$ has a chi-square distribution with $n-1$ degrees of freedom.

D F DISTRIBUTION

If V_1 and V_2 are two independent random variables, where $V_1 \sim \chi^2[\nu_1]$ and $V_2 \sim \chi^2[\nu_2]$, then the random variable

$$\frac{V_1/\nu_1}{V_2/\nu_2}$$

is said to have an F distribution with ν_1 and ν_2 degrees of freedom, respectively. We use the symbol $F[\nu_1, \nu_2]$ to denote an F variable with ν_1 and ν_2 degrees of freedom.

Suppose a random sample of size n_1 is drawn from the $N(\mu_1, \sigma_1^2)$ and an independent random sample of size n_2 is drawn from the $N(\mu_2, \sigma_2^2)$ distribution. If S_1^2 and S_2^2 are the corresponding sample variances, then, from Appendix C, it follows that

$$\frac{(n_1 - 1)S_1^2}{\sigma_1^2} \sim \chi^2[n_1 - 1]$$

and

$$\frac{(n_2 - 1)S_2^2}{\sigma_2^2} \sim \chi^2[n_2 - 1].$$

Therefore, the quotient

$$\frac{S_1^2/\sigma_1^2}{S_2^2/\sigma_2^2} \sim \frac{\chi^2[n_1 - 1]/(n_1 - 1)}{\chi^2[n_2 - 1]/(n_2 - 1)} \sim F[n_1 - 1, n_2 - 1].$$

Fisher (1924) considered the theoretical distribution of $\frac{1}{2}\log_e(S_1^2/S_2^2)$ known as the Z distribution. The distribution of the variance ratio F was derived by Snedecor (1934), who showed that the distribution of F was simply a transformation of Fisher's Z distribution. Snedecor named the distribution of the variance ratio F in honor of Fisher. The distribution has subsequently come to be known as Snedecor's F distribution.

Note that the number of degrees of freedom associated with the chi-square random variable appearing in the numerator of F is always stated first, followed by the number of degrees of freedom associated with the chi-square random variable appearing in the denominator. Thus, the curve of the F distribution depends not only on the two parameters v_1 and v_2, but also on the order in which they occur.

A $100(1 - \alpha)$th percentile value of the $F[v_1, v_2]$ variable is denoted by $F[v_1, v_2; 1 - \alpha]$ and is that point on the $F[v_1, v_2]$ curve for which

$$P\{F[v_1, v_2] \leq F[v_1, v_2; 1 - \alpha]\} = 1 - \alpha.$$

Useful tables of percentiles of the F distribution are given by Hald (1952), Fisher and Yates (1963), and Pearson and Hartley (1970). Mardia and Zemroch (1978) compiled tables of F distributions that include fractional values of v_1 and v_2. The fractional values of the degrees of freedom are useful when an F distribution is used as an approximation. Appendix Table V gives certain selected values of percentiles of $F[v_1, v_2]$ for various sets of values of v_1 and v_2. The mean and variance of an $F[v_1, v_2]$ variable are:

$$E(F[v_1, v_2]) = \frac{v_1}{v_2 - 2}, \quad v_2 > 2$$

and

$$\text{Var}(F[v_1, v_2]) = \frac{2v_2^2(v_1 + v_2 - 2)}{v_1(v_2 - 2)^2(v_2 - 4)}, \quad v_2 > 4.$$

E NONCENTRAL CHI-SQUARE DISTRIBUTION

If $X_1, X_2, \ldots, X_v$ are independently distributed random variables and each X_i has the $N(\mu_i, 1)$ distribution, then the random variable

$$V = \sum_{i=1}^{v} X_i^2$$

is said to have a noncentral chi-square distribution with v degrees of freedom. The quantity $\lambda = (\sum_{i=1}^{v} \mu_i^2)^{\frac{1}{2}}$ is known as the noncentrality parameter of the distribution. We use the symbol $\chi^{2'}[v, \lambda]$ to denote a noncentral chi-square random variable with v degrees of freedom and the noncentrality parameter λ.

Note that the ordinary or central chi-square distribution is the special case of the noncentral distribution when the nonncentrality parameter $\lambda = 0$. In the statistical literature the noncentrality parameter is sometimes defined differently. Some authors use $\lambda = \sum_{i=1}^{v} \mu_i^2$ whereas others use $\lambda = \frac{1}{2} \sum_{i=1}^{v} \mu_i^2$, both using the same symbol λ. The noncentral chi-square variable, like the central chi-square, possesses the reproductive property; that is, the sum of two noncentral chi-square variables is again a noncentral chi-square variable. More specifically, if V_1 and V_2 are independent random variables, such that

$$V_1 \sim \chi^{2'}[\nu_1, \lambda_1]$$

and

$$V_2 \sim \chi^{2'}[\nu_2, \lambda_2],$$

then

$$V_1 + V_2 \sim \chi^{2'}[\nu_1 + \nu_2, \lambda_1 + \lambda_2].$$

The noncentral chi-square distribution can be approximated in terms of a central chi-square distribution. A detailed summary of various approximations including a great deal of other information can be found in Tiku (1985a) and Johnson et al. (1995, Chapter 29). Tables of the noncentral chi-square distribution have been prepared by Hayman et al. (1973). The mean and variance of a $\chi^{2'}[\nu, \lambda]$ variable are:

$$E(\chi^{2'}[\nu, \lambda]) = \nu + \lambda$$

and

$$\mathrm{Var}(\chi^{2'}[\nu, \lambda]) = 2(\nu + 2\lambda).$$

F NONCENTRAL AND DOUBLY NONCENTRAL t DISTRIBUTIONS

If U and V are two independent random variables, where $U \sim N[\delta, 1]$ and $V \sim \chi^2[\nu]$, then the random variable

$$U/\sqrt{V/\nu}$$

is said to have a noncentral t distribution with ν degrees of freedom and the noncentrality parameter δ. We use the symbol $t'[\nu, \delta]$ to denote a noncentral t distribution with ν degrees of freedom and the noncentrality parameter δ. The distribution is useful in evaluating the power of the t test. There are many approximations of the noncentral t distribution in terms of normal and (central) t distributions. A detailed summary of various approximations and other results can be found in Johnson et al. (1995, Chapter 31). A great deal of other information about the noncentral t distribution is given in Owen (1968, 1985). Tables of the noncentral t distribution have been prepared by Resnikoff and Lieberman (1957) and Bagui (1993). The mean and variance of a $t'[\nu, \delta]$

variable are:

$$E(t'[\nu, \delta]) = \frac{\left(\frac{\nu}{2}\right)^{1/2} \Gamma\left(\frac{\nu-1}{2}\right) \delta}{\Gamma\left(\frac{\nu}{2}\right)}$$

and

$$\text{Var}(t'[\nu, \delta]) = \frac{\nu}{\nu-2}(1+\delta^2) - \left[\frac{\left(\frac{\nu}{2}\right)^{1/2} \Gamma\left(\frac{\nu-1}{2}\right) \delta}{\Gamma\left(\frac{\nu}{2}\right)}\right]^2.$$

If V is a noncentral chi-square distribution with the noncentrality parameter λ, then

$$U/\sqrt{V/\nu}$$

has a doubly noncentral t distribution with noncentrality parameters δ and λ, respectively. Tables of the doubly noncentral t distribution are given by Bulgren (1974). Further information including analytic expressions for the distribution function and some computational aspects can be found in Johnson et al. (1995, pp. 533–537) and references cited therein.

G NONCENTRAL F DISTRIBUTION

If V_1 and V_2 are two independent random variables, where $V_1 \sim \chi^{2'}[\nu_1, \lambda]$ and $V_2 \sim \chi^2[\nu_2]$, then the random variable

$$\frac{V_1/\nu_1}{V_2/\nu_2}$$

is said to have a noncentral F distribution with ν_1 and ν_2 degrees of freedom and the noncentrality parameter λ. We use the symbol $F'[\nu_1, \nu_2; \lambda]$ to denote a noncentral F variable with ν_1 and ν_2 degrees of freedom and the noncentrality parameter λ.

It is sometimes useful to approximate a noncentral F distribution in terms of a (central) F distribution. A detailed summary of various approximations including a great deal of other information can be found in Tiku (1985b) and Johnson et al. (1995, Chapter 30). Comprehensive tables of the noncentral F distribution were prepared by Tiku (1967, 1972). Tables and charts of the noncentral F distribution are discussed in Section 2.17. The mean and variance of an $F'[v_1, v_2; \lambda]$ variable are:

$$E(F'[v_1, v_2; \lambda]) = \frac{v_2(v_1 + \lambda)}{v_1(v_2 - 2)}, \quad v_2 > 2$$

and

$$\text{Var}(F'[v_1, v_2; \lambda]) = \frac{2v_2^2\{(v_2 - 2)(v_1 + 2\lambda) + (v_1 + \lambda)^2\}}{v_1^2(v_2 - 2)^2(v_2 - 4)}, \quad v_2 > 4.$$

H DOUBLY NONCENTRAL F DISTRIBUTION

If V_1 and V_2 are two independent random variables, where $V_1 \sim \chi^{2'}[v_1, \lambda_1]$ and $V_2 \sim \chi^{2'}[v_2, \lambda_2]$, then the random variable

$$\frac{V_1/v_1}{V_2/v_2}$$

is said to have a doubly noncentral F distribution. We use the symbol $F''[v_1, v_2; \lambda_1, \lambda_2]$ to denote a doubly noncentral F variable with v_1 and v_2 degrees of freedom and noncentrality parameters λ_1 and λ_2. Thus, the doubly noncentral F distribution is the ratio of two independent variables, each distributed as a noncentral chi-square, divided by their respective degrees of freedom. Tables of the doubly noncentral F distribution were given by Tiku (1974). Further discussions and details about the distribution including applications in contexts other than analysis of variance can be found in Tiku (1974) and Johnson et al. (1995, pp. 499–502).

The doubly noncentral F distribution is related to the doubly noncentral beta in the same way as the (central) beta to the (central) F. It is the distribution of $V_1/(V_1 + V_2)$.

I STUDENTIZED RANGE DISTRIBUTION

Suppose that $(X_1, X_2, \ldots, X_p)$ is a random sample from a normal distribution with mean μ and variance σ^2. Suppose further that S^2 is an unbiased estimate of σ^2 based upon v degrees of freedom. Then the ratio

$$q[p, v] = \{\max(X_i) - \min(X_i)\}/S$$

is called the Studentized range, where the arguments in the square bracket indicate that the distribution of q depends on p and v. In general, the Studentized range arises as the ratio of the range of a sample of size p from a standard normal population to the square root of an independent $\chi^2[v]/v$ variable with v degrees of freedom. In analysis of variance applications, normal samples are usually means of independent samples of the same size, and the denominator is an independent estimate of their common standard error. The sampling distribution of $q[p, v]$ has been tabulated by various workers and good tables are available in Harter (1960), Owen (1962), Pearson and Hartley (1970), and Miller (1981). Perhaps the most comprehensive table of percentiles of the Studentized range distribution is Table B2 given in Harter (1969a), which has $p = 2\,(1)\,20\,(2)\,40\,(10)\,160$; $v = 1\,(1)\,20, 24, 30, 40, 60, 120, \infty$; and upper-tail $\alpha = 0.001, 0.005, 0.01, 0.025, 0.05, 0.1\,(0.1)\,0.9, 0.95, 0.975, 0.99, 0.995, 0.999$. Some selected percentiles of $q[p, v]$ are given in Appendix Table X. The table is fairly easy to use. Let $q[p, v; 1 - \alpha]$ denote the $100(1 - \alpha)$th percentile of the $q[p, v]$ variable. Suppose $p = 10$ and $v = 20$. The 90th percentile of the studentized range distribution is then given by

$$q[10;\ 20;\ 0.90] = 4.51.$$

Thus, with 10 normal observations from a normal population, the probability is 0.90 that their range is not more than 4.51 times as great as an independent sample standard deviation based on 20 degrees of freedom.

J STUDENTIZED MAXIMUM MODULUS DISTRIBUTION

The Studentized maximum modulus is the maximum absolute value of a set of independent unit normal variates which is then Studentized by the standard deviation. Thus, let $X_1, X_2, \ldots, X_p$ be a random sample from the $N(\mu, \sigma^2)$ distribution. Then the Studentized maximum modulus statistic is defined by

$$m[p, v] = \frac{\max |X_i - \bar{X}|}{S},$$

where $\bar{X} = \sum_{i=1}^{p} X_i/p$ and $S^2 = \sum_{i=1}^{p}(X_i - \bar{X})^2/(p-1)$. For the case where S^2 represents an independent estimate of σ^2 such that vS^2/σ^2 has a chi-square distribution, the distribution was first derived and tabulated by Nair (1948). The critical points for the studentized maximum modulus distribution can also be obtained by taking the square roots of the entries in the tables of the Studentized largest chi-square distribution with one degree of freedom for the numerator as given by Armitage and Krishnaiah (1964). Pillai and Ramachandran (1954) gave a table for $\alpha = 0.05$; $p = 1(1)8$; and $v = 5(5)20, 24, 30, 40, 60, 120, \infty$. Dunn and Massey (1965) provided a table for $\alpha = 0.01, 0.025, 0.05, 0.10(0.1)0.50$; $p = 2, 6, 10, 20$; and $v = 4, 10, 30, \infty$. Hahn and Hendrickson (1971) give general tables of percentiles of $m[p, v]$ for $\alpha = 0.01, 0.05, 0.10$; $p = (1)(1)6(2)12, 15, 20$; and $v = 3(1)12, 15, 20, 25, 30, 40, 60$, which also appear in Miller (1981, pp. 277–278). Stoline and Ury (1979) and Ury et al. (1980) give special tables with $p = a(a - 1)/2$ for $a = 2(1)20$ and $v = 2(2)50(5)80(10)100$, respectively. Bechhoefer and Dunnett (1982) have provided tables of the distribution of $m[p, v]$ for $p = 2(1)\,32$; $v = 2(1)12(2)20, 24, 30, 40, 60, \infty$; and $\alpha = 0.10, 0.05$, and 0.01. These tables are abridged versions of more extensive tables given by Bechhoefer and Dunnett (1981). Hochberg and Tamhane (1987) also give tables with $p = a(a-1)/2$ for $a = 2(1)16(2)\ 20$ and $\nu = 2(1)\ 30(5)$ $50(10)\ 60(20)\ 120,\ 200,\ \infty$. Some selected percentiles of $m[p, \nu]$ are given in Appendix Table XV.

K SATTERTHWAITE PROCEDURE AND ITS APPLICATION TO ANALYSIS OF VARIANCE

Many analysis of variance applications involve a linear combination of mean squares. Let S_i^2 ($i = 1, 2, \ldots, p$) be p mean squares such that $\nu_i S_i^2/\sigma_i^2$ has a chi-square distribution with ν_i degrees of freedom. Consider the linear combination $S^2 = \sum_{i=1}^{p} \ell_i S_i^2$ where ℓ_i's are known constants. Satterthwaite (1946) procedure states that $\nu S^2/\sigma^2$ is distributed approximately as a chi-square distribution with ν degrees of freedom where $\sigma^2 = E(S^2)$ and ν is determined by

$$\nu = \frac{\left(\sum_{i=1}^{p} \ell_i S_i^2\right)^2}{\sum_{i=1}^{p} (\ell_i S_i^2)^2/\nu_i}.$$

Satterthwaite procedure is frequently employed for constructing confidence intervals for the mean and the variance components in a random and mixed effects analysis of variance. For example, if a variance component σ^2 is estimated by $S^2 = \sum_{i=1}^{p} \ell_i S_i^2$, then an approximate $100(1-\alpha)\%$ confidence interval for σ^2 is given by

$$\frac{\nu S^2}{\chi^2[\nu, \alpha/2]} < \sigma^2 < \frac{\nu S^2}{\chi^2[\nu, 1-\alpha/2]},$$

where $\chi^2[\nu, \alpha/2]$ and $\chi^2[\nu, 1-\alpha/2]$ are the $100(\alpha/2)$th lower and upper percentiles of the chi-square distribution with ν degrees of freedom and ν is determined by the formula given previously.

Another application of the Satterthwaite procedure involves the construction of a psuedo-F test when an exact F test cannot be found from the ratio of two mean squares. In such cases, one can form linear combinations of mean squares for the numerator, for the denominator, or for both the numerator and the denominator such that their expected values are equal under the null hypothesis. For example, let

$$\text{MS}' = \ell_r S_r^2 + \cdots + \ell_s S_s^2$$

and

$$\text{MS}'' = \ell_u S_u^2 + \cdots + \ell_v S_v^2,$$

where the mean squares are chosen such that $E(\text{MS}') = E(\text{MS}'')$ under the null hypothesis that a particular variance component is zero. Now, an approximate F test of the null hypothesis can be obtained by the statistic

$$F = \frac{\text{MS}'}{\text{MS}''},$$

which has an approximate F distribution with v' and v'' degrees of freedom determined by

$$v' = \frac{\left(\ell_r S_r^2 + \cdots + \ell_s S_s^2\right)^2}{\ell_r^2 S_r^4/v_r + \cdots + \ell_s^2 S_s^4/v_s}$$

and

$$v'' = \frac{\left(\ell_u S_u^2 + \cdots + \ell_v S_v^2\right)^2}{\ell_u^2 S_u^4/v_u + \cdots + \ell_v^2 S_v^4/v_v}.$$

In many situations, it may not be necessary to approximate both the numerator and the denominator mean squares for an approximate F test. However, when both the numerator and the denominator mean squares are constructed, it is always possible to find additive combinations of mean squares, and thereby avoid subtracting mean squares which may result in a poor approximation. For some further discussions of psuedo-F tests, see Anderson (1960) and Eisen (1966).

In many applications of the Satterthwaite procedure, some of the mean squares may involve negative coefficients. Satterthwaite remarked that care should be exercised in applying the approximation when some of the coefficients may be negative. When negative coefficients are involved, one can rewrite the linear combination as $S^2 = S_A^2 - S_B^2$, where S_A^2 contains all the mean squares with positive coefficients and S_B^2 with negative coefficients. Now, the degrees of freedom associated with the approximate chi-square distribution of S^2 are determined by

$$f = \left(S_A^2 - S_B^2\right)^2 / \left(S_A^4/f_A + S_B^4/f_B\right),$$

where f_A and f_B are the degrees of freedom associated with the approximate chi-square distributions of S_A^2 and S_B^2, respectively. Gaylor and Hopper (1969) showed that Satterthwaite approximation for S^2 with f degrees of freedom is an adequate one when

$$S_A^2/S_B^2 > F[f_B, f_A; 0.975] \times F[f_A, f_B; 0.5]$$

if $f_A \leq 100$ and $f_B \geq f_A/2$. The approximation is usually adequate for the differences of mean squares when the mean squares being subtracted are relatively small. Khuri (1995) gives a necessary and sufficient condition for the Satterthwaite approximation to be exact in balanced mixed models.

L COMPONENTS OF VARIANCE

In discussing Models II and III, we have introduced variances corresponding to the random effects terms in the analysis of variance model. These have been designated "components of variance" since they represent the parts of the total variation that can be ascribed to these sources. The variance components are associated with random effects and appear in both random and mixed models.

Variance components were first employed by Fisher (1918) in connection with genetic research on Mendelian laws of inheritance. They have been widely used in evaluating the precision of instruments and, in general, are useful in determining the variables that contribute most to the variability of the process or the different sources contributing to the variation in an observation. This permits corrective actions that can be taken to reduce the effects of these variables. Another use has been described by Cameron (1951), who used variance components in evaluating the precision of estimating the clean content of wool. Kussmaul and Anderson (1967) describe the application of variance components for analyzing composite samples, which are obtained by pooling data from individual samples.

There is a large body of literature related to variance components, which cover results on hypothesis testing, point estimation, and confidence intervals, and fairly complete bibliographies are provided by Sahai (1979), Sahai et al. (1985), and Singhal et al. (1988). Additional works of interest include survey papers by Crump (1951), Searle (1971a, 1995), Khuri and Sahai (1985), and Burdick and Graybill (1988), including texts and monographs by Rao and Kleffe (1988), Burdick and Graybill (1992), Searle et al. (1992), and Rao (1997).

M INTRACLASS CORRELATION

In the random effects model

$$y_{ij} = \mu + \alpha_i + e_{ij}, \quad i = 1, 2, \ldots, a; \; j = 1, 2, \ldots, n,$$

μ is considered to be a fixed constant and α_i's and the e_{ij}'s are independently distributed random variables with mean zero and variances σ_α^2 and σ_e^2, respectively. Thus, as a part of the model,

$$\begin{aligned} E(y_{ij}) &= E(\mu) + E(\alpha_i) + E(e_{ij}) \\ &= \mu + 0 + 0 \\ &= \mu \end{aligned}$$

and

$$\text{Var}(y_{ij}) = \text{Var}(\mu) + \text{Var}(\alpha_i) + \text{Var}(e_{ij})$$
$$= 0 + \sigma_\alpha^2 + \sigma_e^2$$
$$= \sigma_\alpha^2 + \sigma_e^2.$$

The covariance structure of the model may be represented as follows:

$$\text{Cov}(y_{ij}, y_{i'j'}) = \begin{cases} 0, & \text{if } i \ne i' \\ \sigma_e^2 + \sigma_\alpha^2, & \text{if } i = i', \ j = j' \\ \sigma_\alpha^2, & \text{if } i = i', \ j \ne j'. \end{cases}$$

The intraclass correlation is then defined by

$$\rho = \frac{\text{Cov}(y_{ij}, y_{ij'})}{\sqrt{\text{Var}(y_{ij})}\sqrt{\text{Var}(y_{ij})}} = \frac{\sigma_\alpha^2}{\sigma_e^2 + \sigma_\alpha^2}, \quad j \ne j'.$$

Thus, ρ is the correlation between the pair of individuals belonging to the same class and has the range of values from $-1/(n-1)$ to 1. The intraclass correlation was first introduced by Fisher (1918) as a measure of the correlation between the members of the same family, group, or class. It can be interpreted as the proportion of the total variability due to the differences in all possible treatment groups of this type. The intraclass correlation coefficient is a parameter that has been studied classically in statistics (see, e.g., Kendall and Stuart (1961, pp. 302–304)). It has found extensive applications in several different fields of study including use as a measure of the degree of familial resemblance with respect to biological and environmental characteristics. It also plays an important role in reliability theory involving observations on a sample of various judges or raters, and in sensitivity analysis where it has been used to measure the efficacy of an experimental treatment. For a review of inference procedures for the intraclass correlation coefficient in the one-way random effects model, see Donner (1986).

N ANALYSIS OF COVARIANCE

The analysis of covariance is a combination of analysis of variance and regression. In analysis of variance, all the factors being studied are treated qualitatively and in analysis of regression all the factors are treated quantitatively. In analysis of covariance, some factors are treated qualitatively and some are treated quantitatively. The term independent variable often refers to a factor treated quantitatively in analysis of covariance and regression. The term covariate or concomitant variable is also used to denote an independent variable in an analysis of covariance. The analysis of covariance involves adjusting the observed value of the response or dependent variable for the linear effect of the concomitant variable. If such an adjustment for the effects of the concomitant variable is not made, the estimate of the error mean square would be inflated which would make the analysis of variance test less sensitive. The adjustment or elimination of the linear effect of the concomitant variable generally results in a small mean square. The analysis of covariance uses regression analysis techniques for elimination of the linear effect of the concomitant variable. The technique was originally introduced by Fisher (1932) and Cochran (1957) who presented a detailed account of the subject. Analysis of covariance techniques, however, are generally complicated and are often considered to be one of the most misunderstood and misused statistical techniques commonly employed by researchers. A readable account of the subject is given in Snedecor and Cochran (1989, Chapter 18) and Winer et al. (1991, Chapter 10). For a more mathematical treatment of the topic, see Scheffé (1959, Chapter 6). An extended expository review of the analysis of covariance is contained in a set of seven papers which appeared in a special issue of *Biometrics* (Vol. 13, No. 3, 1957). A subsequent issue of the same journal (Vol. 38, No. 3, 1982) includes discussion of complex designs and nonlinear models. Further discussions on analysis of covariance can be found in a series of papers appearing in a special issue of *Communications in Statistics: Part A, Theory and Methods* (Vol. 8, No. 8, 1979). For a book-length account of the subject, see Huitema (1980).

O EQUIVALENCE OF THE ANOVA F AND TWO-SAMPLE t TESTS

In this appendix, it is shown that in a one-way classification, the analysis of variance F test is equivalent to the two-sample t test. Consider a one-way classification with a groups and let the i-th group contain n_i observations with $N = \sum_{i=1}^{a} n_i$. Let y_{ij} be the j-th observation from the i-th group ($i = 1, 2, \ldots, a; j = 1, 2, \ldots, n_i$). The F statistic for testing the hypothesis of the equality of treatment means is defined by

$$F = \frac{\text{MS}_B}{\text{MS}_W} = \frac{\text{SS}_B/df_B}{\text{SS}_W/df_W},$$

where

$$\text{SS}_B = \sum_{i=1}^{a} n_i (\bar{y}_{i.} - \bar{y}_{..})^2,$$

$$\text{SS}_W = \sum_{i=1}^{a} \sum_{j=1}^{n_i} (y_{ij} - \bar{y}_{i.})^2,$$

$$df_B = a - 1 \quad \text{and} \quad df_W = N - a.$$

Appendices

For the case of two groups (i.e., $a = 2$), it follows that

$$SS_B = n_1(\bar{y}_{1.} - \bar{y}_{..})^2 + n_2(\bar{y}_{2.} - \bar{y}_{..})^2,$$
$$SS_W = (n_1 - 1)S_1^2 + (n_2 - 1)S_2^2,$$
$$df_B = 1, \quad \text{and} \quad df_W = n_1 + n_2 - 2,$$

where

$$S_1^2 = \frac{\sum_{j=1}^{n_1}(y_{1j} - \bar{y}_{1.})^2}{n_1 - 1}.$$

and

$$S_2^2 = \frac{\sum_{j=1}^{n_2}(y_{2j} - \bar{y}_{2.})^2}{n_2 - 1}.$$

Furthermore, noting that

$$\bar{y}_{..} = \frac{n_1 \bar{y}_{1.} + n_2 \bar{y}_{2.}}{n_1 + n_2},$$

we have

$$(\bar{y}_{1.} - \bar{y}_{..})^2 = \frac{n_2^2(\bar{y}_{1.} - \bar{y}_{2.})^2}{(n_1 + n_2)^2}$$

and

$$(\bar{y}_{2.} - \bar{y}_{..})^2 = \frac{n_1^2(\bar{y}_{1.} - \bar{y}_{2.})^2}{(n_1 + n_2)^2}.$$

Now, making the substitution, SS_B can be written as

$$SS_B = \frac{n_1 n_2^2(\bar{y}_{1.} - \bar{y}_{2.})^2 + n_1^2 n_2(\bar{y}_{1.} - \bar{y}_{2.})^2}{(n_1 + n_2)^2}$$

$$= \frac{n_1 n_2(\bar{y}_{1.} - \bar{y}_{2.})^2}{n_1 + n_2}$$

$$= \frac{(\bar{y}_{1.} - \bar{y}_{2.})^2}{\frac{1}{n_1} + \frac{1}{n_2}}.$$

Finally, the statistic F can be written simply as

$$F = \frac{(\bar{y}_{1.} - \bar{y}_{2.})^2 / \left(\frac{1}{n_1} + \frac{1}{n_2}\right)}{\{(n_1-1)S_1^2 + (n_2-1)S_2^2\}/(n_1+n_2-2)}$$

$$= \frac{(\bar{y}_{1.} - \bar{y}_{2.})^2}{S_p^2 \left(\frac{1}{n_1} + \frac{1}{n_2}\right)},$$

where

$$S_p^2 = \frac{(n_1-1)S_1^2 + (n_2-1)S_2^2}{n_1+n_2-2}.$$

Since a two-sample t statistic with $n_1 + n_2 - 2$ degrees of freedom is defined by

$$t = \frac{|\bar{y}_{1.} - \bar{y}_{2.}|}{S_p \sqrt{\frac{1}{n_1} + \frac{1}{n_2}}},$$

it follows that $F = t^2$.

P EQUIVALENCE OF THE ANOVA F AND PAIRED t TESTS

In this appendix, it is shown that in a randomized block design with two treatments, the analysis of variance F test is equivalent to the paired t test. Consider a randomized block design with n blocks and t treatments. Let y_{ij} be the observation from the i-th treatment and the j-th block ($i = 1, 2, \ldots, t$; $j = 1, 2, \ldots, n$). The F statistic for testing the hypothesis of the equality of treatment means is defined by

$$F = \frac{MS_\tau}{MS_E} = \frac{SS_\tau / df_\tau}{SS_E / df_E},$$

where

$$SS_\tau = n \sum_{i=1}^{t} (\bar{y}_{i.} - \bar{y}_{..})^2,$$

$$SS_E = \sum_{i=1}^{t} \sum_{j=1}^{n} (y_{ij} - \bar{y}_{i.} - \bar{y}_{.j} + \bar{y}_{..})^2,$$

$$df_\tau = t - 1, \quad \text{and} \quad df_E = (t-1)(n-1).$$

Appendices

For the case of two treatments (i.e., $t = 2$), it follows that

$$SS_\tau = n[(\bar{y}_{1.} - \bar{y}_{..})^2 + (\bar{y}_{2.} - \bar{y}_{..})^2],$$

$$SS_E = \sum_{j=1}^{n} [(y_{1j} - \bar{y}_{1.} - \bar{y}_{.j} + \bar{y}_{..})^2 + (y_{2j} - \bar{y}_{2.} - \bar{y}_{.j} + \bar{y}_{..})^2],$$

$$df_\tau = 1, \quad \text{and} \quad df_E = n - 1.$$

Furthermore, noting that

$$\bar{y}_{.j} = \frac{y_{1j} + y_{2j}}{2}$$

and

$$\bar{y}_{..} = \frac{\bar{y}_{1.} + \bar{y}_{2.}}{2},$$

we have

$$(\bar{y}_{1.} - \bar{y}_{..})^2 = \frac{(\bar{y}_{1.} - \bar{y}_{2.})^2}{4},$$

$$(\bar{y}_{2.} - \bar{y}_{..})^2 = \frac{(\bar{y}_{1.} - \bar{y}_{2.})^2}{4},$$

$$(y_{1j} - \bar{y}_{1.} - \bar{y}_{.j} + \bar{y}_{..})^2 = \frac{[(y_{1j} - y_{2j}) - (\bar{y}_{1.} - \bar{y}_{2.})]^2}{4},$$

and

$$(y_{2j} - \bar{y}_{2.} - \bar{y}_{.j} + \bar{y}_{..})^2 = \frac{[(y_{1j} - y_{2j}) - (\bar{y}_{1.} - \bar{y}_{2.})]^2}{4}.$$

Now, making the substitution, SS_τ and SS_E can be written as

$$SS_\tau = \frac{n}{2}(\bar{y}_{1.} - \bar{y}_{2.})^2$$

and

$$SS_E = \frac{1}{2} \sum_{j=1}^{n} [(y_{1j} - y_{2j}) - (\bar{y}_{1.} - \bar{y}_{2.})]^2.$$

Again, letting $d_j = y_{1j} - y_{2j}$, $\bar{d} = \sum_{j=1}^{n} d_j/n = \bar{y}_{1.} - \bar{y}_{2.}$, we obtain

$$SS_\tau = \frac{n}{2}(\bar{d})^2$$

and

$$SS_E = \frac{1}{2}\sum_{j=1}^{n}(d_j - \bar{d})^2.$$

Finally, the statistic F can be written simply as

$$F = \frac{\frac{n}{2}(\bar{d})^2/1}{\frac{1}{2}\sum_{j=1}^{n}(d_j - \bar{d})^2/(n-1)}$$

$$= \frac{(\bar{d})^2}{S_d^2/n},$$

where

$$S_d^2 = \frac{\sum_{j=1}^{n}(d_j - \bar{d})^2}{n-1}.$$

Since a paired t statistic with $n-1$ degrees of freedom is defined by

$$t = \frac{\bar{d}}{S_d/\sqrt{n}},$$

it follows that $F = t^2$.

Q EXPECTED VALUE AND VARIANCE

If X is a discrete random variable with probability function $p(x)$, the expected value of X, denoted by $E(X)$, is defined as

$$E(X) = \sum_{i=1}^{\infty} x_i p(x_i),$$

provided that $\sum_{i=1}^{\infty} x_i p(x_i) < \infty$. If the series diverges, the expected value is undefined. If X is a continuous random variable with probability density function $f(x)$, then

$$E(X) = \int_{-\infty}^{\infty} x f(x)\,dx,$$

provided that $\int_{-\infty}^{\infty} |x| f(x)\, dx < \infty$. If the integral diverges, the expected value is undefined. $E(X)$ is also referred to as the mathematical expectation or the mean value of X and is often denoted by μ.

The expected value of a random variable is its average value and can be considered as the center of the distribution. For any constants $a, b_1, b_2, \ldots, b_k$, and the random variables $X_1, X_2, \ldots, X_k$, the following properties hold:

$$E(a) = a,$$
$$E(b_i X_i) = b_i E(X_i),$$

and

$$E\left(a + \sum_{i=1}^{k} b_i X_i\right) = a + \sum_{i=1}^{k} b_i E(X_i).$$

If X is a random variable with expected value $E(X)$, the variance of X, denoted by $\text{Var}(X)$, is defined as

$$\text{Var}(X) = E[X - E(X)]^2,$$

provided the expectation exists. The square root of the variance is known as the standard deviation. The variance is often denoted by σ^2 and the standard deviation by σ.

The variance of a random value is the average or expected value of squared deviations from the mean and measures the variation around the mean value. For any constants a and b, and the random variable X, the following properties hold:

$$\text{Var}(a) = 0,$$
$$\text{Var}(bX) = b^2 \text{Var}(X),$$

and

$$\text{Var}(a + bX) = b^2 \text{Var}(X).$$

R COVARIANCE AND CORRELATION

If X and Y are jointly distributed random variables with expected values $E(X)$ and $E(Y)$, the covariance of X and Y, denoted by $\text{Cov}(X, Y)$, is defined as

$$\text{Cov}(X, Y) = E[(X - E(X))(Y - E(Y))].$$

The covariance is the average value of the products of the deviations of the values of X from its mean and the deviations of the values of Y from its mean.

It can be readily shown that

$$\text{Cov}(X, Y) = E(XY) - E(X)E(Y).$$

The variance of a random variable is a measure of its variability and the covariance of two random variables can be considered as a measure of their joint variability or the degree of association. For any constants a, b, c, d, and the random variables X, Y, U, V, the following properties hold:

$$\text{Cov}(a, X) = 0,$$
$$\text{Cov}(aX, bY) = ab\,\text{Cov}(X, Y),$$

and

$$\text{Cov}(aX + bY, cU + dV) = ac\,\text{Cov}(X, U) + ad\,\text{Cov}(X, V)$$
$$+ bc\,\text{Cov}(Y, U) + bd\,\text{Cov}(Y, V).$$

In general, for any constants a_i's, b_j's, and the random variables X_i's and Y_j's ($i = 1, 2, \ldots, k$; $j = 1, 2, \ldots, \ell$), the following relationships hold:

$$\text{Cov}\left(\sum_{i=1}^{k} a_i X_i, \sum_{j=1}^{\ell} b_j Y_j\right) = \sum_{i=1}^{k} \sum_{j=1}^{\ell} a_i b_j\,\text{Cov}(X_i, Y_j),$$

and

$$\text{Var}\left(\sum_{i=1}^{k} a_i X_i\right) = \sum_{i=1}^{k} a_i^2\,\text{Var}(X_i) + 2 \sum_{\substack{i\ i' \\ i < i'}} a_i a_{i'}\,\text{Cov}(X_i, X_{i'}).$$

If X and Y are jointly distributed random variables, the correlation of X and Y denoted by ρ is defined as

$$\rho = \frac{\text{Cov}(X, Y)}{\sqrt{\text{Var}(X)\text{Var}(Y)}}.$$

The correlation can be considered as the standardized covariance. Correlation equals covariance if both variables measure the standardized scores with unit variances. It can be shown that $-1 \leq \rho \leq 1$.

S RULES FOR DETERMINING THE ANALYSIS OF VARIANCE MODEL

In this appendix, we outline rules for determining the analysis of variance model in a balanced experimental layout. The rules are applicable to crossed classifications containing an equal number of observations for each combination of factor levels. They are also applicable to completely nested classifications as well as crossed-nested classifications containing an equal number of levels for each nested factor. We illustrate the rules with a three-factor crossed-nested classification where factors A and C are crossed and factor B is nested within factor A and crossed with factor C. We assume that the factor A has a levels, the factor B has b levels, the factor C has c levels, and there are n replications.

Rule 1. Each model contains a general constant or overall mean to be denoted by μ.

Rule 2. Each model contains a main effect for each factor which is denoted by the corresponding Greek letter with a suffix indicating the level of the factor. If a factor is nested within another factor, the nesting is indicated using the parenthesis notation for its suffix. For the example being considered, the main effects for factors A, B, and C are: α_i, $\beta_j(i)$, and γ_k, $i = 1, \ldots, a$; $j = 1, \ldots, b$; $k = 1, \ldots, c$.

Rule 3. Each model contains interaction terms corresponding to all crossed factors. There are no interaction terms for those factors containing both a nested factor and the factor within which it is nested. For the example being considered, there are $A \times C$ and $B \times C$ interactions. However, there are no $A \times B$ and $A \times B \times C$ interactions since factor B is nested within factor A. The interaction terms in the model are denoted by the combination of the Greek letters enclosed within a pair of parentheses followed by subscripts indicating the levels of the factors being crossed. For the example being considered, the model terms for $A \times C$ and $B \times C$ interactions are: $(\alpha\gamma)_{ik}$ and $(\alpha\gamma)_{jk}$, $i = 1, \ldots, a$; $j = 1, \ldots, b$; $k = 1, \ldots, c$.

Rule 4. Interactions between a nested factor and another factor with which the nested factor is crossed are always themselves nested. In the example being considered, factor B is nested within factor A and is crossed with factor C; thus, the $B \times C$ interaction is considered as nested within factor A. If an interaction term is nested within another, the nesting is indicated by the parenthesis notation for its suffix. For the example being considered, the fact that $(\beta\gamma)_{jk}$ is nested within the levels of factor A is indicated by the parenthesis notation as $(\beta\gamma)_{jk(i)}$, $i = 1, \ldots, a$; $j = 1, \ldots, b$; $k = 1, \ldots, c$.

Rule 5. The final term in the model is the error term which is considered nested within all the factors. For the example being considered, the model term for the error is denoted by $e_{\ell(ijk)}$, $i = 1, \ldots, a$; $j = 1, \ldots, b$; $k = 1, \ldots, c$; $\ell = 1, \ldots, n$.

Rule 6. The final model is written as an algebraic equation between the response variable denoted usually by the Roman letter x or y (with a suffix comprising all the subscripts) in the left-hand side and the sum of all the model terms appearing in the right-hand side. For the example being considered, the model is:

$$y_{ijk\ell} = \mu + \alpha_i + \beta_{j(i)} + \gamma_k + (\alpha\gamma)_{ik} + (\beta\gamma)_{jk(i)} + e_{\ell(ijk)} \quad \begin{cases} i = 1, \ldots, a \\ j = 1, \ldots, b \\ k = 1, \ldots, c \\ \ell = 1, \ldots, n. \end{cases}$$

T RULES FOR CALCULATING SUMS OF SQUARES AND DEGREES OF FREEDOM

In this appendix, we outline rules for calculating sums of squares and degrees of freedom in an analysis of variance model. The rules are applicable to crossed classifications containing an equal number of observations for each combination of factor levels. They are also applicable to completely nested classifications as well as crossed-nested classifications containing an equal number of levels for each nested factor. We illustrate the rules with the three-factor crossed-nested classification considered in Appendix S.

Rule 1. Write the model equation following the rules outlined in Appendix S. For the example being considered, the model equation is

$$y_{ijk\ell} = \mu + \alpha_i + \beta_{j(i)} + \gamma_k + (\alpha\gamma)_{ik} + (\beta\gamma)_{jk(i)} + e_{\ell(ijk)} \quad \begin{cases} i = 1, \ldots, a \\ j = 1, \ldots, b \\ k = 1, \ldots, c \\ \ell = 1, \ldots, n. \end{cases}$$

Rule 2. For each model term (except the general constant) write a symbolic product consisting of the subscripts of the term, using the subscript alone if it is in parentheses and subscript minus 1 if it is not in parentheses. Expand the symbolic product algebraically. For example, the symbolic product for α_i is $i - 1$, for $(\beta\gamma)_{jk(i)}$ it is $i(j - 1)(k - 1) = ijk - ij - ik + i$, and so on.

Rule 3. The typical expression to be squared and summed for obtaining the sum of squares associated with a model term consists of algebraic means indexed by subscripts of the symbolic product determined by Rule 2 and dots for the subscripts missing in the symbolic product. The number 1 is replaced by the suffix containing all the dots and designates the grand mean. In the example being considered, the symbolic product for α_i is $i - 1$ and the algebraic expression to be squared and summed is $\bar{y}_{i\ldots} - \bar{y}_{\ldots\ldots}$. The symbolic product for $(\beta\gamma)_{jk(i)}$ is $i(j - 1)(k - 1) = ijk - ij - ik + i$, and the algebraic expression to be squared and summed is $\bar{y}_{ijk.} - \bar{y}_{ij..} - \bar{y}_{i.k.} + \bar{y}_{i\ldots}$, and so on.

Rule 4. The sum of squares for a model term is obtained by squaring the algebraic expression formed by Rule 3, summing it over the subscripts in the model term, and then multiplying it by the product of the number of levels corresponding to the subscripts not appearing in the suffix of the model term. For the example being considered, the sum of squares for α_i is obtained by squaring $(\bar{y}_{i...} - \bar{y}_{....})$, summing over i, and then multiplying it by bcn; that is, $bcn \sum_{i=1}^{a}(\bar{y}_{i...} - \bar{y}_{....})^2$. Similarly, the sum of squares for $(\beta\gamma)_{jk(i)}$ is obtained by squaring $(\bar{y}_{ijk.} - \bar{y}_{ij..} - \bar{y}_{i.k.} + \bar{y}_{i...})$, summing over (i, j, k), and then multiplying it by n; that is, $n \sum_{i=1}^{a} \sum_{j=1}^{b} \sum_{k=1}^{c}(\bar{y}_{ijk.} - \bar{y}_{ij..} - \bar{y}_{i.k.} + \bar{y}_{i...})^2$, and so on.

Rule 5. The sum of squares for the general constant is obtained by squaring the grand mean $(\bar{y}_{....})$ and then multiplying it by the total number of observations $(abcn)$, that is, $abcn(\bar{y}_{....})^2$. This sum of squares is usually not included in the analysis of variance table.

Rule 6. The total sum of squares is obtained by squaring the deviations of the observations from the grand mean, and then summing it over all the subscripts, that is, $\sum_{i=1}^{a} \sum_{j=1}^{b} \sum_{k=1}^{c} \sum_{\ell=1}^{n}(y_{ijk\ell} - \bar{y}_{....})^2$.

Rule 7. The degrees of freedom corresponding to a sum of squares are calculated by replacing the subscripts in the symbolic product formed by Rule 2 by the number of levels for that subscript. For the example being considered, the degrees of freedom corresponding to the sum of squares for α_i is obtained by replacing i by a in the symbolic product $i - 1$; that is, $a - 1$. Similarly, for $(\beta\gamma)_{jk(i)}$ the symbolic product is $i(j-1)(k-1)$ and the corresponding degrees of freedom are $a(b-1)(c-1)$.

Rule 8. The number of degrees of freedom for the general constant is one, and the total number of degrees of freedom is defined as one less than the total number of observations.

U RULES FOR FINDING EXPECTED MEAN SQUARES

Determination of expected mean squares in an analysis of variance model is essential in order that appropriate mean squares may be used to construct an F statistic for a particular hypothesis of interest. They are also important for finding estimators of the variance components. Although they are not difficult to obtain, it is evident from our previous treatment that the derivation of expected mean squares for various models can be tedious, involving an inordinate amount of time and effort. In this appendix, we outline rules for finding expected mean squares in an analysis of variance model. The rules are applicable to crossed classifications containing an equal number of observations for each combination of factor levels. They are also applicable to completely nested classifications as well as crossed-nested classifications containing an equal number of levels for each nested factor. We illustrate the rules with a three-factor crossed-nested classification considered earlier in Appendices S and T. It should be noted here that in the determination of rules for the analysis of variance model and the calculation of sums of squares and degrees of freedom, it does not matter whether factor effects are fixed or random. However, this is not so in finding expected mean squares and we now assume that factors A and C are fixed whereas factor B is random.

Rule 1. Write the mathematical model following the rules given in Appendix S, including the assumptions for fixed and random effects. For the example being considered, the mathematical model is

$$y_{ijk\ell} = \mu + \alpha_i + \beta_{j(i)} + \gamma_k + (\alpha\gamma)_{ik} + (\beta\gamma)_{jk(i)} + e_{\ell(ijk)} \quad \begin{cases} i = 1, \ldots, a \\ j = 1, \ldots, b \\ k = 1, \ldots, c \\ \ell = 1, \ldots, n, \end{cases}$$

where the α_i's, γ_k's, and $(\alpha\gamma)_{ik}$'s are assumed to be constants subject to the restrictions:

$$\sum_{i=1}^{a} \alpha_i = \sum_{k=1}^{c} \gamma_k = 0,$$

$$\sum_{i=1}^{a} (\alpha\gamma)_{ik} = \sum_{k=1}^{c} (\alpha\gamma)_{ik} = 0.$$

We further assume that the $\beta_{j(i)}$'s, $(\beta\gamma)_{jk(i)}$'s, and $e_{\ell(ijk)}$'s are normally distributed with zero means and variances σ_β^2, $\sigma_{\beta\gamma(\alpha)}^2$, and σ_e^2, respectively; and the three groups of random variables are pairwise independent. The random effects $(\beta\gamma)_{jk(i)}$'s, however, are correlated due to the following restrictions:

$$\sum_{k=1}^{c} (\beta\gamma)_{jk(i)} = 0 \quad \text{for all } j(i).$$

Rule 2. Construct a two-way row × column table where there is a row for each component term in the model including the error term (except the general constant) and there is a column for each of the subscripts that appear in the model. The particular order of rows and columns is immaterial, but it helps to maintain some systematic scheme in order to avoid any mistakes. For the example being considered, the two-way table is constructed as follows:

Appendices

	i	j	k	ℓ
α_i				
$\beta_{j(i)}$				
γ_k				
$(\alpha\gamma)_{ik}$				
$(\beta\gamma)_{jk(i)}$				
$e_{\ell(ijk)}$				

Rule 3. In each row where one or more subscripts are in parentheses, write 1 in the columns corresponding to the subscripts in parentheses. For the example being considered, the two-way table now appears as follows:

	i	j	k	ℓ
α_i	...	...	...	...
$\beta_{j(i)}$	1	...	...	...
γ_k	...	...	...	...
$(\alpha\gamma)_{ik}$	...	...	...	...
$(\beta\gamma)_{jk(i)}$	1	...	...	...
$e_{\ell(ijk)}$	1	1	1	...

Rule 4. In each row where one or more subscripts are not in parentheses:

(i) write 1 in the columns corresponding to subscripts not in parentheses if the subscript represents a random factor;

(ii) write 0 in the columns corresponding to subscripts not in parentheses if the subscript represents a fixed factor.

For the example being considered, the two-way table now appears as follows:

	i	j	k	ℓ
α_i	0	...	...	...
$\beta_{j(i)}$	1	1	...	...
γ_k	...	...	0	...
$(\alpha\gamma)_{ik}$	0	...	0	...
$(\beta\gamma)_{jk(i)}$	1	1	0	...
$e_{\ell(ijk)}$	1	1	1	1

Rule 5. Write the number of levels corresponding to the column subscripts in the remaining cells that are still vacant. For the example being considered, the

two-way table now appears as follows:

	i	j	k	ℓ
α_i	0	b	c	n
$\beta_{j(i)}$	1	1	c	n
γ_k	a	b	0	n
$(\alpha\gamma)_{ik}$	0	b	0	n
$(\beta\gamma)_{jk(i)}$	1	1	0	n
$e_{\ell(ijk)}$	1	1	1	n

Rule 6. Each fixed effect has the effects parameter defined by the sum of squared effects divided by its degrees of freedom. Each random effect has the effects parameter defined by the corresponding variance component. For every model term representing a fixed effect, let $\lambda = \Phi$ designate the effects parameter. For every model term representing a random effect, let $\lambda = \sigma^2$ be the variance component for the random effect. Write all the λ parameters in the last column to the right of the two-way table where each λ parameter appears on the same line as its corresponding model term. The two-way table now appears as follows:

	i	j	k	ℓ	λ
α_i	0	b	c	n	$\Phi(\alpha)$
$\beta_{j(i)}$	1	1	c	n	$\sigma^2_{\beta(\alpha)}$
γ_k	a	b	0	n	$\Phi(\gamma)$
$(\alpha\gamma)_{ik}$	0	b	0	n	$\Phi(\alpha\gamma)$
$(\beta\gamma)_{jk(i)}$	1	1	0	n	$\sigma^2_{\beta\gamma(\alpha)}$
$e_{\ell(ijk)}$	1	1	1	n	σ^2_e

Rule 7. The expected mean square corresponding to any model term is obtained as a linear combination of λ parameters as determined by Rule 6 with the coefficients determined as follows:

(i) The coefficient of the λ parameter is zero if the subscript(s) of the model term in that row (whether in parentheses or not) do not include all of the subscripts (including those in parentheses) in the suffix of the model term whose expected mean square is being evaluated.

(ii) The coefficients of the λ parameters that are not defined as zero by Rule 7(i) are determined by first deleting the columns corresponding to the subscript(s) not in parentheses of the model term whose expected mean square is being evaluated and then multiplying the entries of the remaining columns.

For the example being considered, the coefficients of the λ parameters for different model terms are given as follows:

	\multicolumn{7}{c}{Expected Mean Square of}						
λ	A $[\alpha_i]$	B(A) $[\beta_{j(i)}]$	C $[\gamma_k]$	AC $[(\alpha\gamma)_{ik}]$	BC(A) $[(\beta\gamma)_{jk(i)}]$	Error $[e_{\ell(ijk)}]$
$\Phi(\alpha)$	bn	0	0	0	0	0
$\sigma^2_{\beta(\alpha)}$	cn	cn	n	0	0	0
$\Phi(\gamma)$	0	0	abn	0	0	0
$\Phi(\alpha\gamma)$	0	0	0	bn	0	0
$\sigma^2_{\beta\gamma(\alpha)}$	0	0	0	n	n	0
σ^2_e	1	1	1	1	1	1

Finally, from the preceding table, the expected mean squares are given by

$$E(MS_A) = \sigma^2_e + cn\sigma^2_{\beta(\alpha)} + bn\Phi(\alpha),$$
$$E(MS_{B(A)}) = \sigma^2_e + cn\sigma^2_{\beta(\alpha)},$$
$$E(MS_C) = \sigma^2_e + n\sigma^2_{\beta(\alpha)} + abn\Phi(\gamma),$$
$$E(MS_{AC}) = \sigma^2_e + n\sigma^2_{\beta\gamma(\alpha)} + bn\Phi(\alpha\gamma),$$
$$E(MS_{BC(A)}) = \sigma^2_e + n\sigma^2_{\beta\gamma(\alpha)},$$

and

$$E(MS_E) = \sigma^2_e.$$

V SAMPLES AND SAMPLING DISTRIBUTION

The major objective of any statistical analysis is to make inferences about the parameters of the population(s) under study. If the population is finite and contains only a small number of items or individuals, then it would be ideal to include every member of the population to record or examine the characteristic(s) of interest. However, most populations of interest are either infinite or too large so that it is not feasible in terms of time and money to include every member of the population in the study. Hence, in order to study such a population, the investigator carefully draws a *sample*, which is much smaller than the population, to examine its properties and then generalizes the results of the sample to the population of interest. The process of generalizing the results of a sample to the population is called *statistical inference*.

The basic requirement of a sample is that it should be *representative* of the population under study. However in general, it is difficult to obtain a representative sample. The usual procedure is to select a sample that is *random*.

The concept of *randomness* is intended to ensure that individual biases do not influence the selection of sample values. In addition, the randomness makes it possible to apply the laws of probability in drawing statistical inferences. A random sample is usually drawn with the help of a mechanical process such as throwing a coin or spinning a roulette wheel. The mechanical process generally used to obtain a random sample involves the use of a table of random numbers (see Statistical Tables XXIV). In addition, a great variety of computer programs exists for obtaining a random sample. The standard procedures for obtaining a random sample from a finite population using random numbers are discussed in most introductory statistics textbooks (see, e.g., Snedecor and Cochran (1989, Section 1.9)) and are not described here.

For the purpose of statistical inference discussed in appendix W, we assume that we have a random sample $(x_1, x_2, \ldots, x_n)$ of a given size n where x_i is the observed value of a certain characteristic X on the i-th member of the sample. We then calculate some function $T(x_1, x_2, \ldots, x_n)$ of the random sample, called a *statistic*. We repeat the procedure for every possible samples of size n that can be drawn from the population. Now, the successive samples will differ from one another and will lead to different values of the statistic T. Using a random mechanism, we can draw repeated samples, calculate the value of T for each sample, and derive a *frequency distribution* of the statistic T. Such a distribution of T is known as the *sampling distribution* of T. The following figure shows a schematic representation of the sampling distribution.

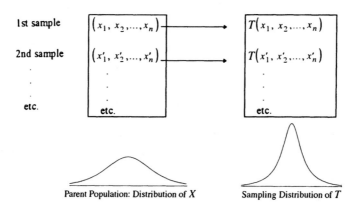

W METHODS OF STATISTICAL INFERENCE

The objective in an analysis of variance procedure is to make statistical inferences about the unknown parameters of the linear model. The main procedures for making inferences are *hypothesis testing* and *point* and *interval estimation*. In this appendix, we briefly summarize basic concepts of each procedure.

HYPOTHESIS TESTING

In *hypothesis testing* the investigator is interested in a particular value of an unknown parameter and wants to employ a statistical test to determine whether the data are consistent with the hypothesized value. The particular hypothesis to be tested is referred to as the *null hypothesis* and is denoted by H_0. In addition, there is another hypothesis which is a complement of the null hypothesis to be concluded if the null hypothesis is found to be false. The complement of the null hypothesis is referred to as the *alternative hypothesis* and is denoted by H_1.

Working provisionally on the assumption that H_0 is true, a *test statistic* is calculated from the data, as an index or measure, which is sensitive to departures from H_0. Extreme values of the test statistic are unlikely to occur if H_0 is true and consequently lead to its rejection as a statement of the true value of the parameter.

A statistical test cannot *prove* that a *hypothesis* is true or false. Even when H_0 is true, sampling variation can produce a very large or very small value of the test statistic and the investigator may be tricked into rejecting a true null hypothesis. The act of rejecting a true null hypothesis is called a *type I error*. The probability of making a type I error is termed the *significance level* and is denoted by α. Similarly, even when H_0 is false (H_1 is true), sampling variation can produce a very small value of the test statistic and the investigator may be tricked into not rejecting (accepting) a false null hypothesis. The act of not rejecting a false null hypothesis is called a *type II error*. The probability of making a type II error is denoted by β and $1 - \beta$ is termed as the *power* of the test.

The *critical region* of a test statistic is made up of extreme values of the test statistic such that H_0 is rejected if the test statistic falls in the critical region. The boundaries of a critical region are determined such that the probability of rejecting a given null hypothesis is just equal to the chosen level of significance. For a given value of α, the boundary values of a critical region are called *critical values*. The value of the level of significance is entirely optional although α-values of 0.05 and 0.01 are frequently used. The *p*-value is defined as the probability of obtaining a value of the test statistic which is more extreme (greater or smaller) than the value calculated from the sample data. The *p*-value being a probability ranges between 0 and 1. If the *p*-value is very small, we prefer alternative hypothesis H_1; that is, we reject H_0 in favor of H_1. Conversely, if the *p*-value is large, we naturally prefer the null hypothesis; that is, we do not reject (accept) H_0. Note that α is the maximum *p*-value at which we decide to reject H_0.

The steps in hypothesis testing can be summarized as follows:

(1) The hypothesis under consideration is formulated by specifying the null and alternative hypotheses.
(2) A value of the level of significance (α) is chosen in advance. The most common values of α are 0.05 and 0.01.

(3) The test statistic for the problem is selected and its value for the sample data is calculated.
(4) The sampling distribution of the test statistic under the assumption of the parent distribution of the study population is determined. The most common sampling distributions of a test statistic are t, χ^2, and F distributions.
(5) The critical value(s) corresponding to the chosen value of α in step 2 is (are) determined from the theoretical values of the sampling distribution of the test statistic identified in step 4 and the associated critical region is defined.
(6) The null hypothesis H_0 is rejected or accepted depending upon whether the value of the statistic calculated in step 3 falls inside or outside the critical region.
(7) The p-value is calculated and reported.

In a more formal statistical procedure, a sample size is chosen in advance that guarantees an acceptably high statistical power (e.g., $1 - \beta = 0.90$) of rejecting H_0 at a given level of significance (e.g., $\alpha = 0.05$).

A statistical test is called exact if its level of significance is exactly equal to a given value of α. Often, it is not possible to obtain a test with a level of significance exactly equal to α, and then the test is referred to as an approximate test. An approximate test with the level of significance less than or equal to α is called a conservative test. Similarly, an approximate test with the level of significance greater than or equal to α is called a liberal test. In general, conservative tests are often preferred when only approximate tests are available. However, if it is known that the actual level of significance of a liberal test is not much greater than α, the liberal test can be recommended.

POINT ESTIMATION

In point estimation of a *parameter*, a selected function of the sample values, known as an *estimator*, is used to make the best guess we can concerning the unknown value of the parameter. The idea of the "best" guess is that the estimator yields a sample value which in some sense is close to the value of the unknown parameter. The observed numerical value obtained by using an estimator for a given sample is called an *estimate*. Since an estimate will assume different values for different samples; it will be close to the true parameter value for some samples and will be far from the parameter value for other samples.

Statistical theory uses various criteria to judge the "goodness" or "merit" of an estimator. One desirable criterion or property used for this purpose is that of *unbiasedness*. An estimator is said to be *unbiased* if its average or expected value is equal to the parameter being estimated. More precisely, an estimator $\hat{\theta}_n$ of a parameter θ is unbiased if

$$E(\hat{\theta}_n) = \theta.$$

An example of an unbiased estimator is the sample variance defined by

$$S^2 = \frac{\sum_{i=1}^{n}(X_i - \bar{X})^2}{n-1},$$

which is an unbiased estimator of the population variance σ^2. To see that S^2 is an unbiased estimator of σ^2, one can write down every possible sample of size n which could be selected from a population, and compute S^2 for each sample. If we calculate the average value of S^2's from all possible samples, we would get σ^2. Obviously, one cannot enumerate every possible sample when the population is infinitely large, but one can derive the property of an unbiased estimator from the sampling distribution of the estimator.

Another desirable property of an estimator is that of *consistency*. An estimator is said to be *consistent* if it approximates more closely the true parameter value with increasing sample size. More precisely, an estimator $\hat{\theta}_n$ is a consistent estimator of θ if for any positive real number ϵ

$$\lim_{n \to \infty} P(|\hat{\theta}_n - \theta| > \epsilon) = 0.$$

A further desirable property of an estimator is that of *efficiency*. An estimator is said to *efficient* if it has minimum variance[1] in the class of all unbiased estimators. A minimum variance unbiased (MVU) estimator is frequently referred to as the *"best"* estimator.

INTERVAL ESTIMATION

In many studies it is generally not enough to obtain just a single value as an estimate for the unknown parameter. It is generally required to specify an index or measure of *reliability* or *uncertainty* associated with the estimate. A point estimate of a parameter provides no such information. A method of estimation known as interval estimation does provide this kind of information. In an interval estimation of an unknown parameter θ, an interval with endpoints $\hat{\theta}_L$ and $\hat{\theta}_U$ is constructed such that

$$P(\hat{\theta}_L < \theta < \hat{\theta}_U) = 1 - \alpha. \tag{W.1}$$

The quantity $1 - \alpha$ in equation (W.1) is known as the *confidence coefficient* or the *level of confidence*. The typical values of a confidence coefficient are 0.99, 0.95, and 0.90, although other values can also be chosen.

---

[1] The variance of an estimator provides a measure of the sampling error that describes the uncertainty of inference based on a particular sample. The square root of a variance estimator is called the standard error of an estimator.

A confidence interval given by equation (W.1) is called exact if the strict equality holds. Often, the equality relationship holds only approximately and then the interval is referred to as an approximate interval. To emphasize the fact that an interval is approximate, equation (W.1) is written as

$$P(\hat{\theta}_L < \theta < \hat{\theta}_U) \doteq 1 - \alpha. \tag{W.2}$$

An approximate interval (W.2) is called conservative if

$$P(\hat{\theta}_L < \theta < \hat{\theta}_U) \geq 1 - \alpha. \tag{W.3}$$

Similarly, an approximate interval (W.2) is called liberal if

$$P(\hat{\theta}_L < \theta < \hat{\theta}_U) \leq 1 - \alpha. \tag{W.4}$$

In general, conservative intervals are preferred when only approximate intervals are available. However, if it is known that the actual confidence coefficient of a liberal interval is not much lower than $1 - \alpha$, the liberal interval can be recommended.

The interval given by equation (W.1) is called a two-sided confidence interval because it has both lower and upper endpoints. In many situations an investigator is interested in an interval with only one endpoint. An interval with only one endpoint is referred to as a one-sided interval. A one-sided interval that satisfies the equation

$$P(\hat{\theta}_L < \theta < \infty) = 1 - \alpha.$$

is called an upper confidence interval. Similarly, an interval that satifies the equation

$$P(-\infty < \theta < \hat{\theta}_U) = 1 - \alpha.$$

is called a lower confidence interval. In this volume, we only consider two-sided confidence intervals. However, one-sided intervals can be readily obtained from two-sided intervals with only a minor modification.

As with estimators statistical theory uses several criteria to judge the goodness or merit of an interval. Again, a desirable criterion or property of an interval is that of *unbiasedness*. A confidence interval is said to be *unbiased* if the probability of containing any value not equal to the true of value of θ is less than or equal to $1 - \alpha$. Another desirable property of an interval is that of *uniformly most accurate* (UMA). A confidence interval is said to be uniformly most accurate if the interval has a smaller probability of containing a value not equal to θ than any other interval with confidence coefficient $1 - \alpha$. A further desirable property of an interval is that of *uniformly most accurate unbiased* (UMAU). A confidence interval that is uniformly most accurate within the class of all unbiased confidence intervals is called a uniformly most accurate unbiased confidence interval. A final desirable property of an interval is that of *uniformly shortest length* (USL). A confidence interval is said to be uniformly shortest length if it has shorter than or shortest expected length of any other interval with confidence coefficient $1 - \alpha$. Generally, if a two-sided confidence interval is UMA (UMAU), then the expected length is shortest within the class of all (unbiased) confidence intervals. For a detailed and rigorous discussion of the properties of confidence intervals, see Graybill (1976, Section 2.9).

X SOME SELECTED LATIN SQUARES

This appendix contains some more representations of Latin Squares from 3×3 to 12×12.

3×3

```
A B C
B C A
C A B
```

4×4

1	2	3	4
A B C D	A B C D	A B C D	A B C D
B A D C	B C D A	B D A C	B A D C
C D B A	C D A B	C A D B	C D A B
D C A B	D A B C	D C B A	D C B A

5×5

```
A B C D E
B A E C D
C D A E B
D E B A C
E C D B A
```

6×6

```
A B C D E F
B F D C A E
C D E F B A
D A F E C B
E C A B F D
F E B A D C
```

7×7

```
A B C D E F G
B C D E F G A
C D E F G A B
D E F G A B C
E F G A B C D
F G A B C D E
G A B C D E F
```

8×8

```
A B C D E F G H
B C D E F G H A
C D E F G H A B
D E F G H A B C
E F G H A B C D
F G H A B C D E
G H A B C D E F
H A B C D E F G
```

9×9

```
A B C D E F G H I
B C D E F G H I A
C D E F G H I A B
D E F G H I A B C
E F G H I A B C D
F G H I A B C D E
G H I A B C D E F
H I A B C D E F G
I A B C D E F G H
```

10×10

```
A B C D E F G H I J
B C D E F G H I J A
C D E F G H I J A B
D E F G H I J A B C
E F G H I J A B C D
F G H I J A B C D E
G H I J A B C D E F
H I J A B C D E F G
I J A B C D E F G H
J A B C D E F G H I
```

11 × 11

A B C D E F G H I J K
B C D E F G H I J K A
C D E F G H I J K A B
D E F G H I J K A B C
E F G H I J K A B C D
F G H I J K A B C D E
G H I J K A B C D E F
H I J K A B C D E F G
I J K A B C D E F G H
J K A B C D E F G H I
K A B C D E F G H I J

12 × 12

A B C D E F G H I J K L
B C D E F G H I J K L A
C D E F G H I J K L A B
D E F G H I J K L A B C
E F G H I J K L A B C D
F G H I J K L A B C D E
G H I J K L A B C D E F
H I J K L A B C D E F G
I J K L A B C D E F G H
J K L A B C D E F G H I
K L A B C D E F G H I J
L A B C D E F G H I J K

Source: Cochran and Cox (1957, pp. 145–146). Used with permission.

Y   SOME SELECTED GRAECO-LATIN SQUARES

This appendix contains some more representations of Graeco-Latin squares from 7×7 to 12×12.

7 × 7

A_α B_ε C_β D_ξ E_γ F_η G_δ
B_β C_ξ D_γ E_η F_δ G_α A_ε
C_γ D_η E_δ F_α G_ε A_β B_ξ
D_δ E_α F_ε G_β A_ξ B_γ C_η
E_ε F_β G_ξ A_γ B_η C_δ D_α
F_ξ G_γ A_η B_δ C_α D_ε E_β
G_η A_δ B_α C_ε D_β E_ξ F_γ

8 × 8

A_α B_ε C_β D_γ E_η F_δ G_θ H_ξ
B_α A_θ G_α F_η H_γ D_ξ C_ε E_δ
C_γ G_δ A_η E_α D_β H_ε B_ξ F_θ
D_δ F_γ E_ξ A_ε C_θ B_α H_η G_β
E_ε H_α D_θ C_δ A_ξ G_γ F_β B_η
F_ξ D_η H_δ B_θ G_ε A_β E_γ C_α
G_η C_ξ B_γ H_β F_α E_θ A_δ D_ε
H_θ E_β F_ε G_ξ B_δ C_η D_α A_γ

9 × 9

A_α B_γ C_β D_η E_τ F_θ G_δ H_ξ I_ε
B_β C_α A_γ E_θ F_η D_τ H_ε I_δ G_ξ
C_γ A_β B_α F_τ D_θ E_η I_ξ G_ε H_δ
D_δ E_ξ F_ε G_α H_γ I_β A_η B_τ C_θ
E_ε F_δ D_ξ H_β I_α G_γ B_θ C_η A_τ
F_ξ D_ε E_δ I_γ G_β H_α C_τ A_θ B_η
G_η H_τ I_θ A_δ B_ξ C_ε D_α E_γ F_β
H_θ I_η G_τ B_ε C_δ A_ξ E_β F_α D_γ
I_τ G_θ H_η C_ξ A_ε B_δ F_γ D_β E_α

Appendices 603

11×11

A_α	B_η	C_β	D_θ	E_γ	F_τ	G_δ	H_κ	I_ε	J_λ	K_ξ
B_β	C_θ	D_γ	E_τ	F_δ	G_κ	H_ε	I_λ	J_ξ	K_α	A_η
C_γ	D_τ	E_δ	F_κ	G_ε	H_λ	I_ξ	J_α	K_η	A_β	B_θ
D_δ	E_κ	F_ε	G_λ	H_ξ	I_α	J_η	K_β	A_θ	B_γ	C_τ
E_ε	F_λ	G_ξ	H_α	I_η	J_β	K_θ	A_γ	B_τ	C_δ	D_κ
F_ξ	G_α	H_η	I_β	J_θ	K_γ	A_τ	B_δ	C_κ	D_ε	F_λ
G_η	H_β	I_θ	J_γ	K_τ	A_δ	B_κ	C_ε	D_λ	E_ξ	F_α
H_θ	I_γ	J_τ	K_δ	A_κ	B_ε	C_λ	D_ξ	E_α	F_η	G_β
I_τ	J_δ	K_κ	A_ε	B_λ	C_ξ	D_α	E_η	F_β	G_θ	H_γ
J_κ	K_ε	A_λ	B_ξ	C_α	D_η	E_β	F_θ	G_γ	H_τ	I_δ
K_λ	A_ξ	B_α	C_η	D_β	E_θ	F_γ	G_τ	H_δ	I_κ	J_ε

12×12

A_α	B_μ	C_ξ	D_η	I_ε	J_δ	K_κ	L_λ	E_ι	F_θ	G_β	H_γ
B_β	A_λ	D_ε	C_θ	J_ξ	I_γ	L_τ	K_μ	F_κ	E_η	H_α	G_δ
C_γ	D_κ	A_θ	B_ε	K_η	L_β	I_μ	J_τ	G_λ	H_ξ	E_δ	F_α
D_δ	C_τ	B_η	A_ξ	L_θ	K_α	J_λ	I_κ	H_μ	G_ε	F_γ	E_β
E_ε	F_δ	G_κ	H_λ	A_τ	B_θ	C_β	D_γ	I_α	J_μ	K_ξ	L_η
F_ξ	E_γ	H_τ	G_μ	B_κ	A_η	D_α	C_δ	J_β	I_λ	L_ε	K_θ
G_η	H_β	E_μ	F_τ	C_λ	D_ξ	A_δ	B_α	K_γ	L_κ	I_θ	J_ε
H_θ	G_α	F_λ	E_κ	D_μ	C_ε	B_γ	A_β	L_δ	K_τ	J_η	I_ξ
I_τ	J_θ	K_β	L_γ	E_α	F_μ	G_ξ	H_η	A_ε	B_δ	C_κ	D_λ
J_κ	I_η	L_α	K_δ	F_β	E_λ	H_ε	G_θ	B_ξ	A_γ	D_τ	C_μ
K_λ	L_ξ	I_δ	J_α	G_γ	H_κ	E_θ	F_ε	C_η	D_β	A_μ	B_τ
L_μ	K_ε	J_γ	I_β	H_δ	G_τ	F_η	E_ξ	D_θ	C_α	B_λ	A_κ

Source: Cochran and Cox (1957, pp. 146–147). Used with permission.

Z PROC MIXED OUTPUTS FOR SOME SELECTED WORKED EXAMPLES

In this appendix, we include some additional outputs using SAS PROC MIXED for some selected worked examples given in Sections 4.16, 7.9, and 10.6. The outputs for these examples using PROC GLM were included in Figures 4.4, 7.4, and 10.9. We did not include these outputs there because the methodology underlying these analyses has not been discussed in this volume. We hope that the readers with adequate background will find these results interesting and useful because they contain estimates of variance components using the maximum likelihood (ML) and the restricted maximum likelihood (REML) procedures. It should be remarked that for balanced designs when the analysis of variance estimates of variance components are nonnegative, they are identical to the REML estimates given here. However, the results of the F tests for the fixed effects are generally not equivalent to significance tests produced by the PROC MIXED procedure.

```
DATA YIELDLOD;                    Ouput for the Worked Example in Section 4.16
INPUT AGING MIX YIELD;                          The SAS System
DATALINES;                                     The MIXED Procedure
1 1 574
1 1 564    .                 REML Estimation Iteration History  Model Fitting Information for YIELD
1 1 550
1 2 524                      Iter.  Eval.   Objective   Criterion     Description          Value
. . .                          0      1    124.13401764               Observations        18.0000
2 3 1055;                      1      1    122.98420101 0.0000000     Res Log Likelihood  -76.1951
PROC MIXED;                                                           Akaike's Inf. Crit. -79.1951
CLASSES AGING MIX;           Convergence criteria met.                Schwarz's Bayes. Crit. -80.3540
MODEL YIELD=AGING;                                                    -2 Res Log Likelihood 152.3902
RANDOM MIX AGING*MIX;
RUN;                         Covariance Parameter Estimates (REML)    Tests of Fixed Effects
Class Level Information      Cov Parm       Estimate        Source   NDF DDF Type III F  Pr > F
Class    Levels  Values      MIX          152.59259259      AGING     1   2   1965.80    0.0005
AGING       2    1 2         AGING*MIX     10.31481481
MIX         3    1 2 3       Residual     532.22222222
```

(i) SAS PROC MIXED Instructions and Output for the Worked Example in Section 4.16

```
DATA GLYCOGEN;                    Ouput for the Worked Example in Section 7.9
INPUT TREATMNT $ RAT                            The SAS System
PREPARAT GLYCOGEN;                             The MIXED Procedure
DATALINES;
CONTROL 1 1 131              REML Estimation Iteration History  Model Fitting Information for GLYCOGEN
. . .
C217SUG 2 3 127;             Iter.  Eval.   Objective   Criterion     Description          Value
PROC MIXED;                    0      1    171.91789976               Observations        36.0000
CLASS TREATMNT RAT             1      1    158.97132463 0.00000000    Res Log Likelihood  -109.811
PREPARAT;MODEL GLY=                                                   Akaike's Inf. Crit. -112.811
TREAT;RANDOM RAT             Convergence criteria met                 Schwarz's Bayes Crit. -115.055
(TREAT) PREPARAT(RAT                                                  -2 Res Log Likelihood 219.6213
TREAT);RUN;
Class Level Information      Covariance Parameter Estimates (REML)    Tests of Fixed Effects
Class  Levels  Values        Cov Parm           Estimate    Source    NDF DDF Type III F  Pr > F
TREATMNT  3    C217          RAT(TREATMNT)     36.06481481  TREATMNT   2   3    2.93     0.1971
               C217SUG CONTROL PREPARAT(TREATMNT*RAT) 14.16666667
RAT       2    1 2           Residual          21.16666667
PREPARAT  3    1 2 3
```

(ii) SAS PROC MIXED Instructions and Output for the Worked Example in Section 7.9

```
DATA SOYBEAN;                     Ouput for the Worked Example in Section 10.6
INPUT BLOCK SPACING                             The SAS System
$VARIETY $YIELD;                               The MIXED Procedure
DATALINES;
1 18" OM 33.6                REML Estimation Iteration History  Model Fitting Information for YIELD
. . .
6 42" B  22.9;               Iter.  Eval.   Objective   Criterion     Description          Value
PROC MIXED;                    0      1    146.43893356               Observations        60.0000
CLASS BLOCK VARIETY            1      3    146.27969658 0.00011090    Res Log Likelihood  -119.083
SPACING;MODEL YIELD=           2      2    146.27230705 0.00000161    Akaike's Inf. Crit. -123.083
VARIETY SPACING VARIETY*       3      1    146.27218808 0.00000000    Schwarz's Bayes. Crit. -126.907
SPACING;RANDOM BLOCK         Convergence criteria met.                -2 Res Log Likelihood 238.1660
VARIETY*BLOCK
SPACING*BLOCK;RUN;           Covariance Parameter Estimates (REML)    Tests of Fixed Effects
Class Level Information      Cov Parm         Estimate      Source   NDF DDF Type III F  Pr > F
Class Levels. Values         BLOCK          0.14028209      VARIETY   1   5    102.33    0.0002
BLOCK   6 1 2 3 4 5 6        BLOCK*VARIETY  0.00000000      SPACING   4  20     11.04    0.0001
VARIETY 2 B OM               BLOCK*SPACING  0.00000000      VAR*SPA   4  20      1.36    0.2820
SPACING 5 18 24 30 36 42     Residual       4.66840543
```

(iii) SAS PROC MIXED Instructions and Output for the Worked Example in Section 10.6

Statistical Tables and Charts

Table I. Cumulative Standard Normal Distribution

This table gives the area under the standard normal curve from $-\infty$ to the indicated values of z. The values of z are provided from 0.00 to 3.99 in increments of 0.01 units.

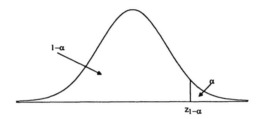

Examples: (i) $P(Z \leq 1.47) = 0.9292$.
  (ii) $P(Z > 2.12) = 1 - P(Z \leq 2.12) = 1 - 0.9830 = 0.0170$.
  (iii) $P(Z \leq -2.51) = 1 - P(Z \leq 2.51) = 1 - 0.9940 = 0.0060$.
  (iv) $P(-1.21 \leq Z \leq 2.68) = P(Z \leq 2.68) - P(Z \leq -1.21) = 0.8832$.

z	P(Z ≤ z)	z	P(Z ≤ z)	z	P(Z ≤ z)	z	P(Z ≤ z)	z	P(Z ≤ z)	z	P(Z ≤ z)	z	P(Z ≤ z)	z	P(Z ≤ z)
0.00	0.5000	0.50	0.6915	1.00	0.8413	1.50	0.9332	2.00	0.9772	2.50	0.9938	3.00	0.9987	3.50	0.9998
0.01	0.5040	0.51	0.6950	1.01	0.8438	1.51	0.9345	2.01	0.9778	2.51	0.9940	3.01	0.9987	3.51	0.9998
0.02	0.5080	0.52	0.6985	1.02	0.8461	1.52	0.9357	2.02	0.9783	2.52	0.9941	3.02	0.9987	3.52	0.9998
0.03	0.5120	0.53	0.7019	1.03	0.8485	1.53	0.9370	2.03	0.9788	2.53	0.9943	3.03	0.9988	3.53	0.9998
0.04	0.5160	0.54	0.7054	1.04	0.8508	1.54	0.9382	2.04	0.9793	2.54	0.9945	3.04	0.9988	3.54	0.9998
0.05	0.5199	0.55	0.7088	1.05	0.8531	1.55	0.9394	2.05	0.9798	2.55	0.9946	3.05	0.9989	3.55	0.9998
0.06	0.5239	0.56	0.7123	1.06	0.8554	1.56	0.9406	2.06	0.9803	2.56	0.9948	3.06	0.9989	3.56	0.9998
0.07	0.5279	0.57	0.7157	1.07	0.8577	1.57	0.9418	2.07	0.9808	2.57	0.9949	3.07	0.9989	3.57	0.9998
0.08	0.5319	0.58	0.7190	1.08	0.8599	1.58	0.9429	2.08	0.9812	2.58	0.9951	3.08	0.9990	3.58	0.9998
0.09	0.5359	0.59	0.7224	1.09	0.8621	1.59	0.9441	2.09	0.9817	2.59	0.9952	3.09	0.9990	3.59	0.9998
0.10	0.5398	0.60	0.7257	1.10	0.8643	1.60	0.9452	2.10	0.9821	2.60	0.9953	3.10	0.9990	3.60	0.9998
0.11	0.5438	0.61	0.7291	1.11	0.8665	1.61	0.9463	2.11	0.9826	2.61	0.9955	3.11	0.9991	3.61	0.9998
0.12	0.5478	0.62	0.7324	1.12	0.8686	1.62	0.9474	2.12	0.9830	2.62	0.9956	3.12	0.9991	3.62	0.9999
0.13	0.5517	0.63	0.7357	1.13	0.8708	1.63	0.9484	2.13	0.9834	2.63	0.9957	3.13	0.9991	3.63	0.9999
0.14	0.5557	0.64	0.7389	1.14	0.8729	1.64	0.9495	2.14	0.9838	2.64	0.9959	3.14	0.9992	3.64	0.9999
0.15	0.5596	0.65	0.7422	1.15	0.8749	1.65	0.9505	2.15	0.9842	2.65	0.9960	3.15	0.9992	3.65	0.9999
0.16	0.5636	0.66	0.7454	1.16	0.8770	1.66	0.9515	2.16	0.9846	2.66	0.9961	3.16	0.9992	3.66	0.9999
0.17	0.5675	0.67	0.7486	1.17	0.8790	1.67	0.9525	2.17	0.9850	2.67	0.9962	3.17	0.9992	3.67	0.9999
0.18	0.5714	0.68	0.7517	1.18	0.8810	1.68	0.9535	2.18	0.9854	2.68	0.9963	3.18	0.9993	3.68	0.9999
0.19	0.5753	0.69	0.7549	1.19	0.8830	1.69	0.9545	2.19	0.9857	2.69	0.9964	3.19	0.9993	3.69	0.9999
0.20	0.5793	0.70	0.7580	1.20	0.8849	1.70	0.9554	2.20	0.9861	2.70	0.9965	3.20	0.9993	3.70	0.9999
0.21	0.5832	0.71	0.7611	1.21	0.8869	1.71	0.9564	2.21	0.9864	2.71	0.9966	3.21	0.9993	3.71	0.9999
0.22	0.5871	0.72	0.7642	1.22	0.8888	1.72	0.9573	2.22	0.9868	2.72	0.9967	3.22	0.9994	3.72	0.9999
0.23	0.5910	0.73	0.7673	1.23	0.8907	1.73	0.9582	2.23	0.9871	2.73	0.9968	3.23	0.9994	3.73	0.9999
0.24	0.5948	0.74	0.7703	1.24	0.8925	1.74	0.9591	2.24	0.9875	2.74	0.9969	3.24	0.9994	3.74	0.9999
0.25	0.5987	0.75	0.7734	1.25	0.8944	1.75	0.9599	2.25	0.9878	2.75	0.9970	3.25	0.9994	3.75	0.9999
0.26	0.6026	0.76	0.7764	1.26	0.8962	1.76	0.9608	2.26	0.9881	2.76	0.9971	3.26	0.9994	3.76	0.9999
0.27	0.6064	0.77	0.7794	1.27	0.8980	1.77	0.9616	2.27	0.9884	2.77	0.9972	3.27	0.9995	3.77	0.9999
0.28	0.6103	0.78	0.7823	1.28	0.8997	1.78	0.9625	2.28	0.9887	2.78	0.9973	3.28	0.9995	3.78	0.9999
0.29	0.6141	0.79	0.7852	1.29	0.9015	1.79	0.9633	2.29	0.9890	2.79	0.9974	3.29	0.9995	3.79	0.9999
0.30	0.6179	0.80	0.7881	1.30	0.9032	1.80	0.9641	2.30	0.9893	2.80	0.9974	3.30	0.9995	3.80	0.9999
0.31	0.6217	0.81	0.7910	1.31	0.9049	1.81	0.9649	2.31	0.9896	2.81	0.9975	3.31	0.9995	3.81	0.9999
0.32	0.6255	0.82	0.7939	1.32	0.9066	1.82	0.9656	2.32	0.9898	2.82	0.9976	3.32	0.9995	3.82	0.9999
0.33	0.6293	0.83	0.7967	1.33	0.9082	1.83	0.9664	2.33	0.9901	2.83	0.9977	3.33	0.9996	3.83	0.9999
0.34	0.6331	0.84	0.7995	1.34	0.9099	1.84	0.9671	2.34	0.9904	2.84	0.9977	3.34	0.9996	3.84	0.9999
0.35	0.6368	0.85	0.8023	1.35	0.9115	1.85	0.9678	2.35	0.9906	2.85	0.9978	3.35	0.9996	3.85	0.9999
0.36	0.6406	0.86	0.8051	1.36	0.9131	1.86	0.9686	2.36	0.9909	2.86	0.9979	3.36	0.9996	3.86	0.9999
0.37	0.6443	0.87	0.8078	1.37	0.9147	1.87	0.9693	2.37	0.9911	2.87	0.9979	3.37	0.9996	3.87	1.0000
0.38	0.6480	0.88	0.8106	1.38	0.9162	1.88	0.9699	2.38	0.9913	2.88	0.9980	3.38	0.9996	3.88	1.0000
0.39	0.6517	0.89	0.8133	1.39	0.9177	1.89	0.9706	2.39	0.9916	2.89	0.9981	3.39	0.9997	3.89	1.0000
0.40	0.6554	0.90	0.8159	1.40	0.9192	1.90	0.9713	2.40	0.9918	2.90	0.9981	3.40	0.9997	3.90	1.0000
0.41	0.6591	0.91	0.8186	1.41	0.9207	1.91	0.9719	2.41	0.9920	2.91	0.9982	3.41	0.9997	3.91	1.0000
0.42	0.6628	0.92	0.8212	1.42	0.9222	1.92	0.9726	2.42	0.9922	2.92	0.9982	3.42	0.9997	3.92	1.0000
0.43	0.6664	0.93	0.8238	1.43	0.9236	1.93	0.9732	2.43	0.9925	2.93	0.9983	3.43	0.9997	3.93	1.0000
0.44	0.6700	0.94	0.8264	1.44	0.9251	1.94	0.9738	2.44	0.9927	2.94	0.9984	3.44	0.9997	3.94	1.0000
0.45	0.6736	0.95	0.8289	1.45	0.9265	1.95	0.9744	2.45	0.9929	2.95	0.9984	3.45	0.9997	3.95	1.0000
0.46	0.6772	0.96	0.8315	1.46	0.9279	1.96	0.9750	2.46	0.9931	2.96	0.9985	3.46	0.9997	3.96	1.0000
0.47	0.6808	0.97	0.8340	1.47	0.9292	1.97	0.9756	2.47	0.9932	2.97	0.9985	3.47	0.9997	3.97	1.0000
0.48	0.6844	0.98	0.8365	1.48	0.9306	1.98	0.9761	2.48	0.9934	2.98	0.9986	3.48	0.9997	3.98	1.0000
0.49	0.6879	0.99	0.8389	1.49	0.9319	1.99	0.9767	2.49	0.9936	2.99	0.9986	3.49	0.9998	3.99	1.0000

Computed Using IMSL* Library Functions.

* IMSL (International Mathematical and Statistical Library) is a registered trade mark of IMSL, Inc.

Table II. Percentage Points of the Standard Normal Distribution

This table is the inverse of Table I. The entries in the table give z values (percentiles) corresponding to a given cumulative probability (i.e., $P(Z \leq z)$), which represents all the area to the left of the z value. The values of $P(Z \leq z)$ are given from 0.0001 to 0.9999.

Examples: (i) $P(Z \leq z) = 0.005, z = -2.57583$.
            (ii) $P(Z \leq z) = 0.200, z = -0.84162$.
            (iii) $P(Z \leq z) = 0.800, z = 0.84162$.
            (iv) $P(Z \leq z) = 0.950, z = 1.64485$.

$P(Z \leq z)$	z	$P(Z \leq z)$	z	$P(Z \leq z)$	z	$P(Z \leq z)$	z	$P(Z \leq z)$	z
0.0001	-3.71902	0.165	-0.97411	0.390	-0.27932	0.615	0.29237	0.8400	0.99446
0.0002	-3.54008	0.170	-0.95417	0.395	-0.26631	0.620	0.30548	0.8450	1.01522
0.0003	-3.43161	0.175	-0.93459	0.400	-0.25335	0.625	0.31864	0.8500	1.03643
0.0004	-3.35279	0.180	-0.91537	0.405	-0.24043	0.630	0.33185	0.8550	1.05812
0.0005	-3.29053	0.185	-0.89647	0.410	-0.22754	0.635	0.34513	0.8600	1.08032
0.0010	-3.09023	0.190	-0.87790	0.415	-0.21470	0.640	0.35846	0.8650	1.10306
0.0020	-2.87816	0.195	-0.85962	0.420	-0.20189	0.645	0.37186	0.8700	1.12639
0.0030	-2.74778	0.200	-0.84162	0.425	-0.18912	0.650	0.38532	0.8750	1.15035
0.0040	-2.65207	0.205	-0.82389	0.430	-0.17637	0.655	0.39886	0.8800	1.17499
0.0050	-2.57583	0.210	-0.80642	0.435	-0.16366	0.660	0.41246	0.8850	1.20036
0.0060	-2.51214	0.215	-0.78919	0.440	-0.15097	0.665	0.42615	0.8900	1.22653
0.0070	-2.45726	0.220	-0.77219	0.445	-0.13830	0.670	0.43991	0.8950	1.25357
0.0080	-2.40892	0.225	-0.75542	0.450	-0.12566	0.675	0.45376	0.9000	1.28155
0.0090	-2.36562	0.230	-0.73885	0.455	-0.11304	0.680	0.46770	0.9050	1.31058
0.0100	-2.32635	0.235	-0.72248	0.460	-0.10043	0.685	0.48173	0.9100	1.34076
0.0150	-2.17009	0.240	-0.70630	0.465	-0.08784	0.690	0.49585	0.9150	1.37220
0.0200	-2.05375	0.245	-0.69031	0.470	-0.07527	0.695	0.51007	0.9200	1.40507
0.0250	-1.95996	0.250	-0.67449	0.475	-0.06271	0.700	0.52440	0.9250	1.43953
0.0300	-1.88079	0.255	-0.65884	0.480	-0.05015	0.705	0.53884	0.9300	1.47579
0.0350	-1.81191	0.260	-0.64335	0.485	-0.03761	0.710	0.55338	0.9350	1.51410
0.0400	-1.75069	0.265	-0.62801	0.490	-0.02507	0.715	0.56805	0.9400	1.55477
0.0450	-1.69540	0.270	-0.61281	0.495	-0.01253	0.720	0.58284	0.9450	1.59819
0.0500	-1.64485	0.275	-0.59776	0.500	0.00000	0.725	0.59776	0.9500	1.64485
0.0550	-1.59819	0.280	-0.58284	0.505	0.01253	0.730	0.61281	0.9550	1.69540
0.0600	-1.55477	0.285	-0.56805	0.510	0.02507	0.735	0.62801	0.9600	1.75069
0.0650	-1.51410	0.290	-0.55338	0.515	0.03761	0.740	0.64335	0.9650	1.81191
0.0700	-1.47579	0.295	-0.53884	0.520	0.05015	0.745	0.65884	0.9700	1.88079
0.0750	-1.43953	0.300	-0.52440	0.525	0.06271	0.750	0.67449	0.9750	1.95996
0.0800	-1.40507	0.305	-0.51007	0.530	0.07527	0.755	0.69031	0.9800	2.05375
0.0850	-1.37220	0.310	-0.49585	0.535	0.08784	0.760	0.70630	0.9850	2.17009
0.0900	-1.34076	0.315	-0.48173	0.540	0.10043	0.765	0.72248	0.9900	2.32635
0.0950	-1.31058	0.320	-0.46770	0.545	0.11304	0.770	0.73885	0.9910	2.36562
0.1000	-1.28155	0.325	-0.45376	0.550	0.12566	0.775	0.75542	0.9920	2.40892
0.1050	-1.25357	0.330	-0.43991	0.555	0.13830	0.780	0.77219	0.9930	2.45726
0.1100	-1.22653	0.335	-0.42615	0.560	0.15097	0.785	0.78919	0.9940	2.51214
0.1150	-1.20036	0.340	-0.41246	0.565	0.16366	0.790	0.80642	0.9950	2.57583
0.1200	-1.17499	0.345	-0.39886	0.570	0.17637	0.795	0.82389	0.9960	2.65207
0.1250	-1.15035	0.350	-0.38532	0.575	0.18912	0.800	0.84162	0.9970	2.74778
0.1300	-1.12639	0.355	-0.37186	0.580	0.20189	0.805	0.85962	0.9980	2.87816
0.1350	-1.10306	0.360	-0.35846	0.585	0.21470	0.810	0.87790	0.9990	3.09023
0.1400	-1.08032	0.365	-0.34513	0.590	0.22754	0.815	0.89647	0.9995	3.29053
0.1450	-1.05812	0.370	-0.33185	0.595	0.24043	0.820	0.91537	0.9996	3.35279
0.1500	-1.03643	0.375	-0.31864	0.600	0.25335	0.825	0.93459	0.9997	3.43161
0.1550	-1.01522	0.380	-0.30548	0.605	0.26631	0.830	0.95417	0.9998	3.54008
0.1600	-0.99446	0.385	-0.29237	0.610	0.27932	0.835	0.97411	0.9999	3.71902

Computed Using IMSL* Library Functions.

---

* IMSL (International Mathematical and Statistical Library) is a registered trade mark of IMSL, Inc.

Table III. Critical Values of the Student's t Distribution

This table gives the critical values of the Student's t distribution for degrees of freedom $\nu = 1\ (1)\ 30, 40, 60, 120, \infty$. The α-values are given corresponding to upper-tail tests of significance. The critical values are given corresponding to one-tail α-levels equal to 0.40, 0.30, 0.20, 0.15, 0.10, 0.025, 0.02, 0.015, 0.01, 0.0075, 0.005, 0.0025, and 0.0005. Since the distribution of t is symmetrical about zero, the one-tailed significance level of α corresponds to the two-tailed significance level of 2α. All the critical values are provided to three decimal places.

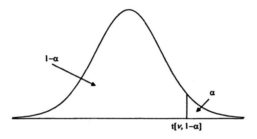

Examples: (i) For $\nu = 15, \alpha = 0.01$, the desired critical value from the table is $t[15, 0.99] = 2.602$.
(ii) For $\nu = 60, \alpha = 0.05$, the desired critical value from the table is $t[60, 0.95] = 1.671$.

ν	0.40	0.30	0.20	0.15	0.10	0.05	α 0.025	0.02	0.015	0.01	0.0075	0.005	0.0025	0.0005
1	0.325	0.727	1.376	1.963	3.078	6.314	12.706	15.895	21.205	31.821	42.434	63.657	127.322	636.590
2	0.289	0.617	1.061	1.386	1.886	2.920	4.303	4.849	5.643	6.965	8.073	9.925	14.089	31.598
3	0.277	0.584	0.978	1.250	1.638	2.353	3.182	3.482	3.896	4.541	5.047	5.841	7.453	12.924
4	0.271	0.569	0.941	1.190	1.533	2.132	2.776	2.999	3.298	3.747	4.088	4.604	5.598	8.610
5	0.267	0.559	0.920	1.156	1.476	2.015	2.571	2.757	3.003	3.365	3.634	4.032	4.773	6.869
6	0.265	0.553	0.906	1.134	1.440	1.943	2.447	2.612	2.829	3.143	3.372	3.707	4.317	5.959
7	0.263	0.549	0.896	1.119	1.415	1.895	2.365	2.517	2.715	2.998	3.203	3.499	4.029	5.408
8	0.262	0.546	0.889	1.108	1.397	1.860	2.306	2.449	2.634	2.896	3.085	3.355	3.833	5.041
9	0.261	0.543	0.883	1.100	1.383	1.833	2.262	2.398	2.574	2.821	2.998	3.250	3.690	4.781
10	0.260	0.542	0.879	1.093	1.372	1.812	2.228	2.359	2.527	2.764	2.932	3.169	3.581	4.587
11	0.260	0.540	0.876	1.088	1.363	1.796	2.201	2.328	2.491	2.718	2.879	3.106	3.497	4.437
12	0.259	0.539	0.873	1.083	1.356	1.782	2.179	2.303	2.461	2.681	2.836	3.055	3.428	4.318
13	0.259	0.537	0.870	1.079	1.350	1.771	2.160	2.282	2.436	2.650	2.801	3.012	3.372	4.221
14	0.258	0.537	0.868	1.076	1.345	1.761	2.145	2.264	2.415	2.624	2.771	2.977	3.326	4.140
15	0.258	0.536	0.866	1.074	1.341	1.753	2.131	2.249	2.397	2.602	2.746	2.947	3.286	4.073
16	0.258	0.535	0.865	1.071	1.337	1.746	2.120	2.235	2.382	2.583	2.724	2.921	3.252	4.015
17	0.257	0.534	0.863	1.069	1.333	1.740	2.110	2.224	2.368	2.567	2.706	2.898	3.222	3.965
18	0.257	0.534	0.862	1.067	1.330	1.734	2.101	2.214	2.356	2.552	2.689	2.878	3.197	3.922
19	0.257	0.533	0.861	1.066	1.328	1.729	2.093	2.205	2.346	2.539	2.674	2.861	3.174	3.883
20	0.257	0.533	0.860	1.064	1.325	1.725	2.086	2.197	2.336	2.528	2.661	2.845	3.153	3.849
21	0.257	0.532	0.859	1.063	1.323	1.721	2.080	2.189	2.328	2.518	2.649	2.831	3.135	3.819
22	0.256	0.532	0.858	1.061	1.321	1.717	2.074	2.183	2.320	2.508	2.639	2.819	3.119	3.792
23	0.256	0.532	0.858	1.060	1.319	1.714	2.069	2.177	2.313	2.500	2.629	2.807	3.104	3.768
24	0.256	0.531	0.857	1.059	1.318	1.711	2.064	2.172	2.307	2.492	2.620	2.797	3.091	3.745
25	0.256	0.531	0.856	1.058	1.316	1.708	2.060	2.167	2.301	2.485	2.612	2.787	3.078	3.725
26	0.256	0.531	0.856	1.058	1.315	1.706	2.056	2.162	2.296	2.479	2.605	2.779	3.067	3.707
27	0.256	0.531	0.855	1.057	1.314	1.703	2.052	2.158	2.291	2.473	2.598	2.771	3.057	3.690
28	0.256	0.530	0.855	1.056	1.313	1.701	2.048	2.154	2.286	2.467	2.592	2.763	3.047	3.674
29	0.256	0.530	0.854	1.055	1.311	1.699	2.045	2.150	2.282	2.462	2.586	2.756	3.038	3.659
30	0.256	0.530	0.854	1.055	1.310	1.697	2.042	2.147	2.278	2.457	2.581	2.750	3.030	3.646
40	0.255	0.529	0.851	1.050	1.303	1.684	2.021	2.123	2.250	2.423	2.542	2.704	2.971	3.551
60	0.254	0.527	0.848	1.045	1.296	1.671	2.000	2.099	2.223	2.390	2.504	2.660	2.915	3.460
120	0.254	0.526	0.845	1.041	1.289	1.658	1.980	2.076	2.196	2.358	2.468	2.617	2.860	3.373
∞	0.253	0.524	0.842	1.036	1.282	1.645	1.960	2.054	2.170	2.326	2.432	2.576	2.807	3.291

From J. Neter, M. H. Kutner, C. J. Nachtsheim and W. Wasserman, *Applied Linear Statistical Models*, Fourth Edition, © 1996 by Richard D. Irwin, Inc., Chicago. Reprinted by permission (from Table B.2).

Table IV. Critical Values of the Chi-Square Distribution

This table gives the critical values of the chi-square (χ^2) distribution for degrees of freedom $\nu = 1\ (1)\ 30\ (10)\ 100$. The critical values are given corresponding to α-levels equal to 0.995, 0.990, 0.975, 0.95, 0.90, 0.75, 0.50, 0.25, 0.10, 0.05, 0.025, 0.01, and 0.005. All the critical values are provided to two decimal places.

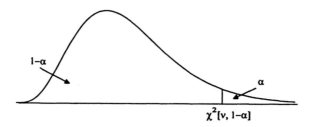

Examples: (i) For $\nu = 15$, $\alpha = 0.05$, the desired critical value from the table is $\chi^2[15, 0.95] = 25.00$.

(ii) For $\nu = 20$, $\alpha = 0.90$, the desired critical value from the table is $\chi^2[20, 0.1] = 12.44$.

					α							
ν	0.995	0.990	0.975	0.950	0.900	0.750	0.500	0.250	0.100	0.050	0.025	0.010	0.005
1	0.0⁴393	0.0³157	0.0²982	0.0²393	0.02	0.10	0.45	1.32	2.71	3.84	5.02	6.63	7.88
2	0.01	0.02	0.05	0.10	0.21	0.58	1.39	2.77	4.61	5.99	7.38	9.21	10.60
3	0.07	0.11	0.22	0.35	0.58	1.21	2.37	4.11	6.25	7.81	9.35	11.34	12.84
4	0.21	0.30	0.48	0.71	1.06	1.92	3.36	5.39	7.78	9.49	11.14	13.28	14.86
5	0.41	0.55	0.83	1.15	1.61	2.67	4.35	6.63	9.24	11.07	12.83	15.09	16.75
6	0.68	0.87	1.24	1.64	2.20	3.45	5.35	7.84	10.64	12.59	14.45	16.81	18.55
7	0.99	1.24	1.69	2.17	2.83	4.25	6.35	9.04	12.02	14.07	16.01	18.48	20.28
8	1.34	1.65	2.18	2.73	3.49	5.07	7.34	10.22	13.36	15.51	17.53	20.09	21.96
9	1.73	2.09	2.70	3.33	4.17	5.90	8.34	11.39	14.68	16.92	19.02	21.67	23.59
10	2.16	2.56	3.25	3.94	4.87	6.74	9.34	12.55	15.99	18.31	20.48	23.21	25.19
11	2.60	3.05	3.82	4.57	5.58	7.58	10.34	13.70	17.28	19.68	21.92	24.72	26.76
12	3.07	3.57	4.40	5.23	6.30	8.44	11.34	14.85	18.55	21.03	23.34	26.22	28.30
13	3.57	4.11	5.01	5.89	7.04	9.30	12.34	15.98	19.81	22.36	24.74	27.69	29.82
14	4.07	4.66	5.63	6.57	7.79	10.17	13.34	17.12	21.06	23.68	26.12	29.14	31.32
15	4.60	5.23	6.27	7.26	8.55	11.04	14.34	18.25	22.31	25.00	27.49	30.58	32.80
16	5.14	5.81	6.91	7.96	9.31	11.91	15.34	19.37	23.54	26.30	28.85	32.00	34.27
17	5.70	6.41	7.56	8.67	10.09	12.79	16.34	20.49	24.77	27.59	30.19	33.41	35.72
18	6.26	7.01	8.23	9.39	10.86	13.68	17.34	21.60	25.99	28.87	31.53	34.81	37.16
19	6.84	7.63	8.91	10.12	11.65	14.56	18.34	22.72	27.20	30.14	32.85	36.19	38.58
20	7.43	8.26	9.59	10.85	12.44	15.45	19.34	23.83	28.41	31.41	34.17	37.57	40.00
21	8.03	8.90	10.28	11.59	13.24	16.34	20.34	24.93	29.62	32.67	35.48	38.93	41.40
22	8.64	9.54	10.98	12.34	14.04	17.24	21.34	26.04	30.81	33.92	36.78	40.29	42.80
23	9.26	10.20	11.69	13.09	14.85	18.14	22.34	27.14	32.01	35.17	38.08	41.64	44.18
24	9.89	10.86	12.40	13.85	15.66	19.04	23.34	28.24	33.20	36.42	39.36	42.98	45.56
25	10.52	11.52	13.12	14.61	16.47	19.94	24.34	29.34	34.38	37.65	40.65	44.31	46.93
26	11.16	12.20	13.84	15.38	17.29	20.84	25.34	30.43	35.56	38.89	41.92	45.64	48.29
27	11.81	12.88	14.57	16.15	18.11	21.75	26.34	31.53	36.74	40.11	43.19	46.96	49.64
28	12.46	13.56	15.31	16.93	18.94	22.66	27.34	32.62	37.92	41.34	44.46	48.28	50.99
29	13.12	14.26	16.05	17.71	19.77	23.57	28.34	33.71	39.09	42.56	45.72	49.59	52.34
30	13.79	14.95	16.79	18.49	20.60	24.48	29.34	34.80	40.26	43.77	46.98	50.89	53.67
40	20.71	22.16	24.43	26.51	29.05	33.66	39.34	45.62	51.80	55.76	59.34	63.69	66.77
50	27.99	29.71	32.36	34.76	37.69	42.94	49.33	56.33	63.17	67.50	71.42	76.15	79.49
60	35.53	37.48	40.48	43.19	46.46	52.29	59.33	66.98	74.40	79.08	83.30	88.38	91.95
70	43.28	45.44	48.76	51.74	55.33	61.70	69.33	77.58	85.53	90.53	95.02	100.42	104.22
80	51.17	53.54	57.15	60.39	64.28	71.14	79.33	88.13	96.58	101.88	106.63	112.33	116.32
90	59.20	61.75	65.65	69.13	73.29	80.62	89.33	98.64	107.56	113.14	118.14	124.12	128.30
100	67.33	70.06	74.22	77.93	82.36	90.13	99.33	109.14	118.50	124.34	129.56	135.81	140.17

From C. M. Thompson, "Table of Percentage Points of the Chi-Square Distribution." *Biometrika*, 32, (1941), 188–189. Reprinted by permission.

Table V. Critical Values of the F Distribution

This table gives the critical values of the F distribution for degrees of freedom $\nu_1 = 1\,(1)\,10, 12, 15, 20, 24, 30, 40, 60, 120, \infty$ arranged across the top of the table and $\nu_2 = 1\,(1)\,30, 40, 60, 120, \infty$ arranged along the left margin of the table. The α-values are given for upper-tail tests of significance. All the critical values are provided to two decimal places. The lower-tailed critical values are not given, but can be obtained using the following relation: $F[\nu_1, \nu_2; 1-\alpha] = 1/F[\nu_2, \nu_1; \alpha]$.

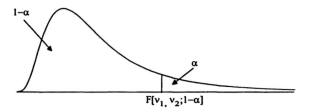

Examples: (i) For $\nu_1 = 6$, $\nu_2 = 30$, $\alpha = 0.05$, the desired critical value from the table is $F[6, 30; 0.95] = 2.42$.

(ii) For $\nu_1 = 10$, $\nu_2 = 60$, $\alpha = 0.10$, the desired critical value from the table is $F[10, 60; 0.90] = 1.71$.

(iii) For $\nu_1 = 8$, $\nu_2 = 24$, $\alpha = 0.95$, the desired critical value is obtained as $F[8, 24, 0.05] = 1/F[24, 8, 0.95] = 1/3.12 = 0.32$.

Statistical Tables and Charts

ν_2	α	1	2	3	4	5	6	7	8	9	10	12	15	20	24	30	40	60	120	∞
1	.100	39.86	49.50	53.59	55.83	57.24	58.20	58.91	59.44	59.86	60.19	60.71	61.22	61.74	62.00	62.26	62.53	62.79	63.06	63.33
	.050	161.4	199.50	215.70	224.60	230.20	234.00	236.80	238.90	240.50	241.90	243.90	245.90	248.00	249.10	250.10	251.10	252.20	253.30	254.30
	.025	647.8	799.50	864.20	899.60	921.80	937.10	948.20	956.70	963.30	968.60	976.70	984.90	993.10	997.20	1001	1006	1010	1014	1018
	.010	4052	4999.50	5403	5625	5764	5859	5928	5982	6022	6056	6106	6157	6209	6235	6261	6287	6313	6339	6366
	.005	16211	20000	21615	22500	23056	23437	23715	23925	24091	24224	24426	24630	24836	24940	25044	25148	25253	25359	25465
2	.100	8.53	9.00	9.16	9.24	9.29	9.33	9.35	9.37	9.38	9.39	9.41	9.42	9.44	9.45	9.46	9.47	9.47	9.48	9.49
	.050	18.51	19.00	19.16	19.25	19.30	19.33	19.35	19.37	19.38	19.40	19.41	19.43	19.45	19.45	19.46	19.47	19.48	19.49	19.50
	.025	38.51	39.00	39.17	39.25	39.30	39.33	39.36	39.37	39.39	39.40	39.41	39.43	39.45	39.46	39.46	39.47	39.48	39.49	39.50
	.010	98.50	99.00	99.17	99.25	99.30	99.33	99.36	99.37	99.39	99.40	99.42	99.43	99.45	99.46	99.47	99.47	99.48	99.49	99.50
	.005	198.5	199.00	199.20	199.20	199.30	199.30	199.40	199.40	199.40	199.40	199.40	199.40	199.40	199.50	199.50	199.50	199.50	199.50	199.50
3	.100	5.54	5.46	5.39	5.34	5.31	5.28	5.27	5.25	5.24	5.23	5.22	5.20	5.18	5.18	5.17	5.16	5.15	5.14	5.13
	.050	10.13	9.55	9.28	9.12	9.01	8.94	8.89	8.85	8.81	8.79	8.74	8.70	8.66	8.64	8.62	8.59	8.57	8.55	8.53
	.025	17.44	16.04	15.44	15.10	14.88	14.73	14.62	14.54	14.47	14.42	14.34	14.25	14.17	14.12	14.08	14.04	13.99	13.95	13.90
	.010	34.12	30.82	29.46	28.71	28.24	27.91	27.67	27.49	27.35	27.23	27.05	26.87	26.69	26.60	26.50	26.41	26.32	26.22	26.13
	.005	55.55	49.80	47.47	46.19	45.39	44.84	44.43	44.13	43.88	43.69	43.39	43.08	42.78	42.62	42.47	42.31	42.15	41.99	41.83
4	.100	4.54	4.32	4.19	4.11	4.05	4.01	3.98	3.95	3.94	3.92	3.90	3.87	3.84	3.83	3.82	3.80	3.79	3.78	3.76
	.050	7.71	6.94	6.59	6.39	6.26	6.16	6.09	6.04	6.00	5.96	5.91	5.86	5.80	5.77	5.75	5.72	5.69	5.66	5.63
	.025	12.22	10.65	9.98	9.60	9.36	9.20	9.07	8.98	8.90	8.84	8.75	8.66	8.56	8.51	8.46	8.41	8.36	8.31	8.26
	.010	21.20	18.00	16.69	15.98	15.52	15.21	14.98	14.80	14.66	14.55	14.37	14.20	14.02	13.93	13.84	13.75	13.65	13.56	13.46
	.005	31.33	26.28	24.26	23.15	22.46	21.97	21.62	21.35	21.14	20.97	20.70	20.44	20.17	20.03	19.89	19.75	19.61	19.47	19.32
5	.100	4.06	3.78	3.62	3.52	3.45	3.40	3.37	3.34	3.32	3.30	3.27	3.24	3.21	3.19	3.17	3.16	3.14	3.12	3.10
	.050	6.61	5.79	5.41	5.19	5.05	4.95	4.88	4.82	4.77	4.74	4.68	4.62	4.56	4.53	4.50	4.46	4.43	4.40	4.36
	.025	10.01	8.43	7.76	7.39	7.15	6.98	6.85	6.76	6.68	6.62	6.52	6.43	6.33	6.28	6.23	6.18	6.12	6.07	6.02
	.010	16.26	13.27	12.06	11.39	10.97	10.67	10.46	10.29	10.16	10.05	9.89	9.72	9.55	9.47	9.38	9.29	9.20	9.11	9.02
	.005	22.78	18.31	16.53	15.56	14.94	14.51	14.20	13.96	13.77	13.62	13.38	13.15	12.90	12.78	12.66	12.53	12.40	12.27	12.14
6	.100	3.78	3.46	3.29	3.18	3.11	3.05	3.01	2.98	2.96	2.94	2.90	2.87	2.84	2.82	2.80	2.78	2.76	2.74	2.72
	.050	5.99	5.14	4.76	4.53	4.39	4.28	4.21	4.15	4.10	4.06	4.00	3.94	3.87	3.84	3.81	3.77	3.74	3.70	3.67
	.025	8.81	7.26	6.60	6.23	5.99	5.82	5.70	5.60	5.52	5.46	5.37	5.27	5.17	5.12	5.07	5.01	4.96	4.90	4.85
	.010	13.75	10.92	9.78	9.15	8.75	8.47	8.26	8.10	7.98	7.87	7.72	7.56	7.40	7.31	7.23	7.14	7.06	6.97	6.88
	.005	18.63	14.54	12.92	12.03	11.46	11.07	10.79	10.57	10.39	10.25	10.03	9.81	9.59	9.47	9.36	9.24	9.12	9.00	8.88
7	.100	3.59	3.26	3.07	2.96	2.88	2.83	2.78	2.75	2.72	2.70	2.67	2.63	2.59	2.58	2.56	2.54	2.51	2.49	2.47
	.050	5.59	4.74	4.35	4.12	3.97	3.87	3.79	3.73	3.68	3.64	3.57	3.51	3.44	3.41	3.38	3.34	3.30	3.27	3.23
	.025	8.07	6.54	5.89	5.52	5.29	5.12	4.99	4.90	4.82	4.76	4.67	4.57	4.47	4.42	4.36	4.31	4.25	4.20	4.14
	.010	12.25	9.55	8.45	7.85	7.46	7.19	6.99	6.84	6.72	6.62	6.47	6.31	6.16	6.07	5.99	5.91	5.82	5.74	5.65
	.005	16.24	12.40	10.88	10.05	9.52	9.16	8.89	8.68	8.51	8.38	8.18	7.97	7.75	7.65	7.53	7.42	7.31	7.19	7.08

TABLE V (continued)

ν_2	α	1	2	3	4	5	6	7	8	9	10	12	15	20	24	30	40	60	120	∞
8	.100	3.46	3.11	2.92	2.81	2.73	2.67	2.62	2.59	2.56	2.54	2.50	2.46	2.42	2.40	2.38	2.36	2.34	2.32	2.29
	.050	5.32	4.46	4.07	3.84	3.69	3.58	3.50	3.44	3.39	3.35	3.28	3.22	3.15	3.12	3.08	3.04	3.01	2.97	2.93
	.025	7.57	6.06	5.42	5.05	4.82	4.65	4.53	4.43	4.36	4.30	4.20	4.10	4.00	3.95	3.89	3.84	3.78	3.73	3.67
	.010	11.26	8.65	7.59	7.01	6.63	6.37	6.18	6.03	5.91	5.81	5.67	5.52	5.36	5.28	5.20	5.12	5.03	4.95	4.86
	.005	14.69	11.04	9.60	8.81	8.30	7.95	7.69	7.50	7.34	7.21	7.01	6.81	6.61	6.50	6.40	6.29	6.18	6.06	5.95
9	.100	3.36	3.01	2.81	2.69	2.61	2.55	2.51	2.47	2.44	2.42	2.38	2.34	2.30	2.28	2.25	2.23	2.21	2.18	2.16
	.050	5.12	4.26	3.86	3.63	3.48	3.37	3.29	3.23	3.18	3.14	3.07	3.01	2.94	2.90	2.86	2.83	2.79	2.75	2.71
	.025	7.21	5.71	5.08	4.72	4.48	4.32	4.20	4.10	4.03	3.96	3.87	3.77	3.67	3.61	3.56	3.51	3.45	3.39	3.33
	.010	10.56	8.02	6.99	6.42	6.06	5.80	5.61	5.47	5.35	5.26	5.11	4.96	4.81	4.73	4.65	4.57	4.48	4.40	4.31
	.005	13.61	10.11	8.72	7.96	7.47	7.13	6.88	6.69	6.54	6.42	6.23	6.03	5.83	5.73	5.62	5.52	5.41	5.30	5.19
10	.100	3.29	2.92	2.73	2.61	2.52	2.46	2.41	2.38	2.35	2.32	2.28	2.24	2.20	2.18	2.16	2.13	2.11	2.08	2.06
	.050	4.96	4.10	3.71	3.48	3.33	3.22	3.14	3.07	3.02	2.98	2.91	2.85	2.77	2.74	2.70	2.66	2.62	2.58	2.54
	.025	6.94	5.46	4.83	4.47	4.24	4.07	3.95	3.85	3.78	3.72	3.62	3.52	3.42	3.37	3.31	3.26	3.20	3.14	3.08
	.010	10.04	7.56	6.55	5.99	5.64	5.39	5.20	5.06	4.94	4.85	4.71	4.56	4.41	4.33	4.25	4.17	4.08	4.00	3.91
	.005	12.83	9.43	8.08	7.34	6.87	6.54	6.30	6.12	5.97	5.85	5.66	5.47	5.27	5.17	5.07	4.97	4.86	4.75	4.64
11	.100	3.23	2.86	2.66	2.54	2.45	2.39	2.34	2.30	2.27	2.25	2.21	2.17	2.12	2.10	2.08	2.05	2.03	2.00	1.97
	.050	4.84	3.98	3.59	3.36	3.20	3.09	3.01	2.95	2.90	2.85	2.79	2.72	2.65	2.61	2.57	2.53	2.49	2.45	2.40
	.025	6.72	5.26	4.63	4.28	4.04	3.88	3.76	3.66	3.59	3.53	3.43	3.33	3.23	3.17	3.12	3.06	3.00	2.94	2.88
	.010	9.65	7.21	6.22	5.67	5.32	5.07	4.89	4.74	4.63	4.54	4.40	4.25	4.10	4.02	3.94	3.86	3.78	3.69	3.60
	.005	12.23	8.91	7.60	6.88	6.42	6.10	5.86	5.68	5.54	5.42	5.24	5.05	4.86	4.76	4.65	4.55	4.44	4.34	4.23
12	.100	3.18	2.81	2.61	2.48	2.39	2.33	2.28	2.24	2.21	2.19	2.15	2.10	2.06	2.04	2.01	1.99	1.96	1.93	1.90
	.050	4.75	3.89	3.49	3.26	3.11	3.00	2.91	2.85	2.80	2.75	2.69	2.62	2.54	2.51	2.47	2.43	2.38	2.34	2.30
	.025	6.55	5.10	4.47	4.12	3.89	3.73	3.61	3.51	3.44	3.37	3.28	3.18	3.07	3.02	2.96	2.91	2.85	2.79	2.72
	.010	9.33	6.93	5.95	5.41	5.06	4.82	4.64	4.50	4.39	4.30	4.16	4.01	3.86	3.78	3.70	3.62	3.54	3.45	3.36
	.005	11.75	8.51	7.23	6.52	6.07	5.76	5.52	5.35	5.20	5.09	4.91	4.72	4.53	4.43	4.33	4.23	4.12	4.01	3.90
13	.100	3.14	2.76	2.56	2.43	2.35	2.28	2.23	2.20	2.16	2.14	2.10	2.05	2.01	1.98	1.96	1.93	1.90	1.88	1.85
	.050	4.67	3.81	3.41	3.18	3.03	2.92	2.83	2.77	2.71	2.67	2.60	2.53	2.46	2.42	2.38	2.34	2.30	2.25	2.21
	.025	6.41	4.97	4.35	4.00	3.77	3.60	3.48	3.39	3.31	3.25	3.15	3.05	2.95	2.89	2.84	2.78	2.72	2.66	2.60
	.010	9.07	6.70	5.74	5.21	4.86	4.62	4.44	4.30	4.19	4.10	3.96	3.82	3.66	3.59	3.51	3.43	3.34	3.25	3.17
	.005	11.37	8.19	6.93	6.23	5.79	5.48	5.25	5.08	4.94	4.82	4.64	4.46	4.27	4.17	4.07	3.97	3.87	3.76	3.65
14	.100	3.10	2.73	2.52	2.39	2.31	2.24	2.19	2.15	2.12	2.10	2.05	2.01	1.96	1.94	1.91	1.89	1.86	1.83	1.80
	.050	4.60	3.74	3.34	3.11	2.96	2.85	2.76	2.70	2.65	2.60	2.53	2.46	2.39	2.35	2.31	2.27	2.22	2.18	2.13
	.025	6.30	4.86	4.24	3.89	3.66	3.50	3.38	3.29	3.21	3.15	3.05	2.95	2.84	2.79	2.73	2.67	2.61	2.55	2.49
	.010	8.86	6.51	5.56	5.04	4.69	4.46	4.28	4.14	4.03	3.94	3.80	3.66	3.51	3.43	3.35	3.27	3.18	3.09	3.00
	.005	11.06	7.92	6.68	6.00	5.56	5.26	5.03	4.86	4.72	4.60	4.43	4.25	4.06	3.96	3.86	3.76	3.66	3.55	3.44

TABLE V (continued)

ν_2	α	\multicolumn{15}{c	}{ν_1}
		1	2	3	4	5	6	7	8	9	10	12	15	20	24	30	40	60	120	∞
15	.100	3.07	2.70	2.49	2.36	2.27	2.21	2.16	2.12	2.09	2.06	2.02	1.97	1.92	1.90	1.87	1.85	1.82	1.79	1.76
	.050	4.54	3.68	3.29	3.06	2.90	2.79	2.71	2.64	2.59	2.54	2.48	2.40	2.33	2.29	2.25	2.20	2.16	2.11	2.07
	.025	6.20	4.77	4.15	3.80	3.58	3.41	3.29	3.20	3.12	3.06	2.96	2.86	2.76	2.70	2.64	2.59	2.52	2.46	2.40
	.010	8.68	6.36	5.42	4.89	4.56	4.32	4.14	4.00	3.89	3.80	3.67	3.52	3.37	3.29	3.21	3.13	3.05	2.96	2.87
	.005	10.80	7.70	6.48	5.80	5.37	5.07	4.85	4.67	4.54	4.42	4.25	4.07	3.88	3.79	3.69	3.58	3.48	3.37	3.26
16	.100	3.05	2.67	2.46	2.33	2.24	2.18	2.13	2.09	2.06	2.03	1.99	1.94	1.89	1.87	1.84	1.81	1.78	1.75	1.72
	.050	4.49	3.63	3.24	3.01	2.85	2.74	2.66	2.59	2.54	2.49	2.42	2.35	2.28	2.24	2.19	2.15	2.11	2.06	2.01
	.025	6.12	4.69	4.08	3.73	3.50	3.34	3.22	3.12	3.05	2.99	2.89	2.79	2.68	2.63	2.57	2.51	2.45	2.38	2.32
	.010	8.53	6.23	5.29	4.77	4.44	4.20	4.03	3.89	3.78	3.69	3.55	3.41	3.26	3.18	3.10	3.02	2.93	2.84	2.75
	.005	10.58	7.51	6.30	5.64	5.21	4.91	4.69	4.52	4.38	4.27	4.10	3.92	3.73	3.64	3.54	3.44	3.33	3.22	3.11
17	.100	3.03	2.64	2.44	2.31	2.22	2.15	2.10	2.06	2.03	2.00	1.96	1.91	1.86	1.84	1.81	1.78	1.75	1.72	1.69
	.050	4.45	3.59	3.20	2.96	2.81	2.70	2.61	2.55	2.49	2.45	2.38	2.31	2.23	2.19	2.15	2.10	2.06	2.01	1.96
	.025	6.04	4.62	4.01	3.66	3.44	3.28	3.16	3.06	2.98	2.92	2.82	2.72	2.62	2.56	2.50	2.44	2.38	2.32	2.25
	.010	8.40	6.11	5.18	4.67	4.34	4.10	3.93	3.79	3.68	3.59	3.46	3.31	3.16	3.08	3.0	2.92	2.83	2.75	2.65
	.005	10.38	7.35	6.16	5.50	5.07	4.78	4.56	4.39	4.25	4.14	3.97	3.79	3.61	3.51	3.41	3.31	3.21	3.10	2.98
18	.100	3.01	2.62	2.42	2.29	2.20	2.13	2.08	2.04	2.00	1.98	1.93	1.89	1.84	1.81	1.78	1.75	1.72	1.69	1.66
	.050	4.41	3.55	3.16	2.93	2.77	2.66	2.58	2.51	2.46	2.41	2.34	2.27	2.19	2.15	2.11	2.06	2.02	1.97	1.92
	.025	5.98	4.56	3.95	3.61	3.38	3.22	3.10	3.01	2.93	2.87	2.77	2.67	2.56	2.50	2.44	2.38	2.32	2.26	2.19
	.010	8.29	6.01	5.09	4.58	4.25	4.01	3.84	3.71	3.60	3.51	3.37	3.23	3.08	3.00	2.92	2.84	2.75	2.66	2.57
	.005	10.22	7.21	6.03	5.37	4.96	4.66	4.44	4.28	4.14	4.03	3.86	3.68	3.50	3.40	3.30	3.20	3.10	2.99	2.87
19	.100	2.99	2.61	2.40	2.27	2.18	2.11	2.06	2.02	1.98	1.96	1.91	1.86	1.81	1.79	1.76	1.73	1.70	1.67	1.63
	.050	4.38	3.52	3.13	2.90	2.74	2.63	2.54	2.48	2.42	2.38	2.31	2.23	2.16	2.11	2.07	2.03	1.98	1.93	1.88
	.025	5.92	4.51	3.90	3.56	3.33	3.17	3.05	2.96	2.88	2.82	2.72	2.62	2.51	2.45	2.39	2.33	2.27	2.20	2.13
	.010	8.18	5.93	5.01	4.50	4.17	3.94	3.77	3.63	3.52	3.43	3.30	3.15	3.00	2.92	2.84	2.76	2.67	2.58	2.49
	.005	10.07	7.09	5.92	5.27	4.85	4.56	4.34	4.18	4.04	3.93	3.76	3.59	3.40	3.31	3.21	3.11	3.00	2.89	2.78
20	.100	2.97	2.59	2.38	2.25	2.16	2.09	2.04	2.00	1.96	1.94	1.89	1.84	1.79	1.77	1.74	1.71	1.68	1.64	1.61
	.050	4.35	3.49	3.10	2.87	2.71	2.60	2.51	2.45	2.39	2.35	2.28	2.20	2.12	2.08	2.04	1.99	1.95	1.90	1.84
	.025	5.87	4.46	3.86	3.51	3.29	3.13	3.01	2.91	2.84	2.77	2.68	2.57	2.46	2.41	2.35	2.29	2.22	2.16	2.09
	.010	8.10	5.85	4.94	4.43	4.10	3.87	3.70	3.56	3.46	3.37	3.23	3.09	2.94	2.86	2.78	2.69	2.61	2.52	2.42
	.005	9.94	6.99	5.82	5.17	4.76	4.47	4.26	4.09	3.96	3.85	3.68	3.50	3.32	3.22	3.12	3.02	2.92	2.81	2.69
21	.100	2.96	2.57	2.36	2.23	2.14	2.08	2.02	1.98	1.95	1.92	1.87	1.83	1.78	1.75	1.72	1.69	1.66	1.62	1.59
	.050	4.32	3.47	3.07	2.84	2.68	2.57	2.49	2.42	2.37	2.32	2.25	2.18	2.10	2.05	2.01	1.96	1.92	1.87	1.81
	.025	5.83	4.42	3.82	3.48	3.25	3.09	2.97	2.87	2.80	2.73	2.64	2.53	2.42	2.37	2.31	2.25	2.18	2.11	2.04
	.010	8.02	5.78	4.87	4.37	4.04	3.81	3.64	3.51	3.40	3.31	3.17	3.03	2.88	2.80	2.72	2.64	2.55	2.46	2.36
	.005	9.83	6.89	5.73	5.09	4.68	4.39	4.18	4.01	3.88	3.77	3.60	3.43	3.24	3.15	3.05	2.95	2.84	2.73	2.61

TABLE V (*continued*)

ν_2	α	1	2	3	4	5	6	7	8	9	10	12	15	20	24	30	40	60	120	∞
22	.100	2.95	2.56	2.35	2.22	2.13	2.06	2.01	1.97	1.93	1.90	1.86	1.81	1.76	1.73	1.70	1.67	1.64	1.60	1.57
	.050	4.30	3.44	3.05	2.82	2.66	2.55	2.46	2.40	2.34	2.30	2.23	2.15	2.07	2.03	1.98	1.94	1.89	1.84	1.78
	.025	5.79	4.38	3.78	3.44	3.22	3.05	2.93	2.84	2.76	2.70	2.60	2.50	2.39	2.33	2.27	2.21	2.14	2.08	2.00
	.010	7.95	5.72	4.82	4.31	3.99	3.76	3.59	3.45	3.35	3.26	3.12	2.98	2.83	2.75	2.67	2.58	2.50	2.40	2.31
	.005	9.73	6.81	5.65	5.02	4.61	4.32	4.11	3.94	3.81	3.70	3.54	3.36	3.18	3.08	2.98	2.88	2.77	2.66	2.55
23	.100	2.94	2.55	2.34	2.21	2.11	2.05	1.99	1.95	1.92	1.89	1.84	1.80	1.74	1.72	1.69	1.66	1.62	1.59	1.55
	.050	4.28	3.42	3.03	2.80	2.64	2.53	2.44	2.37	2.32	2.27	2.20	2.13	2.05	2.01	1.96	1.91	1.86	1.81	1.76
	.025	5.75	4.35	3.75	3.41	3.18	3.02	2.90	2.81	2.73	2.67	2.57	2.47	2.36	2.30	2.24	2.18	2.11	2.04	1.97
	.010	7.88	5.66	4.76	4.26	3.94	3.71	3.54	3.41	3.30	3.21	3.07	2.93	2.78	2.70	2.62	2.54	2.45	2.35	2.26
	.005	9.63	6.73	5.58	4.95	4.54	4.26	4.05	3.88	3.75	3.64	3.47	3.30	3.12	3.02	2.92	2.82	2.71	2.60	2.48
24	.100	2.93	2.54	2.33	2.19	2.10	2.04	1.98	1.94	1.91	1.88	1.83	1.78	1.73	1.70	1.67	1.64	1.61	1.57	1.53
	.050	4.26	3.40	3.01	2.78	2.62	2.51	2.42	2.36	2.30	2.25	2.18	2.11	2.03	1.98	1.94	1.89	1.84	1.79	1.73
	.025	5.72	4.32	3.72	3.38	3.15	2.99	2.87	2.78	2.70	2.64	2.54	2.44	2.33	2.27	2.21	2.15	2.08	2.01	1.94
	.010	7.82	5.61	4.72	4.22	3.90	3.67	3.50	3.36	3.26	3.17	3.03	2.89	2.74	2.66	2.58	2.49	2.40	2.31	2.21
	.005	9.55	6.66	5.52	4.89	4.49	4.20	3.99	3.83	3.69	3.59	3.42	3.25	3.06	2.97	2.87	2.77	2.66	2.55	2.43
25	.100	2.92	2.53	2.32	2.18	2.09	2.02	1.97	1.93	1.89	1.87	1.82	1.77	1.72	1.69	1.66	1.63	1.59	1.56	1.52
	.050	4.24	3.39	2.99	2.76	2.60	2.49	2.40	2.34	2.28	2.24	2.16	2.09	2.01	1.96	1.92	1.87	1.82	1.77	1.71
	.025	5.69	4.29	3.69	3.35	3.13	2.97	2.85	2.75	2.68	2.61	2.51	2.41	2.30	2.24	2.18	2.12	2.05	1.98	1.91
	.010	5.77	5.57	4.68	4.18	3.85	3.63	3.46	3.32	3.22	3.13	2.99	2.85	2.70	2.62	2.54	2.45	2.36	2.27	2.17
	.005	9.48	6.60	5.46	4.84	4.43	4.15	3.94	3.78	3.64	3.54	3.37	3.20	3.01	2.92	2.82	2.72	2.61	2.50	2.38
26	.100	2.91	2.52	2.31	2.17	2.08	2.01	1.96	1.92	1.88	1.86	1.81	1.76	1.71	1.68	1.65	1.61	1.58	1.54	1.50
	.050	4.23	3.37	2.98	2.74	2.59	2.47	2.39	2.32	2.27	2.22	2.15	2.07	1.99	1.95	1.90	1.85	1.80	1.75	1.69
	.025	5.66	4.27	3.67	3.33	3.10	2.94	2.82	2.73	2.65	2.59	2.49	2.39	2.28	2.22	2.16	2.09	2.03	1.95	1.88
	.010	7.72	5.53	4.64	4.14	3.82	3.59	3.42	3.29	3.18	3.09	2.96	2.81	2.66	2.58	2.50	2.42	2.33	2.23	2.13
	.005	9.41	6.54	5.41	4.79	4.38	4.10	3.89	3.73	3.60	3.49	3.33	3.15	2.97	2.87	2.77	2.67	2.56	2.45	2.33
27	.100	2.90	2.51	2.30	2.17	2.07	2.00	1.95	1.91	1.87	1.85	1.80	1.75	1.70	1.67	1.64	1.60	1.57	1.53	1.49
	.050	4.21	3.35	2.96	2.73	2.57	2.46	2.37	2.31	2.25	2.20	2.13	2.06	1.97	1.93	1.88	1.84	1.79	1.73	1.67
	.025	5.63	4.24	3.65	3.31	3.08	2.92	2.80	2.71	2.63	2.57	2.47	2.36	2.25	2.19	2.13	2.07	2.00	1.93	1.85
	.010	7.68	5.49	4.60	4.11	3.78	3.56	3.39	3.26	3.15	3.06	2.93	2.78	2.63	2.55	2.47	2.38	2.29	2.20	2.10
	.005	9.34	6.49	5.36	4.74	4.34	4.06	3.85	3.69	3.56	3.45	3.28	3.11	2.93	2.83	2.73	2.63	2.52	2.41	2.29
28	.100	2.89	2.50	2.29	2.16	2.06	2.00	1.94	1.90	1.87	1.84	1.79	1.74	1.69	1.66	1.63	1.59	1.56	1.52	1.48
	.050	4.20	3.34	2.95	2.71	2.56	2.45	2.36	2.29	2.24	2.19	2.12	2.04	1.96	1.91	1.87	1.82	1.77	1.71	1.65
	.025	5.61	4.22	3.63	3.29	3.06	2.90	2.78	2.69	2.61	2.55	2.45	2.34	2.23	2.17	2.11	2.05	1.98	1.91	1.83
	.010	7.64	5.45	4.57	4.07	3.75	3.53	3.36	3.23	3.12	3.03	2.90	2.75	2.60	2.52	2.44	2.35	2.26	2.17	2.06
	.005	9.28	6.44	5.32	4.70	4.30	4.02	3.81	3.65	3.52	3.41	3.25	3.07	2.89	2.79	2.69	2.59	2.48	2.37	2.25

TABLE V (continued)

v_2	α	1	2	3	4	5	6	7	8	9	10	12	15	20	24	30	40	60	120	∞
29	.100	2.89	2.50	2.28	2.15	2.06	1.99	1.93	1.89	1.86	1.83	1.78	1.73	1.68	1.65	1.62	1.58	1.55	1.51	1.47
	.050	4.18	3.33	2.93	2.70	2.55	2.43	2.35	2.28	2.22	2.18	2.10	2.03	1.94	1.90	1.85	1.81	1.75	1.70	1.64
	.025	5.59	4.20	3.61	3.27	3.04	2.88	2.76	2.67	2.59	2.53	2.43	2.32	2.21	2.15	2.09	2.03	1.96	1.89	1.81
	.010	7.60	5.42	4.54	4.04	3.73	3.50	3.33	3.20	3.09	3.00	2.87	2.73	2.57	2.49	2.41	2.33	2.23	2.14	2.03
	.005	9.23	6.40	5.28	4.66	4.26	3.98	3.77	3.61	3.48	3.38	3.21	3.04	2.86	2.76	2.66	2.56	2.45	2.33	2.21
30	.100	2.88	2.49	2.28	2.14	2.05	1.98	1.93	1.88	1.85	1.82	1.77	1.72	1.67	1.64	1.61	1.57	1.54	1.50	1.46
	.050	4.17	3.32	2.92	2.69	2.53	2.42	2.33	2.27	2.21	2.16	2.09	2.01	1.93	1.89	1.84	1.79	1.74	1.68	1.62
	.025	5.57	4.18	3.59	3.25	3.03	2.87	2.75	2.65	2.57	2.51	2.41	2.31	2.20	2.14	2.07	2.01	1.94	1.87	1.79
	.010	7.56	5.39	4.51	4.02	3.70	3.47	3.30	3.17	3.07	2.98	2.84	2.70	2.55	2.47	2.39	2.30	2.21	2.11	2.01
	.005	9.18	6.35	5.24	4.62	4.23	3.95	3.74	3.58	3.45	3.34	3.18	3.01	2.82	2.73	2.63	2.52	2.42	2.30	2.18
40	.100	2.84	2.44	2.23	2.09	2.00	1.93	1.87	1.83	1.79	1.76	1.71	1.66	1.61	1.57	1.54	1.51	1.47	1.42	1.38
	.050	4.08	3.23	2.84	2.61	2.45	2.34	2.25	2.18	2.12	2.08	2.00	1.92	1.84	1.79	1.74	1.69	1.64	1.58	1.51
	.025	5.42	4.05	3.46	3.13	2.90	2.74	2.62	2.53	2.45	2.39	2.29	2.18	2.07	2.01	1.94	1.88	1.80	1.72	1.64
	.010	7.31	5.18	4.31	3.83	3.51	3.29	3.12	2.99	2.89	2.80	2.66	2.52	2.37	2.29	2.20	2.11	2.02	1.92	1.80
	.005	8.83	6.07	4.98	4.37	3.99	3.71	3.51	3.35	3.22	3.12	2.95	2.78	2.60	2.50	2.40	2.30	2.18	2.06	1.93
60	.100	2.79	2.39	2.18	2.04	1.95	1.87	1.82	1.77	1.74	1.71	1.66	1.60	1.54	1.51	1.48	1.44	1.40	1.35	1.29
	.050	4.00	3.15	2.76	2.53	2.37	2.25	2.17	2.10	2.04	1.99	1.92	1.84	1.75	1.70	1.65	1.59	1.53	1.47	1.39
	.025	5.29	3.93	3.34	3.01	2.79	2.63	2.51	2.41	2.33	2.27	2.17	2.06	1.94	1.88	1.82	1.74	1.67	1.58	1.48
	.010	7.08	4.98	4.13	3.65	3.34	3.12	2.95	2.82	2.72	2.63	2.50	2.35	2.20	2.12	2.03	1.94	1.84	1.73	1.60
	.005	8.49	5.79	4.73	4.14	3.76	3.49	3.29	3.13	3.01	2.90	2.74	2.57	2.39	2.29	2.19	2.08	1.96	1.83	1.69
120	.100	2.75	2.35	2.13	1.99	1.90	1.82	1.77	1.72	1.68	1.65	1.60	1.55	1.48	1.45	1.41	1.37	1.32	1.26	1.19
	.050	3.92	3.07	2.68	2.45	2.29	2.17	2.09	2.02	1.96	1.91	1.83	1.75	1.66	1.61	1.55	1.50	1.43	1.35	1.25
	.025	5.15	3.80	3.23	2.89	2.67	2.52	2.39	2.30	2.22	2.16	2.05	1.94	1.82	1.76	1.69	1.61	1.53	1.43	1.31
	.010	6.85	4.79	3.95	3.48	3.17	2.96	2.79	2.66	2.56	2.47	2.34	2.19	2.03	1.95	1.86	1.76	1.66	1.53	1.38
	.005	8.18	5.54	4.50	3.92	3.55	3.28	3.09	2.93	2.81	2.71	2.54	2.37	2.19	2.09	1.98	1.87	1.75	1.61	1.43
∞	.100	2.71	2.30	2.08	1.94	1.85	1.77	1.72	1.67	1.63	1.60	1.55	1.49	1.42	1.38	1.34	1.30	1.24	1.17	1.00
	.050	3.84	3.00	2.60	2.37	2.21	2.10	2.01	1.94	1.88	1.83	1.75	1.67	1.57	1.52	1.46	1.39	1.32	1.22	1.00
	.025	5.02	3.69	3.12	2.79	2.57	2.41	2.29	2.19	2.11	2.05	1.94	1.83	1.71	1.64	1.57	1.48	1.39	1.27	1.00
	.010	6.63	4.61	3.78	3.32	3.02	2.80	2.64	2.51	2.41	2.32	2.18	2.04	1.88	1.79	1.70	1.59	1.47	1.32	1.00
	.005	7.88	5.30	4.28	3.72	3.35	3.09	2.90	2.74	2.62	2.52	2.36	2.19	2.00	1.90	1.79	1.67	1.53	1.36	1.00

From M. Merrington and C. M. Thompson, "Tables of Percentage Points of the Inverted Beta (F) Distribution," *Biometrika*, 33(1943), 78–87. Reprinted by permission.

Table VI. Power of the Student's t Test

This table gives the values of the noncentrality parameter δ of the noncentral t distribution with degrees of freedom $\nu = 1\ (1)\ 30, 40, 60, 100, \infty$; one-tailed level of significance $\alpha = 0.05, 0.025, 0.01$; and the power $= 1 - \beta = 0.10\ (0.10)\ 0.90, 0.95, 0.99$. Since the distribution of t is symmetrical about zero, the one-tailed levels of significance also represent two-tailed values of $\alpha = 0.10, 0.05,$ and 0.02. The table can be used to determine the power of a test of significance based on the Student's t distribution. For example, the power of the t test corresponding to $\nu = 30, \delta = 3.0,$ and $\alpha = 0.05$ is approximately equal to 0.90.

$\alpha = 0.05$
Power $= 1-\beta$

ν	0.99	0.95	0.90	0.80	0.70	0.60	0.50	0.40	0.30	0.20	0.10
1	16.47	12.53	10.51	8.19	6.63	5.38	4.31	3.35	2.46	1.60	.64
2	6.88	5.52	4.81	3.98	3.40	2.92	2.49	2.07	1.63	1.15	.50
3	5.47	4.46	3.93	3.30	2.85	2.48	2.13	1.79	1.43	1.02	.46
4	4.95	4.07	3.60	3.04	2.64	2.30	1.99	1.67	1.34	.96	.43
5	4.70	3.87	3.43	2.90	2.53	2.21	1.91	1.61	1.29	.92	.42
6	4.55	3.75	3.33	2.82	2.46	2.15	1.86	1.57	1.26	.90	.41
7	4.45	3.67	3.26	2.77	2.41	2.11	1.82	1.54	1.24	.89	.40
8	4.38	3.62	3.21	2.73	2.38	2.08	1.80	1.52	1.22	.88	.40
9	4.32	3.58	3.18	2.70	2.35	2.06	1.78	1.51	1.21	.87	.39
10	4.28	3.54	3.15	2.67	2.33	2.04	1.77	1.49	1.20	.86	.39
11	4.25	3.52	3.13	2.65	2.31	2.02	1.75	1.48	1.19	.86	.39
12	4.22	3.50	3.11	2.64	2.30	2.01	1.74	1.47	1.19	.85	.38
13	4.20	3.48	3.09	2.63	2.29	2.00	1.74	1.47	1.18	.85	.38
14	4.18	3.46	3.08	2.62	2.28	2.00	1.73	1.46	1.18	.84	.38
15	4.17	3.45	3.07	2.61	2.27	1.99	1.72	1.46	1.17	.84	.38
16	4.16	3.44	3.06	2.60	2.27	1.98	1.72	1.45	1.17	.84	.38
17	4.14	3.43	3.05	2.59	2.26	1.98	1.71	1.45	1.17	.84	.38
18	4.13	3.42	3.04	2.59	2.26	1.97	1.71	1.45	1.16	.83	.38
19	4.12	3.41	3.04	2.58	2.25	1.97	1.71	1.44	1.16	.83	.38
20	4.12	3.41	3.03	2.58	2.25	1.97	1.70	1.44	1.16	.83	.38
21	4.11	3.40	3.03	2.57	2.24	1.96	1.70	1.44	1.16	.83	.38
22	4.10	3.40	3.02	2.57	2.24	1.96	1.70	1.44	1.16	.83	.37
23	4.10	3.39	3.02	2.56	2.24	1.96	1.69	1.43	1.15	.83	.37
24	4.09	3.39	3.01	2.56	2.23	1.95	1.69	1.43	1.15	.83	.37
25	4.09	3.38	3.01	2.56	2.23	1.95	1.69	1.43	1.15	.83	.37
26	4.08	3.38	3.01	2.55	2.23	1.95	1.69	1.43	1.15	.82	.37
27	4.08	3.38	3.00	2.55	2.23	1.95	1.69	1.43	1.15	.82	.37
28	4.07	3.37	3.00	2.55	2.22	1.95	1.69	1.43	1.15	.82	.37
29	4.07	3.37	3.00	2.55	2.22	1.94	1.68	1.42	1.15	.82	.37
30	4.07	3.37	3.00	2.54	2.22	1.94	1.68	1.42	1.15	.82	.37
40	4.04	3.35	2.98	2.53	2.21	1.93	1.67	1.42	1.14	.82	.37
60	4.02	3.33	2.96	2.52	2.19	1.92	1.66	1.41	1.13	.81	.37
100	4.00	3.31	2.95	2.50	2.18	1.91	1.66	1.40	1.13	.81	.37
∞	3.97	3.29	2.93	2.49	2.17	1.90	1.64	1.39	1.12	.80	.36

TABLE VI (*continued*)

$\alpha = 0.025$
Power $= 1 - \beta$

ν	0.99	0.95	0.90	0.80	0.70	0.60	0.50	0.40	0.30	0.20	0.10
1	32.83	24.98	20.96	16.33	13.21	10.73	8.60	6.68	4.91	3.22	1.58
2	9.67	7.77	6.80	5.65	4.86	4.21	3.63	3.07	2.50	1.88	1.09
3	6.88	5.65	5.01	4.26	3.72	3.28	2.87	2.47	2.05	1.57	.94
4	5.94	4.93	4.40	3.76	3.31	2.93	2.58	2.23	1.86	1.44	.87
5	5.49	4.57	4.09	3.51	3.10	2.75	2.43	2.11	1.76	1.37	.82
6	5.22	4.37	3.91	3.37	2.98	2.64	2.34	2.03	1.70	1.32	.80
7	5.06	4.23	3.80	3.27	2.89	2.57	2.27	1.98	1.66	1.29	.78
8	4.94	4.14	3.71	3.20	2.83	2.52	2.23	1.94	1.63	1.27	.77
9	4.85	4.07	3.65	3.15	2.79	2.48	2.20	1.91	1.60	1.25	.76
10	4.78	4.01	3.60	3.11	2.75	2.45	2.17	1.89	1.59	1.23	.75
11	4.73	3.97	3.57	3.08	2.73	2.43	2.15	1.87	1.57	1.22	.74
12	4.69	3.93	3.54	3.05	2.70	2.41	2.13	1.85	1.56	1.21	.74
13	4.65	3.91	3.51	3.03	2.69	2.39	2.12	1.84	1.55	1.21	.73
14	4.62	3.88	3.49	3.01	2.67	2.38	2.11	1.83	1.54	1.20	.73
15	4.60	3.86	3.47	3.00	2.66	2.37	2.09	1.82	1.53	1.19	.72
16	4.58	3.84	3.46	2.98	2.65	2.36	2.09	1.81	1.53	1.19	.72
17	4.56	3.83	3.44	2.97	2.64	2.35	2.08	1.81	1.52	1.18	.72
18	4.54	3.82	3.43	2.96	2.63	2.34	2.07	1.80	1.52	1.18	.72
19	4.52	3.80	3.42	2.95	2.61	2.33	2.06	1.80	1.51	1.17	.71
20	4.51	3.79	3.41	2.95	2.61	2.33	2.06	1.79	1.51	1.17	.71
21	4.50	3.78	3.40	2.93	2.60	2.32	2.05	1.79	1.50	1.17	.71
22	4.49	3.77	3.39	2.93	2.60	2.32	2.05	1.78	1.50	1.17	.71
23	4.48	3.77	3.39	2.93	2.59	2.31	2.05	1.78	1.50	1.17	.71
24	4.47	3.76	3.38	2.92	2.59	2.31	2.04	1.78	1.50	1.16	.71
25	4.46	3.75	3.37	2.92	2.58	2.30	2.04	1.77	1.49	1.16	.71
26	4.46	3.75	3.37	2.92	2.58	2.30	2.04	1.77	1.49	1.16	.70
27	4.45	3.74	3.36	2.91	2.58	2.30	2.03	1.77	1.49	1.16	.70
28	4.44	3.73	3.36	2.90	2.57	2.29	2.03	1.77	1.49	1.16	.70
29	4.44	3.73	3.35	2.90	2.57	2.29	2.03	1.77	1.48	1.16	.70
30	4.43	3.73	3.35	2.90	2.57	2.29	2.02	1.76	1.48	1.16	.70
40	4.39	3.69	3.32	2.87	2.55	2.27	2.01	1.75	1.47	1.15	.69
60	4.36	3.66	3.29	2.85	2.53	2.25	1.99	1.73	1.46	1.14	.69
100	4.33	3.64	3.27	2.83	2.51	2.23	1.98	1.73	1.45	1.12	.68
∞	4.29	3.60	3.24	2.80	2.48	2.21	1.96	1.71	1.44	1.12	.68

TABLE VI (*continued*)

$\alpha = 0.01$
Power $= 1 - \beta$

ν	0.99	0.95	0.90	0.80	0.70	0.60	0.50	0.40	0.30	0.20	0.10
1	82.00	62.40	52.37	40.80	33.00	26.79	21.47	16.69	12.27	8.07	4.00
2	15.22	12.26	10.74	8.96	7.73	6.73	5.83	4.98	4.12	3.20	2.08
3	9.34	7.71	6.86	5.87	5.17	4.59	4.07	3.56	3.03	2.44	1.66
4	7.52	6.28	5.64	4.88	4.34	3.88	3.47	3.06	2.63	2.14	1.48
5	6.68	5.62	5.07	4.40	3.93	3.54	3.17	2.81	2.42	1.98	1.38
6	6.21	5.25	4.74	4.13	3.70	3.33	2.99	2.66	2.30	1.88	1.32
7	5.91	5.01	4.53	3.96	3.55	3.20	2.88	2.56	2.22	1.82	1.27
8	5.71	4.85	4.39	3.84	3.44	3.11	2.80	2.49	2.16	1.77	1.24
9	5.56	4.72	4.28	3.75	3.37	3.04	2.74	2.43	2.11	1.74	1.22
10	5.45	4.63	4.20	3.68	3.31	2.99	2.69	2.39	2.08	1.71	1.20
11	5.36	4.56	4.14	3.63	3.26	2.94	2.65	2.36	2.05	1.69	1.18
12	5.29	4.50	4.09	3.58	3.22	2.91	2.62	2.33	2.03	1.67	1.17
13	5.23	4.46	4.04	3.55	3.19	2.88	2.60	2.31	2.01	1.65	1.16
14	5.18	4.42	4.01	3.51	3.16	2.86	2.57	2.29	1.99	1.64	1.15
15	5.14	4.38	3.98	3.49	3.14	2.84	2.56	2.28	1.98	1.63	1.14
16	5.11	4.35	3.95	3.47	3.12	2.82	2.54	2.26	1.97	1.62	1.14
17	5.08	4.33	3.93	3.45	3.10	2.80	2.53	2.25	1.96	1.61	1.13
18	5.05	4.31	3.91	3.43	3.09	2.79	2.52	2.24	1.95	1.60	1.13
19	5.03	4.29	3.89	3.42	3.07	2.78	2.50	2.23	1.94	1.60	1.12
20	5.01	4.27	3.88	3.40	3.06	2.77	2.50	2.22	1.93	1.59	1.12
21	4.99	4.25	3.86	3.39	3.05	2.76	2.49	2.22	1.92	1.59	1.11
22	4.97	4.24	3.85	3.38	3.04	2.75	2.48	2.21	1.92	1.58	1.11
23	4.96	4.23	3.84	3.37	3.03	2.74	2.47	2.20	1.91	1.58	1.11
24	4.94	4.22	3.83	3.36	3.02	2.73	2.47	2.20	1.91	1.57	1.11
25	4.93	4.20	3.82	3.35	3.02	2.73	2.46	2.19	1.90	1.57	1.10
26	4.92	4.19	3.81	3.34	3.01	2.72	2.45	2.19	1.90	1.57	1.10
27	4.91	4.19	3.80	3.34	3.00	2.72	2.45	2.18	1.90	1.56	1.10
28	4.90	4.18	3.79	3.33	3.00	2.71	2.44	2.18	1.89	1.56	1.10
29	4.89	4.17	3.79	3.32	2.99	2.71	2.44	2.17	1.89	1.56	1.10
30	4.88	4.16	3.78	3.32	2.99	2.70	2.44	2.17	1.89	1.55	1.09
40	4.82	4.11	3.74	3.28	2.95	2.67	2.41	2.15	1.86	1.54	1.08
60	4.76	4.06	3.69	3.24	2.92	2.64	2.38	2.12	1.84	1.52	1.07
100	4.72	4.03	3.66	3.21	2.89	2.62	2.36	2.10	1.83	1.51	1.06
∞	4.65	3.97	3.61	3.17	2.85	2.58	2.33	2.07	1.80	1.48	1.04

From D. B. Owen, "The Power of Student's *t* Test," *Journal of the American Statistical Association*, 60 (1965), 320–333. Reprinted by permission.

Table VII. Power of the Analysis of Variance *F* Test

This table gives the values of type II error (β) of a test of significance based on the F distribution corresponding to the numerator degrees of freedom $\nu_1 = 1$ (1) 10 (2) 12; denominator degrees of freedom $\nu_2 = 2$ (2) 30, 40, 60, 120, ∞; standardized noncentrality parameter $\phi = 0.5$ (0.5) 1.0 (0.2) (2.2) (0.4) 3.0; and the level of significance $\alpha = 0.01, 0.05, 0.1$. For example, the power of the F test corresponding to $\nu_1 = 3$, $\nu_2 = 30$, $\phi = 1.4$, and $\alpha = 0.05$ is equal to $1 - 0.4182 = 0.5918$. To obtain power for odd values of ν_2 a linear interpolation in the reciprocal of ν_2 may be used, which generally gives three-decimal-place accuracy. To obtain power for values of ϕ, not given in the table $(0.5 < \phi < 3.0)$, a three-point Lagrangian interpolation may be used, which generally gives an accuracy of at least two decimal places. For $\phi \geq 3$, the values of power are mostly close to one.

$\alpha = 0.01$

ν_2	$\phi = .5$	1.0	1.2	1.4	1.6	1.8	2.0	2.2	2.6	3.0
\multicolumn{11}{c}{$\nu_1 = 1$}
2	.9851	.9705	.9620	.9521	.9408	.9282	.9143	.8991	.8654	.8277
4	.9809	.9492	.9280	.9012	.8682	.8292	.7843	.7341	.6216	.5014
6	.9782	.9340	.9030	.8629	.8131	.7541	.6870	.6136	.4589	.3125
8	.9764	.9236	.8859	.8367	.7759	.7043	.6242	.5387	.3678	.2211
10	.9752	.9163	.8738	.8184	.7501	.6704	.5824	.4904	.3136	.1725
12	.9743	.9109	.8650	.8050	.7314	.6462	.5532	.4574	.2787	.1437
14	.9736	.9068	.8582	.7949	.7174	.6283	.5318	.4336	.2547	.1250
16	.9730	.9036	.8529	.7870	.7066	.6145	.5156	.4158	.2374	.1121
18	.9726	.9010	.8487	.7807	.6979	.6036	.5028	.4020	.2243	.1027
20	.9723	.8989	.8452	.7755	.6908	.5947	.4925	.3910	.2141	.0957
22	.9720	.8971	.8423	.7712	.6850	.5874	.4841	.3820	.2060	.0902
24	.9717	.8956	.8398	.7675	.6801	.5813	.4771	.3746	.1994	.0858
26	.9715	.8943	.8377	.7644	.6758	.5760	.4712	.3683	.1938	.0822
28	.9713	.8931	.8359	.7617	.6722	.5716	.4661	.3630	.1892	.0792
30	.9711	.8922	.8343	.7593	.6690	.5677	.4617	.3584	.1852	.0767
40	.9705	.8886	.8285	.7509	.6578	.5539	.4462	.3424	.1718	.0683
60	.9699	.8850	.8226	.7423	.6463	.5401	.4308	.3267	.1590	.0608
120	.9693	.8812	.8165	.7335	.6347	.5261	.4155	.3113	.1468	.0539
∞	.9687	.8773	.8102	.7244	.6229	.5120	.4003	.2962	.1354	.0478
\multicolumn{11}{c}{$\nu_1 = 2$}
2	.9863	.9753	.9688	.9613	.9527	.9430	.9323	.9207	.8945	.8650
4	.9828	.9567	.9386	.9153	.8862	.8511	.8100	.7635	.6571	.5401
6	.9803	.9409	.9118	.8730	.8237	.7640	.6951	.6191	.4576	.3052
8	.9784	.9288	.8910	.8401	.7754	.6982	.6110	.5182	.3358	.1869
10	.9770	.9196	.8751	.8150	.7393	.6500	.5515	.4498	.2626	.1268
12	.9760	.9124	.8627	.7957	.7118	.6142	.5085	.4022	.2163	.0934
14	.9752	.9067	.8529	.7806	.6905	.5869	.4765	.3678	.1854	.0733
16	.9745	.9021	.8450	.7684	.6736	.5655	.4519	.3420	.1636	.0603
18	.9740	.8983	.8386	.7585	.6600	.5485	.4326	.3221	.1476	.0513
20	.9735	.8951	.8331	.7502	.6486	.5345	.4170	.3063	.1354	.0449
22	.9731	.8924	.8285	.7433	.6392	.5229	.4042	.2936	.1260	.0401
24	.9728	.8901	.8246	.7373	.6312	.5132	.3936	.2830	.1184	.0364
26	.9725	.8881	.8212	.7322	.6243	.5048	.3845	.2742	.1122	.0335
28	.9723	.8863	.8182	.7277	.6182	.4976	.3768	.2667	.1070	.0312
30	.9721	.8848	.8156	.7238	.6130	.4914	.3701	.2603	.1027	.0293
40	.9713	.8791	.8060	.7096	.5943	.4693	.3468	.2382	.0885	.0233
60	.9704	.8731	.7960	.6948	.5749	.4469	.3237	.2170	.0757	.0183
120	.9695	.8668	.7854	.6794	.5551	.4244	.3011	.1968	.0643	.0143
∞	.9686	.8600	.7743	.6634	.5349	.4019	.2789	.1776	.0543	.0111
\multicolumn{11}{c}{$\nu_1 = 3$}
2	.9867	.9769	.9711	.9644	.9567	.9481	.9385	.9280	.9045	.8779
4	.9835	.9592	.9421	.9199	.8919	.8580	.8181	.7726	.6678	.5517
6	.9809	.9427	.9136	.8742	.8237	.7620	.6906	.6117	.4448	.2899
8	.9790	.9291	.8896	.8357	.7665	.6835	.5902	.4917	.3032	.1576
10	.9775	.9181	.8703	.8047	.7214	.5234	.5166	.4085	.2191	.0941
12	.9763	.9093	.8547	.7800	.6861	.5776	.4625	.3504	.1675	.0615
14	.9753	.9021	.8419	.7600	.6580	.5421	.4220	.3086	.1343	.0434
16	.9746	.8961	.8314	.7437	.6354	.5142	.3910	.2776	.1118	.0325
18	.9739	.8910	.8227	.7302	.6169	.4917	.3666	.2540	.0958	.0256
20	.9734	.8868	.8152	.7188	.6016	.4733	.3471	.2355	.0841	.0209
22	.9729	.8831	.8089	.7092	.5886	.4580	.3311	.2207	.0752	.0175
24	.9725	.8799	.8034	.7008	.5776	.4451	.3178	.2087	.0683	.0151
26	.9721	.8772	.7986	.6936	.5681	.4341	.3066	.1987	.0628	.0132
28	.9718	.8747	.7944	.6873	.5599	.4247	.2971	.1903	.0584	.0118
30	.9716	.8725	.7906	.6817	.5526	.4164	.2889	.1831	.0547	.0107
40	.9705	.8645	.7769	.6614	.5266	.3873	.2606	.1590	.0430	.0074
60	.9694	.8558	.7622	.6400	.4997	.3582	.2332	.1367	.0334	.0050
120	.9682	.8464	.7464	.6175	.4721	.3292	.2070	.1163	.0255	.0033
∞	.9669	.8361	.7295	.5938	.4439	.3005	.1821	.0978	.0192	.0022

TABLE VII (continued)

ν_2	$\phi=.5$	1.0	1.2	1.4	$\alpha=0.01$ 1.6	1.8	2.0	2.2	2.6	3.0
					$\nu_1=4$					
2	.9869	.9777	.9723	.9660	.9587	.9506	.9416	.9317	.9096	.8844
4	.9838	.9604	.9438	.9221	.8946	.8612	.8217	.7767	.6725	.5566
6	.9812	.9433	.9139	.8738	.8219	.7585	.6848	.6036	.4330	.2767
8	.9792	.9284	.8873	.8306	.7575	.6697	.5716	.4691	.2776	.1363
10	.9776	.9160	.8650	.7944	.7047	.5996	.4885	.3745	.1867	.0726
12	.9763	.9056	.8464	.7647	.6622	.5452	.4236	.3087	.1330	.0424
14	.9752	.8969	.8309	.7403	.6281	.5027	.3765	.2620	.0998	.0270
16	.9743	.8896	.8178	.7200	.6003	.4691	.3405	.2279	.0783	.0184
18	.9736	.8834	.8068	.7030	.5774	.4420	.3124	.2023	.0636	.0133
20	.9730	.8780	.7974	.6886	.5583	.4199	.2901	.1825	.0533	.0101
22	.9724	.8734	.7892	.6763	.5421	.4015	.2719	.1670	.0457	.0079
24	.9719	.8693	.7821	.6656	.5283	.3861	.2570	.1545	.0400	.0064
26	.9715	.8657	.7759	.6563	.5164	.3730	.2445	.1442	.0355	.0054
28	.9711	.8625	.7704	.6482	.5060	.3617	.2340	.1357	.0320	.0046
30	.9708	.8597	.7655	.6409	.4969	.3519	.2249	.1286	.0292	.0040
40	.9695	.8491	.7473	.6145	.4643	.3176	.1942	.1051	.0207	.0023
60	.9682	.8374	.7275	.5864	.4306	.2836	.1653	.0844	.0143	.0013
120	.9666	.8245	.7060	.5566	.3962	.2504	.1386	.0665	.0096	.0007
∞	.9649	.8103	.6828	.5253	.3614	.2185	.1144	.0513	.0063	.0004
					$\nu_1=5$					
2	.9870	.9782	.9730	.9669	.9600	.9521	.9435	.9340	.9126	.8884
4	.9840	.9611	.9448	.9233	.8961	.8629	.8237	.7789	.6749	.5591
6	.9814	.9435	.9138	.8729	.8199	.7550	.6795	.5965	.4231	.2663
8	.9793	.9276	.8850	.8258	.7494	.6578	.5559	.4504	.2575	.1207
10	.9776	.9138	.8600	.7852	.6899	.5792	.4615	.3471	.1625	.0581
12	.9762	.9021	.8387	.7510	.6413	.5174	.3914	.2757	.1084	.0306
14	.9750	.8920	.8206	.7224	.6017	.4690	.3391	.2257	.0763	.0176
16	.9740	.8835	.8052	.6984	.5692	.4306	.2994	.1898	.0564	.0109
18	.9732	.8761	.7920	.6782	.5423	.3998	.2688	.1633	.0435	.0073
20	.9725	.8696	.7806	.6609	.5198	.3746	.2446	.1433	.0347	.0051
22	.9718	.8640	.7706	.6460	.5007	.3538	.2252	.1278	.0284	.0037
24	.9713	.8590	.7619	.6330	.4844	.3363	.2093	.1156	.0239	.0029
26	.9708	.8546	.7543	.6217	.4704	.3216	.1962	.1057	.0205	.0023
28	.9704	.8507	.7474	.6118	.4581	.3089	.1851	.0976	.0179	.0018
30	.9700	.8472	.7413	.6029	.4474	.2980	.1758	.0909	.0158	.0015
40	.9685	.8339	.7186	.5705	.4090	.2601	.1446	.0695	.0100	.0007
60	.9668	.8189	.6935	.5359	.3696	.2232	.1163	.0516	.0061	.0003
120	.9650	.8023	.6660	.4992	.3298	.1882	.0913	.0372	.0035	.0002
∞	.9628	.7835	.6360	.4606	.2901	.1556	.0699	.0259	.0020	.0000
					$\nu_1=6$					
2	.9871	.9785	.9735	.9675	.9608	.9532	.9447	.9355	.9147	.8910
4	.9841	.9616	.9454	.9241	.8971	.8640	.8250	.7802	.6764	.5605
6	.9815	.9435	.9135	.8720	.8180	.7518	.6749	.5905	.4149	.2578
8	.9794	.9267	.8826	.8216	.7424	.6477	.5428	.4351	.2417	.1090
10	.9776	.9118	.8556	.7770	.6772	.5618	.4406	.3248	.1442	.0480
12	.9761	.8988	.8318	.7388	.6230	.4938	.3647	.2492	.0905	.0230
14	.9748	.8876	.8113	.7064	.5785	.4402	.3083	.1971	.0600	.0120
16	.9737	.8778	.7936	.6790	.5418	.3978	.2659	.1604	.0419	.0068
18	.9728	.8692	.7783	.6556	.5113	.3639	.2335	.1339	.0306	.0042
20	.9720	.8617	.7650	.6356	.4857	.3363	.2082	.1142	.0232	.0027
22	.9713	.8551	.7533	.6183	.4641	.3136	.1881	.0992	.0182	.0019
24	.9707	.8493	.7430	.6032	.4456	.2946	.1719	.0876	.0147	.0013
26	.9701	.8441	.7339	.5900	.4297	.2787	.1586	.0784	.0121	.0010
28	.9696	.8394	.7258	.5783	.4158	.2651	.1476	.0710	.0102	.0008
30	.9692	.8351	.7185	.5679	.4037	.2533	.1383	.0649	.0088	.0006
40	.9675	.8191	.6911	.5299	.3605	.2133	.1080	.0462	.0049	.0002
60	.9655	.8008	.6607	.4891	.3166	.1753	.0816	.0315	.0026	.0001
120	.9633	.7800	.6271	.4459	.2728	.1402	.0595	.0205	.0013	.0000
∞	.9607	.7563	.5901	.4009	.2301	.1089	.0417	.0128	.0006	.0000
					$\nu_1=7$					
2	.9872	.9787	.9738	.9680	.9614	.9539	.9456	.9365	.9161	.8929
4	.9842	.9619	.9459	.9247	.8977	.8648	.8258	.7811	.6773	.5613
6	.9816	.9435	.9132	.8711	.8163	.7490	.6710	.5854	.4082	.2510
8	.9794	.9260	.8810	.8179	.7363	.6390	.5317	.4224	.2290	.1000
10	.9775	.9100	.8516	.7699	.6661	.5469	.4231	.3065	.1299	.0407
12	.9760	.8959	.8256	.7280	.6070	.4735	.3423	.2278	.0770	.0178
14	.9746	.8835	.8029	.6922	.5581	.4156	.2828	.1744	.0482	.0085
16	.9735	.8726	.7831	.6615	.5176	.3698	.2385	.1375	.0319	.0044
18	.9724	.8629	.7658	.6353	.4840	.3333	.2049	.1113	.0221	.0025
20	.9716	.8544	.7506	.6127	.4558	.3038	.1791	.0923	.0160	.0015
22	.9708	.8469	.7373	.5931	.4319	.2796	.1587	.0781	.0120	.0010
24	.9701	.8401	.7254	.5760	.4115	.2596	.1425	.0673	.0093	.0006
26	.9695	.8341	.7149	.5610	.3940	.2429	.1294	.0589	.0074	.0005
28	.9689	.8287	.7055	.5477	.3787	.2286	.1186	.0522	.0060	.0003
30	.9684	.8237	.6971	.5359	.3654	.2165	.1096	.0469	.0050	.0002
40	.9665	.8048	.6651	.4926	.3182	.1754	.0810	.0309	.0025	.0001
60	.9642	.7830	.6294	.4462	.2709	.1375	.0572	.0193	.0011	.0000
120	.9616	.7580	.5896	.3973	.2246	.1038	.0384	.0112	.0005	.0000
∞	.9585	.7290	.5456	.3467	.1807	.0751	.0244	.0062	.0002	.0000

TABLE VII (continued)

ν_2	$\phi=.5$	1.0	1.2	1.4	$\alpha=0.01$ 1.6	1.8	2.0	2.2	2.6	3.0
					$\nu_1=8$					
2	.9872	.9789	.9740	.9683	.9618	.9544	.9463	.9373	.9172	.8943
4	.9843	.9621	.9462	.9251	.8982	.8653	.8264	.7817	.6779	.5619
6	.9817	.9435	.9128	.8703	.8148	.7467	.6676	.5811	.4026	.2454
8	.9794	.9252	.8793	.8147	.7311	.6315	.5223	.4117	.2186	.0928
10	.9775	.9084	.8481	.7635	.6564	.5341	.4082	.2912	.1185	.0352
12	.9758	.8933	.8201	.7184	.5930	.4559	.3234	.2101	.0668	.0142
14	.9744	.8798	.7953	.6794	.5401	.3943	.2614	.1560	.0396	.0063
16	.9732	.8678	.7735	.6458	.4963	.3458	.2157	.1193	.0248	.0030
18	.9721	.8571	.7543	.6169	.4598	.3072	.1815	.0938	.0163	.0016
20	.9711	.8477	.7374	.5919	.4292	.2762	.1554	.0756	.0113	.0009
22	.9703	.8392	.7224	5702	.4034	.2510	.1352	.0624	.0081	.0005
24	.9695	.8316	.7091	.5512	.3814	.2302	.1193	.0524	.0060	.0003
26	.9689	.8247	.6972	.5345	.3625	.2129	.1065	.0449	.0046	.0002
28	9682	.8185	.6866	.5198	.3461	.1983	.0962	.0389	.0036	.0001
30	.9677	.8129	.6770	.5067	.3318	.1859	.0876	.0343	.0029	.0001
40	.9655	.7911	.6406	.4584	.2815	.1447	.0611	.0209	.0012	.0000
60	.9630	.7658	.5995	.4070	.2318	.1079	.0402	.0118	.0005	.0000
120	.9600	.7362	.5536	.3531	.1843	.0765	.0247	.0061	.0002	.0000
∞	.9563	.7016	.5028	.2981	.1406	.0512	.0141	.0029	.0000	.0000
					$\nu_1=9$					
2	.9872	.9790	.9742	.9686	.9621	.9549	.9468	.9380	.9181	.8954
4	.9843	.9623	.9464	.9254	.8986	.8657	.8268	.7821	.6783	.5623
6	.9817	.9434	.9125	.8696	.8135	.7446	.6647	.5774	.3978	.2407
8	.9794	.9246	.8778	.8119	.7265	.6251	.5142	.4025	.2099	.0871
10	.9774	.9070	.8450	.7579	.6479	.5229	.3953	.2782	.1093	.0310
12	.9757	.8909	.8151	.7098	.5806	.4407	.3073	.1955	.0587	.0116
14	.9742	.8764	.7884	.6679	.5242	.3759	.2433	.1410	.0331	.0047
16	.9729	.8634	.7647	.6316	.4774	.3249	1966	.1047	.0197	.0021
18	.9718	.8518	.7438	.6002	.4384	.2847	.1621	.0800	.0124	.0010
20	.9708	.8414	.7252	.5730	.4057	.2525	.1361	.0628	.0081	.0005
22	.9698	.8320	.7086	.5493	.3782	.2265	.1162	.0504	.0056	.0003
24	.9690	.8235	.6939	.5286	.3548	.2053	.1007	.0414	.0040	.0002
26	.9683	.8159	.6807	.5104	.3347	.1877	.0885	.0346	.0029	.0001
28	.9676	.8090	.6689	.4943	.3174	.1730	.0786	.0294	.0022	.0001
30	.9670	.8026	.6581	.4799	.3023	.1606	.0706	.0254	.0017	.0000
40	.9646	.7780	.6174	.4273	.2497	.1199	.0464	.0143	.0006	.0000
60	.9618	.7490	.5712	.3714	.1986	.0848	.0283	.0073	.0002	.0000
120	.9583	.7148	.5194	.3133	.1508	.0562	.0158	.0033	.0001	.0000
∞	.9542	.6745	.4620	.2549	.1085	.0345	.0080	.0014	.0000	.0000
					$\nu_1=10$					
2	.9873	.9791	.9744	.9688	.9624	.9552	.9472	.9385	.9188	.8963
4	.9844	.9625	.9466	.9256	.8989	.8660	.8271	.7825	.6786	.5625
6	.9817	.9433	.9123	.8690	.8124	.7428	.6622	.5742	.3938	.2367
8	.9794	.9240	.8765	.8094	.7225	.6195	.5072	.3947	.2026	.0823
10	.9774	.9057	.8422	.7529	.6404	.5131	.3842	.2672	.1017	.0277
12	.9756	.8888	.8106	.7022	.5696	.4273	.2935	.1831	.0523	.0097
14	.9741	.8734	.7822	.6575	.5101	.3597	.2279	.1285	.0282	.0037
16	.9727	.8594	7567	.6187	.4605	.3068	.1805	.0928	.0160	.0015
18	.9715	.8469	.7341	.5850	.4193	.2652	.1459	.0690	.0096	.0007
20	.9704	.8355	.7139	.5558	.3848	.2322	.1201	.0527	.0060	.0003
22	.9694	.8253	.6959	.5303	.3558	.2057	.1007	.0413	.0039	.0002
24	.9685	.8160	.6798	.5080	.3312	.1841	.0858	.0331	.0027	.0001
26	.9677	.8076	.6653	.4883	.3102	.1665	.0741	.0270	.0019	.0001
28	.9670	.7999	.6523	.4709	.2921	.1518	.0649	.0225	.0014	.0000
30	.9664	.7929	.6405	.4555	.2764	.1395	.0574	.0190	.0010	.0000
40	.9638	.7655	.5955	.3989	.2220	.0998	.0355	.0098	.0003	.0000
60	.9606	.7328	.5443	.3390	.1702	.0668	.0200	.0045	.0001	.0000
120	.9568	.6939	.4869	.2776	.1232	.0411	.0101	.0018	.0000	.0000
∞	.9520	.6475	.4234	.2170	.0832	.0230	.0045	.0007	.0000	.0000
					$\nu_1=12$					
2	.9873	.9793	.9746	.9691	.9628	.9557	.9478	.9392	.9198	.8977
4	.9844	.9627	.9469	.9260	.8993	.8665	.8276	.7829	.6790	.5628
6	.9818	.9432	.9118	.8679	.8104	.7398	.6581	.5689	.3872	.2304
8	.9794	.9231	.8743	.8052	.7158	.6101	.4956	.3819	.1910	.0750
10	.9773	.9035	.8375	.7445	.6277	.4968	.3661	.2494	.0900	.0228
12	.9754	.8850	.8029	.6890	.5509	.4050	.2708	.1635	.0429	.0071
14	.9738	.8680	.7713	.6396	.4860	.3329	.2030	.1092	.0212	.0024
16	.9723	.8523	.7426	.5963	.4318	.2768	.1549	.0749	.0110	.0009
18	.9710	.8380	.7169	.5585	.3868	.2333	.1206	.0529	.0060	.0003
20	.9698	.8250	.6938	.5256	.3493	.1991	.0958	.0384	.0035	.0001
22	.9687	.8131	.6730	.4969	.3179	.1721	.0775	.0286	.0021	.0001
24	.9677	.8023	.6543	.4717	.2914	.1505	.0638	.0219	.0013	.0000
26	.9668	.7924	.6376	.4496	.2690	.1330	.0533	.0171	.0009	.0000
28	.9659	.7833	.6223	.4300	.2498	.1187	.0452	.0136	.0006	.0000
30	.9652	.7750	.6086	.4127	.2333	.1069	.0389	.0110	.0004	.0000
40	.9622	.7419	.5556	.3492	.1770	.0701	.0212	.0048	.0001	.0000
60	.9584	.7019	.4951	.2833	.1256	.0417	.0101	.0018	.0000	.0000
120	.9537	.6534	.4271	.2173	.0819	.0220	.0041	.0005	.0000	.0000
∞	.9476	.5948	.3530	.1552	.0479	.0100	.0015	.0001	.0000	.0000

TABLE VII (continued)

ν_2	$\phi=.5$	1.0	1.2	1.4	$\alpha=0.05$ 1.6	1.8	2.0	2.2	2.6	3.0
					$\nu_1=1$					
2	.9271	.8617	.8256	.7847	.7402	.6927	.6432	.5926	.4915	.3950
4	.9141	.8048	.7415	.6694	.5910	.5095	.4284	.3509	.2169	.1198
6	.9077	.7768	.7010	.6153	.5238	.4315	.3431	.2629	.1374	.0611
8	.9040	.7610	.6784	.5858	.4883	.3916	.3015	.2223	.1054	.0413
10	.9017	.7510	.6642	.5675	.4666	.3680	.2775	.1997	.0890	.0322
12	.9000	.7440	.6544	.5551	.4521	.3524	.2620	.1854	.0793	.0272
14	.8988	.7390	.6474	.5462	.4418	.3414	.2513	.1756	.0728	.0240
16	.8979	.7351	.6420	.5394	.4341	.3333	.2433	.1685	.0683	.0219
18	.8972	.7321	.6379	.5342	.4281	.3270	.2373	.1631	.0649	.0203
20	.8966	.7297	.6345	.5300	.4233	.3220	.2325	.1589	.0623	.0192
22	.8961	.7277	.6317	.5265	.4194	.3180	.2287	.1555	.0603	.0183
24	.8957	.7260	.6294	.5236	.4161	.3146	.2255	.1527	.0586	.0175
26	.8954	.7246	.6274	.5212	.4134	.3118	.2228	.1504	.0573	.0169
28	.8951	.7233	.6258	.5192	.4111	.3094	.2206	.1485	.0561	.0165
30	.8948	.7223	.6243	.5173	.4090	.3073	.2186	.1468	.0551	.0160
40	.8939	.7185	.6192	.5110	.4020	.3001	.2119	.1410	.0518	.0147
60	.8930	.7147	.6140	.5047	.3949	.2930	.2053	.1354	.0487	.0134
120	.8920	.7108	.6087	.4983	.3879	.2859	.1988	.1300	.0457	.0123
∞	.8910	.7070	.6036	.4920	.3810	.2791	.1926	.1248	.0430	.0112
					$\nu_1=2$					
2	.9324	.8814	.8527	.8201	.7840	.7451	.7038	.6608	.5722	.4837
4	.9201	.8239	.7657	.6976	.6219	.5414	.4598	.3804	.2400	.1353
6	.9129	.7891	.7135	.6257	.5303	.4330	.3396	.2554	.1264	.0520
8	.9083	.7672	.6810	.5821	.4769	.3729	.2773	.1955	.0821	.0273
10	.9052	.7523	.6592	.5536	.4430	.3361	.2408	.1624	.0609	.0175
12	.9030	.7417	.6438	.5336	.4197	.3115	.2173	.1419	.0490	.0126
14	.9013	.7337	.6323	.5189	.4028	.2941	.2010	.1281	.0416	.0099
16	.9000	.7274	.6234	.5077	.3901	.2812	.1892	.1183	.0367	.0082
18	.8989	.7225	.6164	.4988	.3802	.2713	.1802	.1110	.0331	.0071
20	.8980	.7184	.6107	.4917	.3723	.2634	.1732	.1054	.0305	.0063
22	.8973	.7150	.6059	.4858	.3658	.2570	.1675	.1009	.0285	.0057
24	.8967	.7122	.6019	.4808	.3603	.2517	.1629	.0973	.0269	.0052
26	.8961	.7097	.5985	.4767	.3558	.2472	.1590	.0943	.0256	.0048
28	.8957	.7076	.5956	.4730	.3518	.2434	.1558	.0918	.0245	.0045
30	.8953	.7058	.5930	.4699	.3484	.2401	.1530	.0896	.0236	.0043
40	.8938	.6992	.5839	.4588	.3365	.2288	.1434	.0824	.0207	.0035
60	.8923	.6924	.5746	.4476	.3247	.2177	.1341	.0756	.0181	.0029
120	.8908	.6855	.5651	.4364	.3129	.2069	.1253	.0692	.0157	.0024
∞	.8892	.6785	.5556	.4251	.3013	.1963	.1168	.0632	.0137	.0019
					$\nu_1=3$					
2	.9342	.8882	.8623	.8327	.7998	.7640	.7260	.6861	.6030	.5187
4	.9221	.8302	.7735	.7064	.6311	.5505	.4683	.3880	.2453	.1384
6	.9144	.7909	.7134	.6226	.5235	.4225	.3264	.2407	.1132	.0435
8	.9092	.7643	.6733	.5683	.4570	.3482	.2504	.1694	.0639	.0184
10	.9056	.7454	.6453	.5314	.4134	.3019	.2059	.1307	.0419	.0098
12	.9028	.7314	.6249	.5050	.3831	.2709	.1776	.1074	.0305	.0061
14	.9007	.7207	.6093	.4853	.3611	.2490	.1583	.0922	.0238	.0042
16	.8990	.7122	.5972	.4701	.3443	.2328	.1444	.0817	.0196	.0031
18	.8976	.7054	.5874	.4581	.3313	.2204	.1340	.0740	.0167	.0025
20	.8965	.6997	.5794	.4483	.3208	.2106	.1259	.0682	.0146	.0020
22	.8955	.6950	.5728	.4402	.3122	.2026	.1195	.0637	.0131	.0017
24	.8947	.6909	.5671	.4333	.3051	.1961	.1143	.0601	.0119	.0015
26	.8940	.6875	.5623	.4275	.2990	.1907	.1100	.0571	.0110	.0013
28	.8934	.6845	.5581	.4225	.2938	.1860	.1064	.0547	.0103	.0012
30	.8928	.6818	.5544	.4182	.2894	.1820	.1033	.0526	.0097	.0011
40	.8909	.6723	.5414	.4028	.2738	.1684	.0930	.0458	.0078	.0008
60	.8888	.6624	.5279	.3872	.2583	.1552	.0833	.0397	.0062	.0006
120	.8866	.6522	.5142	.3716	.2431	.1425	.0743	.0342	.0049	.0004
∞	.8843	.6415	.5000	.3557	.2280	.1304	.0659	.0293	.0038	.0003
					$\nu_1=4$					
2	.9351	.8917	.8672	.8391	.8079	.7738	.7375	.6993	.6193	.5374
4	.9232	.8332	.7771	.7103	.6350	.5542	.4714	.3905	.2466	.1389
6	.9151	.7906	.7112	.6178	.5158	.4122	.3143	.2282	.1030	.0375
8	.9094	.7602	.6649	.5549	.4389	.3271	.2286	.1493	.0515	.0132
10	.9052	.7378	.6315	.5110	.3876	.2736	.1788	.1076	.0301	.0059
12	.9020	.7208	.6066	.4791	.3516	.2380	.1475	.0833	.0199	.0032
14	.8995	.7076	.5875	.4550	.3253	.2129	.1266	.0680	.0143	.0019
16	.8975	.6970	.5723	.4363	.3054	.1945	.1118	.0577	.0109	.0013
18	.8958	.6883	.5600	.4214	.2898	.1804	.1009	.0503	.0087	.0009
20	.8945	.6811	.5498	.4092	.2774	.1695	.0926	.0449	.0073	.0007
22	.8933	.6750	.5413	.3991	.2672	.1607	.0861	.0408	.0062	.0006
24	.8923	.6698	.5341	.3907	.2587	.1535	.0808	.0376	.0054	.0005
26	.8914	.6653	.5279	.3834	.2516	.1475	.0765	.0349	.0049	.0004
28	.8906	.6614	.5225	.3772	.2455	.1424	.0730	.0328	.0044	.0003
30	.8899	.6579	.5178	.3718	.2402	.1381	.0700	.0311	.0040	.0003
40	.8874	.6454	.5009	.3526	.2219	.1234	.0601	.0254	.0029	.0002
60	.8848	.6322	.4833	.3332	.2040	.1095	.0511	.0206	.0021	.0001
120	.8819	.6183	.4652	.3136	.1865	.0965	.0431	.0164	.0015	.0001
∞	.8789	.6038	.4466	.2940	.1695	.0844	.0360	.0130	.0011	.0000

TABLE VII (continued)

ν_2	$\phi=.5$	1.0	1.2	1.4	$\alpha=0.05$ 1.6	1.8	2.0	2.2	2.6	3.0
					$\nu_1=5$					
2	.9356	.8939	.8702	.8431	.8128	.7798	.7445	.7074	.6293	.5490
4	.9238	.8349	.7791	.7124	.6369	.5558	.4727	.3914	.2467	.1386
6	.9154	.7897	.7087	.6131	.5088	.4033	.3044	.2181	.0952	.0333
8	.9093	.7561	.6573	.5432	.4237	.3099	.2115	.1342	.0430	.0100
10	.9048	.7308	.6193	.4933	.3660	.2509	.1579	.0909	.0227	.0038
12	.9012	.7111	.5904	.4566	.3254	.2118	.1249	.0665	.0136	.0018
14	.8983	.6956	.5679	.4287	.2957	.1845	.1033	.0516	.0090	.0010
16	.8960	.6829	.5499	.4069	.2732	.1646	.0883	.0418	.0064	.0006
18	.8941	.6725	.5352	.3895	.2557	.1496	.0774	.0351	.0048	.0004
20	.8924	.6638	.5231	.3753	.2417	.1380	.0693	.0303	.0038	.0003
22	.8910	.6564	.5129	.3635	.2303	.1288	.0630	.0267	.0031	.0002
24	.8898	.6501	.5042	.3536	.2209	.1213	.0580	.0240	.0026	.0001
26	.8888	.6445	.4967	.3451	.2129	.1151	.0540	.0218	.0022	.0001
28	.8878	.6397	.4901	.3379	.2062	.1099	.0507	.0201	.0019	.0001
30	.8870	.6354	.4844	.3315	.2003	.1055	.0479	.0186	.0017	.0001
40	.8840	.6198	.4638	.3091	.1803	.0908	.0390	.0142	.0011	.0000
60	.8807	.6033	.4423	.2864	.1609	.0772	.0313	.0106	.0007	.0000
120	.8771	.5857	.4201	.2638	.1422	.0649	.0247	.0078	.0004	.0000
∞	.8733	.5671	.3971	.2412	.1245	.0538	.0192	.0056	.0003	.0000
					$\nu_1=6$					
2	.9360	.8953	.8722	.8457	.8161	.7839	.7493	.7129	.6361	.5569
4	.9242	.8361	.7803	.7136	.6380	.5567	.4733	.3916	.2464	.1381
6	.9156	.7887	.7063	.6090	.5028	.3959	.2962	.2100	.0893	.0301
8	.9092	.7525	.6506	.5332	.4109	.2958	.1978	.1225	.0369	.0080
10	.9042	.7245	.6086	.4782	.3480	.2325	.1417	.0784	.0177	.0026
12	.9003	.7024	.5761	.4373	.3036	.1908	.1077	.0544	.0097	.0011
14	.8972	.6847	.5506	.4061	.2711	.1619	.0859	.0401	.0059	.0005
16	.8946	.6702	.5301	.3816	.2465	.1412	.0710	.0312	.0039	.0003
18	.8924	.6582	.5132	.3621	.2275	.1257	.0605	.0252	.0028	.0002
20	.8905	.6480	.4992	.3461	.2124	.1139	.0528	.0210	.0021	.0001
22	.8889	.6394	.4874	.3328	.2001	.1045	.0469	.0179	.0016	.0001
24	.8875	.6319	.4773	.3216	.1900	.0970	.0423	.0157	.0013	.0001
26	.8863	.6253	.4686	.3121	.1815	.0908	.0387	.0139	.0011	.0000
28	.8852	.6196	.4610	.3039	.1744	.0857	.0357	.0125	.0009	.0000
30	.8843	.6145	.4543	.2968	.1682	.0814	.0333	.0114	.0008	.0000
40	.8807	.5960	.4302	.2717	.1471	.0672	.0256	.0081	.0004	.0000
60	.8768	.5760	.4050	.2464	.1270	.0545	.0193	.0055	.0002	.0000
120	.8724	.5547	.3789	.2214	.1082	.0434	.0141	.0037	.0001	.0000
∞	.8677	.5319	.3520	.1967	.0907	.0339	.0101	.0024	.0000	.0000
					$\nu_1=7$					
2	.9363	.8963	.8736	.8476	.8185	.7868	.7527	.7168	.6410	.5627
4	.9245	.8368	.7811	.7144	.6387	.5571	.4735	.3916	.2460	.1376
6	.9157	.7878	.7042	.6054	.4978	.3897	.2895	.2035	.0846	.0278
8	.9090	.7492	.6449	.5247	.4002	.2841	.1868	.1133	.0323	.0065
10	.9038	.7189	.5992	.4652	.3328	.2174	.1288	.0689	.0143	.0019
12	.8996	.6947	.5636	.4207	.2852	.1738	.0944	.0454	.0072	.0007
14	.8961	.6750	.5353	.3866	.2505	.1439	.0726	.0320	.0041	.0003
16	.8933	.6588	.5125	.3598	.2243	.1226	.0582	.0238	.0025	.0001
18	.8908	.6452	.4936	.3383	.2041	.1070	.0482	.0185	.0017	.0001
20	.8888	.6336	.4779	.3208	.1882	.0951	.0409	.0149	.0012	.0000
22	.8870	.6238	.4646	.3062	.1753	.0858	.0355	.0123	.0009	.0000
24	.8854	.6152	.4532	.2940	.1647	.0785	.0314	.0105	.0007	.0000
26	.8840	.6077	.4433	.2836	.1559	.0725	.0282	.0091	.0005	.0000
28	.8828	.6011	.4347	.2747	.1485	.0676	.0256	.0080	.0004	.0000
30	.8817	.5952	.4272	.2669	.1421	.0634	.0234	.0071	.0003	.0000
40	.8776	.5737	.3998	.2396	.1206	.0501	.0170	.0046	.0002	.0000
60	.8730	.5504	.3713	.2124	.1005	.0387	.0119	.0029	.0001	.0000
120	.8679	.5253	.3417	.1857	.0821	.0290	.0080	.0017	.0000	.0000
∞	.8622	.4983	.3112	.1597	.0656	.0211	.0052	.0010	.0000	.0000
					$\nu_1=8$					
2	.9365	.8971	.8747	.8490	.8203	.7889	.7553	.7198	.6448	.5671
4	.9274	.8374	.7817	.7149	.6391	.5574	.4735	.3914	.2456	.1371
6	.9158	.7869	.7024	.6023	.4935	.3845	.2839	.1981	.0809	.0259
8	.9088	.7464	.6398	.5173	.3910	.2744	.1777	.1059	.0289	.0055
10	.9033	.7140	.5910	.4540	.3200	.2049	.1184	.0615	.0118	.0014
12	.8989	.6878	.5526	.4063	.2697	.1598	.0838	.0387	.0055	.0005
14	.8951	.6663	.5218	.3696	.2330	.1292	.0624	.0260	.0029	.0002
16	.8920	.6484	.4968	.3407	.2056	.1077	.0485	.0186	.0017	.0001
18	.8894	.6334	.4761	.3176	.1846	.0921	.0390	.0139	.0010	.0000
20	.8871	.6205	.4588	.2988	.1680	.0804	.0323	.0108	.0007	.0000
22	.8851	.6095	.4441	.2832	.1548	.0713	.0274	.0087	.0005	.0000
24	.8834	.5999	.4315	.2700	.1439	.0642	.0237	.0072	.0003	.0000
26	.8819	.5915	.4206	.2589	.1349	.0585	.0208	.0060	.0003	.0000
28	.8805	.5840	.4111	.2493	.1274	.0538	.0186	.0052	.0002	.0000
30	.8793	.5774	.4027	.2410	.1209	.0499	.0168	.0045	.0002	.0000
40	.8746	.5530	.3725	.2120	.0995	.0377	.0114	.0027	.0001	.0000
60	.8694	.5264	.3408	.1834	.0798	.0275	.0074	.0015	.0000	.0000
120	.8635	.4975	.3081	.1556	.0623	.0193	.0046	.0008	.0000	.0000
∞	.8568	.4663	.2745	.1292	.0472	.0130	.0027	.0004	.0000	.0000

TABLE VII (*continued*)

ν_2	$\phi = .5$	1.0	1.2	1.4	$\alpha = 0.05$ 1.6	1.8	2.0	2.2	2.6	3.0
					$\nu_1 = 9$					
2	.9366	.8977	.8756	.8501	.8217	.7906	.7573	.7221	.6477	.5705
4	.9249	.8378	.7821	.7153	.6394	.5575	.4735	.3912	.2452	.1366
6	.9158	.7861	.7007	.5996	.4898	.3800	.2792	.1936	.0778	.0245
8	.9087	.7439	.6354	.5109	.3832	.2661	.1702	.0998	.0262	.0048
10	.9029	.7096	.5838	.4442	.3089	.1944	.1099	.0555	.0100	.0011
12	.8982	.6816	.5428	.3937	.2564	.1481	.0753	.0334	.0043	.0003
14	.8943	.6584	.5097	.3547	.2182	.1171	.0543	.0216	.0021	.0001
16	.8909	.6390	.4827	.3241	.1898	.0956	.0409	.0148	.0011	.0000
18	.8881	.6226	.4604	.2996	.1681	.0801	.0320	.0107	.0007	.0000
20	.8856	.6086	.4416	.2796	.1511	.0686	.0259	.0080	.0004	.0000
22	.8835	.5964	.4257	.2630	.1376	.0599	.0214	.0062	.0003	.0000
24	.8816	.5858	.4120	.2492	.1266	.0531	.0181	.0050	.0002	.0000
26	.8799	.5765	.4002	.2374	.1176	.0477	.0156	.0041	.0001	.0000
28	.8784	.5683	.3898	.2274	.1110	.0433	.0137	.0034	.0001	.0000
30	.8770	.5609	.3807	.2186	.1036	.0397	.0121	.0029	.0001	.0000
40	.8718	.5337	.3477	.1883	.0825	.0286	.0077	.0016	.0000	.0000
60	.8660	.5038	.3133	.1587	.0636	.0197	.0046	.0008	.0000	.0000
120	.8592	.4713	.2778	.1304	.0473	.0129	.0026	.0004	.0000	.0000
∞	.8514	.4361	.2417	.1041	.0337	.0080	.0014	.0002	.0000	.0000
					$\nu_1 = 10$					
2	.9368	.8981	.8762	.8510	.8228	.7920	.7589	.7240	.6500	.5732
4	.9250	.8381	.7825	.7156	.6395	.5575	.4734	.3910	.2448	.1362
6	.9158	.7854	.6992	.5972	.4865	.3762	.2751	.1898	.0752	.0233
8	.9085	.7417	.6315	.5053	.3764	.2591	.1638	.0948	.0241	.0042
10	.9026	.7057	.5774	.4357	.2993	.1854	.1028	.0508	.0086	.0009
12	.8976	.6761	.5340	.3826	.2448	.1383	.0683	.0293	.0035	.0002
14	.8935	.6514	.4990	.3417	.2055	.1070	.0478	.0182	.0016	.0001
16	.8899	.6305	.4702	.3094	.1763	.0856	.0350	.0120	.0008	.0000
18	.8869	.6128	.4463	.2836	.1541	.0704	.0267	.0083	.0004	.0000
20	.8843	.5976	.4262	.2627	.1368	.0592	.0210	.0061	.0003	.0000
22	.8819	.5844	.4091	.2454	.1232	.0508	.0170	.0046	.0002	.0000
24	.8799	.5729	.3944	.2309	.1122	.0443	.0141	.0035	.0001	.0000
26	.8780	.5627	.3817	.2186	.1031	.0393	.0119	.0028	.0001	.0000
28	.8764	.5537	.3705	.2082	.0956	.0352	.0102	.0023	.0001	.0000
30	.8749	.5456	.3607	.1991	.0893	.0319	.0089	.0019	.0000	.0000
40	.8692	.5157	.3253	.1678	.0687	.0218	.0053	.0010	.0000	.0000
60	.8627	.4827	.2884	.1377	.0508	.0142	.0029	.0004	.0000	.0000
120	.8551	.4467	.2506	.1094	.0359	.0086	.0015	.0002	.0000	.0000
∞	.8462	.4047	.2124	.0836	.0240	.0049	.0007	.0000	.0000	.0000
					$\nu_1 = 12$					
2	.9369	.8989	.8772	.8524	.8245	.7941	.7614	.7268	.6536	.5774
4	.9252	.8385	.7829	.7159	.6397	.5575	.4731	.3905	.2441	.1355
6	.9159	.7842	.6968	.5934	.4813	.3700	.2686	.1837	.0713	.0214
8	.9082	.7379	.6250	.4960	.3653	.2476	.1536	.0869	.0208	.0034
10	.9019	.6991	.5666	.4213	.2836	.1711	.0917	.0435	.0067	.0006
12	.8966	.6666	.5192	.3641	.2260	.1227	.0577	.0234	.0024	.0001
14	.8921	.6391	.4805	.3197	.1847	.0913	.0382	.0135	.0010	.0000
16	.8882	.6157	.4485	.2849	.1544	.0703	.0265	.0083	.0004	.0000
18	.8848	.5956	.4219	.2571	.1316	.0557	.0192	.0053	.0002	.0000
20	.8818	.5783	.3994	.2345	.1142	.0452	.0144	.0036	.0001	.0000
22	.8792	.5631	.3803	.2160	.1005	.0376	.0111	.0026	.0001	.0000
24	.8768	.5498	.3638	.2006	.0896	.0318	.0088	.0019	.0000	.0000
26	.8747	.5381	.3496	.1876	.0808	.0274	.0072	.0014	.0000	.0000
28	8728	.5276	.3371	.1766	.0736	.0239	.0059	.0011	.0000	.0000
30	.8710	.5182	.3261	.1671	.0676	.0211	.0050	.0009	.0000	.0000
40	.8643	.4831	.2865	.1347	.0485	.0131	.0026	.0004	.0000	.0000
60	.8565	.4443	.2456	.1044	.0329	.0075	.0012	.0001	.0000	.0000
120	.8472	.4016	.2042	.0770	.0207	.0039	.0005	.0000	.0000	.0000
∞	.8359	.3548	.1632	.0535	.0120	.0018	.0002	.0000	.0000	.0000

TABLE VII (continued)

ν_2	$\phi=.5$	1.0	1.2	1.4	$\alpha=0.10$ 1.6	1.8	2.0	2.2	2.6	3.0
					$\nu_1=1$					
2	.8582	.7443	.6846	.6202	.5534	.4863	.4209	.3588	.2491	.1628
4	.8410	.6773	.5919	.5017	.4118	.3266	.2500	.1846	.0899	.0375
6	.8336	.6498	.5552	.4570	.3613	.2738	.1985	.1373	.0570	.0194
8	.8296	.6353	.5363	.4344	.3367	.2490	.1753	.1172	.0447	.0137
10	.8271	.6265	.5248	.4209	.3223	.2348	.1623	.1063	.0385	.0110
12	.8254	.6205	.5171	.4120	.3128	.2256	.1541	.0996	.0348	.0096
14	.8242	.6162	.5116	.4057	.3062	.2192	.1485	.0949	.0324	.0086
16	.8232	.6130	.5075	.4010	.3012	.2145	.1444	.0916	.0307	.0080
18	.8225	.6104	.5043	.3973	.2974	.2109	.1412	.0891	.0294	.0075
20	.8219	.6084	.5017	.3944	.2944	.2081	.1388	.0871	.0285	.0072
22	.8214	.6068	.4996	.3920	.2920	.2058	.1368	.0856	.0277	.0069
24	.8210	.6054	.4979	.3900	.2900	.2038	.1351	.0843	.0271	.0067
26	.8207	.6042	.4964	.3884	.2882	.2022	.1338	.0832	.0266	.0065
28	.8204	.6032	.4951	.3869	.2868	.2009	.1326	.0823	.0261	.0063
30	.8201	.6023	.4940	.3857	.2855	.1997	.1316	.0815	.0257	.0062
40	.8192	.5993	.4902	.3814	.2811	.1956	.1281	.0788	.0245	.0058
60	.8183	.5962	.4864	.3771	.2768	.1916	.1248	.0762	.0233	.0054
120	.8174	.5932	.4826	.3729	.2726	.1877	.1215	.0737	.0222	.0050
∞	.8165	.5901	.4788	.3686	.2683	.1838	.1183	.0713	.0211	.0047
					$\nu_1=2$					
2	.8669	.7746	.7252	.6707	.6130	.5536	.4939	.4355	.3265	.2333
4	.8486	.6981	.6159	.5268	.4358	.3481	.2680	.1987	.0972	.0405
6	.8392	.6593	.5623	.4595	.3586	.2662	.1876	.1525	.0471	.0140
8	.8335	.6369	.5319	.4228	.3183	.2260	.1508	.0944	.0302	.0073
10	.8298	.6223	.5126	.4000	.2940	.2027	.1305	.0783	.0226	.0048
12	.8272	.6122	.4994	.3846	.2780	.1877	.1179	.0686	.0184	.0035
14	.8252	.6047	.4897	.3734	.2666	.1772	.1093	.0622	.0158	.0028
16	.8237	.5990	.4823	.3651	.2581	.1696	.1031	.0578	.0140	.0024
18	.8225	.5945	.4765	.3586	.2516	.1638	.0984	.0544	.0128	.0021
20	.8215	.5908	.4719	.3533	.2464	.1592	.0948	.0519	.0119	.0019
22	.8207	.5878	.4680	.3490	.2422	.1555	.0919	.0498	.0012	.0017
24	.8200	.5853	.4648	.3455	.2387	.1524	.0895	.0482	.0106	.0016
26	.8194	.5831	.4621	.3425	.2357	.1498	.0875	.0468	.0102	.0015
28	.8190	.5812	.4598	.3399	.2332	.1477	.0859	.0457	.0098	.0014
30	.8185	.5796	.4577	.3376	.2310	.1458	0844	.0447	.0095	.0014
40	.8169	.5739	.4506	.3298	.2235	.1394	.0796	.0415	.0084	.0011
60	.8153	.5681	.4433	.3220	.2160	.1331	.0749	.0384	.0075	.0010
120	.8137	.5622	.4361	.3142	.2087	.1270	.0705	.0355	.0067	.0008
∞	.8120	.5562	.4288	.3064	.2015	.1211	.0662	.0328	.0059	.0007
					$\nu_1=3$					
2	.8700	.7858	.7403	.6899	.6359	.5799	.5231	.4668	.3597	.2655
4	.8513	.7047	.6230	.5336	.4416	.3525	.2709	.2002	.0970	.0399
6	.8406	.6585	.5581	.4514	.3468	.2521	.1729	.1117	.0386	.0103
8	.8338	.6298	.5190	.4040	.2953	.2016	.1281	.0755	.0208	.0041
10	.8291	.6106	.4933	.3738	.2638	.1724	.1038	.0574	.0135	.0022
12	.8257	.5968	.4752	.3531	.2429	.1537	.0890	.0470	.0098	.0014
14	.8232	.5864	.4618	.3380	.2280	.1408	.0792	.0404	.0077	.0010
16	.8211	.5784	.4516	.3266	.2170	.1315	.0723	.0359	.0064	.0007
18	.8195	.5720	.4434	.3177	.2085	.1244	.0671	.0326	.0055	.0006
20	.8182	.5668	.4369	.3105	.2017	.1189	.0632	.0301	.0049	.0005
22	.8170	.5624	.4314	.3047	.1963	.1145	.0601	.0282	.0044	.0004
24	.8161	.5588	.4268	.2998	.1917	.1108	.0576	.0267	.0040	.0004
26	.8153	.5556	.4230	.2956	.1879	.1078	.0555	.0255	.0038	.0003
28	.8145	.5529	.4196	.2921	.1847	.1053	.0537	.0244	.0035	.0003
30	.8139	.5506	.4167	.2890	.1819	.1031	.0523	.0236	.0033	.0003
40	.8117	.5421	.4063	.2782	.1722	.0956	.0473	.0207	.0027	.0002
60	.8093	.5335	.3959	.2674	.1627	.0885	.0427	.0182	.0022	.0002
120	.8069	.5246	.3853	.2567	.1535	.0817	.0384	.0159	.0018	.0001
∞	.8044	.5156	.3745	.2460	.1444	.0752	.0344	.0138	.0015	.0001
					$\nu_1=4$					
2	.8716	.7916	.7482	.6999	.6480	.5939	.5387	.4837	.3781	.2836
4	.8527	.7077	.6259	.5360	.4432	.3532	.2708	.1995	.0958	.0390
6	.8410	.6559	.5527	.4429	.3358	.2399	.1610	.1012	.0327	.0080
8	.8333	.6223	.5066	.3871	.2758	.1821	.1110	.0622	.0151	.0026
10	.8279	.5989	.4754	.3509	.2388	.1489	.0846	.0436	.0086	.0011
12	.8238	.5818	.4531	.3257	.2142	.1279	.0689	.0333	.0056	.0006
14	.8207	.5689	.4365	.3074	.1968	.1136	.0587	.0271	.0040	.0004
16	.8182	.5587	.4236	.2935	.1839	.1033	.0517	.0229	.0031	.0002
18	.8161	.5505	.4133	.2826	.1740	.0957	.0467	.0201	.0025	.0002
20	.8144	.5438	.4050	.2738	.1662	.0898	.0428	.0179	.0021	.0001
22	.8130	.5382	.3981	.2666	.1599	.0851	.0398	.0163	.0018	.0001
24	.8118	.5334	.3922	.2606	.1547	.0812	.0375	.0151	.0016	.0001
26	.8107	.5293	.3872	.2555	.1503	.0781	.0355	.0141	.0014	.0001
28	.8098	.5258	.3829	.2512	.1466	.0754	.0339	.0133	.0013	.0001
30	.8090	.5226	.3792	.2474	.1434	.0732	.0326	.0126	.0012	.0001
40	.8061	.5115	.3659	.2343	.1325	.0655	.0281	.0104	.0009	.0000
60	.8030	.5000	.3524	.2211	.1219	.0584	.0241	.0085	.0007	.0000
120	.7997	.4881	.3387	.2081	.1117	.0518	.0206	.0069	.0005	.0000
∞	.7963	.4758	.3248	.1952	.1019	.0457	.0174	.0056	.0003	.0000

TABLE VII (continued)

$\alpha = 0.10$

ν_2	$\phi = .5$	1.0	1.2	1.4	1.6	1.8	2.0	2.2	2.6	3.0
					$\nu_1 = 5$					
2	.8726	.7952	.7530	.7061	.6555	.6026	.5485	.4943	.3897	.2953
4	.8534	.7093	.6273	.5369	.4435	.3528	.2699	.1983	.0945	.0381
6	.8412	.6532	.5477	.4354	.3266	.2300	.1516	.0933	.0286	.0066
8	.8327	.6154	.4957	.3728	.2599	.1668	.0982	.0528	.0115	.0017
10	.8266	.5885	.4599	.3316	.2187	.1308	.0707	.0343	.0058	.0006
12	.8219	.5685	.4339	.3029	.1913	.1084	.0548	.0245	.0034	.0003
14	.8183	.5532	.4144	.2818	.1720	.0934	.0448	.0188	.0022	.0002
16	.8153	.5410	.3991	.2658	.1578	.0827	.0380	.0152	.0016	.0001
18	.8129	.5311	.3869	.2532	.1470	.0749	.0332	.0128	.0012	.0001
20	.8109	.5229	.3770	.2432	.1385	.0689	.0297	.0110	.0010	.0000
22	.8092	.5161	.3687	.2349	.1316	.0642	.0270	.0097	.0008	.0000
24	.8077	.5103	.3618	.2280	.1260	.0604	.0249	.0087	.0007	.0000
26	.8064	.5053	.3558	.2222	.1213	.0573	.0232	.0080	.0006	.0000
28	.8053	.5009	.3507	.2173	.1174	.0547	.0218	.0074	.0005	.0000
30	.8043	.4971	.3462	.2129	.1140	.0525	.0206	.0068	.0004	.0000
40	.8007	.4833	.3302	.1979	.1024	.0452	.0169	.0053	.0003	.0000
60	.7968	.4689	.3140	.1830	.0914	.0386	.0137	.0040	.0002	.0000
120	.7927	.4539	.2974	.1684	.0810	.0327	.0109	.0030	.0001	.0000
∞	.7883	.4384	.2807	.1540	.0712	.0274	.0086	.0022	.0001	.0000
					$\nu_1 = 6$					
2	.8732	.7976	.7563	.7103	.6606	.6085	.5552	.5016	.3978	.3035
4	.8540	.7102	.6281	.5373	.4434	.3522	.2689	.1971	.0934	.0373
6	.8411	.6507	.5432	.4291	.3189	.2220	.1442	.0872	.0256	.0056
8	.8321	.6094	.4864	.3609	.2469	.1548	.0884	.0459	.0092	.0012
10	.8254	.5794	.4465	.3155	.2023	.1168	.0604	.0278	.0041	.0004
12	.8202	.5568	.4174	.2837	.1728	.0935	.0446	.0187	.0022	.0001
14	.8161	.5392	.3952	.2603	.1522	.0781	.0350	.0136	.0013	.0001
16	.8127	.5252	.3779	.2426	.1371	.0674	.0286	.0104	.0009	.0000
18	.8100	.5137	.3640	.2287	.1257	.0596	.0243	.0084	.0006	.0000
20	.8076	.5042	.3526	.2176	.1167	.0538	.0211	.0070	.0005	.0000
22	.8056	.4962	.3432	.2085	.1096	.0492	.0187	.0060	.0004	.0000
24	.8039	.4894	.3352	.2009	.1038	.0456	.0169	.0052	.0003	.0000
26	.8024	.4835	.3283	.1945	.0989	.0426	.0154	.0046	.0002	.0000
28	.8011	.4783	.3224	.1891	.0949	.0402	.0142	.0042	.0002	.0000
30	.7999	.4738	.3172	.1844	.0914	.0382	.0133	.0038	.0002	.0000
40	.7956	.4574	.2989	.1680	.0797	.0315	.0103	.0027	.0001	.0000
60	.7910	.4403	.2802	.1519	.0688	.0257	.0078	.0019	.0001	.0000
120	.7859	.4223	.2612	.1362	.0587	.0206	.0058	.0013	.0000	.0000
∞	.7805	.4035	.2420	.1210	.0494	.0162	.0042	.0009	.0000	.0000
					$\nu_1 = 7$					
2	.8737	.7994	.7587	.7133	.6643	.6129	.5600	.5069	.4037	.3095
4	.8543	.7108	.6285	.5373	.4431	.3515	.2680	.1960	.0924	.0367
6	.8411	.6485	.5394	.4238	.3126	.2154	.1383	.0825	.0234	.0049
8	.8315	.6042	.4784	.3509	.2362	.1451	.0808	.0407	.0076	.0009
10	.8243	.5714	.4351	.3020	.1890	.1057	.0525	.0231	.0031	.0002
12	.8186	.5465	.4030	.2675	.1578	.0819	.0371	.0146	.0015	.0001
14	.8141	.5269	.3786	.2423	.1362	.0665	.0279	.0101	.0008	.0000
16	.8103	.5111	.3594	.2231	.1204	.0559	.0221	.0074	.0005	.0000
18	.8072	.4982	.3440	.2081	.1086	.0483	.0181	.0057	.0003	.0000
20	.8046	.4874	.3313	.1962	.0995	.0427	.0153	.0046	.0002	.0000
22	.8024	.4784	.3208	.1864	.0922	.0383	.0132	.0038	.0002	.0000
24	.8004	.4705	.3119	.1783	.0863	.0349	.0117	.0032	.0001	.0000
26	.7987	.4638	.3043	.1715	.0815	.0322	.0105	.0028	.0001	.0000
28	.7972	.4579	.2977	.1657	.0774	.0300	.0095	.0024	.0001	.0000
30	.7958	.4527	.2919	.1606	.0740	.0281	.0087	.0022	.0001	.0000
40	.7908	.4339	.2715	.1433	.0625	.0222	.0063	.0014	.0000	.0000
60	.7854	.4140	.2507	.1264	.0520	.0172	.0045	.0009	.0000	.0000
120	.7794	.3932	.2296	.1102	.0426	.0130	.0031	.0006	.0000	.0000
∞	.7728	.3712	.2083	.0948	.0342	.0096	.0021	.0003	.0000	.0000
					$\nu_1 = 8$					
2	.8740	.8006	.7604	.7156	.6670	.6160	.5636	.5109	.4081	.3140
4	.8546	.7112	.6287	.5373	.4427	.3509	.2671	.1950	.0915	.0362
6	.8410	.6466	.5361	.4192	.3072	.2100	.1334	.0786	.0216	.0043
8	.8310	.5996	.4716	.3423	.2273	.1371	.0747	.0367	.0064	.0007
10	.8233	.5645	.4251	.2904	.1778	.0968	.0465	.0196	.0023	.0002
12	.8172	.5373	.3906	.2538	.1454	.0727	.0315	.0117	.0010	.0000
14	.8123	.5159	.3641	.2269	.1230	.0574	.0228	.0077	.0005	.0000
16	.8082	.4986	.3432	.2066	.1069	.0470	.0174	.0054	.0003	.0000
18	.8048	.4843	.3264	.1907	.0949	.0397	.0138	.0040	.0002	.0000
20	.8019	.4724	.3126	.1781	.0857	.0344	.0114	.0031	.0001	.0000
22	.7994	.4622	.3012	.1678	.0784	.0303	.0096	.0025	.0001	.0000
24	.7972	.4535	.2914	.1593	.0726	.0272	.0083	.0020	.0001	.0000
26	.7952	.4460	.2831	.1521	.0678	.0247	.0072	.0017	.0000	.0000
28	.7935	.4394	.2759	.1460	.0638	.0226	.0065	.0015	.0000	.0000
30	.7920	.4335	.2696	.1408	.0604	.0210	.0058	.0013	.0000	.0000
40	.7863	.4123	.2473	.1228	.0494	.0158	.0040	.0008	.0000	.0000
60	.7801	.3899	.2247	.1056	.0395	.0116	.0026	.0004	.0000	.0000
120	.7731	.3663	.2019	.0893	.0309	.0082	.0016	.0002	.0000	.0000
∞	.7653	.3414	.1791	.0740	.0235	.0056	.0010	.0001	.0000	.0000

TABLE VII (continued)

ν_2	$\phi=.5$	1.0	1.2	1.4	$\alpha=0.10$ 1.6	1.8	2.0	2.2	2.6	3.0
					$\nu_1 = 9$					
2	.8743	.8017	.7619	.7174	.6693	.6186	.5666	.5141	.4117	.3177
4	.8548	.7115	.6288	.5372	.4423	.3503	.2663	.1942	.0908	.0357
6	.8409	.6449	.5333	.4153	.3027	.2055	.1294	.0755	.0202	.0039
8	.8305	.5956	.4656	.3350	.2198	.1305	.0698	.0335	.0055	.0006
10	.8224	.5583	.4165	.2805	.1685	.0894	.0417	.0170	.0019	.0001
12	.8160	.5293	.3797	.2420	.1350	.0653	.0271	.0096	.0007	.0000
14	.8107	.5061	.3513	.2137	.1121	.0501	.0189	.0060	.0003	.0000
16	.8063	.4873	.3290	.1924	.0958	.0401	.0139	.0040	.0002	.0000
18	.8025	.4718	.3109	.1758	.0837	.0331	.0107	.0028	.0001	.0000
20	.7994	.4588	.2962	.1627	.0745	.0281	.0086	.0021	.0001	.0000
22	.7966	.4476	.2838	.1520	.0673	.0243	.0071	.0016	.0000	.0000
24	.7942	.4381	.2734	.1432	.0616	.0214	.0059	.0013	.0000	.0000
26	.7921	.4298	.2644	.1358	.0569	.0191	.0051	.0011	.0000	.0000
28	.7902	.4225	.2567	.1295	.0530	.0173	.0045	.0009	.0000	.0000
30	.7885	.4161	.2500	.1241	.0498	.0159	.0040	.0008	.0000	.0000
40	.7821	.3927	.2261	.1058	.0393	.0114	.0025	.0004	.0000	.0000
60	.7750	.3678	.2019	.0885	.0302	.0078	.0015	.0002	.0000	.0000
120	.7671	.3415	.1777	.0724	.0224	.0052	.0009	.0001	.0000	.0000
∞	.7581	.3138	.1537	.0577	.0161	.0033	.0005	.0000	.0000	.0000
					$\nu_1 = 10$					
2	.8746	.8025	.7630	.7188	.6710	.6207	.5689	.5167	.4146	.3206
4	.8550	.7117	.6289	.5370	.4419	.3497	.2656	.1935	.0902	.0353
6	.8408	.6434	.5308	.4119	.2988	.2016	.1260	.0728	.0191	.0036
8	.8301	.5921	.4604	.3287	.2133	.1250	.0657	.0309	.0049	.0005
10	.8216	.5528	.4088	.2719	.1605	.0834	.0378	.0149	.0015	.0001
12	.8148	.5220	.3700	.2317	.1263	.0592	.0237	.0081	.0006	.0000
14	.8092	.4974	.3401	.2023	.1030	.0443	.0160	.0048	.0002	.0000
16	.8045	.4772	.3164	.1802	.0866	.0346	.0114	.0031	.0001	.0000
18	.8005	.4605	.2972	.1630	.0745	.0279	.0085	.0021	.0001	.0000
20	.7971	.4464	.2815	.1494	.0654	.0232	.0066	.0015	.0000	.0000
22	.7941	.4344	.2684	.1384	.0583	.0197	.0053	.0011	.0000	.0000
24	.7914	.4240	.2573	.1294	.0527	.0171	.0044	.0009	.0000	.0000
26	.7891	.4150	.2479	.1219	.0482	.0151	.0037	.0007	.0000	.0000
28	.7870	.4071	.2397	.1155	.0445	.0134	.0031	.0006	.0000	.0000
30	.7852	.4001	.2326	.1101	.0414	.0121	.0027	.0005	.0000	.0000
40	.7782	.3746	.2073	.0916	.0315	.0083	.0016	.0002	.0000	.0000
60	.7703	.3475	.1819	.0745	.0232	.0054	.0009	.0001	.0000	.0000
120	.7613	.3186	.1566	.0588	.0163	.0033	.0005	.0000	.0000	.0000
∞	.7510	.2883	.1319	.0449	.0110	.0019	.0002	.0000	.0000	.0000
					$\nu_1 = 12$					
2	.8749	.8037	.7646	.7210	.6736	.6237	.5723	.5204	.4188	.3250
4	.8552	.7120	.6289	.5368	.4413	.3488	.2645	.1923	.0893	.0348
6	.8406	.6409	.5267	.4064	.2925	.1954	.1207	.0688	.0174	.0032
8	.8293	.5862	.4517	.3182	.2029	.1161	.0594	.0270	.0039	.0003
10	.8203	.5436	.3961	.2577	.1477	.0739	.0320	.0120	.0011	.0001
12	.8129	.5097	.3538	.2149	.1123	.0500	.0187	.0059	.0003	.0000
14	.8067	.4823	.3210	.1836	.0887	.0356	.0118	.0032	.0001	.0000
16	.8014	.4597	.2950	.1603	.0722	.0266	.0079	.0019	.0001	.0000
18	.7969	.4409	.2740	.1423	.0603	.0206	.0056	.0012	.0000	.0000
20	.7930	.4249	.2568	.1281	.0515	.0164	.0041	.0008	.0000	.0000
22	.7896	.4113	.2424	.1167	.0448	.0135	.0031	.0006	.0000	.0000
24	.7865	.3994	.2303	.1074	.0396	.0113	.0025	.0004	.0000	.0000
26	.7838	.3892	.2199	.0998	.0354	.0096	.0020	.0003	.0000	.0000
28	.7814	.3801	.2110	.0933	.0320	.0083	.0016	.0002	.0000	.0000
30	.7792	.3721	.2032	.0879	.0292	.0073	.0014	.0002	.0000	.0000
40	.7709	.3427	.1759	.0697	.0207	.0045	.0007	.0001	.0000	.0000
60	.7614	.3114	.1486	.0532	.0139	.0026	.0003	.0000	.0000	.0000
120	.7503	.2781	.1220	.0389	.0087	.0013	.0001	.0000	.0000	.0000
∞	.7373	.2431	.0967	.0270	.0051	.0006	.0000	.0000	.0000	.0000

From M. L. Tiku, "Tables of the Power of the F-Test," *Journal of the American Statistical Association*, 62 (1967), 525–539 and M. L. Tiku, "More Tables of the Power of the F-test." *Journal of the American Statistical Association*, 67 (1972), 709–710. Abridged and adapted by permission.

Table VIII. Power Values and Optimum Number of Levels for Total Number of Observations in the One-Way Random Effects Analysis of Variance F Test

This table gives power estimates in the one-way random effects analysis of variance F test for specified values of θ_0 (the value of $\sigma_\alpha^2/\sigma_e^2$ under H_0), θ (the value of $\sigma_\alpha^2/\sigma_e^2$ under H_1), $N = an$ (total number of observations), a (the number of treatment groups or levels), and α (the level of significance). For example, in a one-way random effects analysis of variance, consider the simple hypothesis that there are no treatment effects ($\theta_0 = 0$) and the researcher wants to reject the null hypothesis if $\sigma_\alpha^2/\sigma_e^2$ is as large as 1.0 ($\theta = 1.0$) at $\alpha = 0.05$. For 20 treatment groups with 5 subjects per group ($a = 20, n = 5, N = 100$), the power of the test is equal to 0.998.

Statistical Tables and Charts

$\alpha = 0.01$
$\theta_0 = 0.00$

N \ θ	0.2	0.4	0.6	0.8	1.0	2.0	3.0	4.0
10	2, .045	2, .089	2, .132	2, .172	2, .208	2, .341	2, .426	2, .485
20	2, .114	2, .214	4, .302	4, .394	4, .471	5, .706	5, .822	5, .881
30	3, .180	3, .348	5, .474	5, .586	5, .668	6, .875	10, .948	10, .977
40	3, .246	4, .463	5, .615	5, .716	8, .795	10, .954	10, .986	13, .995
50	3, .306	5, .561	5, .708	7, .809	10, .878	10, .981	16, .996	16, .999
60	4, .383	6, .644	6, .788	10, .881	10, .930	15, .994	20, .999	12, 1.00
70	5, .439	7, .713	10, .852	10, .924	14, .960	14, .998	14, 1.00	10, 1.00
80	5, .497	8, .770	10, .896	10, .950	16, .977	20, .999	11, 1.00	9, 1.00
90	5, .547	9, .817	10, .925	15, .970	18, .988	15, 1.00	10, 1.00	8, 1.00
100	5, .591	10, .855	11, .946	20, .980	20, .993	14, 1.00	9, 1.00	8, 1.00
300	15, .965	30, .999	8, 1.00	7, 1.00	6, 1.00	6, 1.00	6, 1.00	5, 1.00
500	15, 1.00	8, 1.00	7, 1.00	7, 1.00	6, 1.00	6, 1.00	6, 1.00	5, 1.00

$\theta_0 = 0.10$

N \ θ	0.3	0.5	0.7	0.9	1.0	2.0	3.0	4.0
10	2, .032	2, .059	2, .089	2, .118	2, .132	2, .250	2, .334	2, .396
20	2, .057	4, .120	4, .194	4, .267	4, .302	5, .567	5, .718	5, .804
30	3, .082	5, .188	5, .302	6, .405	6, .454	6, .751	10, .891	10, .948
40	4, .106	5, .252	8, .395	8, .527	8, .582	10, .878	10, .958	13, .984
50	5, .130	7, .309	10, .484	10, .629	10, .686	10, .931	16, .984	16, .996
60	6, .153	10, .371	10, .569	12, .713	12, .768	15, .971	20, .996	20, .999
70	7, .176	10, .430	14, .636	14, .781	14, .830	23, .985	23, .999	17, 1.00
80	8, .199	10, .479	16, .697	16, .834	16, .877	20, .994	26, 1.00	15, 1.00
90	10, .223	15, .530	15, .751	18, .876	18, .912	30, .997	18, 1.00	14, 1.00
100	10, .245	14, .569	20, .795	20, .907	20, .978	25, .999	16, 1.00	12, 1.00
300	30, .632	50, .967	60, .998	50, 1.00	30, 1.00	17, 1.00	10, 1.00	8, 1.00
500	62, .850	50, .998	22, .100	16, 1.00	15, 1.00	10, 1.00	8, 1.00	8, 1.00

$\theta_0 = 0.50$

N \ θ	0.7	0.8	0.9	1.0	2.0	3.0	4.0	5.0
10	2, .018	2, .023	2, .028	2, .034	2, .095	2, .155	2, .208	3, .259
20	4, .024	4, .034	4, .045	4, .057	5, .218	5, .380	5, .508	5, .604
30	5, .029	6, .043	6, .060	6, .080	10, .329	10, .570	10, .730	10, .828
40	8, .034	8, .053	8, .076	8, .102	10, .445	13, .344	13, .847	13, .919
50	10, .039	10, .062	10, .091	10, .125	16, .526	16, .799	16, .916	16, .963
60	12, .043	12, .071	12, .107	15, .149	20, .634	20, .885	20, .964	20, .988
70	14, .048	14, .081	14, .122	14, .171	23, .703	23, .926	23, .982	23, .995
80	16, .052	16, .090	20, .139	20, .197	20, .762	26, .953	26, .991	40, .998
90	18, .056	18, .099	18, .154	18, .218	30, .823	30, .975	30, .996	45, .999
100	20, .061	20, .109	25, .175	25, .246	33, .856	33, .985	33, .998	30, 1.00
500	125, .255	125, .511	125, .739	125, .884	68, 1.00	20, 1.00	15, 1.00	13, 1.00
1000	250, .503	250, .828	333, .966	190, 1.00	35, 1.00	20, 1.00	15, 1.00	13, 1.00

$\theta_0 = 1.00$

N \ θ	1.2	1.4	1.6	1.8	2.0	3.0	4.0	5.0
10	2, .015	2, .020	2, .025	2, .032	2, .038	2, .074	2, .111	2, .146
20	4, .017	5, .027	5, .039	5, .054	5, .070	5, .166	5, .270	5, .365
30	6, .020	6, .034	6, .052	6, .073	10, .098	10, .260	10, .429	10, .570
40	10, .022	10, .040	10, .065	10, .096	10, .132	10, .344	13, .550	13, .702
50	10, .024	10, .046	10, .076	12, .113	12, .157	16, .425	16, .653	16, .799
60	15, .026	15, .053	15, .091	15, .140	20, .198	20, .525	20, .760	20, .885
70	14, .027	17, .058	23, .102	23, .161	23, .229	23, .592	23, .821	23, .926
80	20, .029	20, .065	20, .118	20, .185	20, .262	26, .651	26, .869	40, .955
90	22, .031	30, .071	30, .132	30, .211	30, .302	30, .721	30, .914	30, .975
100	25, .033	25, .078	25, .145	33, .233	33, .333	33, .765	33, .938	50, .985
500	166, .103	166, .365	166, .678	166, .882	166, .967	94, 1.00	34, 1.00	24, 1.00
1000	333, .202	333, .677	333, .994	250, 1.00	142, 1.00	52, 1.00	34, 1.00	24, 1.00

Table VIII (*continued*)

$\alpha = 0.05$
$\theta_0 = 0.00$

N \ θ	0.2	0.4	0.6	0.8	1.0	2.0	3.0	4.0
10	2, .142	2, .220	2, .282	2, .333	2, .374	3, .518	3, .622	5, .693
20	2, .241	4, .386	4, .507	4, .596	4, .662	5, .847	5, .914	5, .945
30	3, .342	5, .530	5, .666	6, .756	6, .818	10, .949	10, .984	10, .994
40	4, .424	5, .645	8, .768	8, .854	8, .904	10, .984	13, .996	13, .999
50	5, .495	5, .724	10, .843	10, .914	10, .950	16, .994	16, .999	12, 1.00
60	5, .564	6, .792	10, .900	12, .950	12, .974	20, .999	12, 1.00	10, 1.00
70	5, .621	7, .843	10, .934	14, .971	14, .987	23, 1.00	10, 1.00	8, 1.00
80	5, .668	10, .883	10, .955	16, .983	16, .993	13, 1.00	9, 1.00	8, 1.00
90	6, .715	10, .913	15, .965	18, .990	18, .997	11, 1.00	8, 1.00	7, 1.00
100	7, .746	10, .933	14, .980	20, .995	20, .998	10, 1.00	8, 1.00	7, 1.00
300	20, .989	23, 1.00	8, 1.00	7, 1.00	6, 1.00	5, 1.00	5, 1.00	5, 1.00
500	10, 1.00	7, 1.00	7, 1.00	6, 1.00	6, 1.00	5, 1.00	5, 1.00	5, 1.00

$\theta_0 = 0.10$

N \ θ	0.3	0.5	0.7	0.9	1.0	2.0	3.0	4.0
10	2, .112	2, .169	2, .220	2, .263	2, .282	3, .436	3, .547	5, .628
20	4, .163	5, .283	4, .386	5, .473	5, .515	5, .753	5, .854	10, .907
30	5, .211	5, .377	6, .513	6, .618	6, .661	10, .884	10, .962	10, .984
40	5, .255	8, .445	8, .616	8, .727	10, .771	10, .952	13, .988	13, .996
50	7, .290	10, .526	10, .698	10, .806	10, .843	16, .978	16, .996	16, .999
60	10, .325	10, .592	12, .764	15, .863	15, .897	20, .993	20, .999	14, 1.00
70	10, .364	10, .644	14, .816	14, .904	14, .930	23, .997	17, 1.00	13, 1.00
80	10, .397	16, .692	16, .857	20, .934	20, .955	26, .999	15, 1.00	11, 1.00
90	10, .426	15, .738	18, .890	18, .954	18, .969	30, 1.00	14, 1.00	10, 1.00
100	11, .453	20, .771	20, .915	25, .969	25, .981	20, 1.00	12, 1.00	10, 1.00
300	37, .821	50, .991	48, 1.00	26, 1.00	21, 1.00	15, 1.00	8, 1.00	7, 1.00
500	71, .949	29, 1.00	17, 1.00	13, 1.00	12, 1.00	9, 1.00	8, 1.00	7, 1.00

$\theta_0 = 0.50$

N \ θ	0.7	0.8	0.9	1.0	2.0	3.0	4.0	5.0
10	2, .076	2, .090	2, .103	2, .116	3, .239	3, .343	5, .429	5, .509
20	5, .095	5, .121	5, .147	5, .175	5, .429	5, .601	10, .720	10, .809
30	6, .109	6, .144	6, .181	6, .218	10, .578	10, .784	10, .884	10, .934
40	8, .122	10, .167	10, .215	10, .265	13, .677	13, .870	20, .945	20, .978
50	10, .134	10, .186	10, .242	10, .300	16, .755	16, .924	25, .977	25, .993
60	15, .145	15, .207	15, .275	15, .344	20, .832	20, .963	20, .991	30, .998
70	14, .155	14, .224	14, .298	17, .373	23, .875	23, .979	35, .996	35, .999
80	20, .166	20, .245	20, .330	20, .415	26, .907	26, .988	40, .999	24, 1.00
90	18, .175	18, .260	22, .351	30, .442	30, .938	30, .994	45, .999	21, 1.00
100	25, .186	25, .281	25, .381	25, .480	33, .955	33, .997	30, 1.00	20, 1.00
500	125, .499	125, .749	125, .899	166, .967	27, 1.00	16, 1.00	13, 1.00	11, 1.00
1000	250, .745	333, .944	237, 1.00	115, 1.00	27, 1.00	16, 1.00	13, 1.00	11, 1.00

$\theta_0 = 1.00$

N \ θ	1.2	1.4	1.6	1.8	2.0	3.0	4.0	5.0
10	2, .065	2, .081	2, .096	3, .112	3, .129	3, .209	3, .281	5, .350
20	5, .076	5, .105	5, .137	5, .169	5, .203	5, .361	10, .491	10, .610
30	10, .083	10, .123	10, .168	10, .216	10, .267	10, .501	10, .672	10, .784
40	10, .090	10, .139	10, .195	10, .255	13, .317	13, .595	13, .771	20, .878
50	12, .094	16, .152	16, .219	16, .291	16, .365	16, .672	25, .848	25, .935
60	15, .101	20, .170	20, .251	20, .338	20, .424	20, .756	20, .906	30, .966
70	23, .106	23, .183	23, .274	23, .371	23, .466	23, .805	35, .938	35, .982
80	20, .112	20, .196	26, .296	26, .402	26, .505	26, .845	40, .961	40, .991
90	30, .116	30, .212	30, .325	30, .442	30, .533	30, .887	45, .976	45, .996
100	25, .121	33, .224	33, .346	33, .471	33, .587	33, .911	50, .985	50, .998
500	166, .277	166, .625	166, .868	166, .967	166, .993	77, 1.00	25, 1.00	19, 1.00
1000	333, .438	333, .867	333, .992	142, 1.00	100, 1.00	41, 1.00	25, 1.00	19, 1.00

Table VIII (continued)

$\alpha = 0.10$
$\theta_0 = 0.00$

N \ θ	0.2	0.4	0.6	0.8	1.0	2.0	3.0	4.0
10	2, .225	2, .314	2, .380	2, .430	2, .470	3, .633	5, .737	5, .808
20	4, .331	4, .501	4, .615	5, .694	5, .755	5, .897	5, .944	10, .972
30	3, .444	5, .637	6, .754	6, .830	6, .877	10, .973	10, .992	10, .997
40	4, .531	5, .734	8, .843	8, .906	10, .941	10, .992	13, .998	13, 1.00
50	5, .602	7, .801	10, .900	10, .948	10, .971	16, .997	16, 1.00	10, 1.00
60	5, .662	10, .855	10, .938	12, .971	15, .986	20, .999	10, 1.00	8, 1.00
70	7, .711	10, .897	10, .960	14, .984	14, .993	14, 1.00	9, 1.00	7, 1.00
80	8, .753	10, .926	10, .975	16, .991	20, .997	10, 1.00	8, 1.00	7, 1.00
90	6, .791	10, .945	15, .985	18, .995	18, .998	9, 1.00	8, 1.00	6, 1.00
100	9, .821	11, .959	20, .990	20, .998	20, .999	9, 1.00	7, 1.00	6, 1.00
300	20, .994	20, 1.00	7, 1.00	7, 1.00	6, 1.00	5, 1.00	5, 1.00	5, 1.00
500	9, 1.00	7, 1.00	6, 1.00	6, 1.00	6, 1.00	5, 1.00	5, 1.00	5, 1.00

$\theta_0 = 0.10$

N \ θ	0.3	0.5	0.7	0.9	1.0	2.0	3.0	4.0
10	2, .188	2, .258	2, .314	3, .360	3, .385	3, .558	5, .678	5, .758
20	4, .258	4, .396	5, .505	5, .592	5, .628	5, .828	10, .907	10, .953
30	5, .316	6, .496	6, .628	6, .720	6, .754	10, .939	10, .980	10, .992
40	5, .363	5, .558	8, .721	10, .816	10, .850	13, .974	13, .994	20, .998
50	7, .406	10, .646	10, .791	10, .873	10, .900	16, .989	16, .998	25, 1.00
60	10, .448	12, .702	12, .844	15, .918	15, .940	20, .997	20, 1.00	12, 1.00
70	10, .487	14, .750	14, .883	14, .943	14, .960	23, .999	14, 1.00	10, 1.00
80	10, .520	16, .790	16, .913	20, .964	20, .977	26, 1.00	13, 1.00	10, 1.00
90	15, .550	15, .824	18, .935	18, .975	22, .984	18, 1.00	11, 1.00	9, 1.00
100	14, .577	20, .853	20, .952	25, .984	25, .991	16, 1.00	11, 1.00	9, 1.00
300	50, .892	60, .996	33, 1.00	23, 1.00	21, 1.00	10, 1.00	8, 1.00	7, 1.00
500	71, .975	23, 1.00	15, 1.00	12, 1.00	11, 1.00	8, 1.00	7, 1.00	7, 1.00

$\theta_0 = 0.50$

N \ θ	0.7	0.8	0.9	1.0	2.0	3.0	4.0	5.0
10	2, .140	2, .158	3, .177	3, .196	3, .355	5, .483	5, .582	5, .657
20	5, .169	5, .205	5, .241	5, .276	5, .551	10, .720	10, .829	10, .892
30	6, .189	6, .236	10, .284	10, .333	10, .700	10, .863	15, .933	15, .968
40	10, .208	10, .268	10, .328	10, .386	13, .784	13, .925	20, .974	20, .991
50	10, .223	10, .291	10, .358	12, .423	16, .845	25, .961	25, .990	25, .997
60	15, .240	15, .319	15, .399	15, .474	20, .901	20, .982	30, .997	30, .999
70	14, .253	17, .338	17, .424	23, .507	23, .930	23, .990	35, .999	23, 1.00
80	20, .268	20, .365	20, .460	20, .548	26, .950	40, .995	40, 1.00	20, 1.00
90	18, .279	22, .382	30, .485	30, .580	30, .969	30, .998	30, 1.00	18, 1.00
100	25, .294	25, .407	25, .515	25, .612	33, .978	50, .999	25, 1.00	20, 1.00
500	125, .636	125, .845	125, .947	166, .985	23, 1.00	14, 1.00	12, 1.00	10, 1.00
1000	250, .840	333, .976	142, 1.00	93, 1.00	23, 1.00	14, 1.00	12, 1.00	10, 1.00

$\theta_0 = 1.00$

N \ θ	1.2	1.4	1.6	1.8	2.0	3.0	4.0	5.0
10	3, .124	3, .149	3, .173	3, .196	3, .219	3, .330	5, .422	5, .502
20	5, .141	5, .184	5, .227	5, .269	5, .310	10, .492	10, .639	10, .743
30	10, .152	10, .211	10, .272	10, .333	10, .392	10, .632	10, .779	15, .871
40	10, .162	10, .232	13, .305	13, .379	13, .449	13, .716	20, .864	20, .936
50	16, .171	16, .253	16, .340	16, .426	16, .506	16, .787	25, .918	25, .969
60	20, .180	20, .275	20, .375	20, .472	20, .561	20, .846	30, .951	30, .985
70	23, .187	23, .292	23, .402	23, .507	23, .602	23, .883	35, .971	35, .993
80	20, .194	26, .308	26, .427	26, .540	26, .639	26, .913	40, .983	40, .997
90	30, .202	30, .328	30, .459	30, .580	30, .683	30, .938	45, .990	45, .999
100	33, .209	33, .343	33, .482	33, .608	33, .713	50, .953	50, .994	50, .999
500	166, .408	166, .749	166, .929	166, .985	166, .997	69, 1.00	22, 1.00	17, 1.00
1000	333, .580	333, .928	233, 1.00	116, 1.00	82, 1.00	37, 1.00	22, 1.00	17, 1.00

From R. S. Barcikowski, "Optimum Sample Size and Number of Levels in a One-Way Random Effects Analysis of Variance," *The Journal of Experimental Education*, 41 (1973), 10–16. Reprinted by permission.

Table IX. Minimum Sample Size per Treatment Group Needed for a Given Value of p, α, $1 - \beta$, and Effect Size (C) in Sigma Units

This table gives the minimum sample size per treatment group needed in the one-way fixed effects analysis of variance design corresponding to $\alpha = 0.10, 0.05, 0.01$; $1 - \beta = 0.7, 0.8, 0.9, 0.95$; $C = \Delta/\sigma_e = 1.0\,(0.25)\,2\,(0.5)\,3$; and $p = 2\,(1)\,11, 13$. Here, Δ designates the magnitude of the difference between any pair of treatment groups that is meaningful to detect with probability of at least $1 - \beta$. For example, in a one-way fixed effects analysis of variance design, for $p = 3$, $\alpha = 0.05$, $1 - \beta = 0.8$, and $C = 1.0$, the required sample size per treatment group is 21.

		$1-\beta = 0.70$							$1-\beta = 0.80$					
		C							C					
p	α	1.00	1.25	1.50	1.75	2.00	2.50	3.00	1.00	1.25	1.50	1.75	2.00	2.50	3.00
2	.10	11	7	6	4	4	3	3	14	9	7	5	4	3	3
	.05	14	9	7	6	5	4	3	17	12	9	7	6	4	4
	.01	21	15	11	9	7	5	5	26	17	13	10	8	6	5
3	.10	13	9	7	5	4	3	3	17	11	8	6	5	4	3
	.05	17	11	8	7	5	4	3	21	14	10	8	6	5	4
	.01	25	17	12	10	8	6	5	30	20	14	11	9	7	5
4	.10	15	10	7	6	5	4	3	19	13	9	7	6	4	3
	.05	19	13	9	7	6	4	4	23	15	11	9	7	5	4
	.01	28	19	13	10	8	6	5	33	22	16	12	10	7	5
5	.10	17	11	8	6	5	4	3	21	14	10	8	6	4	4
	.05	21	14	10	8	6	5	4	25	17	12	9	7	5	4
	.01	30	20	14	11	9	6	5	35	23	17	13	10	7	6
6	.10	18	12	9	7	5	4	3	22	15	11	8	7	5	4
	.05	22	15	11	8	7	5	4	27	18	13	10	8	6	4
	.01	32	21	15	12	9	7	5	38	25	18	13	11	8	6
7	.10	19	13	9	7	6	4	3	24	16	11	9	7	5	4
	.05	24	16	11	9	7	5	4	29	19	14	10	8	6	5
	.01	34	22	16	12	10	7	5	39	26	18	14	11	8	6
8	.10	20	13	10	7	6	4	3	25	16	12	9	7	5	4
	.05	25	16	12	9	7	5	4	30	20	14	11	9	6	5
	.01	35	23	17	13	10	7	5	41	27	19	15	12	8	6
9	.10	21	14	10	8	6	4	4	26	17	12	9	7	5	4
	.05	26	17	12	9	8	5	4	31	21	15	11	9	6	5
	.01	37	24	17	13	10	7	6	43	28	20	15	12	8	6
10	.10	22	14	10	8	6	5	4	27	18	13	10	8	5	4
	.05	27	18	13	10	8	6	4	33	21	15	12	9	6	5
	.01	38	25	18	14	11	7	6	44	29	21	16	12	8	6
11	.10	23	15	11	8	7	5	4	28	18	13	10	8	6	4
	.05	28	19	13	10	8	6	4	34	22	16	12	9	7	5
	.01	39	26	18	14	11	8	6	46	30	21	16	13	9	7
13	.10	24	16	11	9	7	5	4	30	20	14	11	8	6	4
	.05	30	20	14	11	9	6	5	36	24	17	13	10	7	5
	.01	42	27	19	15	12	8	6	49	32	22	17	13	9	7

Table IX (*continued*)

p	α	\multicolumn{6}{c}{1−β = 0.90, C}						\multicolumn{6}{c}{1−β = 0.95, C}					
		1.00	1.25	1.50	1.75	2.00	2.50	3.00	1.00	1.25	1.50	1.75	2.00	2.50	3.00
2	.10	18	12	9	7	6	4	3	23	15	11	8	7	5	4
	.05	23	15	11	8	7	5	4	27	18	13	10	8	6	5
	.01	32	21	15	12	10	7	6	38	25	18	14	11	8	6
3	.10	22	15	11	8	7	5	4	27	18	13	10	8	6	4
	.05	27	18	13	10	8	6	5	32	21	15	12	9	7	5
	.01	37	24	18	13	11	8	6	43	29	20	16	12	9	7
4	.10	25	16	12	9	7	5	4	30	20	14	11	9	6	5
	.05	30	20	14	11	9	6	5	36	23	17	13	10	7	5
	.01	40	27	19	15	12	8	6	47	31	22	17	13	9	7
5	.10	27	18	13	10	8	5	4	33	22	15	12	9	6	5
	.05	32	21	15	12	9	6	5	39	25	18	14	11	7	6
	.01	43	28	20	15	12	9	7	51	33	23	18	14	10	7
6	.10	29	19	14	10	8	6	4	35	23	16	12	10	7	5
	.05	34	23	16	12	10	7	5	41	27	19	14	11	8	6
	.01	46	30	21	16	13	9	7	53	35	25	19	15	10	8
7	.10	31	20	14	11	9	6	5	37	24	17	13	10	7	5
	.05	36	24	17	13	10	7	5	43	28	20	15	12	8	6
	.01	48	31	22	17	13	9	7	56	36	26	19	15	10	8
8	.10	32	21	15	11	9	6	5	39	25	18	14	11	7	5
	.05	38	25	18	13	11	7	6	45	29	21	16	12	8	6
	.01	50	33	23	17	14	9	7	58	38	27	20	16	11	8
9	.10	33	22	16	12	9	6	5	40	26	19	14	11	8	6
	.05	40	26	18	14	11	8	6	47	30	22	16	13	9	6
	.01	52	34	24	18	14	10	7	60	39	28	21	16	11	8
10	.10	35	23	16	12	10	7	5	42	27	19	15	11	8	6
	.05	41	27	19	14	11	8	6	48	31	22	17	13	9	7
	.01	54	35	25	19	15	10	7	62	40	29	21	17	11	8
11	.10	36	23	17	13	10	7	5	43	28	20	15	12	8	6
	.05	42	28	20	15	12	8	6	50	33	23	17	14	9	7
	.01	55	36	26	19	15	10	8	64	42	29	22	17	12	9
13	.10	38	25	18	13	11	7	5	46	30	21	16	12	8	6
	.05	45	29	21	16	12	8	6	53	34	24	18	14	10	7
	.01	59	38	27	20	16	11	8	68	44	31	23	18	12	9

From T. L. Bratcher, A. M. Moran, and W. J. Zimmer, "Tables of Sample Sizes in the Analysis of Variance," *Journal of Quality Technology*, 2 (1970), 156–164. Abridged and adapted by permission. The adaptation is due to R. E. Kirk, *Experimental Design*, Third Edition, © 1995 by Brooks/Cole, Monterey, CA.

Table X. Critical Values of the Studentized Range Distribution

This table gives the critical values of the Studentized range distribution used in multiple comparisons. The critical values are designated as $q[p, \nu; 1 - \alpha]$ corresponding to a given value of α, p as the total number of treatment groups or r as the number of steps between ordered means, and ν as the number of degrees of freedom for the error. The critical values are given for $\alpha = 0.05, 0.01$; $p = 2$ (1) 20; and $\nu = 2$ (1) 20, 24, 30, 40, 60, 120, ∞. For example, for $\alpha = 0.05$, $p = 4$, and $\nu = 20$, the required critical value is $q\ [4, 20; 0.95] = 3.96$.

Statistical Tables and Charts

		Number of Means (p) or Number of Steps Between Ordered Means (r)								
ν	α	2	3	4	5	6	7	8	9	10	11
2	.05	6.08	8.33	9.80	10.90	11.70	12.40	13.00	13.50	14.00	14.40
	.01	14.00	19.00	22.30	24.70	26.60	28.20	29.50	30.70	31.70	32.60
3	.05	4.50	5.91	6.82	7.50	8.04	8.48	8.85	9.18	9.46	9.72
	.01	8.26	10.60	12.20	13.30	14.20	15.00	15.60	16.20	16.70	17.80
4	.05	3.93	5.04	5.76	6.29	6.71	7.05	7.35	7.60	7.83	8.03
	.01	6.51	8.12	9.17	9.96	10.60	11.10	11.50	11.90	12.30	12.60
5	.05	3.64	4.60	5.22	5.67	6.03	6.33	6.58	6.80	6.99	7.17
	.01	5.70	6.98	7.80	8.42	8.91	9.32	9.67	9.97	10.24	10.48
6	.05	3.46	4.34	4.90	5.30	5.63	5.90	6.12	6.32	6.49	6.65
	.01	5.24	6.33	7.03	7.56	7.97	8.32	8.61	8.87	9.10	9.30
7	.05	3.34	4.16	4.68	5.06	5.36	5.61	5.82	6.00	6.16	6.30
	.01	4.95	5.92	6.54	7.01	7.37	7.68	7.94	8.17	8.37	8.55
8	.05	3.26	4.04	4.53	4.89	5.17	5.40	5.60	5.77	5.92	6.05
	.01	4.75	5.64	6.20	6.62	6.96	7.24	7.47	7.68	7.86	8.03
9	.05	3.20	3.95	4.41	4.76	5.02	5.24	5.43	5.59	5.74	5.87
	.01	4.60	5.43	5.96	6.35	6.66	6.91	7.13	7.33	7.49	7.65
10	.05	3.15	3.88	4.33	4.65	4.91	5.12	5.30	5.46	5.60	5.72
	.01	4.48	5.27	5.77	6.14	6.43	6.67	6.87	7.05	7.21	7.36
11	.05	3.11	3.82	4.26	4.57	4.82	5.03	5.20	5.35	5.49	5.61
	.01	4.39	5.15	5.62	5.97	6.25	6.48	6.67	6.84	6.99	7.13
12	.05	3.08	3.77	4.20	4.51	4.75	4.95	5.12	5.27	5.39	5.51
	.01	4.32	5.05	5.50	5.84	6.10	6.32	6.51	6.67	6.81	6.94
13	.05	3.06	3.73	4.15	4.45	4.69	4.88	5.05	5.19	5.32	5.43
	.01	4.26	4.96	5.40	5.73	5.98	6.19	6.37	6.53	6.67	6.79
14	.05	3.03	3.70	4.11	4.41	4.64	4.83	4.99	5.13	5.25	5.36
	.01	4.21	4.89	5.32	5.63	5.88	6.08	6.26	6.41	6.54	6.66
15	.05	3.01	3.67	4.08	4.37	4.59	4.78	4.94	5.08	5.20	5.31
	.01	4.17	4.84	5.25	5.56	5.80	5.99	6.16	6.31	6.44	6.55
16	.05	3.00	3.65	4.05	4.33	4.56	4.74	4.90	5.03	5.15	5.26
	.01	4.13	4.79	5.19	5.49	5.72	5.92	6.08	6.22	6.35	6.46
17	.05	2.98	3.63	4.02	4.30	4.52	4.70	4.86	4.99	5.11	5.21
	.01	4.10	4.74	5.14	5.43	5.66	5.85	6.01	6.15	6.27	6.38
18	.05	2.97	3.61	4.00	4.28	4.49	4.67	4.82	4.96	5.07	5.17
	.01	4.07	4.70	5.09	5.38	5.60	5.79	5.94	6.08	6.20	6.31
19	.05	2.96	3.59	3.98	4.25	4.47	4.65	4.79	4.92	5.04	5.14
	.01	4.05	4.67	5.05	5.33	5.55	5.73	5.89	6.02	6.14	6.25
20	.05	2.95	3.58	3.96	4.23	4.45	4.62	4.77	4.90	5.01	5.11
	.01	4.02	4.64	5.02	5.29	5.51	5.69	5.84	5.97	6.09	6.19
24	.05	2.92	3.53	3.90	4.17	4.37	4.54	4.68	4.81	4.92	5.01
	.01	3.96	4.55	4.91	5.17	5.37	5.54	5.69	5.81	5.92	6.02
30	.05	2.89	3.49	3.85	4.10	4.30	4.46	4.60	4.72	4.82	4.92
	.01	3.89	4.45	4.80	5.05	5.24	5.40	5.54	5.65	5.76	5.85
40	.05	2.86	3.44	3.79	4.04	4.23	4.39	4.52	4.63	4.73	4.82
	.01	3.82	4.37	4.70	4.93	5.11	5.26	5.39	5.50	5.60	5.69
60	.05	2.83	3.40	3.74	3.98	4.16	4.31	4.44	4.55	4.65	4.73
	.01	3.76	4.28	4.59	4.82	4.99	5.13	5.25	5.36	5.45	5.53
120	.05	2.80	3.36	3.68	3.92	4.10	4.24	4.36	4.47	4.56	4.64
	.01	3.70	4.20	4.50	4.71	4.87	5.01	5.12	5.21	5.30	5.37
∞	.05	2.77	3.31	3.63	3.86	4.03	4.17	4.29	4.39	4.47	4.55
	.01	3.64	4.12	4.40	4.60	4.76	4.88	4.99	5.08	5.16	5.23

Table X (*continued*)

ν	α	Number of Means (*p*) or Number of Steps Between Ordered Means (*r*)								
		12	13	14	15	16	17	18	19	20
2	.05	14.70	15.10	15.40	15.70	15.90	16.10	16.40	16.60	16.80
	.01	33.40	34.10	34.80	35.40	36.00	36.50	37.00	37.50	37.90
3	.05	9.72	10.20	10.30	10.50	10.70	10.80	11.00	11.10	11.20
	.01	17.50	17.90	18.20	18.50	18.80	19.10	19.30	19.50	19.80
4	.05	8.21	8.37	8.52	8.66	8.79	8.91	9.03	9.13	9.23
	.01	12.80	13.10	13.30	13.50	13.70	13.90	14.10	14.20	14.40
5	.05	7.32	7.47	7.60	7.72	7.83	7.93	8.03	8.12	8.21
	.01	10.70	10.89	11.08	11.24	11.40	11.55	11.68	11.81	11.93
6	.05	6.79	6.92	7.03	7.14	7.24	7.34	7.43	7.51	7.59
	.01	9.48	9.65	9.81	9.95	10.08	10.21	10.32	10.43	10.54
7	.05	6.43	6.55	6.66	6.76	6.85	6.94	7.02	7.10	7.17
	.01	8.71	8.86	9.00	9.12	9.24	9.35	9.46	9.55	9.65
8	.05	6.18	6.29	6.39	6.48	6.57	6.65	6.73	6.80	6.87
	.01	8.18	8.31	8.44	8.55	8.66	8.76	8.85	8.94	9.03
9	.05	5.98	6.09	6.19	6.28	6.36	6.44	6.51	6.58	6.64
	.01	7.78	7.91	8.03	8.13	8.23	8.33	8.41	8.49	8.57
10	.05	5.83	5.93	6.03	6.11	6.19	6.27	6.34	6.40	6.47
	.01	7.49	7.60	7.71	7.81	7.91	7.99	8.08	8.15	8.23
11	.05	5.71	5.81	5.90	5.98	6.06	6.13	6.20	6.27	6.33
	.01	7.25	7.36	7.46	7.56	7.65	7.73	7.81	7.88	7.95
12	.05	5.61	5.71	5.80	5.88	5.95	6.02	6.09	6.15	6.21
	.01	7.06	7.17	7.26	7.36	7.44	7.52	7.59	7.66	7.73
13	.05	5.53	5.63	5.71	5.79	5.86	5.93	5.99	6.05	6.11
	.01	6.90	7.01	7.10	7.19	7.27	7.35	7.42	7.48	7.55
14	.05	5.46	5.55	5.64	5.71	5.79	5.85	5.91	5.97	6.03
	.01	6.77	6.87	6.96	7.05	7.13	7.20	7.27	7.33	7.39
15	.05	5.40	5.49	5.57	5.65	5.72	5.78	5.85	5.90	5.96
	.01	6.66	6.76	6.84	6.93	7.00	7.07	7.14	7.20	7.26
16	.05	5.35	5.44	5.52	5.59	5.66	5.73	5.79	5.84	5.90
	.01	6.56	6.66	6.74	6.82	6.90	6.97	7.03	7.09	7.15
17	.05	5.31	5.39	5.47	5.54	5.61	5.67	5.73	5.79	5.84
	.01	6.48	6.57	6.66	6.73	6.81	6.87	6.94	7.00	7.05
18	.05	5.27	5.35	5.43	5.50	5.57	5.63	5.69	5.74	5.79
	.01	6.41	6.50	6.58	6.65	6.73	6.79	6.85	6.91	6.97
19	.05	5.23	5.31	5.39	5.46	5.53	5.59	5.65	5.70	5.75
	.01	6.34	6.43	6.51	6.58	6.65	6.72	6.78	6.84	6.89
20	.05	5.20	5.28	5.36	5.43	5.49	5.55	5.61	5.66	5.71
	.01	6.28	6.37	6.45	6.52	6.59	6.65	6.71	6.77	6.82
24	.05	5.10	5.18	5.25	5.32	5.38	5.44	5.49	5.55	5.59
	.01	6.11	6.19	6.26	6.33	6.39	6.45	6.51	6.56	6.61
30	.05	5.00	5.08	5.15	5.21	5.27	5.33	5.38	5.43	5.47
	.01	5.93	6.01	6.08	6.14	6.20	6.26	6.31	6.36	6.41
40	.05	4.90	4.98	5.04	5.11	5.16	5.22	5.27	5.31	5.36
	.01	5.76	5.83	5.90	5.96	6.02	6.07	6.12	6.16	6.21
60	.05	4.81	4.88	4.94	5.00	5.06	5.11	5.15	5.20	5.24
	.01	5.60	5.67	5.73	5.78	5.84	5.89	5.93	5.97	6.01
120	.05	4.71	4.78	4.84	4.90	4.95	5.00	5.04	5.09	5.13
	.01	5.44	5.50	5.56	5.61	5.66	5.71	5.75	5.79	5.83
∞	.05	4.62	4.68	4.74	4.80	4.85	4.89	4.93	4.97	5.01
	.01	5.29	5.35	5.40	5.45	5.49	5.54	5.57	5.61	5.65

From E. S. Pearson and H. O. Hartley, *Biometrika Tables for Statisticians*, Vol. I, Third Edition, © 1970 by Cambridge University Press, Cambridge. Abridged and adapted by permission (from Table 29).

Table XI. Critical Values of the Dunnett's Test

This table gives the critical values of the Dunnett's test used in comparing all treatment means to a control mean. The critical values are designated as $D[p, v; 1 - \alpha]$ corresponding to a given value of α, p as the number of treatment groups excluding the control, and v as the number of degrees of freedom for the error. The critical values are given for one- and two-tailed tests at $\alpha = 0.05, 0.01$, $p = 1\,(1)\,9$; and $v = 5\,(1)\,20, 24, 30, 40, 60, \infty$. When the researcher is comparing all treatment means to a control, the question often is whether the treatment is better than the control. In this situation, one-tailed critical values should be used. If the researcher wants to test whether the treatment means are simply different from the control, in either direction, two-tailed critical values are more appropriate.

One-Tailed Comparison
Number of Treatment Means, Excluding the Control (p)

ν	α	1	2	3	4	5	6	7	8	9
5	.05	2.02	2.44	2.68	2.85	2.98	3.08	3.16	3.24	3.30
	.01	3.37	3.90	4.21	4.43	4.60	4.73	4.85	4.94	5.03
6	.05	1.94	2.34	2.56	2.71	2.83	2.92	3.00	3.07	3.12
	.01	3.14	3.61	3.88	4.07	4.21	4.33	4.43	4.51	4.59
7	.05	1.89	2.27	2.48	2.62	2.73	2.82	2.89	2.95	3.01
	.01	3.00	3.42	3.66	3.83	3.96	4.07	4.15	4.23	4.30
8	.05	1.86	2.22	2.42	2.55	2.66	2.74	2.81	2.87	2.92
	.01	2.90	3.29	3.51	3.67	3.79	3.88	3.96	4.03	4.09
9	.05	1.83	2.18	2.37	2.50	2.60	2.68	2.75	2.81	2.86
	.01	2.82	3.19	3.40	3.55	3.66	3.75	3.82	3.89	3.94
10	.05	1.81	2.15	2.34	2.47	2.56	2.64	2.70	2.76	2.81
	.01	2.76	3.11	3.31	3.45	3.56	3.64	3.71	3.78	3.83
11	.05	1.80	2.13	2.31	2.44	2.53	2.60	2.67	2.72	2.77
	.01	2.72	3.06	3.25	3.38	3.48	3.56	3.63	3.69	3.74
12	.05	1.78	2.11	2.29	2.41	2.50	2.58	2.64	2.69	2.74
	.01	2.68	3.01	3.19	3.32	3.42	3.50	3.56	3.62	3.67
13	.05	1.77	2.09	2.27	2.39	2.48	2.55	2.61	2.66	2.71
	.01	2.65	2.97	3.15	3.27	3.37	3.44	3.51	3.56	3.61
14	.05	1.76	2.08	2.25	2.37	2.46	2.53	2.59	2.64	2.69
	.01	2.62	2.94	3.11	3.23	3.32	3.40	3.46	3.51	3.56
15	.05	1.75	2.07	2.24	2.36	2.44	2.51	2.57	2.62	2.67
	.01	2.60	2.91	3.08	3.20	3.29	3.36	3.42	3.47	3.52
16	.05	1.75	2.06	2.23	2.34	2.43	2.50	2.56	2.61	2.65
	.01	2.58	2.88	3.05	3.17	3.26	3.33	3.39	3.44	3.48
17	.05	1.74	2.05	2.22	2.33	2.42	2.49	2.54	2.59	2.64
	.01	2.57	2.86	3.03	3.14	3.23	3.30	3.36	3.41	3.45
18	.05	1.73	2.05	2.21	2.32	2.41	2.48	2.53	2.58	2.62
	.01	2.55	2.84	3.01	3.12	3.21	3.27	3.33	3.38	3.42
19	.05	1.73	2.03	2.20	2.31	2.40	2.47	2.52	2.57	2.61
	.01	2.54	2.83	2.99	3.10	3.18	3.25	3.31	3.36	3.40
20	.05	1.72	2.03	2.19	2.30	2.39	2.46	2.51	2.56	2.60
	.01	2.53	2.81	2.97	3.08	3.17	3.23	3.29	3.34	3.38
24	.05	1.71	2.01	2.17	2.28	2.36	2.43	2.48	2.53	2.57
	.01	2.49	2.77	2.92	3.03	3.11	3.17	3.22	3.27	3.31
30	.05	1.70	1.99	2.15	2.25	2.33	2.40	2.45	2.50	2.54
	.01	2.46	2.72	2.87	2.97	3.05	3.11	3.16	3.21	3.24
40	.05	1.68	1.97	2.13	2.23	2.31	2.37	2.42	2.47	2.51
	.01	2.42	2.68	2.82	2.92	2.99	3.05	3.10	3.14	3.18
60	.05	1.67	1.95	2.10	2.21	2.28	2.35	2.39	2.44	2.48
	.01	2.39	2.64	2.78	2.87	2.94	3.00	3.04	3.08	3.12
120	.05	1.66	1.93	2.08	2.18	2.26	2.32	2.37	2.41	2.45
	.01	2.36	2.60	2.73	2.82	2.89	2.94	2.99	3.03	3.06
∞	.05	1.64	1.92	2.06	2.16	2.23	2.29	2.34	2.38	2.42
	.01	2.33	2.56	2.68	2.77	2.84	2.89	2.93	2.97	3.00

Table XI (*continued*)

Two-Tailed Comparison
Number of Treatment Means, Excluding the Control (*p*)

ν	α	1	2	3	4	5	6	7	8	9
5	.05	2.57	3.03	3.29	3.48	3.62	3.73	3.82	3.90	3.97
	.01	4.03	4.63	4.98	5.22	5.41	5.56	5.69	5.80	5.89
6	.05	2.45	2.86	3.10	3.26	3.39	3.49	3.57	3.64	3.71
	.01	3.71	4.21	4.51	4.71	4.87	5.00	5.10	5.20	5.28
7	.05	2.36	2.75	2.97	3.12	3.24	3.33	3.41	3.47	3.53
	.01	3.50	3.95	4.21	4.39	4.53	4.64	4.74	4.82	4.89
8	.05	2.31	2.67	2.88	3.02	3.13	3.22	3.29	3.35	3.41
	.01	3.36	3.77	4.00	4.17	4.29	4.40	4.48	4.56	4.62
9	.05	2.26	2.61	2.81	2.95	3.05	3.14	3.20	3.26	3.32
	.01	3.25	3.63	3.85	4.01	4.12	4.22	4.30	4.37	4.43
10	.05	2.23	2.57	2.76	2.89	2.99	3.07	3.14	3.19	3.24
	.01	3.17	3.53	3.74	3.88	3.99	4.08	4.16	4.22	4.28
11	.05	2.20	2.53	2.72	2.84	2.94	3.02	3.08	3.14	3.19
	.01	3.11	3.45	3.65	3.79	3.89	3.98	4.05	4.11	4.16
12	.05	2.18	2.50	2.68	2.81	2.90	2.98	3.04	3.09	3.14
	.01	3.05	3.39	3.58	3.71	3.81	3.89	3.96	4.02	4.07
13	.05	2.16	2.48	2.65	2.78	2.87	2.94	3.00	3.06	3.10
	.01	3.01	3.33	3.52	3.65	3.74	3.82	3.89	3.94	3.99
14	.05	2.14	2.46	2.63	2.75	2.84	2.91	2.97	3.02	3.07
	.01	2.98	3.29	3.47	3.59	3.69	3.76	3.83	3.88	3.93
15	.05	2.13	2.44	2.61	2.73	2.82	2.89	2.95	3.00	3.04
	.01	2.95	3.25	3.43	3.55	3.64	3.71	3.78	3.83	3.88
16	.05	2.12	2.42	2.59	2.71	2.80	2.87	2.92	2.97	3.02
	.01	2.92	3.22	3.39	3.51	3.60	3.67	3.73	3.78	3.83
17	.05	2.11	2.41	2.58	2.69	2.78	2.85	2.90	2.95	3.00
	.01	2.90	3.19	3.36	3.47	3.56	3.63	3.69	3.74	3.79
18	.05	2.10	2.40	2.56	2.68	2.76	2.83	2.89	2.94	2.98
	.01	2.88	3.17	3.33	3.44	3.53	3.60	3.66	3.71	3.75
19	.05	2.09	2.39	2.55	2.66	2.75	2.81	2.87	2.92	2.96
	.01	2.86	3.15	3.31	3.42	3.50	3.57	3.63	3.68	3.72
20	.05	2.09	2.38	2.54	2.65	2.73	2.80	2.86	2.90	2.95
	.01	2.85	3.13	3.29	3.40	3.48	3.55	3.60	3.65	3.69
24	.05	2.06	2.35	2.51	2.61	2.70	2.76	2.81	2.86	2.90
	.01	2.80	3.07	3.22	3.32	3.40	3.47	3.52	3.57	3.61
30	.05	2.04	2.32	2.47	2.58	2.66	2.72	2.77	2.82	2.86
	.01	2.75	3.01	3.15	3.25	3.33	3.39	3.44	3.49	3.52
40	.05	2.02	2.29	2.44	2.54	2.62	2.68	2.73	2.77	2.81
	.01	2.70	2.95	3.09	3.19	3.26	3.32	3.37	3.41	3.44
60	.05	2.00	2.27	2.41	2.51	2.58	2.64	2.69	2.73	2.77
	.01	2.66	2.90	3.03	3.12	3.19	3.25	3.29	3.33	3.37
120	.05	1.98	2.24	2.38	2.47	2.55	2.60	2.65	2.69	2.73
	.01	2.62	2.85	2.97	3.06	3.12	3.18	3.22	3.26	3.29
∞	.05	1.96	2.21	2.35	2.44	2.51	2.57	2.61	2.65	2.69
	.01	2.58	2.79	2.92	3.00	3.06	3.11	3.15	3.19	3.22

From C. W. Dunnett, "A Multiple Comparison Procedure for Comparing Several Treatments with a Control," *Journal of the American Statistical Association*, 50 (1955), 1096–1121 and C. W. Dunnett, "New Tables for Multiple Comparisons with a Control," *Biometrics*, 20 (1964), 482–491. Reprinted by permission.

Table XII. Critical Values of the Duncan's Multiple Range Test

This table gives the critical values of Duncan's multiple range test which uses protection level α for the collection of all tests. The critical values are designated as $R[r, \nu; 1 - \alpha]$ corresponding to a given level α, the number of means for the range being tested or the number of steps apart of two means in an ordered sequence (r), and the number of degrees of freedom for the error (ν). The critical values are given for $\alpha = 0.05, 0.01; r = 2$ (1) 10 (2) 20, 50, 100; and $\nu = 1$ (1) 20 (2) 30, 40, 60, 100, ∞. For example, for $\alpha = 0.01$, $r = 3$, and $\nu = 13$, the required critical value is obtained as $R[3, 13; 0.99] = 4.48$.

Statistical Tables and Charts

v	α	2	3	4	5	6	7	8	9	10	12	14	16	18	20	50	100
1	.05	18.00	18.00	18.00	18.00	18.00	18.00	18.00	18.00	18.00	18.00	18.00	18.00	18.00	18.00	18.00	18.00
	.01	90.00	90.00	90.00	90.00	90.00	90.00	90.00	90.00	90.00	90.00	90.00	90.00	90.00	90.00	90.00	90.00
2	.05	6.09	6.09	6.09	6.09	6.09	6.09	6.09	6.09	6.09	6.09	6.09	6.09	6.09	6.09	6.09	6.09
	.01	14.00	14.00	14.00	14.00	14.00	14.00	14.00	14.00	14.00	14.00	14.00	14.00	14.00	14.00	14.00	14.00
3	.05	4.50	4.50	4.50	4.50	4.50	4.50	4.50	4.50	4.50	4.50	4.50	4.50	4.50	4.50	4.50	4.50
	.01	8.26	8.50	8.60	8.70	8.80	8.90	8.90	9.00	9.00	9.00	9.10	9.20	9.30	9.30	9.30	9.30
4	.05	3.93	4.01	4.02	4.02	4.02	4.02	4.02	4.02	4.02	4.02	4.02	4.02	4.02	4.02	4.02	4.02
	.01	6.51	6.80	6.90	7.00	7.10	7.10	7.20	7.20	7.30	7.30	7.40	7.40	7.50	7.50	7.50	7.50
5	.05	3.64	3.74	3.79	3.83	3.83	3.83	3.83	3.83	3.83	3.83	3.83	3.83	3.83	3.83	3.83	3.83
	.01	5.70	5.96	6.11	6.18	6.26	6.33	6.40	6.44	6.50	6.60	6.60	6.70	6.70	6.80	6.80	6.80
6	.05	3.46	3.58	3.64	3.68	3.68	3.68	3.68	3.68	3.68	3.68	3.68	3.68	3.68	3.68	3.68	3.68
	.01	5.24	5.51	5.65	5.73	5.81	5.88	5.95	6.00	6.00	6.10	6.20	6.20	6.30	6.30	6.30	6.30
7	.05	3.35	3.47	3.54	3.58	3.60	3.61	3.61	3.61	3.61	3.61	3.61	3.61	3.61	3.61	3.61	3.61
	.01	4.95	5.22	5.37	5.45	5.53	5.61	5.69	5.73	5.80	5.80	5.90	5.90	6.00	6.00	6.00	6.00
8	.05	3.26	3.39	3.47	3.52	3.55	3.56	3.56	3.56	3.56	3.56	3.56	3.56	3.56	3.56	3.56	3.56
	.01	4.74	5.00	5.14	5.23	5.32	5.40	5.47	5.51	5.50	5.60	5.70	5.70	5.80	5.80	5.80	5.80
9	.05	3.20	3.34	3.41	3.47	3.50	3.52	3.52	3.52	3.52	3.52	3.52	3.52	3.52	3.52	3.52	3.52
	.01	4.60	4.86	4.99	5.08	5.17	5.25	5.32	5.36	5.40	5.50	5.50	5.60	5.70	5.70	5.70	5.70
10	.05	3.15	3.30	3.37	3.43	3.46	3.47	3.47	3.47	3.47	3.47	3.47	3.47	3.47	3.48	3.48	3.48
	.01	4.48	4.73	4.88	4.96	5.06	5.13	5.20	5.24	5.28	5.36	5.42	5.48	5.54	5.55	5.55	5.55
11	.05	3.11	3.27	3.35	3.39	3.43	3.44	3.45	3.46	3.46	3.46	3.46	3.46	3.47	3.48	3.48	3.48
	.01	4.39	4.63	4.77	4.86	4.94	5.01	5.06	5.12	5.15	5.24	5.28	5.34	5.38	5.39	5.39	5.39
12	.05	3.08	3.23	3.33	3.36	3.40	3.42	3.44	3.44	3.46	3.46	3.46	3.46	3.47	3.48	3.48	3.48
	.01	4.32	4.55	4.68	4.76	4.84	4.92	4.96	5.02	5.07	5.13	5.17	5.22	5.24	5.26	5.26	5.26
13	.05	3.06	3.21	3.30	3.35	3.38	3.41	3.42	3.44	3.45	3.45	3.46	3.46	3.47	3.47	3.47	3.47
	.01	4.26	4.48	4.62	4.69	4.74	4.84	4.88	4.94	4.98	5.04	5.08	5.13	5.14	5.15	5.15	5.15
14	.05	3.03	3.18	3.27	3.33	3.37	3.39	3.41	3.42	3.44	3.45	3.46	3.46	3.47	3.47	3.47	3.47
	.01	4.21	4.42	4.55	4.63	4.70	4.78	4.83	4.87	4.91	4.96	5.00	5.04	5.06	5.07	5.07	5.07
15	.05	3.01	3.16	3.25	3.31	3.36	3.38	3.40	3.42	3.43	3.44	3.45	3.46	3.47	3.47	3.47	3.47
	.01	4.17	4.37	4.50	4.58	4.64	4.72	4.77	4.81	4.84	4.90	4.94	4.97	4.99	5.00	5.00	5.00

Number of Means for Range Tested (r)

Table XII (*continued*)

		Number of Means for Range Tested (r)														
ν	α	2	3	4	5	6	7	8	9	10	12	14	16	18	20	50	100
16	.05	3.00	3.15	3.23	3.30	3.34	3.37	3.39	3.41	3.43	3.44	3.45	3.46	3.47	3.47	3.47	3.47
	.01	4.13	4.34	4.45	4.54	4.60	4.67	4.72	4.76	4.79	4.84	4.88	4.91	4.93	4.94	4.94	4.94
17	.05	2.98	3.13	3.22	3.28	3.33	3.36	3.38	3.40	3.42	3.44	3.45	3.46	3.47	3.47	3.47	3.47
	.01	4.10	4.30	4.41	4.50	4.56	4.63	4.68	4.72	4.75	4.80	4.83	4.86	4.88	4.89	4.89	4.89
18	.05	2.97	3.12	3.21	3.27	3.32	3.35	3.37	3.39	3.41	3.43	3.45	3.46	3.47	3.47	3.47	3.47
	.01	4.07	4.27	4.38	4.46	4.53	4.59	4.64	4.68	4.71	4.76	4.79	4.82	4.84	4.85	4.85	4.85
19	.05	2.96	3.11	3.19	3.26	3.31	3.35	3.37	3.39	3.41	3.43	3.44	3.46	3.47	3.47	3.47	3.47
	.01	4.05	4.24	4.35	4.43	4.50	4.56	4.61	4.64	4.67	4.72	4.76	4.79	4.81	4.82	4.82	4.82
20	.05	2.95	3.10	3.18	3.25	3.30	3.34	3.36	3.38	3.40	3.43	3.44	3.46	3.46	3.47	3.47	3.47
	.01	4.02	4.22	4.33	4.40	4.47	4.53	4.58	4.61	4.65	4.69	4.73	4.76	4.78	4.79	4.79	4.79
22	.05	2.93	3.08	3.17	3.24	3.29	3.32	3.35	3.37	3.39	3.42	3.44	3.45	3.46	3.47	3.47	3.47
	.01	3.99	4.17	4.28	4.36	4.42	4.48	4.53	4.57	4.60	4.65	4.68	4.71	4.74	4.75	4.75	4.75
24	.05	2.92	3.07	3.15	3.22	3.28	3.31	3.34	3.37	3.38	3.41	3.44	3.45	3.46	3.47	3.47	3.47
	.01	3.96	4.14	4.24	4.33	4.39	4.44	4.49	4.53	4.57	4.62	4.64	4.67	4.70	4.72	4.74	4.74
26	.05	2.91	3.06	3.14	3.21	3.27	3.30	3.34	3.36	3.38	3.41	3.43	3.45	3.46	3.47	3.47	3.47
	.01	3.93	4.11	4.21	4.30	4.36	4.41	4.46	4.50	4.53	4.58	4.62	4.65	4.67	4.69	4.73	4.73
28	.05	2.90	3.04	3.13	3.20	3.26	3.30	3.33	3.35	3.37	3.40	3.43	3.45	3.46	3.47	3.47	3.47
	.01	3.91	4.08	4.18	4.28	4.34	4.39	4.43	4.47	4.51	4.56	4.60	4.62	4.65	4.67	4.72	4.72
30	.05	2.89	3.04	3.12	3.20	3.25	3.29	3.32	3.35	3.37	3.40	3.43	3.44	3.46	3.47	3.47	3.47
	.01	3.89	4.06	4.16	4.22	4.32	4.36	4.41	4.45	4.48	4.54	4.58	4.61	4.63	4.65	4.71	4.71
40	.05	2.86	3.01	3.10	3.17	3.22	3.27	3.30	3.33	3.35	3.39	3.42	3.44	3.46	3.47	3.47	3.47
	.01	3.82	3.99	4.10	4.17	4.24	4.30	4.34	4.37	4.41	4.46	4.51	4.54	4.57	4.59	4.69	4.69
60	.05	2.83	2.98	3.08	3.14	3.20	3.24	3.28	3.31	3.33	3.37	3.40	3.43	3.45	3.47	3.48	3.48
	.01	3.76	3.92	4.03	4.12	4.17	4.23	4.27	4.31	4.34	4.39	4.44	4.47	4.50	4.53	4.66	4.66
100	.05	2.80	2.95	3.05	3.12	3.18	3.22	3.26	3.29	3.32	3.36	3.40	3.42	3.45	3.47	3.53	3.53
	.01	3.71	3.86	3.93	4.06	4.11	4.17	4.21	4.25	4.29	4.35	4.38	4.42	4.45	4.48	4.64	4.65
∞	.05	2.77	2.92	3.02	3.09	3.15	3.19	3.23	3.26	3.29	3.34	3.38	3.41	3.44	3.47	3.61	3.67
	.01	3.64	3.80	3.90	3.98	4.04	4.09	4.14	4.17	4.20	4.26	4.31	4.34	4.38	4.41	4.60	4.68

From D. B. Duncan, "Multiple Range and Multiple *F* Tests," Biometrics, 11 (1955), 1–42. Reprinted by permission.

Table XIII. Critical Values of the Bonferroni t Statistic and Dunn's Multiple Comparison Test

This table gives the critical values of the Bonferroni t statistic and Dunn's multiple comparison procedure. The critical values are given for $\alpha = 0.05, 0.01$; the number of comparisons $p = 1\,(1)\,10\,(5)\,20$ and the error degrees of freedom $v = 2\,(1)\,30\,(5)\,60\,(10)\,120, 250, 500, 1000, \infty$. For example, for $\alpha = 0.05$, $p = 5$, and $v = 10$, the desired critical value is obtained as 3.1693.

$$\alpha_{Bon} = 0.05$$
$$\alpha_{ind} = 0.05/p$$

Number of comparisons (p)

ν \ $100(\alpha/p)$	1 5.0000	2 2.5000	3 1.6667	4 1.2500	5 1.0000	6 0.8333	7 0.7143	8 0.6250	9 0.5556	10 0.5000	15 0.3333	20 0.2500
2	4.3027	6.2053	7.6488	8.8602	9.9248	10.8859	11.7687	12.5897	13.3604	14.0890	17.2772	19.9625
3	3.1824	4.1765	4.8567	5.3919	5.8409	6.2315	6.5797	6.8952	7.1849	7.4533	8.5752	9.4649
4	2.7764	3.4954	3.9608	4.3147	4.6041	4.8510	5.0675	5.2611	5.4366	5.5976	6.2541	6.7583
5	2.5706	3.1634	3.5341	3.8100	4.0321	4.2193	4.3818	4.5257	4.6553	4.7733	5.2474	5.6042
6	2.4469	2.9687	3.2875	3.5212	3.7074	3.8630	3.9971	4.1152	4.2209	4.3168	4.6979	4.9807
7	2.3646	2.8412	3.1276	3.3353	3.4995	3.6358	3.7527	3.8552	3.9467	4.0293	4.3553	4.5946
8	2.3060	2.7515	3.0158	3.2060	3.3554	3.4789	3.5844	3.6766	3.7586	3.8325	4.1224	4.3335
9	2.2622	2.6850	2.9333	3.1109	3.2498	3.3642	3.4616	3.5465	3.6219	3.6897	3.9542	4.1458
10	2.2281	2.6338	2.8701	3.0382	3.1693	3.2768	3.3682	3.4477	3.5182	3.5814	3.8273	4.0045
11	2.2010	2.5931	2.8200	2.9809	3.1058	3.2081	3.2949	3.3702	3.4368	3.4966	3.7283	3.8945
12	2.1788	2.5600	2.7795	2.9345	3.0545	3.1527	3.2357	3.3078	3.3714	3.4284	3.6489	3.8065
13	2.1604	2.5326	2.7459	2.8961	3.0123	3.1070	3.1871	3.2565	3.3177	3.3725	3.5838	3.7345
14	2.1448	2.5096	2.7178	2.8640	2.9768	3.0688	3.1464	3.2135	3.2727	3.3257	3.5296	3.6746
15	2.1314	2.4899	2.6937	2.8366	2.9467	3.0363	3.1118	3.1771	3.2346	3.2860	3.4837	3.6239
16	2.1199	2.4729	2.6730	2.8131	2.9208	3.0083	3.0821	3.1458	3.2019	3.2520	3.4443	3.5805
17	2.1098	2.4581	2.6550	2.7925	2.8982	2.9840	3.0563	3.1186	3.1735	3.2224	3.4102	3.5429
18	2.1009	2.4450	2.6391	2.7745	2.8784	2.9627	3.0336	3.0948	3.1486	3.1966	3.3804	3.5101
19	2.0930	2.4334	2.6251	2.7586	2.8609	2.9439	3.0136	3.0738	3.1266	3.1737	3.3540	3.4812
20	2.0860	2.4231	2.6126	2.7444	2.8453	2.9271	2.9958	3.0550	3.1070	3.1534	3.3306	3.4554
21	2.0796	2.4138	2.6013	2.7316	2.8314	2.9121	2.9799	3.0382	3.0895	3.1352	3.3097	3.4325
22	2.0739	2.4055	2.5912	2.7201	2.8188	2.8985	2.9655	3.0231	3.0737	3.1188	3.2909	3.4118
23	2.0687	2.3979	2.5820	2.7079	2.8073	2.8863	2.9525	3.0095	3.0595	3.1040	3.2739	3.3931
24	2.0639	2.3909	2.5736	2.7002	2.7969	2.8751	2.9406	2.9970	3.0465	3.0905	3.2584	3.3761
25	2.0595	2.3846	2.5660	2.6916	2.7874	2.8649	2.9298	2.9856	3.0346	3.0782	3.2443	3.3606
26	2.0555	2.3788	2.5589	2.6836	2.7787	2.8555	2.9199	2.9752	3.0237	3.0669	3.2313	3.3464
27	2.0518	2.3734	2.5525	2.6763	2.7707	2.8469	2.9107	2.9656	3.0137	3.0565	3.2194	3.3334
28	2.0484	2.3685	2.5465	2.6695	2.7633	2.8389	2.9023	2.9567	3.0045	3.0469	3.2084	3.3214
29	2.0452	2.3638	2.5409	2.6632	2.7564	2.8316	2.8945	2.9485	3.9959	3.0380	3.1982	3.3102
30	2.0423	2.3596	2.5357	2.6574	2.7500	2.8247	2.8872	2.9409	3.9880	3.0298	3.1888	3.2999
35	2.0301	2.3420	2.5145	2.6334	2.7238	2.7966	2.8575	2.9097	2.9554	2.9960	3.1502	3.2577
40	2.0211	2.3289	2.4989	2.6157	2.7045	2.7759	2.8355	2.8867	2.9314	2.9712	3.1218	3.2266
45	2.0141	2.3189	2.4868	2.6021	2.6896	2.7599	2.8187	2.8690	2.9130	2.9521	3.1000	3.2028
50	2.0086	2.3109	2.4772	2.5913	2.6778	2.7473	2.8053	2.8550	2.8984	2.9370	3.0828	3.1840
55	2.0040	2.3044	2.4694	2.5825	2.6682	2.7370	2.7944	2.8436	2.8866	2.9247	3.0688	3.1688
60	2.0003	2.2990	2.4630	2.5752	2.6603	2.7286	2.7855	2.8342	2.8768	2.9146	3.0573	3.1562
70	1.9944	2.2906	2.4529	2.5639	2.6479	2.7153	2.7715	2.8195	2.8615	2.8987	3.0393	3.1366
80	1.9901	2.2844	2.4454	2.5554	2.6387	2.7054	2.7610	2.8086	2.8502	2.8870	3.0259	3.1220
90	1.9867	2.2795	2.4395	2.5489	2.6316	2.6978	2.7530	2.8002	2.8414	2.8779	3.0156	3.1108
100	1.9840	2.2757	2.4349	2.5437	2.6259	2.6918	2.7466	2.7935	2.8344	2.8707	3.0073	3.1018
110	1.9818	2.2725	2.4311	2.5394	2.6213	2.6868	2.7414	2.7880	2.8287	2.8648	3.0007	3.0945
120	1.9799	2.2699	2.4280	2.5359	2.6174	2.6827	2.7370	2.7835	2.8240	2.8599	2.9951	3.0885
250	1.9695	2.2550	2.4102	2.5159	2.5956	2.6594	2.7124	2.7577	2.7972	2.8322	2.9637	3.0543
500	1.9647	2.2482	2.4021	2.5068	2.5857	2.6488	2.7012	2.7460	2.7850	2.8195	2.9494	3.0387
1000	1.9623	2.2448	2.3980	2.5022	2.5808	2.6435	2.6957	2.7402	2.7790	2.8133	2.9423	3.0310
∞	1.9600	2.2414	2.3940	2.4977	2.5758	2.6383	2.6901	2.7344	2.7729	2.8070	2.9352	3.0233

Table XIII (continued)

$$\alpha_{Bon} = 0.01$$
$$\alpha_{ind} = 0.01/p$$

Number of comparisons (p)

ν	$100(\alpha/p)$	1	2	3	4	5	6	7	8	9	10	15	20
		1.0000	0.5000	0.3333	0.2500	0.2000	0.1667	0.1429	0.1250	0.1111	0.1000	0.0667	0.0500
2		9.9248	14.0890	17.2772	19.9625	22.3271	24.4643	26.4292	28.2577	29.9750	31.5991	38.7105	44.7046
3		5.8409	7.4533	8.5752	9.4649	10.2145	10.8668	11.4532	11.9838	12.4715	12.9240	14.8194	16.3263
4		4.6041	5.5976	6.2541	6.7583	7.1732	7.5287	7.8414	8.1216	8.3763	8.6103	9.5679	10.3063
5		4.0321	4.7733	5.2474	5.6042	5.8934	6.1384	6.3518	6.5414	6.7126	6.8688	7.4990	7.9757
6		3.7074	4.3168	4.6979	4.9807	5.2076	5.3982	5.5632	5.7090	5.8399	5.9588	6.4338	6.7883
7		3.4995	4.0293	4.3553	4.5946	4.7853	4.9445	5.0815	5.2022	5.3101	5.4079	5.7954	6.0818
8		3.3554	3.8325	4.1224	4.3335	4.5008	4.6398	4.7590	4.8636	4.9570	5.0413	5.3737	5.6174
9		3.2498	3.6897	3.9542	4.1458	4.2968	4.4219	4.5288	4.6224	4.7058	4.7809	5.0757	5.2907
10		3.1693	3.5814	3.8273	4.0045	4.1437	4.2586	4.3567	4.4423	4.5184	4.5869	4.8547	5.0490
11		3.1058	3.4966	3.7283	3.8945	4.0247	4.1319	4.2232	4.3028	4.3735	4.4370	4.6845	4.8633
12		3.0545	3.4284	3.6489	3.8065	3.9296	4.0308	4.1169	4.1918	4.2582	4.3178	4.5496	4.7165
13		3.0123	3.3725	3.5838	3.7345	3.8520	3.9484	4.0302	4.1013	4.1643	4.2208	4.4401	4.5975
14		2.9768	3.3257	3.5296	3.6746	3.7874	3.8798	3.9582	4.0263	4.0865	4.1405	4.3495	4.4992
15		2.9467	3.2860	3.4837	3.6239	3.7328	3.8220	3.8975	3.9630	4.0209	4.0728	4.2733	4.4166
16		2.9208	3.2520	3.4443	3.5805	3.6862	3.7725	3.8456	3.9089	3.9649	4.0150	4.2084	4.3463
17		2.8982	3.2224	3.4102	3.5429	3.6458	3.7297	3.8007	3.8623	3.9165	3.9651	4.1525	4.2858
18		2.8784	3.1966	3.3804	3.5101	3.6105	3.6924	3.7616	3.8215	3.8744	3.9216	4.1037	4.2332
19		2.8609	3.1737	3.3540	3.4812	3.5794	3.6595	3.7271	3.7857	3.8373	3.8834	4.0609	4.1869
20		2.8453	3.1534	3.3306	3.4554	3.5518	3.6303	3.6966	3.7539	3.8044	3.8495	4.0230	4.1460
21		2.8314	3.1352	3.3097	3.4325	3.5272	3.6043	3.6693	3.7255	3.7750	3.8193	3.9892	4.1096
22		2.8188	3.1188	3.2909	3.4118	3.5050	3.5808	3.6448	3.7000	3.7487	3.7921	3.9589	4.0769
23		2.8073	3.1040	3.2739	3.3931	3.4850	3.5597	3.6226	3.6770	3.7249	3.7676	3.9316	4.0474
24		2.7969	3.0905	3.2584	3.3761	3.4668	3.5405	3.6025	3.6561	3.7033	3.7454	3.9068	4.0207
25		2.7874	3.0782	3.2443	3.3606	3.4502	3.5230	3.5842	3.6371	3.6836	3.7251	3.8842	3.9964
26		2.7787	3.0669	3.2313	3.3464	3.4350	3.5069	3.5674	3.6197	3.6656	3.7066	3.8635	3.9742
27		2.7707	3.0565	3.2194	3.3334	3.4210	3.4922	3.5520	3.6037	3.6491	3.6896	3.8446	3.9538
28		2.7633	3.0469	3.2084	3.3214	3.4082	3.4786	3.5378	3.5889	3.6338	3.6739	3.8271	3.9351
29		2.7564	3.0380	3.1982	3.3102	3.3962	3.4660	3.5247	3.5753	3.6198	3.6594	3.8110	3.9177
30		2.7500	3.0298	3.1888	3.2999	3.3852	3.4544	3.5125	3.5626	3.6067	3.6460	3.7961	3.9016
35		2.7238	2.9960	3.1502	3.2577	3.3400	3.4068	3.4628	3.5110	3.5534	3.5911	3.7352	3.8362
40		2.7045	2.9712	3.1218	3.2266	3.3069	3.3718	3.4263	3.4732	3.5143	3.5510	3.6906	3.7884
45		2.6896	2.9521	3.1000	3.2028	3.2815	3.3451	3.3984	3.4442	3.4845	3.5203	3.6565	3.7519
50		2.6778	2.9370	3.0828	3.1840	3.2614	3.3239	3.3763	3.4214	3.4609	3.4960	3.6297	3.7231
55		2.6682	2.9247	3.0688	3.1688	3.2451	3.3068	3.3585	3.4029	3.4418	3.4764	3.6080	3.6999
60		2.6603	2.9146	3.0573	3.1562	3.2317	3.2927	3.3437	3.3876	3.4260	3.4602	3.5901	3.6807
70		2.6479	2.8987	3.0393	3.1366	3.2108	3.2707	3.3208	3.3638	3.4015	3.4350	3.5622	3.6509
80		2.6387	2.8870	3.0259	3.1220	3.1953	3.2543	3.3037	3.3462	3.3833	3.4163	3.5416	3.6288
90		2.6316	2.8779	3.0156	3.1108	3.1833	3.2417	3.2906	3.3326	3.3693	3.4019	3.5257	3.6118
100		2.6259	2.8707	3.0073	3.1018	3.1737	3.2317	3.2802	3.3218	3.3582	3.3905	3.5131	3.5983
110		2.6213	2.8648	3.0007	3.0945	3.1660	3.2235	3.2717	3.3130	3.3491	3.3812	3.5028	3.5874
120		2.6174	2.8599	2.9951	3.0885	3.1595	3.2168	3.2646	3.3057	3.3416	3.3735	3.4943	3.5783
250		2.5956	2.8322	2.9637	3.0543	3.1232	3.1785	3.2248	3.2644	3.2991	3.3299	3.4462	3.5270
500		2.5857	2.8195	2.9494	3.0387	3.1066	3.1612	3.2067	3.2457	3.2798	3.3101	3.4245	3.5037
1000		2.5808	2.8133	2.9423	3.0310	3.0984	3.1526	3.1977	3.2365	3.2703	3.3003	3.4137	3.4922
∞		2.5758	2.8070	2.9352	3.0233	3.0902	3.1440	3.1888	3.2272	3.2608	3.2905	3.4029	3.4808

From B. J. R. Bailey, "Tables of the Bonferroni t Statistic," *Journal of the American Statistical Association*, 72 (1977), 469–478. Abridged and reprinted by permission.

Table XIV. Critical Values of the Dunn-Šidák's Multiple Comparison Test

This table gives the critical values of the Dunn-Šidák's multiple comparison procedure. The critical values are given for $\alpha = 0.01, 0.05, 0.10, 0.20$; the number of comparisons $p = 2$ (1) 10 (5) 40 (10) 50; and the error degrees of freedom $\nu = 2$ (1) 30, 40, 60, 120, ∞. For example, for $\alpha = 0.05$, $p = 3$, and $\nu = 12$, the required critical value is obtained as 2.770.

Statistical Tables and Charts

v	α	\multicolumn{14}{c}{Number of comparisons (p)}
		2	3	4	5	6	7	8	9	10	15	20	25	30	35	40	50
2	0.01	14.071	17.248	19.925	22.282	24.413	26.372	28.196	29.908	31.528	38.620	44.598	49.865	54.626	59.004	63.079	70.526
	0.05	6.164	7.582	8.774	9.823	10.769	11.639	12.449	13.208	13.927	17.072	19.721	22.054	24.163	26.103	27.908	31.206
	0.10	4.243	5.243	6.081	6.816	7.480	8.090	8.656	9.188	9.691	11.890	13.741	15.371	16.845	18.199	19.459	21.761
	0.20	2.828	3.531	4.116	4.628	5.089	5.512	5.904	6.272	6.620	8.138	9.414	10.537	11.552	12.484	13.351	14.936
3	0.01	7.447	8.565	9.453	10.201	10.853	11.436	11.966	12.453	12.904	14.796	16.300	17.569	18.678	19.670	20.570	22.167
	0.05	4.156	4.626	5.355	5.799	6.185	6.529	6.842	7.128	7.394	8.505	9.387	10.129	10.778	11.357	11.883	12.815
	0.10	3.149	3.690	4.115	4.471	4.780	5.055	5.304	5.532	5.744	6.627	7.326	7.914	8.427	8.886	9.301	10.038
	0.20	2.294	2.734	3.077	3.363	3.610	3.829	4.028	4.209	4.377	5.076	5.628	6.091	6.495	6.855	7.181	7.759
4	0.01	5.594	6.248	6.751	7.166	7.520	7.832	8.112	8.367	8.600	9.556	10.294	10.902	11.424	11.884	12.297	13.017
	0.05	3.481	3.941	4.290	4.577	4.822	5.036	5.228	5.402	5.562	6.214	6.714	7.127	7.480	7.790	8.069	8.554
	0.10	2.751	3.150	3.452	3.699	3.909	4.093	4.257	4.406	4.542	5.097	5.521	5.870	6.169	6.432	6.667	7.076
	0.20	2.084	2.434	2.697	2.911	3.092	3.250	3.391	3.518	3.635	4.107	4.468	4.763	5.015	5.237	5.435	5.779
5	0.01	4.771	5.243	5.599	5.888	6.133	6.346	6.535	6.706	6.862	7.491	7.968	8.355	8.684	8.971	9.226	9.668
	0.05	3.152	3.518	3.791	4.012	4.197	4.358	4.501	4.630	4.747	5.219	5.573	5.861	6.105	6.317	6.506	6.831
	0.10	2.549	2.882	3.129	3.327	3.493	3.638	3.765	3.880	3.985	4.403	4.718	4.972	5.187	5.374	5.540	5.826
	0.20	1.973	2.278	2.503	2.683	2.834	2.964	3.079	3.182	3.275	3.649	3.928	4.153	4.343	4.508	4.654	4.906
6	0.01	4.315	4.695	4.977	5.203	5.394	5.559	5.704	5.835	5.954	6.428	6.782	7.068	7.308	7.516	7.701	8.018
	0.05	2.959	3.274	3.505	3.690	3.845	3.978	4.095	4.200	4.296	4.675	4.956	5.182	5.372	5.536	5.682	5.930
	0.10	2.428	2.723	2.939	3.110	3.253	3.376	3.484	3.580	3.668	4.015	4.272	4.477	4.649	4.798	4.930	5.155
	0.20	1.904	2.184	2.387	2.547	2.681	2.795	2.895	2.985	3.066	3.385	3.620	3.808	3.965	4.100	4.220	4.424
7	0.01	4.027	4.353	4.591	4.782	4.941	5.078	5.198	5.306	5.404	5.791	6.077	6.306	6.497	6.663	6.809	7.058
	0.05	2.832	3.115	3.321	3.484	3.620	3.736	3.838	3.929	4.011	4.336	4.574	4.764	4.923	5.059	5.180	5.385
	0.10	2.347	2.618	2.814	2.969	3.097	3.206	3.302	3.388	3.465	3.768	3.990	4.167	4.314	4.441	4.552	4.741
	0.20	1.858	2.120	2.309	2.457	2.579	2.684	2.775	2.856	2.929	3.214	3.423	3.588	3.725	3.842	3.946	4.121
8	0.01	3.831	4.120	4.331	4.498	4.637	4.756	4.860	4.953	5.038	5.370	5.613	5.807	5.969	6.107	6.230	6.437
	0.05	2.743	3.005	3.193	3.342	3.464	3.569	3.661	3.743	3.816	4.105	4.316	4.482	4.621	4.740	4.844	5.021
	0.10	2.289	2.544	2.726	2.869	2.987	3.088	3.176	3.254	3.324	3.598	3.798	3.955	4.086	4.198	4.296	4.462
	0.20	1.824	2.075	2.254	2.393	2.508	2.605	2.690	2.765	2.832	3.095	3.286	3.435	3.559	3.665	3.758	3.914
9	0.01	3.688	3.952	4.143	4.294	4.419	4.526	4.619	4.703	4.778	5.072	5.287	5.457	5.598	5.720	5.826	6.006
	0.05	2.677	2.923	3.099	3.237	3.351	3.448	3.532	3.607	3.675	3.938	4.129	4.280	4.405	4.512	4.605	4.763
	0.10	2.246	2.488	2.661	2.796	2.907	3.001	3.083	3.155	3.221	3.474	3.658	3.802	3.921	4.023	4.112	4.262
	0.20	1.799	2.041	2.212	2.345	2.454	2.546	2.627	2.698	2.761	3.008	3.185	3.324	3.438	3.536	3.621	3.765
10	0.01	3.580	3.825	4.002	4.141	4.256	4.354	4.439	4.515	4.584	4.852	5.046	5.199	5.326	5.434	5.529	5.690
	0.05	2.626	2.860	3.027	3.157	3.264	3.355	3.434	3.505	3.568	3.813	3.989	4.128	4.243	4.341	4.426	4.571
	0.10	2.213	2.446	2.611	2.739	2.845	2.934	3.012	3.080	3.142	3.380	3.552	3.686	3.796	3.891	3.973	4.112
	0.20	1.799	2.014	2.180	2.308	2.413	2.501	2.578	2.646	2.706	2.941	3.108	3.239	3.346	3.438	3.517	3.651
11	0.01	3.495	3.726	3.892	4.022	4.129	4.221	4.300	4.371	4.434	4.682	4.860	5.001	5.117	5.216	5.303	5.450
	0.05	2.586	2.811	2.970	3.094	3.196	3.283	3.358	3.424	3.484	3.715	3.880	4.010	4.117	4.208	4.288	4.422
	0.10	2.186	2.412	2.571	2.695	2.796	2.881	2.955	3.021	3.079	3.306	3.468	3.595	3.699	3.788	3.865	3.995
	0.20	1.763	1.993	2.154	2.279	2.380	2.465	2.539	2.605	2.663	2.888	3.048	3.172	3.274	3.361	3.436	3.583
12	0.01	3.427	3.647	3.804	3.927	4.029	4.114	4.189	4.256	4.315	4.547	4.714	4.845	4.953	5.045	5.125	5.260
	0.05	2.553	2.770	2.924	3.044	3.141	3.224	3.296	3.359	3.416	3.636	3.793	3.916	4.017	4.103	4.178	4.304
	0.10	2.164	2.384	2.539	2.658	2.756	2.838	2.910	2.973	3.029	3.247	3.402	3.522	3.621	3.705	3.779	3.901
	0.20	1.750	1.975	2.133	2.254	2.353	2.436	2.508	2.571	2.628	2.845	2.999	3.118	3.216	3.299	3.371	3.491

Table XIV (continued)

ν	α	2	3	4	5	6	7	8	9	10	15	20	25	30	35	40	50
13	0.01	3.371	3.582	3.733	3.850	3.946	4.028	4.099	4.162	4.218	4.438	4.595	4.718	4.819	4.906	4.981	5.108
	0.05	2.526	2.737	2.886	3.002	3.096	3.176	3.245	3.306	3.361	3.571	3.722	3.839	3.935	4.017	4.088	4.207
	0.10	2.146	2.361	2.512	2.628	2.723	2.803	2.872	2.933	2.988	3.198	3.347	3.463	3.557	3.638	3.708	3.825
	0.20	1.739	1.961	2.116	2.234	2.331	2.412	2.482	2.544	2.599	2.809	2.958	3.074	3.168	3.248	3.317	3.433
14	0.01	3.324	3.528	3.673	3.785	3.878	3.956	4.024	4.084	4.138	4.347	4.497	4.614	4.710	4.792	4.863	4.982
	0.05	2.503	2.709	2.854	2.967	3.058	3.135	3.202	3.261	3.314	3.518	3.662	3.775	3.867	3.946	4.014	4.128
	0.10	2.131	2.342	2.489	2.603	2.696	2.774	2.841	2.900	2.953	3.157	3.301	3.413	3.504	3.582	3.649	3.761
	0.20	1.730	1.949	2.101	2.217	2.312	2.392	2.460	2.520	2.574	2.779	2.924	3.036	3.128	3.205	3.272	3.384
15	0.01	3.285	3.482	3.622	3.731	3.820	3.895	3.961	4.019	4.071	4.271	4.414	4.526	4.618	4.696	4.764	4.877
	0.05	2.483	2.685	2.827	2.937	3.026	3.101	3.166	3.224	3.275	3.472	3.612	3.721	3.810	3.885	3.951	4.060
	0.10	2.118	2.325	2.470	2.582	2.672	2.748	2.814	2.872	2.924	3.122	3.262	3.370	3.459	3.534	3.599	3.708
	0.20	1.722	1.938	2.088	2.203	2.296	2.374	2.441	2.500	2.553	2.754	2.896	3.005	3.094	3.169	3.234	3.343
16	0.01	3.251	3.443	3.579	3.684	3.771	3.844	3.907	3.963	4.013	4.206	4.344	4.451	4.540	4.614	4.679	4.788
	0.05	2.467	2.665	2.804	2.911	2.998	3.072	3.135	3.191	3.241	3.433	3.569	3.675	3.761	3.834	3.897	4.003
	0.10	2.106	2.311	2.453	2.563	2.652	2.726	2.791	2.848	2.898	3.092	3.228	3.334	3.420	3.493	3.556	3.662
	0.20	1.715	1.929	2.077	2.190	2.282	2.359	2.425	2.483	2.535	2.732	2.871	2.978	3.064	3.138	3.201	3.307
17	0.01	3.221	3.409	3.541	3.664	3.278	3.799	3.860	3.914	3.963	4.150	4.284	4.387	4.472	4.544	4.607	4.712
	0.05	2.452	2.647	2.783	2.889	2.974	3.046	3.108	3.163	3.212	3.399	3.532	3.634	3.718	3.789	3.851	3.954
	0.10	2.096	2.298	2.439	2.547	2.634	2.708	2.771	2.826	2.876	3.066	3.199	3.303	3.387	3.458	3.519	3.622
	0.20	1.709	1.921	2.068	2.179	2.270	2.346	2.411	2.468	2.519	2.713	2.849	2.954	3.039	3.111	3.173	3.276
18	0.01	3.195	3.379	3.508	3.609	3.691	3.760	3.820	3.872	3.920	4.102	4.231	4.332	4.414	4.484	4.544	4.646
	0.05	2.439	2.631	2.766	2.869	2.953	3.024	3.085	3.138	3.186	3.370	3.499	3.599	3.681	3.750	3.810	3.910
	0.10	2.088	2.287	2.426	2.532	2.619	2.691	2.753	2.808	2.857	3.043	3.174	3.275	3.358	3.427	3.487	3.587
	0.20	1.704	1.914	2.059	2.170	2.259	2.334	2.399	2.455	2.505	2.696	2.830	2.933	3.017	3.087	3.148	3.249
19	0.01	3.173	3.353	3.479	3.578	3.658	3.725	3.784	3.835	3.881	4.059	4.185	4.283	4.363	4.430	4.489	4.588
	0.05	2.427	2.617	2.750	2.852	2.934	3.004	3.064	3.116	3.163	3.343	3.470	3.569	3.649	3.716	3.775	3.872
	0.10	2.080	2.277	2.415	2.520	2.605	2.676	2.738	2.791	2.839	3.023	3.152	3.251	3.332	3.400	3.459	3.557
	0.20	1.699	1.908	2.052	2.161	2.250	2.324	2.388	2.443	2.493	2.682	2.813	2.915	2.997	3.066	3.126	3.225
20	0.01	3.152	3.329	3.454	3.550	3.629	3.695	3.752	3.802	3.848	4.021	4.144	4.239	4.317	4.383	4.441	4.536
	0.05	2.417	2.605	2.736	2.836	2.918	2.986	3.045	3.097	3.143	3.320	3.445	3.541	3.620	3.686	3.743	3.839
	0.10	2.073	2.269	2.405	2.508	2.593	2.663	2.724	2.777	2.824	3.005	3.132	3.229	3.309	3.376	3.433	3.530
	0.20	1.695	1.902	2.045	2.154	2.241	2.315	2.378	2.433	2.482	2.668	2.798	2.898	2.979	3.048	3.106	3.204
21	0.01	3.134	3.308	3.431	3.525	3.602	3.667	3.724	3.773	3.817	3.987	4.108	4.201	4.277	4.342	4.397	4.491
	0.05	2.408	2.594	2.723	2.822	2.903	2.970	3.028	3.080	3.125	3.300	3.422	3.517	3.594	3.659	3.715	3.809
	0.10	2.067	2.261	2.396	2.498	2.581	2.651	2.711	2.764	2.810	2.989	3.114	3.210	3.288	3.354	3.411	3.505
	0.20	1.691	1.897	2.039	2.147	2.234	2.306	2.369	2.424	2.472	2.656	2.785	2.884	2.964	3.031	3.089	3.185
22	0.01	3.118	3.289	3.410	3.503	3.579	3.643	3.698	3.747	3.790	3.957	4.075	4.166	4.241	4.304	4.359	4.450
	0.05	2.400	2.584	2.712	2.810	2.889	2.956	3.014	3.064	3.109	3.281	3.402	3.495	3.571	3.634	3.690	3.782
	0.10	2.061	2.254	2.387	2.489	2.572	2.641	2.700	2.752	2.798	2.974	3.098	3.193	3.270	3.334	3.390	3.484
	0.20	1.688	1.892	2.033	2.141	2.227	2.299	2.361	2.415	2.463	2.646	2.773	2.871	2.950	3.016	3.073	3.168
23	0.01	3.103	3.272	3.392	3.483	3.558	3.621	3.675	3.723	3.766	3.930	4.046	4.135	4.208	4.270	4.324	4.413
	0.05	2.392	2.574	2.701	2.798	2.877	2.943	3.000	3.050	3.094	3.264	3.383	3.475	3.550	3.613	3.667	3.757
	0.10	2.056	2.247	2.380	2.481	2.563	2.631	2.690	2.741	2.787	2.961	3.083	3.177	3.253	3.317	3.372	3.464
	0.20	1.685	1.888	2.028	2.135	2.221	2.292	2.354	2.407	2.455	2.636	2.762	2.859	2.937	3.002	3.059	3.153

Number of comparisons (p)

Table XIV (continued)

ν	α	\multicolumn{14}{c	}{Number of comparisons (p)}
		2	3	4	5	6	7	8	9	10	15	20	25	30	35	40	50
24	0.01	3.089	3.257	3.375	3.465	3.539	3.601	3.654	3.702	3.744	3.905	4.019	4.107	4.179	4.240	4.292	4.380
	0.05	2.385	2.566	2.692	2.788	2.866	2.931	2.988	3.037	3.081	3.249	3.366	3.457	3.531	3.593	3.646	3.735
	0.10	2.051	2.241	2.373	2.473	2.554	2.622	2.680	2.731	2.777	2.949	3.070	3.162	3.238	3.301	3.355	3.446
	0.20	1.682	1.884	2.024	2.130	2.215	2.286	2.347	2.400	2.448	2.627	2.752	2.848	2.925	2.990	3.046	3.139
25	0.01	3.077	3.243	3.359	3.449	3.521	3.583	3.635	3.682	3.723	3.882	3.995	4.081	4.152	4.212	4.263	4.350
	0.05	2.379	2.558	2.683	2.779	2.856	2.921	2.976	3.025	3.069	3.235	3.351	3.440	3.513	3.574	3.627	3.715
	0.10	2.047	2.236	2.367	2.466	2.547	2.614	2.672	2.722	2.767	2.938	3.058	3.149	3.224	3.286	3.340	3.430
	0.20	1.679	1.881	2.020	2.125	2.210	2.280	2.341	2.394	2.441	2.619	2.743	2.838	2.914	2.979	3.034	3.126
26	0.01	3.066	3.230	3.345	3.433	3.505	3.566	3.618	3.664	3.705	3.862	3.972	4.058	4.128	4.186	4.237	4.322
	0.05	2.373	2.551	2.675	2.770	2.847	2.911	2.966	3.014	3.058	3.222	3.337	3.425	3.497	3.558	3.610	3.697
	0.10	2.043	2.231	2.361	2.460	2.540	2.607	2.664	2.714	2.759	2.928	3.047	3.137	3.211	3.273	3.326	3.415
	0.20	1.677	1.878	2.012	2.121	2.205	2.275	2.335	2.388	2.435	2.612	2.735	2.829	2.905	2.968	3.023	3.114
27	0.01	3.056	3.218	3.332	3.419	3.491	3.550	3.602	3.647	3.688	3.843	3.952	4.036	4.105	4.163	4.213	4.297
	0.05	2.368	2.545	2.668	2.762	2.838	2.902	2.956	3.004	3.047	3.210	3.324	3.411	3.483	3.542	3.594	3.680
	0.10	2.039	2.227	2.356	2.454	2.534	2.600	2.657	2.707	2.751	2.919	3.036	3.126	3.199	3.261	3.313	3.401
	0.20	1.675	1.875	2.012	2.117	2.201	2.270	2.330	2.383	2.429	2.605	2.727	2.820	2.896	2.959	3.013	3.103
28	0.01	3.046	3.207	3.320	3.407	3.477	3.536	3.587	3.632	3.672	3.825	3.933	4.017	4.084	4.142	4.191	4.274
	0.05	2.363	2.539	2.661	2.755	2.830	2.893	2.948	2.995	3.039	3.199	3.312	3.399	3.469	3.528	3.579	3.664
	0.10	2.036	2.222	2.351	2.449	2.528	2.594	2.650	2.700	2.744	2.911	3.027	3.116	3.188	3.249	3.301	3.388
	0.20	1.672	1.872	2.009	2.113	2.196	2.266	2.326	2.378	2.424	2.599	2.720	2.812	2.887	2.950	3.004	3.093
29	0.01	3.037	3.197	3.309	3.395	3.464	3.523	3.574	3.618	3.658	3.809	3.916	3.998	4.065	4.122	4.171	4.252
	0.05	2.358	2.534	2.655	2.748	2.823	2.886	2.940	2.987	3.029	3.189	3.301	3.387	3.457	3.515	3.566	3.650
	0.10	2.033	2.218	2.346	2.444	2.522	2.588	2.644	2.693	2.737	2.903	3.018	3.107	3.178	3.239	3.291	3.377
	0.20	1.671	1.869	2.006	2.110	2.193	2.262	2.321	2.373	2.419	2.593	2.713	2.805	2.880	2.942	2.995	3.084
30	0.01	3.029	3.188	3.298	3.384	3.453	3.511	3.561	3.605	3.644	3.794	3.900	3.981	4.048	4.103	4.152	4.232
	0.05	2.354	2.528	2.649	2.742	2.816	2.878	2.932	2.979	3.021	3.180	3.291	3.376	3.445	3.503	3.553	3.637
	0.10	2.030	2.215	2.342	2.439	2.517	2.582	2.638	2.687	2.731	2.895	3.010	3.098	3.169	3.229	3.280	3.366
	0.20	1.669	1.867	2.003	2.106	2.189	2.258	2.317	2.369	2.414	2.587	2.707	2.798	2.872	2.934	2.987	3.076
40	0.01	2.970	3.121	3.225	3.305	3.370	3.425	3.472	3.513	3.549	3.689	3.787	3.862	3.923	3.975	4.019	4.093
	0.05	2.323	2.492	2.608	2.696	2.768	2.827	2.878	2.923	2.963	3.113	3.218	3.298	3.363	3.418	3.464	3.542
	0.10	2.009	2.189	2.312	2.406	2.481	2.544	2.597	2.644	2.686	2.843	2.952	3.036	3.103	3.160	3.208	3.289
	0.20	1.656	1.850	1.983	2.083	2.164	2.231	2.288	2.338	2.382	2.548	2.663	2.751	2.821	2.880	2.931	3.015
60	0.01	2.914	3.056	3.155	3.230	3.291	3.342	3.386	3.425	3.459	3.589	3.679	3.749	3.805	3.852	3.893	3.961
	0.05	2.294	2.456	2.568	2.653	2.721	2.777	2.826	2.869	2.906	3.049	3.148	3.223	3.284	3.336	3.379	3.452
	0.10	1.989	2.163	2.283	2.373	2.446	2.506	2.558	2.603	2.643	2.793	2.897	2.976	3.040	3.093	3.139	3.214
	0.20	1.643	1.834	1.963	2.061	2.139	2.204	2.259	2.308	2.350	2.511	2.621	2.705	2.772	2.828	2.876	2.956
120	0.01	2.859	2.994	3.087	3.158	3.215	3.263	3.304	3.340	3.372	3.493	3.577	3.641	3.693	3.736	3.774	3.836
	0.05	2.265	2.422	2.529	2.610	2.675	2.729	2.776	2.816	2.852	2.987	3.081	3.152	3.209	3.257	3.298	3.366
	0.10	1.968	2.138	2.254	2.342	2.411	2.469	2.519	2.562	2.600	2.744	2.843	2.918	2.978	3.029	3.072	3.143
	0.20	1.631	1.817	1.944	2.039	2.115	2.178	2.231	2.278	2.319	2.474	2.580	2.660	2.724	2.778	2.824	2.899
∞	0.01	2.806	2.934	3.022	3.089	3.143	3.188	3.226	3.260	3.289	3.402	3.480	3.539	3.587	3.627	3.661	3.718
	0.05	2.237	2.388	2.491	2.569	2.631	2.683	2.727	2.766	2.800	2.928	3.016	3.083	3.137	3.182	3.220	3.284
	0.10	1.949	2.114	2.226	2.311	2.378	2.434	2.482	2.523	2.560	2.697	2.791	2.862	2.920	2.967	3.008	3.075
	0.20	1.618	1.801	1.925	2.018	2.091	2.152	2.204	2.249	2.289	2.438	2.540	2.617	2.678	2.729	2.773	2.844

From P. A. Games, "An Improved t Table for Simultaneous Control on Contrasts," *Journal of the American Statistical Association*, 72 (1977), 531–534. Reprinted by permission.

Table XV. Critical Values of the Studentized Maximum Modulus Distribution

This table gives the critical values of the Studentized maximum modulus distribution used in multiple comparisons. The critical values are given for $\alpha = 0.10, 0.05, 0.01$; the number of comparisons $p = 2$ (1) 5; and the error degrees of freedom $\nu = 2$ (1) 12 (2) 20, 24, 30, 40, 60, ∞. For example, for $\alpha = 0.05$, $p = 3$, and $\nu = 12$, the required critical value is obtained as 2.75.

ν	α	\multicolumn{14}{c}{Number of comparisons (p)}
		2	3	4	5	6	7	8	9	10	11	12	13	14	15
2	0.10	3.83	4.38	4.77	5.06	5.30	5.50	5.67	5.82	5.96	6.08	6.18	6.28	6.37	6.45
	0.05	5.57	6.34	6.89	7.31	7.65	7.93	8.17	8.38	8.57	8.74	8.89	9.03	9.16	9.28
	0.01	12.73	14.44	15.65	16.59	17.35	17.99	18.53	19.01	19.43	19.81	20.15	20.46	20.75	21.02
3	0.10	2.99	3.37	3.64	3.84	4.01	4.15	4.27	4.38	4.47	4.55	4.63	4.70	4.76	4.82
	0.05	3.96	4.43	4.76	5.02	5.23	5.41	5.56	5.69	5.81	5.92	6.01	6.10	6.18	6.26
	0.01	7.13	7.91	8.48	8.92	9.28	9.58	9.84	10.06	10.27	10.45	10.61	10.76	10.90	11.03
4	0.10	2.66	2.98	3.20	3.37	3.51	3.62	3.72	3.81	3.89	3.96	4.02	4.08	4.13	4.18
	0.05	3.38	3.74	4.00	4.20	4.37	4.50	4.62	4.72	4.82	4.90	4.98	5.04	5.11	5.17
	0.01	5.46	5.99	6.36	6.66	6.90	7.10	7.27	7.43	7.57	7.69	7.80	7.91	8.00	8.09
5	0.10	2.49	2.77	2.96	3.12	3.24	3.34	3.43	3.51	3.58	3.64	3.69	3.75	3.79	3.84
	0.05	3.09	3.40	3.62	3.79	3.93	4.04	4.14	4.23	4.31	4.38	4.45	4.51	4.56	4.61
	0.01	4.70	5.11	5.40	5.63	5.81	5.97	6.11	6.23	6.33	6.43	6.52	6.60	6.67	6.74
6	0.10	2.39	2.64	2.82	2.96	3.07	3.17	3.25	3.32	3.38	3.44	3.49	3.54	3.58	3.62
	0.05	2.92	3.19	3.39	3.54	3.66	3.77	3.86	3.94	4.01	4.07	4.13	4.18	4.23	4.28
	0.01	4.27	4.61	4.86	5.05	5.20	5.33	5.45	5.55	5.64	5.72	5.80	5.86	5.93	5.99
7	0.10	2.31	2.56	2.73	2.86	2.96	3.05	3.13	3.19	3.25	3.31	3.35	3.40	3.44	3.48
	0.05	2.80	3.06	3.24	3.38	3.49	3.59	3.67	3.74	3.80	3.86	3.92	3.96	4.01	4.05
	0.01	4.00	4.30	4.51	4.68	4.81	4.93	5.03	5.12	5.20	5.27	5.33	5.39	5.45	5.50
8	0.10	2.26	2.49	2.66	2.78	2.88	2.97	3.04	3.10	3.16	3.21	3.26	3.30	3.34	3.37
	0.05	2.72	2.96	3.13	3.26	3.36	3.45	3.53	3.60	3.66	3.71	3.76	3.81	3.85	3.89
	0.01	3.81	4.08	4.27	4.42	4.55	4.65	4.74	4.82	4.89	4.96	5.02	5.07	5.12	5.17
9	0.10	2.22	2.45	2.60	2.72	2.82	2.90	2.97	3.03	3.09	3.13	3.18	3.22	3.26	3.29
	0.05	2.66	2.89	3.05	3.17	3.27	3.36	3.43	3.49	3.55	3.60	3.65	3.69	3.73	3.77
	0.01	3.67	3.92	4.10	4.24	4.35	4.45	4.53	4.61	4.67	4.73	4.79	4.84	4.88	4.92
10	0.10	2.19	2.41	2.56	2.68	2.77	2.85	2.92	2.98	3.03	3.08	3.12	3.16	3.20	3.23
	0.05	2.61	2.83	2.98	3.10	3.20	3.28	3.35	3.41	3.47	3.52	3.56	3.60	3.64	3.68
	0.01	3.57	3.80	3.97	4.10	4.20	4.29	4.37	4.44	4.50	4.56	4.61	4.66	4.70	4.74
11	0.10	2.17	2.38	2.53	2.64	2.73	2.81	2.88	2.93	2.98	3.03	3.07	3.11	3.15	3.18
	0.05	2.57	2.78	2.93	3.05	3.14	3.22	3.29	3.35	3.40	3.45	3.49	3.53	3.57	3.60
	0.01	3.48	3.71	3.87	3.99	4.09	4.17	4.25	4.31	4.37	4.42	4.47	4.51	4.55	4.59
12	0.10	2.15	2.36	2.50	2.61	2.70	2.78	2.84	2.90	2.95	2.99	3.03	3.07	3.10	3.14
	0.05	2.54	2.75	2.89	3.00	3.09	3.17	3.24	3.29	3.34	3.39	3.43	3.47	3.51	3.54
	0.01	3.42	3.63	3.78	3.90	4.00	4.08	4.15	4.21	4.26	4.31	4.36	4.40	4.44	4.48
14	0.10	2.12	2.32	2.46	2.57	2.65	2.72	2.79	2.84	2.89	2.93	2.97	3.01	3.04	3.07
	0.05	2.49	2.69	2.83	2.94	3.02	3.09	3.16	3.21	3.26	3.30	3.34	3.38	3.41	3.45
	0.01	3.32	3.52	3.66	3.77	3.85	3.93	3.99	4.05	4.10	4.15	4.19	4.23	4.26	4.30
16	0.10	2.10	2.29	2.43	2.53	2.62	2.69	2.75	2.80	2.85	2.89	2.93	2.96	2.99	3.02
	0.05	2.46	2.65	2.78	2.89	2.97	3.04	3.10	3.15	3.20	3.24	3.28	3.31	3.35	3.38
	0.01	3.25	3.43	3.57	3.67	3.75	3.82	3.88	3.94	3.99	4.03	4.07	4.11	4.14	4.17

Table XV (*continued*)

ν	α	\multicolumn{14}{c}{Number of comparisons (p)}													
		2	3	4	5	6	7	8	9	10	11	12	13	14	15
18	0.10	2.08	2.27	2.41	2.51	2.59	2.66	2.72	2.77	2.81	2.85	2.89	2.92	2.96	2.99
	0.05	2.43	2.62	2.75	2.85	2.93	3.00	3.05	3.11	3.15	3.19	3.23	3.26	3.29	3.32
	0.01	3.19	3.37	3.50	3.60	3.68	3.74	3.80	3.85	3.90	3.94	3.98	4.01	4.04	4.07
20	0.10	2.07	2.26	2.39	2.49	2.57	2.63	2.69	2.74	2.79	2.83	2.86	2.90	2.93	2.96
	0.05	2.41	2.59	2.72	2.82	2.90	2.96	3.02	3.07	3.11	3.15	3.19	3.22	3.25	3.28
	0.01	3.15	3.32	3.45	3.54	3.62	3.68	3.74	3.79	3.83	3.87	3.91	3.94	3.97	4.00
24	0.10	2.05	2.23	2.36	2.46	2.53	2.60	2.66	2.70	2.75	2.79	2.82	2.85	2.88	2.91
	0.05	2.38	2.56	2.68	2.77	2.85	2.91	2.97	3.02	3.06	3.10	3.13	3.16	3.19	3.22
	0.01	3.09	3.25	3.37	3.46	3.53	3.59	3.64	3.69	3.73	3.77	3.80	3.83	3.86	3.89
30	0.10	2.03	2.21	2.33	2.43	2.50	2.57	2.62	2.67	2.71	2.75	2.78	2.81	2.84	2.87
	0.05	2.35	2.52	2.64	2.73	2.80	2.87	2.92	2.96	3.00	3.04	3.07	3.11	3.13	3.16
	0.01	3.03	3.18	3.29	3.38	3.45	3.51	3.55	3.60	3.64	3.67	3.70	3.73	3.76	3.78
40	0.10	2.01	2.18	2.30	2.40	2.47	2.53	2.58	2.63	2.67	2.71	2.74	2.77	2.80	2.82
	0.05	2.32	2.49	2.60	2.69	2.76	2.82	2.87	2.91	2.95	2.99	3.02	3.05	3.08	3.10
	0.01	2.97	3.12	3.22	3.30	3.37	3.42	3.47	3.51	3.54	3.58	3.61	3.63	3.66	3.68
60	0.10	1.99	2.16	2.28	2.37	2.44	2.50	2.55	2.59	2.63	2.67	2.70	2.73	2.76	2.78
	0.05	2.29	2.45	2.56	2.65	2.72	2.77	2.82	2.86	2.90	2.93	2.96	2.99	3.02	3.04
	0.01	2.91	3.05	3.15	3.23	3.29	3.34	3.38	3.42	3.46	3.49	3.51	3.54	3.56	3.59
∞	0.10	1.95	2.11	2.23	2.31	2.38	2.43	2.48	2.52	2.56	2.59	2.62	2.65	2.67	2.70
	0.05	2.24	2.39	2.49	2.57	2.63	2.68	2.73	2.77	2.80	2.83	2.86	2.88	2.91	2.93
	0.01	2.81	2.93	3.02	3.09	3.14	3.19	3.23	3.26	3.29	3.32	3.34	3.36	3.38	3.40

From R. E. Bechhofer and C. W. Dunnett, "Comparisons for Orthogonal Contrasts: Examples and Tables," *Technometrics*, 24 (1982), 213–222. Abridged and reprinted by permission.

Table XVI. Critical Values of the Studentized Augmented Range Distribution

This table gives the critical values of the Studentized augmented range distribution used in multiple comparisons. The critical values are given for $\alpha = 0.20, 0.10, 0.05, 0.01$; the number of comparisons $p = 2\,(1)\,8$; and the error degrees of freedom $\nu = 5, 7, 10, 12\,(4)\,24, 30, 40, 60, 120, \infty$. For example, for $\alpha = 0.05$, $p = 4$, and $\nu = 16$, the desired critical value is obtained as 4.050.

		Number of comparisons (p)						
ν	α	2	3	4	5	6	7	8
5	.20	2.326	2.935	3.379	3.719	3.991	4.215	4.406
	.10	3.060	3.772	4.282	4.671	4.982	5.239	5.458
	.05	3.832	4.654	5.236	5.680	6.036	6.331	6.583
	.01	5.903	7.030	7.823	8.429	8.916	9.322	9.669
7	.20	2.213	2.783	3.195	3.508	3.757	3.963	4.137
	.10	2.848	3.491	3.943	4.285	4.556	4.781	4.972
	.05	3.486	4.198	4.692	5.064	5.360	5.606	5.816
	.01	5.063	5.947	6.551	7.008	7.374	7.679	7.939
10	.20	2.133	2.676	3.066	3.359	3.592	3.783	3.944
	.10	2.704	3.300	3.712	4.021	4.265	4.466	4.636
	.05	3.259	3.899	4.333	4.656	4.913	5.124	5.305
	.01	4.550	5.284	5.773	6.138	6.428	6.669	6.875
12	.20	2.103	2.636	3.017	3.303	3.530	3.715	3.872
	.10	2.651	3.230	3.628	3.924	4.157	4.349	4.511
	.05	3.177	3.791	4.204	4.509	4.751	4.950	5.119
	.01	4.373	5.056	5.505	5.837	6.101	6.321	6.507
16	.20	2.066	2.587	2.958	3.235	3.453	3.632	3.782
	.10	2.588	3.146	3.526	3.806	4.027	4.207	4.360
	.05	3.080	3.663	4.050	4.334	4.557	4.741	4.897
	.01	4.169	4.792	5.194	5.489	5.722	5.915	6.079
20	.20	2.045	2.558	2.923	3.195	3.408	3.582	3.729
	.10	2.551	3.097	3.466	3.738	3.950	4.124	4.271
	.05	3.024	3.590	3.961	4.233	4.446	4.620	4.768
	.01	4.055	4.644	5.019	5.294	5.510	5.688	5.839
24	.20	2.031	2.539	2.900	3.168	3.378	3.549	3.694
	.10	2.527	3.065	3.427	3.693	3.901	4.070	4.213
	.05	2.988	3.542	3.904	4.167	4.373	4.541	4.684
	.01	3.982	4.549	4.908	5.169	5.374	5.542	5.685
30	.20	2.017	2.521	2.877	3.142	3.348	3.517	3.659
	.10	2.503	3.034	3.389	3.649	3.851	4.016	4.155
	.05	2.952	3.496	3.847	4.103	4.320	4.464	4.602
	.01	3.912	4.458	4.800	5.048	5.242	5.401	5.536
40	.20	2.003	2.502	2.855	3.116	3.319	3.485	3.624
	.10	2.480	3.003	3.352	3.605	3.803	3.963	4.099
	.05	2.918	3.450	3.792	4.040	4.232	4.389	4.521
	.01	3.844	4.370	4.696	4.931	5.115	5.265	5.392
60	.20	1.990	2.484	2.833	3.090	3.290	3.453	3.589
	.10	2.457	2.927	3.315	3.563	3.755	3.911	4.042
	.05	2.884	3.406	3.738	3.978	4.163	4.314	4.441
	.01	3.778	4.284	4.595	4.818	4.991	5.133	5.253
120	.20	1.976	2.466	2.811	3.064	3.261	3.421	3.554
	.10	2.434	2.943	3.278	3.520	3.707	3.859	3.987
	.05	2.851	3.362	3.686	3.917	4.096	4.241	4.363
	.01	3.714	4.201	4.497	4.709	4.872	5.005	5.118
∞	.20	1.963	2.448	2.789	3.039	3.232	3.389	3.520
	.10	2.412	2.913	3.243	3.479	3.661	3.808	3.931
	.05	2.819	3.320	3.634	3.858	4.030	4.170	4.286
	.01	3.653	4.121	4.403	4.603	4.757	4.882	4.987

From M. R. Stoline, "Tables of the Studentized Augmented Range and Applications to Problems of Multiple Comparisons," *Journal of the American Statistical Association*, 73 (1978), 656–660. Adapted and reprinted by permission.

Table XVII (a). Critical Values of the Distribution of $\hat{\gamma}_1$ for Testing Skewness

This table gives the upper-tailed critical values of the sample estimate of the coefficient of skewness ($\hat{\gamma}_1$). The critical values are given for $\alpha = 0.05, 0.01$, and the sample size $n = 25\ (5)\ 50\ (10)\ 100\ (25)\ 200\ (5)\ 500$. Since the distribution of the statistic $\hat{\gamma}_1$ is symmetrical about zero, the one-tailed critical values also represent two-tailed values of 0.10 and 0.02. For example, for $\alpha = 0.05$ and $n = 30$, the desired critical value is obtained as 0.661.

	Critical Value (α)				Critical Value (α)		
n	0.05	0.01	Standard Deviation	n	0.05	0.01	Standard Deviation
25	0.711	1.061	0.4354	100	0.389	0.567	0.2377
30	0.661	0.982	0.4052	125	0.350	0.508	0.2139
35	0.621	0.921	0.3804	150	0.321	0.464	0.1961
40	0.587	0.869	0.3596	175	0.298	0.430	0.1820
45	0.558	0.825	0.3418	200	0.280	0.403	0.1706
50	0.533	0.787	0.3264	250	0.251	0.360	0.1531
60	0.492	0.723	0.3009	300	0.230	0.329	0.1400
70	0.459	0.673	0.2806	350	0.213	0.305	0.1298
80	0.432	0.631	0.2638	400	0.200	0.285	0.1216
90	0.409	0.596	0.2498	450	0.188	0.269	0.1147
100	0.389	0.567	0.2377	500	0.179	0.255	0.1089

Table XVII (b). Critical Values of the Distribution of $\hat{\gamma}_2$ for Testing Kurtosis

This table gives upper- and lower-tailed critical values of the sample estimate of the coefficient of kurtosis ($\hat{\gamma}_2$). The critical values are given for $\alpha = 0.05, 0.01$, and the sample size $n = 50\ (25)\ 150\ (50)\ 1000\ (200)\ 2000$. For example, for $\alpha = 0.05$ and $n = 50$, the upper-tailed critical value is obtained as 3.99.

	Critical Value (α)					Critical Value (α)			
	Upper		Lower			Upper		Lower	
n	0.01	0.05	0.05	0.01	n	0.01	0.05	0.05	0.01
50	4.88	3.99	2.15	1.95	600	3.54	3.34	2.70	2.60
75	4.59	3.87	2.27	2.08	650	3.52	3.33	2.71	2.61
100	4.39	3.77	2.35	2.18	700	3.50	3.31	2.72	2.62
125	4.24	3.71	2.40	2.24	750	3.48	3.30	2.73	2.64
150	4.13	3.65	2.45	2.29	800	3.46	3.29	2.74	2.65
					850	3.45	3.28	2.74	2.66
200	3.98	3.57	2.51	2.37	900	3.43	3.28	2.75	2.66
250	3.87	3.52	2.55	2.42	950	3.42	3.27	2.76	2.67
300	3.79	3.47	2.59	2.46	1000	3.41	3.26	2.76	2.68
350	3.72	3.44	2.62	2.50					
400	3.67	3.41	2.64	2.52	1200	3.37	3.24	2.78	2.71
450	3.63	3.39	2.66	2.55	1400	3.34	3.22	2.80	2.72
500	3.60	3.37	2.67	2.57	1600	3.32	3.21	2.81	2.74
550	3.57	3.35	2.69	2.58	1800	3.30	3.20	2.82	2.76
600	3.54	3.34	2.70	2.60	2000	3.28	3.18	2.83	2.77

From E. S. Pearson and H. O. Hartley, *Biometrika Tables for Statisticians*, Vol. I, Third Edition, © 1970 by Cambridge University Press, Cambridge. Reprinted by permission (from Table 34B and 34C).

Table XVIII. Coefficients of Order Statistics for the Shapiro-Wilk's W Test for Normality

The table gives coefficients $\{a_{n-i+1}\}$ ($i = 1, 2, \ldots, n$) of the order statistics for determining the Shapiro-Wilk's W statistic. The coefficients are given for $n = 2$ (1) 30. Shapiro and Wilk (1965) used approximations for $n > 20$. The values given here are exact upto $n = 30$.

$i \backslash n$	2	3	4	5	6	7	8	9	10
1	0.7071	0.7071	0.6872	0.6646	0.6431	0.6233	0.6052	0.5888	0.5739
2	–	0.0000	0.1668	0.2413	0.2806	0.3031	0.3164	0.3244	0.3290
3	–	–	–	0.0000	0.0875	0.1401	0.1743	0.1976	0.2141
4	–	–	–	–	–	0.0000	0.0561	0.0947	0.1224
5	–	–	–	–	–	–	–	0.0000	0.0399

$i \backslash n$	11	12	13	14	15	16	17	18	19	20
1	0.5601	0.5475	0.5359	0.5251	0.5150	0.5056	0.4968	0.4886	0.4808	0.4734
2	0.3315	0.3325	0.3325	0.3318	0.3306	0.3290	0.3273	0.3253	0.3232	0.3211
3	0.2260	0.2347	0.2412	0.2460	0.2495	0.2521	0.2540	0.2552	0.2561	0.2565
4	0.1429	0.1586	0.1707	0.1803	0.1878	0.1939	0.1988	0.2027	0.2059	0.2085
5	0.0695	0.0922	0.1100	0.1240	0.1354	0.1447	0.1523	0.1587	0.1641	0.1686
6	0.0000	0.0303	0.0538	0.0727	0.0880	0.1005	0.1109	0.1197	0.1271	0.1334
7	–	–	0.0000	0.0240	0.0434	0.0593	0.0725	0.0837	0.0932	0.1013
8	–	–	–	–	0.0000	0.0196	0.0359	0.0496	0.0612	0.0712
9	–	–	–	–	–	–	0.0000	0.0164	0.0303	0.0422
10	–	–	–	–	–	–	–	–	0.0000	0.0140

$i \backslash n$	21	22	23	24	25	26	27	28	29	30
1	0.4664	0.4598	0.4536	0.4476	0.4419	0.4364	0.4312	0.4262	0.4214	0.4168
2	0.3189	0.3167	0.3144	0.3122	0.3100	0.3078	0.3056	0.3035	0.3014	0.2993
3	0.2567	0.2566	0.2564	0.2560	0.2554	0.2548	0.2541	0.2533	0.2525	0.2516
4	0.2106	0.2122	0.2136	0.2146	0.2154	0.2160	0.2164	0.2167	0.2168	0.2169
5	0.1724	0.1756	0.1783	0.1806	0.1826	0.1842	0.1856	0.1868	0.1878	0.1886
6	0.1388	0.1435	0.1475	0.1510	0.1540	0.1567	0.1590	0.1610	0.1628	0.1643
7	0.1083	0.1144	0.1197	0.1243	0.1284	0.1320	0.1351	0.1380	0.1404	0.1427
8	0.0798	0.0873	0.0938	0.0997	0.1047	0.1092	0.1132	0.1167	0.1200	0.1228
9	0.0525	0.0615	0.0693	0.0759	0.0823	0.0878	0.0926	0.0969	0.1008	0.1044
10	0.0261	0.0366	0.0457	0.0540	0.0610	0.0673	0.0730	0.0781	0.0827	0.0869
11	0.0000	0.0121	0.0227	0.0321	0.0403	0.0476	0.0542	0.0601	0.0654	0.0702
12	–	–	0.0000	0.0107	0.0201	0.0284	0.0359	0.0426	0.0486	0.0541
13	–	–	–	–	0.0000	0.0094	0.0179	0.0254	0.0322	0.0383
14	–	–	–	–	–	–	0.0000	0.0084	0.0160	0.0229
15	–	–	–	–	–	–	–	–	0.0000	0.0076

From T. J. Lorenzen and V. L. Anderson, *Design of Experiments: A No-Name Approach*, © 1993 by Marcel Dekker, New York. Reprinted by permission.

Table XIX. Critical Values of the Shapiro-Wilk's W Test for Normality

This table gives critical values of the Shapiro-Wilk's W test for normality. The critical values are given for $\alpha = 0.01, 0.02, 0.05, 0.10, 0.50$, and $n = 3\,(1)\,50$.

	Critical Value (α)				
n	0.01	0.02	0.05	0.10	0.50
3	0.753	0.756	0.767	0.789	0.959
4	0.687	0.707	0.748	0.792	0.935
5	0.686	0.715	0.762	0.806	0.927
6	0.713	0.743	0.788	0.826	0.927
7	0.730	0.760	0.803	0.838	0.928
8	0.749	0.778	0.818	0.851	0.932
9	0.764	0.791	0.829	0.859	0.935
10	0.781	0.806	0.842	0.869	0.938
11	0.792	0.817	0.850	0.876	0.940
12	0.805	0.828	0.859	0.883	0.943
13	0.814	0.837	0.866	0.889	0.945
14	0.825	0.846	0.874	0.895	0.947
15	0.835	0.855	0.881	0.901	0.950
16	0.844	0.863	0.887	0.906	0.952
17	0.851	0.869	0.892	0.910	0.954
18	0.858	0.874	0.897	0.914	0.956
19	0.863	0.879	0.901	0.917	0.957
20	0.868	0.884	0.905	0.920	0.959
21	0.873	0.888	0.908	0.923	0.960
22	0.878	0.892	0.911	0.926	0.961
23	0.881	0.895	0.914	0.928	0.962
24	0.884	0.898	0.916	0.930	0.963
25	0.888	0.901	0.918	0.931	0.964
26	0.891	0.904	0.920	0.933	0.965
27	0.894	0.906	0.923	0.935	0.965
28	0.896	0.908	0.924	0.936	0.966
29	0.898	0.910	0.926	0.937	0.966
30	0.900	0.912	0.927	0.939	0.967
31	0.902	0.914	0.929	0.940	0.967
32	0.904	0.915	0.930	0.941	0.968
33	0.906	0.917	0.931	0.942	0.968
34	0.908	0.919	0.933	0.943	0.969
35	0.910	0.920	0.934	0.944	0.969
36	0.912	0.922	0.935	0.945	0.970
37	0.914	0.924	0.936	0.946	0.970
38	0.916	0.925	0.938	0.947	0.971
39	0.917	0.927	0.939	0.948	0.971
40	0.919	0.928	0.940	0.949	0.972
41	0.920	0.929	0.941	0.950	0.972
42	0.922	0.930	0.942	0.951	0.972
43	0.923	0.932	0.943	0.951	0.973
44	0.924	0.933	0.944	0.952	0.973
45	0.926	0.934	0.945	0.953	0.973
46	0.927	0.935	0.945	0.953	0.974
47	0.928	0.928	0.946	0.954	0.974
48	0.929	0.937	0.947	0.954	0.974
49	0.929	0.937	0.947	0.955	0.974
50	0.930	0.938	0.947	0.955	0.974

From S. S. Shapiro and M. B. Wilk, "An Analysis of Variance Test for Normality (Complete Samples)," *Biometrika*, 52 (1965), 591–611. Reprinted by permission.

Table XX. Critical Values of the D'Agostino's D Test for Normality

This table gives critical values of the D'Agostino's D test for normality. The critical values are given for $\alpha = 0.20, 0.10, 0.05, 0.02, 0.01$; and $n = 10$ (2) 50 (10) 100 (20) 200 (50) 1000 (250) 2000.

	Critical Value (α)				
n	0.20	0.10	0.05	0.02	0.01
10	0.2632, 0.2835	0.2573, 0.2843	0.2513, 0.2849	0.2436, 0.2855	0.2379, 0.2857
12	0.2653, 0.2841	0.2598, 0.2849	0.2544, 0.2854	0.2473, 0.2859	0.2420, 0.2862
14	0.2669, 0.2846	0.2618, 0.2853	0.2568, 0.2858	0.2503, 0.2862	0.2455, 0.2865
16	0.2681, 0.2848	0.2634, 0.2855	0.2587, 0.2860	0.2527, 0.2865	0.2482, 0.2867
18	0.2690, 0.2850	0.2646, 0.2855	0.2603, 0.2862	0.2547, 0.2866	0.2505, 0.2868
20	0.2699, 0.2852	0.2657, 0.2857	0.2617, 0.2863	0.2564, 0.2867	0.2525, 0.2869
22	0.2705, 0.2853	0.2670, 0.2859	0.2629, 0.2864	0.2579, 0.2869	0.2542, 0.2870
24	0.2711, 0.2853	0.2675, 0.2860	0.2638, 0.2865	0.2591, 0.2870	0.2557, 0.2871
26	0.2717, 0.2854	0.2682, 0.2861	0.2647, 0.2866	0.2603, 0.2870	0.2570, 0.2872
28	0.2721, 0.2854	0.2688, 0.2861	0.2655, 0.2866	0.2612, 0.2870	0.2581, 0.2873
30	0.2725, 0.2854	0.2693, 0.2861	0.2662, 0.2866	0.2622, 0.2871	0.2592, 0.2872
32	0.2729, 0.2854	0.2698, 0.2862	0.2668, 0.2867	0.2630, 0.2871	0.2600, 0.2873
34	0.2732, 0.2854	0.2703, 0.2862	0.2674, 0.2867	0.2636, 0.2871	0.2609, 0.2873
36	0.2735, 0.2854	0.2707, 0.2862	0.2679, 0.2867	0.2643, 0.2871	0.2617, 0.2873
38	0.2738, 0.2854	0.2710, 0.2862	0.2683, 0.2867	0.2649, 0.2871	0.2623, 0.2873
40	0.2740, 0.2854	0.2714, 0.2862	0.2688, 0.2867	0.2655, 0.2871	0.2630, 0.2874
42	0.2743, 0.2854	0.2717, 0.2861	0.2691, 0.2867	0.2659, 0.2871	0.2636, 0.2874
44	0.2745, 0.2854	0.2720, 0.2861	0.2695, 0.2867	0.2664, 0.2871	0.2641, 0.2874
46	0.2747, 0.2854	0.2722, 0.2861	0.2698, 0.2866	0.2668, 0.2871	0.2646, 0.2874
48	0.2749, 0.2854	0.2725, 0.2861	0.2702, 0.2866	0.2672, 0.2871	0.2651, 0.2874
50	0.2751, 0.2853	0.2727, 0.2861	0.2705, 0.2866	0.2676, 0.2871	0.2655, 0.2874
60	0.2757, 0.2852	0.2737, 0.2860	0.2717, 0.2865	0.2692, 0.2870	0.2673, 0.2873
70	0.2763, 0.2851	0.2744, 0.2859	0.2726, 0.2864	0.2708, 0.2869	0.2687, 0.2872
80	0.2768, 0.2850	0.2750, 0.2857	0.2734, 0.2863	0.2713, 0.2868	0.2698, 0.2871
90	0.2771, 0.2849	0.2755, 0.2856	0.2740, 0.2862	0.2721, 0.2866	0.2707, 0.2870
100	0.2774, 0.2849	0.2759, 0.2855	0.2745, 0.2860	0.2727, 0.2865	0.2714, 0.2869
120	0.2779, 0.2847	0.2765, 0.2853	0.2752, 0.2858	0.2737, 0.2863	0.2725, 0.2866
140	0.2782, 0.2846	0.2770, 0.2852	0.2758, 0.2856	0.2744, 0.2862	0.2734, 0.2865
160	0.2785, 0.2845	0.2774, 0.2851	0.2763, 0.2855	0.2750, 0.2860	0.2741, 0.2863
180	0.2787, 0.2844	0.2777, 0.2850	0.2767, 0.2854	0.2755, 0.2859	0.2746, 0.2862
200	0.2789, 0.2843	0.2779, 0.2848	0.2770, 0.2853	0.2759, 0.2857	0.2751, 0.2860
250	0.2793, 0.2841	0.2784, 0.2846	0.2776, 0.2850	0.2767, 0.2855	0.2760, 0.2858
300	0.2796, 0.2840	0.2788, 0.2844	0.2781, 0.2848	0.2772, 0.2853	0.2766, 0.2855
350	0.2798, 0.2839	0.2791, 0.2843	0.2784, 0.2847	0.2776, 0.2851	0.2771, 0.2853
400	0.2799, 0.2838	0.2793, 0.2842	0.2787, 0.2845	0.2780, 0.2849	0.2775, 0.2852
450	0.2801, 0.2837	0.2795, 0.2841	0.2789, 0.2844	0.2782, 0.2848	0.2778, 0.2851
500	0.2802, 0.2836	0.2796, 0.2840	0.2791, 0.2843	0.2785, 0.2847	0.2780, 0.2849
600	0.2804, 0.2835	0.2799, 0.2839	0.2794, 0.2842	0.2788, 0.2845	0.2784, 0.2847
700	0.2805, 0.2834	0.2800, 0.2838	0.2796, 0.2840	0.2791, 0.2844	0.2787, 0.2846
800	0.2806, 0.2833	0.2802, 0.2837	0.2798, 0.2839	0.2793, 0.2842	0.2790, 0.2844
900	0.2807, 0.2833	0.2803, 0.2836	0.2799, 0.2838	0.2795, 0.2841	0.2792, 0.2843
1000	0.2808, 0.2832	0.2804, 0.2835	0.2800, 0.2838	0.2796, 0.2840	0.2793, 0.2842
1250	0.2809, 0.2831	0.2806, 0.2834	0.2803, 0.2836	0.2799, 0.2839	0.2797, 0.2840
1500	0.2810, 0.2830	0.2807, 0.2833	0.2805, 0.2835	0.2801, 0.2837	0.2799, 0.2839
1750	0.2811, 0.2830	0.2808, 0.2832	0.2806, 0.2834	0.2803, 0.2836	0.2801, 0.2838
2000	0.2812, 0.2829	0.2809, 0.2831	0.2807, 0.2833	0.2804, 0.2835	0.2802, 0.2837

From R. B. D'Agostino and M. A. Stephens, *Goodness-of-Fit Techniques*, © 1986 by Marcel Dekker, Inc., New York. Reprinted by permission.

Statistical Tables and Charts

Table XXI. Critical Values of the Bartlett's Test for Homogeneity of Variances

This table gives critical values of the Bartlett's test for homogeneity of variances having equal sample sizes in each group. Bartlett's test statistic is the ratio of the weighted geometric mean of the sample variances to their weighted arithmetic mean (the weights are relative degrees of freedom). The critical values are given for $\alpha = 0.01, 0.05, 0.10$; the number of groups $p = 2$ (1) 10; and the sample size in each group $n = 3$ (1) 30 (10) 60 (20) 100. We reject the hypothesis of homogeneity of variances at the α-level of significance if $B < B_p(n, \alpha)$, where B is the calculated value of the Bartlett's statistic and $B_p(n, \alpha)$ is the critical value having an area of size α in the left-tail of the Bartlett's distribution. The critical values for equal sample sizes given in this table can also be used to obtain a highly accurate approximation of the critical values in the unequal sample size case by employing the following relation:

$$B_p(n_1, n_2, \ldots, n_p; \alpha) \cong (n_1/N)B_p(n_1, \alpha) + (n_2/N)B_p(n_2, \alpha)$$
$$+ \cdots + (n_p/N)B_p(n_p, \alpha),$$

where $N = \sum_{i=1}^{p} n_i$, $B_p(n_1, n_2, \ldots, n_p; \alpha)$ denotes the α-level critical value of the Bartlett's test statistic with p groups having $n_1, n_2, \ldots, n_p$ observations, respectively, and $B_p(n_i, \alpha)$, $i = 1, 2, \ldots, p$, denotes the α-level critical value in the equal sample size case with n_i observations in all p groups. For a given p, where $p = 2$ (1) 10, and for any combination of sample sizes from 5 (1) 100, the absolute error of this approximation is less than 0.005 (the percentage relative error is less than one-half of one percent) when $\alpha = 0.05, 0.10$, or 0.25. When $\alpha = 0.01$, the absolute error is approximately 0.015 in the extreme case and less than 0.005 when min $(n_1, n_2, \ldots, n_p) \geq 10$. The approximation can be improved with the help of correction factors given in Dyer and Keating (1980, Table 2) and the absolute error of the corrected approximation is as small as for any other α values. To illustrate, suppose $p = 4$ and $n_1 = 5$, $n_2 = 6$, $n_3 = 10$, $n_4 = 50$. Using the relation given previoulsy, $B_4(5, 6, 10, 50; 0.01) \cong (5/71) (0.4607) + (6/71) (0.5430) + (10/71) (0.7195) + (50/71) (0.9433) = 0.8440$. Using the correction factors given in Dyer and Keating (1980, Table 2) it can be shown that $B_4(5, 6, 10, 50; 0.01) \cong 0.8364$. The exact value is $B_4(5, 6, 10, 50; 0.01) = 0.8359$.

	Number of Groups (p)								
n	2	3	4	5	6	7	8	9	10
					$\alpha = 0.01$				
3	.1411	.1672	–	–	–	–	–	–	–
4	.2843	.3165	.3475	.3729	.3937	.4110	–	–	–
5	.3984	.4304	.4607	.4850	.5046	.5207	.5343	.5458	.5558
6	.4850	.5149	.5430	.5653	.5832	.5978	.6100	.6204	.6293
7	.5512	.5787	.6045	.6248	.6410	.6542	.6652	.6744	.6824
8	.6031	.6282	.6518	.6704	.6851	.6970	.7069	.7153	.7225
9	.6445	.6676	.6892	.7062	.7197	.7305	.7395	.7471	.7536
10	.6783	.6996	.7195	.7352	.7475	.7575	.7657	.7726	.7786
11	.7063	.7260	.7445	.7590	.7703	.7795	.7871	.7935	.7990
12	.7299	.7483	.7654	.7789	.7894	.7980	.8050	.8109	.8160
13	.7501	.7672	.7832	.7958	.8056	.8135	.8201	.8256	.8303
14	.7674	.7835	.7985	.8103	.8195	.8269	.8330	.8382	.8426
15	.7825	.7977	.8118	.8229	.8315	.8385	.8443	.8491	.8532
16	.7958	.8101	.8235	.8339	.8421	.8486	.8541	.8586	.8625
17	.8076	.8211	.8338	.8436	.8514	.8576	.8627	.8670	.8707
18	.8181	.8309	.8429	.8523	.8596	.8655	.8704	.8745	.8780
19	.8275	.8397	.8512	.8601	.8670	.8727	.8773	.8811	.8845
20	.8360	.8476	.8586	.8671	.8737	.8791	.8835	.8871	.8903
21	.8437	.8548	.8653	.8734	.8797	.8848	.8890	.8926	.8956
22	.8507	.8614	.8714	.8791	.8852	.8901	.8941	.8975	.9004
23	.8571	.8673	.8769	.8844	.8902	.8949	.8988	.9020	.9047
24	.8630	.8728	.8820	.8892	.8948	.8993	.9030	.9061	.9087
25	.8684	.8779	.8867	.8936	.8990	.9034	.9069	.9099	.9124
26	.8734	.8825	.8911	.8977	.9029	.9071	.9105	.9134	.9158
27	.8781	.8869	.8951	.9015	.9065	.9105	.9138	.9166	.9190
28	.8824	.8909	.8988	.9050	.9099	.9138	.9169	.9196	.9219
29	.8864	.8946	.9023	.9083	.9130	.9167	.9198	.9224	.9246
30	.8902	.8981	.9056	.9114	.9159	.9195	.9225	.9250	.9271
40	.9175	.9235	.9291	.9335	.9370	.9397	.9420	.9439	.9455
50	.9339	.9387	.9433	.9468	.9496	.9518	.9536	.9551	.9564
60	.9449	.9489	.9527	.9557	.9580	.9599	.9614	.9626	.9637
80	.9586	.9617	.9646	.9668	.9685	.9699	.9711	.9720	.9728
100	.9669	.9693	.9716	.9734	.9748	.9759	.9769	.9776	.9783

Table XXI (continued)

	\multicolumn{9}{c}{Number of Groups (p)}
n	2	3	4	5	6	7	8	9	10
					$\alpha = 0.05$				
3	.3123	.3058	.3173	.3299	–	–	–	–	–
4	.4780	.4699	.4803	.4921	.5028	.5122	.5204	.5277	.5341
5	.5845	.5762	.5850	.5952	.6045	.6126	.6197	.6260	.6315
6	.6563	.6483	.6559	.6646	.6727	.6798	.6860	.6914	.6961
7	.7075	.7000	.7065	.7142	.7213	.7275	.7329	.7376	.7418
8	.7456	.7387	.7444	.7512	.7574	.7629	.7677	.7719	.7757
9	.7751	.7686	.7737	.7798	.7854	.7903	.7946	.7984	.8017
10	.7984	.7924	.7970	.8025	.8076	.8121	.8160	.8194	.8224
11	.8175	.8118	.8160	.8210	.8257	.8298	.8333	.8365	.8392
12	.8332	.8280	.8317	.8364	.8407	.8444	.8477	.8506	.8531
13	.8465	.8415	.8450	.8493	.8533	.8568	.8598	.8625	.8648
14	.8578	.8532	.8564	.8604	.8641	.8673	.8701	.8726	.8748
15	.8676	.8632	.8662	.8699	.8734	.8764	.8790	.8814	.8834
16	.8761	.8719	.8747	.8782	.8815	.8843	.8868	.8890	.8909
17	.8836	.8796	.8823	.8856	.8886	.8913	.8936	.8957	.8975
18	.8902	.8865	.8890	.8921	.8949	.8975	.8997	.9016	.9033
19	.8961	.8926	.8949	.8979	.9006	.9030	.9051	.9069	.9086
20	.9015	.8980	.9003	.9031	.9057	.9080	.9100	.9117	.9132
21	.9063	.9030	.9051	.9078	.9103	.9124	.9143	.9160	.9175
22	.9106	.9075	.9095	.9120	.9144	.9165	.9183	.9199	.9213
23	.9146	.9116	.9135	.9159	.9182	.9202	.9219	.9235	.9248
24	.9182	.9153	.9172	.9195	.9217	.9236	.9253	.9267	.9280
25	.9216	.9187	.9205	.9228	.9249	.9267	.9283	.9297	.9309
26	.9246	.9219	.9236	.9258	.9278	.9296	.9311	.9325	.9336
27	.9275	.9249	.9265	.9286	.9305	.9322	.9337	.9350	.9361
28	.9301	.9276	.9292	.9312	.9330	.9347	.9361	.9374	.9385
29	.9326	.9301	.9316	.9336	.9354	.9370	.9383	.9396	.9406
30	.9348	.9325	.9340	.9358	.9376	.9391	.9404	.9416	.9426
40	.9513	.9495	.9506	.9520	.9533	.9545	.9555	.9564	.9572
50	.9612	.9597	.9606	.9617	.9628	.9637	.9645	.9652	.9658
60	.9677	.9665	.9672	.9681	.9690	.9698	.9705	.9710	.9716
80	.9758	.9749	.9754	.9761	.9768	.9774	.9779	.9783	.9787
100	.9807	.9799	.9804	.9809	.9815	.9819	.9823	.9827	.9830

Table XXI (*continued*)

				Number of Groups (*p*)					
n	2	3	4	5	6	7	8	9	10
					$\alpha = 0.10$				
3	.4359	.3991	.3966	.4006	.4061	.4116	—	—	—
4	.5928	.5583	.5551	.5582	.5626	.5673	.5717	.5759	.5797
5	.6842	.6539	.6507	.6530	.6566	.6605	.6642	.6676	.6708
6	.7429	.7163	.7133	.7151	.7182	.7214	.7245	.7274	.7301
7	.7834	.7600	.7572	.7587	.7612	.7640	.7667	.7692	.7716
8	.8130	.7921	.7895	.7908	.7930	.7955	.7978	.8000	.8021
9	.8356	.8168	.8143	.8154	.8174	.8196	.8217	.8236	.8254
10	.8533	.8362	.8339	.8349	.8367	.8386	.8405	.8423	.8439
11	.8676	.8519	.8498	.8507	.8523	.8540	.8557	.8574	.8589
12	.8794	.8649	.8629	.8637	.8652	.8668	.8683	.8698	.8712
13	.8892	.8758	.8740	.8746	.8760	.8775	.8789	.8803	.8816
14	.8976	.8851	.8833	.8840	.8852	.8866	.8879	.8892	.8904
15	.9048	.8931	.8914	.8920	.8932	.8944	.8957	.8969	.8980
16	.9110	.9000	.8985	.8990	.9001	.9013	.9025	.9036	.9046
17	.9165	.9061	.9046	.9051	.9062	.9073	.9084	.9094	.9104
18	.9214	.9115	.9101	.9106	.9115	.9126	.9137	.9146	.9156
19	.9257	.9163	.9150	.9154	.9163	.9174	.9183	.9193	.9201
20	.9295	.9206	.9194	.9198	.9207	.9216	.9226	.9234	.9243
21	.9330	.9245	.9233	.9237	.9245	.9255	.9263	.9272	.9280
22	.9362	.9281	.9269	.9273	.9281	.9289	.9298	.9306	.9313
23	.9390	.9313	.9302	.9305	.9313	.9321	.9329	.9337	.9344
24	.9417	.9342	.9332	.9335	.9342	.9350	.9358	.9365	.9372
25	.9441	.9369	.9359	.9362	.9369	.9377	.9384	.9391	.9398
26	.9463	.9394	.9384	.9387	.9394	.9401	.9408	.9415	.9421
27	.9484	.9417	.9408	.9410	.9417	.9424	.9431	.9437	.9443
28	.9503	.9439	.9429	.9432	.9438	.9445	.9452	.9458	.9464
29	.9520	.9458	.9449	.9452	.9458	.9464	.9471	.9477	.9483
30	.9537	.9477	.9468	.9471	.9476	.9483	.9489	.9495	.9500
40	.9655	.9610	.9603	.9605	.9609	.9614	.9619	.9623	.9627
50	.9725	.9689	.9683	.9685	.9688	.9692	.9696	.9699	.9703
60	.9771	.9741	.9737	.9738	.9741	.9744	.9747	.9750	.9753
80	.9829	.9806	.9803	.9804	.9806	.9808	.9811	.9813	.9815
100	.9864	.9845	.9843	.9843	.9845	.9847	.9849	.9851	.9852

From D. D. Dyer and J. P. Keating, "On the Determination of Critical Values for Bartlett's Test," *Journal of the American Statistical Association*, 75 (1980), 313-319. Abridged and reprinted by permission.

Table XXII. Critical Values of the Hartley's Maximum F Ratio Test for Homogeneity of Variances

This table gives critical values of the Hartley's maximum F ratio test for homogeneity of variances having equal sample sizes in each group. The critical values are given for $\alpha = 0.05, 0.01$; the number groups $p = 2\ (1)\ 12$; and the number of degrees of freedom for variance estimate $\nu = 2\ (1)\ 10, 12, 15, 20, 30, 60, \infty$.

		Number of Groups (p)									
ν	α	2	3	4	5	6	7	8	9	10	11	12
2	.05	39.00	87.50	142.00	202.00	266.00	333.00	403.00	475.00	550.00	626.00	704.00
	.01	199.00	448.00	729.00	1036.00	1362.00	1705.00	2063.00	2432.00	2813.00	3204.00	3605.00
3	.05	15.40	27.80	39.20	50.70	62.00	72.90	83.50	93.90	104.00	114.00	124.00
	.01	47.50	85.00	120.00	151.00	184.00	216.00	249.00	281.00	310.00	337.00	361.00
4	.05	9.60	15.50	20.60	25.20	29.50	33.60	37.50	41.40	44.60	48.00	51.40
	.01	23.20	37.00	49.00	59.00	69.00	79.00	89.00	97.00	106.00	113.00	120.00
5	.05	7.15	10.80	13.70	16.30	18.70	20.80	22.90	24.70	26.50	28.20	29.90
	.01	14.90	22.00	28.00	33.00	38.00	42.00	46.00	50.00	54.00	57.00	60.00
6	.05	5.82	8.38	10.40	12.10	13.70	15.00	16.30	17.50	18.60	19.70	20.70
	.01	11.10	15.50	19.10	22.00	25.00	27.00	30.00	32.00	34.00	36.00	37.0
7	.05	4.99	6.94	8.44	9.70	10.80	11.80	12.70	13.50	14.30	15.10	15.80
	.01	8.89	12.10	14.50	16.50	18.40	20.00	22.00	23.00	24.00	26.00	27.00
8	.05	4.43	6.00	7.18	8.12	9.03	9.78	10.50	11.10	11.70	12.20	12.70
	.01	7.50	9.90	11.70	13.20	14.50	15.80	16.90	17.90	18.90	19.80	21.00
9	.05	4.03	5.34	6.31	7.11	7.80	8.41	8.95	9.45	9.91	10.30	10.70
	.01	6.54	8.50	9.90	11.10	12.10	13.10	13.90	14.70	15.30	16.00	16.60
10	.05	3.72	4.85	5.67	6.34	6.92	7.42	7.87	8.28	8.66	9.01	9.34
	.01	5.85	7.40	8.60	9.60	10.40	11.10	11.80	12.40	12.90	13.40	13.90
12	.05	3.28	4.16	4.79	5.30	5.72	6.09	6.42	6.72	7.00	7.25	7.48
	.01	4.91	6.10	6.90	7.60	8.20	8.70	9.10	9.50	9.90	10.20	10.60
15	.05	2.86	3.54	4.01	4.37	4.68	4.95	5.19	5.40	5.59	5.77	5.93
	.01	4.07	4.90	5.50	6.00	6.40	6.70	7.10	7.30	7.50	7.80	8.00
20	.05	2.46	2.95	3.29	3.54	3.76	3.94	4.10	4.24	4.37	4.49	4.59
	.01	3.32	3.80	4.30	4.60	4.90	5.10	5.30	5.50	5.60	5.80	5.90
30	.05	2.07	2.40	2.61	2.78	2.91	3.02	3.12	3.21	3.29	3.36	3.39
	.01	2.63	3.00	3.30	3.40	3.60	3.70	3.80	3.90	4.00	4.10	4.20
60	.05	1.67	1.85	1.96	2.04	2.11	2.17	2.22	2.26	2.30	2.33	2.36
	.01	1.96	2.20	2.30	2.40	2.40	2.50	2.50	2.60	2.60	2.70	2.70
∞	.05	1.00	1.00	1.00	1.00	1.00	1.00	1.00	1.00	1.00	1.00	1.00
	.01	1.00	1.00	1.00	1.00	1.00	1.00	1.00	1.00	1.00	1.00	1.00

From H. A. David, "Upper 5 and 1% Points of the Maximum F-Ratio" *Biometrika*, 39 (1952), 422–424. Reprinted by permission.

Table XXIII. Critical Values of the Cochran's C Test for Homogeneity of Variances

This table gives critical values of Cochran's C test for homogeneity of variances having equal sample sizes in each group. The critical values are given for $\alpha = 0.05, 0.01$; the number of groups $p = 2$ (1) 10, 12, 15, 20, 24, 30, 40, 60, 120; and the number of degrees of freedom for variance estimate $\nu = 1$ (1) 10, 16, 36, 144, ∞.

							Number of Groups (p)									
ν	α	2	3	4	5	6	7	8	9	10	12	15	20	24	30	40	60	120
1	.05	.9985	.9669	.9065	.8412	.7808	.7271	.6798	.6385	.6020	.5410	.4709	.3894	.3434	.2929	.2370	.1737	.0998
	.01	.9999	.9933	.9676	.9279	.8828	.8376	.7945	.7544	.7175	.6528	.5747	.4799	.4247	.3632	.2940	.2151	.1225
2	.05	.9750	.8709	.7679	.6838	.6161	.5612	.5157	.4775	.4450	.3924	.3346	.2705	.2354	.1980	.1567	.1131	.0632
	.01	.9950	.9423	.8643	.7885	.7218	.6644	.6152	.5727	.5358	.4751	.4069	.3297	.2871	.2412	.1915	.1371	.0759
3	.05	.9392	.7977	.6841	.5981	.5321	.4800	.4377	.4027	.3733	.3264	.2758	.2205	.1907	.1593	.1259	.0895	.0495
	.01	.9794	.8831	.7814	.6957	.6258	.5685	.5209	.4810	.4469	.3919	.3317	.2654	.2295	.1913	.1508	.1069	.0585
4	.05	.9057	.7457	.6287	.5441	.4803	.4307	.3910	.3584	.3311	.2880	.2419	.1921	.1656	.1377	.1082	.0765	.0419
	.01	.9586	.8335	.7212	.6329	.5635	.5080	.4627	.4251	.3934	.3428	.2882	.2288	.1970	.1635	.1281	.0902	.0489
5	.05	.8772	.7071	.5895	.5065	.4447	.3974	.3595	.3286	.3029	.2624	.2195	.1735	.1493	.1237	.0968	.0682	.0371
	.01	.9373	.7933	.6761	.5875	.5195	.4659	.4226	.3870	.3572	.3099	.2593	.2048	.1759	.1454	.1135	.0796	.0429
6	.05	.8534	.6771	.5598	.4783	.4184	.3726	.3362	.3067	.2823	.2439	.2034	.1602	.1374	.1137	.0887	.0623	.0337
	.01	.9172	.7606	.6410	.5531	.4866	.4347	.3932	.3592	.3308	.2861	.2386	.1877	.1608	.1327	.1033	.0722	.0387
7	.05	.8332	.6530	.5365	.4564	.3980	.3535	.3185	.2901	.2666	.2299	.1911	.1501	.1286	.1061	.0827	.0583	.0312
	.01	.8988	.7335	.6129	.5259	.4608	.4105	.3704	.3378	.3106	.2680	.2228	.1748	.1495	.1232	.0957	.0668	.0357
8	.05	.8159	.6333	.5175	.4387	.3817	.3384	.3043	.2768	.2541	.2187	.1815	.1422	.1216	.1002	.0780	.0552	.0292
	.01	.8823	.7107	.5897	.5037	.4401	.3911	.3522	.3207	.2945	.2535	.2104	.1646	.1406	.1157	.0898	.0625	.0334
9	.05	.8010	.6167	.5017	.4241	.3682	.3259	.2926	.2659	.2439	.2098	.1736	.1357	.1160	.0958	.0745	.0520	.0279
	.01	.8674	.6912	.5702	.4854	.4229	.3751	.3373	.3067	.2813	.2419	.2002	.1567	.1388	.1100	.0853	.0594	.0316
10	.05	.7880	.6025	.4884	.4118	.3568	.3154	.2829	.2568	.2353	.2020	.1671	.1303	.1113	.0921	.0713	.0497	.0266
	.01	.8539	.6743	.5536	.4697	.4084	.3616	.3248	.2950	.2704	.2320	.1918	.1501	.1283	.1054	.0816	.0567	.0302
16	.05	.7341	.5466	.4366	.3645	.3135	.2756	.2462	.2226	.2032	.1737	.1429	.1108	.0942	.0771	.0595	.0411	.0218
	.01	.7949	.6059	.4884	.4094	.3529	.3105	.2779	.2514	.2297	.1961	.1612	.1248	.1060	.0867	.0668	.0461	.0242
36	.05	.6602	.4748	.3720	.3066	.2612	.2278	.2022	.1820	.1655	.1403	.1144	.0879	.0743	.0604	.0462	.0316	.0165
	.01	.7067	.5153	.4057	.3351	.2858	.2494	.2214	.1992	.1811	.1535	.1251	.0960	.0810	.0658	.0503	.0344	.0178
144	.05	.5813	.4031	.3093	.2513	.2119	.1833	.1616	.1446	.1308	.1100	.0889	.0675	.0567	.0457	.0347	.0234	.0120
	.01	.6062	.4230	.3251	.2644	.2229	.1929	.1700	.1521	.1376	.1157	.0934	.0709	.0595	.0480	.0363	.0245	.0125
∞	.05	.5000	.3333	.2500	.2000	.1667	.1429	.1250	.1111	.1000	.0833	.0667	.0500	.0417	.0333	.0250	.0167	.0083
	.01	.5000	.3333	.2500	.2000	.1667	.1429	.1250	.1111	.1000	.0833	.0667	.0500	.0417	.0333	.0250	.0167	.0083

From C. Eisenhart, M. W. Hastay, and W. A. Wallis, *Techniques of Statistical Analysis*, Chapter 15, pp. 390–391, © 1947 by McGraw-Hill, New York. Reprinted by permission.

Table XXIV. Random Numbers

This table gives computer-generated pseudorandom digits that may be drawn in any direction: horizontal, left-to-right; vertical, up or down. The numbers may be read in single digits, double digits, or digits of any size. For ease of reading, the numbers are arranged in groups of five digits which should be ignored when reading the table. The table should be employed using a random start and crossing out the digits as they are used so that each portion of the table is used only once in a given experiment.

	00-04	05-09	10-14	15-19	20-24	25-29	30-34	35-39	40-44	45-49	50-54	55-59	60-64	65-69	70-74	75-79	80-84	85-89	90-94	95-99
00	54463	22662	65905	70639	79365	67382	29085	69831	47058	08186	59391	58030	52098	82718	87024	82848	04190	96574	90464	29065
01	15389	85205	18850	39226	42249	90669	96325	23248	60933	26927	99567	76364	77204	04615	27062	96621	43918	01896	83991	51141
02	85941	40756	82414	02015	13858	78030	16269	65978	01385	15345	10363	97518	51400	25670	98342	61891	27101	37855	06235	33316
03	61149	69440	11286	88218	58925	03638	52862	62733	33451	77455	86859	19558	64432	16706	99612	59798	32803	67708	15297	28612
04	05219	81619	10651	67079	92511	59888	84502	72095	83463	75577	11258	24591	36863	55368	31721	94335	34936	02566	80972	08188
05	41417	98326	87719	92294	46614	50948	64886	20002	97365	30976	95068	88628	35911	14530	33020	80428	39936	31855	34334	64865
06	28357	94070	20652	35774	16249	75019	21145	05217	47286	76305	54463	47237	73800	91017	36239	71824	83671	39892	60518	37092
07	17783	00015	10806	83091	91530	36466	39981	62481	49177	75779	16874	62677	57412	13215	31389	62233	80827	73917	82802	84420
08	40950	84820	29881	85966	62800	70326	84740	62660	77379	90279	92494	63157	76593	91316	03505	72389	96363	52887	01087	66091
09	82995	64157	66164	41180	10089	41757	78258	96488	88629	37231	15669	56689	35682	40844	53256	81872	35213	09840	34471	74441
10	96754	17676	55659	44105	47361	34833	86679	23930	53249	27083	99116	75486	84989	23476	52967	67104	39495	39100	17217	74073
11	34357	88040	53364	71726	45690	66334	60332	22554	90600	71113	15696	10703	65178	90637	63110	17622	53988	71087	84148	11670
12	06318	37403	49927	57715	50423	67372	63116	48888	21505	80182	97720	15369	51269	69620	03388	13699	33423	67453	43269	56720
13	62111	52820	07243	79931	89292	84767	85693	73947	22278	11551	11666	13841	71681	98000	35979	39719	81899	07449	47985	46967
14	47534	09243	67879	00544	23410	12740	02540	54440	32949	13491	71628	73130	78783	75691	41632	09847	61547	18707	85489	69944
15	98614	75993	84460	62846	59844	14922	48730	73443	48167	34770	40501	51089	99943	91843	41995	88931	73631	69361	05375	15417
16	24856	03648	44898	09351	98795	18644	39765	71058	90368	44104	22518	55576	98215	82068	10798	86211	36584	67466	69373	40054
17	96887	12479	80621	66223	86085	78285	02432	53342	42846	94771	75112	30485	62173	02132	14878	92879	22281	16783	86352	00077
18	90801	21472	42815	77408	37390	76766	52615	32141	30268	18106	80327	02671	98191	84342	90813	49268	95441	15496	20168	09271
19	55165	77312	83666	36028	28420	70219	81369	41943	47366	41067	60251	45548	02146	05597	48228	81366	34598	72856	66762	17002
20	75884	12952	84318	95108	72305	64620	91318	89872	45375	85436	57430	82270	10421	00540	43648	75888	66049	21511	47676	33444
21	16777	37116	58550	42958	21460	43910	01175	87894	81378	10620	73528	39559	34434	88596	54086	71693	43132	14414	79949	85193
22	46230	43877	80207	88877	89380	32992	91380	03164	98656	59337	25991	65959	70769	64721	86413	33475	42740	06175	82758	66248
23	42902	66892	46134	01432	94710	23474	20423	60137	60609	13119	78388	16638	09134	59980	63806	48472	39318	35434	24057	74739
24	81007	00333	39693	28039	10154	95425	39220	19774	31782	49037	12477	09965	96657	57994	59439	76330	24596	77515	09577	91871
25	68089	01122	51111	72373	06902	74373	96199	97017	41273	21546	83266	32883	42451	15579	38155	29793	40914	65990	16255	17777
26	20411	67081	89950	16944	93054	87687	96693	87236	77054	33848	76970	80876	10237	39515	79152	74798	39357	09054	73579	92359
27	58212	13160	06468	15718	82627	76999	05999	58680	96739	63700	37074	65198	44785	68624	98336	84481	97610	78735	46703	98265
28	70577	42866	24969	61210	76046	67699	42054	12696	93758	03283	83712	06514	30101	78295	54656	85417	43189	60048	72781	72606
29	94522	74358	71659	62038	79643	79169	44741	05437	39038	13163	20287	56862	69727	94443	64936	08366	27227	05158	50326	59566

Table XXIV (continued)

	00-04	05-09	10-14	15-19	20-24	25-29	30-34	35-39	40-44	45-49	50-54	55-59	60-64	65-69	70-74	75-79	80-84	85-89	90-94	95-99
30	42626	86819	85651	88678	17401	03252	99547	32404	17918	62880	74261	32592	86538	27041	65172	85532	07571	80609	39285	65340
31	16051	33763	57194	16752	54450	19031	58580	47629	54132	60631	64081	49863	08478	96001	18888	14810	70545	89755	59064	07210
32	08244	27647	33851	44705	94211	46716	11738	55784	95374	72655	05617	75818	47750	67814	29575	10526	66192	44464	27058	40467
33	59497	04392	09419	89964	51211	04894	72882	17805	21896	83864	26793	74951	95466	74307	13330	42664	85515	20632	05497	33625
34	97155	13428	40293	09985	58434	01412	69124	82171	59058	82859	65988	72850	48737	54719	52056	01596	03845	35067	03134	70322
35	98409	66162	95763	47420	20792	61527	20441	39435	11859	41567	27366	42271	44300	73399	21105	03280	73457	43093	05192	48657
36	45476	84882	65109	96597	25930	66790	65706	61203	53634	22557	56760	10909	98147	34736	33863	95256	12731	66598	50771	83665
37	89300	69700	50741	30329	11658	23166	05400	66669	48708	03887	72880	43338	93643	58904	59543	23943	11231	83268	65938	81581
38	50051	95137	91631	66315	91428	12275	24816	68091	71710	33258	77888	38100	03062	58103	47961	83841	25878	23746	55903	44115
39	31753	85178	31310	89642	98364	02306	24617	09609	83942	22716	28440	07819	21580	51459	47971	29882	13990	29226	23608	15873
40	79152	53829	77250	20190	56535	18760	69942	77448	33278	48805	63525	94441	77033	12147	51054	49955	58312	76923	96071	05813
41	44560	38750	83635	56540	64900	42912	13953	79149	18710	68618	47606	93410	16359	89033	89696	47231	64498	31776	05383	39902
42	68328	83378	63369	71381	39564	05615	42451	64559	97501	65747	52669	45030	96279	14709	52372	87832	02735	50803	72744	88208
43	46939	38689	58625	08342	30459	85863	20781	09284	26333	91777	16738	60159	07425	62369	07515	82721	37875	71153	21315	00132
44	83544	86141	15707	96256	23068	13782	08467	89469	93842	55349	59348	11695	45751	15865	74739	05572	32688	20271	65128	14551
45	91621	00881	04900	54224	46177	55309	17852	27491	89415	23466	12900	71775	29845	60774	94924	21810	38636	33717	67598	82521
46	91896	67126	04151	03795	59077	11848	12630	98375	52068	60142	75086	23537	49939	33595	13484	97588	28617	17979	70749	35234
47	55751	62515	21108	80830	02263	29303	37204	96926	30506	09808	99495	51434	29181	09993	38190	42553	68922	52125	91077	40197
48	85156	87689	95493	88842	00664	55017	55539	17771	69448	87530	26075	31671	45386	36583	93459	48599	52022	41330	60651	91321
49	07521	56898	12236	60277	39102	62315	12239	07105	11844	01117	13636	93596	23377	51133	95126	61496	42474	44141	46660	42338
50	64249	63664	39652	40646	97306	31741	07294	84149	46797	82487	32847	31282	03345	89593	69214	70381	78285	20054	91018	16742
51	26538	44249	04050	48174	65570	44072	40192	51153	11397	58212	16916	00041	30236	55023	14253	76582	12092	86533	92426	37655
52	05845	00512	78630	55328	18116	69296	91705	86224	29503	57071	66176	34037	21005	27137	03193	48970	64625	22394	39622	79085
53	74897	68373	67359	51014	33510	83048	17056	72506	82949	54600	46299	13335	12180	16861	38043	59292	62675	63631	37020	78195
54	20872	54570	35017	88132	25730	22626	86723	91691	13191	77212	22847	47839	45385	23289	47526	54098	45683	55849	51575	64689
55	31432	96156	89177	75541	81355	24480	77243	76690	42507	94362	41851	54160	92320	69936	34803	92479	33399	71160	64777	83378
56	66890	61505	01240	00660	05873	13568	76082	79172	57913	93448	28444	59497	91586	95917	68553	28639	06455	34174	11130	91994
57	41894	57790	79970	33106	86904	48119	52503	24130	72824	21627	47520	62378	98855	83174	13088	16561	68559	26679	06238	51254
58	11303	87118	81471	52936	08555	28420	49416	44448	04269	27029	34978	63271	13142	82681	05271	08822	06490	44984	49307	61717
59	54374	57325	16947	45356	78371	10563	97191	53798	12693	27928	37404	80416	69035	92980	49486	74378	75610	74976	70056	15478

Table XXIV (continued)

	00-04	05-09	10-14	15-19	20-24	25-29	30-34	35-39	40-44	45-49	50-54	55-59	60-64	65-69	70-74	75-79	80-84	85-89	90-94	95-99
60	64852	34421	61046	90849	13966	39810	42699	21753	76192	10508	32400	65482	52099	53676	74648	94148	65095	69597	52771	71551
61	16309	20384	09491	91588	97720	89846	30376	76970	23063	35894	89262	86332	51718	70663	11623	29834	79820	73002	84886	03591
62	42587	37065	24526	72602	57589	98131	37292	05967	26002	51945	86866	09127	98021	03871	27789	58444	44832	36505	40672	30180
63	40177	98590	97161	41682	84533	67588	62036	49967	01990	72308	90814	14833	08759	74645	05046	94056	99094	65091	32663	73040
64	82309	76128	93965	26743	24141	04838	40254	26065	07938	76236	19192	82756	20553	58446	55376	88914	75096	26119	83898	43816
65	79788	68243	59732	04257	27084	14743	17520	95401	55811	76099	77585	52593	55612	95766	10019	29531	73064	20953	53523	58136
66	40538	79000	89559	25026	42274	23489	34502	75508	06059	86682	23757	16364	05096	03192	62386	45389	85332	18877	55710	96459
67	64016	73598	18609	73150	62463	33102	45205	87440	96767	67042	45989	96257	23850	26216	23309	21526	07425	50254	19455	29315
68	49767	12691	17903	93871	99721	79109	09425	26904	07419	76013	92970	94243	07316	41467	64837	52406	25225	51553	31220	14032
69	76974	55108	29795	08404	82684	00497	51126	79935	57450	55671	74346	59596	40088	98176	17896	86900	20249	77753	19099	48885
70	23854	08480	85983	96025	50117	64610	99425	62291	86943	21541	87646	41309	27636	45153	29988	94770	07255	70908	05340	99751
71	68973	70551	25098	78033	98573	79848	31778	29555	61446	23037	50099	71038	45146	06146	55211	99429	43169	66259	97786	59180
72	36444	93600	65350	14971	25325	00427	52073	64280	18847	24768	10127	46900	64984	75348	04115	33624	68774	60013	35515	62556
73	03003	87800	07391	11594	21196	00781	32550	57158	58887	73041	67995	81977	18984	64091	02785	27762	42529	97144	80407	64524
74	17540	26188	36647	78386	04558	61463	57842	90382	77019	24210	26304	80217	84934	82657	69291	35397	98714	35104	08187	48109
75	38916	55809	47982	41968	69760	79422	80154	91486	19180	15100	81994	41070	56642	64091	31229	02595	13513	45148	78722	30144
76	64288	19843	69122	42502	48508	28820	59933	72998	99942	10515	59537	34662	79631	89403	65212	09975	06118	86197	58208	16162
77	86809	51564	38040	39418	49915	19000	58050	16899	79952	57849	51228	10937	62396	81460	47331	91403	95007	06047	16846	64809
78	99800	99566	14742	05028	30033	94889	53381	23656	75787	59223	31089	37995	29577	07828	42272	54016	21950	86192	99046	84864
79	92345	31890	95712	08279	91794	94068	49337	88674	35355	12267	38207	97938	93459	75174	79460	55436	57206	87644	21296	43393
80	90363	65162	32245	82279	79256	80834	06088	99462	56705	06118	88666	31142	09474	89712	63153	62333	42212	06140	42594	43671
81	64437	32242	48431	04835	39070	59702	31508	60935	22390	52246	53365	56134	67582	92557	89520	33452	05134	70628	27612	33738
82	91714	53662	28373	34333	55791	74758	51144	18827	10704	76803	89807	74530	38004	90102	11693	90257	05500	79920	62700	43325
83	20902	17646	31391	31459	33315	03444	55743	74701	58851	27427	18682	81038	85662	90915	91631	22223	91588	80774	07716	12548
84	12217	86007	70371	52281	14510	76094	96579	54853	78339	20839	63571	32579	63942	25371	09234	94592	98475	76884	37635	33608
85	45177	02863	42307	53571	22532	74921	17735	42201	80540	54721	68927	56492	67799	95398	77642	54913	91583	08421	81450	76229
86	28325	90814	08804	52746	47913	54577	47525	77705	95330	21866	56401	63186	39389	88798	31356	89235	97036	32341	33292	73757
87	29019	28776	56116	54791	64604	08815	46049	71186	34650	14994	24333	95603	02359	72942	46287	95382	08452	62862	97869	71775
88	84979	81353	56219	67062	26146	82567	33122	14124	46240	92973	17025	84202	95199	62272	06366	16175	97577	99304	41587	03686
89	50371	26347	48513	63915	11158	25563	91915	18431	92978	11591	02804	08253	52133	20224	68034	50865	57868	22343	55111	03607

From G. W. Snedecor and W. G. Cochran, *Statistical Methods*, Eighth Edition, © 1989 by Iowa State University Press, Ames, Iowa. Adapted and reprinted by permission (from Table A1).

Chart I. Power Functions of the Two-Sided Student's t Test

This chart shows graphs of power functions of the two-sided Student's t test. There are two graphs of power functions corresponding to two levels of significance $\alpha = 0.01$ and 0.05. Since the distribution of t is symmetrical about zero, the two-tailed levels of significance also represent one-tailed values of 0.005 and 0.025. For each graph, curves are drawn for values of $df = 1, 2, 3, 4, 6, 12, 24,$ and ∞, the degrees of freedom associated with the variance estimate. The horizontal scale of the graph represents the noncentrality parameter δ and the vertical scale the corresponding power. For example, to test the hypothesis $H_0: \mu = \mu_0$ against the alternative $H_1: \mu = \mu_1$, the statistic $t = (\bar{X} - \mu_0)/(S/\sqrt{n})$ is used. The distribution of t when H_0 is true is the t distribution with $df = n - 1$ degrees of freedom, and the critical region $t > t[df, 1 - \alpha]$ would have a significance level α. Now, under H_1, the distribution of $t(\delta) = [(\bar{X} - \mu_1) + \delta\sigma/\sqrt{n}]/(S/\sqrt{n})$ is noncentral t with the noncentrality parameter $\delta = (\mu_1 - \mu_0)/(\sigma/\sqrt{n})$. The power of the test is given by $P\{t(\delta) > t[df, 1 - \alpha]\}$. For example, for a two-tailed test with $\alpha = 0.05$, $df = 6$, and $\delta = 3$, the corresponding power read from the graph is approximately 0.70.

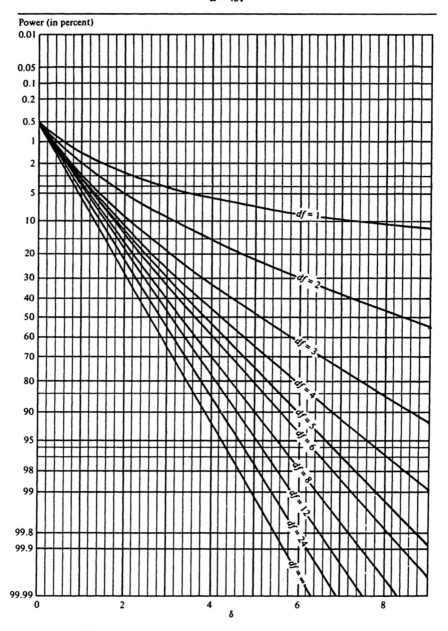

Statistical Tables and Charts

Chart I (*continued*)

$\alpha = .05$

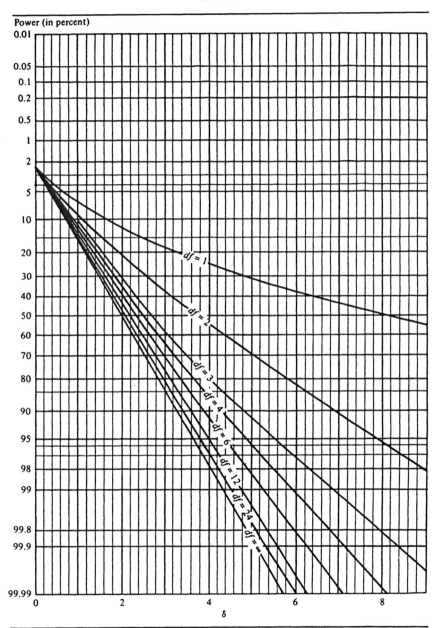

From D. B. Owen, *Handbook of Statistical Tables*, © 1962 by Addison-Wesley. Reprinted by permission.

Chart II. Power Functions of the Analysis of Variance F Tests (Fixed Effects Model): Pearson-Hartley Charts

This chart shows a set of graphs of power functions of the analysis of variance F tests in a fixed effects analysis of variance model. There are eight graphs for eight values of the numerator degrees of freedom $v_1 = 1\ (1)\ 8$. For each value of v_1, there are several values of the denominator degrees of freedom $v_2 = 6\ (1)\ 10,\ 12,\ 15,\ 20,\ 30,\ 60,\ \infty$. Each graph depicts two groups of power functions corresponding to two levels of significance, $\alpha = 0.01$ and 0.05. The horizontal scale of the graph represents the normalized noncentrality parameter (ϕ) and the vertical scale the corresponding power $(1 - \beta)$. There are two x-scales depending on which level of significance is employed. For example, for $v_1 = 2,\ v_2 = 6,\ \alpha = 0.05$, and $\phi = 2$, the corresponding power read from the graph is approximately $1 - \beta = 0.66$.

Statistical Tables and Charts

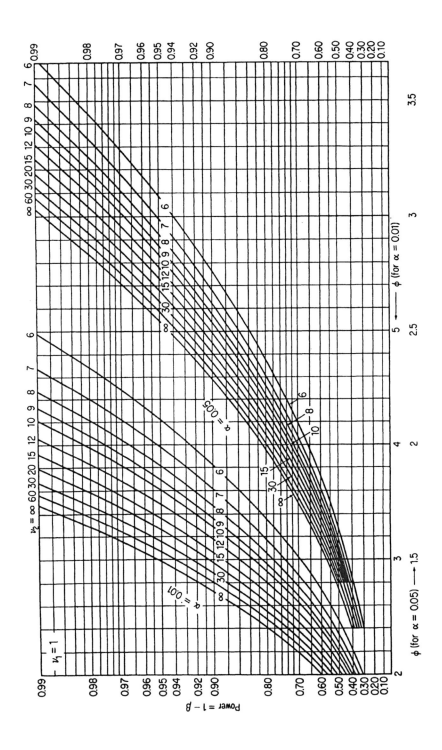

Chart II (*continued*)

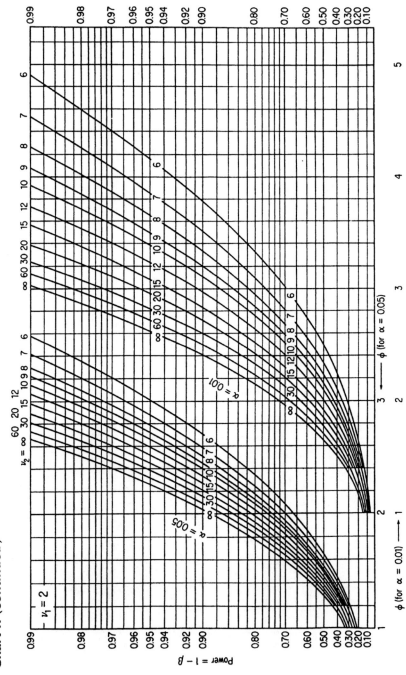

Statistical Tables and Charts

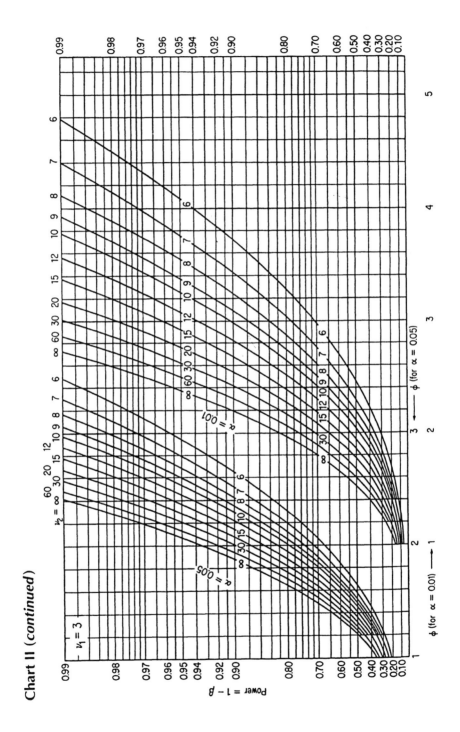

Chart II (continued)

Chart II (continued)

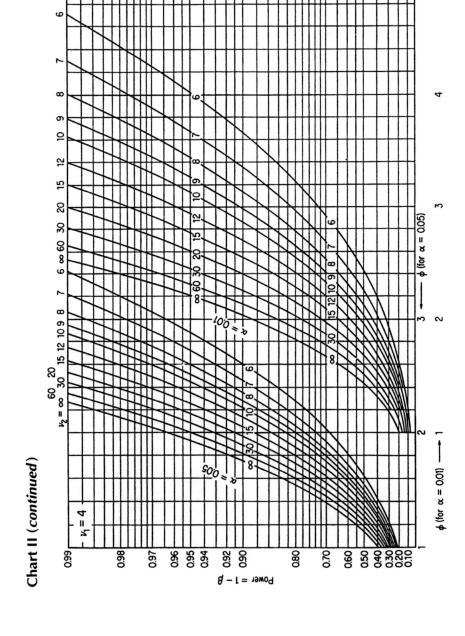

Chart II (*continued*)

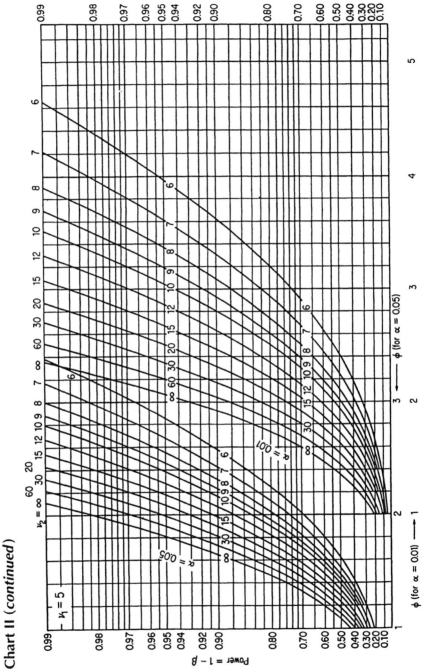

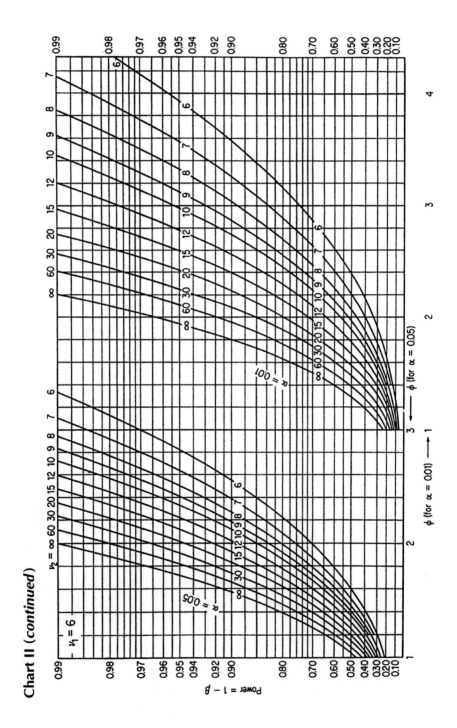

Chart II (continued)

Statistical Tables and Charts

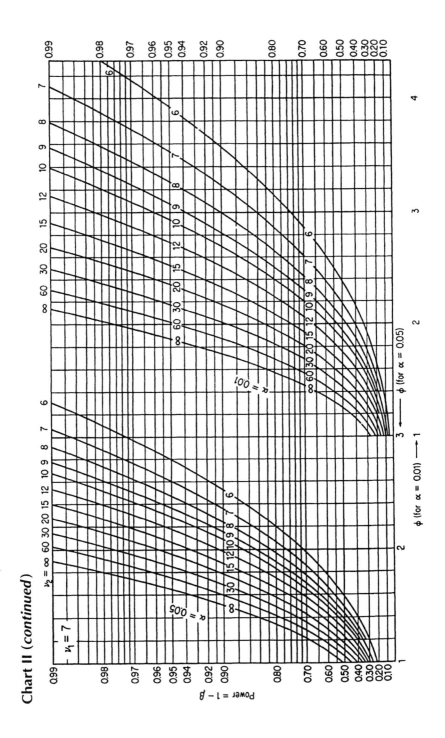

Chart II (continued)

Chart II (continued)

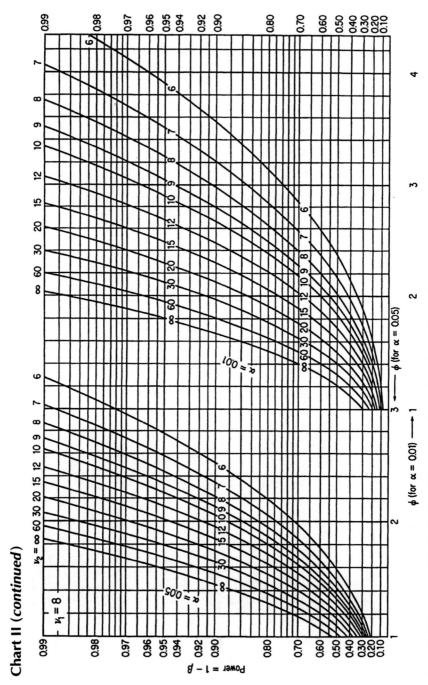

From E. S. Pearson and H. O. Hartley, "Charts of the Power Function of the Analysis of Variance Tests, Derived from the Non-central F-Distribution," *Biometrika*, 38 (1951), 112–130. Reprinted by permission.

Chart III. Operating Characteristic Curves for the Analysis of Variance F Tests (Random Effects Model)

This chart shows graphs of curves giving 1 − power for the analysis of variance F tests in a random effects analysis of variance model. There are eight graphs for eight values of the number of degrees of freedom $v_1 = 1\,(1)\,8$. For each value of v_1, there are several values of the denominator degrees of freedom $v_2 = 6\,(1)$ 10, 12, 15, 20, 30, 60, ∞. Each graph depicts two groups of operating characteristic curves corresponding to two levels of significance $\alpha = 0.01$ and 0.05. The horizontal scale of each graph represents the parameter $\lambda = \sqrt{1/(1 + n\sigma_\alpha^2/\sigma_e^2)}$ and the vertical scale the corresponding probability of accepting the hypothesis (1 − power). There are two x-scales depending on which level of significance is employed. For example, for $v_1 = 2$, $v_2 = 6$, $\alpha = 0.01$, and $\lambda = 7$, the corresponding power read from the graph is approximately $1 - 0.20 = 0.80$

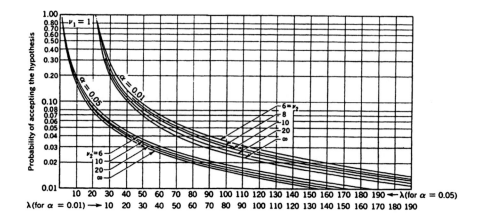
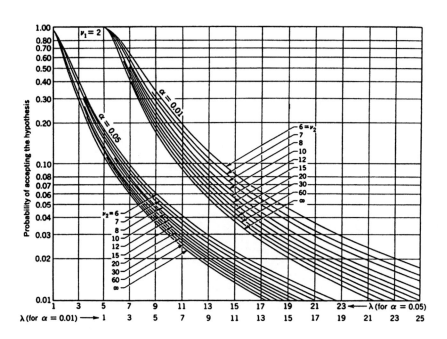

Chart III (*continued*)

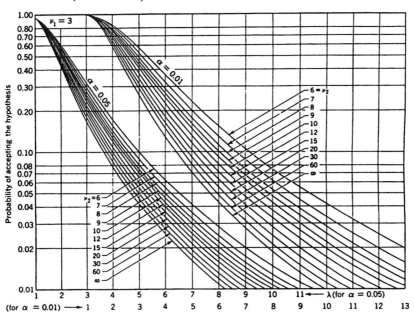

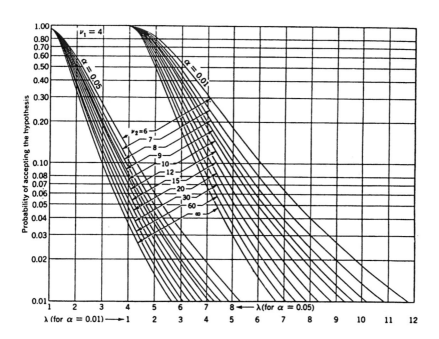

Chart III (*continued*)

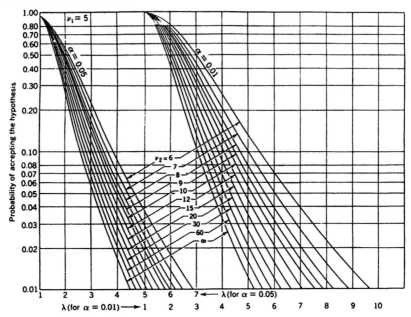

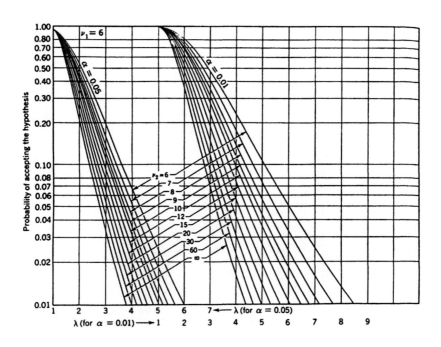

Chart III (*continued*)

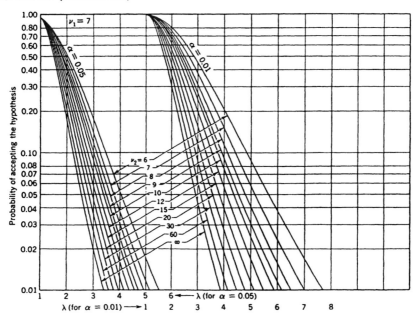

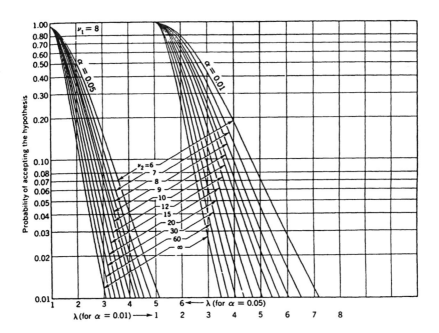

From A. H. Bowker and G. J. Lieberman, *Engineering Statistics*, 2nd ed., ©1972 by Prentice-Hall, Englewood Cliffs, New Jersey. Reprinted by permission.

Chart IV. Curves of Constant Power for Determination of Sample Size in a One-Way Analysis of Variance (Fixed Effects Model): Feldt-Mahmoud Charts

This chart shows graphs of curves of constant power for the analysis of variance F tests in a fixed effects analysis of variance model. The graphs give the values of n (y-scale) as a function of $\phi' = \frac{1}{\sigma_e}\sqrt{\sum_{i=1}^{r} \alpha_i^2/r}$ for specified values of the number of groups r, the level of significance α, and the power $P(1-\beta)$. Each graph depicts two groups of curves corresponding to two levels of significance, $\alpha = 0.05$ and 0.01. The graphs are given for $r = 2, 3, 4, 5$; and for each value of r, the values of P used in drawing the curves are equal to $0.5, 0.7, 0.8, 0.9,$ and 0.95. There are two x-scales depending on which level of significance is employed. For a given set of values of $r, \alpha, P,$ and ϕ', the sample size n may be read from the ordinate of the graph. For example, for $r = 3, \alpha = 0.05, P = 0.7,$ and $\phi' = 0.3$, the value of n read from the chart is approximately equal to 29.

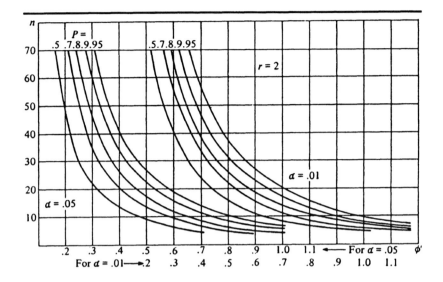

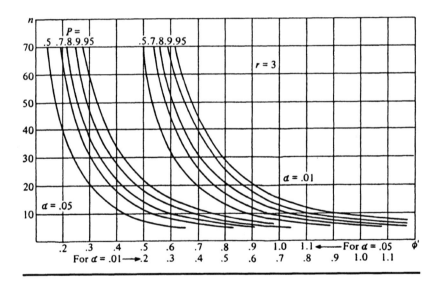

Chart IV (*continued*)

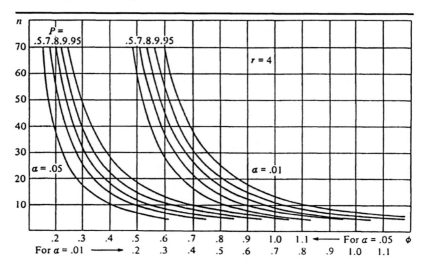

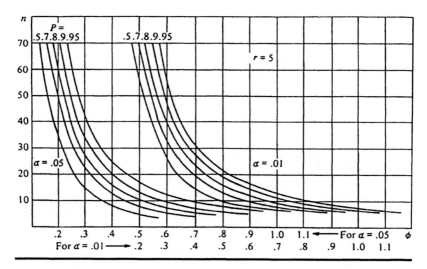

From L. S. Feldt and M. W. Mahmoud, "Power Function Charts for Specifying Number of Observations in Analysis of Variance of Fixed Effects," *Annals of Mathematical Statistics*, 29 (1958), 871–877. Reprinted by permission.

References

Abraham, J. K. (1960). Note 154: On an alternative method of computing Tukey's statistic for the Latin square model. *Biometrics*, 16, 686–691.

Afifi, A. A. and Elashoff, R. M. (1966). Missing observations in multivariate statistics, I. Review of the literature. *J. Amer. Stat. Assoc.*, 61, 595–604.

Afifi, A. A. and Elashoff, R. M. (1967). Missing observations in multivariate statistics, II. Point estimation in simple linear regression. *J. Amer. Stat. Assoc.*, 62, 10–29.

Akutowicz, F. and Traux, H. M. (1956). Establishing control of tire cord testing laboratories. *Indus. Qual. Contr.*, 13 (2), 4–5.

Alexander, R. A. and Govern, D. M. (1994). A new and simple approximation for ANOVA under variance heterogeneity. *J. Educ. Stat.*, 19, 91–101.

Algina, J., Blair, R. C., and Coombs, W. T. (1995). A maximum test for scale: Type I error rates and power. *J. Educ. Beh. Stat.*, 20, 27–39.

Allen, R. E. and Wishart, J. (1930). A method of estimating the yield of a missing plot in field experimental work . *J. Agri. Sci.*, 20, 399–406.

Anderson, R. L. (1946). Missing-plot techniques. *Biom. Bull.*, 2, 41–47.

Anderson, R. L. (1954). Components of variance and mixed models. In: *Quality Control Convention Papers: Proceedings Eighth Annual Convention*, pp. 633–645. American Society for Quality Control, Milwaukee, Wisconsin.

Anderson, R. L. (1960). Use of variance component analysis in the interpretation of biological experiments. *Bull. Inter. Stat. Inst.*, 37, 1–22.

Anderson, R. L. and Bancroft, T. A. (1952). *Statistical Theory in Research*. McGraw-Hill, New York.

Anderson, R. L. and Houseman, E. E. (1942). Tables of orthogonal polynomial values extended to $N = 104$. *Res. Bull.*, 297, Ames, Iowa.

Anderson, T. W. (1984). *An Introduction to Multivariate Statistical Analysis*, 2nd ed. John Wiley, New York. (1st ed., 1958.)

Anderson, T. W. (1985). Components of variance in MANOVA. In: *Multivariate Analysis IV*, pp. 1–8 (Ed. P. R. Krishnaiah). North-Holland, Amsterdam.

Anderson, V. L. and McLean, R. A. (1974). *Design of Experiments: A Realistic Approach*. Marcel Dekker, New York.

Andrews, D. M. and Herzberg, A. (1985). *Data: A Collection of Problems from Many Fields for the Students and Research Workers*. Springer-Verlag, New York.

Anionwu, E., Watford, D., Brozovic, M., and Kirkwood, B. (1981). Sickle cell disease in a British urban community. *Brit. Med. J.*, 282, 283–286.

Anscombe, F. J. (1948). The transformation of Poisson, binomial and negative binomial data. *Biometrika*, 35, 246–254.

Armitage, J. V. and Krishnaiah, P. R. (1964). *Tables for the Studentized Largest Chi-square Distribution and their Applications*. Tech. Rep. No. ARL 64-188, Aerospace Research Laboratories, Wright-Patterson Air Force Base, Dayton, Ohio.

Arteaga, C., Jeyaratnanam, S., and Graybill, F. A. (1982). Confidence intervals for proportions of total variance in the two-way cross component of variance model. *Commun. Stat., A: Theo. & Meth.*, 11, 1643–1658.
Arvesen, J. N. and Layard, M. W. J. (1975). Asymptotically robust tests in unbalanced variance component models. *Ann. Stat.*, 3, 1122–1134.
Arvesen, J. N. and Schmitz, T. H. (1970). Robust procedures for variance component problems using the jackknife. *Biometrics*, 26, 677–686.
Aster, R. (1994). *SAS Foundations: From Installation to Operation.* McGraw-Hill, New York.
Atkinson, A. C. (1982). Developments in the design of experiments. *Inter. Stat. Rev.*, 50, 161–177.
Bagui, S. C. (1993). *CRC Handbook of Percentiles of Noncentral t-Distribution.* CRC Press, Boca Raton, Florida.
Bailey, B. J. R. (1977). Tables of the Bonferroni t statistic. *J. Amer. Stat. Assoc.*, 72, 469–478.
Bancroft, T. A. (1968). *Topics in Intermediate Statistical Methods*, Vol. I. Iowa State University Press, Ames, Iowa.
Bankier, J. D. (1960a). Operators and the r-way crossed classification. *Amer. Math. Month.*, 67, 841–846.
Bankier, J. D. (1960b). An operational approach to the r-way crossed classification. *Ann. Math. Stat.*, 31, 16–22.
Barnett, V. D. (1962). Large sample tables of percentage points for Hartley's correction to Bartlett's criterion for testing the homogeneity of a set of variances. *Biometrika*, 49, 487–494.
Barnett, V. D. and Lewis, T. (1994). *Outliers in Statistical Data*, 3rd ed. John Wiley, New York.
Bartlett, M. S. (1936). The square-root transformation in the analysis of variance. *J. R. Stat. Soc., Suppl.*, 3, 68–78.
Bartlett, M. S. (1937a). Properties of sufficiency and statistical tests. *Proc. R. Soc. London, Ser. A*, 160, 268–282.
Bartlett, M. S. (1937b). Some examples of statistical methods of research in agriculture and applied biology. *J. R. Stat. Soc., Suppl.*, 4, 137–183.
Bartlett, M. S. (1947). The use of transformations. *Biometrics*, 3, 39–52.
Bartlett, M. S. and Kendal, D. G. (1946). The statistical analysis of variance: Heterogeneity and logarithmic transformation. *J. R. Stat. Soc., Ser. B*, 8, 128–150.
Beall, G. (1942). The transformation of data from entomological field experiments so that the analysis of variance becomes applicable. *Biometrika*, 29, 243–262.
Bechhoefer, R. E. and Dunnett, C. W. (1981). *Multiple Comparisons for Orthogonal Contrasts: Examples and Tables.* Tech. Rep. No. 495, School of Operations Research and Industrial Engineering, Cornell University, Ithaca, New York.
Bechhoefer, R. E. and Dunnett, C. W. (1982). Comparisons for orthogonal contrasts: Examples and tables. *Technom.*, 24, 213–222.
Beckman, R. J. and Cook, R. D. (1983). Outlier..........s. *Technom.*, 25, 119–149.
Bennett, C. A. and Franklin, N. L. (1954). *Statistical Analysis in Chemistry and the Chemical Industry.* John Wiley, New York.
Berry, D. A. (1987). Logarithmic transformations in ANOVA. *Biometrics*, 43, 439–456.
Birch, N. J., Burdick, R. K., and Ting, N. (1990). Confidence intervals and bounds for a ratio of summed expected mean squares. *Technom.*, 32, 437–444.

Bishop, D. J. and Nair, U. S. (1939). A note on certain methods of testing for the homogeneity of a set of estimated variances. *J. R.. Stat. Soc., Suppl.*, 6, 89–99.

Bishop, T. A. and Dudewicz, E. J. (1978). Exact analysis of variance with unequal variances: Test procedures and tables. *Technom.*, 20, 419–424.

Blackwell, T., Brown, C., and Mosteller, F. (1991). Which denominator? In: *Fundamentals of Exploratory Analysis of Variance*, pp. 252–294 (Eds. D. C. Hoaglin, F. Mosteller, and J. W. Tukey). John Wiley, New York.

Blischke, W. R. (1966). Variances of estimates of variance components in a three-way classification. Biometrics, 22, 553–565.

Blischke, W. R. (1968). Variances of moment estimators of variance components in the unbalanced r-way classification. Biometrics, 24, 527–540.

Bliss, C. (1967). *Statistics in Biology*, Vol. 1. McGraw-Hill, New York.

Boardman, T. J. (1974). Confidence intervals for variance components-A comparative Monte Carlo study. *Biometrics*, 30, 251–262.

Bock, R. D. (1963). Programming univariate and multivariate analysis of variances. *Technom.*, 5, 95–117.

Bolk, R. J. (1993). Testing additivity in two-way classifications with no replications: The locally best invariant test. *J. Appl. Stat.*, 20, 41–55.

Boneau, C. A. (1960). The effects of violation of assumptions underlying the t-test. *Psychol. Bull.*, 57, 49–64.

Boneau, C. A. (1962). A comparison of the power of the U and t tests. *Psychol. Rev.*, 59, 246–256.

Bose, R. C. and Shrikhande, S. S. (1959). On the falsity of Euler's conjecture about the nonexistence of two orthogonal Latin squares of order $4t + 2$. *Proc. Nat. Acad. Sci.*, 45, 734–737.

Bowker, A. H. and Lieberman, G. J. (1972). *Engineering Statistics*, 2nd ed. Prentice Hall, Englewood Cliffs, New Jersey.

Bowman, K. O. (1972). Tables of the sample size requriments. *Biometrika*, 59, 234.

Bowman, K. O. and Kastenbaum, M. (1975). Sample size determination in single and double classification experiments. In: *Selected Tables in Mathematical Statistics*, Vol. 3, pp. 1–23 (Eds. H. L. Harter and D. B. Owen). American Mathematical Society, Providence, Rhode Island.

Box, G. E. P. (1953). Nonnormality and tests on variances. *Biometrika*, 40, 318–335.

Box, G. E. P. (1954a). Some theorems on quadratic forms applied in the study of analysis of variance problems, I. Effect of inequality of variance in the one-way classification. *Ann. Math. Stat.*, 25, 290–302.

Box, G. E. P. (1954b). Some theorems on quadratic forms applied in the study of analysis of variance problems, II. Effect of inequality of variance and of correlation errors in the two-way classification. *Ann. Math. Stat.*, 25, 484–498.

Box, G. E. P. and Anderson, S. L. (1955). Permutation theory in the derivation of robust criteria and the study of departures from assumption. *J. R. Stat. Soc., Ser. B*, 17, 1–26.

Box, G. E. P. and Cox, D. R. (1964). An analysis of transformations. *J. R. Stat. Soc., Ser. B*, 26, 211–243.

Box, G. E. P. and Cox, D. R. (1982). An analysis of transformations revisited, rebutted. *J. Amer. Stat. Assoc.*, 77, 209–210.

Box, G. E. P. and Draper, N. (1987). *Empirical Model Building and Response Surfaces*. John Wiley, New York.

Box, G. E. P., Hunter, W. G., and Hunter, J. S. (1978). *Statistics for Experimenters*. John Wiley, New York.

Box, G. E. P. and Tiao, G. C. (1973). *Bayesian Inference in Statistical Analysis*. Addison-Wesley, Reading, Massachusetts. (Wiley Classic Edition, 1992.)

Bozivich, H., Bancroft, T. A., and Hartley, H. O. (1956). Power of analysis of variance test procedures for certain incompletely specified models. *Ann. Math. Stat.*, 27, 1017–1043.

Bradley, J. V. (1964). *Studies in Research Methodology, VI. The Central Limit Effect for a Variety of Populations and the Robustness of Z, t, and F*. Tech. Rep. No. AMRL-54-123, Aerospace Medical Research Laboratories, Wright-Patterson Air Force Base, Dayton, Ohio.

Bratcher, T. L., Moran M. A., and Zimmer, W. J. (1970). Tables of sample sizes in the analysis of variance. *J. Qual. Tech.*, 2, 156–164.

Broemeling, L. D. (1985). *Bayesian Analysis of Linear Models*. Marcel Dekker, New York.

Brown, M. B. and Forsythe, A. B. (1974a). The small size sample behavior of some statistics which test the equality of several means. *Technom.*, 16, 129–132.

Brown, M. B. and Forsythe, A. B. (1974b). The ANOVA and multiple comparisons for data with heterogeneous variances. *Biometrics*, 30, 719–724.

Brown, M. B. and Forsythe, A. B. (1974c). Robust tests for the equality of variances. *J. Amer. Stat. Assoc.*, 69, 364–367.

Brown, R. A. (1974). Robustness of the Studentized range statistic. *Biometrika*, 61, 171–175.

Brownlee, K. A. (1953). *Industrial Experimentation*. Chemical Publishing Co., New York.

Brownlee, K. A. (1965). *Statistical Theory and Methodology in Science and Engineering*, 2nd ed. John Wiley, New York. (1st ed. 1960.)

Budescu, D. V. and Applebaum, M. I. (1981). Variance stabilizing transformations and the power of the F test. *J. Educ. Stat.*, 6, 55–74.

Bulgren, W. G. (1974). Probability integral of the doubly noncentral t distribution with degrees of freedom n and noncentrality parameters δ and λ. In: *Selected Tables in Mathematical Statistics*. Vol. 2, pp. 1–138 (Eds. H. L. Harter and D. B. Owen). American Mathematical Society, Providence, Rhode Island.

Bulmer, M. G. (1957). Approximate confidence limits for components of variance. *Biometrika*, 44, 159–167.

Burch, L. and King, S. J. (1994). *SAS Software Roadmaps: Your Guide to Discovering the SAS System*. SAS Institute, Cary, North Carolina.

Burdick, R. K. (1994). Using confidence intervals to test variance components. *J. Qual. Tech.*, 26, 30–38.

Burdick, R. K., Birch, N. J., and Graybill, F. A. (1986a). Confidence intervals on measures of variability in an unbalanced two-fold nested design with equal subsampling. *J. Stat. Comp. Simul.*, 25, 259–272.

Burdick, R. K. and Eickman, J. (1986). Confidence intervals on the among group variance component in the unbalanced one-fold nested design. *J. Stat. Comp. Simul.*, 26, 205–219.

Burdick, R. K. and Graybill, F. A. (1984). Confidence intervals on linear combinations of variance components in the unbalanced one-way classification. *Technom.*, 26, 131–136.

Burdick, R. K. and Graybill, F. A. (1985). Confidence intervals on the total variance in an unbalanced two-fold nested classification with equal sub-sampling. *Commun. Stat., A: Theo. & Meth.*, 14, 761–774.

Burdick, R. K. and Graybill, F. A. (1988). The present status of confidence interval estimation on variance components in balanced and unbalanced random models. *Commun. Stat., A: Theo. & Meth.*, 17, 1165–1195.

Burdick, R. K. and Graybill, F. A. (1992). *Confidence Intervals on Variance Components*. Marcel Dekker, New York.

Burdick, R. K., Maqsood, F., and Graybill, F. A. (1986b). Confidence intervals on the intraclass correlation in the unbalanced one-way classification. *Commun. Stat., A: Theo. & Meth.*, 15, 3353–3378.

Cameron, J. M. (1951). Use of components of variance in preparing schedules for the sampling of baled wool. *Biometrics*, 7, 83–96.

Chambers, J. M., Cleveland, W. S., Kleiner, B., and Tukey, P. A. (1983). *Graphical Methods for Data Analysis*. Wadsworth, Pacific Grove, California.

Chao, M.-T. and Glaser, R. E. (1978). The exact distribution of Bartlett's test statistic for homogeneity of variances with unequal sample sizes. *J. Amer. Stat. Assoc.*, 73, 422–426.

Chew, V. (1976a). *Comparisons Among Treatment Means in Analysis of Variance*. Tech. Bull., Agricultural Research Service, U.S. Department of Agriculture, Washington, D.C.

Chew, V. (1976b). Uses and abuses of Duncan's multiple range test. *Hort. Sci.*, 11, 251–253.

Chew, V. (1976c). Comparing treatment means: A compendium. *Hort. Sci.*, 11, 348–357.

Christensen, R. (1996). *Plane Answers to Complex Questions: The Theory of Linear Models*, 2nd ed. Springer-Verlag, New York.

Cleveland, W. S. (1985). *The Elements of Graphing Data*. Wadsworth. Pacific Grove, California.

Clinch, J. J. and Keselman, H. J. (1982). Parametric alternatives to the analysis of variance. *J. Educ. Stat.*, 7, 207–214.

Coakes, S. J. and Steed, L. G. (1997). *SPSS: Analysis without Anguish*. John Wiley, Chichester.

Cochran, W. G. (1937). Catalogue of uniformity trial data. *J. R. Stat. Soc., Suppl.*, 4, 233–253.

Cochran, W. G. (1940). The analysis of variance when experimental errors follow the Poisson or binomial laws. *Ann. Math. Stat.*, 11, 335–347.

Cochran, W. G. (1941). The distribution of the largest of a set of estimated variances as a fraction of their total. *Ann. Eug.*, 11, 47–52.

Cochran, W. G. (1951). Testing a linear relation among variances. *Biometrics*, 7, 17–32.

Cochran, W. G. (1954). Some methods for strengthening the common χ^2 tests. *Biometrics*, 10, 417–451.

Cochran, W. G. (1957). Analysis of covariance: Its nature and uses. *Biometrics*, 13, 261–281.

Cochran, W. G. (1964). Approximate significance levels of the Behrens-Fisher test. *Biometrics*, 20, 191–195.

Cochran, W. G. and Cox, G. M. (1957). *Experimental Designs*, 2nd ed. John Wiley, New York.

Cody, R. P. and Smith, J. K. (1997). *Applied Statistics and the SAS Programming Language*, 4th ed. SAS Institute, Cary, North Carolina.

Cohen, A. and Strawderman, W. E. (1971). Unbiasedness of tests for homogeneity of variances. *Ann. Math. Stat.*, 42, 355–360.
Cohen, J. (1988). *Statistical Power Analysis for the Behavioral Sciences*, 2nd ed. Lawrence Erlbaum, Hillsdale, New Jersey.
Collyer, C. E. and Enns, J. T. (1987). *Analysis of Variance: The Basic Designs.* Nelson Hall, Chicago.
Conover, W. J. (1971). *Practical Nonparametric Statistics.* John Wiley, New York.
Conover, W. J., Johnson, M. E., and Johnson, M. M. (1981). A comparative study of tests for homogeneity of variances with applications to outer continental shelf bidding data. *Technom.*, 23, 351–361. (Corrigendum *ibid.*, 26, 302.)
Cooley, W. W. and Lohnes, P. R. (1962). *Multivariate Procedures for the Behavioral Sciences.* John Wiley, New York.
Coombs, W. T., Algina, J., and Oltman, D. O. (1996). Univariate and multivariate omnibus hypothesis tests selected to control type I error rates when population variances are not necessarily equal. *Rev. Educ. Res.*, 66, 137–179.
Cornfield, J. and Tukey, J. W. (1956). Average values of mean squares in factorials. *Ann. Math. Stat.*, 27, 907–949.
Cox, D. R. (1958a). *Planning of Experiments.* John Wiley, New York. (Wiley Classic Edition, 1992.)
Cox, D. R. (1958b). The interpretation of the effects of non-additivity in the Latin square. *Biometrika*, 45, 69–73.
Cox, D. R. (1977). Nonlinear models, residuals and transformations. *Math. Operationsforsch. Stat., Ser. Stat.*, 8, 3–22.
Cox, D. R. (1984). Interaction. *Inter. Stat. Rev.*, 52, 1–31.
Crisler, L. (1991). *Computer Based Data Analysis: Using SPSS-X in the Social and Behavioral Sciences.* Nelson-Hall, Chicago.
Crowder, M. J. and Hand, D. J. (1990). *Analysis of Repeated Measures.* Chapman & Hall, London.
Crump, S. L. (1946). The estimation of variance components in analysis of variance. *Biom. Bull.*, 2, 7–11.
Crump, S. L. (1951). The present status of variance component analysis. *Biometrics*, 7, 1–16.
Cummings, W. B. and Gaylor, D. W. (1974). Variance component testing in unbalanced nested designs. *J. Amer. Stat. Assoc.*, 69, 765–771.
Curtiss, J. H. (1943). On transformations used in the analysis of variance. *Ann. Math. Stat.*, 14, 107–122.
D'Agostino, R. B. (1971). An omnibus test for normality for moderate and large samples. *Biometrika*, 58, 341–348.
D'Agostino, R. B. (1972). Small sample probability points for the D test of normality. *Biometrika*, 59, 219–221.
Damon, R. A., Jr. and Harvey, W. R. (1987). *Experimental Design, ANOVA, and Regression.* Harper & Row, New York.
Daniel, W. W. (1990). *Applied Nonparametric Statistics*, 2nd ed. Brooks/Cole, Belmont, California. (1st ed., 1978.)
Daniels, H. E. (1939). The estimation of components of variance. *J. R. Stat. Soc., Suppl.*, 6, 186–197.
Das, M. N. and Giri, N. C. (1979). *Design and Analysis of Experiments.* John Wiley (Eastern), New Delhi.
Davenport, J. M. (1975). Two methods of estimating the degrees of freedom of an approximate F. *Biometrika*, 62, 682–684.

Davenport, J. M. and Webester, J. T. (1973). A comparison of some approximate F-tests. *Technom.*, 15, 779–789.
David, H. A. (1952). Upper 5 and 1% points of the maximum F-ratio. *Biometrika*, 39, 422–424.
Davies, O. L. (Ed.) (1954, 1956, 1960). *The Design and Analysis of Industrial Experiments*. 1st, 2nd, & 3rd eds. Oliver and Boyd, Edinburgh.
Davies, O. L. and Goldsmith, P. L. (Eds.) (1972). *Statistical Methods in Research and Production*, 4th ed. Oliver and Boyd, Edinburgh.
Day, S. J. and Graham, D. F. (1991). Sample size estimation for comparing two or more treatment groups in clinical trials. *Stat. Med.*, 10, 33–43.
DeLury, D. B. (1946). The analysis of Latin squares when some observations are missing. *J. Amer. Stat. Assoc.*, 41, 370–389.
Dempster, A. P., Lord, N. M., and Rubin, D. B. (1977). Maximum likelihood from incomplete data via the E. M. algorithm (with discussion). *J. R. Stat. Soc., Ser. B*, 39, 1–38.
Dénes, J. and Keedwell, A. D. (1974). *Latin Squares and Their Applications*. Akadémia Kiad. English Universities Press, Budapest/Academic Press, London.
Desmond, D. J. (1954). Quality control on the setting of voltage regulators. *Appl. Stat.*, 3, 9–15.
DiIorio, F. C. (1991). *SAS Applications Programming: A Gentle Introduction*. PWS-KENT, Boston.
DiIorio, F. C. and Hardy, K. A. (1996). *Quick Start to Data Analysis with SAS*. Duxbury Press, Belmont, California.
Dijkstra, J. B. and Werter, S. P. J. (1981). Testing the equality of several means when the population variances are unequal. *Commun. Stat., B: Simul. & Comp.*, 10, 557–569.
Dixon, W. J. (Ed.) (1992). *BMDP Statistical Software Manual*, Vols. 1, 2, & 3. University of California Press, Los Angeles.
Dobson, A. J. (1990). *An Introduction to Generalized Linear Models*. Chapman & Hall, London.
Dodge, Y. (1985). *Analysis of Experiments with Missing Data*. John Wiley, New York.
Dodge, Y. and Shah, K. R. (1977). Estimation of parameters in Latin squares and Graeco-Latin squares with missing observations. *Commun. Stat., A: Theo. & Meth.*, 6, 1465–1472.
Donaldson, T. S. (1968). Robustness of the F-test to errors of both kinds and the correlation between the numerator and denominator of the F-ratio. *J. Amer. Stat. Assoc.*, 63, 660–676.
Donner, A. (1986). A review of inference procedures for the intraclass correlation coefficient in the one-way random effects model. *Inter. Stat. Rev.*, 54, 67–82.
Donner, A. and Koval, J. J. (1989). The effect of imbalance on significance testing in one-way model II analysis of variance. *Commun. Stat., A: Theo. & Meth.*, 18, 1239–1250.
Donner, A. and Wells, G. (1986). A comparison of confidence interval methods for the intraclass correlation coefficient. *Biometrics*, 42, 401–412.
Donoghue, J. R. and Collins, L. M. (1990). A note on the unbiased estimation of the intraclass correlation. *Psychom.*, 55, 159–164.
Draper, N. R. and Hunter, W. G. (1969). Transformations: Some examples revisited. *Technom.*, 11, 23–40.
Draper, N. R. and Smith, H. (1981). *Applied Regression Analysis*, 2nd ed. John Wiley, New York. (3rd ed., 1998.)

Duncan, A. J. (1957). Charts of the 10% and 50% points of the operating characteristic curves for fixed effects analysis of variance F tests, $\alpha = .01$ and $.05$. *J. Amer. Stat. Assoc.*, 52, 345–349.

Duncan, D. B. (1952). On the properties of multiple comparison test. *Virg. J. Sci.*, 3, 49–67.

Duncan, D. B. (1955). Multiple range and multiple F-tests. *Biometrics*, 11, 1–42.

Dunlop, G. (1933). Methods of experimentation in animal nutrition. *J. Agri. Sci.*, 23, 580–614.

Dunn, O. J. (1958). Estimation of the means of dependent variables. *Ann. Math. Stat.*, 29, 1095–1111.

Dunn, O. J. (1959). Confidence intervals for the means of dependent, normally distributed variables. *J. Amer. Stat. Assoc.*, 54, 613–621.

Dunn, O. J. (1961). Multiple comparisons among means. *J. Amer. Stat. Assoc.*, 56, 52–64.

Dunn, O. J. and Clark, V. A. (1974, 1987). *Applied Statistics: Analysis of Variance and Regression*. 1st & 2nd eds. John Wiley, New York.

Dunn, O. J. and Massey, F. J. (1965). Estimation of multiple contrasts using t distribution. *J. Amer. Stat. Assoc.*, 60, 573–583.

Dunnett, C. W. (1955). A multiple comparison procedure for comparing several treatments with a control. *J. Amer. Stat. Assoc.*, 50, 1096–1121.

Dunnett, C. W. (1980a). Pairwise multiple comparisons in the homogeneous variance, unequal sample size case. *J. Amer. Stat. Assoc.*, 75, 789–795.

Dunnett, C. W. (1980b). Pairwise multiple comparisons in the unequal variance case. *J. Amer. Stat. Assoc.*, 75, 796–800.

Dunnett, C. W. (1982). Robust multiple comparisons. *Commun. Stat., A: Theo. & Meth.*, 11, 2611–2629.

Dyer, D. D. and Keating, J. P. (1980). On the determination of critical values for Bartlett's test. *J. Amer. Stat. Assoc.*, 75, 313–319.

Efron, B. (1982). Transformation theory: How normal is a family of distributions? *Ann. Stat.*, 10, 323–339.

Eisen, E. J. (1966). The quasi-F test for an unnested fixed factor in an unbalanced hierarchical design with a mixed model. *Biometrics*, 22, 937–942.

Eisenhart, C. (1947a). The assumptions underlying the analysis of variance. *Biometrics*, 3, 1–21.

Eisenhart, C. (1947b). Inverse sine transformation of proportion. In: *Selected Techniques of Statistical Analysis*, Chapter 16, pp. 395–416 (Eds. C. Eisenhart, M. W. Hastay, and W. A. Wallis). McGraw-Hill, New York.

Eisenhart, C. and Solomon, H. (1947). Significance of the largest of a set of sample estimates of variance. In: *Selected Techniques of Statistical Analysis*, Chapter 15, pp. 383–394 (Eds. C. Eisenhart, M. W. Hastay, and W. A. Wallis). McGraw-Hill, New York.

Eisenhart, C., Hastay, M. W., and Wallis, W. A. (Eds.) (1947). *Selected Techniques of Statistical Analysis*. McGraw-Hill, New York.

Elliott, R. J. (1995). *Learning SAS in the Computer Lab*. Duxbury Press, Belmont, California.

Euler, L. (1782). Recherches sur une nouvelle espéce de quarrés magiques. *Verh. Zeeu. Genoot. Wetens. Vlissen.*, 9, 85–239.

Everitt, B. and Derr, G. (1996). *A Handbook of Statistical Analyses Using SAS*. Chapman & Hall, London.

References

Federer, W. T. (1955). *Experimental Design: Theory and Applications*. MacMillan, New York.

Federer, W. T. (1980). Some recent results in experimental design with a bibliography, I. *Inter. Stat. Rev.*, 48, 337–368.

Federer, W. T. (1981a). Some recent results in experimental design with a bibliography, II: A-K. *Inter. Stat. Rev.*, 49, 95–109.

Federer, W. T. (1981b). Some recent results in experimental design with a bibliography, III: L-Z. *Inter. Stat. Rev.*, 49, 185–197.

Federer, W. T. and Balaam, L. N. (1972). *Bibliography on Experiment and Treatment Design: Pre-1968*. Oliver and Boyd, Edinburgh (for the International Statistical Institute).

Federer, W. T. and Federer, A. J. (1973). A study of design publications: 1968 through 1971. *Amer. Stat.*, 27, 160–163.

Federer, W. T. and Zelen, M. (1966). Analysis of multifactor classifications with unequal number of observations. *Biometrics*, 22, 525–552.

Feldt, L. S. and Mahmoud, M. W. (1958a). Power function charts for specification of sample size in analysis of variance. *Psychom.*, 23, 201–210.

Feldt, L. S. and Mahmoud, M. W. (1958b). Power function charts for specifying numbers of observations in analysis of variance of fixed effects. *Ann. Math. Stat.*, 29, 871–877.

Fisher, L. and McDonald, J. (1978). *Fixed Effects Analysis of Variance*. Academic Press, New York.

Fisher, R. A. (1918). The correlation between relatives on the supposition of Mendelian law on inheritance. *Trans. R. Soc., Edin.*, 52, 399–433.

Fisher, R. A. (1924). On a distribution yielding the error functions of several well-known statistics. *Proc. Inter. Math. Cong.* (*Toronto*), pp. 805–813.

Fisher, R. A. (1925). *Statistical Methods for Research Workers*, 1st ed. Oliver and Boyd, Edinburgh and London.

Fisher, R. A. (1926). The arrangements of field experiments. *J. Ministry Agri.*, 33, 503–513.

Fisher, R. A. (1932). *Statistical Methods for Research Workers*, 4th ed. Oliver and Boyd, Edinburgh.

Fisher, R. A. (1935). *The Design of Experiments*, 1st ed. Oliver & Boyd, Edinburgh & London. (7th ed., 1960; 8th ed., 1966.)

Fisher, R. A. (1958). *Statistical Methods for Research Workers*, 13th ed. Hafner, New York. (19th ed., 1997.)

Fisher, R. A. and Mackenzie, W. A. (1923). Studies in crop variation, II. The manurial response of different potato varieties. *J. Agri. Sci.*, 13, 311–320.

Fisher, R. A. and Yates, F. (1963). *Statistical Tables for Biological, Agriculture and Medical Research*, 6th ed. Hafner, New York. (4th ed., 1953; 5th ed., 1957.)

Fleiss, J. L. (1986). *The Design and Analysis of Clinical Experiments*. John Wiley, New York.

Fox, M. (1956). Charts of the power of the F-test. *Ann. Math. Stat.*, 27, 484–497.

Freeman, M. F. and Tukey, J. W. (1950). Transformations related to the angular and the square root. *Ann. Math. Stat.*, 21, 607–611.

Freund, R. J. and Littell, R. C. (1991). *SAS System for Regression*, 2nd ed. SAS Institute, Cary, North Carolina.

Friendly, M. (1991). *SAS System for Statistical Graphics*. SAS Institute, Cary, North Carolina.

Frude, N. (1993). *A Guide to SPSS-PC+*. Springer-Verlag, New York.
Gabriel, K. R. (1964). Procedure for testing the homogeneity of all sets of means in analysis of variance. *Biometrics*, 20, 459–477.
Gallo, J. and Khuri, A. I. (1990). Exact tests for the random and fixed effects in an unbalanced mixed two-way cross-classification model. *Biometrics*, 46, 1087–1095.
Games, P. A. (1977). An improved t table for simultaneous control on g contrasts. *J. Amer. Stat. Assoc.*, 72, 531–534.
Games, P. A. and Howell, J. F. (1976). Pairwise multiple comparison procedures with unequal n's and/or variances. *J. Educ. Stat.*, 1, 113–125.
Games, P. A., Winkler, H. B., and Probert, D. A. (1972). Robust tests for homogeneity of variance. *Educ. Psychol. Meas.*, 32, 887–909.
Ganguli, M. (1941). A note on nested sampling. *Sankhyā*, 5, 449–452.
Gartside, P. S. (1972). A study of methods for comparing several variances. *J. Amer. Stat. Assoc.*, 67, 342–346.
Gates, C. E. and Shiue, C. (1962). The analysis of variance of the s-stage hierarchical classification. *Biometrics*, 25, 427–430.
Gayen, A. K. (1950). The distribution of the variance ratio in random samples of any size drawn from non-normal universes. *Biometrika*, 37, 236–255.
Gaylor, D. W. and Hartwell, T. D. (1969). Expected mean squares for nested classifications. *Biometrics*, 25, 427–430.
Gaylor, D. W. and Hopper, F. N. (1969). Estimating the degrees of freedom for linear combinations of mean squares by Satterthwaite's formula. *Technom.*, 11, 691–706.
Geary, R. C. (1935). The ratio of the mean deviation to the standard deviation as a (new) test of normality. *Biometrika*, 27, 310–332.
Geary, R. C. (1936). Moments of the ratio of the mean deviation to the standard deviation for normal samples. *Biometrika*, 28, 298–305.
Geary, R. C. (1947). Testing for normality. *Biometrika*, 34, 209–242.
Ghosh, M. N. and Sharma, D. (1963). Power of Tukey's test for non-additivity. *J. R. Stat. Soc., Ser. B*, 25, 213–219.
Gibbons, J. D. and Pratt, J. W. (1975). P-values: Interpretations and methodology. *Amer. Stat.*, 29, 20–25.
Gill, J. L. (1978). *Design and Analysis of Experiments in the Animal and Medical Sciences*, Vols. 1, 2, & 3. Iowa State University Press, Ames, Iowa.
Glaser, R. E. (1976). Exact critical values for Bartlett's test for homogeneity of variances. *J. Amer. Stat. Assoc.*, 71, 488–490.
Glass, G. V., Peckham, P. D., and Sanders, J. R. (1972). Consequences of failure to meet assumptions underlying the fixed effects analysis of variance and covariance. *Rev. Educ. Res.*, 42, 239–288.
Glen, W. A. and Kramer, C. Y. (1958). Analysis of variance of a randomized block design with missing observations. *Appl. Stat.*, 7, 173–185.
Gosslee, D. G. and Lucas, H. L. (1965). Analysis of variance of disproportionate data when interaction is present. *Biometrics*, 21, 115–133.
Gower, J. C. (1962). Variance component estimation for unbalanced hierarchical classifications. *Biometrics*, 18, 427–430.
Graybill, F. A. (1954). On quadratic estimation of variance components. *Ann. Math. Stat.*, 25, 367–372.
Graybill, F. A. (1961). *An Introduction to Linear Statistical Models*, Vol. I. McGraw-Hill, New York.
Graybill, F. A. (1976). *Theory and Applications of Linear Models*. Duxbury Press, North Scituate, Massachusetts.

References

Graybill, F. A. and Hultquist, R. A. (1961). Theorems concerning Eisenhart's Model II. *Ann. Math. Stat.*, 32, 261–269.

Graybill, F. A. and Wortham, A. W. (1956). A note on uniformly best unbiased estimators for variance components. *J. Amer. Stat. Assoc.*, 51, 266–268.

Greenberg, B. G. (1951). Why randomize? *Biometrics*, 7, 309–322.

Groggel, D. J., Wackerly, D. D., and Rao, P. V. (1988). Nonparametric estimation in one-way random effects models. *Commun. Stat., B: Simul. & Comp.*, 17, 887–903.

Guenther, W. C. (1964). *Analysis of Variance*. Prentice-Hall, Englewood Cliffs, New Jersey.

Hahn, G. J. (1982). Design of experiments: An annotated bibliography. In: *Encyclopedia of Statistical Sciences*, Vol. 2, pp. 359–366 (Eds. S. Kotz and N. L. Johnson). John Wiley, New York.

Hahn, G. J. and Hendrickson, R. W. (1971). A table of percentage points of the distribution of the largest absolute value of k Student t variate and its applications. *Biometrika*, 58, 323–332.

Hald, A. (1952). *Statistical Tables and Formulas*. John Wiley, New York.

Hald, A. and Sinkbaek, S. A. (1950). A table of percentage points of the chi-square distribution. *Skand. Aktuar.*, 33, 168–175.

Hall, I. J. (1972). Some comparisons of tests for equality of variances. *J. Stat. Comp. Simul.*, 1, 183–194.

Halvorsen, K. T. (1991). Value splitting involving more factors. In: *Fundamentals of Exploratory Analysis of Variance*, pp. 114–145 (Eds. D.C. Hoaglin, F. Mosteller, and J. W. Tukey). John Wiley, New York.

Hampel, F. R., Rochetti, E. M., Rousseeuw, P. J., and Stahel, W. A. (1986). *Robust Statistics: The Approach Based on Influence Functions*. John Wiley, New York.

Hand, D. J., Daly, F., McConway, K., Lunn, D., and Ostrowski, E. (1993). *A Handbook of Small Data Sets*. Chapman & Hall, London.

Harsaae, E. (1969). On the computation and use of a table of percentage points of Bartlett's M. *Biometrika*, 56, 273–281.

Harter, H. L. (1960). Tables of range and Studentized range. *Ann. Math. Stat.*, 31, 1122–1147.

Harter, H. L. (1961). Expected values of normal order statistics. *Biometrika*, 48, 151–165.

Harter, H. L. (1964a). A new table of percentage points of the χ^2 distribution. *Biometrika*, 51, 231–239.

Harter, H. L. (1964b). *New Tables of the Incomplete Gamma Function Ratio and of Percentage Points of the Chi-square and Beta Distributions*. U.S. Government Printing Office, Washington, D.C.

Harter, H. L. (1969a). *Order Statistics and Their Use in Testing and Estimation*, Vol. 1. *Tests Based on Range and Studentized Range of Samples from a Normal Population*. U.S. Government Printing Office, Washington, D.C.

Harter, H. L. (1969b). *Order Statistics and Their Use in Testing and Estimation*, Vol. 2. *Estimates Based on Order Statistics of Samples from Various Populations*. U.S. Government Printing Office, Washington, D.C.

Hartley, H. O. (1940). Testing the homogeneity of a set of variances. *Biometrika*, 31, 249–255.

Hartley, H. O. (1950). The maximum F-ratio as a short-cut test for heterogeneity of variance. *Biometrika*, 37, 308–312.

Hartley, H. O. (1956). A plan for programming analysis of variance for general purpose computers. *Biometrics*, 12, 110–122.

Hartley, H. O. (1962). Analysis of variance. In: *Mathematical Methods for Digital Computers*, Vol. 1, pp. 221–230 (Eds. A. Ralston and H. S. Wilf). John Wiley, New York.

Hartung, J. and Voet, B. (1987). An asymptotic χ^2-test for variance components. In: *Contributions to Stochastics*, pp. 153–163 (Ed. W. Sendler). Physica-Verlag, Heidelberg.

Harville, D. A. (1969). *Variance Components Estimation for the Unbalanced One-Way Random Classification–A Critique*. Tech. Rep. No. ARL-69-0180, Aerospace Research Laboratories, Wright-Patterson Air Force Base, Dayton, Ohio.

Harville, D. A. (1977). Maximum-likelihood approaches to variance component estimation and to related problems. *J. Amer. Stat. Assoc.*, 72, 320–340.

Harville, D. A. (1978). Alternative formulations and procedures for the two-way mixed model. *Biometrics*, 34, 441–454.

Hatcher, L. and Stepanski, E. J. (1994). *A Step-by-Step Approach to Using the SAS System for Univariate and Multivariate Statistics*. SAS Institute, Cary, North Carolina.

Hawkins, D. M. (1980). *Identification of Outliers*. Chapman & Hall, London.

Hayman, G. E., Govindarajulu, Z., and Leone, F. C. (1973). Tables of the cumulative noncentral chi-square distribution. In: *Selected Tables in Mathematical Statistics*, Vol. 1, pp. 1–78 (Eds. H. L. Harter and D.B. Owen). American Mathematical Society, Providence, Rhode Island.

Hayter, A. J. (1984). A proof of the conjecture that the Tukey-Kramer multiple comparisons procedure is conservative. *Ann. Stat.*, 12, 61–75.

Healy, M. J. R. and Westmacott, M. (1956). Missing values in experiments analyzed on automatic computers. *Appl. Stat.*, 5, 203–206.

Hedayat, A. and Afsarinejad, K. (1975). Repeated measurements designs, I. In: *A Survey of Statistical Design and Linear Models*, pp. 229–242 (Ed. J. N. Srivastava). North-Holland, Amsterdam.

Hedayat, A. and Afsarinejad, K. (1978). Repeated measurements design, II. *Ann. Stat.*, 6, 619–628.

Hedderson, J. (1991). *SPSS-PC Plus Made Simple*. Wadsworth, Belmont, California.

Hedderson, J. and Fisher, M. (1993). *SPSS-X Made Simple*, 2nd ed. Wadsworth, Belmont, California.

Hegemann, V. and Johnson, D. E. (1976). The power of two tests for additivity. *J. Amer. Stat. Assoc.*, 71, 945–948.

Heiberger, R. M. (1989). *Computation for the Analysis of Designed Experiment*. John Wiley, New York.

Hemmerle, W. J. (1964). Algebraic specification of statistical models for analysis of variance computations. *Assoc. Comput. Mach.*, 11, 234–239.

Henderson, C. R. (1953). Estimation of variance and covariance components. *Biometrics*, 9, 226–252.

Henderson, C. R. (1959). Design and analysis of animal husbandry experiments. In: *Techniques and Procedures in Animal Production Research*, Chapter 1, pp. 2–56 (Ed. C. R. Henderson). American Society of Animal Production, Beltsville, Maryland.

Henderson, C. R. (1969). Design and analysis of animal husbandry experiments. In: *Techniques and Procedures in Animal Science Research*, 2nd ed., Chapter 1, pp. 1–35 (Ed. C. R. Henderson). American Society of Animal Science Monograph, Quality Corporation, Albany, New York.

Hendy, M. F. and Charles, J. A. (1970). The production techniques, silver content and circulation history of the twelfth century Byzantine Trachy. *Archaeometry*, 12, 13–21.

Herbach, L. H. (1959). Properties of Model II type analysis of variance tests, A: Optimum nature of the F-test for Model II in balanced case. *Ann. Math. Stat.*, 30, 939–959.

Hernandez, R. P. and Burdick, R. K. (1993). Confidence intervals on the total variance in an unbalanced two-fold nested design. *Biomet. J.*, 35, 515–522.

Hernandez, R. P., Burdick, R. K., and Birch, N. J. (1992). Confidence intervals and tests of hypotheses on variance components in an unbalanced two-fold nested design. *Biomet. J.*, 34, 387–402.

Herr, D. G. and Gaebelin, J. (1978). Nonorthogonal two-way analysis of variance. *Psychol. Bull.*, 85, 207–216.

Herzberg, A. M. and Cox, D. R. (1959). Recent work in the design of experiments: A bibliography and a review. *J. R. Stat. Soc., Ser. A*, 132, 29–67.

Herzberg, P. A. (1994). *How SAS Works: A Comprehensive Introduction.* Springer-Verlag, New York.

Hicks, C. R. (1956). Fundamentals of analysis of variance, Part III. *Indust. Qual. Contr.*, 13 (4), 13–16.

Hicks, C. R. (1987). *Fundamental Concepts in the Design of Experiments*, 3rd ed. Holt, Rinehart and Winston, New York.

Hinkelmann, K. and Kempthorne, O. (1994). *Design and Analysis of Experiments*, Vol. I, *Introduction to Experimental Design*. John Wiley, New York.

Hinkley, D. V., Reid, N., and Snell, E. J. (Eds.) (1991). *Statistical Theory and Modelling*. Chapman & Hall, London.

Hirotsu, C. (1968). An approximate test for the case of random effects model in a two-way layout with unequal cell frequencies. *Rep. Stat. Appl. Res. (JUSE)*, 15, 13–26.

Hirotsu, C. (1973). Multiple comparisons in two-way layout. *Rep. Stat. Appl. Res. (JUSE)*, 20, 1–10.

Hirotsu, C. (1983). An approach to defining the pattern of interaction effects in a two-way layout. *Ann. Inst. Stat. Math. (Japan), Ser. A*, 35, 77–90.

Hoaglin, D. C. (1988). Transformations in everyday experience. *Chance*, 1 (4), 40–45.

Hoaglin, D. C., Mosteller, F., and Tukey, J. W. (Eds.) (1983). *Understanding Robust and Exploratory Data Analysis*. John Wiley, New York.

Hoaglin, D. C., Mosteller, F., and Tukey, J. W. (Eds.) (1991). *Fundamentals of Exploratory Analysis of Variance*. John Wiley, New York.

Hochberg, Y. (1974). Some conservative generalizations of the T-method in simultaneous inference. *J. Multivar. Anal.*, 4, 224–234.

Hochberg, Y. (1976). A modification of the T-method of multiple comparisons for one-way layout with unequal variances. *J. Amer. Stat. Assoc.*, 71, 200–203.

Hochberg, Y. and Tamhane, A. C. (1983). Multiple comparisons in a mixed model. *Amer. Stat.*, 37, 305–307.

Hochberg, Y. and Tamhane, A. C. (1987). *Multiple Comparison Procedures.* John Wiley, New York.

Hocking, R. R. (1973). A discussion of the two-way mixed model. *Amer. Stat.*, 27, 148–152.

Hocking, R. R. (1985). *The Analysis of Linear Model.* Brooks/Cole, Monterey, California.

Hocking, R. R. (1993). Variance component estimation in mixed linear models. In: *Applied Analysis of Variance in Behavioral Science*, pp. 541–571 (Ed. L. K. Edwards). Marcel Dekker, New York.
Hocking, R. R. (1996). *The Analysis of Linear Models: Regression and Analysis of Variance*. John Wiley, New York.
Hodges, J. L., Jr. and Lehmann, E. L. (1970). *Basic Concepts of Probability and Statistics*, 2nd ed. Holden-Day, San Francisco.
Hogg, R. V. and Craig, A. T. (1995). *Introduction to Mathematical Statistics*, 5th ed. Prentice-Hall, Englewood Cliffs, New Jersey.
Hollander, M. and Wolfe, D. A. (1998). *Nonparametric Statistical Methods*, 2nd ed. John Wiley, New York. (1st ed. 1973.)
Holm, S. (1979). A simple sequentially rejective multiple test procedure. *Scand. J. Stat.*, 6, 65–70.
Horsnell, G. (1953). The effect of unequal group variances on the F-test for the homogeneity of group means. *Biometrika*, 40, 128–136.
Howell, J. F. and Games, P. A. (1973). The effects of variance heterogeneity on simultaneous multiple-comparison procedures with equal sample size. *Brit. J. Math. Stat. Psychol.*, 27, 72–81.
Hoyle, M. H. (1971). Spoilt data — An introduction and a bibliography. *J. R. Stat. Soc., Ser. A*, 134, 429–439.
Hoyle, M. H. (1973). Transformations — An introduction and a bibliography. *Inter. Stat. Rev.*, 41, 203–223.
Hsu, J. (1996). *Multiple Comparisons: Theory and Methods*. Chapman & Hall, London.
Huber, P. J. (1981). *Robust Statistics*. John Wiley, New York.
Huck, S. W. and Layne, B. H. (1974). Checking for proportional n's in factorial ANOVA's. *Educ. Psychol. Meas.*, 34, 281–287.
Hudson, J. D. and Krutchkoff, R. G. (1968). A Monte Carlo investigation of the size and power of tests employing Satterthwaite's synthetic mean squares. *Biometrika*, 55, 431–433.
Huitema, B. E. (1980). *The Analysis of Covariance and Alternatives*. John Wiley, New York.
Huitson, A. (1971). *The Analysis of Variance*. Griffin, London.
Hussein, M. and Milliken, G. A. (1978a). An unbalanced two-way model with random effects having unequal variances. *Biomet. J.*, 20, 203–213.
Hussein, M. and Milliken, G. A. (1978b). An unbalanced nested model with random effects having unequal variances. *Biomet. J.*, 20, 329–338.
Imhof, J. P. (1958). *Contributions to the Theory of Mixed Models in the Analysis of Variance*. Ph.D. Thesis, Department of Statistics, University of California, Berkeley, California.
Imhof, J. P. (1960). A mixed model for the complete three-way layout with two random effects factors. *Ann. Math. Stat.*, 31, 906–928.
Jackson, R. W. B. (1939). Reliability of mental tests. *Brit. J. Psychol.*, 29, 267–287.
Jaffe, J. A. (1994). *Mastering the SAS System*, 2nd ed. Van Nostrand, Reinhold, New York.
James, G. S. (1951). The comparison of several groups of observations when the ratios of population variances are unknown. *Biometrika*, 38, 324–329.
Japanese Standards Association (1972). *Selected Tables and Formulas with Computer Applications*. Japanese Standards Association, Tokyo.

Jeyaratnam, S. and Graybill, F. Y. (1980). Confidence intervals on variance components in three-factor cross-classification models. *Technom.*, 22, 375–380.
John, J. A. and Quenouille, M. H. (1977). *Experiments: Design and Analysis.* Griffin, London.
John, P. W. M. (1971). *Statistical Design and Analysis of Experiments.* McMillan, New York. (Reprinted 1998, SIAM, Philadelphia.)
Johnson, D. E. and Graybill, F. A. (1972a). The estimation of σ^2 in a two-way classification with interaction. *J. Amer. Stat. Assoc.*, 67, 388–394.
Johnson, D. E. and Graybill, F. A. (1972b). An analysis of a two-way model with interaction and no replication. *J. Amer. Stat. Assoc.*, 67, 862–868.
Johnson, N. L. (1948). Alternative systems in the analysis of variance. *Biometrika*, 35, 80–87.
Johnson, N. L., Kotz, S., and Balakrishnan, N. (1995). *Continuous Univariate Distributions*, Vol. 2, 2nd ed. John Wiley, New York.
Johnson, N. L. and Leone, F. C. (1964, 1977). *Statistics and Experimental Design in Engineering and the Physical Sciences*, Vol. 2. 1st & 2nd eds. John Wiley, New York.
Jones, B. and Kenward, M. G. (1989). *Design and Analysis of Cross-Over Trials.* Chapman & Hall, London.
Kafadar, K. and Tukey, J. W. (1988). A bidec t table. *J. Amer. Stat. Assoc.*, 83, 532–539.
Kastenbaum, M. A., Hoel, D. G., and Bowman, K. O. (1970a). Sample size requirements: One-way analysis of variance. *Biometrika*, 57, 421–430.
Kastenbaum, M. A., Hoel, D. G., and Bowman, K. O. (1970b). Sample size requirements: Randomized block designs. *Biometrika*, 57, 573–577.
Kempthorne, O. (1952). *The Design and Analysis of Experiments.* John Wiley, New York.
Kempthorne, O. (1955). The randomization theory of experimental inference. *J. Amer. Stat. Assoc.*, 50, 946–967.
Kempthorne, O. (1976). The analysis of variance and factorial design. In: *On the History of Statistics and Probability*, pp. 29–54 (Ed. D. B. Owen). Marcel Dekker, New York.
Kempthorne, O. (1977). Why randomize? *J. Stat. Plann. Inf.*, 1, 1–25.
Kempthorne, O. and Folks, L. (1971). *Probability, Statistics and Data Analysis.* The Iowa State University Press, Ames, Iowa.
Kendall, M. G. and Stuart, A. (1961). *The Advanced Theory of Statistics*, Vol. 2. *Inference and Relationship*, 3rd ed. Griffin, London.
Kendall, M. G., Stuart, A., and Ord, J. K. (1983). *The Advanced Theory of Statistics*, Vol. 3. *Design and Analysis, and Time-Series*, 4th ed. MacMillan, New York.
Keselman, H. J., Games, P. A., and Clinch, J. J. (1979). Tests for homogeneity of variance. *Commun. Stat., B: Simul. & Comp.*, 88, 113–139.
Keselman, H. J. and Rogan, J. C. (1978). A comparison of modified-Tukey and Scheffé methods of multiple comparisons for pairwise contrasts. *J. Amer. Stat. Assoc.*, 73, 47–51.
Keselman, H. J., Toothaker, L. E., and Shooter, M. (1975). An evaluation of two unequal n_k forms of the Tukey multiple comparison statistic. *J. Amer. Stat. Assoc.*, 70, 584–587.
Keuls, M. (1952). The use of the "Studentized range" in connection with an analysis of variance. *Euphytica*, 1, 112–122.

Khargonkar, S. A. (1948). The estimation of missing plot values in split-plot and strip trials. *J. Ind. Soc. Agri. Stat.*, 1, 147–161.

Khuri, A. I. (1987). An exact test for the nesting effect's variance component in an unbalanced two-fold nested model. *Stat. Prob. Lett.*, 5, 305–311.

Khuri, A. I. (1990). Exact tests for random models with unequal cell frequencies in the last stage. *J. Stat. Plann. Inf.*, 24, 177–193.

Khuri, A. I. (1995). A test to detect inadequacy of Satterthwaite's approximation in balanced mixed models. *Statistics*, 27, 45–54.

Khuri, A. I. and Cornell, J. (1996). *Response Surfaces: Designs and Analyses*, 2nd ed. Marcel Dekker, New York.

Khuri, A. I. and Littell, R. C. (1987). Exact tests for the main effects variance components in an unbalanced random two-way model. *Biometrics*, 43, 545–560.

Khuri, A. I., Mathew, T., and Sinha, B. K. (1998). *Statistical Tests for Mixed Linear Models*. John Wiley, New York.

Khuri, A. I. and Sahai, H. (1985). Variance components analysis: A selective literature survey. *Inter. Stat. Rev.*, 53, 279–300.

Kihlberg, J. K., Herson, J. H., and Schutz, W. E. (1972). Square root transformation revisited. *Appl. Stat.*, 21, 76–81.

Kimball, A. W. (1951). On dependent tests of significance in the analysis of variance. *Ann. Math. Stat.*, 22, 600–602.

Kirk, R. E. (1995). *Experimental Design: Procedures for the Behavioral Sciences*, 3rd ed. Brooks/Cole, Belmont, California. (1st ed., 1968; 2nd ed., 1982.)

Klotz, J. H. (1969). A simple proof of Scheffé's multiple comparison theorem for contrasts in the one-way layout. *Amer. Stat.*, 23, 44–45.

Koch, G. G., Elashoff, J. D., and Amara, I. A. (1988). Repeated measurements — Design and Analysis. In: *Encyclopedia of Statistical Sciences*, Vol. 8, pp. 46–73 (Eds. S. Kotz and N. L. Johnson). John Wiley, New York.

Koehler, K. J. (1983). A simple approximation for the percentiles of the t distribution. *Technom.*, 25, 103–105.

Koehler, K. J. and Larntz, K. (1980). An empirical investigation of goodness-of-fit statistics for sparse multinomials. *J. Amer. Stat. Assoc.*, 75, 336–344.

Kohr, R. L. and Games, P. A. (1974). Robustness of the analysis of variance, the Welch procedure, and a Box procedure to heterogeneous variances. *J. Exper. Educ.*, 43, 61–69.

Kramer, C. Y. (1956). Extension of multiple range test to group means with unequal numbers of replications. *Biometrics*, 12, 307–310.

Kramer, C. Y. (1957). Extension of multiple range tests to group correlated adjusted means. *Biometrics*, 13, 13–18.

Kramer, C. Y. and Glass, S. (1960). Analysis of variance of a Latin square design with missing observations. *Appl. Stat.*, 9, 43–50.

Krishnaiah, P. R. (1979). Some developments on simultaneous test procedures: A review. In: *Development in Statistics*, Vol. 2, pp. 157–201 (Ed. P. R. Krishnaiah). Academic Press, New York.

Krishnaiah, P. R. (Ed.) (1980). *Handbook of Statistics*, Vol. 1: *Analysis of Variance*. North-Holland, Amsterdam.

Krishnaiah, P. R. and Yochmowitz, M. G. (1980). Inference on the structure of interaction in two-way classification model. In: *Handbook of Statistics*, Vol. 1: *Analysis of Variance*, pp. 973–994 (Ed. P. R. Krishnaiah). North-Holland, Amsterdam.

References

Krutchkoff, R. G. (1988). One-way fixed effects analysis of variance when the error variances may be unequal. *J. Stat. Comp. Simul.*, 30, 259–271.

Krutchkoff, R. G. (1989). Two-way fixed effects analysis of variance when the error variances may be unequal. *J. Stat. Comp. Simul.*, 32, 177–183.

Kurtz, T. E., Link, R. F., Tukey, J. W., and Wallace, D. L. (1965). Short-cut multiple comparisons for balanced single and double classifications, Part 1. Results. *Technom.*, 7, 95–165. (Authors' reply to Anscombe's comment, *Technom.*, 7, 169.)

Kussmaul, K. and Anderson, R. L. (1967). Estimation of variance components in two-stage nested designs with composite samples. *Technom.*, 9, 373–389.

LaMotte, L. R. (1973). Quadratic estimation of variance components. *Biometrics*, 29, 311–330.

Larntz, K. (1978). Small sample comparison of exact levels of chi-squared goodness-of-fit statistics. *J. Amer. Stat. Assoc.*, 73, 253–263.

Laubscher, N. F. (1965). Interpolation in F tables. *Amer. Stat.*, 19, 28 and 40.

Layard, M. W. J. (1973). Robust large sample tests for homogeneity of variance. *J. Amer. Stat. Assoc.*, 68, 195–198.

Lee, P. M. (1997). *Bayesian Statistics: An Introduction*, 2nd ed. Arnold, London.

Lehmann, E. L. (1975). *Nonparametric Statistical Methods Based on Ranks*. Holden-Day, San Francisco.

Lehmer, E. (1944). Inverse tables of probabilities of errors of the second kind. *Ann. Math. Stat.*, 15, 338–398.

Lentner, M., Arnold, J., and Hinklemann, K. (1989). The efficiency of blocking: How to use MS(block)/MS(error) correctly. *Amer. Stat.*, 43, 106–111.

Levene, H. (1960). Robust tests for equality of variances. In: *Contributions to Probability and Statistics*, pp. 278–292 (Eds. I. Olkin, S. G. Ghurye, W. Hoeffding, W. G. Madow, and H. B. Mann). Stanford University Press, Stanford, California.

Levine, G. (1991). *A Guide to SPSS for Analysis of Variance*. Lawrence Erlbaum, Hillsdale, New Jersey.

Levy, K. J. (1978a). An empirical comparison of the ANOVA F-test with alternatives which are more robust against heterogeneity of variance. *J. Stat. Comp. Simul.*, 8, 49–57.

Levy, K. J. (1978b). An empirical study of the cube root test for homogeneity of variances with respect to the effects of a non-normality and power. *J. Stat. Comp. Simul.*, 8, 71–78.

Lindman, H. R. (1992). *Analysis of Variance in Experimental Design*. Springer-Verlag, New York.

Lindsey, J. K. (1993). *Models for Repeated Measurements*. Clarendon Press, Oxford, U.K.

Littell, R. C., Freund, R. J., and Spector, P. C. (1991). *SAS Systems for Linear Models*. SAS Institute, Cary, North Carolina.

Littell, R. C., Milliken, G. A., Stroup, W. W., and Wolfinger, R. D. (1996). *SAS System for Mixed Models*. SAS Institute, Cary, North Carolina.

Lix, L. M., Keselman, J. C., and Keselman, H. J. (1996). Consequences of assumption violations revisited: A quantitative review of alternatives to the one-way analysis of variance F test. *Rev. Educ. Res.*, 66, 579–619.

Lorenzen, T. J. (1977). *Derivation of Expected Mean Squares and F-tests in Statistical Experimental Design*. Res. Publ., GMR-2442, Mathematics Department, General Motors Research Laboratories, Warren, Michigan.

Lorenzen, T. J. (1984). Randomization and blocking in the design of experiments. *Commun. Stat., A: Theo. & Meth.*, 13, 2601–2623.

Lorenzen, T. J. (1987). *A Comparison of Approximate F' Tests Under Pooling Rules*. Res. Publ., GMR-5928, Mathematics Department, General Motors Research Laboratories, Warren, Michigan.

Lorenzen, T. J. and Anderson, V. L. (1993). *Design of Experiments: A No-Name Approach*. Marcel Dekker, New York.

Lurigio, A., Seng, M., Sinecore, J., and Dantzker, M. (1995). *Computer Applications Using SPSS for Windows*. Butterworth/Heinemann, Stoneham, Massachusetts.

Mahalanobis, P. C. (1964). Professor Ronald Aylmer Fisher. *Biometrics*, 20, 238–251. (Reprinted from Sankhyā, Vol. 4, 1938, pp. 265–272.)

Mahamunulu, D. M. (1963). Sampling variances of the estimates of variance components in the unbalanced 3-way nested classification. *Ann. Math. Stat.*, 34, 521–527.

Mandel, J. (1971). A new analysis of variance model for non-additive data. *Technom.*, 13, 1–18.

Marcuse, S. (1949). Optimum allocation and variance components in nested sampling with an application to chemical analysis. *Biometrics*, 5, 189–206.

Mardia, K. V. and Zemroch, P. J. (1978). *Tables of the F and Related Distributions with Algorithms*. Academic Press, New York.

Maurais, J. and Quimet, R. (1986). Exact critical values of Bartlett's test of homogeneity of variances for unequal sample sizes for two populations and power of the test. *Metrika*, 33, 275–289.

Maxwell, S. E. and Delaney, H. D. (1990). *Designing Experiments and Analyzing Data: A Model Comparison Perspective*. Wadsworth, Belmont, California.

McCullagh, P. and Nelder, J. A. (1983, 1989). *Generalized Linear Models*, 1st & 2nd eds. Chapman & Hall, London.

McHugh, R. B. and Mielke, P. W., Jr. (1968). Negative variance estimates and statistical dependence in nested sampling. *J. Amer. Stat. Assoc.*, 63, 1000–1003.

McLean, R. A., Sanders, W. L., and Stroup, W. W. (1991). A unified approach to mixed linear models. *Amer. Stat.*, 45, 54–63.

Mead, R., Bancroft, T. A., and Han, C. P. (1975). Power of analysis of variance test procedures for incompletely specified mixed models. *Ann. Stat.*, 3, 797–808.

Micceri, T. (1989). The unicorn, the normal curve, and other improbable creatures. *Psychol. Bull.*, 105, 156–166.

Miller, J. J. (1977). Asymptotic properties of maximum likelihood estimates in the mixed model of the analysis of variance. *Ann. Stat.*, 5, 746–762.

Miller, R. G., Jr. (1966). *Simultaneous Statistical Inference*. McGraw-Hill, New York.

Miller, R. G., Jr. (1968). Jackknifing variances. *Ann. Math. Stat.*, 39, 567–582.

Miller, R. G., Jr. (1977). Developments in multiple comparisons, 1966–1976. *J. Amer. Stat. Assoc.*, 72, 779–788.

Miller, R. G., Jr. (1981). *Simultaneous Statistical Inference*, 2nd ed. Springer-Verlag, New York.

Miller, R. G., Jr. (1985). Multiple comparisons. In: *Encyclopedia of Statistical Sciences*, Vol. 5, pp. 679–689 (Eds. S. Kotz and N. L. Johnson). John Wiley, New York.

Miller, R. G., Jr. (1986). *Beyond ANOVA: Basics of Applied Statistics*. John Wiley, New York. (Reprinted 1996, Chapman & Hall, New York.)

Milliken, G. A. and Graybill, F. A. (1970). Extensions of the general linear hypothesis model. *J. Amer. Stat. Assoc.*, 65, 797–807.

References

Milliken, G. A. and Graybill, F. A. (1971). Tests for interaction in the two-way model with missing data. *Biometrics*, 27, 1079–1083.

Milliken, G. A. and Graybill, F. A. (1972). Interaction models for Latin square. *Aust. J. Stat.*, 14, 129–138.

Milliken, G. A. and Johnson, D. E. (1992). *Analysis of Messy Data*, Vol. 1. Chapman & Hall, London.

Millman, J. and Glass, G. V. (1967). Rules of thumb for writing the ANOVA table. *J. Educ. Meas.*, 4, 41–51.

Miron, T. (1993). *SAS Software Solutions: Basic Data Processing.* SAS Institute, Cary, North Carolina.

Miyakawa, M. (1993). An interpretation of the interaction terms in Mandel's ANOVA Model from Hirotsu's interaction elements. *Rep. Stat. Appl. Res. (JUSE)*, 20, 1–10.

Montgomery, D. C. (1991). *Design and Analysis of Experiments*, 3rd ed. John Wiley, New York. (1st ed., 1976; 2nd ed., 1984).

Morrison, D. F. (1990). *Multivariate Statistical Methods*, 3rd ed. McGraw-Hill, New York. (1st ed., 1967; 2nd ed., 1976.)

Moses, L. E. (1978). Charts for finding upper percentage points of Student's t in the range .01 to .00001. *Commun. Stat., B: Simul. & Comp.*, 7, 479–490.

Mosteller, F. and Tukey, J. W. (1977). *Data Analysis and Regression.* Addison-Wesley, Reading, Massachusetts.

Murdock, G. R. and Williford, W. O. (1977). Tables for obtaining optimal confidence intervals involving the chi-square distribution. In: *Selected Tables in Mathematical Statistics*, Vol. 5, pp. 205–230 (Eds. D. B. Owen and R. E. Odeh). American Mathematical Society, Providence, Rhode Island.

Myers, R. H. (1976). *Response Surface Methodology.* Allyn and Bacon, Boston.

Myers, R. H. and Howe, R. B. (1971). On alternative approximate F tests for hypotheses involving variance components. *Biometrika*, 58, 393–396.

Myers, R. H. and Montgomery, D. C. (1995). *Response Surface Methodology.* John Wiley, New York.

Nagasenkar, P. B. (1984). On Bartlett's test for homogeneity of variances. *Biometrika*, 71, 405–407.

Naik, U. D. (1974). On tests of main effects and interactions in higher-way layouts in the analysis of variance random effects model. *Technom.*, 16, 17–25.

Nair, K. R. (1940). The application of the technique of analysis of covariance to field experiments with several missing or mixed-up plots. *Sankhyā*, 4, 581–588.

Nair, K. R. (1948). Distribution of the extreme deviate from the sample mean. *Biometrika*, 35, 118–144.

Natrella, M. G. (1963). *Experimental Statistics.* John Wiley, New York. (Reprint of the original edition published by the National Bureau of Standards, Washington, D.C. as Handbook No. 91.)

Nelson, L. S. (1983). A comparison of sample sizes for the analysis of means and the analysis of variance. *J. Qual. Tech.*, 15, 33–39.

Nelson, L. S. (1985). Sample size tables for analysis of variance. *J. Qual. Tech.*, 17, 167–169.

Neter, J., Kutner, M. H., Nachtsheim, C. J., and Wasserman, W. (1996). *Applied Linear Statistical Models.* 4th ed. Irwin, Burr Ridge, Illinois.

Neter, J., Wasserman, W., and Kutner, M. H. (1990). *Applied Linear Statistical Models: Regression, Analysis of Variance, and Experimental Designs*, 3rd ed. Irwin, Burr Ridge, Illinois. (1st ed. 1974; 2nd ed 1985).

Newman, D. (1939). The distribution of the range in samples from a normal population, expressed in terms of an independent estimate of standard deviation. *Biometrika*, 31, 20–30.

Norton, H. W. (1939). The 7 × 7 squares. *Ann. Eug.*, 9, 269–307.

O'Brien, R. G. (1979). An improved ANOVA method for robust tests of additive models for variances. *J. Amer. Stat. Assoc.*, 74, 877–880.

O'Brien, R. G. (1981). A simple test for variance effects in experimental designs. *Psychol. Bull.*, 89, 570–574.

O'Neil, R. and Wetherill, G. B. (1971). The present state of multiple comparison methods. *J. R. Stat. Soc., Ser. B*, 33, 218–250.

Olejnik, S. F. and Algina, J. (1987). Type I error rates and power estimates of selected parametric and nonparametric tests of scales. *J. Educ. Stat.*, 12, 45–61.

Olkin, I. and Pratt, J. W. (1958). Unbiased estimation of certain correlation coefficients. *Ann. Math. Stat.*, 29, 201–211.

Ostle, B. (1952). Answer to query no. 95. *Biometrics*, 8, 264–266.

Ostle, B. and Malone, L. C. (1988). *Statistics in Research: Basic Concepts and Techniques for Research Workers*, 4th ed. Iowa State University Press, Ames, Iowa.

Ostle, B. and Mensing, R. W. (1975). *Statistics in Research: Basic Concepts and Techniques For Research Workers*, 3rd ed. Iowa State University Press, Ames, Iowa.

Owen, D. B. (1962). *Handbook of Statistical Tables*. Addison-Wesley, Reading, Massachusetts.

Owen, D. B. (1968). A survey of properties and applications of the noncentral t distribution. *Technom.*, 10, 445–478.

Owen, D. B. (1985). Noncentral t distribution. In: *Encyclopedia of Statistical Sciences*, Vol. 6, pp. 286–290 (Eds. S. Kotz and N. L. Johnson). John Wiley, New York.

Parker, E. T. (1959). Orthogonal Latin squares. *Proc. Nat. Acad. Sci.*, 45, 459–462.

Paull, A. E. (1950). On a preliminary test for pooling mean squares in the analysis of variance. *Ann. Math. Stat.*, 21, 539–556.

Pearson, E. S. (1931). The analysis of variance in cases of non-normal variation. *Biometrika*, 23, 114–133.

Pearson, E. S. and Hartley, H. O. (1951). Charts of the power function for analysis of variance tests, derived from the non-central F-distribution. *Biometrika*, 38, 112–130.

Pearson, E. S. and Hartley, H. O. (1970). *Biometrika Tables for Statisticians*, Vol. 1, 3rd ed. Cambridge University Press, Cambridge, U.K.

Pearson, E. S. and Hartley, H. O. (1973). *Biometrika Tables for Statisticians*, Vol. II, 3rd ed. Cambridge University Press, Cambridge, U.K.

Pearson, E. S., D'Agostino, R. B., and Bowman, K. O. (1977). Tests for departure from normality: Comparison of powers. *Biometrika*, 64, 231–246.

Peng, K. C. (1967). *The Design and Analysis of Scientific Experiments*. Addison Wesley, Reading, Massachusetts.

Perry, J. N., Wall, C., and Greenway, A. R. (1980). Latin square designs in field experiments involving sex attractants. *Ecol. Entomol.*, 5, 385–396.

Petrinovich, L. F. and Hardyck, C. D. (1969). Error rates for multiple comparison methods. *Psychol. Bull.*, 71, 43–54.

Pillai, K. C. S. and Ramachandran, K. V. (1954). On the distribution of the ratio of the i-th observation in an ordered sample from a normal population to an independent estimate of the standard deviation. *Ann. Math. Stat.*, 25, 565–572.

References

Pitman, E. J. G. (1938). Significance tests which may be applied to samples from any population, III. The analysis of variance test. *Biometrika*, 29, 322–335.
Plackett, R. L. (1960). Models in analysis of variance (with discussion). *J. R. Stat. Soc., Ser. B*, 22, 195–217.
Pratt, J. W. and Gibbons, J. D. (1981). *Concepts of Nonparametric Theory*. Springer-Verlag, New York.
Pukelsheim, F. (1981). On the existence of unbiased nonnegative estimates of variance and covariance components. *Ann. Stat.*, 9, 293–299.
Ramsey, P. H. (1994). Testing variances in psychological and educational research. *J. Educ. Stat.*, 19, 23–42.
Ramsey, P. H. and Brailsford, E. A. (1989). Robustness and power of tests of variability on two independent groups. *J. Math. Stat. Psychol.*, 43, 113–130.
Rankin, N. O. (1974). The harmonic mean method for one-way and two-way analysis of variance. *Biometrika*, 61, 117–122.
Rao, C. R. (1971). Estimation of variance and covariance components — MINQUE theory. *J. Multivar. Anal.*, 1, 257–275.
Rao, C. R. (1972). Estimation of variance and covariance components in linear models. *J. Amer. Stat. Assoc.*, 67, 112–115.
Rao, C. R. (1973). *Linear Statistical Inference and Its Applications*, 2nd. ed. John Wiley, New York. (1st ed., 1965.)
Rao, C. R. and Kleffe, J. (1988). *Estimation of Variance Components and Applications*. North-Holland, Amsterdam.
Rao, C. R. and Toutenburg, H. (1995). *Linear Models: Least Squares and Alternatives*. Springer-Verlag, New York.
Rao, P. S. R. S. (1997). *Variance Components Estimation: Mixed Models, Methodologies and Applications*. Chapman & Hall, London.
Ratkowski, D. A., Evans, M. A., and Alldredge, J. R. (1993). *Cross-Over Experiments: Design, Analysis and Application*. Marcel Dekker, New York.
Resnikoff, G. J. and Lieberman, G. J. (1957). *Tables of the Noncentral t Distribution*. Stanford University Press, Stanford, California.
Ringland, J. T. (1983). Robust multiple comparisons. *J. Amer. Stat. Assoc.*, 78, 145–151.
Robertson, A. (1962). Weighting in the estimation of variance components in the unbalanced single classification. *Biometrics*, 18, 413–417.
Rogan, J. C. and Keselman, H. J. (1977). Is the ANOVA F-test robust to variance heterogeneity when sample sizes are equal? : An investigation via a coefficient of variation. *Amer. Educ. Res. J.*, 14, 493–498.
Rosenthal, R. and Rosnow, R. L. (1985). *Contrast Analysis: Focused Comparison in the Analysis of Variance*. Cambridge University Press, Cambridge, U.K.
Royston, J. P. (1982a). An extension of Shapiro and Wilk's W test for normality to large samples. *Appl. Stat.*, 31, 115–124.
Royston, J. P. (1982b). The W test for normality. *Appl. Stat.*, 31, 176–180.
Royston, J. P. (1983). A simple method for evaluating the Shapiro-Francia W' test for non-normality. *Statistician*, 32, 297–300.
Royston, J. P. (1991). Estimating departures from normality. *Stat. Med.*, 10, 1283–1291.
Royston, J. P. (1993a). Graphical detection of non-normality by using Michael's statistic. *Appl. Stat.*, 42, 153–158.
Royston, J. P. (1993b). A toolkit for non-normality in complete and censored samples. *Statistician*, 42, 37–43.

Royston, J. P. (1993c). A pocket calculator algorithm for the Shapiro-Francia test for non-normality: An application to medicine. *Stat. Med.*, 12, 181–184.

Royston, J. P., Flecknell, P. A., and Wootton, R. (1982). New evidence that the intrauterine growth retarded piglet is a member of a discrete subpopulation. *Biol. Neon.*, 42, 100–104.

Rubin, D. B. (1972). A non-iterative algorithm for least squares estimation of missing values in any analysis of variance design. *Appl. Stat.*, 21, 136–141.

Sahai, H. (1974a). On negative estimates of variance components under finite population models. *S. Afric. Stat. J.*, 8, 157–166.

Sahai, H. (1974b) Non-negative maximum likelihood and restricted maximum likelihood estimators of variances components in two simple linear models. *Util. Math.*, 5, 151–160.

Sahai, H. (1976). A comparison of estimators of variance components in the balanced three-stage nested random effects model using mean squared error criterion. *J. Amer. Stat. Assoc.*, 71, 435–444.

Sahai, H. (1979). A bibliography on variance components. *Inter. Stat. Rev.*, 47, 177–222.

Sahai, H. (1988). Two-way mixed model: A brief review. *N. Zealand Statist.*, 23, 58–65.

Sahai, H., Khuri, A. I., and Kapadia, C. H. (1985). A second bibliography on variance components. *Commun. Stat., A: Theo. & Meth.*, 14, 63–115.

Sahai, H. and Khurshid, A. (1992). A comparison of estimators of variance components in a two-way balanced crossed classification random effects model. *Statistics*, 23, 128–143.

Sahai, H. and Thompson, W. O. (1973). Non-negative maximum likelihood estimators of variance components in a simple linear model. *Amer. Stat.*, 27, 112–113.

Samuels, M. L., Casella, G., and McCabe, G. P. (1991). Interpreting blocks and random factors. *J. Amer. Stat. Assoc.*, 86, 798–808.

SAS Institute (1989). *SAS Language and Procedures: Usage.* Version 6.0. SAS Institute, Cary, North Carolina.

SAS Institute (1990a). *SAS Language: Reference.* Version 6.0. SAS Institute, Cary, North Carolina.

SAS Institute (1990b). *SAS Procedures Guide.* Version 6.0, 3rd ed. SAS Institute, Cary, North Carolina.

SAS Institute (1990c). *SAS/STAT User's Guide.* Version 6.0, Vols. I & II, 4th ed. SAS Institute, Cary, North Carolina.

SAS Institute (1991). *SAS Language and Procedures: Usage 2.* Version 6.0. SAS Institute, Cary, North Carolina.

SAS Institute (1992). *SAS Introductory Guide for PC's.* Version 6.03. SAS Institute, Cary, North Carolina.

SAS Institute (1997). *SAS/STAT Software: Changes and Enhancements through Release 6.12.* SAS Institute, Cary, North Carolina.

Satterthwaite, F. E. (1946). An approximate distribution of estimates of variance components. *Biom. Bull.*, 2, 110–114.

Scheffé, H. (1953). A method for judging all contrasts in the analysis of variance. *Biometrika*, 40, 87–104.

Scheffé, H. (1956a). A "mixed model" for the analysis of variance. *Ann. Math. Stat.*, 27, 23–36.

Scheffé, H. (1956b). Alternative models for the analysis of variance. *Ann. Math. Stat.*, 27, 251–271.

References

Scheffé, H. (1959). *The Analysis of Variance*. John Wiley, New York.
Schervish, M. L. (1992). Bayesian analysis of lineal models. In: *Bayesian Statistics*, IV, pp. 419–434 (Eds. J. M. Bernardo, J. V. Berger, A. P. David, and A. F. M. Smith). Oxford University Press, New York.
Schlotzhauer, S. D. and Littell, R. C. (1997). *SAS System for Elementary Statistical Analysis*, 2nd ed. SAS Institute, Cary, North Carolina. (1st ed., 1987.)
Schultz, E. F., Jr. (1954). Answer to query no. 110. *Biometrics*, 10, 407–411.
Schultz, E. F., Jr. (1955). Rules of thumb for determining expectations of mean squares in analysis of variance. *Biometrics*, 11, 123–148.
Searle, S. R. (1961). Variance components in the unbalanced two-way nested classification. *Ann. Math. Stat.*, 32, 1161–1166.
Searle, S. R. (1968). Another look at Henderson's methods of estimating variance components. *Biometrics*, 24, 749–778.
Searle, S. R. (1971a). Topics in variance components estimation. *Biometrics*, 27, 1–76.
Searle, S. R. (1971b). *Linear Models*. John Wiley, New York. (Wiley Classic Edition, 1997.)
Searle, S. R. (1987). *Linear Models for Unbalanced Data*. John Wiley, New York.
Searle, S. R. (1988). Mixed models and unbalanced data: Wherefrom, whereat, and whereto? *Commun. Stat., A: Theo.& Meth.*, 17, 935–968.
Searle, S. R. (1995). An overview of variance component estimation. *Metrika*, 42, 215–230.
Searle, S. R., Casella, G., and McCulloch, C. E. (1992). *Variance Components*. John Wiley, New York.
Searle, S. R. and Fawcett, R. F. (1970). Expected mean squares in variance components models having finite populations. *Biometrics*, 26, 243–254.
Seely, J. F. and El-Bassiouni, Y. (1983). Applying Wald's variance component test. *Ann. Stat.*, 11, 197–201.
Seifert, B. (1981). Explicit formulae of exact tests in mixed balanced ANOVA models. *Biomet. J.*, 23, 535–550.
Sen, B., Graybill, F. A., and Ting, N. (1992). Confidence intervals on ratios of variance components for the unbalanced two-factor nested model. *Biomet. J.*, 34, 259–274.
Senn, S. (1993). *Cross-Over Trials in Clinical Research*. John Wiley, Chichester, U.K.
Shaffer, J. P. (1977). Multiple comparison emphasizing selected contrasts: An extension and generalization of Dunnett's procedure. *Biometrics*, 33, 293–303.
Shaffer, J. P. (1986). Modified sequentially rejective multiple test procedures. *J. Amer. Stat. Assoc.*, 81, 826–831.
Shapiro, S. S. and Francia, R. S. (1972). An approximate analysis of variance test for normality. *J. Amer. Stat. Assoc.*, 67, 215–216.
Shapiro, S. S. and Wilk, M. B. (1965). An analysis of variance test for normality (complete samples). *Biometrika*, 52, 591–611.
Shapiro, S. S., Wilk, M. B., and Chen, H. J. (1968). A comparative study of various tests for normality. *J. Amer. Stat. Assoc.*, 63, 1343–1372.
Šidák, A. (1967). Rectangular confidence regions for the means of multivariate normal distributions. *J. Amer. Stat. Assoc.*, 62, 626–633.
Singh, B. (1987). On the non-null distribution of ANOVA F-ratio in one-way unbalanced random model. *Calcutta Stat. Assoc. Bull.*, 36, 57–62.
Singhal, R. A. (1987). Confidence limits on heritability under nonnormal variations. *Biomet. J.*, 29, 571–578.

Singhal, R. A. and Sahai, H. (1992). Sampling distribution of the ANOVA estimator of between variance component in samples from a non-normal universe. *J. Stat. Comp. Simul.*, 43, 19–30.

Singhal, R. A. and Sahai, H. (1994). Effects of non-normality on the power function in a one-way random model. *Stat. Papers*, 35, 113–125.

Singhal, R. A. and Singh, C. (1984). Distribution of the variance ratio test in a non-normal random effects model. *Sankhyā, Ser. B*, 46, 29–35.

Singhal, R. A., Tiwari, C. B., and Sahai, H. (1988). A selected and annotated bibliography on the robustness studies to non-normality in variance components models. *J. Jap. Stat. Soc.*, 18, 195–206.

Smith, D. W. and Murray, L. W. (1984). An alternative to Eisenhart's Model II and mixed model in the case of negative variance estimates. *J. Amer. Stat. Assoc.*, 79, 145–151.

Smith, H. F. (1951). Analysis of variance with unequal but proportional numbers of observations in the sub-classes of a two-way classification. *Biometrics*, 7, 70–74.

Snedecor, G. W. (1934). *Analysis of Variance and Covariance*. Iowa State University Press, Ames, Iowa.

Snedecor, G. W. (1955). Query no. 113. *Biometrics*, 11, 111–113.

Snedecor, G. W. and Cochran, W. G. (1967, 1989). *Statistical Methods*, 6th & 8th eds. Iowa State University Press, Ames, Iowa.

Snee, R. D. (1985). Graphical display of results of three-treatment randomized block experiments. *Appl. Stat.*, 34, 71–77.

Snell, E. J. (1987). *Applied Statistics: A Handbook of BMDP Analyses*. Chapman & Hall, London.

Sokal, R. R. and Rohlf, F. J. (1995). *Biometry*, 3rd ed. W. H. Freeman, New York. (1st ed. 1969; 2nd ed. 1981.)

Spector, P. E. (1993). *SAS Programming for Researchers and Social Scientists*. Sage, Thousand Oaks, California.

Spjotvoll, E. (1967). Optimum invariant tests in unbalanced variance components models. *Ann. Math. Stat.*, 38, 422–428.

Spjotvoll, E. (1968). Confidence intervals and tests for variance ratios in unbalanced variance components models. *Inter. Stat. Rev.*, 36, 37–42.

Spjotvoll, E. and Stoline, M. R. (1973). An extension of the T-method of multiple comparisons to include the cases with unequal sample sizes. *J. Amer. Stat. Assoc.*, 68, 975–978.

Sprent, P. (1997). *Applied Nonparametric Statistical Methods*, 2nd ed. Chapman & Hall, London. (1st ed., 1989.)

SPSS, Inc. (1997a). *SPSS Base 7.5 for Windows User's Guide*. SPSS, Inc., Chicago, Illinois.

SPSS, Inc. (1997b). *SPSS Professional Statistics 7.5*. SPSS, Inc., Chicago, Illonios.

SPSS, Inc. (1997c). *SPSS Advanced Statistics 7.5*. SPSS, Inc., Chicago, Illonois.

Srivastava, A. B. L. (1959). Effects of non-normality on the power of the analysis of variance test. *Biometrika*, 46, 114–122.

Srivastava, S. R. and Bozivich, H. (1962). Power of certain analysis of variance test procedures involving preliminary test. *Bull. Inter. Stat. Inst.*, 39, 133.

Steel, G. D. and Torrie, J. H. (1980). *Principles and Procedures of Statistics: A Biometrical Approach*, 2nd ed. McGraw-Hill, New York. (1st ed., 1960.)

Steel, G. D., Torrie, J. H., and Dickey, D. A. (1997). *Principles and Procedure of Statistics: A Biometrical Approach*, 3rd ed. McGraw-Hill, New York.

References

Stoline, M. R. (1978). Tables of the Studentized augmented range and applications to problems of multiple comparison. *J. Amer. Stat. Assoc.*, 73, 656–660.

Stoline, M. R. (1981). The status of multiple comparisons: Simultaneous estimation of all pairwise comparisons in one-way ANOVA designs. *Amer. Stat.*, 35, 134–141.

Stoline, M. R. and Ury, H. K. (1979). Tables of the Studentized distribution and an application to multiple comparisons among means. *Technom.*, 21, 87–93.

Street, A. P. and Street, D. J. (1987). *Combinatorics of Experimental Design*. Clarendon Press, Oxford.

Street, A. P. and Street, D. J. (1988). Latin squares and agriculture: The other bicentennial. *Math. Scient.*, 13, 48–55.

Stroup, W. W. (1989). Why mixed models? Applications of mixed models in agriculture and related disciplines. *South. Coop. Ser. Bull. No.* 343, 183–201.

Szatrowski, T. H. and Miller, J. J. (1980). Explicit maximum likelihood estimates for balanced data in the mixed model of the analysis of variance. *Ann. Stat.*, 8, 811–819.

Tabachnick, B. G. and Fidell, L. S. (1991). Software for advanced ANOVA courses: A survey. *Beh. Res. Meth., Instrum. & Comp.*, 23, 208–211.

Tamhane, A. C. (1979). A comparison of procedures for multiple comparisons of means with unequal variances. *J. Amer. Stat. Assoc.*, 74, 471–480.

Tan, W. Y. (1981). The power function and an approximation for testing variance components in the presence of interaction in two-way random effects models. *Canad. J. Stat.*, 9, 91–99.

Tan, W. Y. and Cheng, S. S. (1984). On testing variance components in three-stages unbalanced nested random effects models. *Sankhyā, Ser. B*, 46, 188–200.

Tan, W. Y. and Tabatabai, M. A. (1986). Some Monte Carlo studies on the comparison of several means under heteroscedasticity and robustness with respect to departure from normality. *Biomet. J.*, 28, 801–814.

Tan, W. Y., Tabatabai, M. A., and Balakrishnan, N. (1988). Harmonic mean approach to unbalanced random effects model under heteroscedasticity. *Commun. Stat., A: Theo. & Meth.*, 17, 1261–1286.

Tan, W. Y. and Wong, S. P. (1980). On approximating the null and non-null distributions of the F ratio in unbalanced random effects models from non-normal universes. *J. Amer. Stat. Assoc.*, 75, 655–662.

Tang, P. C. (1938). The power function of the analysis of variance tests with tables and illustrations for their use. *Stat. Res. Mem.*, 2, 126–149.

Tate, R. F. and Klett, G. W. (1959). Optimal confidence intervals for the variance of a normal distribution. *J. Amer. Stat. Assoc.*, 54, 674–682.

Theune, J. A. (1973). Comparison of power for the D'Agostino and the Wilk-Shapiro tests of normality for small and moderate samples. *Stat. Neerland.*, 27, 163–169.

Thomas, J. D. and Hultquist, R. A. (1978). Interval estimation for the unbalanced case of the one-way random effects model. *Ann. Stat.*, 6, 582–587.

Thomsen, I. B. (1975). Testing hypotheses in unbalanced variance components models for two-way layouts. *Ann. Stat.*, 3, 257–265.

Thöni, H. (1967). *Transformation of Variables Used in the Analysis of Experimental and Observational Data: A Review*. Tech. Rep. No. 7, Statistical Laboratory, Iowa State University, Ames, Iowa.

Tietjen, G. L. (1974). Exact and approximate tests for unbalanced random effects designs. *Biometrics*, 30, 573–581.

Tiku, M. L. (1964). Approximating the general non-normal variance-ratio sampling distributions. *Biometrika*, 51, 83–95.

Tiku, M. L. (1967). Tables of the power of the F-test. *J. Amer. Stat. Assoc.*, 62, 529–539. (Corrigenda 63, 1551.)

Tiku, M. L. (1971). Power function of F-test under non-normal situations. *J. Amer. Stat. Assoc.*, 66, 913–916.

Tiku, M. L. (1972). More tables of the power of the F-test. *J. Amer. Stat. Assoc.*, 67, 709–710.

Tiku, M. L. (1974). Doubly noncentral F distribution – Tables and applications. In: *Selected Tables in Mathematical Statistics*, Vol. 2, pp. 139–178 (Eds. H. L. Harter and D. B. Owen). American Mathematical Society, Providence, Rhode Island.

Tiku, M. L. (1985a). Noncentral chi-square distribution. In: *Encyclopedia of Statistical Sciences*, Vol. 6, pp. 276–280 (Eds. S. Kotz and N. L. Johnson). John Wiley, New York.

Tiku, M. L. (1985b). Noncentral F-distribution. In: *Encyclopedia of Statistical Sciences*, Vol. 6, pp. 280–284 (Eds. S. Kotz and N. L. Johnson). John Wiley, New York.

Ting, N., Burdick, R. K., Graybill, F. A., Jeyaratnam, S., and Lu, T. F. C. (1990). Confidence intervals on linear combinations of variance components that are unrestricted in sign. *J. Stat. Comp. Simul.*, 35, 135–143.

Tippett, L. H. C. (1931). *The Methods of Statistics*, 1st ed. William and Norgate, London. (4th ed., John Wiley, New York.)

Tippett, L. H. C. (1934). *Applications of Statistical Methods to the Control of Quality in Industrial Production*. Manchester Statistical Society, Manchester, U.K.

Tocher, K. D. (1952). The design and analysis of block experiments. *J. R. Stat. Soc., Ser. B*, 14, 45–100.

Toothaker, L. E. (1991). *Multiple Comparison for Researchers*. Sage, Thousand Oaks, California.

Tukey, J. W. (1949a). Dyadic ANOVA, an analysis of variance for vectors. *Hum. Biol.*, 21, 65–110.

Tukey, J. W. (1949b). One degree of freedom for nonadditivity. *Biometrics*, 5, 232–242.

Tukey, J. W. (1949c). *Interaction in a Row-by-Column Design*. Mem. Rep. 18, Statistical Research Group, Princeton University, Princeton, New Jersey.

Tukey, J. W. (1950). *Finite Sampling Simplified*. Mem. Rep. 45, Statistical Research Group, Princeton University, Princeton, New Jersey.

Tukey, J. W. (1953). *The Problem of Multiple Comparisons* (Mimeographed manuscript of 396 pages). Department of Mathematics, Princeton University, Princeton, New Jersey.

Tukey, J. W. (1955). Answer to query no. 113. *Biometrics*, 11, 111–113.

Tukey, J. W. (1957). On the comparative anatomy of transformations. *Ann. Math. Stat.*, 28, 602–632.

Tukey, J. W. (1977). *Exploratory Data Analysis*. Addison-Wesley, Reading, Massachusetts.

Tukey, J. W. (1991). The philosophy of multiple comparisons. *Stat. Sci.*, 6, 100–116.

Ury, H. K. (1976). A comparison of four procedures for multiple comparisons among means (pairwise contrast) for arbitrary sample sizes. *Technom.*, 18, 89–97.

Ury, H. K., Stoline, M., and Mitchell, B. T. (1980). Further tables of the Studentized maximum modulus distribution. *Commun. Stat., B: Simul. & Comp.*, 9, 167–178.

Vanderbeck, J. P. and Cook, J. R. (1961). *Extended Table of Percentage Points of the*

Chi-Square Distribution. Nau. Rep. No. 7770, U.S. Naval Ordinance Test Station, China Lake, California.
Verdooren, L. R. (1988). Exact tests and confidence intervals for ratio of variance components in unbalanced two- and three-stage nested designs. *Commun. Stat., A: Theo. & Meth.*, 9, 1197–1230.
Vidmar, T. J. and Brunden, M. N. (1980). Optimal allocation with fixed power in a completely randomized design with levels of subsampling. *Commun. Stat., A: Theo. & Meth.*, 9, 757–763.
Walker, H. M. (1940). Degrees of freedom. *J. Educ. Psychol.*, 31, 253–269.
Wang, S.-G. and Chow, S.-C. (1994). *Advanced Linear Models: Theory and Application.* Marcel Dekker, New York.
Weekes, A. J. (1983). *A Genstat Primer.* Arnold, London.
Weerahandi, S. (1995). ANOVA under unequal error variances. *Biometrics*, 51, 589–599.
Welch, B. L. (1936). The specification of rules for rejecting too-variable a product, with particular reference to an electric lamp problem. *J. R. Stat. Soc., Suppl.*, 3, 29–48.
Welch, B. L. (1937). On the Z-test in randomized blocks and Latin squares. *Biometrika*, 29, 21–52.
Welch, B. L. (1956). On linear combinations of several variances. *J. Amer. Stat. Assoc.*, 51, 132–148.
Wheeler, R. E. (1974). Portable power. *Technom.*, 16, 193–201.
Wilcox, R. R. (1988). A new alternative to the ANOVA F and new results on James' second-order method. *Brit. J. Math. Stat. Psychol.*, 41, 109–117.
Wilcox, R. R. (1993). Robustness in ANOVA. In: *Applied Analysis of Variance in Behavioral Science*, pp. 345–374 (Ed. L. K. Edwards). Marcel Dekker, New York.
Wilk, M. B. (1955). *Linear Models and Randomized Experiments.* Ph.D. Thesis, Iowa State College, Ames, Iowa.
Wilk, M. B. and Kempthorne, O. (1955). Fixed, mixed and random models. *J. Amer. Stat. Assoc.*, 50, 1144–1167 (Corrigenda 51, 652.)
Wilk, M. B. and Kempthorne, O. (1956). Some aspects of the analysis of factorial experiments in a completely randomized design. *Ann. Math. Stat.*, 27, 950–985.
Wilk, M. B. and Kempthorne, O. (1957). Non-additivities in a Latin square design. *J. Amer. Stat. Assoc.*, 52, 218–236.
Williams, J. S. (1962). A confidence interval for variance components. *Biometrika*, 49, 278–281.
Winer, B. J. (1962, 1971). *Statistical Principles in Experimental Design.* 1st & 2nd eds. McGraw-Hill, New York.
Winer, B. J., Brown, D. R., and Michels, K. M. (1991). *Statistical Principles in Experimental Design*, 3rd ed. McGraw-Hill, New York.
Wolach, A. H. (1983). *BASIC Analysis of Variance Programs for Microcomputers.* Brooks/Cole, Montery, California.
Yates, F. (1933). The analysis of replicated experiments when the field results are incomplete. *Empor. J. Exper. Agri.*, 1, 129–142.
Yates, F. (1934). The analysis of multiple classification with unequal numbers in the different classes. *J. Amer. Stat. Assoc.*, 29, 51–66.
Yates, F. (1936a). Incomplete randomized blocks. *Ann. Eug.*, 7, 121–140.
Yates, F. (1936b). Incomplete Latin squares. *J. Agri. Sci.*, 26, 301–315.
Yates, F. (1937a). A further note on the arrangement of variety trials: Quasi-Latin squares. *Ann. Eug.*, 7, 319–332.

Yates, F. (1937b). *The Design and Analysis of Factorial Experiments.* Imperial Bureau of Soil Science, Harpenden, England.

Yates, F. and Hale, R. W. (1939). The analysis of Latin squares when two or more rows, columns, or treatments are missing. *J. R. Stat. Soc., Suppl.,* 6, 67–69.

Youden, W. J. (1937). Use of incomplete block replications in estimating tobacco-mosaic virus. *Contr. Boyce Thompson Inst.,* 9, 41–48.

Youden, W. J. (1940). Experimental designs to increase accuracy of greenhouse studies. *Contr. Boyce Thompson Inst.,* 11, 219–228.

Youden, W. J. (1951). *Statistical Methods for Chemists.* John Wiley, New York.

Zar, J. H. (1996). *Biostatistical Analysis,* 3rd ed. Prentice-Hall, Upper Saddle River, New Jersey. (1st ed., 1974; 2nd ed., 1984.)

Author Index

Abraham, J. K., 503, 689
Afifi, A. A., 147, 689
Afsarinejad, K., 522, 700
Akutowicz, F., 444, 689
Alexander, R. A., 86, 689
Algina, J., 86, 108, 689, 694, 708
Alldredge, J. R., 521, 709
Allen, R. E., 146, 689
Amara, I. A., 523, 704
Anderson, D. L., 368
Anderson, R. L., 66, 147, 314, 316, 428, 429, 456, 457, 494, 514, 530, 536, 579, 580, 689, 705
Anderson, S. L., 85, 106, 691
Anderson, T. W., 9, 689
Anderson, V. L., 307, 523, 527, 528, 656, 689, 706
Andrews, D. M., xi, 689
Anionwu, E., 121, 122, 689
Anscombe, F. J., 110, 111, 689
Applebaum, M. E., 111, 692
Armitage, J. V., 577, 689
Arnold, J., 492, 705
Arteaga, C., 143, 690
Arvesen, J. N., 33, 86, 690
Aster, R., 544, 690
Atkinson, A. C., 485, 690

Bagui, S. C., 574, 690
Bailey, B. J. R., 79, 647, 690
Balaam, L. N., 485, 697
Balakrishnan, N., 59, 60, 224, 574, 575, 576, 703, 713
Bancroft, T. A., 200, 222, 223, 314, 316, 428, 429, 530, 536, 689, 690, 692, 706
Bankier, J. D., 307, 690
Barcikowski, R. S., 633
Barnett, V. D., 97, 100, 690
Bartlett, M. S., vii, 98, 109, 110, 111, 690
Beall, G., 110, 690

Bechhoefer, R. E., 577, 653, 690
Beckman, R. J., 97, 690
Bennett, C. A., 8, 147, 307, 435, 461, 464, 470, 474, 690
Berry, D. A., 109, 174, 690
Birch, N. J., 291, 365, 366, 690, 692, 701
Bishop, D. J., 100, 691
Bishop, T. A., 86, 87, 691
Blackwell, T., 237, 238, 307, 691
Blair, R. C., 108, 689
Blischke, W. R., 314, 691
Bliss, C., 365, 427, 428, 691
Boardman, T. J., 32, 691
Bock, R. D., 307, 691
Bolk, R. J., 263, 691
Boneau, C. A., 85, 691
Bose, R. C., 507, 691
Bowker, A. H., 61, 685, 691
Bowman, K. O., 62, 64, 93, 228, 691, 703, 708
Box, G. E. P., 9, 85, 86, 87, 106, 109, 112, 168, 278, 311, 371, 372, 519, 524, 691, 692
Bozivich, H., 200, 692, 712
Bradley, J. V., 85, 692
Brailsford, E. A., 107, 709
Bratcher, T. L., 64, 635, 692
Broemeling, L. D., 9, 692
Brown, C., 237, 238, 307, 691
Brown, D. R., 29, 30, 485, 521, 582, 715
Brown, M. B., 86, 87, 107, 692
Brown, R. A., 86, 692
Brownlee, K. A., 407, 408, 453, 454, 464, 692
Brozovic, M., 121, 122, 689
Brunden, M. N., 400, 715
Budescu, D. V., 111, 692
Bulgren, W. G., 575, 692
Bulmer, M. G., 32, 692
Burch, L., 544, 692

Burdick, R. K., 32, 33, 38, 143, 144, 208, 209, 226, 244, 245, 291, 295, 297, 353, 358, 365, 366, 367, 400, 403, 406, 407, 580, 690, 692, 693, 701, 714

Cameron, J. M., 580, 693
Casella, G., 9, 29, 193, 225, 226, 403, 492, 580, 710, 711
Chambers, J. M., 89, 693
Chao, M.-T., 100, 693
Charles, J. A., 121, 701
Chen, H. J., 93, 711
Cheng, S. S., 365, 713
Chew, V., 64, 693
Chow, S.-C., 8, 715
Christensen, R., 8, 693
Clark, V. A., ix, 4, 696
Cleveland, W. S., 89, 693
Clinch, J. J., 86, 108, 693, 703
Coakes, S. J., 550, 693
Cochran, W. G., 84, 85, 90, 105, 106, 109, 147, 148, 223, 238, 290, 378, 379, 484, 485, 493, 497, 503, 507, 509, 516, 519, 520, 521, 524, 582, 596, 603, 676, 693, 712
Cody, R. P., 544, 693
Cohen, A., 98, 694
Cohen, J., 62, 694
Collins, L. M., 30, 695
Collyer, C. E., 543, 694
Conover, W. J., 9, 108, 694
Cook, J. R., 571, 714
Cook, R. D., 97, 690
Cooley, W. W., 307, 694
Coombs, W. T., 86, 108, 689, 694
Cornell, J., 311, 704
Cornfield, J., 181, 307, 461, 464, 471, 694
Cox, D. R., 109, 112, 259, 278, 484, 485, 503, 519, 521, 691, 694, 701
Cox, G. M., 147, 148, 290, 484, 485, 497, 507, 509, 516, 519, 520, 521, 522, 524, 603, 693
Craig, A. T., 572, 702
Crisler, L., 550, 694
Crowder, M. J., 522, 694
Crump, S. L., 4, 277, 580, 694

Cummings, W. B., 365, 694
Curtiss, J. H., 109, 694

D'Agostino, R. B., 93, 96, 97, 658, 694, 708
Daly, F., xi, 699
Damon, R. A., Jr., 343, 368, 411, 412, 516, 537, 694
Daniel, W. W., 9, 694
Daniels, H. E., 4, 694
Dantzker, M., 550, 706
Das, M. N., 516, 694
Davenport, J. M., 291, 694, 695
David, H. A., 105, 663, 695
Davies, O. L., 151, 279, 280, 345, 504, 509, 520, 695
Day, S. J., 60, 62, 695
Delaney, H. D., 521, 706
DeLury, D. B., 503, 695
Dempster, A. P., 147, 695
Dénes, J., 505, 695
Derr, G., 544, 696
Desmond, D. J., 455, 695
Dickey, D. A., 485, 712
DiIorio, F. C., 544, 695
Dijkstra, J. B., 86, 695
Dixon, W. J., 307, 558, 695
Dobson, A. J., 8, 695
Dodge, Y., 147, 509, 695
Donaldson, T. S., 85, 695
Donner, A., 31, 38, 581, 695
Donoghue, J. R., 30, 695
Draper, N. R., vii, 109, 223, 311, 691, 695
Dudewicz, E. J., 86, 87, 691
Duncan, A. J., 58, 696
Duncan, D. B., 81, 644, 696
Dunlop, G., 509, 696
Dunn, O. J., ix, 4, 79, 80, 577, 696
Dunnett, C. W., 81, 83, 84, 86, 577, 641, 690, 696
Dyer, D. D., 100, 102, 103, 659, 662, 696
Dykstra, Jr., O., 344

Efron, B., 109, 696
Eickman, J., 38, 692
Eisen, E. J., 579, 696

Eisenhart, C., 4, 106, 109, 664, 696
Elashoff, J. D., 523, 704
Elashoff, R. M., 147, 689
El-Bassiouni, Y., 366, 711
Elliott, R. J., 544, 696
Enns, J. T., 543, 694
Euler, L., 507, 696
Evans, M. A., 521, 709
Everitt, B., 544, 696

Fawcett, R. F., 8, 475, 711
Federer, A. J., 485, 697
Federer, W. T., 222, 485, 516, 520, 522, 697
Feldt, L. S., 62, 686, 688, 697
Fidell, L. S., 543, 713
Fisher, L., ix, 527, 531, 697
Fisher, M., 550, 700
Fisher, R. A., vii, 1, 2, 3, 4, 66, 77, 112, 484, 487, 488, 497, 504, 507, 537, 538, 569, 572, 573, 580, 581, 582, 697
Flecknell, P. A., 95, 710
Fleiss, J. L., 60, 79, 485, 516, 521, 697
Folks, L., 5, 22, 703
Forsythe, A. B., 86, 87, 107, 692
Fox, M., 58, 697
Francia, R. S., 94, 95, 711
Franklin, N. L., 8, 147, 307, 435, 461, 464, 470, 474, 690
Freeman, M. F., 109, 110, 111, 697
Freund, R. J., 8, 544, 697, 705
Friendly, M., 544, 697
Frude, N., 550, 698

Gabriel, K. R., 73, 698
Gaebelin, J., 213, 701
Gallo, J., 227, 698
Galton, F. F., vii
Games, P. A., 80, 82, 84, 86, 87, 108, 651, 698, 702, 703, 704
Ganguli, M., 401, 698
Gartside, P. S., 106, 698
Gates, C. E., 407, 698
Gayen, A. K., 85, 698
Gaylor, D. W., 8, 290, 365, 475, 579, 694, 698

Geary, R. C., 85, 92, 698
Ghosh, M. N., 263, 698
Gibbons, J. D., 23, 698, 709
Gill, J. L., 485, 698
Giri, N. C., 516, 694
Glaser, R. E., 100, 693, 698
Glass, G. V., 85, 86, 307, 698, 707
Glass, S., 503, 704
Glen, W. A., 147, 698
Goldsmith, P. L., 279, 280, 695
Gosset, W. S., 569
Gosslee, D. G., 219, 698
Govern, D. M., 86, 689
Govindarajulu, Z., 574, 700
Gower, J. C., 407, 698
Graham, D. F., 60, 62, 695
Graybill, F. A., 8, 22, 27, 28, 29, 32, 33, 37, 38, 57, 135, 143, 144, 182, 193, 208, 209, 225, 226, 227, 244, 245, 263, 264, 291, 295, 297, 353, 358, 363, 366, 367, 374, 375, 400, 403, 406, 407, 503, 580, 601, 690, 692, 693, 698, 699, 703, 706, 707, 711, 714
Greenberg, B. G., 484, 699
Greenway, A. R., 509, 708
Groggel, D. J., 34, 699
Guenther, W. C., ix, 699

Hahn, G. J., 485, 577, 699
Hald, A., 569, 571, 573, 699
Hale, R. W., 503, 716
Hall, I. J., 108, 699
Halvorsen, K. T., 328, 699
Hampel, F. R., 97, 699
Han, C. P., 200, 706
Hand, D. J., xi, 522, 694, 699
Hardy, K. A., 544, 695
Hardyck, C. D., 86, 708
Harsaae, E., 100, 699
Harter, H. L., 95, 571, 577, 699
Hartley, H. O., 58, 79, 91, 92, 100, 104, 105, 106, 200, 307, 569, 571, 573, 577, 638, 655, 680, 692, 699, 700, 708
Hartung, J., 353, 700
Hartwell, T. D., 8, 475, 698

Harvey, W. R., 343, 368, 411, 412, 516, 537, 694
Harville, D. A., 29, 181, 700
Hastay, M. W., 664, 696
Hatcher, L., 544, 700
Hawkins, D. M., 97, 700
Hayman, G. E., 574, 700
Hayter, A. J., 82, 700
Healy, M. J. R., 147, 700
Hedayat, A., 522, 700
Hedderson, J., 550, 700
Hegemann, V., 263, 700
Heiberger, R. M., 543, 700
Hemmerle, W. J., 307, 700
Henderson, C. R., 224, 307, 700
Hendrickson, R. W., 577, 699
Hendy, M. F., 121, 701
Herbach, L. H., 206, 701
Hernandez, R. P., 365, 366, 701
Herr, D. G., 213, 701
Herson, J. H., 111, 704
Herzberg, A., xi, 689
Herzberg, A. M., 485, 701
Herzberg, P. A., 544, 701
Hicks, C. R., 390, 391, 485, 701
Hinkelmann, K., 146, 168, 485, 492, 503, 516, 701, 705
Hinkley, D. V., 8, 701
Hirotsu, C., 150, 223, 224, 263, 701
Hoaglin, D. C., 89, 109, 701
Hochberg, Y., 64, 83, 84, 268, 577, 701
Hocking, R. R., 8, 181, 226, 701, 702
Hodges, J. L., Jr., 23, 702
Hoel, D. G., 62, 703
Hogg, R. V., 572, 702
Hollander, M., 9, 702
Holm, S., 79, 702
Hopper, F. N., 290, 579, 698
Horsnell, G., 87, 702
Houseman, E. E., 66, 689
Howe, R. B., 291, 707
Howell, J. F., 82, 84, 698, 702
Hoyle, M. H., 109, 147, 702
Hsu, J., 64, 702
Huber, P. J., 97, 702
Huck, S. W., 215, 702
Hudson, J. D., 290, 702
Huitema, B. E., 582, 702

Huitson, A., ix, 224, 702
Hultquist, R. A., 28, 38, 699, 713
Hunter, J. S., 371, 372, 519, 524, 692
Hunter, W. G., 109, 223, 371, 372, 519, 524, 692, 695
Hussein, M., 365, 702

Imhof, J. P., 268, 285, 702

Jackson, R. W. B., 4, 702
Jaffe, J. A., 544, 702
James, G. S., 86, 87, 702
Japanese Standards Association, 106, 702
Jeyaratnam, S., 33, 143, 291, 690, 703, 714
John, J. A., 521, 532, 703
John, P. W. M., 154, 531, 541, 542, 703
Johnson, D. E., 252, 263, 264, 700, 703, 707
Johnson, M. E., 108, 694
Johnson, M. M., 108, 694
Johnson, N. L., 59, 60, 322, 323, 510, 574, 575, 576, 703
Jones, B. 521, 703

Kafadar, K., 79, 703
Kapadia, C. H., 580, 710
Kastenbaum, M. A., 62, 64, 228, 691, 703
Keating, J. P., 100, 102, 103, 659, 662, 696
Keedwell, A. D., 505, 695
Kempthorne, O., 3, 5, 22, 146, 168, 223, 264, 484, 485, 503, 516, 520, 524, 701, 703, 715
Kendall, D. G., 109, 690
Kendall, M. G., 3, 14, 37, 311, 581, 703
Kenward, M. G., 521, 703
Keselman, H. J., 82, 86, 87, 108, 693, 703, 705, 709
Keselman, J. C., 87, 705
Keuls, M., 80, 703
Khargonkar, S. A., 514, 704
Khuri, A. I., 29, 224, 226, 227, 311, 365, 407, 580, 698, 704, 710
Khurshid, A., 141, 710
Kihlberg, J. K., 111, 704
Kimball, A. W., 200, 704
King, S. J., 544, 692
Kirk, R. E., 485, 521, 641, 704

Kirkwood, B., 121, 122, 689
Kleffe, J., 580, 709
Kleiner, B., 89, 693
Klett, G. W., 32, 713
Klotz, J. H., 73, 704
Koch, G. G., 523, 704
Koehler, K. J., 79, 90, 704
Kohr, R. L., 86, 87, 704
Kotz, S., 59, 60, 574, 575, 576, 703
Koval, J. J., 38, 695
Kramer, C. Y., 82, 147, 503, 698, 704
Krishnaiah, P. R., 9, 64, 263, 577, 689, 704
Krutchkoff, R. G., 86, 87, 269, 290, 702, 705
Kurtz, T. E., 64, 705
Kussmaul, K., 580, 705
Kutner, M. H., 8, 503, 609, 707

Lambert, J. W., 516
LaMotte, L. R., 224, 705
Larntz, K., 90, 704, 705
Larson, R., 411
Laubscher, N. F., 100, 705
Layard, M. W. J., 86, 168, 690, 705
Layne, B. H., 215, 702
Lee, P. M., 9, 705
Lehmann, E. L., 9, 23, 702, 705
Lehmer, E., 57, 705
Lentner, M., 492, 705
Leone, F. C., 322, 323, 510, 574, 700, 703
Levene, H., 107, 705
Levine, G., 590, 705
Levy, K. J., 86, 108, 705
Lewis, T., 97, 690
Lexis, W. H. R. A., 3
Lieberman, G. J., 61, 574, 685, 691, 709
Lindman, H. R., 9, 705
Lindsey, J. K., 522, 705
Link, R. F., 64, 705
Linnerud, A. C., 233
Linthurst, R. A., 533
Littell, R. C., 8, 224, 522, 544, 545, 697, 704, 705, 711
Lix, L. M., 87, 705
Lohnes, P. R., 307, 694
Lord, N. M., 147, 695
Lorenzen, T. J., 291, 307, 484, 527, 528, 656, 705, 706
Lu, T. F. C., 33, 714

Lucas, H. L., 219, 698
Lunn, D., xi, 699
Lurigio, A., 550, 706

Mackenzie, W. A., 4, 697
Mahalanobis, P. C., 3, 706
Mahamunulu, D. M., 403, 706
Mahmoud, M. W., 62, 686, 688, 697
Malone, L. C., 4, 708
Mandel, J., 263, 706
Maqsood, F., 38, 693
Marcuse, S., 391, 392, 400, 706
Mardia, K. V., 573, 706
Massey, F. J., 79, 577, 696
Mathew, T., 226, 704
Maurais, J., 100, 706
Maxwell, S. E., 521, 706
McCabe, G. P., 492, 710
McConway, K., xi, 699
McCullagh, P., 8, 706
McCulloch, C. E., 9, 29, 193, 225, 226, 403, 580, 711
McDonald, J., ix, 527, 531, 697
McHugh, R. B., 8, 706
McLean, R. A., 226, 523, 689, 706
Mead, R., 200, 706
Mensing, R. W., 263, 708
Merrington, M., 617
Micceri, T., 108, 706
Michels, K. M., 29 30, 485, 521, 582, 715
Mielke, P. W., Jr., 8, 706
Miller, J. J., 64, 206, 706, 713
Miller, R. G., Jr., 23, 33, 64, 79, 82, 85, 107, 268, 577, 706
Milliken, G. A., 252, 263, 264, 365, 503, 522, 545, 702, 705, 706, 707
Millman, J., 307, 707
Miron, T., 544, 707
Mitchell, B. T., 577, 714
Miyakawa, M., 263, 707
Montgomery, D. C., 146, 311, 707
Moran, M. A., 64, 635, 692
Morrison, D. F., 9, 707
Moses, L. E., 79, 707
Mosteller, F., 89, 172, 237, 238, 307, 691, 701, 707
Murdock, G. R., 32, 707
Murray, L. W., 264, 712
Myers, R. H., 291, 311, 707

Nachtsheim, C. J., 8, 609, 707
Nagasenkar, P. B., 100, 707
Naik, U. D., 291, 295, 707
Nair, K. R., 509, 577, 707
Nair, U. S., 100, 691
Natrella, M. G., 109, 520, 707
Nelder, J. A., 8, 706
Nelson, L. S., 64, 707
Neter, J., 8, 503, 609, 707
Newman, D., 80, 708
Nie, N. H., 550
Norton, H. W., 495, 708

O'Brien, R. G., 108, 708
Olejnik, S. F., 108, 708
Olkin, I., 30, 708
Oltman, D. O., 86, 694
O'Neil, R., 64, 708
Ord, J. K., 14, 37, 311, 703
Ostle, B., 4, 263, 341, 342, 708
Ostrowski, E., xi, 699
Owen, D. B., 105, 112, 569, 574, 577, 620, 671, 708

Parker, E. T., 507, 708
Paull, A. E., 200, 708
Pearson, E. S., 57, 58, 79, 85, 91, 92, 93, 100, 105, 106, 569, 570, 571, 573, 577, 644, 655, 680, 708
Pearson, K., 3, 570
Peckham, P. D., 85, 86, 698
Peng, K. C., 146, 307, 708
Perry, J. N., 509, 708
Petrinovich, L. F., 86, 708
Pillai, K. C. S., 577, 708
Pitman, E. J. G., 168, 709
Plackett, R. L., 4, 481, 709
Pratt, J. W., 23, 30, 698, 708, 709.
Probert, D. A., 108, 698
Pukelsheim, F., 224, 709

Quenouille, M. H., 521, 532, 703
Quimet, R., 100, 706

Ramachandran, K. V., 577, 708
Ramsey, P. H., 107, 108, 709
Rankin, N. O., 219, 709
Rao, C. R., 8, 224, 263, 580, 709

Rao, P. S. R. S., 225, 580, 709
Rao, P. V., 34, 699
Ratkowski, D. A., 521, 709
Reid, N., 8, 701
Resnikoff, G. J., 574, 709
Ringland, J. T., 86, 709
Robertson, A., 37, 709
Rochetti, E. M., 97, 699
Rogan, J. C., 82, 86, 87, 703, 709
Rohlf, F. J., 122, 123, 391, 392, 394, 417, 418, 712
Rosenthal, R., 64, 709
Rosnow, R. L., 64, 709
Rousseeuw, P. J., 97, 699
Royston, J. P., 93, 95, 96, 709, 710
Rubin, D. B., 147, 695, 710

Sahai, H., 8, 28, 29, 61, 86, 141, 181, 356, 580, 704, 710, 712
Samuels, M. L., 492, 710
Sanders, J. R., 85, 86, 698
Sanders, W. L., 226, 706
SAS Institute, 82, 520, 544, 545, 710
Satterthwaite, F. E., 289, 308, 578, 710
Scheffé, H., ix, 4, 22, 32, 58, 71, 73, 85, 87, 88, 135, 181, 193, 263, 264, 267, 268, 278, 279, 299, 307, 344, 363, 399, 401, 582, 710, 711
Schervish, M. L., 9, 711
Schlotzhauer, S. D., 544, 711
Schmitz, T. H., 33, 86, 690
Schultz, E. F., Jr., 286, 305, 307, 439, 441, 456, 711
Schutz, W. E., 111, 704
Searle, S. R., 8, 9, 29, 33, 37, 135, 193, 218, 221, 223, 224, 225, 226, 363, 366, 403, 475, 545, 580, 711
Seely, J. F., 366, 711
Seifert, B., 291, 711
Sen, B., 366, 711
Seneca, E. D., 533
Seng, M., 550, 706
Senn, S., 521, 711
Shaffer, J. P., 79, 82, 711
Shah, K. R., 509, 695
Shapiro, S. S., 93, 94, 95, 656, 657, 711
Sharma, D., 263, 698
Shiue, C., 407, 698

Shooter, M., 82, 703
Shrikhande, S. S., 507, 691
Šidák, A., 80, 711
Sinecore, J., 550, 706
Singh, B., 38, 711
Singh, C., 86, 712
Singhal, R. A., 34, 61, 86, 580, 711, 712
Sinha, B. K., 226, 704
Sinkbaek, S. A., 571, 699
Smith, D. W., 264, 712
Smith, H., vii, 109, 223, 695
Smith, H. F., 226, 712
Smith, J. K., 544, 693
Snedecor, G. W., 48, 85, 147, 223, 238, 378, 379, 503, 505, 516, 572, 582, 596, 676, 712
Snee, R. D., 532, 533, 712
Snell, E. J., 8, 558, 701, 712
Sokal, R. R., 122, 123, 391, 392, 394, 417, 418, 712
Solomon, H., 106, 696
Spector, P. C., 8, 544, 705
Spector, P. E., 544, 712
Spjotvoll, E., 25, 83, 224, 712
Sprent, P., 8, 712
SPSS, Inc., 550, 551, 558, 565, 566, 712
Srivastava, A. B. L., 85, 200, 712
Srivastava, S. R., 200, 712
Stahel, W. A., 97, 699
Steed, L. G., 550, 693
Steel, G. D., 147, 233, 234, 393, 485, 503, 516, 517, 526, 533, 534, 538, 712
Stepanski, E. J., 544, 700
Stephens, M. A., 658
Stoline, M. R., 64, 83, 84, 577, 654, 712, 713, 714
Strawderman, W. E., 98, 694
Street, A. P., 504, 505, 713
Street, D. J., 505, 713
Stroup, W. W., 226, 522, 545, 705, 706, 713
Stuart, A., 3, 14, 37, 311, 581, 703
Szatrowski, T. H., 206, 713

Tabachnick, B. G., 543, 713
Tabatabai, M. A., 87, 713

Tamhane, A. C., 64, 83, 84, 268, 577, 701, 713
Tan, W. Y., 61, 86, 87, 224, 264, 365, 713
Tang, P. C., 57, 713
Tate, R. F., 32, 713
Theune, J. A., 97, 713
Thiele, T. N., 3
Thomas, J. D., 38, 713
Thompson, C. M., 611, 617
Thompson, W. O., 28, 710
Thomsen, I. B., 224, 713
Thöni, H., 109, 713
Tiao, G. C., 9, 692
Tietjen, G. L., 365, 713
Tiku, M. L., 57, 59, 85, 574, 575, 576, 629, 714
Ting, N., 33, 291, 366, 690, 711, 714
Tippett, L. H. C., 4, 504, 509, 714
Tiwari, C. B., 86, 580, 712
Tocher, K. D., 147, 714
Toothaker, L. E., 64, 82, 703, 714
Torrie, J. H., 147, 233, 234, 393, 485, 503, 516, 517, 526, 533, 534, 538, 712
Toutenburg, H., 8, 709
Traux, H. M., 444, 689
Tukey, J. W., 5, 64, 71, 79, 82, 89, 109, 110, 111, 172, 181, 262, 264, 307, 461, 464, 471, 481, 503, 694, 697, 701, 703, 705, 707, 714
Tukey, P. A., 89, 693

Urey, F. R., 526
Ury, H. K., 83, 577, 713, 714

Vanderbeck, J. P., 571, 714
Verdooren, L. R., 366, 715
Vidmar, T. J., 400, 715
Voet, B., 353, 700

Wackerly, D. D., 34, 699
Walker, H. M., 16, 715
Wall, C., 509, 708
Wallace, D. L., 64, 705
Wallis, W. A., 664, 696
Wang, S.-G., 8, 715
Wasserman, W., 8, 503, 609, 707
Watford, D., 121, 122, 689

Webester, J. T., 291, 695
Weekes, A. J., 173, 715
Weerahandi, S., 84, 98, 715
Welch, B. L., 86, 87, 168, 289, 715
Wells, G., 38, 695
Werter, S. P. J., 86, 695
Westmacott, M., 147, 700
Wetherill, G. B., 64, 708
Wheeler, R. E., 61, 715
Wilcox, R. R., 86, 715
Wilk, M. B., 93, 181, 223, 264, 503, 656, 657, 711, 715
Williams, J. S., 32, 715
Williford, W. O., 32, 707
Winer, B. J., 29, 30, 82, 485, 521, 582, 715
Winkler, H. B., 108, 698
Wishart, J., 146, 689

Wolach, A. H., 543, 715
Wolfe, D. A., 9, 702
Wolfinger, R. D., 522, 545, 705
Wong, S. P., 61, 86, 713
Wootton, R., 95, 710
Wortham, A. W., 28, 699

Yates, F., 3, 66, 112, 146, 217, 220, 368, 497, 503, 507, 509, 519, 520, 541, 569, 573, 697, 715, 716
Yochmowitz, M. G., 263, 704
Youden, W. J., 247, 248, 520, 716

Zar, J. H., 86, 716
Zelen, M., 222, 697
Zemroch, P. J., 573, 706
Zimmer, W. J., 64, 635, 692

Subject Index

Additivity. *See* Nonadditivity *entries*
Aggregate variation, 2
Agricultural experiments, 4
Alternate mixed models, 264–268
Alternative hypothesis, defined, 597
Analysis of covariance, 581–582
Analysis of variance (ANOVA),
   for completely randomized design,
     485–486
   for Graeco-Latin square design,
     507–509
   historical developments in, 3–4
   for Latin square design, 498–500
   methodology, 1
   nonparametric, 9
   for partially nested classifications,
     433–436
   for randomized block design, 490–493
   for split-plot design, 513–516
   for three-way nested classification,
     396–399
   for two-way nested (hierarchical)
     classification, 350–351
   for unequal numbers of observations,
     in one-way classification, 36–39
   for various other designs or models.
     See under a specific design or
     model
   term, 2
   using BMDP, 558–559
   using SAS, 543–550
   using SPSS, 550–558
   using statistical computing packages,
     543–567
Analysis of variance F test. *See* F test
Analysis of variance (ANOVA) models,
   defined, 4–5
   finite population, 7–8, 461–482. *See
     also* finite population models
     entries
   fixed effects. *See* Fixed effects model
     (Model I)
   linear, 4–5
   mixed. *See* Mixed model (Model III)
   Model I. *See* Fixed effects model
     (Model I)
   Model II. *See* Random effects model
     (Model II)
   Model III. *See* Mixed model
     (Model III)
   with multivariate response, 9
   one-way. *See* One-way classification
   random effects. *See* Random effects
     model (Model II)
   rules for determining, 588–590
   univariate, 8
   variance components. *See* Random
     effects model (Model II), Mixed
     model (Model III)
Analysis of variance table, defined, 26
ANOVA. *See* Analysis of variance
   (ANOVA) *entries*
ANOVA procedure, in SPSS, 164, 252,
   333, 551, 552
   program and output, 165, 489
Approximate confidence interval,
   defined, 600
   in two-way crossed finite population
     model, 470
Approximate test, defined, 598
Arcsine transformation, 110, 112
Assumption of normality, 20, 135, 193,
   286, 351, 399, 407

Balanced incomplete block design, 519
Balanced lattice design, 520
Balancing experimental units, 484
Bartlett's test for homogeneity of
   variances, 98–104, 106–107
   table of critical values of, 659–662

Bayesian inference, 9
Behrens-Fisher problem, 84
Best linear unbiased estimates (BLUE),
    in one-way classification, 27
    in two-way crossed classification with
        interaction, 204
    in two-way crossed classification
        without interaction, 140
Beta statistic, 168
Between group sum of squares, 15, 35–36
Biomedical Programs. See BMDP
Block design
    incomplete, 516, 519
    randomized. See Randomized block
        design
Blocking experimental units, 484
Blocks fixed
    treatments fixed and, 490–492
    treatments random and, 493
Blocks random
    treatments fixed and, 492–493
    treatments random and, 492
BLUE. See Best linear unbiased estimates
BMD (Biomedical) package, 558
BMDP (Biomedical Programs), 543. See
    also Statistical computing
    packages
    analysis of variance using, 558–559
BMDP 3D, 1V, 4V, and 5V, 559
BMDP 7D, 52, 164, 252, 253, 559, 565,
    567
    program and output, 53, 54, 254, 489
BMDP 2V, 164, 252, 253, 334, 381, 451,
    558, 559
    program and output, 165, 255, 335,
        496, 506, 511
BMDP 3V, 52, 164, 252, 381, 421, 451,
    559
    program and output, 56, 385, 424
BMDP 8V, 52, 55, 164, 252, 253, 381,
    421, 451, 559
    program and output, 55, 166, 167, 257,
        259, 336, 338, 382, 383, 386, 422,
        425, 451, 519
Bonferroni inequality, 78
Bonferroni t statistic, 78–79, table of
    critical values of, 645–647
Bonferroni's test/method/interval
    in one-way classification, 78–79, 80

    in three-way crossed classification,
        321–322
    in two-way crossed classification with
        interaction, 233
    in two-way crossed classification
        without interaction, 150
    in two-way nested classification,
        361–362, 365
    using statistical computing packages,
        561–563
Boole inequality, 78
Box-plots, 89
Brown-Forsythe procedure, 107–108,
    566
BY keyword, in a SPSS procedure,
    551–552

C test for homogeneity of variances. See
    Cochran's C test for homogeneity
    of variances.
Calculators, electronic, x
Cells, 126
Central limit theorem, 368, 569
Chi-square distribution, definition and
    properties of, 570–571
    table of critical values of, 610–611
    noncentral. See Noncentral chi-square
        distribution
Chi-square goodness-of-fit test for
    normality, 89–90
CLASS keyword/statement, in a SAS
    procedure, 524, 546–549
Classification, term, 125–126
Cochran's C test for homogeneity of
    variances, 98, 105–107
    table of critical values of, 664
Coefficient of kurtosis, 91–92
    table of critical values of sample
        estimate of, 655
Coefficient of skewness, 90–91
    table of critical values of sample
        estimate of, 655
Coefficients of order statistics for the
    Shapiro-Wilk's W test for
    normality, table of, 656
Column efficiency, of a Latin square
    design, 503
Completely nested design, definition and
    examples of, 347–348, 395

Subject Index

Completely randomized design (CRD), 485–488, 489
　analysis of variance for, 486–487
　mathematical model of, 486
　worked example for, 487, 488
Components of variance. *See* Variance components
Computational formulae and procedure for sums of squares
　in Latin square design, 502
　in one-way classification, 35–36
　in partially nested classifications, 437
　in three-way crossed classification, 297–298
　in three- and four-way nested classifications, 396–397, 401, 403–404
　in two-way crossed classification with interaction, 210–212
　in two-way crossed classification without interaction, 144–145
　in two-way nested (hierarchical) classification, 359, 362–363
Computing power, use of statistical packages for, 560–561
Concomitant variable, term, 581
Confidence coefficient, defined, 599
Confidence intervals for variance components, in one-way random effects model, 31–34
Conservative confidence interval, defined, 600
Conservative test, defined, 598
Consistency, defined, 599
Consistent estimator, defined, 599
CONTRAST command/subcommand, in SAS ANOVA/GLM and SPSS ONEWAY/GLM procedures, 562–565
Contrasts, defined, 65
　test of hypothesis involving, 67–69
Control, in experimental design, 484–485
Correlation, defined, 588
Covariance, defined, 587–588
　analysis of, 581–582
Covariate, term, 581
CRD. *See* Completely randomized design

CRITERIA subcommand, in SPSS GLM and VARCOMP procedures, 558, 561
Critical range values, Studentized, 80
Critical region, defined, 597
Critical values, defined, 597
　of Bartlett's test for homogeneity of variances, table of, 659–662
　of chi-square distribution, table of, 610–611
　of Cochran's C test for homogeneity of variances, table of, 664
　of D'Agostino's D test for normality, table of, 658
　of Duncan's multiple range test, table of, 642–644
　of Dunn-Šidák's multiple comparison test, table of, 648–651
　of Dunnett's test, table of, 639–641
　of F distribution, table of, 612–617
　of Hartley's maximum F ratio test for homogeneity of variances, table of, 663
　of sample estimate of coefficient of kurtosis, table of, 655
　of sample estimate of coefficient of skewness, table of, 655
　of Shapiro-Wilk's W test for normality, table of, 657
　of Studentized augmented range distribution, table of, 654
　of Studentized maximum modulus distribution, table of, 652–653
　of Studentized range distribution, table of, 636–638
　of Student's t distribution, table of, 608–609
Cross-nested design, defined, 431
Cross-over design, 520–521
Crossed classification, contrasted with a nested classification, 347
　term, 125–126
Cumulative standard normal distribution, table of critical values of, 605–606
Curves of constant power for F tests in fixed effects model (Model I) for determination of sample size in a one-way classification, 686–688

D'Agostino's D test for normality, 96–97
  table of critical values of, 658
Degrees of freedom, 2, 569, 570, 572–576
  concept of, 15–17
  for error variance in a one-way classification, 125
  in partially nested classifications, 435
  rules for calculating, 590–591
DESIGN keyword/subcommand, in SPSS GLM and MANOVA procedures, 552–557
Detecting outliers, 97
Doubly noncentral beta, related to doubly noncentral F, 576
Doubly noncentral F distribution, 262
  definition and properties of, 576
Doubly noncentral t distribution, definition and properties of, 575
Duncan's multiple range test, 81, 561–563
  table of critical values of, 642–644
Dunn-Šidák's multiple comparison test, 79–80, 562–653
  table of critical values of, 648–651
Dunnett's multiple comparison test, 81–82, 563
  table of critical values of, 639–641
Dunn's multiple comparison test, 79
  table of critical values of, 645–647

Effects parameter, 594
Efficiency, defined, 599
Efficient estimator, defined, 599
Electronic calculators, x
EMS (error mean square), its custom-made use for multiple comparisons using statistical packages, 561
Equal variances, departures from, 86–88
Error mean square (EMS). See EMS (error mean square)
Error sum of squares, 129, 183, 286, 351, 397, 435, 464, 475, 486, 491, 499, 508, 513
Error terms, 5, 589
  departures from independence of, 88–89
Error variance, degrees of freedom in one-way classification for, 125

Estimate, defined, 598
Estimated total variance, in one-way random effects model, 29
Estimator, defined, 598
Exact confidence interval, defined, 600
Exact statistical test, defined, 598
EXAMINE procedure, in SPSS, 567
Expectations of mean squares
  in fixed effects models (Model I),
    two-way crossed classification with interaction, 186–188
    two-way crossed classification without interaction, 131–132
  in mixed models (Model III),
    in two-way crossed classification with interaction, 190–193
    in two-way crossed classification without interaction, 133
  in one-way classification, 17–20
  in random effects model (Model II)
    in two-way crossed classification with interaction, 188–190
    in two-way crossed classification without interaction, 132–133
  in three-way crossed classification, 286
  in two-way crossed classification with interaction, 184–193
  in two-way crossed classification without interaction, 130–134
  in various other designs or models. See under a specific design or model
Expected mean squares, rules for finding, 591–595
Expected subclass numbers, method of, 222
Expected value, defined, 586–587
Experimental data, collecting, 483
Experimental designs,
  in agriculture and other sciences, 4
  literature on, 485
  principles of, 483–485
  some simple, 483–542
Experimental unit, defined, 484

F distribution, 23, 138, 198, 199, 201, 202, definition and properties of, 572–573
  table of critical values of, 612–617
  doubly noncentral, 262, definition and properties of, 576

Subject Index

F distribution (cont.)
  noncentral. See Noncentral
    F distribution
F test(s)
  equivalent to paired t test, 584–586
  equivalent to two-sample t test, in
    one-way classification, 582–584
  in fixed effects model (Model I)
    curves of constant power for
      determination of sample size in
      one-way classification, 686–688
    in three-way crossed classification,
      288, 289
    in two-way crossed classification
      with interaction, 198–200
    in two-way crossed classification
      without interaction, 137–139
    power function charts of, 672–680
  in mixed model (Model III)
    in three-way crossed classification,
      291–292
    in two-way crossed classification
      with interaction, 202–203
    in two-way crossed classification
      without interaction, 139
  in one-way classification, 22–25
  in random effects model (Model II)
    in one-way classification, table of
      power and optimum number of
      levels in, 630–633
    in three-way crossed classification,
      288–291, 292
    in two-way crossed classification
      with interaction, 200–202
    in two-way crossed classification
      without interaction, 139
  operating characteristic curves for,
    61, 681–685
  in three-way crossed classification,
    286, 288–292
  in two-way crossed classification with
    interaction, 197–203
  in two-way crossed classification
    without interaction, 137–139
  in two-way crossed finite population
    model, 466–467
  in two-way nested (hierarchical)
    classification, 351, 353, 363–365
  in various other designs or models. See
    under a specific design or model

  power of. See Power of F test
  F value in an analysis of variance
    table, 26
Feldt-Mahmoud charts, 62, 686–688
Finite population models, 461–482
  four-way crossed, 472–473
  more complex, 481
  nested, 474–475
  one-way, 461–462
  statistical computing packages in, 481
  three-way crossed, 470–472
  two-way crossed, 462–470
  unablanced, 475
  worked example for, 475–481
Finite population theory, 8
Finite populations, 7–8, 461
Fisher's Z distribution, related to F
  distribution, 572
Fixed effects, 4, 5, concept of, 6–7
Fixed effects analysis, in an unbalanced
  two-way crossed classification
  model, 215–223
  general case of unequal frequencies
    for, 217–223
  proportional frequencies for,
    215–217
Fixed effects model (Model I), 4, 5,
  481
  curves of constant power for F tests for
    determination of sample size in
    one-way classification, 686–688
  effects of violations of assumptions of
    two-way crossed classification
    with interaction, 269
  expectations of mean squares in,
    in two-way crossed classification
      with interaction, 186–188
    in two-way crossed classification
      without interaction, 131–132
  F tests
    in one-way classification, 22–24
    in three-way crossed classification,
      288, 289
    in two-way crossed classification
      with interaction, 198–200
    in two-way crossed classification
      without interaction, 137–139
    in various other designs or models.
      See under a specific design or
      model

Fixed effects model (Model I) (*cont.*)
  interval estimation
    in two-way crossed classification with interaction, 207–208
    in two-way crossed classification without interaction, 142–143
    in two-way nested (hierarchical) classification, 356–357
  one-way classification, table of minimum sample size per treatment group needed in, 634–635
  point estimation in
    in two-way crossed classification with interaction, 203–205
    in two-way crossed classification without interaction, 139–141
    in two-way nested (hierarchical) classification, 354–355
    in various other designs or models. *See under* a specific design or model
  power function charts of F tests in, 672–680
  power of F test in
    in one-way classification, 57–60
    in two-way crossed classification with interaction, 227–228
    in various other designs or models. *See under* a specific design or model
  sampling distribution of mean squares in
    in two-way crossed classification with interaction, 193, 195–196
    in two-way crossed classification without interaction, 135–136
    in various other designs or models. *See under* a specific design or model
  worked examples for
    in one-way classification, 39–43
    in three-way crossed classification, 314–322
    in two-way crossed classification with interaction, 233–237
      for unequal sample sizes per cell, 237–244
    in two-way crossed classification without interaction, 151–154
    in two-way nested (hierarchical) classification, 368–371
    in various other design or models. *See under* a specific design or model
Four-factor partially nested classification, 438–439, 440
Four-way crossed classification, 302, 304–307
Four-way crossed finite population model, 472–473
Four-way nested classification, 403–406
Fox charts, 58
Fractional replications, 523–524

General case of unequal frequencies
  for fixed effects analysis, 217–223
  for random effects analysis, 223–226
General constant, in a linear model, 5
General linear models (GLM), 5, 8
General q-way nested classification, 406–407
Generalized linear models, 8
Generalized randomized block design (GRBD), 494
GLM (General linear models), 5, 8
GLM procedure, in SPSS, 52, 164, 252, 253, 333, 334, 381, 421, 451, 551–554, 556, 557
  program and output, 55, 56, 166, 167, 256, 258, 336, 337, 383, 384, 386, 422, 423, 425, 450, 518
Graeco-Latin square design, 507–512
  analysis of variance for, 507–509
  mathematical model for, 507
  worked example in, 510–512
Graeco-Latin squares, 507
  some more examples of, 602–603
GRBD (generalized randomized block design), 494
Grouping experimental units, 484

Hartley's maximum F ratio test for homogeneity of variances, 98, 104–105, 106–107
  table of critical values of, 663
Hierarchical models, partially, 431–460

Subject Index

Hierarchically nested design. *See* completely nested design
Higher-order crossed classifications, 307–311
Homogeneity of variances
　Bartlett's test for, 98–104, 106–107
　　table of critical values of, 659–662
　Cochran's C test for, 98, 105–107
　　table of critical values of, 664
　Hartley's maximum F ratio test for, 98, 104–105, 106–107
　　table of critical values of, 663
　other tests for, 107–108
HOMOGENEITY option, in SPSS procedures, 566
Homoscedasticity
　tests for, 97–108. *See also* Homogeneity of variances *entries*
　use of statistical packages for, 565–567
　transformations to correct lack of, 110–113
HOVTEST option, in SAS GLM procedure, 566
Hyper-Graeco-Latin square design, 522
Hypersquares, 522
Hypothesis testing, general procedure of, 597–598

Incomplete block design, 516, 519
Independence of error terms, departures from, 88–89
Independent variable, term, in analysis of covariance, 581
Infinite population theory, 461
Infinite populations, 7–8, 461
Interaction, defined, 177
　meaning and interpretation of, 253, 256–259
　with one observation per cell, 259, 261–264
　significant, 257
　two-way crossed classification with. *See* Two-way crossed classification with interaction
　two-way crossed classification without. *See* Two-way crossed classification without interaction

Interaction effect, in a two-way crossed classification model, 180
Interaction sum of squares, in a two-way crossed classification model, 183
Interaction terms, in an analysis of variance model, 589
Interval estimation, general method of, 599–601
　in fixed effects model (Model I)
　　in two-way crossed classification with interaction, 207–208
　　in two-way crossed classification without interaction, 142–143
　　in two-way nested (hierarchical) classification, 356–357
　　in various other designs or models. *See under* a specific design or model
　in Latin square design, 500–501
　in mixed model (Model III)
　　in two-way crossed classification with interaction, 209–210
　　in two-way crossed classification without interaction, 143–144
　　in two-way nested (hierarchical) classification, 359
　　in various other designs or models. *See under* a specific design or model
　in random effects model (Model II)
　　in two-way crossed classification with interaction, 208–209
　　in two-way crossed classification without interaction, 143
　　in two-way nested (hierarchical) classification, 358
　　in various other designs or models. *See under* a specific design or model
　in three-way crossed classification, 292–297
　in two-way crossed classification with interaction, 207–210
　in two-way crossed classification without interaction, 142–144
　in two-way crossed finite population model, 468–470
　in two-way nested (hierarchical) classification, 356–359, 365–367

Interval estimation (*cont.*)
  in various other designs or models. *See under* a specific design or model
Intraclass correlation, definition and properties of, 580–581
Intraclass correlation coefficient, defined, 581
intraclass correlations
  in one-way classification, 29, 30, 34, 38
  in two-way crossed classification with interaction, 182
  in two-way crossed classification without interaction, 128
Inverse hyperbolic sine transformation, 110n

Jackknife technique, in finding confidence interval of a variance component 33n, in testing homogeneity of variances, 107–108

Kruskal-Wallis test, 87
K test, 87, 269
Kurtosis, 85, 91
  coefficient of. *See* Coefficient of kurtosis
  test for, 91–93

Lagrangian interpolation, three-point, to calculate power of an F test, 621
Latin square design, 495, 497–507
  analysis of variance for, 498–500
  computational formulae and procedure for sums of squares in, 502
  interval estimation in, 500–501
  mathematical model of, 498
  missing observations in, 502–503
  multiple comparisons in, 501
  point estimation in, 500–501
  power of F test in, 501
  relative efficiency of, 503–504
  replications in, 504–505
  worked example in, 505–507
Latin squares, 495, 497
  some more examples of, 601–602
Lattice design, 519–520
Least significant difference test, 77–78
Least squares estimators
  in two-way crossed classification with interaction, 203
  in two-way crossed classification without interaction, 139
  in two-way nested classification, 354
Level of confidence, definition and interpretation of, 599
Levene's test for homogeneity of variances, 107, 566–567
Liberal interval, defined, 600
Liberal test, defined, 598
Linear combination of means, defined, 65
Linear (statistical) models, defined, 8
Logarithmic transformation, 109, 111
Lower confidence interval, defined, 600
LSD (least significance difference), 77–78

Magic Latin squares, 522
Main effects, in two-way crossed classification with interaction, 180
MANOVA procedure, in SPSS, 164, 252, 253, 334, 381, 451, 551–555
  program and output, 254, 255, 334, 382, 496, 506, 511
Mathematical expectation, defined, 586–587
Maximum likelihood estimators
  in one-way classification, 28
  in two-way crossed classification with interaction, 203n, 206n
  in two-way crossed classification without interaction, 139n, 141
  in two-way nested classification, 356
Maximum test for scale, for homogeneity of variances, 108
MAXORDERS subcommand, in SPSS ANOVA procedure, 553
Mean squares, defined, 2
  expectations of. *See* Expectations of mean squares
  expected, rules for finding, 591–595
  sampling distribution of. *See* Sampling distribution of mean squares
Mean value, defined, 587
Means
  linear combination of, defined, 65
  range of the set of, defined, 80
MEANS procedure, in SPSS, 550

Subject Index

MEANS statement, in SAS GLM procedure, 52, 562, 565–566
Method of expected subclass numbers, 222
Method of unweighted means, 217–220
Method of weighted-squares-of-means, 220–222
METHOD subcommand, in SPSS VARCOMP procedure, 557–558
Minimum norm quadratic unbiased estimation (MINQUE), 224
Minimum sample size per treatment group needed in one-way fixed effects design, table of, 634–635
Minimum variance quadratic unbiased estimators (MIVQUE), 224
Minimum variance unbiased (MVU) estimator, defined, 599
MINQUE (minimum norm quadratic unbiased estimation), 224
Missing observations or values
  in Latin square design, 502–503
  in split-plot design, 514, 516
  in two-way crossed classification without interaction, 145–148
  worked example for, in two-way crossed classification without interaction, 161–164
MIVQUE (minimum variance quadratic unbiased estimators), 224
Mixed-classification design, defined, 431
Mixed effects analysis, in an unbalanced two-way crossed classification model, 226–227
Mixed model (Model III), 4, 5, 481
  alternate, 264–268
  effects of violations of assumptions of two-way crossed classification with interaction, 270
  expectations of mean squares in
    in two-way crossed classification with interaction, 190–193
    in two-way crossed classification without interaction, 133
  F tests in
    in three-way crossed classification, 291–292
    in two-way crossed classification with interaction, 202–203
    in two-way crossed classification without interaction, 139
    in various other designs or models. See under a specific design or model
  interval estimation in
    in two-way crossed classification with interaction, 209–210
    in two-way crossed classification without interaction, 143–144
    in two-way nested (hierarchical) classification, 359, 367
    in various other designs or models. See under a specific design or model
  point estimation in
    in two-way crossed classification with interaction, 206–207
    in two-way crossed classification without interaction, 141
    in two-way nested (hierarchical) classification, 356, 357
    in various other designs or models. See under a specific design or model
  power of F test in, in two-way crossed classification with interaction, 229–230
    in various other designs or models. See under a specific design or model
  sampling distribution of mean squares in
    in two-way crossed classification with interaction, 196–197
    in two-way crossed classification without interaction, 136–137
  Scheffé's, 267–268, 285n
  worked examples for
    in partially nested classifications, 444–448, 449
    in three-way crossed classification, 328–333
    in three-way nested classification, 417–420
    in two-way crossed classification with interaction, 247–252
    in two-way crossed classification without interaction, 157–161
    in two-way nested (hierarchical) classification, 378–380

Model I. *See* Fixed effects model (Model I)
Model II. *See* Random effects model (Model II)
Model III. *See* Mixed model (Model III)
MODEL keyword/statement, in SAS procedures, 524–525, 546–549
Modified sequentially rejective Bonferroni (MSRB) test, 79
MSRB (modified sequentially rejective Bonferroni) test, 79
Multifactor layouts, 281
Multiple comparisons
  in Latin square design, 501
  in one-way classification, 64–84
    Bonferroni's test for. *See* Bonferroni's test/method interval
    Dunn-Šidák test for, 79–80
    Dunnett's test for, 81–82
    Dunn's procedure of, 78–79
    least significant difference test for, 77–78
    MSRB test for, 79
    Newman-Keul's test for, 80–81
    Scheffé's method of. *See* Scheffé's method of multiple comparison
    SRB test for, 79
    Tukey's method of. *See* Tukey's method of multiple comparison
    Various other methods of, 82–84
  in three-way crossed classification, 299–301
  in two-way crossed classification with interaction, 230–233
  in two-way crossed classification without interaction, 149–151
  in two-way nested (hierarchical) classification, 360–362
  use of statistical packages for, 561–565
Multiple regression analysis, method based on, 222–223
Multivariate response, analysis of variance (ANOVA) models with, 9
Mutual orthogonality of contrasts, 66
MVU (minimum variance unbiased) estimator, defined, 599

Negative estimates of variance components, 28, 141, 206, 356

Nested classifications
  four-way, 403–406
  general q-way, 406–407
  partially. *See* Partially nested classifications
  three-way. *See* Three-way nested classification
Nested design, completely or hierarchically, 348–349
Nested-factorial design, defined, 431
Nested finite population models, 474–475
  two-way, 474–475
Newman-Keul's test, 80–81, 561–563, 565
Nonadditivity, sum of squares for, 262
  Tukey's one degree of freedom test for, 262–264, 503
Noncentral chi-square distribution, 21, 135, 195, 197, definition and properties of, 573–574
Noncentral F distribution, 59, 60, 138, 139, 148, 227, 229, 262, definition and properties of, 575–576
Noncentral t distribution, 251, 618, 669, definition and properties of, 574–575
Noncentrality parameter, 21, 57–59, 135, 136, 138, 195, 251, 298, 618, 669, defined, 573
Nonnegative maximum likelihood estimators of variance components, $28n$, $141n$, $206n$, $356n$
Nonparametric analysis of variance, 9
Normality
  assumption of, 20, 135, 193, 286, 351, 399, 407
  chi-square goodness of-fit test for, 89–90
  D'Agostino's D test for. *See* D'Agostino's D test for normality
  effects of departures from assumption of, 85–86
  Shapiro-Francia's test for, 94–96
  Shapiro-Wilk's W test for. *See* Shapiro-Wilk's W test for normality

Subject Index

Normality (*cont.*)
   tests for, 89–97
   use of statistical packages for tests of, 567
NPAR TESTS procedure, in SPSS, 567
Null hypothesis, defined, 597

O'Brien procedure for testing homogeneity of variances, 108
Observations, physical, variation among, 1
One observation per cell, three-way classification with, 301–302, 303
One-sided confidence interval, defined, 600
One-way classification, 11–123
   advantages and disadvantages of, 125
   assumptions of, 11–12
   computational formulae and procedure for sums of squares in, 35–36
   confidence intervals for variance components in, 31–34
   corrections for departures from assumptions of, 108–113
   effects of departures from assumptions underlying, 84–89
   F test equivalent to two-sample t test in, 582–584
   F tests in, 22–25
   mathematical model of, 11
   point estimation in, 26–31
   power of F test
      in fixed effects model (Model I), 57–60
      in random effects model (Model II), 60–61
   statistical computing packages in, 52
   worked examples using, 52, 53–56
   tests for departures from assumptions of, 89–108
One-way finite population model, 461–462
One-way fixed effects design, table of minimum sample size per treatment group needed in, 634–635
One-way random effects design, table of power and optimum number of levels in, 630–633

ONEWAY procedure, in SPSS, 52, 550–552
   program and output, 53, 54, 489
Operating characteristic curves for F tests in random effects model (Model II), 61, 681–685
Orthogonal contrasts, defined, 65
Outliers, detecting, 97

p-value, defined, 23, in hypothesis testing, 597
Paired t test, F test equivalent to, 584–586
Parameter, defined, 598
Partially hierarchical models, 434
Partially nested classifications, 431–460
   analysis of variance for, 433–436
   computational formulae and procedure for sums of squares in, 437
   degrees of freedom in, rule for finding, 435
   four-factor, 438–439
   mathematical model of, 431–433
   statistical computing packages in, 448, worked example using, 450–451
Partition of the total sum of squares
   in one-way classification, 14–15
   in three-way crossed classification, 285–286
   in two-way crossed classification with interaction, 182–183
   in two-way crossed classification without interaction, 128–129
Pearson-Hartley charts, 58, 59, 61, 148, 251, 672–680
Percentage points of the standard normal distribution, table of, 607
Percentiles of the chi-square distribution, table of, 610–611
Physical observations, variation among, 1
Point estimation, general method of, 598–599
   in fixed effects model (Model I)
      in two-way crossed classification with interaction, 203–205
      in two-way crossed classification without interaction, 139–141
      in two-way nested (hierarchical) classification, 354–355

Point estimation (*cont.*)
  in Latin square design, 500–501
  in mixed model (Model III)
    in two-way crossed classification with interaction, 206–207
    in two-way crossed classification without interaction, 141
    in two-way nested (hierarchical) classification, 356, 357
  in one-way classification, 26–31
  in random effects model (Model II)
    in two-way crossed classification with interaction, 205–206
    in two-way crossed classification without interaction, 141
    in two-way nested (hierarchical) classification, 355–356
  in three-way crossed classification, 292–297
  in two-way crossed classification with interaction, 203–207
  in two-way crossed classification without interaction, 139–141
  in two-way crossed finite population model, 467–468
  in two-way nested (hierarchical) classification, 354–356, 365–366
Population variances, multiple comparison for unequal, 84
Populations, finite and infinite, 7–8
POSTHOC subcommand, in SPSS ONEWAY and GLM procedures, 563–564
Power and optimum number of levels in one-way random effects F test, table of, 630–633
Power function charts
  of F test in fixed effects model (Model I), 672–680
  of two-sided Student's t test, 669–671
Power of a test, defined, 597
Power of F test
  in fixed effects model (Model I), table of, 621–629
  in Latin square design, 501
  in one-way classification, 52, 57–61
    in random effects model (Model II), table of optimum number of levels and, 630–633
  in three-way crossed classification, 298–299
  in two-way crossed classification with interaction, 227–230
    in fixed effects model (Model I), 227–228
    in mixed model (Model III), 229–230
    in random effects model (Model II), 228–229
  in two-way crossed classification without interaction, 148–149
  in two-way nested (hierarchical) classification, 360
Power of Student's t test, table of, 618–620
POWER subcommand, in SPSS MANOVA procedure, 560
Power transformation, 112–113
PRINT subcommand, in SPSS VARCOMP procedure, 558
PROC ANOVA, in SAS, 52, 164, 252, 253, 333, 334, 381, 448, 524, 544–547, 549
  program and output, 53, 54, 165, 254, 334, 335, 489, 496, 506, 511
PROC GLM, in SAS, 52, 164, 252, 253, 333, 334, 381, 421, 448, 451, 524, 544, 549
  program and output, 55, 56, 166, 167, 255, 256, 258, 337, 382, 384, 385, 421, 423, 424, 450, 518
PROC LATTICE, in SAS, 520
PROC MIXED, in SAS, 164, 252, 333, 381, 448, 525, 545, 549
  program and output, 604
  worked examples using, 603–604
PROC MULTTEST, in SAS, 562
PROC NESTED, in SAS, 381, 448, 544, 545, 547, 548
  program and output, 383
PROC UNIVARIATE, in SAS, 52, 567
PROC VARCOMP, in SAS, 164, 252, 333, 381, 448, 545, 549, 557
Proportional frequencies in an unbalanced three-way crossed classification, 311–314
  unbalanced two-way crossed classification,

Subject Index 737

Proportional frequencies (*cont.*)
   for fixed effects analysis, 215–217
   for random effects analysis, 223
Protected least significant difference, 77–78
Pseudo-F test
   in higher order crossed classifications, 308
   in partially nested classifications, 442–443
   in three-way crossed classification, 290–292, 293, 327
   in two-way crossed finite population model, 466–467, 477–479
   use of Satterthwaite procedure in constructing, 578–580

Quasi-factorials, 519

Random effects, 4, 5, concept of, 6–7
Random effects analysis, in an unbalanced two-way crossed classification model, 223–226
   general case of unequal frequencies for, 223–226
   proportional frequencies for, 223
Random effects model (Model II), 4, 5, 481
   effects of violations of assumptions of two-way crossed classification with interaction, 269
   expectations of mean squares in
      in two-way crossed classification with interaction, 188–190
      in two-way crossed classification without interaction, 132–133
   F tests in
      in one-way classification, 24–25
      in three-way crossed classification, 288–291
      in two-way crossed classification with interaction, 200–202
      in two-way crossed classification without interaction, 139
      in various other designs or models. *See under* a specific design or model
   interval estimation in
      in two-way crossed classification with interaction, 208–209
      in two-way crossed classification without interaction, 143
      in two-way nested (hierarchical) classification, 358
      in various other designs or models. *See under* a specific design or model
   operating characteristic curves for F tests in, charts of, 61, 681–685
   point estimation in
      in two-way crossed classification with interaction, 205–206
      in two-way crossed classification without interaction, 141
      in two-way nested (hierarchical) classification, 355–356
      in various other designs or models. *See under* a specific design or model
   power and optimum number of levels of F test in one-way, table of, 630–633
   power of F test in
      in one-way classification, 60–61
      in two-way crossed classification with interaction, 228–229
      in various other designs or models. *See under* a specific design or model
   sampling distribution of mean squares in
      in two-way crossed classification with interaction, 196
      in two-way crossed classification without interaction, 136
      in various other designs or models. *See under* a specific design or model
   worked examples for
      in one-way classification, 43–52
      in partially nested classifications, 439–443
      in three-way crossed classification, 322–328
      in three-way nested classification, 407–411
      in two-way crossed classification with interaction, 244–247
      in two-way crossed classification without interaction, 154–157

Random effects model (Model II) (*cont.*)
  in two-way nested (hierarchical) classification, 371–374
  for unequal numbers in subclasses
    in three-way nested classification, 411–417
    in two-way nested (hierarchical) classification, 374–378
RANDOM keyword/statement/subcommand, in SAS and SPSS GLM procedures, 252, 333, 381, 448, 525, 546, 547–548, 556
Random numbers, table of, 665–668
Random sample, defined, 595–596
Randomization, in experimental design, 483–484, 488, 493, 497
Randomized block design (RBD), 488, 490–495, 496
  analysis of variance for, 490–491
  generalized (GRBD), 494
  mathematical model of, 490
  missing observations in, 493
  relative efficiency of, 493–494
  replications in, 494
  worked example for, 494–495
Randomized design, completely. *See* Completely randomized design
Randomness, concept of, 596
Range of the set of means, defined, 80
RANGES subcommand, in SPSS ONEWAY procedure, 562–563
RBD. *See* Randomized block design
RE. *See* Relative efficiency
Reciprocal transformation, 111–112
Relative efficiency (RE), defined, 493n
  of randomized block design, 493–494
  of Latin square design, 503–504
Reliability, of an estimate, 599
REML, option for computing restricted maximum likelihood estimates in SAS PROC MIXED and SPSS VARCOMP procedures, 549, 557
Repeated measures design, 521–522
Replications, in experimental design, 483
  in Latin square design, 504
  in randomized block design, 494
Representative sample, concept of, 595

Restricted maximum likelihood. *See* REML
Row efficiency, of a Latin square design, 504

Sample size, power and determination of, 61–64
Sample size determination using smallest detectable difference, 63–64
Sample, defined, 595
Sampling distribution, defined, 596
Sampling distribution of mean squares
  in fixed effects model (Model I)
    in two-way crossed classification with interaction, 193, 195–196
    in two-way crossed classification without interaction, 135–136
  in mixed model (Model III)
    in two-way crossed classification with interaction, 196–197
    in two-way crossed classification without interaction, 136–137
  in one-way classification, 20–22
  in random effects model (Model II)
    in two-way crossed classification with interaction, 196
    in two-way crossed classification without interaction, 136
  in two-way crossed classification with interaction, 193–197
  in two-way crossed classification without interaction, 135–137
SAS (Statistical Analysis System), 543–550. *See also* Statistical computing packages and *entries* following PROC
SAS PROBMC function, 82
Satterthwaite procedure, 144, 209, 210, 365, 367, 403, 468, 578–580
Scheffé's mixed model, 267–268, 285n
Scheffé's method of multiple comparison
  in Latin square design, 501
  in one-way classification, 73–76
    effects of departures from assumptions in, 86, 87
    interpretation of, 76–77
    relative merits and drawbacks of, 77
  in three-way crossed classification, 300–301

Subject Index

Scheffé's method (*cont.*)
  in two-way crossed classification with interaction, 230, 231, 232, 233, 251
  in two-way crossed classification without interaction, 150, 153–154, 160–161
  in two-way nested (hierarchical) classification, 361, 365
  using statistical computing packages, 561–565
Sequentially rejective Bonferroni (SRB) procedure, 79
Shapiro-Francia's test for normality, 94–96
Shapiro-Wilk's W test for normality, 93–94, 102–103
  table of coefficients of order statistics for, 656
  table of critical values of, 657
Significance level, concept of, 597
Skewness, 85, 90
  coefficient of. *See* Coefficient of skewness
  test for, 90–91
Smallest detectable difference, sample size determination using, 63–64
Snedecor's F distribution. *See* F distribution
SPECIAL keyword, in SPSS GLM procedure, 564
Split-plot design, 512–516
  analysis of variance for, 513–515
  mathematical model of, 513
  missing values in, 514, 516
  worked example for, 516, 517–519
Split-split-plot design, 523
SPSS, 543, 550–558. *See also* Statistical computing packages and *entries* following a particular procedure
Square-root transformation, 109–110, 111
Square transformation, 112
SRB (sequentially rejective Bonferroni) procedure, 79
Standard deviation, defined, 587
Standard error of an estimator, defined, 599*n*

Standard normal distribution
  cumulative, table of, 605–606
  percentage points of, table of, 607
Statistic, term, 596
Statistical Analysis System. *See* SAS
Statistical computing packages, 543
  analysis of variance using, 543–567
  in finite population models, 481
  in one-way classification, 52
    worked examples using, 52, 53–56
  in partially nested classifications, 448
    worked example using, 450–451
  in three-way crossed classification, 333–334
    worked examples using, 334–338
  in three-way nested classifications, 421
    worked examples using, 421–425
  in two-way crossed classification with interaction, 252
    worked examples using, 253, 254–256, 257, 258, 259
  in two-way crossed classification without interaction, 164
    worked examples using, 164–167
  in two-way nested (hierarchical) classification, 381
    worked examples using, 381–386
  in various other designs or models. *See under* a specific design or model
Statistical inference, methods of, 596–601
Statistical packages, use of
  for computing power, 560–561
  for performing multiple comparisons, 561–565
  for performing tests of homoscedasticity, 565–567
Statistical Product and Service Solutions, 543
Statistical tables and charts, 605–688. *See also entries* under a specific table or chart
Studentized augmented range distribution 83, table of critical values of, 654
Studentized critical range values, 80
Studentized maximum modulus distribution, 84
  definition and properties of, 577–578
  table of critical values of, 652–653

Studentized range distribution, 71, 80
  definition and properties of, 576–577
  table of critical values of, 636–638
Student's t distribution. See t distribution
Student's t test. See t test
Sum(s) of squares, defined, 2
  for Tukey's test of nonadditivity, 262
  rules for calculating, 590–591
  Types I, II, III, and IV, 545
Super magic Latin squares, 522
Systemic effects, 4. See also entries under Fixed effects

t distribution, 68, 74, 76, 618, 669,
  definition and properties of, 569–570
  doubly noncentral. See Doubly noncentral t distribution
  noncentral. See Noncentral t distribution
  table of critical values of, 608–609
t test, 68, 69, 78, 252, 618, 669
  paired, F test equivalent to, 584–586
  power function charts of the two-sided, 669–671
  two-sample, F test equivalent to, in one-way classification, 582–584
TEST statement/option/subcommand, in SAS and SPSS GLM procedures, 252, 333, 381, 448, 547–548, 556–557
Test statistic, defined, 597
Three- and higher-order crossed classifications, 281–345, unequal sample sizes in, 311–314
Three-way classification with one observation per cell, 301–302, 303
Three-way crossed classification, 281–302, 311–345
  assumptions of, 284–285
  computational formulae and procedure for sums of squares in, 297–298
  expectations of mean squares in, 286
  F tests in, 286, 288–292
  interval estimation in, 292–297
  mathematical model of, 281–283
  multiple comparisons in, 299–301
  partition of the total sum of squares in, 285–286
  point estimation in, 292–297
  power of F test in, 298–299
  statistical computing packages in, 333–334
  worked examples using, 334–338
Three-way crossed finite population model, 470–472
Three-way nested classification, 395–429
  analysis of variance of, 396–399
  mathematical model of, 395–396
  statistical computing packages in, 421
  worked examples using, 421–425
  tests of hypotheses and estimation in, 399–400
  unequal numbers in subclasses in, 400–403
  worked example for, 411–417
Transformations, 109–113
  to correct lack of normality, 109–110
  to correct lack of homoscedasticity, 110–113
Treatments fixed
  blocks fixed and, 490–492
  blocks random and, 492–493
Treatments random
  blocks fixed and, 493
  blocks random and, 492
Tukey-Kramer intervals, 82
Tukey-Kramer-Miller-Winer procedure, 82
Tukey's method of multiple comparison
  in Latin square design, 501
  in one-way classification, 70–72
    effects of departures from assumptions in, 86
    interpretation of, 76–77
    relative merits and drawbacks of, 77
  in three-way crossed classification, 300–301
  in two-way crossed classification with interaction, 230, 231, 232, 233, 250
  in two-way crossed classification without interaction, 150, 152–153, 159–160
  in two-way nested (hierarchical) classification, 361
  using statistical computing packages, 561–563, 565

Tukey's one degree of freedom test for
nonadditivity, 262–264, 503
Two-sample t test, F test equivalent to, in
one-way classification, 582–584
Two-sided confidence interval, defined,
600
Two-sided Student's t test, power
function charts of, 669–671
Two-stage nested design, 13
Two-way crossed classification, defined,
125–126
Two-way crossed classification with
interaction, 177–280
assumptions of, 180–182
best linear unbiased estimation
(BLUE) in, 204
computational formulae and procedure
for sums of squares in, 210–212
effects of violations of assumptions of,
268–270
expectations of mean squares in,
184–193
F tests in, 197–203
interval estimation in, 207–210
mathematical model of, 177–180
multiple comparisons in, 230–233
partition of the total sum of squares in,
182–183
point estimation in, 203–207
power of F test in, 227–230
sampling distribution of mean squares
in, 193, 195–197
statistical computing packages in, 252
worked examples using, 253,
254–256, 257, 258, 259
with unequal sample sizes per cell,
212–227
worked example for, 237–244
Two-way crossed classification without
interaction, 125–175
assumptions of, 127–128
best linear unbiased estimates (BLUE)
in, 140
computational formulae and procedure
for sums of squares in, 144–145
effects of violations of assumptions of,
168
expectations of mean squares in,
130–134
F tests in, 137–139

interval estimation in, 142–144
mathematical model of, 126
missing observations in, 145–148
multiple comparisons in, 149–150
partition of the total sum of squares in,
128–129
point estimation in, 139–141
power of F test in, 148–149
sampling distribution of mean squares
in, 135–137
statistical computing packages in, 164
worked examples using, 164–167
Two-way crossed finite population
model, 462–470
F tests in, 466–467
interval estimation in, 468–470
point estimation in, 467–468
Two-way nested (hierarchical)
classification, 347–394
analysis of variance of, 350–351
assumptions of, 350
computational formulae and procedure
for sums of squares in, 359,
362–363
F tests in, 351, 353, 363, 365
interval estimation in, 356–359,
365–367
mathematical model of, 349–350
multiple comparisons in, 360–362
point estimation in, 354–356, 365–366
power of F test in, 360
statistical computing packages in, 381
worked examples using, 381–386
unequal numbers in subclasses in,
362–368
worked example for, 374–378
Two-way nested finite population model,
474–475
2^p design, 523–524
Type I error, defined, 597
Type II error, 36, defined, 597
Types I, II, III, and IV sums of squares,
252, 545

UMA (uniformly most accurate) interval,
defined, 600
UMAU (uniformly most accurate
unbiased) interval, defined, 601
UMVU estimator, of the ratio of two
variance components, 29

Unbalanced finite population models, 475
Unbiased confidence interval, defined, 600
Unbiased estimator, definition and example of, 598–599
Unbiasedness, 598, 600
Uncertainty, of an estimate, 599
Unequal numbers of observations in one-way classification, 36–39
Unequal numbers in subclasses
 in three-way nested classification, 400–403
  worked example for, 411–417
 in two-way nested (hierarchical) classification, 362–368
  worked example for, 374–378
Unequal sample sizes and population variances
 in multiple comparisons, 82–84
Unequal sample sizes per cell
 in two-way crossed classification with interaction, 212–227
 in three- and higher-order classifications, 311–314
UNIANOVA procedure, in SPSS, 551, 566
Uniformly minimum variance unbiased estimator. *See* UMVU estimator
Uniformly most accurate (UMA) interval, defined, 600
Uniformly most accurate unbiased (UMAU) interval, defined, 601
Uniformly shortest length (USL) interval, defined, 601
Univariate analysis of variance (ANOVA) models, defined, 8
Unweighted means, method of, 217–220
Upper confidence interval, defined, 600

USL (uniformly shortest length) interval, defined, 601

VARCOMP procedure, in SPSS, 164, 252, 333, 451, 557–558
Variance(s), defined, 587
 error, degrees of freedom in one-way classification for, 125
 homogeneity of. *See* Homogeneity of variances
 unequal population, in multiple comparisons, 84
Variance components, defined, 12
 literature on, 580
 confidence intervals for, in one-way classification, 31–34
 estimation of. *See* Point estimation and Interval estimation *entries*
Variance components model. *See* Random effects model (Model II), Mixed effects model (Model III)
VS keyword, in SPSS GLM and MANOVA procedures, 554–557

W test. *See* Shapiro-Wilk's W test for normality
Weighted-squares-of-means analysis, 220–222
WELCH option, in SAS GLM MEANS statement, 566
Within group sum of squares, 15, 35–36
WITHIN keyword, in SPSS GLM and MANOVA procedures, 553–555
Worked examples. *See under* a specific model or design

Youden squares, 520
Z distribution. *See* Fisher's Z distribution

Printed in the United States
61035LVS00001B/19-21